Entertainmer

"Thorsten Hennig-Thurau and Mark B. Houston—two of our finest scholars in the area of entertainment marketing—have produced a definitive research-based compendium that cuts across various branches of the arts to explain the phenomena that provide consumption experiences to capture the hearts and minds of audiences."

—Morris B. Holbrook, W. T. Dillard *Professor Emeritus of Marketing, Columbia University*

"*Entertainment Science* is a must-read for everyone working in the entertainment industry today, where the impact of digital and the use of big data can't be ignored anymore. Hennig-Thurau and Houston are the scientific frontrunners of knowledge that the industry urgently needs."

—Michael Kölmel, *media entrepreneur and Honorary Professor of Media Economics at University of Leipzig*

"*Entertainment Science*'s winning combination of creativity, theory, and data analytics offers managers in the creative industries and beyond a novel, compelling, and comprehensive approach to support their decision making. This ground-breaking book marks the dawn of a new Golden Age of fruitful conversation between entertainment scholars, managers, and artists."

—Allègre Hadida, *Associate Professor in Strategy, University of Cambridge*

"Finally, a pioneering step from conventional power-political and gut-feel decision-making to a research-based guide that can readily be applied to all segments of the entertainment and media universe. More than anything that has been previously available, *Entertainment Science* provides readers with a deep understanding of what makes the industry tick and what raises the probability of profitability. Media executives, scholars, students, and buffs will find this book an invaluable reference."

—Harold (Hal) L. Vogel, *author of "Entertainment Industry Economics: A Guide for Financial Analysis".*

"Hennig-Thurau and Houston have done a terrific job organizing, summarizing, and articulating lucidly cumulative scholarly research on the entertainment industry. *Entertainment Science* challenges convincingly the "Nobody-Knows-Anything" old mantra, making empirical-based knowledge and findings accessible to a wide range of audiences."

—Jehoshua Eliashberg, *Professor of Marketing, The Wharton School.*

"*Entertainment Science* offers a new dimension of how statistical analysis can be applied to the intersection of art and science."

—Jason E. Squire, *editor of "The Movie Business Book" and Associate Professor of Practice, USC School of Cinematic Arts.*

Thorsten Hennig-Thurau Mark B. Houston

For the latest developments in *Entertainment Science* and to engage with the authors and other readers, please visit our website http://entertainment-science.com and our Facebook page at https://www.facebook.com/EntertainmentScience.

#EntertainmentScience

Thorsten Hennig-Thurau · Mark B. Houston

Entertainment Science

Data Analytics and Practical Theory for Movies, Games, Books, and Music

Thorsten Hennig-Thurau
Marketing Center Münster
University of Münster
Münster, Germany

Mark B. Houston
The Neeley School of Business
Texas Christian University
Fort Worth, TX, USA

ISBN 978-3-030-07733-4   ISBN 978-3-319-89292-4 (eBook)
https://doi.org/10.1007/978-3-319-89292-4

Softcover re-print of the Hardcover 1st edition 2019

Printed on acid-free paper

This Springer imprint is published by the registered company Springer International Publishing AG part of Springer Nature
The registered company address is: Gewerbestrasse 11, 6330 Cham, Switzerland

To Claudia and the boys, for their love and support and inspiration, as well as their sheer infinite patience with me during the writing of this book. And to Sergio and Clint, for making me fall in love with the movies and the art (and science) of entertainment.

T. H.-T.

To Nancy, Jon, Elise, Wil, and Shane for their love and laughter, regardless of circumstance.

M. B. H.

And to Bruce Mallen, without whose inspiration and efforts this book and a large part of the research from which it arose would never exist. We will always be thankful for your contributions to Entertainment Science and hope that there will be a way through which our words will reach you.

T. H.-T. & M. B. H.

Preface and Acknowledgements

The entertainment industry, enlightening billions of people with movies, games, books, and music, is often characterized by its "Nobody-Knows-Anything" mantra. This mantra, coined more than 30 years ago by screenwriter legend William Goldman, argues that survival and success is a function of managerial intuition and instinct *only* and refuses the existence of economic rules and laws for entertainment products.

The Goldman adage strongly collides with today's production and marketing budgets for entertainment products which often exceed $100 million and can reach up to $500 million—for a single new movie or video game. This book introduces *Entertainment Science* as an alternative, and more timely, paradigm. *Entertainment Science* builds on the assumption that in the era of almost unlimited data and computer power, the combination of smart analytics and powerful theories can provide valuable insights to those who have room for them in their decision making. Our aim to retire the Goldman mantra must not be confused with any desire to retire creativity and intuition—*Entertainment Science* considers data analytics and theory as *complementary* resources to these basic skills, not as their substitutes.

Entertainment Science (the book) offers a systematic investigation of the knowledge that has been accumulated by scholars in various fields such as marketing and economics regarding the factors that make entertainment products successful—or let them flop. This knowledge has gone unnoticed by many who manage entertainment products and determine the industry's course. But the knowledge has also suffered from a lack of integration, with most studies being relatively isolated scholarly endeavors of particular aspects of the entertainment business.

A main contribution of this book is that we open a unique vault of more than 35 years of high-quality scholarly research on the entertainment industry and make it accessible to future and current decision-makers. In other words, we link the practical skills of Hollywood with the intellectual powers of Harvard, UCLA, Wharton, TCU, and Münster University, just to name a few of the many places around the globe where scholars have contributed to the development of *Entertainment Science*. But our goals are even more ambitious—the idea is to offer our readers a comprehensive approach toward what defines "good" marketing and management in the entertainment industry, something that requires an integration of the many different scholarly pieces into a coherent puzzle. This integrative nature makes this book an attempt in theory-building itself—in its totality, this book can be considered the first draft of a theory of *Entertainment Science*, at whose core is the explanation of what makes an entertainment product successful (and what doesn't work in entertainment). You will note that it is a theory with many remaining gaps and blank spaces, and some parts of it are supported by stronger arguments, richer data, and more rigorous statistical methods than others. But this developing character is typical for theories of any kind, which are, by definition, hardly definite and final.

At the core of *Entertainment Science* (the book and the theory) is a *probabilistic* worldview. The book substitutes the deterministic perspective, which—often unnoticed—underlies the Goldman adage, with the argument that success in the entertainment industries is all about probability, not determination. Whereas the devotees of Mr. Goldman have been right that nobody will ever be *sure* that a new entertainment product will succeed in the marketplace, the insights about industry mechanisms, consumer patterns, and marketing instruments compiled in this book will increase every manager's *probability* to be successful with his or her next offering if taking them to heart.

It's this probabilistic perspective that adds value to the findings that have been generated when scholars apply rigorous analytical methods to big data sets on the performance of movies, games, books, or music. But we argue that a theory of the entertainment industry and its actions cannot really work if it does not take into account the powerful theories that scholars of entertainment and other fields have already developed regarding the actions of firms and its customers. This is why we have configured *Entertainment Science* as the combination of creativity, data analytics, and good theories. In other words, and paraphrasing a classic saying, we argue here that for being successful in today's entertainment industry, there is nothing so practical as the combination of powerful analytics and good theory to complement entertainment's traditional elements of creativity and intuition.

Our book is targeted to various groups of readers. It is targeted to those who are, as students of business or the creative arts, or as employees in another industry, fascinated by entertainment and the firms that provide it. We hope that *Entertainment Science* helps you to deepen your fascination with this unique industry and to better understand its economic mechanisms. Arranging the many fragments of scientific studies on entertainment into a holistic theory of *Entertainment Science* should inspire scholars' next explorations and help to identify exciting "unknowns," but also frame scholars' work and help to interpret new findings. We also have written our book with those current decision-makers in entertainment firms in mind who like to have their thinking challenged, who are looking for ways to grow as a decision-maker, and who seek to improve their firm's ability to pick winners and avoid losers. Our digital times provide us with the tools that enable and facilitate a lively exchange between all these groups about the many fascinating facets of *Entertainment Science*. We invite you to join the community at our website http://entertainment-science.com and on Facebook at https://www.facebook.com/EntertainmentScience. This is also where we will keep you updated on new discoveries and developments.

There are some more things we want to explain before you, our reader, dive into the world of *Entertainment Science*. We (THT and MBH) are both marketing scholars by training, and so it should not come as a surprise that our book takes a market-centered perspective that focuses on winning customers in a competitive world. It deals to a much lesser degree with the internal organizational processes which are required to ensure that customers are actually won over, and that market-related goals are achieved (one exception is the organization of the innovation process, because here organizational and procedural issues are literally inseparable from crafting powerful new offers for customers). The market-centered approach implies decisions that refer to the overall firm strategy, and we highlight how success, or failure, can result from market-centered decision making. We believe that there is enormous value in taking a market-centered perspective, and we feel that a lot of the changes that are happening in the entertainment industry these days are caused by firms that share our line of thinking—Amazon and Netflix, for example.

We also want to note that writing a book that aims at bringing together the often-separate worlds of entertainment practice and academia sometimes requires the use of terminology which runs counter to either one party's or the other's standard language. One of the more drastic examples is that we will refer to movie theaters as distributors of movies, not as "exhibitors," as a result of our value-chain analysis of the entertainment industry: studios/labels are those who "sell" their products to those who distribute it to the consumers, just like retailers distribute products to them. Distribution still remains an important part of the marketing mix (we assign a full chapter to it), but from an *industry* perspective, it's the theaters that distribute entertainment to those who are longing for it. It is important for us to stress that we by no means dispute the adequacy of the industry's choice of terms. Our ambition though is to disseminate knowledge also from other markets and industries, and doing so requires some harmonization of concepts and terms, as general theories and models occasionally run counter to established "industry language." We hope that the industry representatives among our readers (who are used to refer to showing films to audiences via theaters as "theatrical distribution" themselves) will pardon us this procedure, for the sake that both parties will be able to learn from each other.

Moreover, the combination of two authors from two parts of the entertainment world carries additional value. We both have an international perspective, but combine our "geocentric" approach with our knowledge of regional specifics in North America (MBH) and the European continent (THT). This also enabled us to cover entertainment insights unveiled in different parts of the world and in different languages. Of course, the world of entertainment is, in our globalized times, one that is much larger than just those two continents, and readers will certainly note that although we aim to bring in insights from other parts of the entertainment universe, as well, those parts have ended up somewhat underrepresented in the book. Please forgive this bias—at least we've tried.

It is often heard that writing a book is a collective effort, and there couldn't be more truth to this phrase. We are deeply grateful to the various groups of people who have shaped our thinking—often challenging our ideas—and inspired our work. Let us begin by saying that *Entertainment Science* would not exist without the direct and indirect contributions by many, many colleagues. Specifically, we would like to thank our co-authors on various entertainment industry research projects, namely (in strict alphabetical order) Suman Basuroy, Sabine Best, Matthias Bode, Björn Bohnenkamp, Subimal Chatterjee, Haipeng "Allan" Chen, Michel Clement, Dominik Dallwitz-Wegner, Felix Eggers, Jehoshua "Josh" Eliashberg, Fabian Feldhaus, Stefan Fuchs, Tim(othy) Heath, Torsten Heitjans, Victor Henning, Barbara Hiller, Julian Hofmann, Ram Janakiraman, Shane Johnson, Alegra Kaczinski, Ann-Kristin Kupfer (Knapp), Bruno Kocher, Raoul Kübler, André Marchand, Paul Marx, Juliane Mathys, Sangkil Moon, Eunho Park, Nora Pähler vor der Holte, Rishika Rishika, Henrik Sattler, Ricarda Schauerte, Reo Song, Shrihari "Hari" Sridhar, Franziska Völckner,

Gianfranco "Johnny" (a.k.a. Frank) Walsh, Charles "Chuck" Weinberg, Berend Wierenga, Caroline Wiertz, and Oliver "Olli" Wruck. Whenever we refer to "our" work in this book, we've got you in mind too. Ann-Kristin and also Björn deserve special credit for co-developing the entertainment and media lectures which provided the foundation of this book, in addition to earlier attempts by scholars to structure the field of entertainment research (Hadida 2009 and Peltoniemi 2015 were among the works which inspired us the most). And Ronny Behrens deserves a very special mention—he co-authored the book's innovation management chapter with us, provided feedback, and also helped with numerous editorial issues.

Further, it is literally impossible to express how much we have benefitted from interactions with the collection of scholars who congregate each year for the Mallen Economics of Filmed Entertainment Conference, which will take place for the 20th time in this book's release year. Bruce Mallen was the driving force who motivated and created this conference, and we will always be thankful to him for his initiative and long-term support—we dedicate this book to him. S. Abraham "Avri" Ravid and Olav Sorenson have kept it rolling full speed ahead, and many individuals have contributed through the years to make the gathering such an amazing incubator for ideas, including Darlene Chisholm, Art De Vany, Anita Elberse, Natasha Foutz, Allègre Hadida, the one-and-only Morris Holbrook, Amit Joshi, Yong Liu, Jordi McKenzie, Jamal Shamsie, Michael (D.) Smith, Jason Squire, and Harold "Hal" Vogel, as well as several of our co-authors we have mentioned above, such as Josh and Chuck and Suman and Ann-Kristin.

We also owe a debt of intellectual gratitude to our current and former faculty colleagues. For Thorsten, this includes Thorsten Wiesel and Manfred Krafft from University of Münster, Armin Rott, Wolfgang Kissel, and Tom Gross from his inspiring time at Bauhaus University of Weimar, and Caroline Wiertz, but also Vince Mitchell and Joe Lampel from his City University London days. Mark's thanks go to his colleagues at Texas Christian University (including Bob Leone, Eric Yorkston, and Chris White), the faculty of Texas A&M University's marketing department, and Comic-Con aficionado Peter Bloch from his time at University of Missouri. Our universities have provided the resources and the bandwidth to invest time and effort into this stream of research that made writing the book possible.

But our book could not succeed in building bridges between scholars and the industry without the help from all the great people who are part of the entertainment industry in so many different roles and functions and who have shared their insights and cooperated with us on projects, providing data and context. We cannot list them all by name here, but among those who we benefitted from mostly (and enjoyed interacting with!) are Andreas Bareiss, Malte Probst, Michael Kölmel, Jan Rickers, Wilfried Berauer, Dietmar Güntsche, Andreas Kramer, Caroline Bernhardt, and Jannis Funk, who co-chaired the Big Data, Big Movies Conference with Thorsten in Potsdam and Berlin in 2016. We thank Michael and Dietmar's Weltkino team, Bernhard Glöggler, Christine Weber, Roger Grotti, and Robert Rossberg at Disney Germany, Fox Germany's Volker Lauster, Germar Tetzlaff, and colleagues, Kalle Fritz and his StudioCanal Germany team, and Marcel Lenz and Guido Schwab from Weimar-based ostlicht ("meer is immer!"). Of course, if there should be anything in the book with what you, the reader, do not agree, we, the authors, are the sole and only ones to blame.

In addition, let us stress that lots of the insights we report in this book would not exist without the help of several generations of Bachelor's, Master's, and doctoral students at Bauhaus University of Weimar and University of Münster, who have helped us shaping ideas and theories and have provided great help, as research assistants, by compiling the databases which were the sources for several key findings and which we make heavy use of in this book. Several of them are listed above, as they have become entertainment scholars or managers themselves. Nora Pähler and Ricarda Schauerte, along with Tanja Geringhoff who manages Thorsten's *Lehrstuhl* in Münster, helped also with the finalization of the manuscript and with the proofreading, as did Alegra Kaczinski and also Utz Riehl. Further, Jack Grimes and Preyan Choudhuri from Texas A&M helped us with some technical aspects of the book, and Kira Schlender from Münster so marvelously crafted the website. Kai Pohlkamp created the title photo with the "data analytics" and "theory" chairs (and its variation inside the book), and Maris Hartmanis and his Studio Tense team have helped us with the design of several figures and also turned the chairs into the book's logo.

Finally, we have to express our gratitude to our families and close friends. Thorsten thanks his longtime pals Olli, Alex Deseniss, and Ronnie Zietz for numerous hours of joint "entertainment action research" which have shaped his entertainment taste in countless and ongoing debates over the decades. But most of all, he is grateful to his wife Claudia (of 20+ years) and the most entertaining of all boy groups, featuring Frederick (watching The Walking Dead's first season together certainly is a bonding experience…), Patrick (we'll always have EA's FIFA!), and Tom ("to infinity and beyond!") for the unlimited love and support he has received. Mark is thankful to his wife (of 30+ years) Nancy, his partners-in-crime at too many movies to count, Jon, Elise, and Wil, and his gamer-extraordinaire Shane (any proceeds that Mark gets from this book will have long been spent by Shane on the latest MARIO or SUPER SMASH BROS title). It was our families' inspiration and also their sheer infinite patience that made this book become a reality. We owe each and every one of you—big time.

Münster, Germany — Thorsten Hennig-Thurau
Fort Worth, USA — Mark B. Houston

References

Hadida, A. L. (2009). Motion picture performance: A review and research agenda. *International Journal of Management Review, 11*, 297–335.

Peltoniemi, M. (2015). Cultural industries: Product-market characteristics, management challenges and industry dynamics. *International Journal of Management Reviews, 17*, 41–68.

Contents

1

Forget the "Nobody-Knows-Anything" Mantra: It's Time for *Entertainment Science*!

We love—and often take for granted—the steady stream of movies, video games, books, and music that entertain us and enrichen our lives. These products vie for our time, attention, and money. Take a moment to stop and think about the managerial decisions that affect the success of an entertainment product, whether it is the newest installment of STAR WARS, a Nicholas Sparks novel, the current GRAND THEFT AUTO or SUPER MARIO game, or Adele's latest track.

Among many other choices, managers decide …

- whether to "greenlight" the product,
- whether to pay for the big star or franchise or go with unknown artists or characters,
- how the product is to be financed … distributed … sold,
- how many screens or how much shelf-space to devote, and
- how to communicate and engage with potential customers.

These decisions determine the fate of new entertainment products—and sometimes even the fate of the company that produces them and the careers of the managers who make them. So, anyone who is in charge of making entertainment products wants to use an approach that helps him or her to make the "right" decisions. As we lay out in this book, the traditional approach to deal with such decisions is to rely on "gut feeling." The entertainment industry has established this approach several decades ago and made it its mantra—more specifically, the "Nobody-Knows-Anything" mantra, which

T. Hennig-Thurau and M. B. Houston, *Entertainment Science*,
https://doi.org/10.1007/978-3-319-89292-4_1

draws on a phrase from Hollywood legend William Goldman. Anything suggestive of a scientific approach has often been ridiculed or dismissed as naïve by the proponents of "Nobody Knows Anything." In recent years, however, several scholars and progressive managers have argued that big data and complex analytics can function as an alternative to such gut-feeling-based decision making for entertainment products. Spurred by great promise, and encouraged by analogies from other industries, quantitative statistics are hailed by some as the "new way" forward in entertainment.

We argue that both approaches, the traditional "Nobody-Knows-Anything" mantra and this data-driven "new way," are troublesome, though in quite different ways. *Entertainment Science*, as we present it in this book, provides a way to manage entertainment products that is superior to both the "traditional" and the "new" way. The reason is that both approaches are tied to a specific "trap" that entertainment managers should avoid, with *Entertainment Science* helping to do exactly that. The first trap associated with the "Nobody-Knows-Anything" mantra stems from the argument that gut-feeling-based decision making is no longer an adequate way to address the challenges of today's entertainment industry (we're not sure that it *ever* was). In a competitive digital environment, with so much information being available, managers can no longer justify making important decisions solely based on their personal feelings—doing so would be short-sighted and lead them to become victims of what we refer to as the "Nobody-Knows-Anything" trap. The second trap reflects our credo that the use of big data carries some risks of its own. A manager can grab data and model them until some significant empirical relationship between a variable and his or her product's success emerges. But our own experiences have shown us that it is likely that such an analysis will yield impressive-looking-but-idiosyncratic results that are of short-term value, at best, or blatantly misleading and even counterproductive at worst. We call this the "False-Precision" trap of entertainment decision making.

This book introduces *Entertainment Science* as a response to both the "Nobody-Knows-Anything" trap and the "False-Precision" trap, as we illustrate in Fig. 1.1. *Entertainment Science* suggests an approach to managing entertainment products that acknowledges that data analytics have the potential for incredible value, but assigns a crucial role to another fundamental element of good science—theory. Theory is the foundation for useful science—as for good decisions of *any* kind, including those by entertainment managers. We discuss below what we mean by "theory" and argue that applying a theory perspective to entertainment, when done right, does not rule out creativity and gut feelings, but instead embraces them.

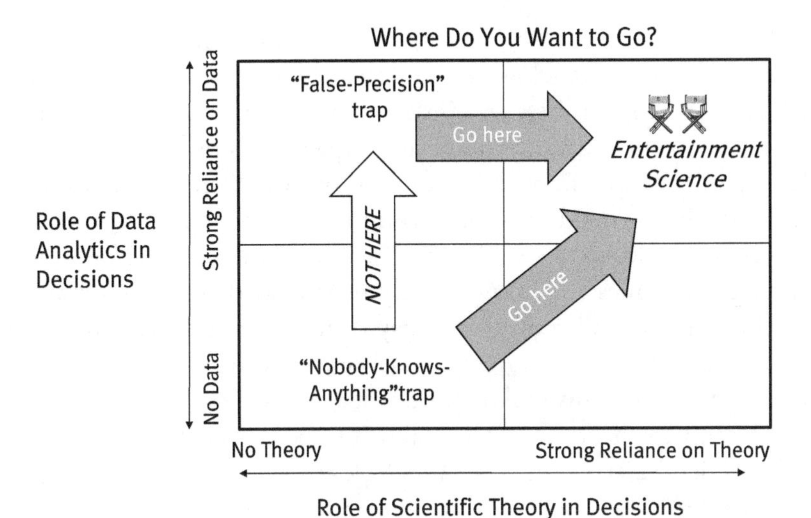

Fig. 1.1 Avoiding managerial traps with *Entertainment Science*
Note: Authors' own illustration.

So, *Entertainment Science* combines the use of data analytics with powerful theories. In the following paragraphs, we lay out this idea in more detail and describe why such a "scientific" approach is not in conflict with the requirements of day-to-day decision making in the entertainment industry. But before we do so, let us elaborate our criticism of the quasi-holy "Nobody-Knows-Anything" mantra more fully.

The "Nobody-Knows-Anything" Mantra in Entertainment

> "A high-ranking executive at a major studio … insisted that successfully choosing which movies to send into production was still primarily 'a gut game'."
> —Zafirau *(2009, p. 196)*

When Goldman (1983) wrote that "Not a single person in the entire motion picture field knows for a certainty what's going to work" (p. 39) some 35 years ago, it is hard to believe that he anticipated that his words would become the economic foundation of decision making for several generations of entertainment managers. Goldman is an acclaimed screenwriter and novelist who won numerous awards for his work, including Oscars for his wonderful screenplays for the Robert Redford–Paul Newman collaboration Butch Cassidy and the Sundance Kid and the Watergate movie

All the President's Men. His academic credentials include Bachelor and Master of Arts degrees. However, Goldman has never studied business or economics, and his business credentials consist of a single producer notion, for an unsuccessful and largely-forgotten sequel to his Butch Cassidy movie.

All of this has not hurt the credibility of Goldman's "Nobody-Knows" mantra—probably because it only summed up what industry leaders in entertainment have considered the essence of their work from the very beginning. But the mantra has strongly reverberated since Goldman's original writing and is still present in today's entertainment business. Consider the movie business, where Adam Fogelson, CEO of STX Entertainment (and previously of Universal Pictures), lets filmmakers "describe the film they .. intend to make, then trusts his gut about whether it sounds commercial" (Friend 2016). Universal CEO Donna Langley tells journalists that the "movie business is special because it is so irrational and risky" (quoted in Beier 2016). And DreamWorks co-founder and former Disney chairman Jeffrey Katzenberg states that the "crazy thing about the movie business is that there is absolutely no recipe for success" (*Mediabiz* 2016). Similar stories abound for music, games, and books.

When ethnographer Zafirau (2009) enrolled in classes taught by senior entertainment producers and executives and also conducted several interviews with entertainment leaders, he drew from his research that "the notion of 'instinct' continues to be an important one in explanations that many .. executives have for why they make the decisions that they do" (Zafirau 2009, p. 196). Even some of our academic colleagues have fallen in love with Goldman's claim, arguing that "it is impossible to identify hits in advance" (Peltoniemi 2015, p. 43) and that "'Nobody knows' is the core problem" of creative industries (De Vany 2006, p. 619).[1]

The Goldman mantra emphasizes the importance of managerial "instincts" and "gut feelings" for determining idiosyncratic characteristics that drive success—nobody *can* know anything, despite how hard he or she tries. When we talked to German producer Stefan Arndt of X Filme, he told us that "you have to have the right instinct and be up-to-date, there isn't much more to [being successful]" (Arndt 2009, p. 59). It is the instinct, that "golden touch," that constitutes *the* main driver of success (along with luck). Such perceptions are usually reinforced by an executive's personal life experiences,

[1]Art De Vany's reference to the Goldman idea is sometimes used as an academic fig leaf by industry managers who despise the use of analytical approaches. It, however, contrasts strongly with De Vany's role as one of the "founding fathers" of empirical entertainment research, who studied the creative industry's patterns and "rules" extensively with ambitious mathematical and statistical tools (De Vany [2004] offers a summary of his work). De Vany also used his insights as a consultant for movie producers via his firm, Extremal Film Partners (see, e.g., *Indiewire Team* 2011).

such as reactions from other executives and friends, as well as his or her own consumption of other entertainment products and news (Zafirau 2009).

Does this mean that the entertainment industry utterly despises logical thinking? Of course not. Mr. Arndt also stated that some producers such as Jerry Bruckheimer, in addition to being equipped with a high degree of intuition, have a "deep understanding into the economics of movies which many others lack" (Arndt 2009, p. 59). And after reflecting for a few moments, he even conceded that he *might* have a personal "formula" for success. However, he then quickly stressed that he is not consciously aware of this formula—and could never articulate it. It is an essential element of the Goldman mantra that such knowledge is highly tacit and thus cannot be shared with others, or even be formalized. Analytical or scientific methods stand in stark contrast to the Goldman mantra. As Zafirau (2009) notes, the executive who referred to industry success as a "gut game" in the opening quotation also dismissed the possibility that "scientific" research could be of substantial value.

The Goldman mantra is not a mannerism but has quite far-reaching implications. The biggest problem is that its underlying "uniqueness assumption" (Austin 1989, p. 2) systematically inhibits progress. If every product is a unique artifact and needs to be developed from scratch by gut feeling, there is no room for generalizable knowledge, and learning is just not possible—and thus deserves no resources to be spent. No insights gathered from the study of any single entertainment product and its audience can be generalized to the next one. Moreover, the mantra implies that failing is "natural" and "just happens," rather than being the result of erroneous managerial processes and decisions, so that any attempts to systematically improve decision-making processes seem almost illegitimate (Thompson et al. 2007, p. 630). Traditional marketing and business theories do not apply. It is through this mechanism that the Goldman mantra prevents industry progress. Why are all movies priced the same at the box office, regardless of whether they attract millions of consumers by featuring major stars and attractions or not? Blame "Nobody-Knows-Anything": because "in the absence of an ability to know much about the demand for any given movie …, the best price for any particular movie is, more or less, a total guess" (McKenzie 2008; see also our chapter on entertainment pricing).

The absence of learning and progress causes problems for consumers and audiences, because the entertainment products and the ways they are marketed, which are the consequence of these constrained managerial decisions, do not live up to the potentials that marketing offers. Stagnation dominates in what is greenlighted and how it is offered to consumers. If the mantra is wrong, and there *is* indeed something "out there" that can be learned, a problem also exists for entertainment firms themselves, because such hidden treasure can be unearthed by others *who know*, providing them with a competitive advantage (and putting the Goldman followers in a troublesome position).

Why *Entertainment Science* Should Be the New Mantra of the Entertainment Industry

We live in a world where a single entertainment product can cost a studio or game designer hundreds of millions of dollars to build, distribute, and promote. In this world, a single flop can threaten a firm's future viability. We argue that living by the mystique of the "guru" can threaten the competitiveness of any company, as it leads to inferior decisions compared to combining hard-won industry experience and creativity with practical theory and data analytics, the two major power sources at the core of *Entertainment Science*. Let us explain now what value data analytics has to offer (i.e., what it means to avoid the "Nobody-Knows-Anything" trap), and what managers can gain by combining data analytics with insights that practical theory can provide (i.e., what it means to avoid the "False-Precision" trap).

Avoiding the "Nobody-Knows-Anything" Trap with Data Analytics

> "Big Data helps us gauge potential audience size better than others."
> —Ted Sarandos, *as Chief Content Officer at Netflix (quoted in* Nocera *2016)*

The first power source, and an integral element of *Entertainment Science*, is data. Today, an abundance of data exists in the field of entertainment, along with people who, for the right price, are eager to analyze these data for entertainment producers. Today, several data providers, such as CinemaScore and Rentrak, conduct extensive consumer surveys that collect information about consumers' awareness and perceptions of new products, sometimes even months before their releases (Moon et al. 2015). But equally (if not more) valuable is the data that is generated in the normal course of business and in the conduct of consumers' lives. Often this data is "owned" by entertainment firms, particularly social media data that happens on their websites, as well as daily sales and pre-sales information. Many analytics firms have made entertainment their specialty and add to the troves of data that are now available. Most of these analytics firms are new start-ups, such as RelishMix (a social analytics firm that delivers marketing intelligence and data visualizations for entertainment and consumer brands), Moviepilot (an entertainment website that covers news and editorial for films, television, popular culture, and video gaming), and Next Big Sound (which "scours the Web for Spotify listens, Instagram mentions, and other traces of digital fandom to forecast breakouts," Thompson 2014). Others offer early success predictions based on a data-driven decomposition of product ideas (e.g., Worldwide Motion Picture Group—see Barnes [2013], and Epagogix—see Gladwell [2006], both of which analyze movie screenplays). Several established industry players also are compiling new data, such as Variety's "Vscore," a social media-driven measure of actors' starpower (*Variety* 2014).

When it comes to demonstrating the power of data and analytics in creative contexts, the success of "sabermetrics" in baseball is an obvious reference—the econometric analysis of in-game activity.[2] Though criticized by many baseball "purists," the concept gained wide publicity through the Brad Pitt–starring movie MONEYBALL, which chronicled the Oakland Athletics' rise from also-ran to division champion by assembling a carefully constructed hodgepodge of cast-off players and mid-level stars who were overlooked by other teams, but identified by Oakland's data analytics efforts. In Fig. 1.2, analytics expert Peter Brand, portrayed by Jonah Hill, introduces manager Billy Beane, played by Brad Pitt, to the logic of "sabermetrics."

Fig. 1.2 MONEYBALL, or applying data analytics to baseball
Notes: Freeze frame from the trailer for the movie MONEYBALL. Courtesy of Columbia Pictures.

[2] "Sabermetrics" was brought to prominence in the landmark book "Historical Baseball Abstract" by James (1985), in which the author uses historical data from the Society for American Baseball Research (SABR, hence "saber") and applied advanced statistical analyses to identify the 100 top players in history at each baseball position.

If it is helpful, think of *Entertainment Science* as "moneyflick," "moneytune," "moneygame," and "moneybook"—that is, the MONEYBALL equivalent for entertainment products. Critically, analytics, including *Entertainment Science*, is all about probability, not determination. Whereas the devotees to the Goldman mantra have been right that nobody will ever be *sure* that a new entertainment product will succeed in the marketplace, the insights into industry rules and processes shared in this book will increase every manager's *probability* of being successful with his or her next offering.

Let's take another look at the sports business: when Mercedes lost a Formula 1 race because of an erroneous algorithm, its team chief, Toto Wolff, stressed the long-term, multi-race perspective of his team, stating that "to win races *regularly*, it is better to rely on data than on gut feeling alone" (Sturm 2015). Consider this the essence of the probabilistic worldview that underlies analytics and *Entertainment Science*: whereas algorithmic insights cannot guarantee that a single product will become a hit (which would require deterministic knowledge), it raises the probability that products will be successful. This pays across a slate of products, where random deviations from the "average" cancel each other out.

Netflix and Amazon, like most other major challengers of traditional entertainment companies, have made data analytics a core element of their business models, although in unique ways. Consider how Neil Hunt, as Netflix's then-chief product officer, described how the firm makes use of these resources: "[in the case of HOUSE OF CARDS,] David Fincher came to us with a story and a star. We then calculated how many people this would appeal to and found: this series is worth a pile of money. … [And for the film BEASTS OF NO NATION,] we knew upfront that it would be about child soldiers in Africa, that Idris Elba plays the lead role, and that the atmosphere will be bleak. Based on this information, we built a model, using viewers' behavior in the past. How many people have watched similarly dramatic films, such as HOTEL RUANDA and SCHINDLER's LIST? Using such data we extrapolated the potential number of viewers. Our creative team then tells us how much this number is worth in monetary terms. … Our estimates are pretty accurate. … My data helps me to make a well-informed decision" (quoted in Brodnig 2015). Netflix also uses data for other purposes, such as for providing their customers with powerful recommendations (Lohr 2009) and to learn about popular trends (for the latter, the company has been reported to monitor illegal downloads on sites like BitTorrent; *tickld* 2016). Many in the entertainment industry believe that the decisions that Netflix and, to a certain degree, Amazon have been making are, on average, better than those of most established players in the industry (e.g., Bart 2017).

At least some of the established players have also gotten the message and have begun investing in analytical resources. Examples include Vivendi SA's Universal Music Group, which has been reported to judge their employees partly on their interest in data and analytics and has created an "Artist Portal" which "allows users to track and compare artists' sales, streaming, social-media buzz and airplay globally in real time, while offering insight into the driving factors for spikes and dips in each metric" (Karp 2014). Disney experiments with audience recognition technologies to forecast their films' success better. And Sony has been said to work on an analytical platform that is intended to shed light on the drivers of costs and revenues of their movies and TV offerings, enabling the firm to make decisions based on a proactive, data-driven model. The alleged title for this initiative is properly chosen: the "Moneyball initiative."

Avoiding the "False-Precision" Trap with Theory

The second power source, and an equally essential element of *Entertainment Science*, is theory. Whereas data analytics serves as antidote to the Goldman mantra and can help entertainment managers avoid the "Nobody-Knows-Anything" trap, such analytics can be dangerous when applied in a "theory-free" way, leading managers into the "False-Precision" trap. Again consider the MONEYBALL case. Many baseball teams failed to find success relying solely on analytics. For example, the Boston Red Sox invested heavily in the sabermetrics approach introduced so successfully by the Oakland Athletics but performed poorly in 2014 and 2015. The formulas used were sophisticated but failed to accurately capture or predict reality. Why? Because analytics requires not only big data and smart algorithms, but also a profound understanding of market mechanisms and the actors in the market, such as the consumers of entertainment products and a firm's competitors (Walker 2016).

In the entertainment context, we have also come across several cases that fall into this "False-Precision" trap. The journalists of *The Economist* (2016) were so fascinated by the mere possibilities of applying big data to the movie business that they ran their own analyses, which led them to conclude that "the strongest predictor of absolute box-office receipts is a film's budget" and that "a movie would generate an average of 80 cents at American and Canadian cinemas for every dollar a studio promises to spend on it." If you think this sounds somewhat simplistic, if not downright obscure, you are right—it is because the journalists overlooked or ignored the complexities behind their data. These results would basically suggest that producers increase the budget of films indefinitely (there are many other revenue sources for film producers other than the North American box office). This would be a very dangerous thing to do—because these estimates were not based on sound theory. We will later discuss the reasons empirical budget parameters should not be treated as being of a causal nature.

Part of the "False-Precision" trap is that theory-free data-driven management can be easily (mis-)used in opportunistic ways. When the industry treats visits of a music band's Wikipedia page as a predictor of the band's future success purely based on empirical correlations reported by analytics firm Next Big Sound (Thompson 2014), without reflecting on the mechanisms and logic behind those correlations, this creates an incentive for musicians to systematically manipulate that metric: send all your friends to Wikipedia! The realities of entertainment products require complex theories to serve as the underpinning of ambitious empirical analyses, as this book will prove. Sound theories can show *why* a metric (e.g., Wikipedia visits) is empirically linked to success, and by doing so they enable a management approach that is more robust against misinterpretation and manipulation. Let us say that several of the approaches by commercial data analysts we came across during the writing of this book did not impress us for similar reasons.

But what does "theory" mean in this context? When scientists speak of theory, they don't mean an opinion or loose idea (as in "that's Mark's theory—everyone is entitled to their opinion"). Instead, to cite a distinguished biology colleague, "a theory is a system of explanations that ties together a whole bunch of facts. It not only explains those facts, but predicts what you ought to find from other observations and experiments" (Kenneth R. Miller, quoted in Zimmer 2016). Theory, in the scholarly sense, implies a careful definition of the respective question, a precise use of terminology, and explanations of causes and effects in very clear language (often in the form of formal hypotheses), as well as rigid tests of which claims can be supported with data and which cannot.

By applying a scientific approach to entertainment, scholars have compiled extensive evidence in the past 35 years that rules and patterns that determine the economic success of entertainment products *do* indeed exist. This work comes from an international cadre of scholars who work across many academic fields, including, but not limited to, marketing, finance, economics, management, information systems, and media science, and who represent some of the world's highest-ranked academic institutions. Their insights have often been published in science's finest and most rigorous journals, after having been scrutinized thoroughly by other leading experts in the field.

Theory needs testing, and as we report in this book, often scholars have tested their arguments empirically, ruling out alternative explanations (was the *zeitgeist* the true reason behind a movie's success, not its star or director?). Sometimes they have used experiments or questionnaires, but mostly the tests are run against large databases of real-word information on entertainment product success and consumption. Meaningful theory needs to mirror the non-linear and multifaceted nature of the real world, and so it is not surprising that most of this entertainment research focuses on so-called "contingencies," rather than on simple direct (often termed "main") effects. Contingencies describe the *conditions* under which certain decisions are successful. For example, whether stars matter is a much less interesting question than why stars are *sometimes* the most important thing to have—but irrelevant, if not counterproductive, in other situations. Why do some sequels and remakes work better than others? Under what circumstances does 3D offer a commercial advantage for a movie? We share the answers to these and many other contingency-related questions in this book.

If the goal is to make a good decision, then theory-based decisions supported by empirical findings are the way to go, as powerful theory has been the foundation for good decisions in all parts of the economy (as well as in most, if not all, other parts of life). We believe it is about time for the entertainment industry to benefit from it as well. Some of the industry's members are already doing so, with Netflix certainly being among those that not only use data analytics, but also have close ties to theory and science (e.g., Amatriain and Basilico 2012a, b; Gomez-Uribe and Hunt 2015; Liu et al. 2018). Much less is known about whether, and how, Amazon's entertainment decisions are shaped by theory. Some reported controversy suggests that the firm is putting less emphasis on this aspect, more resembling traditional entertainment conglomerates in terms of people and focus (Littleton and Holloway 2017)—something that might compromise the powers of their data analytical skills, preventing the firm from exploiting the full potential of *Entertainment Science*.

And what about the traditional producers, the studios and labels, when it comes to making use of theory? The outstanding track record of Pixar, certainly not a "traditional" institution, comes hand in hand with its leaders stressing the need to "stay close to innovations happening in the academic community" (Catmull 2008, p. 71), but their interest is tied more closely to technological than economic scholarly advances. The above-mentioned efforts of Pixar's parent company Disney to "read" movie audiences' emotional reactions also imply some ties with information technology science. But our very own experiences when crafting this book paint a picture that is far from euphoric—introducing ourselves as "entertainment scholars" raised numerous eyebrows and was often met by blank stares. We hoped to include more movie and game posters, song lyrics, and quotes from novels, but those in charge of property rights either did not respond to our requests at all, despite numerous attempts, or asked for fees that resembled those a firm would pay to create licensed toys or other commercial products. These (non-)responses signal a lack of understanding and interest in scholarly work by some of the industry's most traditional firms, a reaction that we have not experienced in our research in other industries. But in a way it all makes sense: in a universe dominated by "Nobody-Knows-Anything" thinking, theory and its creators shall have a not-so-easy time. Dealing with entertainment studios and labels became the ultimate confirmation that we had picked the right topic.

The vast repertoire of scholarly studies that have been conducted in the field of entertainment provides us with a generalizable understanding of the entertainment industry. In other words, the wide-ranging scholarly studies contribute to a larger theory on its own that we present with this book—a highly fragmented and incomplete, but holistic theory of *Entertainment Science*, or at least a first draft of it. This theory explains how and why some entertainment products are hugely successful, whereas others fail commercially. This *Entertainment Science* theory tells managers to overcome the "Nobody-Knows-Anything" trap but also helps them avoid the "False-Precision" trap that misguides managers through big (but dumb) data and impressive looking (but irrelevant) analyses. We are convinced that this holistic theory of entertainment can benefit every firm that deals with such content, and even several outside of the entertainment industry.

By combining smart statistics with smart thinking, *Entertainment Science* is also the antithesis to Anderson's (2008) proclamation of the "end of theory," a title story in Wired magazine. Anderson stated that "[w]ith enough data, the numbers speak for themselves." Though potentially true for limited, specific questions, we could not oppose this view more strongly for many of the important decisions the entertainment manager must make. A key assumption of our work is that theory has never been as useful as in the age of abundant big data. In other words, it is the combination with theory that turns big data into smart information.

It is important to stress that what we call *Entertainment Science* theory is not in conflict with intuition. Instead, a key finding of scholarly research in entertainment is that great entertainment always depends on creativity as one of its key characteristics. So, *Entertainment Science* acknowledges this critical role of creative processes rather than denying it—borrowing from Netflix's Ted Sarandos, *Entertainment Science* is "definitely art and science mixing" (quoted in Vance 2013). This perspective also clarifies the role of data analytics—as a complement and multiplier of creativity and intuition, not as their substitute.

As we explain in this book, creativity is essential for offering value to consumers of entertainment. As Walker (2016) states in his insightful analysis of why the Boston Red Sox were not successful by applying "sabermetrics," "if the company pursues the wrong creative, then all bets are off." And humans still outperform algorithms when it comes to creating something new and unexpected, unless artificial intelligence *really* takes off. Can computers write a great screenplay? We show that whereas analytics has a lot to tell about the general structure of great narratives, dialogue and *mise-en-scène* require original decisions artificial intelligence simply cannot provide. If you are not convinced, just watch the experimental short film SUNSPRING, which was written entirely by a recurrent neural network that was trained with the screenplays of several science fiction classics (Newitz 2016), and you will most probably agree. The inability of algorithms to generate creative content was further demonstrated when a group of computer scientists used artificial intelligence for crafting a theater play in London—the result lacked sensation, context, and a longer-term structure (Jordanous 2016). Google aims to unlock the creativity potential of algorithms with their Magenta project (Eck 2016), but we classify such efforts as science fiction, at least in the period in which we wrote this first edition of our book.

Thus, *Entertainment Science* is intended to serve as a decision-support approach, not a standalone one. This decision-support perspective is also in line with today's most successful uses of analytics in the entertainment industry. It is a widespread misunderstanding that Netflix tells creatives how to write screenplays based on their data-crunching efforts. In reality, their actions are informed by powerful data analyses in combination with theoretical thinking, very much in line with *Entertainment Science*—but data analytics does not substitute their *creative* decisions. As CEO Reed Hastings phrases it: "We don't tell creatives, 'Add a dog. Series with dogs do better…' Creatives would not want to work under such conditions. Results would be devastating. It would be Frankenstein" (quoted in Reinartz 2016).

Scholars Seifert and Hadida (2013) even provide initial empirical evidence of the power of such a decision-support function of *Entertainment Science*. They asked a sample of music managers to predict the future chart performances of 40 singles between one and two weeks before their releases (in

2007); they also used a parsimonious, but theory-inspired linear regression model to predict each single's performance.[3] The algorithm was as good as or better than the managers in predicting the songs' success when artists had an established track record, but less accurate for new artists. However, the essence of Seifert & Hadida's study was a different one: across a number of conditions (managers from major and independent labels, songs of established artists and newcomers), the *combination* of human judgment and econometric estimation clearly outperformed either the algorithm or the managers' personal judgment alone, with an improvement in accuracy of up to 25%.

In sum, *Entertainment Science* enables us (and, more important, you, the reader) to take a radically new, systematic look at the entertainment industry and to shed light on the mechanisms underlying the economic success of entertainment products. *Entertainment Science* offers theoretical arguments that will help you to better understand the subtle, but critical nuances of entertainment product decisions, and often provides empirical evidence for these arguments generated through analytical techniques and models, often using "big data." Be aware though that *Entertainment Science* is far from being a fully harmonic and homogenous approach—instead, given the variety of backgrounds of its scholarly contributors, its value stems from it being a highly diverse concept, encompassing insights generated through many different ways of thinking, techniques, and models.

For any readers who are afraid that "science" sounds too time-consuming—we have done much of the work for you. Because managers make decisions in a time-pressed world, one in which there is simply not enough room to experimentally pretest every alternative, we present the key insights from the main models and empirical studies for the entertainment industry in a way that should be digestible without requiring readers to dive overly deep into the statistical methods and hard core psychological theories which have been used by scientists.

However, for readers interested in taking a deeper look, the book contains a short section on the methods used by *Entertainment Science* scholars for deriving their insights (see "Before We Get Started: Some Words on the Empirical Methods Employed by Entertainment Scientists"). We will also name the methods and approaches that have led scholars to arrive at their insights, when appropriate, to enable a better framing of the reported findings and potential limitations.

---

[3]Specifically, the econometric model was fed with information on advertising spending, the artist's previous performance (a kind of "star power"), and the song itself.

Understanding (and Overcoming) the Persistent Forces of the "Nobody-Knows-Anything" Mantra

Even if you agree with us that *Entertainment Science* is the way to go, there are several persistent forces that managers and their firms need to overcome on their transition to the new, data-and-theory-based approach. First, there is often a fundamental misunderstanding of the nature of the concept of uncertainty as it pertains to business decisions and outcomes. This is because probabilistic thinking differs fundamentally from deterministic thinking.

At the core of probabilistic thinking is the recognition that one unexpected hit or miss does not confirm or reject a theory, in the same way that one heavy smoker's 100th birthday party does not prove the preponderance of cancer research wrong. Consider the exemplary statement "The performance of any one movie is unpredictable" by a journalist, which he or she invokes as clear proof of the impossibility of infusing managerial decisions with data and algorithms (Wallace 2016). Here, the term "unpredictable" implies a binary understanding of predictions: it is either completely "right" or fully "wrong." However, in reality, as well as in probabilistic statistics, predictions are always *more* or *less* right and *more* or *less* wrong.

What gives a lot of people headaches (apparently including at least some decision makers in entertainment) is that something that is almost intuitive when applying it to *many* cases (the percentage of people with lung cancer among those who smoke is higher than among those who do not) needs to be applied to a *single* case. Even if the probability is very low that a film for which everything is done wrong will become a huge hit, it *can* happen (and has). What a probabilistic view recommends is that a manager should not *bet* that this is going to happen.

Managing the probability of "rightness," or working to reduce the magnitude of prediction error, is what characterizes a probabilistic approach. Deterministic thinking does not go well with *Entertainment Science*, so adopting a probabilistic perspective is essential for recognizing the value it can offer. William Goldman and his followers apply a deterministic worldview, as expressed with statements such as that nobody knows with "certainty" what will work. Do we need to have such certainty to make good decisions? No. Any smart manager knows that *nothing* is certain in life. As we demonstrate in this book, *Entertainment Science* can help reduce the uncertainty of entertainment-related decisions (or at least determine their certainty), which makes it a powerful source of substantive competitive advantage.

Second, there is somewhat of an agency problem for powerful decision makers in entertainment. What is best for a firm may not align perfectly with what is best for the individual manager, so some subconscious self-protection can occur. In the "Nobody-Knows-Anything" environment, the professional reputation of an entertainment manager is often based on his or her idiosyncratic "tacit" knowledge that cannot be articulated (think of Mr. Arndt's statement we quoted earlier in this book). With "Nobody Knows Anything," it is only the manager's singular ability to spot a winner or a dud that makes the difference. Making this knowledge available and transparent as part of a transformation toward scientific models and algorithms may make the manager less indispensable—which might not be in the manager's best *personal* interest. Thus, companies need to think of how such concerns can be overcome, such as through incentive systems that reduce the gap between what is best for the individual manager and what is best for the firm.

Third, there is what could be termed "devaluation *angst*." Though working in what is an economic industry at its core, many entertainment product managers were (and still are) attracted to movies, books, music, or games by their cultural underpinnings. Being part of a cultural environment is often a focal element of an industry participant's self-concept (e.g., "Music is part of who I am"). Stressing the economic elements of entertainment production over entertainment's artistic character seems crass, taboo, or unenlightened and can threaten the identities and motivation of these deeply committed aficionados-turned-managers. As MINORITY REPORT producer Gary Goldman once told us during a panel discussion (see also Dehn 2007): "With your statistical approach, you guys are destroying the magic." As we told him then, the idea of *Entertainment Science* is a not in conflict with the creation of art per se: it is only in conflict of creating art that loses money. Learning about the commercial potential of new ideas and projects from data and theory, in addition to feelings, might even help get radical ideas for which conventional managerial thinking (conservatism is a core characteristic of "Nobody Knows Anything") has no room off the ground. It's the quality of the data, the analyses, and the theories that makes the difference.

Finally, probably the deepest, most rigid persistent force impeding the adaptation of *Entertainment Science* is the industry's long history of self-mythology (see Austin 1989, p. 5). Trade stories and sagas claim that success in the industry rests on people's ability to manage relationships with trade partners or colleagues rather than on educational achievements or intellectual skills (Caldwell 2008). The one big myth of entertainment is that what is mainly required are characteristics such as being hardy, tough traveling, confident, and aggressive. Only those who are equipped with these characteristics can work (and succeed) in the industry, only those who work there are equipped with them, and only they possess the authority to judge "how the industry works, what it means, and what 'really' goes on 'behind the scenes'" (Caldwell 2008, p. 10). In other words, only those who are part of it can understand it.

The result of this myth is a strong separation between the industry and everything outside it (certainly including science), which legitimates ignorance ("If someone within the industry doesn't know or understand it, how can it be good or even important?") and can border on hubris or narcissism ("I am part of the industry, so I *must* be great"). The myth that relationships are what it takes is also the foundation for entertainment's reputation as a "party industry." When movie producer Mario Kassar looked back on his career, festivities deserved a prominent place: "And, of course, there were the parties. … Everybody was dancing and drinking and eating like crazy … It was unbelievable" (quoted in Jaafar 2016).

If there is nothing to be learned but everything to be gained by managing relationships, spending time and money partying big time is a better investment than studying and running econometric models. Reed Hastings and other proponents of an analytical approach do not adhere to this myth, so their successes are also questioning its credibility—and probably the best chance to retire it at some point.

Adopting *Entertainment Science* requires firms and their managers to fight and overcome all these forces. This is definitely not an easy and short-term task, as broad-scale change implies nothing less than a redefinition of the industry's core values and identity. But because there is no reason to believe that any of the four arguments above is what drives business performance (at least in a positive way), we argue that adopting the *Entertainment Science* approach will be worth the effort. If done right, *Entertainment Science*, in a world full of rich data and powerful theory, is the right alternative to the Goldman adage, a mantra whose time has passed. So, let's begin to unlock the potential of *Entertainment Science*!

How This Book is Organized: *Entertainment Science* as a Cross-Product Approach to Knowledge Generation

Almost any other source of information or guidance for entertainment managers focuses on a specific kind of entertainment product—be it film *or* games *or* novels *or* music. This book takes a different approach by looking at all four forms of entertainment, taking what we call a "cross-product perspective" on the entertainment industry. Why do we do this? Although we clearly recognize (and highlight throughout the book) that some unique drivers of success exist for each type of entertainment product, the research we have assembled here provides strong evidence that these products have more in common than not in terms of the fundamental principles that underlie success.

Furthermore, and equally importantly, there are synergies and opportunities for learning that come from examining all four end-product businesses together. Extensive experience in a specific part of the entertainment industry, such as music, offers insights into the subtle nuances of that field. However, operating solely within one part, with its standards and "rules of thumb," can also constrain creative and analytical thinking. We argue that there is much to be gained by backing away and looking at problems from a more generalized perspective, to truly understand a problem's essence. For example, when acknowledging piracy as a cross-industry challenge, managers of movies and books who adopt this perspective can learn much from the disastrous outcomes of the music industry's initial anti-piracy efforts, but also from how game producers have navigated these troubled waters much more successfully. By putting things into a more general perspective, we offer entertainment managers insights that can help them address fundamental issues in new and inspired ways. Similar learning opportunities exist in many areas of *Entertainment Science*, ranging from how brands are (and should be, according to scholarly investigations) managed by those who run movie franchises to the distribution of a product across various channels, as practiced by film managers (and as studied by movie scholars).

For this reason, we have organized this book not around different entertainment products, but around key issues, key managerial decisions, and key success drivers. By focusing on common underlying principles that capture the true nature of entertainment industry challenges, the majority of our insights apply to all the different entertainment products covered in this book. At the same time, we acknowledge industry specifics when called for. So, do not misread us: our approach is not intended to substitute our readers' hard-earned expertise and wisdom from their specific field, but rather to help them broaden that expertise by learning from releases of other forms of entertainment.

Specifically, we have structured our following exploration of *Entertainment Science* in two major parts. Part I provides an overview of essential concepts, theories, and practices that are intended to contribute to a rich understanding of the fundamentals of entertainment and key entertainment markets. We have come across a bipolar view of how similar entertainment products and other offerings are: whereas several industry outsiders argue that entertainment products and markets are "just like any other," suggesting that what works with fast-moving consumer goods and cars will also be effective for movies, games, books, and music, those inside the industry often take a diametric position—managing entertainment is so unique that *nothing* could be learned from other industries.

We show that, like so often, the truth lies somewhere in the middle. On the one hand, our book provides evidence that substantial similarities exist between entertainment and other products and industries, with these similarities constituting the basis for the transfer of outside knowledge to the business of entertainment. The way Disney, Warner, and other entertainment conglomerates have adapted the concept of branding, in a way that has created some of the globe's most valuable and well-known trademarks (think AVENGERS! Think STAR WARS!) provides a pretty good example for such learning potential. Several *Entertainment Science* scholars have often drawn on concepts and approaches developed for other products when tackling entertainment-related questions. Examples can be found throughout this book; the chapters on branding and the management of innovation (and diffusion modeling in particular) contain some of the most striking examples.

On the other hand, we highlight several unique characteristics that, as scholars have shown, differentiate entertainment from other parts of the economy—these characteristics refer to the products and the way they are produced and consumed (e.g., entertainment products are "cultural" products), as well as to the markets in which they are exchanged (which, for example, require specific resources). Any manager who is in charge of entertainment products should be aware of these particularities, as they affect the performance of management approaches, as we point out, based on scholarly evidence. For example, the cultural aspect of entertainment products assigns importance to the role of taste in consumers' decision making, but it also, in combination with entertainment's hedonic nature, offers the potential for the enormous pre-release buzz on which the industry has built its "blockbuster strategy." *Entertainment Science* scholars have developed a deep understanding of these specifics and the implications they have for running a business.

In addition to discussing the specifics of entertainment products and markets (and building on this discussion), we investigate the essential business models of the industry in the first part of the book. Specifically, we show the general ways revenues can be generated with entertainment and also how the risk of doing so can be addressed. We frame our analysis of business models by an in-depth analysis of how value is generated in the entertainment industry, dissecting and linking the various roles that are involved in turning content into economic value. Knowing these roles and their interplay is essential for understanding how a company can enhance its value by extending (or reducing) its role in the market, for example, by strategic integration or disintegration decisions. Our value-creation analysis of the entertainment industry helps assess recent business decisions by Disney, Amazon, Netflix, and others, as well as determine what might be the best way forward for any current (or future) player in entertainment. We also provide an overview of the main entertainment producers' assets and activities. Here and in the rest of the book, our focus is on entertainment content and its producers, but our value-creation analysis clarifies the linkages between such content and other value-creation activities.

The final section of Part I then deals with those who ultimately determine the economic success of all entertainment activities—the consumers. Under the assumption that every good management decision requires a thorough understanding of the target group, we offer rich insights into how, and why, consumers actually behave with regard to entertainment products. We dive into the motivations of entertainment consumption and also investigate the cognitive and emotional processes that are associated with such consumption experiences, including focal concepts that scholars refer to as "narrative transportation," "immersion," and the state of "flow."

In Part II, we build on these general insights and present powerful theory-inspired frameworks and analytical approaches regarding the managerial decisions that affect the success potential of a new entertainment product. As marketing scholars, it seemed almost intuitive for us to build on the classic "Four P" typology of "product," "promotion," "place," and "price" activities. This typology has, in the 50+ years since its introduction by McCarthy (1960), established itself as a highly instructive framework for how business decisions affect the market success of products. We argue and show that the four "Ps" are as instrumental for entertainment products as for any other product. Entertainment must be created and offered (the "product"), communicated to the customers (the "promotion"), distributed (the "place"), and priced (the "price"), even if the price for the product is (or *appears* to be) zero for consumers in entertainment markets that have more than one side (e.g., mobile games such as CLASH OF CLANS).

The "Four P" framework has rarely been employed in the entertainment context,[4] and we believe that its use is a strength of this book. By adopting it, we are able to identify several industry "rules" that have captured the attention of scholars, but been ignored by managers (who have often operated under the "Nobody-Knows" mantra). Rest assured though that our book is certainly not one about "Marketing 101." Instead, it applies the "Four Ps" framework to entertainment and focuses on the specific

[4]Michel Clement's paper (2004) marks a notable exception—in German though.

decisions that are required for entertainment products and that stem from their unique characteristics.

Because the product is at the heart of all entertainment activities, we have dedicated four chapters to product-related decisions. In these chapters, we begin by discussing the product experience itself as a key success driver. But in a context in which many purchase decisions are made by consumers without knowing a product's "true" quality, we also devote substantial room to the factors that *signal* such quality to consumers, such as a film's genre or country of origin. We also discuss how different brands (a term that overlaps with what the industry often calls "IP," or intellectual property), such as sequels and stars, influence consumers' entertainment decisions. Our analysis of entertainment brands includes approaches that help assess a brand's economic value, and what is required for effectively managing "brandscapes," as Disney has been doing so well with their "Marvel cinematic universe." The fourth and final product-related chapter is then dedicated to the theories and analytical approaches that can help firms effectively design the innovation process for entertainment products.

The next chapters are dedicated to the other Ps: promotion (alias communication), place (alias distribution), and price. We split our discussion of entertainment communication across two chapters. In the first chapter, we discuss how entertainment should be communicated through "paid" (e.g., advertising) and "owned" (e.g., Facebook pages) media, considering questions such as what prospective customers should know about a new product (and what they should better *not* know!), and how much should be spent on each media and when. The experience nature of entertainment products, along with their cultural role, also assigns importance to "earned" media that comes from customers (through word-of-mouth and "herding" cascades), algorithms (through automated recommendations), experts (through professional reviews), and various kinds of awards. In the second communication chapter, we summarize what research reveals about how and under which conditions these different kinds of "earned" media affect the success of entertainment products and how they might be managed.

In our distribution chapter, we focus on timing decisions, the complex coordination of the multiple channels that entertainment can use these digital days, and the competition from illegal sources that has shaped the industry so strongly in recent decades. Pricing has received less attention by entertainment firms than the other marketing instruments, being an obvious victim of the "Nobody-Knows-Anything" mantra. However, scholars have highlighted several opportunities for the pricing of entertainment products, including empirical field tests of differential pricing between products; we summarize these findings in our chapter on pricing decisions. Thus, entertainment thinkers and producers who are open to adopting *Entertainment Science* might get particularly interesting and innovative insights from this part of the book.

At the end of the book, we integrate all the scientific evidence we have laid out in front of you, the reader, over several hundred pages. We do this by presenting a scholar's eye view of the two key integrative marketing strategies that have evolved into the dominating ones in entertainment in the last decades—the blockbuster concept and the niche concept. Informed by the knowledge that theoretical and analytical studies have produced, we distill which concept works best under which conditions. We also issue some warnings: as things tend to become imbalanced, that imbalance threatens some forms of entertainment, and maybe even the traditional producers of films and series, games, books, and music, as a whole. We then leave it up to you, our reader, to apply what we have put together—and unlock the power of *Entertainment Science* to enhance your own decision making in practical ways. Figure 1.3 summarizes this structure—the parts and the chapters and how they all work together to create an "*Entertainment Science* perspective" on the successful management of movies (and series), games, books, and music.

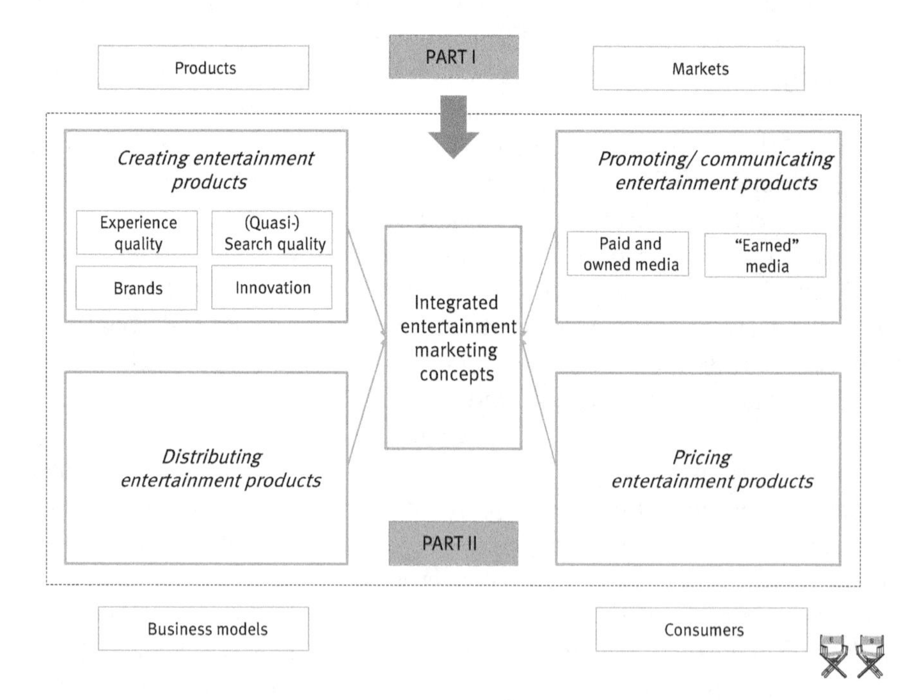

Fig. 1.3 The structure of *Entertainment Science*
Note: Authors' own illustration.

Before We (Really) Get Started: Some Words on the Empirical Methods Employed by *Entertainment Science* Scholars

Theories are vital because they guide *Entertainment Science* scholars in asking the right questions and provide explanations to patterns found in data. Equally essential to *Entertainment Science* are the statistical methods that help researchers identify those patterns and to extract knowledge from an otherwise random-looking set of numbers. Both theories and methods are often challenging to the non-scientist because of their mere complexity. But our understanding of empirical research methods (and their results) is impeded by one additional characteristic—they are expressed in a formal "language" of numbers, notations, and signs. This language is very different from our usual style of communication and is also based on a set of statistical assumptions that are not readily familiar to anyone who is a few years removed from his or her last statistics course (or might have decided to instantly forget them…).

Our purpose with this book is not to turn our readers into statistical-methods gurus, with expertise across the entire range of methods that are used in *Entertainment Science*. Instead, we want to spread valuable insights from *Entertainment Science* among scholars, but also share it with those who are either future or current decision makers in the entertainment industry. To make this possible, we have worked hard to translate the research into a language that can be decoded without the need for a master's or doctoral degree in economics, business, or math. At the same time, because the logic of *Entertainment Science* is a probabilistic one, following the general, underlying logic of its key methods and owning a certain "scholarly" vocabulary is definitely helpful. Such skills will also help readers to interpret research results and interact with the data scientists who crunch the numbers. Thus, we use this section to offer a "crash course" in research methods to facilitate the benefits our readers derive from *Entertainment Science*.

Regression Analysis as the Econometric "Mother" of *Entertainment Science* Methods

If you are someone who feels overwhelmed by statistics, here is some good news. Although scholars use a wide variety of research methods, the clear majority of these methods share a common statistical foundation: regression analysis. The main intention of regression analysis is straightforward: to identify the relationship between two or more phenomena (or "variables") based on how the phenomena have behaved in the past, either over time or across a number of observations (e.g., products, consumers). For example, a sociologist might gather historical data on individuals' education levels and income and use regression analysis to determine how these two variables relate to each other. Essentially, the question is how changes in one variable relate to changes in the other. Of note, regression analysis assumes such effects to be causal, but it does not formally *prove* causality—at least not in regression's basic form.

To bring this discussion back to our focus on entertainment products, Fig. 1.4 shows the North American box-office performance and the amount of advertising spent before the release date for five sample movies: Under the Same Moon, Wild (the 2014 version), K-Pax, Atlas Shrugged: Part I, and The Devil Wears Prada.

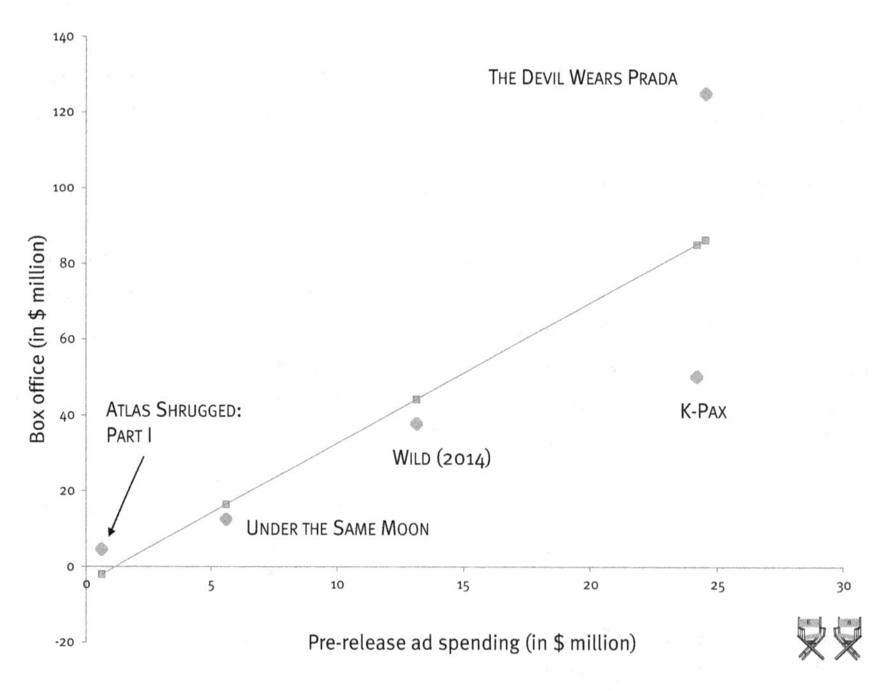

Fig. 1.4 A simple ordinary least squares regression analysis of movie box office on movie advertising

Notes: Authors' own illustration. Orange markers are the actual values for the movies in terms of pre-release advertising spending and box-office revenues, and blue markers are the values predicted by the regression function. The function's R^2 is 0.69. Advertising data was provided by Kantar Media and box-office data by The Numbers.

In the figure, we plot the five movies according to their actual values in terms of ad spending and the revenues they earned at the box office; the ad spending, in millions of dollars, is the horizontal axis of the plot (the *x*-axis), and the box office, also in $ million, is the vertical axis (or *y*-axis). From the positions of the five orange dots in this two-dimensional space, you can quickly see the levels of the ad spending and box office for each film. For example, the producers of K-PAX spent a little more than $24 million on advertising to support the film's release in North American theaters in October 2001, and their movie generated approximately $50 million at the North American box office.

Why have we chosen to use the *x*-axis for advertising, instead of the *y*-axis? Because it is a convention in statistics to use the *x*-axis for the variable that is thought to exert an impact on the other variable. This variable is called the "independent" variable, or IV (because it is assumed not to be influenced by other variables; it is "independent"). The *y*-axis is for the variable that is believed to be affected by the other (it is called the "dependent" variable, or DV—because its values depend on those of the IVs). In our case, we assume that advertising spending (x) influences the box-office results (y), not vice versa. But does it have to be that way? If producers adjust their advertising levels based on a movie's early performance, this would violate our causality assumption, and the results might not be trusted. But we get back to this matter shortly.

Regression analysis considers the IV and DV for the entire set of members (or cases) of the research sample (in our example, ad spending and box office for each of the five movies) and mathematically solves for the relationship between the two variables. Depending on this relationship you can then plot a predicted value of the DV for each level of the IV. If you connect these predicted values, you end up with a line that cuts through the scatterplot of the observed variables in a two-dimensional space, like the one in Fig. 1.4 which shows how the IV is linked to the DV. The most basic form of regression assumes a linear relationship between the two and places the line, or function, in a way that meets some optimization criterion. In the case of ordinary least squares (or OLS) regression, which is the standard type and the one we use for our example here, the function is placed so that it minimizes the sum of the squared differences between the actual values of the dependent variable and the values the function has estimated; these differences are called "residuals."

In Fig. 1.4, this criterion is met best by the blue line, and the five blue markers are the box-office values predicted by the regression function for the five movies, based on the films' ad spending (and *only* on ad spending). Knowing the course of the line (i.e., the regression function) provides us with several useful features. First, it helps us to know if the IV (advertising) is indeed associated with the DV (box office) and how strong that association is. The strength of an IV's "impact" is quantified in the form of a coefficient or "parameter" that is calculated during the regression process. Note that we use the term "impact" with care—because we are not completely sure about whether the relationship is indeed a *causal* one. The shape of the function is crucial; in the example above, the function's slope tells us that spending an additional \$1 million on advertising would, on average and all else held constant, generate \$3.7 million in additional box-office revenues. Keep in mind that these results are based on a small sample of only five movies that we picked solely for didactical reasons—we discuss advertising effects in greater detail in our chapter on paid entertainment communication.

Next, the parameters of the regression function allow us to predict the effect of any level of the IV on the DV simply by inserting that value in the function. The function takes the form of a constant "intercept" value (i.e., the value of the DV if the IV = 0; the intercept is usually called α, or alpha) and a parameter (usually named β, or beta) that is the coefficient of the IV and expresses how much the DV changes for each unit of change in the IV. In our example, the estimated function is as follows:

$$BoxOffice_m = -4.26 + 3.70 \times Advertising_m$$

This means that a hypothetical movie m that has an advertising budget of \$10 million could be expected to generate (−4.26 + 3.70 × \$10 million =) \$32.74 million in North American movie theaters, based on the regression results.

Finally, we can also determine whether the results can be generalized beyond the data set from which we have derived them. Such information is provided by the statistical significance of a parameter, which is a measure of the error the researcher must accept when determining whether regression results are systematic, instead of being the result of arbitrary forces. A convention widely shared among scholars is that a 5% error separates findings that are "significant" from those that are not. Our five-movie sample is not sufficiently large to produce significant results—a larger sample of movies would be required. Note, however, that in very large samples, it is often *too* easy to find significant parameters because significance is affected by the sample size itself.

How meaningful are the results of a regression? Meaningfulness depends on the distance between the actual values and those predicted by the analysis: the smaller the distance, the more accurate the model and the more meaningful the results. In our example, the analysis is able to predict the values of three movies fairly well (ATLAS SHRUGGED: PART I, UNDER THE SAME MOON, and WILD), but the residuals are much larger for the other two movies, K-PAX and THE DEVIL WEARS PRADA. Whereas the analysis underestimates the box office for DEVIL, it overestimates K-PAX's performance; using the amount of ad spending the latter film had received, the analysis predicts a box office of about $85 million, instead of the $50 million the movie actually made. This is fairly typical—almost no regression is *perfectly* fitted with market reality.

A widely used measure of meaningfulness, or "fit," of a regression analysis is the "coefficient of determination," or R^2, which ranges from 0 (when the IV explains *none* of the variation in the DV across all cases) to 1 (when the variation is perfectly explained and predicted values match the actual values exactly). In our example, the regression function has an R^2 of 0.69, meaning that about two-thirds of the variation in box office between the five movies is explained by their advertising spending.

Is this "good" enough? The answer depends on what the analyst wants to achieve. If there is solid theory that predicts that ad spending influences box office, the results make sense and explain quite a bit of variance. We can see the value of theory by choosing other IVs for which such impact on box office is not suggested by theoretical arguments. For example, an alternative model that uses the films' *running time* as the IV to explain box office has an R^2 of only 0.12 for our five movies. Several other fit measures also exist: the Mean Absolute Percentage Error (or MAPE), for example, reports the average percentage deviations of the predicted from the actual values. It is particularly useful when prediction (versus explanation) is the main goal of running a regression. The MAPE of our model is 58% in our example and 37% if the smallest of the films is excluded.[5]

Some Challenges—and a Quick Glance at Methodological Approaches to Master Them

One obvious limitation of the movie example above is that we treat advertising as the *only* determinant of box office. This is of course an oversimplification of a much more complex reality (which is the reason why this book

[5]One of the problems of deviation metrics such as MAPE is that percentage deviations are systematically higher for smaller values of the dependent variable than for greater values. One approach to correct this is to weigh the cases. When we do so in our example, using the ad spending per movie as weight, the MAPE shrinks to 43% (from 58%). See also our discussion of prediction measures in the context of innovation management for entertainment.

is so thick, by the way!). Every entertainment manager and student could easily name other factors that might also play a role. Part II of our book is dedicated to these multiple determinants, and we discuss them one by one. But this is not a serious problem for regression analysis: the method allows us to include more independent variables, or determinants, in the estimation—it can handle multiple IVs and determine their relative levels of influence simultaneously.

Among the movies in our example, we do the worst job of predicting the performance of The Devil Wears Prada—our regression function estimates $39 million less than the film actually made. Why? The theory of *Entertainment Science*, as we lay it out in this book, suggests that films' performance is also influenced by the brands they involve. Devil, in particular, was based on a bestselling novel (as was Atlas Shrugged, by the way). So when we add whether the films in our sample were based on a bestseller or not as a second IV, we find that the R^2 increases to 0.88, or by almost 30%, and the prediction error for Devil shrinks to $14 million.[6]

Our prediction for K-Pax is also not good. In this case, an explanation could be that the movie was released only a month after the September 11 terror attacks in 2001, an extreme time in history when few people were in the mood to go to the movies. We could address this by adding a "released-during-crisis" variable to the model (which would be 1 for K-Pax, but 0 for the four others). Doing so would result in an almost perfect model fit: the share of explained box office variance increases to 0.99, and the prediction error is *very* small now for all movies. But this result is an artifact that needs to be avoided—if the number of IVs in relation to the number of cases exceeds a critical level, the results become meaningless because of so-called "overfitting." Thus, the number of cases one has on hand might pose a limitation for how many IVs can be considered. Our "crisis" variable is also problematic for another reason: it is tied to a *single* film only, which enables the regression model to assign all success deviation for this film to this idiosyncratic factor.[7]

Another issue that deserves attention is that our DV here, box office, is a "continuous" variable—it can take on virtually an infinite number of

[6]This remains the case, by the way, when we adjust the R^2 for the number of IVs—this adjusted R^2 value rises from 0.59 to 0.76.

[7]The same would happen if you would try to explain the extraordinary success of James Cameron's Titanic as part of a sample of less successful films and include "Lead-actor-says-'I-am-the-king-of-the-world'-while-standing-on-large-ship" as an IV.

possible values. In some cases, managers might be interested in other DVs that are not continuous, but can only take on one of a *discrete* number of possible values (e.g., a "binary" DV has only two possible values). What are the differences between video games that break even and those that do not? Which songs win Grammy awards? Basic linear regression cannot be used with DVs that are discrete, not continuous. However, with slight modifications, there are other kinds of analyses such as logit or probit regressions that are well-suited to handle them.

Other challenges are less easy to address—and also less easy to spot. Running a regression today is a fairly simple task, as statistical packages with menu-driven interfaces exist (even Excel can do it!). But such simplicity brings along some pitfalls because it is almost as easy to get *wrong* empirical results as to get any results at all. These pitfalls are not exclusive to statistical novices; they also limit the meaningfulness of numerous otherwise-scientific studies on *Entertainment Science* (we do our best in this book to only report reliable studies and mention some limitations when appropriate). Let us list some of the most pressing issues and describe how scholars work to avoid or deal with them.

The database. The phrase that "any study is only as good as its data" is particularly true for regression analyses. Regressions always use past information, making predictions about the future based on what happened before. So always take a close look at from which *part* of the past researchers have derived their findings. Is their data set old or new, is it North American or German, is it comprehensive or does it systematically leave out certain products (such as low- or high-budget products), or is it from a unique period of time (e.g., collected during a recession, or before the Internet existed)? The studies cited in this book differ quite extensively in the data they use. Often generalizations are possible and legitimate, but you should always carefully consider the foundation on which conclusions are based before relying on those findings.

The spuriousness challenge. The statistical explanation of something with regression may or may not be causal. In several cases, the correlations of factors underlying a regression exist simply because of a quirk of fate, or because of an "omitted variable" (see below). Vigen (2015) provides a full book of examples for meaningless, non-causal empirical links: statistically, Brad Pitt's annual earnings explain about 84% of the average American's ice cream consumption between 2001 and 2009 (p. 29), and the percentage of Argentina's GDP that is spent by the government explains a whopping 97% of the North American viewers of the TV comedy BIG BANG THEORY (p. 115). Does this mean that ice cream consumption in the U.S. could be reduced by lowering Mr. Pitt's fees? Or that the producers of BIG BANG should stimulate Argentina's economy? Of course not—because these correlations lack a causal character. Many spurious effects are less obvious but equally misleading. The essential problem of non-causal relations is that they suggest the use of remedies that will turn out to be ineffective. Thus, thorough tests should rule out the spurious nature of a link, but we consider having a powerful explanation at hand as even more essential. While there can be some value in exploring data in a theory-free way to let it "speak," empirical correlations should never be trusted without sound theory.

Non-linear relationships. Sometimes, the relationships between determinants and outcomes are not linear—not every change in the IV leads to the same level of change in the DV. There are numerous reasons for such non-linearity: satiation levels or thresholds may exist, or an effect may develop its own dynamics. Satiation is a common phenomenon for entertainment consumption, and dynamics exist in the form of cascades and feedback effects (e.g., when buzz creates more buzz and when charts benefit those who are already successful; see our chapter on "earned" entertainment communication). Thus, failing to find an impact of an IV in a linear model does not mean that the variable exerts no effect *at all*—the relationship could be non-linear. To test for non-linearity, researchers add squared terms (for U- or inverted-U-shaped relationships) and cubed terms to their regression models, or they use specific techniques, such as quantile regression.

Interplay between variables. A particular case of non-linearity is when two or more IVs affect an outcome *jointly*. Consider the case of a movie sequel: as we discuss later, the popularity of the previous film increases the sequel's success potential, but this popularity effect is not the same for all sequels—it depends on whether the previous installment's star returns in the sequel. To account for this type of interplay, some researchers add "interaction terms" to their models by multiplying together the variables that are assumed to exert a joint effect, while other researchers run "subsample," or "multigroup," analyses in which they compare the regression results for subsets of the data.

Omitted variable biases. Regressions yield robust results only to the degree that the right IVs are included in the model to adequately reflect actual market realities. Particularly if the goal of a regression analysis is to establish causality, *all* variables that influence the DV need to be included in the model to control for their effects. If key variables are omitted, however, spurious correlations can show up—the variables in the model claim shares of the dependent variable's variance that, in reality, belong to the other IVs that are not accounted for in the analysis. In our five-movie example above, the lack of other factors, such as the films' genre and star power, leads to an exaggeration of the impact of advertising on box-office revenues—and causes its misspecification.[8]

Multicollinearity bias. In regressions with more than one IV ("multiple regressions"), results can also be distorted by high correlations between the IVs. For example, in regressions aiming to explain the box-office success of films, the simultaneous inclusion of advertising, the number of theaters in which a film is shown, and the production budget often leads to biased results, because all three IVs are similarly distributed ("collinear") across different movies—low for independent films, high for normal studio films, and *very* high for blockbusters. High (multi-)collinearity inflates statistical significance levels and makes the results nearly impossible to interpret. In addition to measuring and reporting formal metrics such as "variance inflation factors," researchers can also consider using the *residuals* of an IV, instead of the raw information that overlaps with other IVs. Residuals isolate the unique information that is not covered by other IVs.

Endogeneity bias. Regression analysis assumes that all independent variables are truly "independent." Econometricians use the more precise term "exogenous" for such independence, meaning that the IVs need to be created autonomously and are neither unduly determined by other factors in the model, nor by the factors outside of it. If this exogeneity condition is not met, a variable is considered "endogenous." In practice, the exogeneity condition is almost always violated because it is hard to find *anything* in our world that is not at least slightly influenced by other factors. But in some constellations, endogeneity leads to misleading results in a regression, which must be avoided. When an OLS regression finds that the presence of stars increases a movie's revenues, on average, by $13 million (Litman and Kohl 1989), does this mean stars do have this impact? Not necessarily—because stars might systematically sign on to films that have higher success potential, and these films will probably also receive advantageous treatment by their producers because of that higher potential. In other words, because stars may be endogenous, estimations of their impact might be inflated unless this endogeneity is accounted for. To do so, various approaches and methods

[8]Vigen (2015, p. 29) offers a colorful example when citing very high correlations between murder rates and ice cream consumption. So which ice cream ingredient turns us into killers? None, at least as far as we (and food scientists) know. Instead, the *season* is the omitted variable here: murders are more common in the summer, which is when people usually eat ice cream. Thus, any attempt to ban ice cream to reduce murder rates would turn out to be quite ineffective…

have been developed for use by researchers, such as statistical matching, two-stage least squares regression (2SLS), and three-stage least squares regression (3SLS). (To see what stars are *really* worth, go to our chapter on entertainment product brands.)

Heterogeneity. Another source for bias in regression is the heterogeneity that characterizes entertainment products and exists among consumers of entertainment. Keep in mind that regression parameters, by definition, are average effects (because *one* parameter is created from *all* cases in a sample), and thus their validity depends on how meaningful an overall mathematical average is for the data set in general. If you hold one foot in ice water and the other in boiling water, the average temperature of your feet might appear to be cozy, but this average in no way accurately reflects how you actually feel. Because of similar considerations, researchers need to be aware that findings might differ between forms of entertainment and/or between consumer segments. Methods exist to account for heterogeneity, including latent class regressions and the estimation of regression models for subsamples.

There are two main takeaways from knowing about these pitfalls. The first of them links the task of data analysis back to *theory* and underscores the importance of this key element of *Entertainment Science*. Powerful theory is necessary to design regression models in a proper way—one that reflects the realities in which consumers experience entertainment products and managers make decisions. Theory helps you avoid the problems and biases mentioned here, namely endogenous, non-linear, heterogeneous, and interactive relationships by selecting the relevant variables and data. In other words, if a regression model is ill-defined and in conflict with theory, its results will only make things worse.

The second main takeaway is that we ask our readers to pay attention to these requirements not only when conducting studies themselves, but also when hearing about new research, journalistic discoveries, and particularly when confronted with commercial data consulting offers. The pitfalls and complexities of data analytics offer room for misinterpretations or, worse, manipulation. Consider, for example, how Relativity Media reportedly (mis)used algorithms to get more promising financial projections to impress investors. When the company wanted to determine the success potential of a prequel movie to The Untouchables, it did so by calculating the average box-office results of previous films starring lead-actor candidate Nicholas Cage. While the general lack of sophistication of this approach could be criticized, a more fundamental problem was that Relativity was found to have "massaged" the results by leaving Cage's flop Snake Eyes out of the analysis (Wallace 2016).

We have also learned that analytics companies often hide their methodological approaches, substituting information with mystique. In the case of screenplay analyzer Epagogix, employee credentials are hidden by cool-sounding pseudonyms such as "Mr. Pink" and "Mr. Brown," references to Quentin Tarantino's RESERVOIR DOGS movie (Gladwell 2006). We passionately believe that any lack of transparency in econometric methods should be a major reason for concern, just as is the case when our scholarly submissions to a scientific journal are reviewed by our colleagues. If one needs to rely on findings to make important decision, he or she must be able to judge the quality of the methods and theory that have led to those findings. Transparency, particularly regarding data, method-related information, and goodness-of-fit statistics, is an essential requirement for good *Entertainment Science*.

This discussion has provided some basics on the methods and data used by the studies on which we build our theory of *Entertainment Science*. Many details are missing, of course, but we wanted to provide some background on how the insights we report herein have been generated. And you never have to take our word as gospel; throughout the book, we always name the respective scholars behind a finding and encourage you to review their respective works. All the information you need to do so is provided in the book's reference section.[9]

Concluding Comments

In this initial chapter, we confronted the "Nobody-Knows-Anything" mantra that has pervaded the entertainment industry and shaped its management decisions head-on, offering *Entertainment Science* as a timely alternative. We clarified that our goal is *not* to abolish managerial intuition, and that it is also *not* to worship at the altar of Big Data. We show that while a "Nobody-Knows" approach ignores the learning potentials that data and its analysis offer managers, a "data-only" management approach leads to a "False-Precision" trap because quantitative analyses look so authoritative, even when conducted in a thoughtless way, such as leaving out key explanatory

[9]For readers who want to dive deeper into the method dimension of *Entertainment Science*, there is an abundance of good books about regression analysis and its extensions. For beginners, we recommend Hair Jr. et al.'s (2014) book, which covers the fundamentals of regression analysis, along with other statistical methods, in a way that is both highly competent and readable. For advanced topics and questions on regression analysis, we suggest the work of Angrist and Pischke (2009), who devote detailed attention to most of the pitfall issues we have listed here—as well as many others.

variables. Instead, *Entertainment Science* pairs intuition not only with data analytics, but also with practical scientific theory—an approach which, as we show, captures the benefits of each approach, while compensating for the weaknesses of each.

This chapter also outlined the structure of the book. In the five chapters of Part I, we cover the fundamentals of entertainment, including critical examinations of what makes entertainment products and markets different from other contexts, how entertainment firms make money, and how consumers make decisions regarding these hedonic, creative products. The insights lay the groundwork for Part II of the book in which, over the course of nine chapters, we look at the specific decisions that entertainment managers make when creating, promoting, distributing, and pricing their products, as well as the power of thoroughly integrating them in a strategic way. We wrapped up this chapter with a short primer on research methods—foundations and pitfalls—to give readers a general understanding, or reminder, of the approaches underlying the findings which, in their totality, constitute *Entertainment Science*. Now let's move on to the good stuff.

References

Amatriain, X., & Basilico, J. (2012a). Netflix recommendations: Beyond the 5 stars (Part 2). *The Netflix Tech Blog*, June 20, https://goo.gl/N6QNK9.

Amatriain, X., & Basilico, J. (2012b). Netflix recommendations: Beyond the 5 stars (Part 1). *The Netflix Technology Blog*, April 5, https://goo.gl/6ur2Qs.

Anderson, C. (2008). The end of theory: The data deluge makes the scientific method obsolete. *Wired*, June 23, https://goo.gl/usVwdK.

Angrist, J. D., & Pischke, J.-S. (2009). *Mostly harmless econometrics: An empiricist's companion*. Princeton: Princeton University Press.

Arndt, S. (2009). Marken aus der Manufaktur: Filme für das 21. Jahrhundert. In T. Hennig-Thurau & V. Henning (Eds.), *Guru Talk – Die deutsche Filmindustrie im 21. Jahrhundert* (pp. 47–63). Marburg: Schüren Verlag.

Austin, B. A. (1989). *Immediate seating: A look at movie audiences*. California: Wadsworth Pub. Co.

Barnes, B. (2013). Solving equation of a hit film script, with data. *The New York Times*, May 5, https://goo.gl/S5RvQt.

Bart, P. (2017). Peter Bart: Amazon raises bet on movie business, but rivals still baffled about long-term strategy. *Deadline*, March 10, https://goo.gl/t46Vvj.

Beier, L.-O. (2016). Instinkt. Geschmack. Eier. *Der Spiegel*, February 27, https://goo.gl/r5dKPH.

Brodnig, I. (2015). Netflix-Produktchef Neil Hunt: 'Ich weiß das alles über Sie'. *profil*, December 11, https://goo.gl/TSHFgD.
Caldwell, J. T. (2008). *Production culture: Industrial reflexivity and critical practice in film and television*. Durham and London: Duke University Press.
Catmull, Ed. (2008). How Pixar fosters collective creativity. *Harvard Business Review, 86*, 64–72.
Clement, M. (2004). Erfolgsfaktoren von Spielfilmen im Kino – Eine Übersicht der empirischen betriebswirtschaftlichen Literatur. *Medien & Kommunikationswissenschaft, 52*, 250–271.
Dehn, P. (2007). InsightOut: Blick auf digitales Kino und Verwertungskette insgesamt. *Film-TV-Video.de*, April 2, https://goo.gl/p432yW.
De Vany, A. (2004). *Hollywood economics: How extreme uncertainty shapes the film industry*. London: Routledge.
De Vany, A. (2006). The movies. In V. A. Ginsburgh & D. Throsby (Eds.), *Handbook of the economics of art and culture* (pp. 615–665). Amsterdam: Elsevier.
Eck, D. (2016). Welcome to Magenta! *Magenta*, June 1, https://goo.gl/gWNt8B.
Friend, T. (2016). The mogul of the middle. *The New Yorker*, January 11, https://goo.gl/8hYXxT.
Gladwell, M. (2006). Annals of entertainment: The formula. *The New Yorker*, October 9, https://goo.gl/x9e7qD.
Goldman, W. (1983). *Adventures in the screen trade*. New York: Warner Books.
Gomez-Uribe, C. A., & Hunt, N. (2015). The Netflix recommender system: Algorithms, business value, and innovation. *ACM Transactions on Management Information Systems, 6,* 13–19.
Hair Jr., J. F., Black, W. C., Babin, B. J., & Anderson, R. E. (2014). *Multivariate data analysis* (7th ed.). Harlow: Pearson.
Indiewire Team (2011). Show me the money: Extremal Film Partners pitches a new financing product. September 7, https://goo.gl/U3NLkE.
Jaafar, A. (2016). The king of Cannes Mario Kassar on the glory days of Carolco, why buying Arnie a plane made sense and talking vaginas. *Deadline Hollywood*, May 12, https://goo.gl/eoW3fz.
James, B. (1985). *The Bill James historical baseball abstract*. New York: Villard Books.
Jordanous, A. (2016). Has computational creativity successfully made it 'Beyond the Fence' in musical theatre? Conference Paper, University of Kent.
Karp, H. (2014). Music business plays to big data's beat. *The Wall Street Journal*, December 14, https://goo.gl/Rc3Tv7.
Litman, B. R., & Kohl, L. S. (1989). Predicting financial success of motion pictures: The 80's experience. *Journal of Media Economics, 2*, 35–50.
Littleton, C., & Holloway, D. (2017). Jeff Bezos mandates programming shift at Amazon Studios. *Variety*, September 8, https://goo.gl/GFtV37.
Liu, X., Shi, S., Teixeira, T., & Wedel, M. (2018). Video content marketing: The making of clips. *Journal of Marketing, 82*, 86–101.

Lohr, S. (2009). Netflix awards $1 million prize and starts a new contest. *The New York Times*, September 21, https://goo.gl/ZX38Up.

McCarthy, E. J. (1960). *Basic marketing: A managerial approach*. Homewood, IL: Richard D. Irwin.

McKenzie, R. B. (2008). *Why popcorn costs so much at the movies*. Leipzig: Springer.

Mediabiz (2016). Jeffrey Katzenbergs Gedankenspiele um Paramount. March 3, https://goo.gl/Fn1TNA.

Moon, S., & Song, R. (2015). The roles of cultural elements in international retailing of cultural products: An application to the motion picture industry. *Journal of Retailing, 91*, 154–170.

Newitz, A. (2016). Movie written by algorithm turns out to be hilarious and intense. *Ars Technica*, June 9, https://goo.gl/3RrB5n.

Nocera, J. (2016). Can Netflix survive in the new world it created? *The New York Times Magazine*, June 15, https://goo.gl/e1d2Zu.

Peltoniemi, M. (2015). Cultural industries: Product-market characteristics, management challenges and industry dynamics. *International Journal of Management Reviews, 17*, 41–68.

Reinartz, P. (2016). Nimm einen Hund rein. *Die Zeit*, April 22, https://goo.gl/pKVQ7o.

Seifert, M., & Hadida, A. L. (2013). On the relative importance of linear model and human judge(s) in combined forecasting. *Organizational Behavior and Human Decision Processes, 120*, 24–36.

Sturm, K. (2015). Mercedes-Rechenfehler in Monaco: 'Wie zum Teufel konnte das passieren?' *Spiegel Online*, May 25, https://goo.gl/K8HLRy.

The Economist (2016). Silver-screen playbook: How to make a hit film. February 27, https://goo.gl/J7daa7.

Thompson, D. (2014). The Shazam effect. *The Atlantic*, December, https://goo.gl/LasX8e.

Thompson, P., Jones, M., & Warhurst, C. (2007). From conception to consumption: Creativity and the missing managerial link. *Journal of Organizational Behavior, 28*, 625–640.

Tickld (2016). 31 Fascinating things most people don't know about Netflix. https://goo.gl/8pNw7X.

Vance, A. (2013). Netflix, Reed Hastings survive missteps to join Silicon Valley's elite. *Business Week*, May 10, https://goo.gl/ss45xZ.

Variety (2014). Variety launches Vscore to measure actors' value. August 6, https://goo.gl/9W6tE7.

Vigen, T. (2015). *Spurious correlations*. New York: Hachette.

Walker, J. (2016). OMG! Analytics didn't work! *LinkedIn*, February 25 https://goo.gl/2s2rNW.

Wallace, B. (2016). The epic fail of Hollywood's hottest algorithm. *Vulture*, https://goo.gl/QbEYiK.

Zafirau, S. (2009). Audience knowledge and the everyday lives of cultural producers in Hollywood. In V. Mayer, M. J. Banks, & J. T. Caldwell (Eds.), *Production studies: Cultural studies of media industries* (pp. 190–202).

Zimmer, C. (2016). In science, it's never 'Just a Theory'. *The New York Times*, April 8, https://goo.gl/dker2P.

Part I

Products, Markets, & Consumers—The Business and Economics of Entertainment

This first part of our book is intended to lay out the foundations for successfully marketing and managing entertainment products. We assume that you, the reader, command a sound general managerial knowledge already, and we want to enrich this knowledge with key insights on the characteristics of entertainment that make the business of movies, books, games, and music such a fascinating matter.

Understanding entertainment's characteristics is essential for developing marketing strategies that do entertainment products justice and help to avoid not only costly missteps, but also *schadenfreude* by some industry traditionalists who consider *Entertainment Science* (with its data and analytics) as a threat rather than an opportunity. As we have argued before, learning about what kind of marketing strategies are effective in the context of entertainment is possible, but not by simply transferring knowledge from other industry contexts—only by carefully adapting it to the specifics of entertainment.

In the following chapters, we will thus shed light on such specifics with regard to the products that are intended to entertain people, the economic markets on which they are offered, and the business models through which financial value can be generated with entertainment. We will then also distil the core insights that consumer researchers have gathered over the last decades. Building on the pioneering work by Morris Holbrook and Elisabeth Hirschman on hedonic consumption, we integrate what is known from diverse fields and advance it toward a comprehensive understanding of entertainment consumer behavior.

But before we begin, give us a moment to investigate the subject of our investigation, our "labor of love": what is entertainment after all, and why is it worth spending much time studying and managing it?

2

The Fundamentals of Entertainment

What's Entertainment?

Entertainment is *big*. If you google "entertainment" these days, you will get 2.4 billion websites. And people search the term "entertainment" on the Internet far more often than they search for fundamental concepts such as "economy," "politics," and even "happiness." But what exactly *is* entertainment, the topic we study in this book? When it comes to developing a common understanding about a subject matter between authors (us) and their readers (you), definitions beyond the anecdotal ones that Fred Astaire and his co-singers offer in their famous 1953 song THAT'S ENTERTAINMENT! (Mean villains! Romantic dreams! Fighting! Clowns! Sex!) are essential. So, what do we mean when we write about "entertainment" in this book?

Our approach to define entertainment is pretty straightforward and hopefully in alignment with our readers' intuitive understanding of the concept. We take a producer-sided perspective and consider entertainment as *any market offering whose main purpose is to offer* pleasure *to consumers, versus offering primarily functional utility*. As we discuss in more detail later in this book, pleasure, which Drake (1919, p. 666) defined almost a century ago as the finding of "a certain quality in our experience," is one of the fundamental states which we, as consumers, strive for in our lives; it may even be the most essential of all psychological end states (see our chapter on entertainment consumption for more details if you just can't wait). This "certain quality" encompasses a broad spectrum of consumer states, ranging from frivolous amusement to sensual gratification to distraction to mental challenge—entertainment products span this entire range.

T. Hennig-Thurau and M. B. Houston, *Entertainment Science*,
https://doi.org/10.1007/978-3-319-89292-4_2

There are many ways that pleasure can be offered to consumers and achieved by them. Consistent with the traditional understanding of an "industry" (such as in "the entertainment industry"), we focus on entertainment which is *pre-produced* and delivered via media, instead of "live" services. Such pre-produced entertainment can offer consumers pleasure by providing them access to one or more of the following forms of content:

- Filmed content (such as fiction movies and series, documentaries, video clips),
- written content (such as novels and poems),
- recorded content (such as pop songs, classical compositions, movie soundtracks), and
- programmed content (such as console games, massively multiplayer online games or MMOGs, and smartphone games).

Because our perspective is an economic one, with the firms that create such content at the center of our analysis, we usually have entertainment *products* in mind. Such products are the offerings which consumers can acquire, either for a fee or for free, and either for a limited time period or forever, and which provide the consumers with access to entertaining content. Entertainment products can be either of a material or immaterial type. Material entertainment products are physical tools (such as a DVD or a CD) that contain entertainment content which is transferred to consumers; immaterial entertainment products, in contrast, transmit content through (technical) channels to consumers,[10] such as the Internet, cable, or satellite. Figure 2.1 names the different forms of entertainment content and the corresponding products we study in this book, listing popular examples for each combination of content and format.

And what about advertising—doesn't it also qualify as entertainment? There is hardly any doubt that advertising *can* be entertaining, as it surely offers pleasure to consumers, as evidenced through millions of clicks for some advertising spots on YouTube. But the main purpose of advertising is not to please consumers, but something else—namely to make consumers remember *other* brand names and to encourage them to buy *other* products. So in the case of advertising, consumers' pleasure experience is only a means

[10]We have added the term "technical" here to differentiate the channels we are talking about here from the channels we discuss in the context of managerial distribution decisions.

Content form	*Exemplary material entertainment products*	*Exemplary immaterial entertainment products*
Filmed content	THE HUNGER GAMES Blu-ray	THE HUNGER GAMES streamed via Amazon Video THE HUNGER GAMES as MP4 download
Written content	THE HUNGER GAMES book	THE HUNGER GAMES ebook
Recorded content	Lady Gaga CD	Lady Gaga songs aired via (Internet) radio Lady Gaga MP3 songs Lady Gaga songs streamed via Spotify
Programmed content	ASSASSIN'S CREED PlayStation 3 disc	ASSASSIN'S CREED file from PlayStation Store WORLD OF WARCRAFT online game

Fig. 2.1 A typology of entertainment products
Notes: Authors' own illustration. Brands are trademarked.

to a separate end.[11] This is why we do not treat advertising as entertainment in this book (but, of course, discuss it as an important element of the marketing mix for other entertainment products such as movies or games). This does certainly not mean that producers of advertising can't benefit from the insights we report herein, such as the motivational underpinnings of entertainment consumption. We assume that this should be even more so the case for those who are in the growing business of "content marketing." Content marketers' task is to produce "products" which are, like advertising, still means to enhancing another product's success, but whose effectiveness relies even more strongly on the pleasure which consumers associate with experiencing them.

Why is Entertainment Important After All?

Students who enroll in a business school often do not get exposure to entertainment as a study subject, as the focus of academic curricula is usually heavily biased toward "more important" products and industries, such

[11]One might say that pleasure is also tied to different goals in the case of movies, music etc., such as making their creators popular and wealthy—but here the "other" goal is an *immediate* result from consumers' pleasure or at least their anticipation of it, whereas for advertising consumers' pleasure and its economic effectiveness are simply two separate things. Also, the quality of an entertainment product does not only have an instrumental function, but is almost always also an end state by itself. We discuss this as an inherent characteristic of entertainment products.

as fast-moving goods (soft drinks!) and durables (cars!). In reality, there are a range of reasons for considering entertainment as much more than a fun topic. As we show in this section, entertainment creates substantial economic value. But it can also offer extensive insights for other industries (beyond the content marketing business) as a result of the pioneering role that entertainment products have played with regard to several business concepts. In addition, and of at least equal importance to us, entertainment is an unparalleled source of personal identification and value for consumers.

Entertainment Generates Substantial Economic Value!

In total, we estimate the annual revenues produced by the entertainment products we analyze in this book to be close to $750 billion globally, without taking into account consumer expenditures on entertainment hardware such as TVs, Kindles, and smartphones. Filmed entertainment content accounts for the largest share, at least if indirect revenues are considered; whereas films generate close to $100 billion these days through theatrical and home entertainment channels, they are also responsible for roughly twice that dollar amount of TV advertising. Filmed content also generates pay-TV subscription fees of a similar size. Programmed content in the form of electronic games accumulates about $100 billion from consumers and advertisers across all platforms and channels, written entertainment content in the form of recreational books (not counting other publishing categories, such as scholarly and educational) is responsible for about $75 billion in global consumer spending, and recorded (musical) content revenues are close to $15 billion, while also serving as the backbone for advertising and other earnings by radio stations of more than three time that amount.[12]

Adding in the hardware that is required for consuming such content (such as TV sets, computers, and gaming consoles) makes the entertainment and media industry one of the largest in economic terms. In the U.S., only housing, health, and food and beverages receive more from consumers' budgets than entertainment and media; spending on automobiles, furnishings,

[12]All numbers given in this section are our own calculations and should be treated as rough estimates, building on publically available information by, among others, McKinsey, PricewaterhouseCoopers, Statista, Datamonitor, and IFPI, as well as various conversations with industry experts. Please note that these numbers in general reflect the "retail" value paid by consumers or advertisers, not the share of money that flows back to entertainment producers. We shed some more light on the latter in our chapter on value creation for the different forms of entertainment.

and education each trail behind, according to data from the Bureau of Economic Analysis.

The economic substance of entertainment also becomes evident when looking at the success potential of single products. When the Walt Disney company released the movie The Force Awakens in December 2015, the seventh entry in the Star Wars movie series for which the firm had bought the rights from Lucasfilm for $4 billion, it needed just *three days* for the film to generate theatrical revenues of $248 million in North America and $528 million globally, not including China (where it was released later). When the film crossed the $1 billion mark, only 12 days had passed since its market entry, and by its 53rd day, The Force Awakens became the third film in history that cracked the $2 billion revenue mark. Although Disney had reportedly invested a whopping $259 million in the film's production and an additional $185 million to support and promote its global release, experts estimate that the film will provide its producing studio a net profit of $780 million, an amount that does not even include merchandising revenues and the film's long-term revenue potential, such as further sequels and spin-offs (Fleming 2016).

In sum, one should make no mistake by confusing the fun and light-heartedness that often characterize great entertainment with the economic seriousness of the industry itself. Entertainment is among the largest industries, involves products that require extreme financial investments, and can compete with almost any other product category in terms of return of investment and also absolute gains.

Entertainment is a Pioneering Industry!

Studying the entertainment industry not only benefits a reader's career in this particular field, but can also boost understanding of critical aspects of other industries outside of entertainment. When BusinessWeek asked our colleague Anita Elberse why her entertainment marketing course was so popular among Harvard students though relatively few took a job in the industry, she said: "Even if [my students] are not going into entertainment, it's a very useful course to understand the world of marketing. I think many sectors are adopting some of the concepts we see in entertainment, and … things have changed so much" (quoted in Zlomek 2013).

Several key issues have been on entertainment managers' agendas for quite some time, but are only now becoming critical for success in other business areas. Managers in those other industries can benefit from what has been learned in entertainment and what we have accumulated in this book. Let us look more closely at seven of these issues:

- *Pre-release buzz.* Entertainment marketing has become very often *"front-loaded,"* with the focus of advertising, distribution, and other activities being on the period prior to a new product's release. This is partly a result of entertainment's attributes, but also the result of entertainment managers' embrace of the "blockbuster" concept and pre-release new product buzz (see our chapter on "earned" entertainment communication). The advantages that are associated with a focus on buzz have begun to inspire other industry leaders, such as Apple and Tesla; today, it is hard to think of any major product launch or IPO that is not accompanied by live-streamed media coverage of waiting lines and fan boys' exuberant excitement.
- *Social networks.* The social networks of consumers in society are a key element of the entertainment industry, and leveraging such networks, either directly or indirectly, has been an essential challenge for the marketing of entertainment products. Today, however, managing networks is no longer a niche subject left to entertainment managers; instead, with almost every manager now recognizing the value potential of social media (and also the difficulties associated with realizing this value), network-focused strategies have become an essential part of marketing, in general. Because of the high intrinsic interest consumers have for new movies, music, etc., progressive entertainment managers provide powerful examples of the "pinball" approach of marketing communication, where "owned" media such as Facebook brand pages are used to get consumers engaged (see our chapter on paid and owned entertainment communication).
- *Continuous innovation.* In an industry which features an unparalleled number of innovations every year, organizing the innovation process effectively is a *conditio sine qua non*—you simply cannot survive in entertainment if your innovation management does not work. Thus, visionary managers, along with scholars, have given extensive consideration to how such effective organization for innovation (e.g., forecasting success, establishing an "innovation culture") looks like (see this book's chapter on entertainment product innovation). There is hardly any debate that such innovation-related knowledge is valuable for *any* firm in these times of ever-shrinking product life cycles, mega-competition, and relentless technological advances.
- *Creativity* and the management of creatives (e.g., artists, actors, authors, directors, etc.) are essential for entertainment products and for any firm that is part of what scholars such as Richard Caves (2000) call the "creative industries" (see our chapter on entertainment product characteristics). Today,

when most product markets outside of entertainment are characterized by limited functional differentiation, and market share is often gained based on psychological and social benefits offered to consumers, it is very often the creative element that makes the difference; in some ways, all industries are migrating toward the creative ones. So, why not learn from those who make a living from creative ideas and their management in the first place?

- *Storytelling*. Related to creativity is a growing interest in *storytelling* capabilities. Creating a convincing narrative has long been recognized as an important prerequisite for acclaimed movies, video games, or novels. As traditional, informative advertising is considered by many to be losing its power, and "content marketing" is seen as a valid alternative, managers' attention shifts toward storytelling skills when designing communication campaigns. The success (or failure) of Coca Cola's "content factory" follows a similar logic as the stories told with films, TV series, or novels. Further, scholars have provided empirical evidence that the most impactful consumer reviews on TripAdvisor are those that follow a narrative pattern (van Laer et al. 2017). If there is a place to learn about storytelling, it is certainly entertainment, and the works of *Entertainment Science* scholars have shed theoretical and empirical light on its mechanisms (see our chapter on entertainment product quality).
- *Building brands and brand alliances*. Although the term brand was "discovered" by entertainment managers only recently, it is hard to find industries these days that make use of the powerful potential of branding in such versatile ways, working with ambitious "brandscapes" and applying complex strategies (such as Marvel's Cinematic Universe of comic heroes, with The Avengers at its core). *Entertainment Science* scholars have shed extensive light on how to brand entertainment products, testing the effectiveness of entertainment branding approaches, but also developing new branding frameworks and methods (see the chapter on entertainment brands). It is obvious that insights on the dynamic management of multi-faceted components of entertainment brands can now inspire brand managers in other industries in their development and innovative management of brand universes.
- *Dealing with digital disruption*. Because the content of entertainment products essentially consists of information, the entertainment industry has, although certainly not always intentionally, adapted a pioneering role in dealing with the challenges of digitalization. Some industry segments within entertainment have suffered grievously for their decisions (such as not making music legally available through digital channels when the MP3 format and broadband connections arrived). But today, entertainment firms have largely accepted the role of thought leaders for digital ideas and concepts

(see examples in our chapter on entertainment distribution). As digitalization now affects *all* industries in a fundamental way, similar challenges exist, or are about to arrive, for managers in other fields beyond entertainment.

Entertainment managers have gained extensive experience and developed initial strategies in all these areas, and entertainment scholars have supported their learning expeditions, often by providing empirical insight into what works (and what doesn't) in the world of entertainment. Much pain can be avoided if a manager is willing to learn from the treasure trove of detailed scholarly investigations into the successes and failures of *actual* decisions by *actual* entertainment product managers, as well as carefully conducted experiments.

As our references above suggest, we provide such lessons throughout this book; they are embedded in our discussion of how effective marketing approaches need to be designed in light of specific conditions and contexts that differ across products and customers, as well as over time. We discuss some of the issues above in specific chapters (such as brand management and innovation), whereas others, such as digitalization, have impacted our thoughts across many aspects of *Entertainment Science* and can thus be found in several parts of the book, from distribution to communication. In essence, the insights reported in *Entertainment Science* are for those interested in entertainment, but certainly not limited to them.

Entertainment Defines Our (and Your) World!

"[Great entertainment provides] the voices, soundtracks, and stories of our personal lives and memories."
—David M. Rubenstein *(quoted in* Viagas *2015)*

Consumers all over the world love to spend time with entertainment products. Harold Vogel (2015, p. XIX) reports that Americans now dedicate about 160 billion hours per year to different forms of entertainment. In 2013, the average American spent 11.4 hours of every day consuming entertainment and media products, an increase of 86% compared to usage rates from 1970. About one-third of this consumption takes now place over the Internet, where every type of entertainment content is available. For offline entertainment product consumption, filmed content takes the lion's share of consumer time (still mostly watched on TV), followed by recorded content (mostly listened to on the radio), programmed content, and written content (Fig. 2.2). We assume that entertainment time shares look similar in other developed parts of the world.

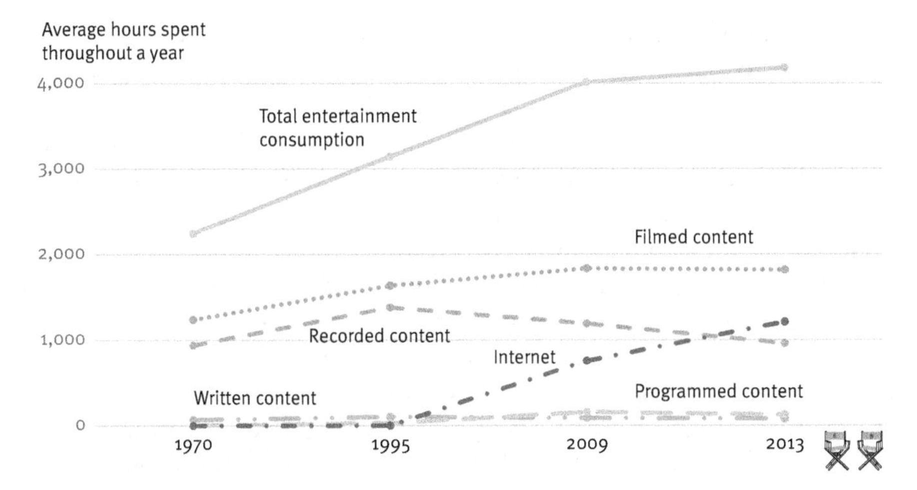

Fig. 2.2 Average hours spent for different forms of entertainment per adult
Note: Authors' own illustration based on data reported in Vogel (2015 and previous editions).

But the impact of entertainment on consumers reaches far beyond the sheer number of hours we invest in its consumption. Entertainment resonates throughout our culture, shapes our view of what exists around us, and influences our actions, as well as our vocabulary. Often, entertainment even inspires within consumers what might be considered among the rarest of resources: motivation and personal meaning.

Entertainment Shapes Our View of the World (and of Dogs)

With its ability to offer information and pseudo-experiences, entertainment is a valuable source of knowledge for consumers about many aspects of their world, including people, historical events, and cultural and political institutions (Kolker 1999). This learning, regardless of precise historical accuracy, occurs because entertainment helps consumers to "experience" vicariously a certain version of reality that the consumer has not experienced in real life (Pautz 2015). Our perception of Indian revolutionary Mahatma Gandhi is shaped by Ben Kingsley's portrayal of him (in the movie GANDHI), and we visualize British officer T. E. Lawrence as he was characterized by Peter O'Toole on the screen in LAWRENCE OF ARABIA. Falco's pop hit ROCK ME

AMADEUS and Tom Hulce's depiction in AMADEUS (the movie) add a "rock star" tinge to our perceptions of classical composer Wolfgang Amadeus Mozart.

Similarly, peoples' views of John F. Kennedy's assassination have been influenced by Oliver Stone's storytelling in the movie JFK (which "reopened the case and put the conspiracy back on the table"; Tiefenthäler and Scott 2017) as well as Stephen King's time-travel bestseller 11/22/63 (and its TV series adaptation 11.22.63 by J. J. Abrams). Germany's public debate of its Nazi past was triggered in the late 1970s by the airing of the American TV series HOLOCAUST that presented a tale of woe of the fictional Weiss family; our collective images of the Allied invasion of Normandy on D-Day are inseparable from its depiction in Steven Spielberg's movie SAVING PRIVATE RYAN.

Are such entertainment portrayals accurate? Not necessarily, because the entertainment industry, and we as human beings, have a tendency to favor the legend over the fact, as famously coined by John Ford in his western, THE MAN WHO SHOT LIBERTY VALANCE. Or, as the screenwriters of the movie JACKIE let the lead character phrase it: "[T]he characters we read about on the page end up being more real than the men who stand beside us." Sometimes reality is even modeled after the legend, such as in the case of the famous Parisian Notre Dame cathedral. Those readers who have visited the cathedral might have been impressed by how closely Victor Hugo captured the church's Galerie des Chimères and its mythical and fantastic creatures in his classic novel THE HUNCHBACK OF NOTRE-DAME. In reality though, it was Hugo's work that inspired the creatures which were added as part of a 19th century restoration program by architect Eugène Viollet-le-Duc—who was a dedicated devotee of the author!

These types of entertainment-based perceptions and knowledge can become the foundation for an individual's attitudes, preferences, and, eventually, behaviors. Pautz (2015) demonstrates the influence of movies on what we think of our governments and how we judge their work. In a classical experimental design, she showed the movies ARGO and ZERO DARK THIRTY to a sample of 69 students and found that about 25% of them had a more positive view of American politics and the country's government after having watched one of the films. Glas and Taylor (2018) provide evidence that watching films which carry authoritarian (the movie 300) or antiauthoritarian themes (V FOR VENDETTA) activates the respective dispositions in consumers, at least in the short run.

Related, others have argued that the high level of trust British citizens have in their security agencies, compared to people in other countries, might be the result of the agencies' portrayals in popular British entertainment.

These include novels and films by or based on John le Carré, but most prominently those featuring legendary agent James Bond: "with James Bond and the Enigma codebreakers as our heroes, we've always believed the intelligence agencies protect us" (Freedland 2015). For a sample of 367 Americans who have never been to Paris, Gkritzali et al. (2016) show that the Hollywood movies they have watched about the French capital influence their image of the city.

Research also suggests that the entertainment icons of our childhood influence our development. For a sample of 198 children around the age of five years, Coyne et al. (2016) linked the children's engagement with Disney princesses with their gender-stereotypical behaviors (such as playing "dressing up" and liking "pretty things") one year later. Even when controlling in a structural equations model for initial levels of such behaviors, the authors find statistical evidence that the children's level of engagement with Disney princesses was associated with higher levels of female gender-stereotypical behavior.

Finally, there is proof that entertainment choices can also have a much more heart-warming effect on our everyday life. Ghirlanda et al. (2014) have demonstrated empirically that entertainment can even affect the dog we choose to be our best friend. Drawing from a large dog registry database, the scholars analyzed whether the release of 29 movies that featured a dog as a main character influenced the respective dog breed's popularity, as indicated by *actual registration trends* in the U.S. Their results provide evidence that movies can have a long-lasting influence on preferences for dog breeds: over the course of 1–10 years, registration trends increased by between 3% and almost 10% on average (the longer the post-release period, the higher the increase). The number of movies' opening weekend viewers correlate significantly with the increase in registration trends, informing us that the more people are drawn into a theater by a movie featuring a dog, the stronger will be the movie's impact on dog-related preferences and behaviors on a societal level. Figure 2.3 shows this empirical effect for four exemplary movies.

Entertainment Gives Us Language

Entertainment not only influences what we think of the world and its institutions, where we travel, and which dogs we take out for a walk, but also helps us to express ourselves verbally. Many specific words and phrases from the entertainment repertoire have become standard vocabulary, being used by many of us in our own lives and in many different situations. Often, we

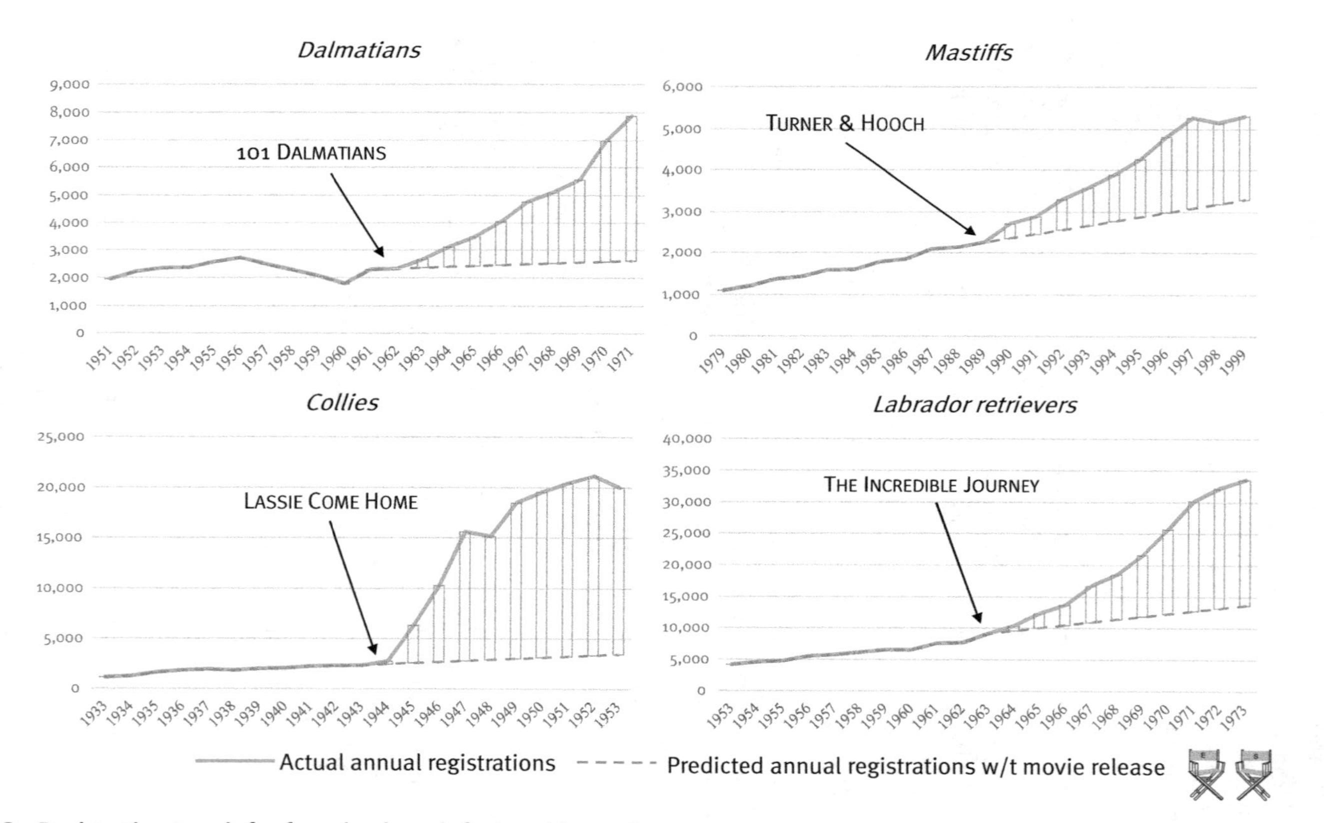

Fig. 2.3 Registration trends for four dog breeds featured in movies
Note: Authors' own illustration based on data and ideas reported in Ghirlanda et al. (2014), which is distributed under the terms of the Creative Commons Attribution License (https://goo.gl/n6FkkT). The data itself is available at https://goo.gl/Gr9ctX.

quote a favorite line to bring some levity and to share a laugh over a shared experience with friends. But these socially meaningful words and phrases can also help us form a point of connection with strangers or even to make an important point in serious occasions in public discourse. For example, when U.S. President Ronald Reagan threatened to veto legislation raising taxes in 1985, he stated: "I have only one thing to say to the tax increasers: *Go ahead, make my day.*" These words, which became a phrase that defined Reagan's career (Curry 2004), were borrowed from Clint Eastwood, who had used them to cow a villain in his Dirty Harry movie SUDDEN IMPACT the year before.

Language is given us by movies as well as novels—think of Shakespeare's HAMLET famous quote "To be, or not to be, that is the question" or Goethe's dictum "[G]ray are all theories, and green alone Life's golden tree" in his magnum opus FAUST. It is also provided by songs: let us just think of John Lennon's words from IMAGINE that encouraged so many to envision, and jointly strive for, a more peaceful world. And is there any better way to tell someone you love her or him than by quoting great song lyrics, such as the Beatles' ALL YOU NEED IS LOVE? Even some video games have added to our language, as evidenced in numerous "best video game quotes" lists on the Internet. And some of us may have, in a situation where progress has been made, but more work is still needed, said "Thanks" to an imaginary Mario, while at the same time stressing the need to head on to another castle to finally free the princess, just like it happened to all players of the legendary SUPER MARIO BROS.

If you want to know if your own favorite entertainment quotes are also the ones that had the most lasting impact on other people's dictionary, you might enjoy taking a look at what the American Film Institute considers the "100 greatest movie quotes of all time"—phrases that "circulate through popular culture" and have become "part of the [American] lexicon" (*AFI* 2005). Michael Curtiz' wonderful CASABLANCA alone accounts for six of them—how many of them come to your mind (think of kids, friendships, time, suspects, gin joints, and of Paris, of course!), and how many have you used when chatting with friends?

Entertainment Provides Us Meaning and Motivation

One of the most existential pursuits of mankind is the self-discovery of personal meaning. Personal meaning, when discovered, is intertwined with motivation to fulfill that meaning in a sustained manner throughout life.

Personal meaning and motivation are some of the rarest and most valuable "resources" that a person can acquire. Entertainment can be an important source for these rare resources, something that has been shown to be true for all forms of entertainment we discuss in this book.

We often ask ourselves questions such as "Who am I?" and "What are my values?" There are many stories of people who found out about themselves by consuming entertainment. Take the example of Joseph Winkler, who grew up as an orthodox Jew, but found that his true values are those of a bleeding-heart liberal (Winkler 2013). How did that discovery happen? He located "the seedlings of the answer in, of all things, THE SIMPSONS." The numerous stories about the chaos, suffering, and also happiness in an ordinary family's life, about the struggles between the workers and the magnates gave him the stories he needed to find out who he actually was.

Many of us have been similarly influenced by stories from entertainment. Entertainment products can provide guidance and identification on our winding journey through life, such as both THE BREAKFAST CLUB and RISKY BUSINESS have done for many struggling teenagers, although in very different ways. The introvert starts to value his intellectual capabilities. The assimilated discovers the extraordinary. Watching THE SECRET LIFE OF WALTER MITTY put a once-reclusive child and teenager "into positions of leadership," who now often makes "the first move in instigating actions" (Marshall 2016). For her, the movie was "more liberating than I can put into words." Others have confessed that they stopped looking down on weaker people after falling in love with FORREST GUMP, whereas some have begun to value relationships more as a result of INTO THE WILD ("Happiness is only real when shared").

Just as we are all individuals (remember LIFE OF BRIAN!), the entertainment offerings that tell us the answers or guide us toward them are different for each of us, as different as the answers themselves. So, it might have been THE SIMPSONS for Mr. Winkler, but chances are that it will be some alternate series for you, our reader. Or an alternative movie. Or game. Or novel. Or album.

And what is it that you want to do in life? Entertainment can also offer a helping hand in this regard, both as a general source of inspiration and as a personal career counselor. Juan Gallardo, a Latino who now works as a web developer, took the motivation to attend college and to "think bigger" from watching the movie STAND AND DELIVER, which tells the story of Hispanic students and their inspirational teacher who frees their potentials. Movies such as ROCKY and THE PURSUIT OF HAPPYNESS inspire people less about *what* course to follow, but about *how* to reach whatever goal they set—about being determined. Or, to cite the latter film, "Don't ever let somebody tell you that you can't do something."

But entertainment can also offer much more specific hints about what career to pursue. We speculate that the record number of astronaut applications NASA received in 2015 for its Mars missions are not uncorrelated with entertainment hits such as The Martian, which took consumers by storm, both as book and film. As the book's author Andy Weir argues: "There's a virtuous cycle in progress. People are fascinated by space again, causing the entertainment industry to make more space fiction, which causes more people to be fascinated by space" (quoted in Berger 2016). Similarly, when Top Gun ruled the box office in 1986, applications for the armed services in the U.S. skyrocketed (Rigby 2015). And Hollywood also plays a role when it comes to explain why Adam Rutherford, now a renowned geneticist, pursued a career in the natural sciences: It was the time-traveling DeLorean in the movie Back to the Future and the parapsychological adventures of the Ghostbusters team that excited him and pointed him toward the field (Rigby 2015). We found others who became scientists after having encountered the entertaining universe of Star Trek.

Sometimes such career inspirations derive from unexpected sources. The movie Wall Street, intended to be a tale of the dangers of unrestricted capitalism, has "exuded an almost hypnotic attraction on scores of would-be bankers and traders" (Guerrera 2010). Michael Douglas' portrayal of iconic evil banker Gordon Gekko made a lasting impression on many students, who were impressed by his glamorous amoral cool and copied his one-liners ("Lunch is for wimps!"), his literature sources (the Chinese military treatise "The Art of War"), and his dress code (yep, the suspenders!). The film's director, Oliver Stone, revealed that *many* have told him over the years: "I went to Wall Street because of that movie" (quoted in Wise 2009). One might even argue that the film's motivational impact was so strong that, in the end, it not only influenced the lives of some students, but the institution as a whole: it turned banking into the shiny platform for corporate raiders and speculative deals that, 20 years later, brought the world to the edge of financial collapse. Ironically, the financial crisis very much resembled the situation against which director Oliver Stone wanted to warn society in the first place.

Finally, entertainment products can also ignite excitement for the medium itself. It was during a showing of Arizona Junior when the then 14-year-old Jay Duplass and his younger brother Mark were first propelled into a joint career as filmmakers (Metz 2012). After the experience, they "started watching movies from a completely different perspective—that this is a gigantic piece of art that people work on for years, and there are ...

people who are creatively in charge. It opened up the idea that making movies might even be a possibility" (quoted in Metz 2012). Mr. Rutherford, the scientist whose passion for science was sparked by entertainment, not only fell in love with science through Hollywood's works, but also fell in love with entertainment itself, as he today advises the industry on the science-side of movies such as WORLD WAR Z. And it has been said that some marketing professors might have become close observers and analysts of the entertainment industry as a response to their own enthusiasm for its creations. But that, of course, remains pure speculation.

Concluding Comments

What justifies the reading of a book as thick as ours? The studying of entertainment? Dedicating a career (and life) to it? Entertainment is an important economic domain—big enough that every reader of this book can make a fortune in it. It is also a pioneering industry that offers numerous lessons for those who prefer to make their living in other parts of the economy. And even if this would not be enough to merit your interest, it is also an industry whose products have a deep meaning, not only as means to more important ends, but as ends themselves, both on the societal level (where entertainment helps us to understand the world) and on the individual level (where entertainment can inspire us). So, we conclude—and hope you agree—that it is hard, if possible at all, to overstate the importance of the world of entertainment.

References

AFI (2005). AFI'S 100 Years…100 Movie Quotes. *American Film Institute*, June 21, https://goo.gl/ogd2hW.

Berger, E. (2016). NASA just smashed its record for astronaut applications—18,000+. *ARS Technica*, February 19, https://goo.gl/C7kJJD.

Caves, R. E. (2000). *Creative industries: Contracts between art and commerce*. Cambridge, MA: Harvard University Press.

Coyne, S. M., Linder, J. R., Rasmussen, E. E., Nelson, D. A., & Birkbeck, V. (2016). Pretty as a princess: Longitudinal effects of engagement with Disney princesses on gender stereotypes, body esteem, and prosocial behavior in children. *Child Development, 87*, 1909–1925.

Curry, T. (2004). Phrases that defined a career. *NBC News*, June 5, https://goo.gl/TzFNWT.

Drake, D. (1919). Is pleasure objective? *The Journal of Philosophy, Psychology and Scientific Methods, 16*, 665–668.
Fleming, Jr., M. (2016). No. 1 'Star Wars: The Force Awakens'—2015 most valuable movie blockbuster tournament. *Deadline*, March 28, https://goo.gl/Rtj96X.
Freedland, J. (2015). The spooks will keep spying on us Brits: We clearly don't care. *The Guardian*, November 6, https://goo.gl/nY471m.
Ghirlanda, S., Acerbi, A., & Herzog, H. (2014). Dog movie stars and dog breed popularity: A case study in media influence on choice. *PLOS ONE, 9*, 1–5.
Gkritzali, A., Lampel, J., & Wiertz, C. (2016). Blame it on Hollywood: The influence of films on Paris as product location. *Journal of Business Research, 69*, 2363–2370.
Glas, J. M., & Benjamin Taylor, J. (2018). The silver screen and authoritarianism: How popular films activate latent personality dispositions and affect American political attitudes. *American Politics Research, 46*, 246–275.
Guerrera, F. (2010). How 'Wall Street' changed Wall Street. *Financial Times*, September 24, https://goo.gl/ptq8fG.
Kolker, R. P. (1999). *Film, form, and culture* (4th ed.). New York: Routledge.
Marshall, E. (2016). The secret life of Walter Mitty. *Quora*, March 31, https://goo.gl/8iaRxy.
Metz, N. (2012). Duplasses found a career via Coens. *Chicago Tribune*, July 6, https://goo.gl/9bV1Sm.
Pautz, M. C. (2015). Argo and Zero Dark Thirty: Film, government, and audiences. *Political Science and Politics, 48*, 120–128.
Rigby, R. (2015). How Hollywood films inspire careers. *Financial Times*, February 18, https://goo.gl/MTRR6n.
Tiefenthäler, A., & Scott, A. O. (2017). Why do we love J.F.K. conspiracy theories? Blame the movies. *The New York Times*, October 26, https://goo.gl/q5KyH2.
Van Laer, T., Escalas, J. E., Ludwig, S., & van den Hende, E. A. (2017). What happens in vegas stays on tripadvisor? Computerized text analysis of narrativity in online consumer reviews. Working Paper, Vanderbilt Owen Graduate School of Management.
Viagas, R. (2015). Stephen Colbert hosts 2015 Kennedy Center Honors tonight. *Playbill*, December 6, https://goo.gl/PRkGGN.
Vogel, H. L. (2015). *Entertainment industry economics: A guide for financial analysis* (9th ed.). Cambridge: Cambridge University Press.
Winkler, J. (2013). Everything I know I learned from watching 'The Simpsons'. *Medium*, October 24, https://goo.gl/W8Xi8Y.
Wise, Z. (2009). A conversation with Oliver Stone. *The New York Times*, September 8, https://goo.gl/3tRzZa.
Zlomek, E. (2013). HBS Professor brings Jay-Z, Lady Gaga to the classroom. *Business Week*, April 25, https://goo.gl/ZAoBP2.

3

Why Entertainment Products are Unique: Key Characteristics

Entertainment products differ from many other products in ways that affect not only consumers and their decision making, but also the companies that make a living out of producing these products and providing them to audiences. Being able to develop effective marketing strategies for entertainment products requires a solid understanding of the general marketing canon; but to avoid a miscalibration of marketing instruments, the manager in charge must also grapple with these unique characteristics of entertainment products. Whereas entertainment managers do not have to re-invent the wheel of marketing, they have to be aware how the vehicle to which they are attaching this marketing wheel differs from other vehicles which are carrying, let's say, fast moving consumer packaged goods or industrial products. If you drive a Tesla, don't fill up at the standard gas station.

In this book, we argue that each entertainment product possesses up to eight unique characteristics. As shown in Fig. 3.1, four of them (which we label "consumer-sided characteristics") relate to consumers and their entertainment-related attitudes and behaviors: the provision of hedonic benefits, the existence of satiation effects in consumption, a cultural character, and the difficulties of judging product quality. The other four are "producer-sided characteristics": they bear directly on major decisions facing managers. They are the "information good" character of entertainment products, their creative character, their short life cycles, and the existence of "externalities." But make no mistake: the consumer characteristics are also absolutely critical for managers to grasp, because a product's success (and thus the manager's success) depends on understanding how consumers make decisions. Not all eight characteristics must apply to

T. Hennig-Thurau and M. B. Houston, *Entertainment Science*,
https://doi.org/10.1007/978-3-319-89292-4_3

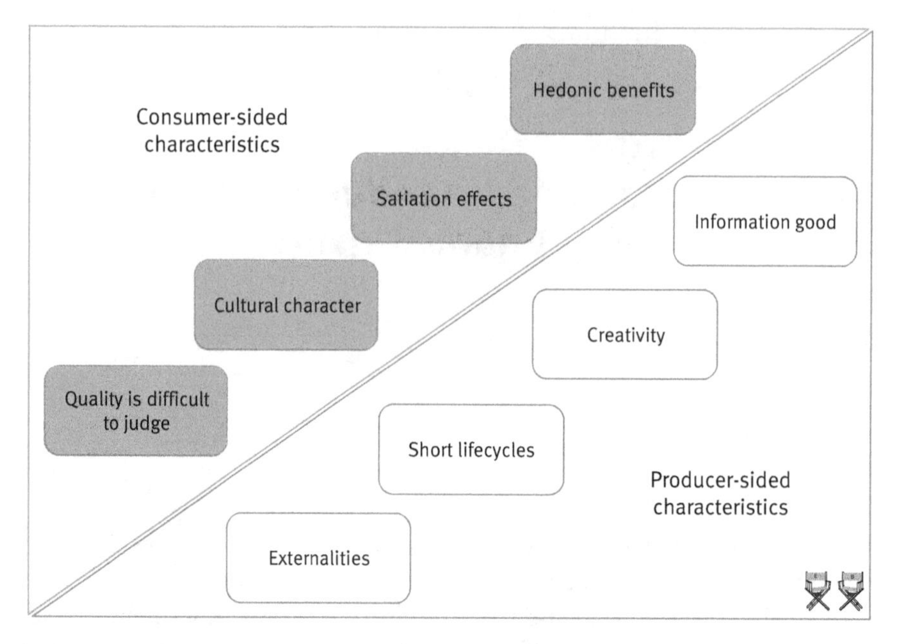

Fig. 3.1 Key characteristics of entertainment products
Note: Authors' own illustration.

each entertainment product, and their importance varies between different products (which span widely from short texts to big-budgeted games), but all are common and typical for entertainment products, shaping our understanding (and, hopefully, handling) of them.

Before we discuss these eight characteristics of entertainment, let us take a moment to bring up two other aspects which some of our readers might have expected to appear in the figure, but we have decided to *not* include: (a) the riskiness of entertainment products and (b) their adoption patterns, marked by exponential decay.

Numerous entertainment executives and also some entertainment scholars have argued that entertainment products are characterized by an abnormally high level of risk, or uncertainty (e.g., Caves 2006; De Vany 2006). Indeed, some of the entertainment characteristics we discuss in this section (such as entertainment's creative nature) impede management control. But instead of considering high risk as a natural "given," we look at the *sources* of the riskiness of entertainment products and believe that managers should do the same. We do so because risk is not a product-level concept, but one that takes place on the level of the firm, where decisions have to be made to counter the sources of risk. Thus, we dedicate a part of the later business model section of this book to risk-related business strategies.

But our perspective conflicts even more fundamentally with those who consider producing entertainment as inherently riskier than producing most other products. Over the last centuries, the industry has developed strategies (such as the "blockbuster concept") to mitigate product risk by addressing its sources (see also Elberse 2013 for a similar perspective). Those strategies have been quite effective, proving that at least for those who are exclusive providers of scarce content, producing entertainment does not need to entail an extra level of business risk. Such effectiveness is reflected in the impressive, and quite stable, profits that several entertainment conglomerates have been able to accumulate over the last decades (not ruling out failures that result from mismanagement, of course[13])—even during periods of economic turmoil like the late 2000s. Suggestive empirical evidence comes from industry expert Stephen Follows (2016), whose analyses of a rich data set of 279 high-budgeted (i.e., $100 million and above) Hollywood films let him reject the often cited "20-80" rule, which claims that out of ten films, only two are profitable on average. Instead, he finds that more than half of the films in his database were profitable, and out of those with budgets of $200+ million, even three-quarters generated a profit.[14]

In a similar analysis, Sparviero (2015) calculated the profitability of 191 motion pictures released in 2007 by the leading studios, complementing available data with "case scenarios" of fees and cost streams. His results indicate that more than 70% of the films were "likely to have generated a positive return for the producers, if the revenue from the secondary windows is taken into account," and that this number becomes even more positive on the level of the conglomerates (at which the different distribution branches are also considered). He argues that, at least for movies, the 20–80 rule and, more generally, the claim of entertainment being extraordinarily risky "is a pernicious narrative that in numerous ways serves the interests of major conglomerates." In sum, while we certainly do not ignore the existence of substantial risk for entertainment, we focus on understanding its sources and discuss managerial consequences in the context of business models. By offering insights into the effectiveness of various marketing approaches for entertainment, we hope that *Entertainment Science* (the theory *and* the book) might further alleviate the riskiness of the industry.

---

[13]Our book points at quite a few of those failures. See for example the fallacy of the once-ambitious Cannon Group in the 1980s which we summarize in our chapter on entertainment product innovation.

[14]Follows' analysis even tends to be conservative, as he does not consider potential profits made through what the industry refers to as "distribution fees."

Moreover, some scholars have suggested that an exponentially decaying adoption pattern, in which product revenues peak immediately after release and decline afterward at a rate that is consistent over time (so that the function decline becomes less steep when time passes), is a standard element of entertainment products (e.g., Jedidi et al. 1998; Luan and Sudhir 2010).[15] Although such patterns for entertainment products can indeed be observed quite often, we argue that they are not a "given," in contrast to the characteristics of entertainment we discuss in this chapter. Instead, we consider them the result of a particular strategic treatment given to the entertainment product—a treatment that is summarized by the "blockbuster concept," an ambitious integrated marketing strategy that we review toward the end of this book. Indeed, entertainment products that are not treated as blockbusters by the industry usually show very different adoption patterns, which often resemble those known for non-entertainment products. If there is indeed an inherent element in the decay function, then it results from the particular value patterns and short life cycles of entertainment products, which we address separately in this chapter on entertainment characteristics. Thus, even while we do not dedicate a separate section to decaying adoption pattern, we discuss the *reasons* for them.

Let us now move on to what we consider as the eight key characteristics of entertainment products. For each of the characteristics, we also point out its respective impact for the marketing of entertainment, which we will then discuss in more detail in this book's second half.

Entertainment Products Offer Hedonic Benefits

The Pleasure Principle

> "Writing 2,300 years ago, Aristotle concluded that, above all else, people seek personal happiness and pleasure."
> —Chen *(2007, p. 31)*

The first consumer-sided characteristic of entertainment products relates to the types of benefits consumers receive from them. As we expressed in

---

[15]In a formal perspective, in such a model revenues y are the result of $a \times b^t$, where a is the starting value (such as the revenues generated by a movie in its first week of release), b is the decay factor (or "multiplier") which determines the slope of the curve, and t is the time period (with 1 being the first week of the movie's release, 2 the second week, etc.).

our definition of entertainment, at the core of entertainment products is a unique motivation that sets entertainment activities apart from most everything else we do as consumers. *Entertainment Science* scholars call this particular motivation "hedonic" motivation, and they often contrast it with "utilitarian" motives that drive most of our consumption of other products (Alba and Williams 2013). This hedonic nature of entertainment is so foundational that experts often refer to movies, music, and novels as "hedonic products" (e.g., Strahilevitz and Myers 1998).

Morris Holbrook and Elizabeth Hirschman are two of the world's leading experts on this topic. In one of their seminal articles on hedonic consumption (Holbrook and Hirschman 1982), they explained the essence of the hedonic concept by linking it to Sigmund Freud's fundamental psychoanalytical concepts. They argue that for most other (i.e., utilitarian) products, such as fast-moving consumer goods or home appliances, people act as problem solvers who consume a product in order to reach a functional goal; i.e., the goal is accessed *through* the product. The product serves as a tool, or a means to a separate aspired end.

For example, we choose a low-fat food not because we love the taste, but to pursue the goal of losing weight. You buy a calendar app as a tool to help you gain control of your time or organize your busy life. We don't desire to consume these products per se; instead we want the outcomes they produce. The benefit comes from reaching the goal—not from consuming the product. When choosing a product like this, consumers carry out cognitive activities such as searching for information, rummaging their memory, and weighing arguments—activities which reflect "secondary process thinking" in that they acknowledge the requirements of a complex outside reality with which the consumer has to arrange his or her inner desires (Holbrook and Hirschman 1982).

Things are fundamentally different in the case of hedonic consumption. Instead of using a product to reach some separate goal, *it is the consumption of the product itself* that provides the consumer with gratification. That is, the product is not simply consumed as a tool or means to another goal; instead, it is the desire to experience the product which motivates the consumption. And why do customers desire to consume a product simply for the goal of experiencing that product? Because hedonic products offer, as we noted in our earlier definition of entertainment, a "certain quality" of experience—one that is often labeled *pleasure* (or enjoyment). Remember that pleasure is not only frivolous amusement, but can range from sensual gratification (in all its forms, e.g., desire, surprise, and the fear resulting from an exciting horror movie) to distraction to moral or cognitive challenge. Hedonic consumption is all about the experience of enjoying a product, or in our case, the experience of being entertained.

This "pleasure principle" which is at the core of hedonic consumption corresponds with Freud's idea of "primary process thinking"—a behavior which is carried out to immediately gratify the consumer's inner desires and needs. It is "'primary' in the sense that it hearkens back to the way a baby pursues immediate pleasure or gratification" (Holbrook and Hirschman 1982, p. 135). The entertainment product is an instrument that *directly* produces pleasure, whereas non-hedonic products only *indirectly* produce gratification by enabling the accomplishment of a separate goal. So in sum, entertainment is all about directly offering pleasure experiences to consumers. And, thus, entertainment products are judged by consumers solely on their ability to offer such pleasure (Peltoniemi 2015).

How "valuable" are such pleasure experiences, by the way? We have already shown that the provision of pleasure has substantive economic effects (in its totality) and can provide meaning and motivation for those who experience it. But isn't the mere act of consuming a product to experience pleasure a childish and cheap one, compared to being productive (such as by studying a book like ours)? It is not only Aristotle who disagrees, as expressed in our initial quote. In his Pulitzer prize winning novel ALL THE LIGHT WE CANNOT SEE,Doerr (2014) lets his protagonist answer this question: when he, a young German soldier of the Wehrmacht who has been indoctrinated by the Nazis and was involved in brutal war scenes, sees a young girl on a swing in war-torn Vienna, he considers her behavior not infantile, but quite the opposite: "This is life, … this is why we live, to play like this…" (p. 366). Scholarly support for this answer comes from social-psychologist Shalom Schwartz and his colleagues, who, in an extensive investigation of people's motivation, concluded that enjoyment is one of a small set of "universal human values," corresponding to a basic human requirement (e.g., Schwartz and Bilsky 1990).

Understanding the hedonic character of entertainment consumption is an important prerequisite to powerful marketing decisions for entertainment products, as consumer decisions aiming at pleasurable experiences differ substantially from those in which consumers are pursuing other goals.[16] As

[16]Let us add that utilitarian products can possess certain hedonic elements. All else equal, we will choose a better-tasting low-fat food while pursuing weight loss and you will choose a calendar app that has more attractive graphics while striving to organize your life. However, the essential motivation for consumption of the food or the app is to accomplish another goal—*not* the desire to experience the product itself.

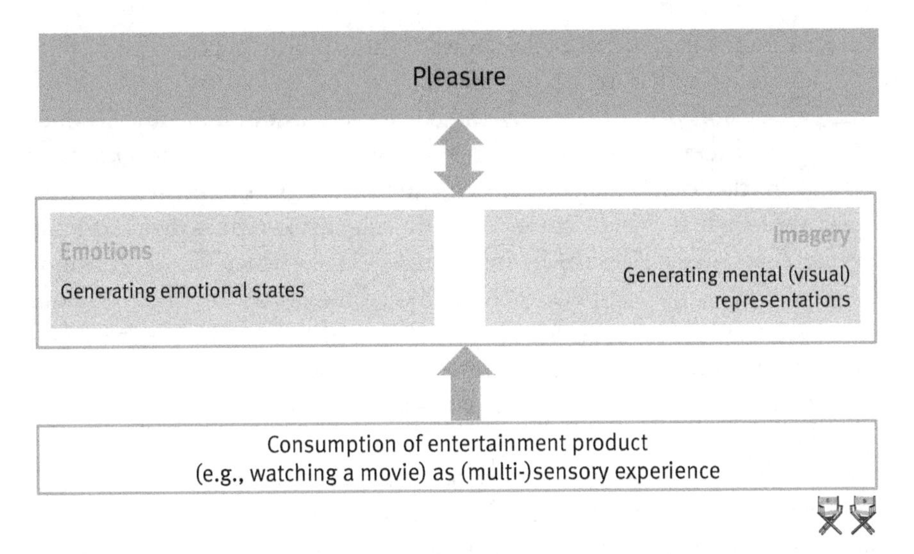

Fig. 3.2 Emotions, imagery, and sensory processes as key facets of hedonic consumption
Note: Authors' own illustration based on ideas by Hirschman and Holbrook (1982).

focal elements of hedonic consumption processes, Hirschman and Holbrook (1982) have highlighted the critical role of *consumer emotions* and so-called *imagery*. The latter describes a cognitive process that involves the mental, and often visual, representation of sensual experiences—think about the pictures that come to your mind when you hear the word "Jedi," or listen to the STAR WARS musical theme when someone else uses it as ringtone, as a result of having experienced the movie saga yourself many times. Figure 3.2 illustrates our thinking; for creating pleasurable experiences, the consumption process must generate emotional states and mental representations (e.g., images). The figure also points to the importance of sensory processes for triggering such emotions and images, and eventually, pleasure (Hirschman and Holbrook 1982).

Holistic Judgment

In addition to the particular roles of emotions, imagery, and sensory experiences, the pleasure principle that guides hedonic consumption also causes consumers' decision making and judgment—the very ways in which information is processed—to differ between hedonic and utilitarian products.

Extensive research in marketing shows that decision making for utilitarian products is predominantly focused on product characteristics, with the purchase decision being driven by the consumer's assessment of a product's multiple attributes. The weighted integration of these attribute judgments produce in the consumer an overall, and cognitive-dominated, attitude toward the product. This kind of decision making is reflected in expectancy-value theory (e.g., Fishbein and Ajzen 1975), which views consumer choices as the result of a person's implicit estimation of the likely outcome that would result from choosing that product and how attractive or valuable that outcome is to the consumer; scholars have developed various so-called multi-attribute models of consumer attitudes that operationalize this theory (e.g., Mazis et al. 1975).

For hedonic products that people consume as they strive for pleasure (versus striving for functional, utilitarian benefits), decision making does not follow the same logic. Specifically, individual product attributes are *not* at the center of a consumer's judgment, nor are multiple individual attributes integrated through "cognitive algebra" processes. Instead, consumers judge hedonic products more holistically and emotionally; e.g., does a product (or its advertising—such as trailers, samples, etc.) trigger the desired emotional states and reactions? Does the product spark *desire* within the consumer?

Thus, the consumer's judgment of a hedonic product's quality is not the (weighted) sum of its parts/attributes, but is their overall reaction to the product *as a whole*. Product attributes, such as the availability of a star actor or a prominent director, are not quality dimensions per se, but are instead used by consumers to infer whether a product will likely cause the anticipated pleasure state for which they would consume it. Accordingly, Belk et al. (2000) offer a "desire paradigm" as a hedonic alternative to the attribute-based, cognitively dominated information-processing utility paradigm.

Hedonic Does Not Rule Out Utilitarian

But if you have been trained as a marketer of fast-moving consumer goods, do not throw the baby out with the bath water. Products usually contain both hedonic and utilitarian elements, so that the idea of entertainment products as *purely* hedonic is an oversimplification to some extent. Just as people sometimes drive cars, cook meals, and use their computer to

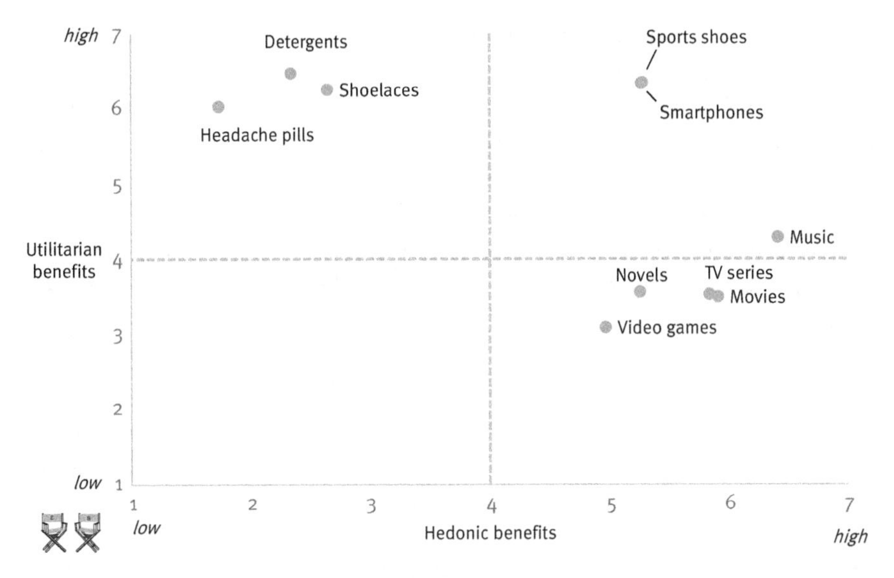

Fig. 3.3 The hedonic and utilitarian benefits consumers derive from selected entertainment and non-entertainment products

Notes: Authors' own illustration based on a survey of 359 undergraduate students; the number of ratings per product ranges between 313 (video games) and 358. We measured hedonic and utilitarian benefits with three items each on a 1–7 agreement scale and then used the mean scores for positioning the products.

experience pleasure, people also consume entertainment products to obtain utilitarian benefits in certain situations—such as when someone reads a classic novel to increase his or her *Bildung*, when s/he plays a game to remain part of a social group of friends, or watches a documentary or drama film to learn about a specific part of history.

Scholars have provided evidence for this non-exclusive existence of hedonic and utilitarian elements in products. For example, Voss et al. (2003) developed scales for measuring products' hedonic and utilitarian levels and then classified 16 products in a two-dimensional space, based on ratings by up to 380 students; whereas their results place video games in the high hedonic-low utilitarian quadrant, they find TV sets to be both highly hedonic *and* highly utilitarian. Figure 3.3 shows the results of a classification we conducted ourselves, based on 359 Münster business students' ratings of the hedonic and utilitarian benefits they derive from the entertainment products that this book is about (movies, music, video games, novels, and TV series), along with five diverse non-entertainment products (sports shoes, smartphones, detergents, shoelaces, and headache pills).[17]

Consistent with our arguments in this chapter, all five entertainment products offer high levels of hedonic benefits to consumers. Their utilitarian benefits are comparably lower; for the average respondent, music offers the highest level of hedonic benefits, followed by movies and TV series, whereas video games are perceived as somewhat less beneficial. At the same time, they are clearly distinct from zero—which supports our arguments that entertainment consumption, to a certain degree, can be motivated by utilitarian interests as well.[18] Detergents, shoelaces, and also headache pills, in contrast, are used almost solely for utilitarian purposes by consumers, and sports shoes and smartphones offer hedonic and utilitarian benefits to similar degrees (scoring high on both criteria), further stressing the two-dimensional nature of these concepts.

Managerial Consequences of the Hedonic Character

The hedonic character of entertainment products has several implications for their management and marketing. We address them in different parts of this book:

- *Consumer behavior.* As the degree of pleasure that consumers draw from entertainment depends heavily on the generation of emotions and imagery, entertainment scientists have dedicated substantial effort toward understanding these facets of entertainment consumption; we summarize their findings in our chapter on entertainment consumer behavior. We also introduce "sensations" and "familiarity" as critical concepts when it comes to providing pleasure to entertainment consumers—and develop a theoretical framework that puts these concepts at the center of *Entertainment Science.*
- *Product decisions.* In four separate chapters, we discuss a variety of issues that impact product decisions. For example, the importance of consumers'

[17]Specifically, we used the following statements for measuring utilitarian benefits: (1) "In general, [product] are very practical [praktisch] for me."; (2) "In general, I think of [product] as very useful [zweckmäßig]."; (3) "In most cases, I perceive [product] as very functional [funktional]." For hedonic benefits, we used the following items: (1) "In most cases, using [product] gives me a lot of pleasure [Vergnügen]."; (2) "In general, I really enjoy [Freude haben] using [product]."; (3) "Usually, I have a lot of fun [Spaß] when using [product]."

[18]Another reason might be entertainment's ability to support consumers' need for managing their mood, a motive that contains a certain instrumental (i.e., utilitarian) element.

sensual experiences leads us to pay particular attention to the role of technologies (such as Virtual Reality and 3D projections), which can influence the level of sensual stimulation in entertainment. Further, the holistic judgment that is typical for hedonic consumption assigns a different function to individual product attributes in entertainment—these attributes are used by consumers as "inferential cues" to help form conclusions about the overall quality of an entertainment product. Also, the holistic judgment means that traditional innovation techniques developed for consumer-packaged goods, which focus on functional attributes, are less appropriate in an entertainment context.

- *Communication decisions*. The hedonic character, with its focus on pleasure instead of functional performance, also carries important implications for marketing communications for entertainment products. Whereas the marginal utility of additional information about a product is nearly always positive when consumers are striving for utilitarian benefits, for entertainment products a threshold exists for adding new knowledge—when you reveal the identity of the murderer in the ads for your new thriller novel, the consumer's excitement might go *down* instead of up. Managers have to understand that too much information can hamper pleasure, so that determining the optimal amount of information provision can be crucial for an entertainment product's success. Further, we will show that pre-release "buzz" plays a huge role in building consumer anticipation for a new entertainment product, with fine-tuning marketing communications to stimulate consumer-to-consumer excitement being key.
- *Distribution decisions*. The pleasure-seeking motivation of entertainment also affects the demand for entertainment products over time. Demand for entertainment among consumers varies with economic conditions and consumer sentiment in unconventional, counter-cyclical ways, which should be considered when making distribution decisions.

Entertainment Products Are Prone to Satiation Effects

"People usually like experiences less as they repeat them: they satiate."
—Redden *(2008, p. 624)*

Of Utilities and Satiation

This second consumer-sided characteristic of entertainment relates to patterns in the amount of benefits that customers gain from repeated consumption of an entertainment product. Consumers' utility functions for entertainment products differ from those for other products: whereas the utility a consumer derives from using a "normal" product remains largely constant over time (e.g., using a washing machine or computer delivers the same utility on day 1, 2,..., *n*), the consumer's utility function for entertainment products is often "single-peaked." Specifically, the utility of entertainment usage peaks early and then declines with the number of usages, often in an escalating way. Consider the example of a TV series, whose initial season was aired by a nationwide broadcaster in Germany for the first time in 1997. When the identical series was aired again by the same station seven years later, each of the series' episodes attracted substantially fewer viewers (see Fig. 3.4).

Coombs and Avrunin (1977) have analyzed the psychological processes underlying this pattern, linking the single-peaked course to entertainment's hedonic character. They argue that when products are consumed for pleasure, consumers always face "good" consequences (such as new and exciting hedonic benefits), but they also face "bad" consequences (such as the opportunity costs that stem from not engaging in something "productive" and useful). For the observed utility pattern, it is crucial how these "good" and "bad" consequences of consumption develop over time when one consumes the entertainment product repeatedly.

Coombs and Avrunin argue that whereas the "good" consequences accumulate over time, they do so more and more slowly because the stimulation

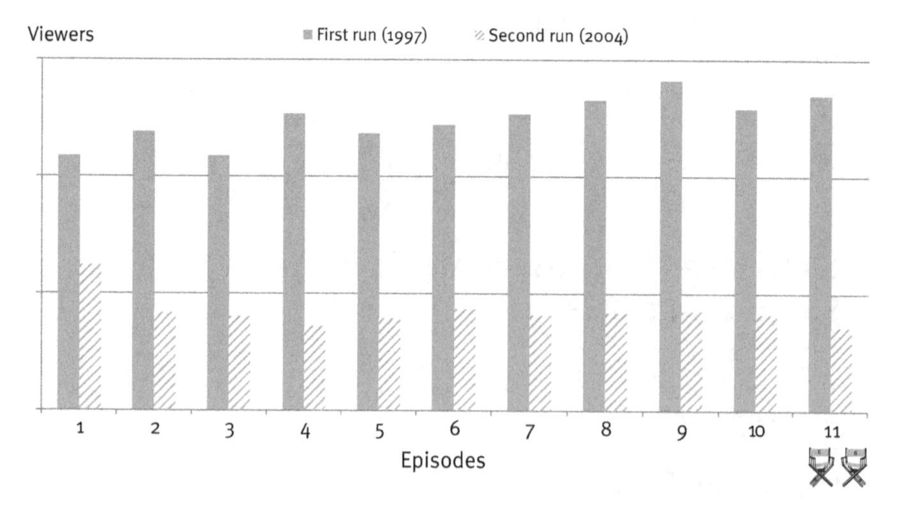

Fig. 3.4 First and second run ratings for a TV series in Germany
Note: Authors' own illustration based on data reported in various media.

consumers get from an entertainment product *satiates*—watching a specific TV show or movie, reading a specific novel, or playing a specific game over and over reduces the fun and can become a tedious experience, even if we like the product. The sensations that we enjoyed when watching a movie the first time do not repeat. The existence of satiation has led other researchers to compare life with a "hedonic treadmill," in which consumers constantly try new experiences in their pursuit of happiness (Brickman and Campbell 1971; Redden 2008).

In addition, the bad consequences not only increase with the passing of time and repeated consumption, but do so in an accelerated way. So, when it comes to the repeated consumption of an entertainment product over time, "[g]ood things satiate and bad things escalate" (Coombs and Avrunin 1977, p. 224). And because the total (net) utility consumers receive from consuming entertainment is the sum of the good and bad consequences, this net utility peaks early and falls off afterward, as we illustrate it in the right panel of Fig. 3.5.

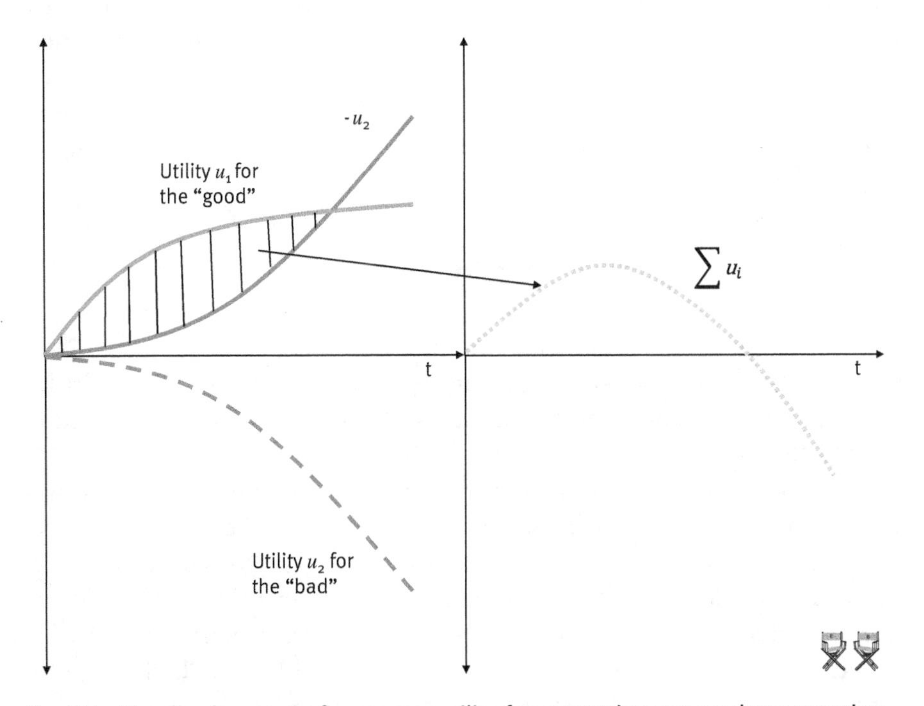

Fig. 3.5 The development of consumer utility for entertainment products over time
Notes: Authors' own illustration based on Coombs and Avrunin (1977). In the figure, u_1 are the "good" consequences of consumption, u_2 are the "bad" consequences of consumption (and $-u_2$ is their additive inverse), u is the sum of the "good" and the "bad" consequences. A consumer's total net utility from a product is the sum of u over a period of time t.

Reality, as is often the case, is more complicated—essentially because sensations are not the sole driver of entertainment utility. As we discuss in detail in the context of our sensations-familiarity framework of entertainment (see our entertainment consumption chapter), our familiarity with the material can cause us to enjoy a favored song or movie time and time again (if you have children who insist on watching Toy Story every evening, you know what we are talking about) or to repeatedly play a video game because of the skills that come with higher familiarity. Familiarity is the stuff that cult classics are made of: when Pretty Woman was shown by German broadcaster ZDF in 2013 (i.e., 23 years after its theatrical premiere), and after having numerous previous airings on TV, more than 5 million people tuned in, constituting one of the year's highest audiences for a movie shown on TV.

Because of this additional utility source, the *net* utility of an entertainment product does not necessarily have to be at its maximum when consuming the product for the first time for *every* entertainment product and consumer. But nevertheless, satiation is real and will reduce each entertainment product's sensation utility, as a major part of the total utility we derive from it, over time. That's why our kids eventually lose interest in watching a beloved film, and that is why we eventually stop playing a game if we have mastered it: blame satiation. Cult classics like Pretty Woman are, by definition, rare exceptions (or outliers, if you prefer statistical vocabulary), and we urge our readers to focus on norms, not exceptions.

Levels of Satiation in Entertainment

In the context of entertainment consumption, satiation can exist on different conceptual levels, as we illustrate in Fig. 3.6. First, satiation can exist for a single particular entertainment product—this is what we have discussed in the previous section. Because consuming an entertainment product repeatedly provides decreasing stimulation, the utility of doing so decreases over time. Please note that in Panel A of the figure, we let the consumer derive some utility from a new product even before s/he has consumed it for the first time—we do so because consumers often "anticipate" positive emotions from an entertainment product prior to actually consuming it.

Kahn et al. (1997) provide empirical evidence for this "basic" type of satiation: they conducted a laboratory experiment in which students had to choose a selection of songs that they had already rated earlier. Results clearly showed that the *repetition* of a sequence of three favorite songs reduced the

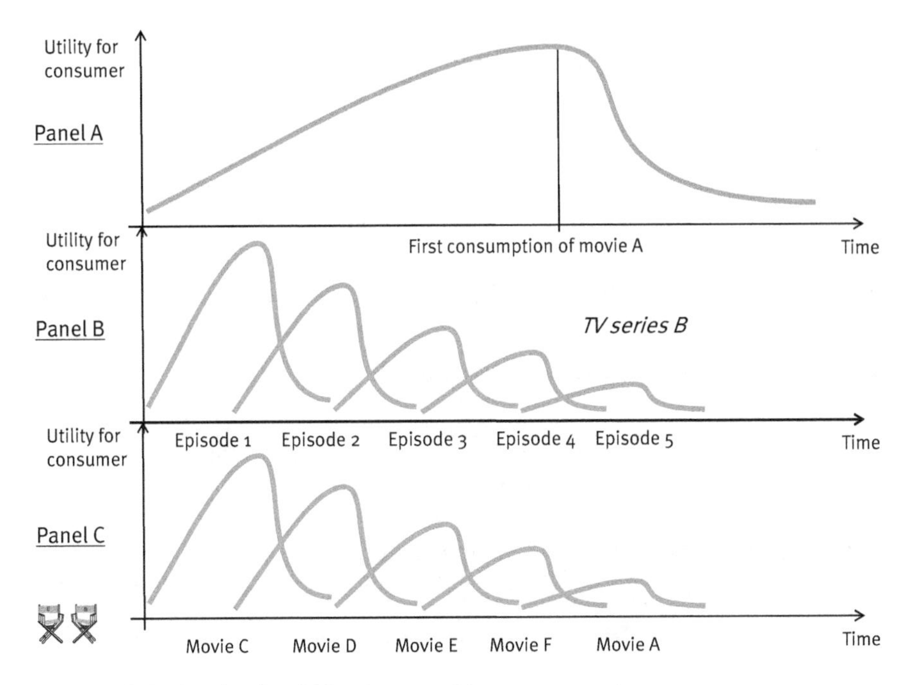

Fig. 3.6 Three levels of satiation in entertainment
Note: Authors' own illustration.

listeners' enjoyment level. When the same scholars asked students to listen to a sequence of fifteen 45-second clips of songs in a follow-up experiment (Ratner et al. 1999), they found that the students' enjoyment of listening to a song they liked "very much" was quite drastically lower when they had to listen to the same song repeatedly; their enjoyment scores dropped from 80 to 20 on a 0–100 scale. Figure 3.7 shows that this reduction in the listeners' enjoyment was less strong when the repeated playing of the favored song was interspersed with a second song ("alternation condition")—despite the fact that the other song was the one which participants had named their *least-favored* choice among all available songs in the experiment. Obviously, the variation (or reduction in satiation) that the second song added was valuable enough to overcome the deficit in liking it.

Second, satiation can also exist between iterated offerings of a joint product or brand. Consider the different episodes of a TV series such as House of Cards, the different issues of a comic like Spider-Man, or the different sequels to the Rocky movie or the Call of Duty game: in all these cases, the consumption utility of a new offering (such as the newest House of Cards episode) can suffer from a lack of stimulation (because it closely

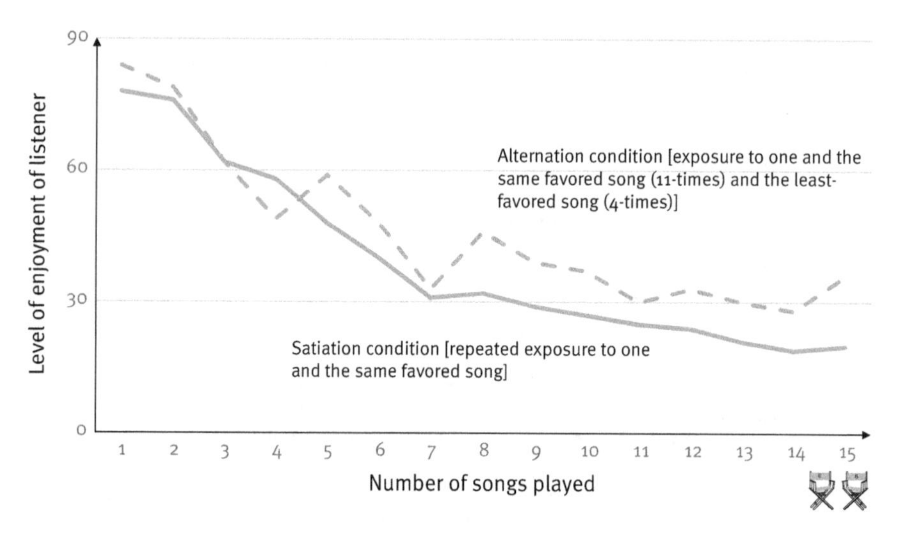

Fig. 3.7 Satiation effects for repeated music consumption
Notes: Authors' own illustration based on information reported in Ratner et al. (1999). In the alternation condition, the least-favored song was played at positions 4, 7, 11, and 14.

resembles previous episodes of the series), causing satiation with the product or brand. We illustrate this kind of satiation in Fig. 3.6's Panel B for the episodes of a TV series. Again, note that this pattern is not set in stone as familiarity can also cause utility to increase, at least for a certain period of time.

Third, satiation can even be caused by the existence of other entertainment products which are not part of the same series or brand, but are perceived as similar by consumers because of their content, style, etc. (see Panel C of Fig. 3.6). Barroso et al. (2016) provide empirical evidence for the existence of this kind of satiation effects that takes place beyond a single entertainment brand. They analyze the survival rates of all 2,245 television series aired in the United States from 1946 to 2003, coding their content based on descriptions by historians and subsequently determining clusters of similar shows. Using probit regressions they show that the more series address a similar theme as other shows have done before, the higher the probability that a series is canceled, with no further seasons being produced. Based on their results, the scholars conclude that "[r]epeated consumption of a particular product appears to reduce consumer interest not just for the product itself but also for other similar products (those in the same niche)" (Barroso et al. 2016, p. 576).

And in the context of music, Askin and Mauskapf (2017) analyze how a song's "typicality" (measured as its similarity to all other chart hits during the year before the song's release across 11 musical features, such as "dance-bility") influences it sales potential. For about 25,000 songs which reached the Billboard charts between 1958 and 2016, they find (via ordered logit regressions and negative binomial regressions, respectively) that from a certain point on, a higher level of such typicality/similarity goes along with a *lower* peak position and also *fewer* weeks on the charts.

We suspect that this kind of generic satiation hurt Kevin Costner's movie WYATT EARP, which was released (too) closely behind Val Kilmer's popular portrayal of this real-life lawman from the American "Wild West" in the movie TOMBSTONE. And we find it likely that consumers' tepid reactions to later installments of the DIVERGENT and the MAZE RUNNER movie series were at least in part due to consumers' satiation with the films' setting—after four HUNGER GAMES films, people were not eagerly waiting for more "teen-focused, dystopian future" movies.

Managerial Consequences of the Satiation Effect

The satiation effect inherent in entertainment products also has a number of implications for the way they should be marketed and managed. We address these implications in different parts of *Entertainment Science*:

- *Consumer behavior.* As managers try to win over customers, the satiation effect highlights a major challenge that needs to be overcome. This is clearly not a trivial matter, as satiation is closely tied to the sensations-familiarity framework which we argue is so fundamental for entertainment consumption. The framework points managers to the high "value potential" of familiar elements in entertainment for consumers, while also stressing that those familiar elements can cause satiation—and thus *reduce* the attractiveness of an entertainment product. This represents a delicate, but crucial, balancing act. We discuss the framework and the general requirement to balance familiarity with new sensations in our consumer behavior chapter.
- *Product decisions.* Addressing the satiation threat is a key topic when designing entertainment products. Satiation is relevant for original products, but is particularly key for those which are part of a product series or brand franchise. In our analysis of product decisions, we discuss the satiation that is inherent in brand extensions (such as sequels, remakes, and screen adaptations of books or games) and the consequences of satiation (such as for naming entertainment products).

- *Communication decisions.* The fact that consumer utility can be reduced by satiation affects also the information that an entertainment manager should share with consumers about a new product. Although enough information has to be revealed to attract customers and build anticipation, providing consumers access to (too) much of the product via trailers or samples might reduce the attractiveness of that product for them. They might feel that they have already experienced the new product, seeing little value in doing it again. Thus, managers must carefully consider what part of an entertainment product should be made available as part of the communication campaign.
- *Distribution decisions.* Entertainment products often reach consumers through a sequence of distribution channels (e.g., movies being released in theaters first, followed by physical home entertainment, streaming, TV, etc.). Whether a channel sequence is lucrative depends on how consumption in one channel influences the demand for a product in other channels. Satiation is a major factor in such inter-channel constellations, with high satiation reducing the likelihood that a person consumes a product multiple times across different channels.
- *Integrated decisions.* Finally, we believe there's a possibility that generic satiation might result from an overly narrow interpretation of the blockbuster concept of entertainment marketing by its biggest producers, something that could hurt the demand for the forms of entertainment we discuss in this book *in toto*.

Entertainment Products are Cultural Products

> "Hollywood movies are key cultural artifacts that offer a window into American cultural and social history."
>
> —Ibbi *(2013, p. 96f.)*

A third consumer-sided characteristic arises because popular entertainment products are so inherently prominent within cultures. "Culture" is a complex and amorphous concept that has been the subject of many academic studies and even more informal debates. Despite the existence of a multitude of perspectives, most culture scholars tend to agree that the core of any culture is a set of attitudes, values, and beliefs that is shared by a group of people, whether this group is a segment of consumers, the employees of a firm, the members of a tribe, or the citizens of a nation. Culture provides members of the group with norms for their behavior (e.g., Deshpande and Webster 1989).

A key to understanding how entertainment products perform economically is recognizing that they are "cultural" products; they represent important elements of a culture's existence, development, and content. Specifically, entertainment products "transport" their creators' attitudes and values which have the potential to influence people's perception of the world. This is why access restrictions have been erected for certain works of entertainment, either for all members of a culture or for a selection of them (such as minors). On the consumer side, it is these attitudes and values that entertainment transports, along with its aesthetics and their symbolic potential, that can shape a culture's identity and influence its entertainment consumption patterns.

Entertainment Products Express Attitudes and Values

Entertainment products are key transportation vehicles for a wide range of cultural attitudes and corresponding values and beliefs that are held by the artists who create them. Such values might be subtly embedded in a story, song text, or a product's aesthetics, as the movie RAIN MAN might be celebrated not only as a dramatic family story, but as a plea for treating people with handicaps with respect and appreciating their specific abilities. But values can also be more overtly presented in entertainment products, and they can be explicitly political. For example, movies have expressed right-wing ideology (see RED DAWN, a saga of a Russian invasion of the U.S. directed by John Milius, who named himself "an extreme right-wing reactionary," as quoted in White 2000), as they have transported socialist ideology (see most films by Sergei Eisenstein, Pier Paolo Pasolini, and Jean-Luc Godard).

Very similar arguments can be made for all other forms of entertainment featured in this book. The ability of music to transport values is evidenced by countless politicians who used, or wanted to use, songs as part of their campaign efforts. Examples include Bill Clinton (who featured Fleetwood Mac's DON'T STOP) and also Donald Trump—who used Aerosmith's DREAM ON in 2016 campaign rallies until the band's frontman Steven Tyler threatened to sue the "America-First" candidate for unauthorized use of the song (Kreps 2015). A popular video game brand that transports specific ("patriotic") political values is HOMEFRONT, a first-person shooter which, like RED DAWN, follows the narrative of a foreign invasion on U.S. soil (in this case by North Korean forces)—and for whose initial installment Mr. Milius served as consultant. Figure 3.8 shows an occupied America and a patriotic "Resistance" fighter in graphic art for the game's "reboot" HOMEFRONT: THE REVOLUTION.

Fig. 3.8 Transporting political values in entertainment
Notes: Graphical art for the game HOMEFRONT: THE REVOLUTION.

Because entertainment products carry societal attitudes and values, they also violate others. Thus, multiple regulations have been installed, which reflect a society's (or its leaders') perception of whether and how its members should be exposed to, or protected from, certain entertainment content and its depiction of events, institutions, and people. The most obvious of such regulations are restrictions on access, which either apply to a country's population as a whole or only to subgroups within it.

General access restrictions to *certain* kinds of entertainment are in place almost everywhere, but differ strongly between countries. In Germany, entertainment works are prohibited if they are believed to carry the potential to "incite hatred" (an echo of anti-Jewish Nazi propaganda such as the infamous JUD SÜß, which "encourage[s] a dislike of all Jews"—Culbert 2003, p. 205—and is still banned) or "injure human dignity" (for this reason, THE EVIL DEAD was banned until 2016; in September 2017, a total of 445 horror films and games are prohibited).

In China and Russia, foreign films and other entertainment content are only allowed to be distributed when narratives align with the ruling political ideology. Take the example of the Hollywood film CHILD 44 about a serial killer in the 1950s Soviet Union, which was pulled from release in

Russia because the ministry considered it "unacceptable to show this kind of film on the eve of the 70th anniversary of victory" (Walker 2015). In these countries, restrictions are also not limited to explicit political or moral elements, but also affect much more ordinary aspects of stories, scripts, and settings, following the idea that entertainment shapes consumers' perceptions about real-world phenomena of almost any kind. An insider reported that in China the "censorship always goes back to the Communist Party. They're in charge and they're always looking at how China is portrayed" (T. J. Green, CEO of Apex Entertainment, quoted in Langfitt 2015). When in the James Bond movie SKYFALL an assassin walked into a skyscraper in Shanghai and shoots a security guard, censors believed this resulted in China "looking weak"—and thus required the scene to be removed.[19]

And in countries with orthodox religious views, entertainment has been banned for "controversial" religious themes and blasphemy (e.g., THE LAST TEMPTATION OF CHRIST was banned in Israel, Mexico, and Turkey, among others). When Saudi Arabia adopted "ultraconservative religious standards" (Cowell and Kirkpatrick 2017) in 1979, this meant the end of *any* public screening of filmed entertainment because they were considered a "source of depravity" by religious leaders. When the country is lifting the ban in 2018, films will still be subject to moral censorship and require to be in line with "Sharia laws and ethical values of the kingdom," an official announcement stated.

Other, somewhat less radical restrictions are in place to prevent *certain parts* of a population, often minors, from accessing entertainment products that are not considered suitable. In the U.S., for example, games are rated by the Entertainment Software Rating Board (ESRB), with categories between eC ("early childhood") and A ("adults only 18+"). Music can be saddled with a "Parental Advisory Label," and different rating models are in order for comic books. Movies are classified by the Motion Picture Association of America (MPAA), based on their handling of sensitive issues and topics ranging from G (all ages are admitted) to the rarely assigned NC-17 ("no one under 17").[20]

[19]The political role of entertainment is further demonstrated by a recent Chinese initiative which precedes movie screenings with video messages, in which movie stars such as Jackie Chan promote "socialist core values," quoting from Mao and other national leaders (Qin 2017).

[20]In-between rating categories for movies are PG ("parental guidance" is suggested for children under 13), PG-13 (parents of children of 12 or younger are "strongly cautioned"), and R (under 17 year olds require adults to accompany them). For games, additional categories are E ("everyone"), E10+("everyone 10+"), T ("teen"), and M ("mature 17+"). We find it interesting that no such restrictions exists for books, in general. Obviously watching character Anastacia Steele having intercourse in 50 SHADES OF GREY—the movie—is considered to have more impact than reading its description in 50 SHADES OF GREY—the novel.

Other countries have similar rating systems in place, but they differ in terms of their rigidity (i.e., whereas parents have the last word in the U.S., ratings have law-like status in Germany) as well as their judgments, with the discrepancies in ratings reflecting the countries' respective cultural values and norms. For example, U.S. ratings often are highly sensitive toward sexual depictions and verbal indecency (reflecting the nation's puritanical roots). In contrast, Western European countries such as Germany, France, and Spain focus more strongly on the display of violence (whereas the U.S., due to its historical origins, has a more relaxed attitude toward weapons and fighting; Bellesiles 1996).

The movie SHAKESPEARE IN LOVE provides a quite drastic example of the consequences of the differing values when it comes to age ratings. Whereas the film is available for children of 6 years and above in Germany and for even younger children in Spain (because of its almost complete lack of violence), it received an R-rating in the U.S., restricting access for people under the age of 17. The MPAA blamed the film's "sexuality" for its restrictive judgment, but the raters' concern about mild nudity and implied love-making was probably heightened by the adulterous affair between the two lead characters, which caused moral conflict by violating conservative sexual mores.[21] In contrast, the horror comedy movie GREMLINS was considered appropriate for children by the MPAA, but Germany's jurors restricted access to the film to audiences of 16 years or older due to explicit scenes of gore and violence.

Entertainment Science scholars Leenders and Eliashberg (2011) provide empirical evidence regarding the systematic nature of these differences in rating systems. They studied age ratings of 227 U.S.-produced movies that were internationally released between 1996 and 2000, comparing ratings by the MPAA with those in eight other countries (mostly European, but also Australia and Hong Kong). Their results show that U.S. ratings are systematically more restrictive than in *all* other countries studied by the authors (with those in France and Italy being the most lenient). Leenders and Eliashberg also find that these differences can be partly explained by the culture dimensions identified by social psychologist Geert Hofstede (e.g., Hofstede 1991), in that *more* restrictive ratings are associated with higher

[21]This moral conflict is obvious in many comments on religious websites about the film. For example, Prins (1999), in his review for Christian Answers Network, judges the film as "extremely morally offensive" and writes: "Many Christians will no doubt be disturbed by the fact that Shakespeare and Viola are in a sexual relationship despite Shakespeare being married and Viola being engaged to another man."

levels of masculinity (a preference for achievements and material rewards) and individualism (people take care of themselves only) in a culture. In contrast, higher levels of a culture's uncertainty avoidance (i.e., people feel uncomfortable with and attempt to avoid uncertainty and ambiguity) go hand in hand with *less* restrictive movie ratings.

Restrictions on entertainment content not only reflect a culture's underlying values, but are also driven by the culture's beliefs regarding how being exposed to such elements affects consumers' well-being. Scholars have been working to shed light into these effects, particularly those of *violence* depicted in entertainment. Their debate of whether such violence has a lasting effect on consumers is a fervent one, with findings being "complex and multifaceted" (Marchand and Hennig-Thurau 2013). Several experimental and also correlational studies report increased physiological arousal and aggressive behavior, as well as decreased "prosocial" behaviors (such as showing empathy for others) in association with consumers' playing of violent video games (see Anderson's 2003 summary and his meta-analysis from 2010)[22]. Also, when Bushman (2016) meta-analyzes the link between a person's consumption of violent entertainment media and his or her subsequent perception of *others'* actions as aggressive ("hostile appraisals"), he finds a "small to moderate," but robust correlation.

However, other scholars such as Ferguson (2013) consider such findings misleading because of methodological problems. They argue that the use of invalid measures and the failure to include important control variables (which might offer alternative explanations) leave room for spurious effects and alternative explanations; like in so many fields of *Entertainment Science*, one has to be careful to not confuse correlations with causal effects.[23] The research design and data clearly matters, as several studies do not find that

[22]Meta analysis is a research technique that does not use original data, but combines data from many studies on a topic, trying to determine a "true" average effect.

[23]See also the critique of Anderson et al.'s (2010) meta-analysis results by Hilgard et al. (2017), who, after re-analyzing the same studies, question some of the authors' key findings. In a separate work, DeCamp (2017) provides rich insights into factors that determine people's *playing* of violent games, which potentially also influence real-life violence—a potential source of an endogeneity bias that might underlie the empirically measured correlations between playing violent games and real-life violence. Using a variety of surveys conducted in public and public-charter schools in Delaware in 2015, his cross-sectional OLS regressions show that playing violent games differs with gender (males play *way* more), several family factors (e.g., students play more when family members are in the military, the father has lost his job, or a family member was recently in prison), and social variables (those who feel safe at school or their neighborhood and find support from teachers all play less violent games). They also found impacts for health issues (e.g., those who take medication for bipolar disorder and/or are around people who smoke play more) and psychological states and attitudes (those who feel worried play more, as do those who have been bullied in their neighborhood).

violent entertainment increases aggression levels within consumers (e.g., McCarthy et al. 2016, who compare "aggressive inclinations" between players of a violent game and a "non-violent" game for a sample of 386 consumers). Some studies even find *reduced* aggression for some consumers who watch violent content (e.g., Unsworth et al. 2007)—a result that is consistent with psychology's catharsis theory, which states that violent media consumption can replace the need for aggressive behavior in real life (Feshbach and Singer 1971).

But like many other experiments on violence effects, even these studies suffer from a serious design limitation: most violence research has a very short-term focus, dealing with *immediate* effects only. Thus, it remains unclear how consumers' longer-term, real-life behavior might be affected by violent entertainment consumption. Maybe the work by Szycik et al. (2016) offers an insightful contribution: they examine brain reactions of 28 heavy users of violent shooter games toward positive, negative and neutral pictures, comparing them to a control group of the same size. They find *no* differences that would point to players' emotional desensitization, a precondition for violence spreading from entertainment to the streets (see also Szycik et al. 2017). Let us add that we are not alone in our skeptical view of the supposed scholarly evidence for a causal link of media violence and real-life behaviors. When the U.S. Supreme Court was asked to prevent minors from accessing violent games, it spoke quite critically about the state of research on violence effects, and particularly the work by Anderson and his colleagues—the judges refused to take their findings into consideration.[24]

Entertainment Products Constitute Cultures and Influence Their Choices

Because entertainment products transport meaning, they also shape and define any culture. Consumers judge entertainment based on the messages and values they transport, but even more so on "aesthetic criteria"

---

[24]Specifically, the judges stated that "[Anderson's studies] do not prove that violent video games *cause* minors to *act* aggressively (which would at least be a beginning). Instead, '[n]early all of the research is based on correlation, not evidence of causation, and most of the studies suffer from significant, admitted flaws in methodology.' (Video Software Dealers Assn. 556 F. 3d, at 964). They show at best some correlation between exposure to violent entertainment and minuscule real-world effects, such as children's feeling more aggressive or making louder noises in the few minutes after playing a violent game than after playing a nonviolent game. [But even] those effects are both small and indistinguishable from effects produced by other media [such as Bugs Bunny TV shows]" (*Supreme Court* 2011). For those of our readers who want to dive deeper into potential antisocial, but also prosocial effects of entertainment, and video games in particular, we recommend the book edited by Kowert and Quandt (2016).

(Thompson et al. 2007, p. 630). As such, entertainment products are important representations of almost any group's identity and their achievements and status, something that applies to a local scene, a society, or even a civilization. Consumers' entertainment choices tell others their "personal values, ambitions, beliefs, and perceptions of the world and themselves" (Schäfer and Sedlmeier 2009). Take the example of the Star Trek sub-culture: based on a 20 month-long fieldwork investigation, Kozinets (2001) reports qualitative evidence that, for "Trekkies," the Star Trek brand is essential to "construct a sense of self and what matters in life" (p. 67).

The aesthetics of entertainment play a particular role in this: they serve as a "signifying system" (Markusen et al. 2008) and offer consumers "sign-value" (e.g., DeFillippi et al. 2007). The attitudes one holds in regard to aesthetics incarnate the group's "taste" (for a detailed discussion of the taste concept, see our chapter on entertainment product characteristics). Think about the smartphones ringtones and jingles that you use: why you use them and whether your choice was influenced by what they might tell others about you (Audley 2015). "[I]t's all identity now," writes Abebe (2017) about music, and that applies to all kinds of entertainment choices.

Thus, a group's entertainment consumption choices send a strong symbolic message to both their members and to others. They establish "symbolic borders" against other groups, particularly for consumption choices that are publically visible. When we tell others that we are going to watch a movie or announce via Facebook that we are "listening to" a song or playing a game, we send a message that we belong to, or *would like* to belong to, a certain cultural group. In essence, cultures often define themselves via the entertainment products they value, the ones they ignore, and the ones they despise. Consistent with this signaling role of entertainment, Iannone et al. (2018) found, in a series of experiments, that consumers experience negative psychological consequences when they feel "out of the cultural loop in everyday life," lacking knowledge regarding, for example, musicians, movies, and books. The consumers reported less satisfaction with fundamental human needs—namely, they felt more disconnected, less good about themselves, and less in control of things.

Figure 3.9 shows that the link between a culture and entertainment products is actually a reciprocal one. A specific set of films, TV shows, music etc. help to define a group's culture through their "sign-value," which goes along with the development of preferences for such forms of entertainment. It is these preferences which then influence the future entertainment choices of the culture's members, who will watch movies and listen to music which are in line with these cultural values.

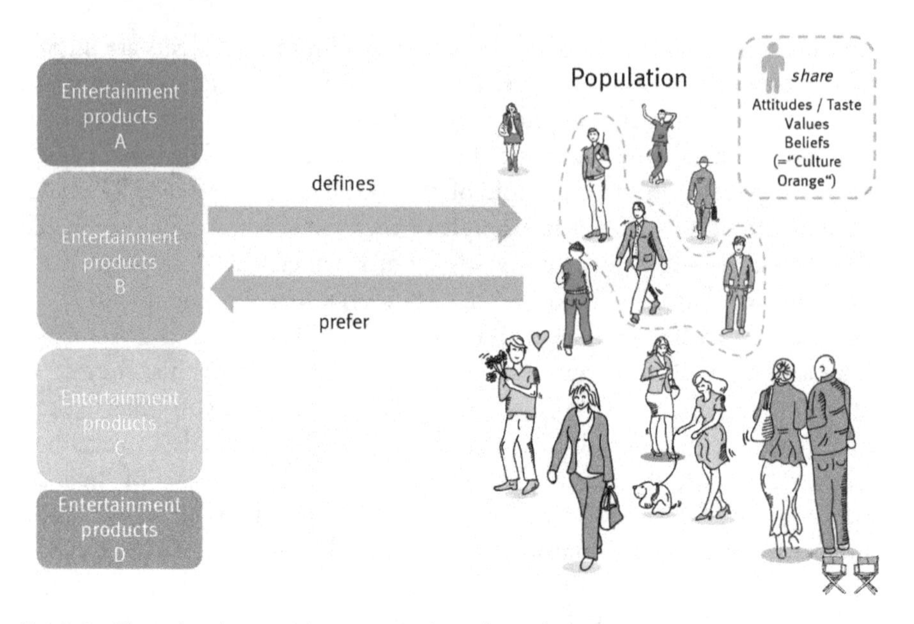

Fig. 3.9 The role of entertainment products for cultures
Notes: Authors' own illustration. With graphical contributions by Studio Tense.

Entertainment products can define and reinforce a culture and its taste, but they can also *challenge* the culture's taste: that's because new entertainment experiences require new quality judgments from the group, and the very nature of entertainment requires a lot of those judgments to be made in relatively short time. Is Bernardo Bertolucci's LAST TANGO IN PARIS a progressive masterpiece—or smut disguised as art? And is Clint Eastwood's AMERICAN SNIPER an anti-war movie, or a right-wing fantasy that trivializes the complexities and victims of war? If these directors, the involved stars, or the kinds of films are important for a group and it cannot find consensus regarding such controversies, even the group's very existence can be threatened.

Consider the example of the "poppers," a (sub-)culture formed by a group of young German consumers that rose in the 1980s as a counter-movement to the then-soaring "punk" scene. Poppers were mostly upscale kids who celebrated consumption and were conformist, a kind of protest against the then prevalent "anti-consumption" youth movement. Like any culture, poppers needed an aesthetic identity, which the group found in the hedonic cool of movie characters such as Tom Cruise's Joel Goodson (in RISKY BUSINESS) and in glossy pop music from bands such as Roxy Music and Spandau Ballet, whose brands and dress styles were largely influential for the poppers.

The group lost coherence and eventually dissolved with its members joining other (sub-)cultures, when the poppers' entertainment idols adopted new trends, such as Tom Cruise moving on and starring in ambitious dramas (such as RAIN MAN) and historic anti-war movies (BORN ON THE FOURTH OF JULY).

Managerial Consequences of the Cultural Character

As with the hedonic and satiation aspects, the cultural character of entertainment products has implications for how they should be marketed and managed. Specifically, the following parts of this book are affected by it:

- *Product decisions.* Managers have to account for how an entertainment product's content and aesthetics will resonate with different members of the culture that constitutes the product's target markets. For consumers, an entertainment product's cultural fit is mainly determined by its genre and content themes. Here, cultural differences are strong *between* countries, which requires a thorough understanding of international entertainment markets when a global release strategy is scheduled. Moreover, managers need to be aware that the attitudes and beliefs that define a culture are not constant over time, but often vary with political, social, and economic occurrences, thereby affecting the demand for corresponding entertainment products. As a result, the success potential for an entertainment product can be affected by a *zeitgeist* factor. In addition to the consumers who make up the potential audience for the product, managers must also consider reactions by governments and other authorities who decide about accessibility for consumers; something that sheds light on why invasion tales, such as the HOMEFRONT game, feature attackers of obscure origin. Also, because the cultural character of entertainment implies age ratings, managers need to know about how such ratings, as well as the content elements that drive the rating, influence a product's success potential by restricting, as well as attracting, potential customers.
- *Communication decisions.* Because of their symbolic value for consumers, entertainment products can hold a strong personal resonance for different groups of consumers which serves as a source for anticipation of a new product. If such anticipation is expressed in observable behaviors, it can stimulate massive consumer buzz for a product (you might remember the frenzy surrounding the release of STAR WARS: THE FORCE AWAKENS!) We

discuss how such buzz can trigger self-enhancing loops, attracting even more consumers, and what managers need to do to stimulate and harvest such buzz.

Entertainment Products are Difficult to Judge

The fourth consumer-sided characteristic that makes entertainment products stand out and requires specific marketing and management solutions is the difficulty which consumers face in judging their quality. As we will discuss in this section, this difficulty is particularly pronounced *prior* to consuming a new entertainment product for the first time, but lingers on even after the consumption act. It is fed by two sources that have strong theoretical bases: entertainment products' "experience good" character (meaning that consumers have to choose with less-than-complete product knowledge) and these products being works of art (with appreciation of art being a matter of "taste").

The Experience (and Quasi-Search) Good Character of Entertainment

For the consumer, the quality of an entertainment product is largely unobservable prior to consumption due to its experience good character. As De Vany and Walls (1999, p. 288) have so concisely phrased it for the context of filmed entertainment, "No one knows they like a movie until they see it." The same can be said for all other forms of entertainment that we discuss in this book—customers do not truly know if they will like a video game until they have played it, a novel until they have read it, or a song until they have listened to it.

The concept of "experience goods" was introduced by economics scholar Nelson (1970), who argued that for certain products, the information that a consumer can gain about a product's quality via experiencing the product is *far* superior to the information about its quality he or she can gain through pre-consumption search. Nelson's theory recognizes that, for many products, thorough and informative inspections of the product can happen prior to purchase. For example, trying on a dress in a store or test-driving a car enables you to more accurately assess a product's quality before buying it. And even without trial, one can judge the color and design of a car before having driven it. In the case of experience products however, such pre-purchase judgments are either not possible (e.g., it's difficult to "try on" a haircut) or can happen only by the consumer incurring prohibitive costs.

In reality, products are unlikely to be classified as 100% search or 100% experience, but instead consist of combinations in differing proportions (e.g., Ekelund et al. 1995). Thus, let's not consider the classification as a dichotomy, but instead as a continuum, on which products can be placed based on the share of quality information that consumers can access prior to purchasing it and the share they can only determine thereafter. We name product elements as "search attributes" if their quality can be judged by the customer without purchasing the product (e.g., the color of a car, the material of a chair, the size and resolution of a smartphone screen), whereas "experience attributes" are those aspects of a product that have to be experienced after purchase to be evaluated (e.g., Hennig-Thurau et al. 2001).

But why do we refer to entertainment as experience goods (or dominant on experience attributes, to be precise)? Isn't there quite a lot of information available for entertainment products which *can* be used to make informed judgments about a new movie or game prior to paying for it? There is information about a movie's running time, the bonus features on the Blu-ray, and its cover design or the movie's poster. It will be clear prior to consumption whether the film is shown in 2D or 3D and the game has a multi-player mode. And there is information about the movie's genre, its stars, director, and producers, as well as about its central characters (particularly if it is a sequel or adaptation). Heck, on the Internet you can even learn about a product's development budget! So, why aren't entertainment products search goods then?

The answer to this question is three-fold. First, entertainment products indeed possess some search attributes, such as the movie's runtime, the bonus features, and the cover design in the listing above. However, these bits of information are usually not central for the consumer's overall quality judgment. Very few people love a film for its runtime or a game for its cover. Search attributes exist, but their informational value is modest.

Second, some other search attributes refer to technological aspects of the entertainment product (such as a 3D presentation, the availability on a popular e-reader, or a multiplayer mode). These objective features can be important to consumers, but they are usually *generic* in nature—they describe a type of products rather than a single, specific product. Thus, they seem to complement the product-specific quality judgment of a consumer, but do not form it. Very few people watch a movie *because* it is shown in 3D. Instead, they might prefer to watch the 3D version of the film over its 2D version (or vice versa).

Third and most important, the other objective attributes mentioned above (e.g., artists, sequels, budgets) are not reasons for a consumer to love an entertainment product; instead, these attributes serve as *signals* of the product's quality. Consumers use such signals to make pre-consumption *inferences* (based on prior own experiences with other entertainment products), but final quality judgments *after* consumption do not rely on these earlier inferences. Take the example of a movie star like Johnny Depp: Although consumers will never be quite sure whether they are going to like a movie in which Mr. Depp stars (everyone has hated at least one movie featuring his or her favorite star or despised a song by a beloved singer!), they might *infer* from his participation, and their own personal history with other movies in which Mr. Depp appeared, that this new movie will have a similar level of appeal to them (and make a ticket purchase decision accordingly).

We have coined such signals "quasi-search attributes" (Hennig-Thurau et al. 2001); the degree to which they produce high quality cannot be truly known by the consumer a priori (unlike the case with "true" search attributes), but they provide a basis from which consumers can make *assumptions* about how the attributes might affect consumers' pleasure levels. If someone who tends to love Johnny Depp movies hears about a new movie starring the actor, she (or he) might infer from her (or his) own experiences with prior Depp movies that this new movie will also offer the desired benefits.[25]

However, these inferences clearly do not *guarantee* that the consumer will eventually like the new movie—a quasi-search attribute is only a signal, and like all other signals a product sends, consumers' inferences based on them can be (and often are) erroneous. Other than perhaps by priming the customer's expectations, quasi-search attributes do not influence the consumer's consumption-based quality judgment during and after the experience. Let's also keep in mind that the inferential character of any single quasi-search attribute is reconciled during the holistic way consumers judge hedonic entertainment products; as we have discussed earlier, consumers rarely like a movie or a book because of their individual attributes (such as Mr. Depp's participation), but because of the total consumption experience.

In summary, experience attributes dominate consumers' quality judgments for entertainment products, whereas search attributes play only a limited role. Quasi-search attributes are vehicles that are used by consumers to transform what are experience attributes into information that can be used

[25]Let us note that there are specific constellations in which a particular quasi-search attribute functions as a "true" search quality for *some* consumers. In the case of Mr. Depp, a particularly dedicated fan may find him so appealing that he or she goes to a movie simply to *look at* Mr. Depp, not for his contribution to the film—and thus indeed knows what he or she will get in advance.

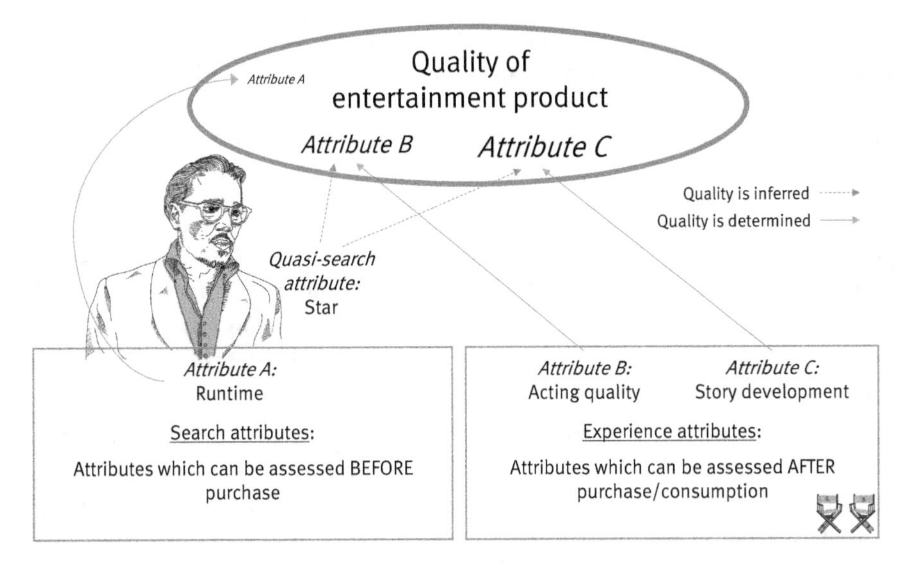

Fig. 3.10 Search, experience, and quasi-search attributes of entertainment products
Notes: Authors' own illustration. With graphical contributions by Studio Tense.

to make a better judgement about a new product's quality prior to purchase and consumption. A good way to think of them is as *proxies* for experience attributes. Figure 3.10 illustrates the different attribute types for entertainment products. A new movie's runtime can be determined in advance and thus acts as a search attribute, but it plays only a marginal role for the overall quality assessment, as expressed in the small font in the figure. In contrast, the quality of the acting and the story are of greater importance to overall quality judgments, but they require the consumer to actually watch the film; thus, these are experience attributes. A known movie star provides a proxy for these quality aspects prior to consumption, and the consumer generates inferences from this quasi-search attribute based on previous experiences and associations.

Entertainment Choices Depend on Taste

The Trouble with Taste Judgments

But the experience character isn't the whole story when it comes to understanding the difficulties consumers have with judging an entertainment product's quality. Instead, judging the quality of an experience is much more delicate for entertainment than for other experience goods. The reason is entertainment's cultural character, which implies that aesthetic and artistic aspects of the product play a focal role for its appreciation. How much benefit a consumer draws from an entertainment experience essentially depends on the consumer's taste—and taste judgments always involve a notorious level of subjectivity, because after all, "Beauty lies in the eye of the beholder." Because taste judgments are both subjective and holistic, they have to be generated from scratch every time a consumer experiences a new entertainment product and cannot be determined by simple heuristics or formal rules. Do I really like the song I just heard on the radio? And how much? And why exactly? Justifying our tastes vis á vis our friends can be a tremendous challenge.

This uncertainty surrounding taste is increased by the fact that consumer judgments about the quality of an entertainment product are also affected by some external standards of "artistic excellence." Such standards have developed in the form of cultural and aesthetic criteria, over time; they have been defined by experts, such as media theorists and philosophers. Knowledge regarding such artistic criteria of excellence varies immensely across consumers, but the artistry of entertainment products forces consumers to make taste judgments—even when they lack the expertise that would be required.

Consumers who read a novel after being told that it is "a true classic" will probably rate the book's quality more highly than they would have without any taste-related foreknowledge. Even if we are not *sure* whether what we read, watch, play, or listen to is of high quality, our quality perception is influenced by the artistic excellence we *assume* the product has, with this perception then having some impact on our pleasure experience, or consumption enjoyment. Scholars have stressed the coexistence of an "objective" facet of taste with a subjective, personal facet—Zenatti (1994, p. 177), for example, defines taste as the assessment of a work's "[artistic] value on the one hand, and the perception that [it] pleases or displeases us on the other. [Taste] is … expressed by judgments that mainly concern either the [artistic] value granted to the work or personal feelings of enjoyment."

Do Consumers Have Taste, After All?

A focal question for understanding consumers' taste judgments and their uncertainty regarding such judgments is determining to what extent "ordinary people" actually possess taste. Let's begin with looking at a cultural experiment that was conducted in January 2007 in Washington, D.C. To find an answer to the question whether beauty "would transcend" (Weingarten 2007), an ordinary looking man in jeans, T-shirt, and baseball cap played six classical pieces on the violin during morning rush hour for three-quarters of an hour at the city's main traffic hub. The incognito violinist was Joshua Bell, one of the world's most acclaimed classical musicians, who performed some of the finest tunes ever written, such as Schubert's AVE MARIA, on a Stradivari violin valued at $3.5 million. In the end, out of the more than 1,000 people passing by, only seven stopped and listened to what was indisputably of artistic excellence, for a minute or longer, and 27 gave some money. There was never a crowd. Clearly, the beauty of Mr. Bell's performance did not transcend his street attire, at least not in this specific setting.[26]

Entertainment Science scholars have also have tackled the same question in more systematic ways. Their generalizations draw a somewhat differentiated picture of people's taste—most empirical investigations confirm that taste differences exist between experts and lay consumers, but do so only to a certain extent. The seminal study of the topic dates back almost half a century when Getzels and Csíkszentmihályi (1969) compared how small groups of experts (artists and art instructors) and non-experts (math and business students) rated the aesthetic value of 31 drawings. The judgments of the two expert groups were highly correlated ($r=0.73$), as were those of the two non-expert groups. But when comparing the experts to the non-experts, the overlap between judgments was, though still clearly different from zero and positive, significantly lower (correlations ranged from 0.11 to 0.43).

Their findings foreshadowed future investigations that center on this book's focal entertainment products, despite methodological variations. For example, Wanderer (1970) compared ratings for 5,644 movies by (up to nine) professional critics and an unreported number of members of Consumers Union ("lay audiences") that were published in CONSUMER REPORTS between 1947 and 1968. Classifying movie ratings as "equal" (when both groups rated a movie identically on a 10-point scale) or "different," Wanderer finds the overall average of "equal" movies to be 53%, ranging from 27 to 71% for individual years. Separately, for a sample of 1,000 films from pre-1986 that had all been recognized for high aesthetic qualities (by awards or inclusion in best-films lists), Holbrook (1999) reported a significant, but limited correlation of 0.25 between the ratings by six movie

[26]Mr. Weingarten's (2007) coverage of the experiment in the Washington Post is as fascinating as the experiment itself; although our book refers to a large number of awarded pieces of entertainment, it is probably the only item in our reference list that itself was honored with a Pulitzer Prize. You might also watch a time-lapse version of Mr. Bell's performance at https://goo.gl/MmwRBi.

guides (his expert measure) and the subscribers of the pay-TV channel HBO (his measure of ordinary consumers).[27]

Interestingly, when he used secondary data from 219 movies released in the U.S. in the year 2000 to compare expert and consumer tastes, he found a much stronger correlation of 0.84 (Holbrook 2005). Why is this? Rather than disproving taste differences, his result point at the existence of substantive heterogeneity among different consumer groups. Instead of ordinary HBO subscribers, Holbrook this time used IMDb users as raters: consumers who tend to be "enthusiasts" and thus will, on average, have a much higher-than-average involvement and expertise than "normal" consumers. More systematic evidence for such taste heterogeneity is offered by Debenedetti and Larcenieux (2011), who calculate correlations between experts (professional critics) and ordinary moviegoers (surveyed via exit polls) as well as the users of allocine.fr, a French film site comparable to IMDb. Whereas the ratings of experts and moviegoers for 622 popular films released between 2005 and 2009 in France correlate with a small value of 0.19 (which compares nicely with Holbrook's 1999 study), the correlation between experts and film site users is 2.5 times as high ($r = 0.49$), being closer to the results from Holbrook's 2005 study. Interestingly, the overlap between experts and film-site users is even higher than between the two customer groups (which correlate at 0.40). So, taste is not (solely) determined by occupation.

With these results in mind, we could not resist taking a look at taste differences ourselves. Figure 3.11 lists the correlations between movie ratings of four groups—professional reviewers, IMDb users, opening night moviegoers, and subscribers of an international movie/series streaming service ("subscription video-on-demand," or SVOD). We analyzed the groups' quality judgments for a random sample of 200 films representing ten genres (20 films per genre).[28] Our findings confirm those from prior studies: whereas the correlations between the different groups are all statistically significant (i.e., higher than from zero), they are also far from perfect overlap, with the average correlation across groups being only 0.56 (which equals a shared variance, or R^2, of only about 30%).

The largest differences in group taste exist between professional reviewers and "ordinary" consumers (as captured by both opening night mov-

[27] As the squared correlation coefficient equals the shared variance of the two ratings, professional critics' evaluations explain only 6% of the ordinary consumers' preferences, and vice versa in Holbrook's study.

[28] The genres were: action, comedy, drama, horror, independent, international, romance, science fiction/fantasy, sports, and thriller.

	Opening night moviegoers	*SVOD subscribers*	*IMDb users*	*Professional reviewers*
Opening night moviegoers	1			
SVOD subscribers	.66	1		
IMDb users	.45	.71	1	
Professional reviewers	.34	.43	.74	1

Fig. 3.11 Correlations among movie quality ratings by four different groups
Notes: Authors' own illustration based on data published by various sources, including Metacritic (for professional reviews), CinemaScore (for ratings by opening night moviegoers), and IMDb. Because opening night ratings were only available for films that were released widely in North America, correlations for this variable are only based on ratings of 180 films. We converted all ratings into numerical scores.

iegoers and SVOD subscribers), whereas we also find film "enthusiasts" to have a similar taste as professional reviewers. One fresh insight is that we find that taste standards differ between *product types*—whereas ratings are fairly homogeneous for some genres (e.g., the average correlation for science fiction/fantasy is an impressive 0.81), taste judgments vary quite strongly between groups for others. For example, for independent films the average correlation is only 0.34.

And how about taste for other entertainment products? In a study on musical taste, Holbrook et al. (2006) compared experts' (faculty and graduate students in music) and ordinary consumers' (students from non-music colleges) ratings of 200 musical variations of the jazz/pop song My Funny Valentine. The experts' judgment of the "aesthetic excellence" of a song version (which the authors define as "artistic creativity and technical precision") and consumers' "excellence judgments" correlate somewhat stronger than in most movie studies, but still far from perfect ($r=0.55$). However, Holbrook et al.'s results also provide evidence for something else: that for us, as consumers, the perception of something being of "great art" does only partially translate into great personal enjoyment (the correlation between both is 0.59). We will get back to this (and its managerial consequences) when debating the link between quality judgments and the financial success of entertainment products in our chapter on entertainment product quality.

Where Do Differences in Taste Stem From?

Why do such differences in taste exist? Research has stressed three factors as determinants of persons' taste: their "cultural capital," their age, and their national culture. Let's take a quick look at their respective roles for shaping consumers' tastes.

Cultural capital. The concept of "cultural capital" was introduced by French sociologist Pierre Bourdieu (e.g., Bourdieu 2002) as one of three resources that determine a person's status in a society, in addition to economic capital (financial resources) and social capital (relationships, affiliations, and networks). Cultural capital consists of "a set of socially rare and distinctive tastes, skills, knowledge and practices" (Holt 1998), which relate to the arts, but also to politics, education, etc. This cultural capital is acted out not only through visiting galleries and owning works of art, but also through consumption activities in entertainment, sports, and other parts of life.

According to Bourdieu, a consumer's "social milieu" is crucial for the level of his or her cultural capital. It is the upbringing, the formal education, and the socialization that convey and continuously refine the consumer's cultural skills and resources (e.g., Holt 1998). In other words, cultural skills are taught by parents and teachers; they are mainly a function of the parents' education and occupation, in addition to a consumer's own education and interactions with others.[29]

Accordingly, the level of cultural capital a consumer possesses should determine his or her ability to decode the "innate excellence" embedded in an entertainment product and also to alter the degree to which the consumer enjoys such entertainment. Based on a sample of 1,005 Israeli consumers, Yaish and Katz-Gerro (2010) use factor analysis to classify various forms of entertainment into "high-brow" versus "popular" (i.e., mainstream) products. The scholars then provide evidence (via structural equation modeling) that consumers' preferences regarding these kinds of products (i.e., the consumers' entertainment tastes) differ with their respective cultural capital. For example, a "high-brow" taste was influenced by a consumer's education as well as his or her parents' consumption of high-brow choices.

Rössel and Bromberger (2009) find similar results for the movie-related taste of 590 German consumers; regression results show that "high" cultural tastes (expressed by a consumer's preference for "arthouse" films) are positively associated with the level of the father's education. And for a sample of 1,860 German consumers, we show that the extent with which people watch new "high quality" drama series such as Breaking Bad and Mad Men can be linked to higher levels of cultural capital (Pähler vor der Holte and Hennig-Thurau 2016).

[29]Holt (1998, p. 23) also suggests a pragmatic measure of cultural capital, which essentially weights a consumer's and his/her parents' education and occupation.

Whereas all these studies implicitly assign consumers into two general segments that each have a distinct taste (low cultural capital consumers who prefer "low-brow" genres versus "elite" consumers who prefer only "high-brow" genres), scholars have also pointed to the existence of a third group, the so-called "cultural omnivores": consumers who possess a high level of cultural capital, but are at the same time open to experiencing a wider range of genres and products (e.g., Petersen and Kern 1996). Related, using a combination of methods and data sets we found evidence that an entertainment product's *lack of* artistic excellence can sometimes be the very reason for high cultural-capital consumers enjoying the product. For example, liking "media trash" can allow these consumers to challenge the societal norms which led to the product's classification as being of low quality in the first place (e.g., Bohnenkamp et al. 2012).

Consumer age. A separate stream of taste research has studied the role of a consumer's age for understanding taste. A key finding here is that consumers have "impressionable years" (Peltoniemi 2015, p. 44) in their lives during which their taste is determined. Specifically, Holbrook and Schindler (1989) analyzed the preferences of 108 consumers for musical recordings they had heard at different points of their lives and found them to follow an inverted U-shape pattern: we like those songs the most that had been popular when we were in our early 20s (see Panel A of Fig. 3.12).

Janssen et al. (2007) then extended these insights by asking an (international) sample of 2,161 consumers to name their favorites in three entertainment categories: music, books, and movies. They also found patterns that resemble inverted U-shapes, with most favorites in all three product categories stemming from the period when consumers were between 16 and 20 years old (Panel B of Fig. 3.12). The curves' patterns are pretty similar across the different entertainment products, with novels we read at a later age tending to have a slightly stronger influence on us than songs and movies. (So if you agree with THT that Michael Jackson's music is the greatest, we have an idea how old you are…)

A potential psychological explanation for these findings is the "reminiscence bump" effect. It means that once we reach a certain age, we recall early-life memories most readily, because these memories often represent first-time events which are more vivid, and thus more easily retrievable from our brain's "long-term storage" (Jansari and Parkin 1989). This recall-based effect can be heightened by some consumers' beliefs that things were better back "then." They discount modern entertainment offers accordingly—a consumer characteristic that Holbrook (1993) coined "nostalgia proneness." His results from a principal component analysis based on a sample of 170

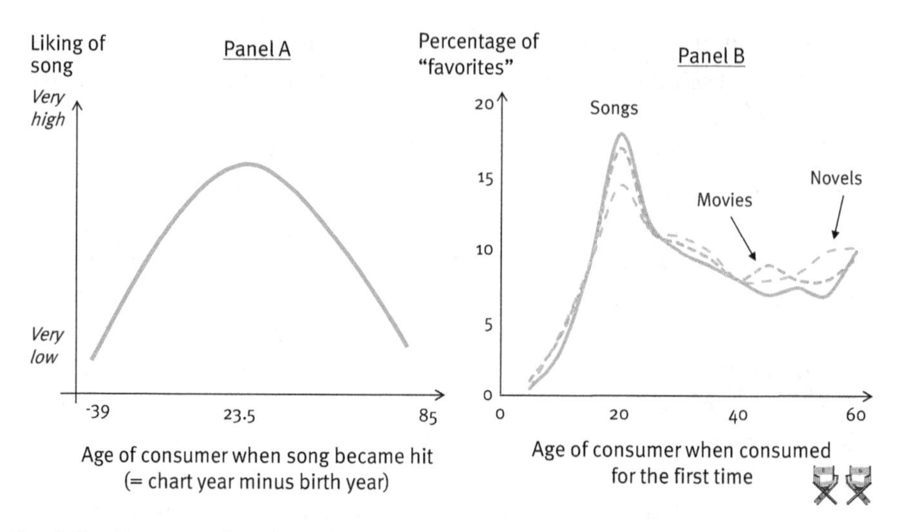

Fig. 3.12 Taste as a function of consumer age
Notes: Panel A is the authors' own illustration based on results reported in Holbrook and Schindler (1989). Panel B is the authors' own illustration based on results reported in Janssen et al. (2007). The depicted courses are stylized.

business students show that in the case of movies, this nostalgia trait goes along with a preference for more tenderhearted films, whereas violent content is devalued by those of us who score high on nostalgia.

National culture. Finally, taste has also been shown to be influenced by the culture that characterizes the nation in which consumers live. A nation transmits the values of what is to be appreciated (and what is not) to its members, as part of a continuous and often lifelong socialization process. With entertainment product's cultural nature, one might argue that quality perceptions of entertainment are embedded in a nation's culture.

We used regression analysis to analyze consumers' evaluations for 260 movies in 25 countries (the films that received a wide release in the U.S. in 2007–2008) with a focus on a film's use of "cultural elements" that are familiar to a particular country's consumers: for example, martial arts is familiar in China, but much less so in Germany. The results showed that such "cultural congruence" between a movie and the watching audience influences how much consumers like the movie (Song et al. 2018). Harvesting consumer reviews on IMDb to identify culture-specific elements and then computing a "culture score" for each film, we found that an increase of the congruence between a movie's cultural content and the culture of his/her region of one standard deviation goes hand in hand with a 0.5 points higher quality rating of the movie on the IMDb's 10-point scale.

We also learned that this effect is even higher for "culturally loaded" products—products for which the cultural content plays a more central role, such as is the case with independent films. So it is with entertainment as with other parts of life: we like what have learned to like.[30]

Managerial Consequences of the Difficulty to Judge Entertainment

In this section, we have shown that the quality of entertainment products is notoriously difficult to judge for consumers for two reasons: because entertainment products lack search attributes and also contain an inert artistic excellence that requires certain specific skills to recognize and value. This judgment difficulty carries a number of implications for entertainment marketing and management, which we address in this book.

- *Product decisions.* A focal consequence of the "experience good" character of entertainment products is that the consumer's decision whether to spend time and money on an entertainment product carries a serious amount of risk for him or her. To overcome such risk, entertainment producers must develop powerful strategies to lower consumers' uncertainty perceptions. In the context of product decisions, firms must provide consumers with "cues" that are informative enough to customers to reduce their uncertainty. Managers have to understand the signaling power of different product attributes, ranging from "unbranded" attributes (such as a novel's genre or a movie's country of origin) to the various kinds of brands an entertainment manager can employ, including well-known artists and producers. Both unbranded and branded signals can be used by consumers as quasi-search characteristics. But because of the artistic nature of entertainment products, understanding what consumers like when they have experienced it is far from trivial; we discuss what *Entertainment Science* scholars have found regarding the *experience* quality of entertainment in the first of four chapters on entertainment product decisions.
- *Communication decisions.* Consumers' inherent uncertainty about the quality of an entertainment product prior to purchase assigns particular importance to a manager's *pre-release* communication activities. Such

[30]The taste-forming role of a nation's culture also affects preferences for elements and forms of entertainment, such as genre.

communication must provide prospect customers with convincing arguments if a new entertainment product is to succeed in wide release. Samples such as trailers, beta versions, preview chapters, and song previews might help to overcome uncertainty barriers, but at what costs? Can samples spoil the consumption experience itself? Pre-purchase uncertainty also points to the importance of all kinds of what we call "earned" media. In addition to being exposed to paid advertisements by firms, entertainment consumers can usually access large amounts of information about a new entertainment product that is earned by its quality, not purchased by a manager. Such "earned" media includes professional reviews by "taste experts," word of mouth from other consumers who have already consumed a product, and even "success-breeds-success" cascades (that provide "social proof" of quality via bestseller charts and the like). At the same time, the artistic nature of entertainment somewhat limits the power of such (non-individual) information, because we have shown that taste contains idiosyncratic elements. This is where automated recommender systems show great promise as they find an individual consumer's "taste neighbors" out of a large base of other consumers.

Entertainment Products are Creative Products

> "The trouble with movies as a business is that it's an art, and the trouble with movies as an art is that it's business."
> —*Hollywood adage, often attributed to writer* Garson Kanin

Let us now turn to the *producer*-sided characteristics of entertainment and begin our investigation with the creative nature of entertainment products. To successfully craft appealing offerings for consumers, producers must address entertainment's consumer-sided characteristics, from their hedonic and aesthetic nature to their satiation tendency. Doing so requires producers to offer "creative" content which is equipped with the two "hallmark" dimensions of creativity (Amabile 1983):

- an "originality" dimension that requires creative products to be novel, surprising, or even shocking, and
- a "value" or "appropriateness" dimension, which expresses the idea that creativity is linked to a certain goal or objective (such as to develop an "exciting new thriller movie") (e.g., Runco and Charles 1993).

Many of the terms that consumers use to describe great entertainment products, such as "imaginative," "beautiful," and "touching," require both originality and appropriateness to be present in a product.

To develop a creative product that meets these requirements, one needs a certain type of people—people who have the right artistic instincts. These "creatives," such as novelists, actors, and directors, have an artistic creativity that differs from other types of creativity possessed by employees who develop new products in other industries (e.g., consumer packaged goods, business-to-business technologies). Specifically, people involved in the creation of new entertainment products must certainly have foundational "creativity-relevant skills" (e.g., keeping response options open when facing a task, aesthetic skills) as well as a unique task motivation (Amabile 1983). But they also must combine this general creativity with the soul, eye, and/or ear of an artist (i.e., "artistic logic," which we discuss below).

The required skills are closely tied to certain personality characteristics which are not found in most of us: the lack of a need to conform in thinking, a high level of independence, being impervious to social approval, and a capacity to think imaginatively (Bryant and Throsby 2006). In addition, the task motivation essential for creative tasks implies an intrinsic interest in the entertainment product itself, instead of seeing the product's development as a means to some extrinsic goal (Amabile 1983).

It is this "human side" of creativity that sets creative products apart from more analytical creations. It is reflected in three specific properties of entertainment products (Caves 2006): an "art-for-art's-sake" property, a "motley crew" property, and an "infinite variety" property.

The "Art-for-Art's-Sake" Property of Entertainment

> "Suits suck."
> —*Text on T-shirt worn by movie director* Billy Walsh, *a character in the TV series* ENTOURAGE

Were you among those who watched Ridley Scott's movie BLADE RUNNER when it debuted in theaters in the early 1980s? If you were, you may still remember being appalled by the dull voice-over that the movie's star Harrison Ford delivered—something that almost ruined a movie that is now considered one of the medium's greatest artistic achievements. So, what was going on? Warner Bros, the studio that was providing financing, was worried that the film might be incomprehensible to potential audiences and ordered the addition of a voice-over. Mr. Ford was contractually compelled to oblige, despite serious misgivings about the wisdom of the voice-over. The star fulfilled his contract, but did so in a way that even the producers considered "an insult" (Merchant 2013)—he "simply read" (Harrison Ford, quoted in *Empire* 1999) the lines, hoping the producers would never use it. But they did.

The history of entertainment is full of stories about conflicts featuring creatives versus executives in charge of business decisions. Actor Peter Sellers left the filming of the James Bond film CASINO ROYALE after a fight with the studio head (Koski 2014). Despite playing the lead character, Edward Norton, after quarreling with Marvel Studios over how the story of THE INCREDIBLE HULK should be told on screen, embarked on a month-long trip to Africa rather than participating in the promotion of the film (Lee 2013). And numerous artists have released music tracks that make it more or less clear to their listeners that the songs exist only for contractual reasons.[31]

An underlying theme in such conflicts is that the creatives felt that their artistic integrity (and/or freedom) was threatened. And make no mistake about how deeply artists care about their work: about the originality of the ideas and compositions, the technical skills and their execution, and the artistic achievement that finally results from the creative act (Caves 2006). Often the creatives are driven by a desire for the art itself, rather than any economic interest; their reward is derived from the aesthetic or cultural value of the product (Bryant and Throsby 2006).

Eikhof and Haunschild (2007) refer to such a perspective as the "artistic logic of practice," which they contrast with the "economic logic of practice." Whereas the latter emphasizes the market (or financial) value of a product, the artistic logic "is marked by the desire to produce *l'art pour l'art.* Art itself is seen as an abstract quality that surfaces, for example, in specific aesthetics or individual reactions by the recipient, and needs no external legitimization" (p. 526). Market value does not play a focal role in this logic—instead, the principal legitimization for producing entertainment is the contribution to art as a greater good (whether self-fulfillment for the artist, full realization of a beautiful creation, or "changing the world" with movie, song or even a rock show—just think of character Dewey Finn's mantra in the film SCHOOL OF ROCK).

---

[31]Forde (2013) has compiled a list of some of these cases where musicians such as The Mamas & Papas, Van Morrison, and Marvin Gaye produced songs and albums not because of an intrinsic motivation, but solely to avoid lawsuits. Probably the most drastic case of such "artistic disobedience" might be the 1970 song SCHOOLBOY BLUES by the Rolling Stones, whose quite explicit references to certain sexual "techniques" were so radical that they prevented the managers at Decca from releasing it (in line with the creators' intentions).

Take the example of acclaimed film director Steven Spielberg and his motivation to make a multi-million dollar film adaptation of the TINTIN comics by Hergé. Instead of being driven by commercial interest, Spielberg admits that he was "struck by Hergé's illustrations. They were so evocative of storytelling, plot and character relationships that by the end, without knowing one word of the language, I understood the whole story. [...] I said to Kathy (Kennedy), my fellow producer, 'We've got to make this into a movie'" (Spielberg 2012). The film was, thus, in Spielberg's own words, "a bet of $135 million on a cartoon reporter and his dog"—using resources mainly provided by two major Hollywood studios.

A consequence of the "artistic logic" of creatives arises when business decisions have to be made about an entertainment product. Rather than cooperate with managers ("Don't be a sell-out!"), creatives tend to forswear compromise ("One *must not* compromise art!"). Discussions between creatives and managers about the quality of entertainment products are further complicated by the vagueness of whether something is of artistic value or not. Finally, because art is the result of creatives' visions rather than logical arguments, artists tend to resist making commitments to specific courses of action or specific timelines—tomorrow's vision might be superior to today's. After the singer Prince had recorded his BLACK ALBUM, he encountered a spontaneous spiritual epiphany that the whole album was "evil"—and requested Warner Music to not publish it, despite having already manufactured half a million discs (Hahn 2004).

The "Motley Crew" Property of Entertainment

> "Teamwork makes the dream work. Together we can make something ... dare I say it? Bravura."
>
> —*Character* Jimmy McGill *alias* Saul Goodman *in the series BETTER CALL SAUL [Courtesy of Sony Pictures Television]*

In entertainment, many finished products require a highly diverse set of creative skills that depend on inputs from various people. The quality of a movie is determined by the contributions of actors and actresses, writers, directors, and composers, but also on creative contributions by people with skills in special effects, editing, photography, and sound design. Musical works require the creative inputs from composers, lyricists, arrangers, singers, and various instrumentalists, but also stage managers, sound engineers, etc. How do all these creative contributions affect the overall quality of an entertainment product, as judged by the consumer? The overall quality is *not* the simple "sum" of the individual performances. Instead, it is determined by the complex interplay of the different performances, which has two characteristics: it is (a) non-compensatory and (b) non-linear.

By non-compensatory, we mean that the presence of one individual component—even at high levels of quality—cannot compensate for the absence of another component. So, across the "cast of characters" required to produce an entertainment product, if one performance is mediocre, excellent performances by other people usually cannot fully make up for this weakness. Instead, a movie can be spoiled by a single scene or even a bad soundtrack, despite an otherwise excellent script and strong acting performances. Scrawler (2016) has assembled a list of films he feels are ruined "by just one bad scene," with Peter Parker's dancing in SPIDER-MAN 3 taking first rank.[32] In the same way, a game can be ruined by the voice of its lead character, although the graphics and storyline are top-notch, and great vocals and instrumentals cannot overcome a less-than-optimal song arrangement. If it helps, you can think of the production function of entertainment products as being of a *multiplicative* kind—the overall quality of the entertainment product is the result of a multiplication (not addition) of the different performances that make up the product. If one performance fails, it impacts the value of others. Or, as Caves (2006, p. 5), phrases it: "a large number multiplied by zero is still zero."

Another nuance to understand about this motley crew of contributors is that, in addition to being non-compensatory, the quality of each individual performance is altered by the quality of other performances, as individual performances combine together to create overall quality. In other words, performances are interdependent and integrate in a non-linear way. Whereas a singer's voice can be impressive because of its clarity and range, the voice quality that listeners perceive will be influenced by other aspects of the entertainment product, such as the respective song material or the fit of the singer's voice with a duet partner.

Take the example of musicians Paul Simon and Art Garfunkel, both amazing talents with impressive musical skills that range from song writing to vocalizing. These skills translated into critical acclaim and success for their respective solo projects. However, it was the *combination* of their talents that created something unique and magical: "Simon's whispering, almost in falsetto and Garfunkel's seraphim harmonizing produced something … ethereal, even spiritual" (Scaruffi 1999). The way these two singers *jointly*

[32]We will revisit several of the scenes Scrawler has included in his list in our consumer behavior chapter when discussing the need for verisimilitude for great storytelling.

vocalized their material had a synergistic effect on quality, which econometricians refer to as a *positive interaction* that drives entertainment quality. Negative interactions also exist, for example if two musical voices are incompatible, or, to cite Frank Price, former president of Universal Pictures: "if you are making a romance and the chemistry isn't there between the leading woman and man. You're dead" (quoted in Fleming 2015).

Often, because entertainment is created by people, the quality of a creative product is influenced by the *social* match between the different creatives involved in the project—or, worst case, the lack of such match. A social match can boost the motivation of creatives and subsequently enhance performance quality, but also produce interaction effects between individual performers that can be clearly seen or heard in the final product (such as the transcendent personal liking between actors Ethan Hawke and Julie Delpy in Richard Linklater's BEFORE trilogy). Given the unique personality profiles of creatives, it is not surprising that the absence of such social fit has left its mark on entertainment history in the form of low-quality performances (take the reported lack of *off-screen* chemistry between on-screen lovers Julia Roberts and Nick Nolte in I LOVE TROUBLE as an example; Brennan 1994).

Lack of social match has also prevented the future creation of works of high quality. For example, whereas Simon and Garfunkel's professional creative skills harmonized beautifully, their social skills didn't mesh so well ("he was getting on my nerves. The jokes had run dry;" Garfunkel referring to Simon, quoted in Farndale 2015). This interpersonal—not creative—tension caused regular, ongoing fights between the two musicians, with only short periods of productivity in between, clearly limiting their output and economic value creation. The fans of The Beatles, Pink Floyd, and ABBA will also longingly understand the importance of social match.

The "Infinite Variety" Property of Entertainment

A final property of entertainment products resulting from their creative nature is that an endless array of design and gestalt options exists for each product. Creatives can choose from a "universe of possibilities" (Caves 2006, p. 6) when making decisions about a new entertainment product. The number of novels that can be written is essentially infinite, as is the number of movies and TV series that can be filmed, the number of games that can be programmed, and the number of songs that can be composed. In economic terms, entertainment options are horizontally differentiated (i.e., qualitatively different in other dimensions than price) to a maximal degree (Caves 2006).

Focusing on just one of many elements, for example, creatives can vary the number of characters in a novel, their moves, and their arguments in more ways than we can think of in a lifetime. "The same screenplay can generate countless end products, based on all the possible combinations of directors, actors, sound tracks, editing, and so on" (Troilo 2015, p. 7). The number of potential attributes of a creative product is endless, and, to phrase it more analytically, these attributes cannot be reduced to a lower-dimensional space of "design alternatives" or solutions. Instead, the solution space for entertainment is always multidimensional, and each dimension of this space (such as a movie's narrative, or any of its characters) is essentially infinite itself, as well as highly fuzzy and dynamic. As a result, there exists a continuously evolving, virtually endless room of alternatives. Even more complexity is added by the possible combinations of the infinite product dimensions.

This complexity of the solution space and the infinite variety of product options is also closely linked to the heuristic nature of creative processes (Amabile 1983). Infinite variety limits algorithmic solutions, demanding a heuristic approach to the creation of entertainment (which, however, is not the same as saying that algorithms can't help us to understand what makes a great story, as we discuss later).

Managerial Consequences of the Creative Character

The creative character of entertainment products also carries a number of implications for managers that we address in upcoming sections of our book.

- *Product decisions: innovation.* One area that is particularly affected by creativity is innovation management. To address the crucial role of creatives (each of whom has a unique perspective) for the development of successful products, entertainment producers must ensure that the innovation context—including strategy, organization, culture, and the design of the innovation process itself—simultaneously addresses artistic considerations and economic goals. The luminous presence and reputation of creatives can also create a danger for managers who work closely with them—they can fall in love with the art itself, or can be deluded in mistaking their close association with art(ists) for actual artistic talent; we name this danger the "artistic fallacy" (see the case of the Cannon Group that we describe in our innovation chapter). The "motley crew" property has implications for how entertainment producers design the innovation process, and particularly how to craft the ensemble of creatives that are involved in the creation of a new entertainment product. And the "infinite variety" of design alternatives invalidates many traditional, attribute-based market research techniques for entertainment products, so different research approaches are needed that leave sufficient room for creativity decisions.

- *Product decisions: quality.* How can producers address the "infinite variety" property of entertainment when trying to create a high-quality final product, finding the "right" combination of performances among a myriad of options? With regard to storyline, one way is to develop an understanding of general narrative/storytelling patterns and how these patterns influence the quality of the resulting entertainment work. Such knowledge can help producers to better frame detailed decision making by at least reducing the set of design alternatives that are clearly inappropriate.
- *Product decisions: branding.* Creatives play a key role for the development of a new entertainment product, but they play just as key a role in building consumer demand for the product upon release. Well-known creatives function as brands and influence consumers' judgments—movie actors, musicians, game designers, or novelists can be an important driver for consumers' interest in and eventual liking of a new entertainment product. We discuss this role of creatives as brands in various contexts in this book; for example, we show how creatives are strategic resources for entertainment firms and discuss their instrumental roles as "human ingredient brands" and "parasocial relationship partners" for consumers.
- *Communication decisions.* The roles that creatives play in the minds of consumers also give guidance for the firm's communication mix for entertainment products. Creatives often have strong presences on social media; thus, their own communication can impact consumers' anticipation of entertainment products over and above the usual impact of star power (stars as "influencers").

Finally, the creative character of entertainment—and particularly its multiplicative production function—provides some general insights for managers: they cannot cherry-pick certain product elements or performances at the costs of others.

Entertainment Products are Information Goods

The "First-Copy Cost" Property of Entertainment

This second producer-sided characteristic arises because the core benefit for which people consume an entertainment product does *not* come from its tangible delivery method (e.g., the disk, the paper, the cartridge). Instead, the true value of entertainment comes from the product's intangible information content. Thus, scholars refer to entertainment media products as "information goods"—economic offerings that are valued by customers mostly because of the information they carry (Wang and Zhang 2009). Please note that we do not use the term "information" to refer to only a certain kind of content, but instead to connote a rather technical meaning—information is "anything that can be digitized" (Varian 1998, p. 3). Any novel, game, movie or TV show, or song can be envisioned as a combination of a large number of 0s and 1s and, hence, stored in digital form. Amazon's Kindle book store, Netflix's movie library, Spotify's music playlists, and Sony's PlayStation store are examples that bring to life this information character of entertainment.

How do the economics of information goods differ from other market offerings? The main consequence is that information has a different cost structure, and it does so in two ways. First, and most importantly, the information character affects the allocation of fixed and marginal costs. The fixed costs of producing information can be very high, but once the information product is finalized, the marginal costs of reproducing and distributing additional copies are low (e.g., Varian 1998). Although there can be variations across specific information products, there is a general pattern in the *relative* height of fixed versus marginal costs—the dominant share of costs is fixed in the case of information goods, something that applies particularly for entertainment. This sets information goods apart from industrial goods, for which "the unit costs of production and distribution are often dominant" (Jones and Mendelson 2011, p. 164), even if fixed costs can be absolutely high for these products, too (e.g., for products that require specialized tooling or production facilities).

A second aspect of the cost structure of information (entertainment) products is that the majority of costs are essentially *sunk*—the fixed costs are incurred completely before the market entry of the product. These early fixed costs are largely irrecoverable at a later point, regardless of the outcome of the production process, unlike investments in machines or other tangible assets for which some value can be reclaimed as these assets are auctioned off or redeployed. Hal Varian (1998, p. 3) described this sunk nature of entertainment costs via an example from the film industry: “If the movie bombs, there isn’t much of a market for its script, no matter how much it cost to produce.” Or, using the words of legendary Hollywood producer Robert Evans: “[T]here’s no closeout value [for a film]. Unlike a car, which you can close out if it doesn’t sell, a film is like a parachute: if it doesn’t open, you’re dead” (quoted in Grobel 1993).

When you combine a dominant proportion of fixed costs and the early “sunk” nature of these costs, it becomes clear that entertainment products have a highly asymmetric cost structure. We refer to this as the *high first-copy cost* characteristic of entertainment (e.g., Varian 1995). Take the example of a major Hollywood film such as SPIDER-MAN 2, which was produced by Sony in 2004. Young et al. (2008) report that the studio spent about $30 million for pre-filming work ($10 million for the script and $20 million for licensing the SPIDER-MAN brand from Marvel), $100 million for the filming itself ($55 million for actors, director, and producers, $45 million for logistics, equipment, and other “below-the-line” crew members), and about another $70 million for post-filming/pre-release work on the film ($65 million for special effects and $5 million for music). We expect the studio to have added a global advertising budget of $60 million or higher, most of which was spent before the release of the film.

Of these total costs of $260 million, at least the $200 million spent prior to advertising are the first copy costs of the film; they would have accumulated if only *a single copy* was to be produced. Compare this with the costs for the second, third, etc. copies of the film which industry sources have shared with us: an (analogue) film print costs about $1,500 (or $0.0015 million), a digital copy transferred to theaters about $20 to $25 (or $0.000025 million), a Blu-ray about $3 (or $0.000003 million), and a digital file for download by consumers quickly approaches zero marginal costs for the producer. These amounts are clearly negligible for economic decisions in comparison to the costs of the first copy (Peltoniemi 2015). Similar calculations apply for the other entertainment products we feature in this book.

Let us finish by saying that the cost structure of an information good is impacted very little by whether they are actually sold as digital files (such as music as MP3 file) or in analogue form (the same music on a physical CD)—the key element is that digitization is a mere *possibility*. The economic logic is the same regardless of final product format, because the logic is tied to the information character of the content offered. And this content remains unaltered by the transformation of information into analogue forms. If anything, though, the logic might be somewhat more radical for purely digital entertainment offers because, as the numbers from SPIDER-MAN above illustrate, the marginal distribution costs for a digital Spidey are effectively—*zero*.

Managerial Consequences of the Information Good Character

The information good character of entertainment products and the resulting cost structure have a number of implications for entertainment marketing and management, which we address at different places in our book.

- *Business model decisions.* The asymmetric allocation of production costs requires substantial upfront investments that carry financial risk. As a result, it is critical that managers craft effective ways for addressing such risk with their entertainment firms' business models. We discuss and evaluate, in a separate chapter, the core business models that the entertainment industry has developed.
- *Pricing decisions.* The strongest managerial impact the information-based cost structure has is on the pricing of entertainment. With marginal costs being negligible in many cases, profit maximization becomes equal to revenue maximization, consistent with neoclassical economic theory (Shapiro and Varian 1999). The irrelevance of marginal costs for determining the "right" price for a product provides a lot of leeway for the manager who is making pricing decisions for entertainment, including price discrimination strategies such as versioning and bundling. Acknowledging the huge differences in fixed costs between entertainment alternatives, *Entertainment Science* scholars also question that most new-release movies are offered at the same price (as are similar new songs, similar new books, and first-run games), wondering whether prices should also vary between products.
- *Distribution decisions.* The information good character enables the distribution of products via various different (digital) channels. A major challenge for distribution managers is to coordinate the timing of these channels, as we discuss in our chapter on entertainment distribution. The information good character of entertainment products also is the foundation for the important role that piracy plays for entertainment marketing—by its very nature, information is "easy to copy and share" (Varian 1998, p. 16—and the "Piracy Challenge" section in the distribution chapter in this book).

- *Product decisions*. The front-loaded cost structure of information puts emphasis on the importance of effectively forecasting the demand for a new entertainment product very early in its development process. Such early forecasting is essential for avoiding the wasting of otherwise sunk resources.
- *The blockbuster concept*. The near irrelevance of marginal costs for most entertainment products carries an important strategic implication: it enables managers to leverage a successful entertainment product on a global scale. If a new game, movie, or song is well received by certain audiences, extending its availability is not limited by production nor distribution costs. However, this strategy does not work equally with all content, but requires specific content that is attractive to global audiences and a marketing strategy that ensures global awareness. In our chapter on integrated entertainment marketing, we will discuss the blockbuster concept as one particular approach for exploiting this scalability of entertainment as information.

Entertainment Products Have Short Life Cycles

> "An important characteristic of entertainment products … is that they have remarkably short life cycles… the majority of demand occurs within a few weeks."
> —Luan *and* Sudhir *(2010, p. 445)*

The "Perishability" Property of Entertainment

Short life cycles are the third producer-sided characteristic of entertainment products. The product life cycle is a fundamental management concept, which describes the sales pattern of a product over time, from the moment it is introduced until its removal (e.g., Rink and Swan 1979). Although originally developed to describe the trajectory of an entire category of products, the general notion of the product life cycle has also been usefully applied to individual products within a category. The concept has triggered extensive research in marketing and management; the distinction between life cycle stages such as introduction, maturity, and decline being among the most prominent descriptive insights (Anderson and Zeithaml 1984). The life cycle concept is the conceptual underpinning for diffusion models, which are essential for understanding how a new entertainment product is adopted by consumers.

We don't want to discuss the stages of the life cycle here, but its *length*. This is because life cycles for entertainment products are systematically shorter than those for many other products. When we say shorter, we do not mean the *absolute* length of a product's life cycle, as measured by its mere availability (entertainment products are often available for decades, and some even for centuries). Instead, we are talking about the time frame in which products generate *substantial* revenues.

What time period do we have in mind when we say that entertainment life cycles are "short"? Please take a look at how cumulative revenues for movies are distributed over time. Of all revenues earned by 240 movies released by a Hollywood studio in Germany between 1999 and 2009, about 80% of the box office was generated within just *four weeks* (see Panel A of Fig. 3.13). A similar pattern exists for films' home entertainment sales—David Walls (2010) reports, based on a data set of almost 1,000 films released in the U.S. between 2006 and 2009, that only about 5% of hit films are still listed in the DVD charts after ten weeks.

Similarly, Clements and Ohashi (2005, p. 523) state that for video games "[m]ore than 50% of the revenues for a particular game title were typically made in the first year after the game release." Our analysis of the distribution of North American sales for all 100 Xbox games released between October 2011 and October 2012 shows an even more radical life cycle (Panel B of Fig. 3.13): 80% of the total unit sales were generated within a time frame of less than seven weeks after a game's release. And for music, sales patterns for songs and albums that reached the peak position in Japan in 2005 show that hit singles earn 80% of their total revenues with just only nine weeks. In the case of music albums, the time frame is even shorter: they make 80% in only four weeks (Asai 2009; Panel C of Fig. 3.13).

The pattern exist also for books (where the average time for a book on the German bestseller list is less than six weeks, for example) and even for TV series. Barroso et al. (2016), in their analysis of all 2,245 fictional television series broadcasted in the U.S. between 1946 and 2003, calculate an average run of 1.5 years, with the median being just a single season. For all these entertainment products, the lion's share of revenues is usually generated within a few weeks or months.

This shortness of entertainment products' life cycles is the result of a number of systematic forces, some of which we have discussed in the context of the consumer-sided characteristics of entertainment. Entertainment's cultural role is certainly a main driver: entertainment consumption is symbolic consumption, with the symbolism not being limited to what a consumer chooses, but also when he or she does so. It tells us something about

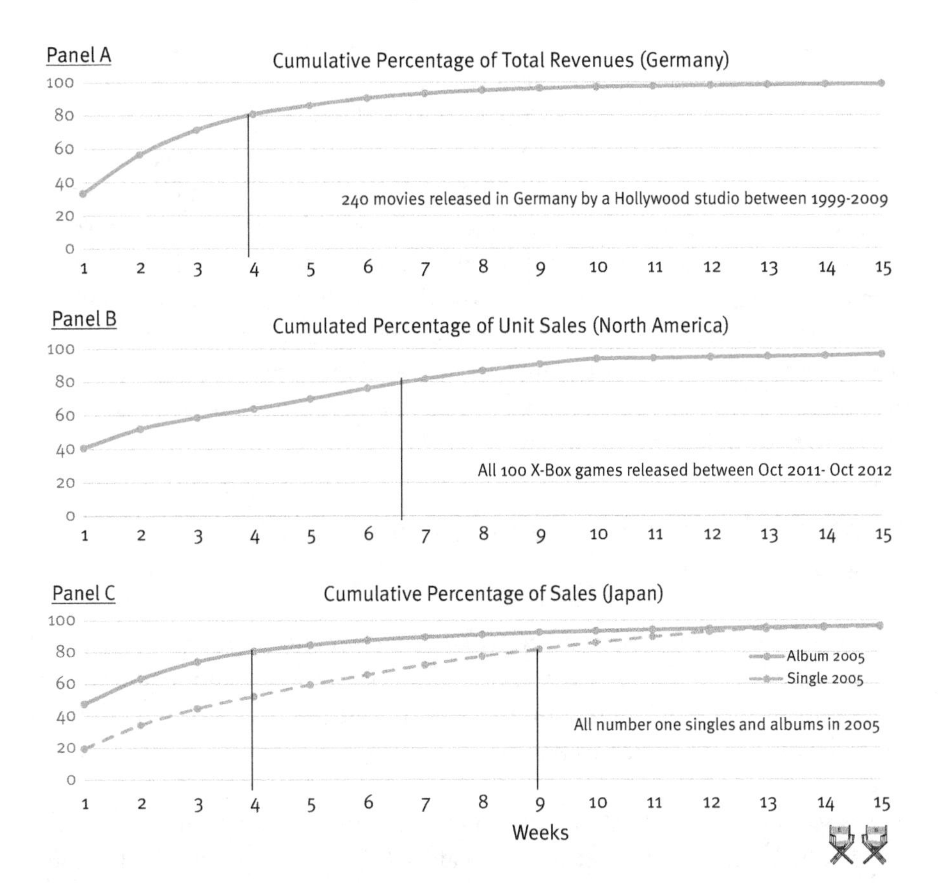

Fig. 3.13 Empirical life cycles for movies, games, and music
Notes: Authors' own illustration based on our own data (Panels A and B) and data reported by Asai (2009) (Panel C).

a person if he or she has not seen the new movie everyone is talking about on its opening weekend. Also, as entertainment transports societal attitudes, the value of a product is closely tied to how well these attitudes resonate with the interests of the target group. Such resonance is notoriously dynamic and fast changing—whereas the next addition to Marvel's cinematic universe might be what everyone is looking for today, few of us care much once the buzz has evaporated.

As a result, the Google search volume for the STAR WARS movie THE FORCE AWAKENS was 11% of its peak just three month after the release, that for Coldplay's album A HEAD FULL OF DREAMS 12%, for the CALL OF DUTY: INFINITE WARFARE game 17%, and just 10% searched for the book version of HARRY POTTER AND THE CURSED CHILD three months later—and those were all big *hits*! To transfer one of the news industry's key axioms into the context of entertainment, there is hardly anything as dead in the world as last month's movie release, hit song, or bestselling novel.

In addition, the length of entertainment life cycles is also restricted by entertainment's satiation character. Satiation decreases consumers' desire for experiencing an entertainment product again, whether in the same form or in some other version (such as through a new channel). This limits further demand for the product after the target group has been reached. Finally, competition is also a factor—as we will discuss later, entertainment markets are notorious for a high frequency of innovations. This ongoing stream of new products distracts consumers from discovering products that were introduced to the market months or years ago.[33]

Managerial Consequences of Entertainment's Short Life Cycles

Like with the other characteristics of entertainment products, managerial decision making must take the short life cycles of entertainment products into account. Let us highlight a number of implications which we address in the following chapters of this book.

- *Product decisions.* The short time frame in which an entertainment product can be expected to generate meaningful revenue points to the managerial need to engage in constant innovation to assure a continuous cash flow. We dedicate a chapter to how such innovation activity should be managed. Also, the need to ensure that an entertainment product achieves maximal resonance with consumers in the first days and weeks following launch has consequences for product design; some product elements are better suited for stimulating consumer awareness and interest than others. For example, we have devoted one chapter to "branded" product features, which include stars and well-known characters (e.g., Superman).
- *Communication decisions.* The immediacy of consumer response also depends on a firm's communication activities. One popular approach implies a focus on pre-launch communication. Such an approach can be

[33]Some argue that the development of automated recommender systems and other digital innovations will cause a shift toward the "long tail," which would go hand in hand with a reduction of this perishability property, extending the length of entertainment's life cycle. We discuss this potential effect later in the book.

particularly effective when it generates pre-release buzz for the new product; it can also help to extend the product life cycle by triggering success-breeds-success cascades.

- *Distribution decisions.* The life cycle's shortness requires managers to pay particular attention to the timing of a new product's release—if you get off on the wrong foot, you often have no chance to counter this later. We will discuss that entertainment timing matters both in absolute terms (e.g., should a new movie be released in summer?) and in relation to competing offers or events during a given release weekend or season.
- *The blockbuster concept.* Finally, the entertainment industry has developed the blockbuster concept as an integrated marketing response to the short life cycle challenge. The strategy has exhibited strong potential, but it also exacerbates the problem: because its central idea is to allocate resources in a way that facilitates a strong opening, a blockbuster release strategy contributes to even shorter entertainment life cycles!

Entertainment Products (Potentially) Have Externalities

The "Two-Sided" Property of Entertainment

The final producer-sided characteristic of entertainment products highlights what economists call "externalities." What does this mean? Most (other) products are targeted at a single group of customers which are part of a specific market: a premium car such as Tesla's model S is aimed at wealthy people with a predilection for modern, even ecologically clean vehicles. Here, the seller (Tesla) and the customers (the wealthy people) constitute the market. Sometimes products have multiple target group of customers (think of a printer which is sold by HP to consumers as well as small business owners), but each of them is dealt with in isolation by the seller (HP). Externalities exist in neither case.

However, other products encompass different benefits that are offered to separate groups of customers—in distinct markets. Take the case of a newspaper for which valuable journalistic content is required to attract one or more groups of readers (e.g., single copy purchasers and subscribers). However, the newspaper's revenues also come from those who pay the newspaper to run advertisements. The value proposition for this advertiser audience is that the newspaper provides access to a large number of eyeballs attached to the right kind of readers, i.e., exposure to a desired target audience. In other words: the newspaper's success with readers affects its value for advertisers. That's the externality.

Now, entertainment products have more in common with newspapers that one might realize at first glance. In addition to offering enjoyment to consumers, they also offer room for firms to promote their brands and products via the entertainment product (let us call the latter "advertisers," for simplicity's sake). Examples include product placement in movies and shows and in-game advertising. Like in the case of the newspaper, the two customer groups of entertainment consumers and "advertisers" are not acting in isolation, but their actions affect each other. The entertainment producer operates the product, which serves as a "platform" for the actions of the "advertiser": "the benefit enjoyed by a member of one group depends upon how well the platform does in attracting customers from the other group" (Armstrong 2006, p. 668). The degree to which the entertainment product addresses the respective needs of each group impacts *both* groups simultaneously.

Externalities can be both negative and positive, and both types are found in entertainment. In the case of negative externalities, the better the product appeals to the needs of the one group, the less attractive it is for the other. If there are too many advertisements in a newspaper, readers often feel annoyed. Similarly, when consumers turn off a TV because they find the advertiser's actions a nuisance that reduces the enjoyment of watching the program, or stop playing a game or listening to music for the same reasons, it is a negative externality (Anderson and Gabszewicz 2006). Clear evidence of the nuisance factor is offered by the avoidance strategies that consumers employ to avoid commercial communications when experiencing entertainment—from zapping to a different program during the advertising break on TV, to using digital video recorders that allow the consumer to skip commercials, to using ad blockers when browsing the Internet.[34]

In contrast, positive externalities describe a synergistic relationship between the two groups: the success of the product with one group increases its attractiveness for the other. If a producer of a game console sells licenses

[34]Although beyond the scope of this book, it is an interesting question how such avoidance strategies impact the value of consumers' product usage for *advertisers*. Bronnenberg et al. (2010) study how the ownership of a TiVo digital video recorder (DVR) changed the shopping behaviors of 819 Texas households, comparing purchases in several product categories for 13 months prior and 26 months after the adoption of the DVR with a large control sample. Analyzing the differences in spending behavior show no significant effects—in other words, obtaining a TiVo DVR and the associated ad skipping does not impact purchase behaviors, at least not in Bronnenberg et al.'s setting.

to many game producers, the influx of new games increases the console's value for consumers, thus representing a positive externality. And, vice versa, the higher the number of consumers who own the console, the more attractive it is for producers for developing new games for the console (the platform).

We illustrate this two-sided property of entertainment products in Fig. 3.14 with entertainment content in mind (versus hardware such as consoles), in line with our book's focus. The "other customer" in the figure often has its own product to sell, using the entertainment product as a communication vehicle to carry a message to the entertainment consumers. We have already mentioned films, TV shows, and games as platforms for advertising and placements. Music can serve as a communication vehicle for other products when distributed via music videos (shown on YouTube or similar channels), streaming providers, and of course also radio. Amazon sells advertising space on its "Kindle with Special Offers" ebook readers, but in general, communication for advertisers' products is less prominent so far in books. Some, however, have argued that this might change (Adner and Vincent 2010).

Please note that instead of speaking of revenues when describing the contributions that content consumers or other customers make to the success of an entertainment product, we replace that term with "usage" in the figure. The reason is that several business models we discuss in this book require usage of the product by consumers, but they do not necessarily require that customers *pay* for them with their own money. In many situations, such as

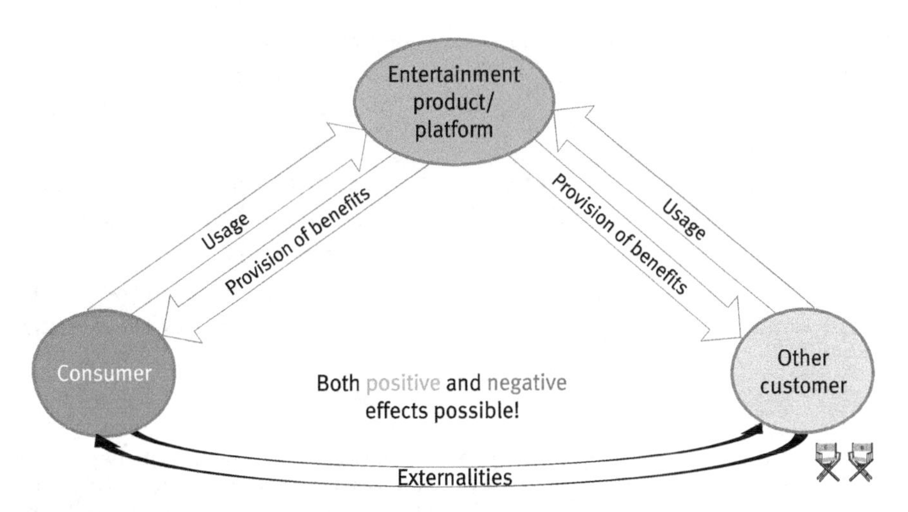

Fig. 3.14 The two-sided property of entertainment products
Note: Authors' own illustration.

the website of a newspaper, but also a game that consumers can play for free, if one side "subsidizes" the other, the other may not need to actually pay for product usage—at least not "directly." This is an important aspect of the two-sided property of entertainment markets.

Because this book deals mainly with content, we focus on negative externalities as a characteristic of entertainment products—we have named examples above. Positive externalities are more relevant for providers of platform products such as producers of gaming *consoles* (e.g., Sony's PlayStation), but are somewhat less so for producers of entertainment products themselves (e.g., a game developer). Nevertheless, we discuss positive externalities in the context of *market* characteristics in entertainment, which producers should be well aware of also.

As is the case for some of the other characteristics we discuss in this chapter, externalities are also present in other industries besides entertainment (see Rochet and Tirole 2006 for a list of examples). However, entertainment products are ripe for them: the information good character of entertainment allows the inclusion of communicative messages in the product *itself* in many ways. Adding an advertisement to a film or game is much easier than integrating it in a tangible product such as a car or washing machine.[35]

Managerial Consequences of Entertainment Products' Externalities

Like the previously discussed characteristics of entertainment products, the two-sided property has consequences for managing and marketing entertainment products. It affects both the fundamental business model for a product and how this business model is translated to the consumer via marketing instruments.

The main implication is that producers of entertainment must consider the potential commercial value of their product to more than one market: those who *consume* the entertainment product to get its hedonic benefits and those who want to *use* the entertainment product to reach the hedonic consumers. At the same time, managers must be carefully aware of the externalities that

[35]Please note that one can also argue that (positive) externalities exist between different groups of customers in the printer example above: consumers might consider a printer as more powerful if it is adopted by business customers. However, usually the link between the two customer groups is weak and is not related to *separate* benefits, as is the case with advertising—that is why we are hesitant to use the term "two-sided market" for such constellations.

go along with such a two-sided structure. We find it particularly important to avoid, or at least to account for, the negative effects on consumers' enjoyment levels that can result from addressing advertisers' needs. In the next chapter, we discuss the conditions under which integrating communication messages into an entertainment product can hurt its success with consumers.

Some entertainment products are solely supported by advertising, but in some cases, managers also offer a "paid" version (i.e., one without advertising) of the same product. For example, many video games are available in both versions ("free" and paid), so that customers can self-select between them. Here, determining the "optimal" amount of advertising, along with the "right" design of the paid variant, can be a difficult challenge for the manager.

Finally, a general implication is that, if more than one customer group exists, an entire range of artistic and economic decisions must be made in ways that account for (1) the interests of the different customer groups and (2) the degree to which the different groups affect each other. This challenge is heightened by the fact that the different customer groups are not necessarily equally important for the economic success of the entertainment product. Thus, a potential concern is that an entertainment product may cater too much to the desires of one group, with the product becoming unappealing to other groups.

Concluding Comments

Understanding the critical characteristics of entertainment products—four that are consumer-sided and another four that are producer-sided—is essential for fully appreciating the managerial issues that we will discuss in Part II of the book. Decisions that make sense for other types of products—especially those that are tangible, functional/utilitarian, long-life-cycle products—won't work effectively for managers of entertainment without modification. Failing to account for the unique aspects of entertainment products will, in the absence of "luck," lead to failure.

In the past, the failure to acknowledge and address these particularities of entertainment has prevented several firms who entered the entertainment industry to prevail (we name some of them later in the next chapter of this book). Their ignorance, or inability, has also fueled the "Nobody-Knows-Anything" myth. The failing of those who applied established business rules to entertainment products created the illusion of a "rule-free" environment—although it was the lack of *adapting* the rules to the unique characteristics of entertainment that caused the failure, rather than the general unfitness of rules per se.

But even considering these special characteristics that set entertainment products apart from other market offerings, producing and marketing an entertainment product does not take place in a context-free space. Entertainment *markets* also have unique characteristics that must be understood for success. We take a look at those markets in the next chapter.

References

Abebe, N. (2017). 25 Songs that tell us where music is going. *The New York Times Magazine*, March 9, https://goo.gl/LYWgBY.

Adner, R., & Vincent, W. (2010). Get ready for ads in books. *The Wall Street Journal*, August 19, https://goo.gl/otEf9J.

Alba, J. W., & Williams, E. F. (2013). Pleasure principles: A review of research on hedonic consumption. *Journal of Consumer Psychology, 23*, 2–18.

Amabile, T. M. (1983). The social psychology of creativity: A componential conceptualization. *Journal of Personality and Social Psychology, 45*, 357–377.

Anderson, C. A. (2003). Violent video games: Myths, facts, and unanswered questions. *Psychological Science Agenda*, October, https://goo.gl/iwfXiH.

Anderson, C. R., & Zeithaml, C. P. (1984). Stage of product life cycle, business strategy, and business performance. *The Academy of Management Journal, 27*, 5–24.

Anderson, S. P., & Gabszewicz, J. J. (2006). The media and advertising: A tale of two-sided markets. *Handbook of the Economics of Art and Culture, 1*, 567–614.

Anderson, C. A., Akiko, S., Nobuko, I., Swing, E. L., Bushman, B. J., Sakamoto, A., Rothstein, H. R., & Saleem, M. (2010). Violent video game effects on aggression, empathy, and prosocial behavior in eastern and western countries: A meta-analytic review. *Psychological Bulletin, 136*, 151–73.

Armstrong, M. (2006). Competition in two-sided markets. *The RAND Journal of Economics, 37*, 668–691.

Asai, S. (2009). Sales patterns of hit music in Japan. *Journal of Media Economics, 22*, 81–101.

Askin, N., & Mauskapf, M. (2017). What makes popular culture popular? Product features and optimal differentiation in music. *American Sociological Review, 82*, 910–944.

Audley, A. (2015). What does your ringtone say about you? *The Telegraph*, January 15, https://goo.gl/YWSgYF.

Barroso, A., Giarratana, M. S., Reis, S., & Sorenson, O. (2016). Crowding, satiation, and saturation: The days of television series' lives. *Strategic Management Journal, 37*, 565–585.

Belk, R. W., Ger, G., & Askegaard, S. (2000). The missing streetcar named desire. In S. Ratneshwar, D. Glen Mick, & C. Huffman (Eds.), *The why of consumption: Contemporary perspectives on consumer motives, goals, and desires* (pp. 98–199). London: Routledge.

Bellesiles, M. A. (1996). The origins of gun culture in the United States, 1760–1865. *Journal of American History, 83*, 425–455.

Bohnenkamp, B., Wiertz, C., & Hennig-Thurau, T. (2012). Consuming 'Media Trash:' When "Bad" becomes "Good". In Z. Gürhan-Canli, C. Otnes, & R. (Juliet) Zhu (Eds.), Advances in consumer research (Vol. 40, pp. 1035–1036). Duluth, MN: Association for Consumer Research.

Bourdieu, P. (2002). The forms of capital. In N. Woolsey Biggart (Ed.), *Readings in economic sociology* (pp. 280–291). Blackwell: Malden.

Brennan, J. (1994). Trouble on 'Trouble' set? Take your pick: (a) co-stars Julia Roberts and Nick Nolte got on each other's nerves; (b) the filmmakers got on their nerves; (c) snoopy questions are getting on everyone's nerves. *Los Angeles Times*, July 3, https://goo.gl/mY6NSe.

Brickman, P., & Campbell, D. T. (1971). Hedonic relativism and planning the good society. In M. H. Appley (Ed.), *Adaptation-level theory: A symposium* (pp. 287–302). New York: Academic Press.

Bronnenberg, B. J., Dub, J., & Mela, C. F. (2010). Do digital video recorders influence sales? *Journal of Marketing Research, 47*, 998–1010.

Bryant, W. D. A., & Throsby, D. (2006). Creativity and the behavior of artists. In V. A. Ginsburgh & D. Throsby (Eds.), *Handbook of the economics of art and culture* (pp. 507–529).

Bushman, B. J. (2016). Violent media and hostile appraisals: A meta-analytic review. *Aggressive Behavior, 42*, 605–613.

Caves, R. E. (2006). Chapter 17 Organization of arts and entertainment industries. In *Handbook of economics of art and culture* (Vol. V).

Chen, J. (2007). Flow in games (and everything else). *Communications of the ACM, 50*, 31–34.

Clements, M. T., & Ohashi, H. (2005). Indirect network effects and the product cycle: Video games in the U.S., 1994–2002. *Journal of Industrial Economics, 53*, 515–542.

Coombs, C. H., & Avrunin, G. S. (1977). Single-peaked functions and the theory of preference. *Psychological Review, 84*, 216–230.

Cowell, A., & Kirkpatrick, D. D. (2017). Saudi Arabia to allow movie theaters after 35-year ban. *The New York Times*, December 11, https://goo.gl/hv36xp.

Culbert, D. (2003). Jud Süss. In N. J. Cull, D. Culbert, & D. Welch (Eds.), *Propaganda and mass persuasion* (p. 205). Santa Barbara: ABC-Clio.

Debenedetti, S., & Larcenieux, F. (2011). 'The Taste of Others': Divergences in tastes between professional experts and ordinary consumers of movies in France. *Recherche et Applications en Marketing, 26*, 71–88.

DeCamp, W. (2017). Who plays violent video games? An exploratory analysis of predictors of playing violent games. *Personality and Individual Differences, 117*, 260–266.

DeFillippi, R., Grabher, G., & Jones, C. (2007). Introduction to paradoxes of creativity: Managerial and organizational challenges in the cultural economy. *Journal of Organizational Behavior, 28*, 511–521.

Deshpande, R., & Webster Jr., F. E. (1989). Organizational culture and marketing: Defining the research agenda. *Journal of Marketing, 53*, 3–15.

De Vany, A. (2006). The movies. In V. A. Ginsburgh & D. Throsby (Eds.), *Handbook of the economics of art and culture* (pp. 615–665). Amsterdam: Elsevier.

De Vany, A., & David Walls, W. (1999). Uncertainty in the movie industry: Does star power reduce the terror of the box office? *Journal of Cultural Economics, 23*, 285–318.

Doerr, A. (2014). *All the light we cannot see*. New York: Scribner.

Eikhof, D. R., & Haunschild, A. (2007). For art's sake! Artistic and economic logics in creative production. *Journal of Organizational Behavior, 28*, 523–538.

Ekelund Jr., R. B., Mixon Jr., F. G., & Ressler, R. (1995). Advertising and information: An empirical study of search, experience and credence goods. *Journal of Economic Studies, 22*, 33–43.

Elberse, A. (2013). *Blockbusters: Hit-making, risk-taking, and the big business of entertainment*. New York: Henry Holt and Company.

Empire (1999). Harrison Ford's Blade Runner gripe, October 7, https://goo.gl/BKkbU7.

Farndale, N. (2015). Art Garfunkel on Paul Simon: 'I created a monster'. *The Telegraph*, May 24, https://goo.gl/Rg4XgR.

Ferguson, C. J. (2013). Violent video games and the supreme court. *American Psychologist, 68*, 57–74.

Feshbach, S., & Singer, R. D. (1971). *Television and aggression*. San Francisco: Jossey-Bass Inc.

Fishbein, M., & Ajzen, I. (1975). *Belief, attitude, intention, and behavior: An introduction to theory and research*. Reading, MA: Addison-Wesley. [An online version of the book can be found at Ajzen's website at https://goo.gl/Re6HGP].

Fleming Jr., M. (2015). Blast from the past on 'Back To The Future': How Frank Price rescued Robert Zemeckis' classic from obscurity. *Deadline*, October 21, https://goo.gl/8MvUpd.

Follows, S. (2016). How movies make money: $100 m + Hollywood blockbusters, July 10, https://goo.gl/uYwnJe.

Forde, K. (2013). Top 10 albums only recorded because of contractual obligations. *Top Tenz*, April 16, https://goo.gl/qWLGXB.

Getzels, J. W., & Csíkszentmihályi, M. (1969). Aesthetic opinion: An empirical study. *Public Opinion Quarterly, 33*, 34–45.

Grobel, L. (1993). The dark side of fame: Robert Evans Pt. II. *Movieline*, September 1, https://goo.gl/RD1J34.

Hahn, A. (2004). *Possessed: The rise and fall of Prince*. New York: Billboard Books.

Hennig-Thurau, T., Walsh, G., & Wruck, O. (2001). An investigation into the factors determining the success of service innovations: The case of motion pictures. *Academy of Marketing Science Review, 1*, 1–23.

Hilgard, J., Engelhardt, C. R., & Rouder, J. N. (2017). Overstated evidence for short-term effects of violent games on affect and behavior: A reanalysis of Anderson et al. (2010). *Psychological Bulletin, 143*, 757–774.

Hirschman, E. C., & Holbrook, M. B. (1982). Hedonic consumption: Emerging concepts, methods and propositions. *Journal of Marketing, 46*, 92–101.

Hofstede, G. (1991). *Cultures and organizations: Software of the mind*. London: McGraw-Hill.

Holbrook, M. B. (1993). Nostalgia and consumption preferences: Some emerging patterns of consumer tastes. *Journal of Consumer Research, 20*, 245–256.

Holbrook, M. B. (1999). Popular appeal versus expert judgments of motion pictures. *Journal of Consumer Research, 26*, 144–155.

Holbrook, M. B. (2005). The role of ordinary evaluations in the market for popular culture: Do consumers have 'Good Taste'? *Marketing Letters, 16*, 75–86.

Holbrook, M. B., & Hirschman, E. C. (1982). The experiential aspects of consumption: Consumer fantasies, feelings, and fun. *Journal of Consumer Research, 9*, 132–140.

Holbrook, M. B., & Schindler, R. M. (1989). Some exploratory findings on the development of musical tastes. *Journal of Consumer Research, 16*, 119–124.

Holbrook, M. B., Lacher, K. T., & LaTour, M. S. (2006). Audience judgments as potential missing link between expert judgments and audience appeal: An illustration based on musical recordings of 'My Funny Valentine'. *Journal of the Academy of Marketing Science, 34*, 8–18.

Holt, D. B. (1998). Does cultural capital structure american consumption? *Journal of Consumer Research, 25*, 1–25.

Iannone, N. E., Kelly. J. R., & Williams, K. D. (2018). 'Who's That?' The negative consequences of being out of the loop on pop culture. *Psychology of Popular Media Culture, 7*, 113–129.

Ibbi, A. A. (2013). Hollywood, The American image and the global film industry. *CINEJ Cinema Journal, 3*, 93–106.

Jansari, A., & Parkin, A. J. (1989). Things that go bump in your life: Explaining the reminiscence bump in autobiographical memory. *Psychology and Aging, 11*, 85–91.

Janssen, S. M. J., Chessa, A. G., & Murre, J. M. J. (2007). Temporal distribution of favourite books, movies, and records: Differential encoding and re-sampling. *Memory, 15*, 755–767.

Jedidi, K., Krider, R., & Weinberg, C. (1998). Clustering at the movies. *Marketing Letters, 9*, 393–405.

Jones, R., & Mendelson, H. (2011). Information goods vs. industrial goods: Cost structure and competition. *Management Science, 57*, 164–176.

Kahn, B., Ratner, R., & Kahnemann, D. (1997). Patterns of hedonic consumption over time. *Marketing Letters, 8*, 85–96.

Koski, D. (2014). 10 movies sabotaged by their own creators. *Listverse*, September 25, https://goo.gl/J6wpur.

Kowert, R., & Quandt, T. (Eds.). (2016). *The video game debate: Unravelling the physical, social, and psychological effects of digital games*. New York: Routledge.

Kozinets, R. V. (2001). Utopian enterprise: Articulating the meanings of Star Trek's culture of consumption. *Journal of Consumer Research, 28*, 67–88.

Kreps, D. (2015). Aerosmith warns Donald Trump over 'Dream On' use, October 11, https://goo.gl/N7ZTEo.

Langfitt, F. (2015). How China's censors influence Hollywood. *NPR*, May 18, https://goo.gl/fNtSM9.

Lee, C. (2013). A history of flexing his muscles. *Los Angeles Times*, June 13, https://goo.gl/7ktRF1.

Leenders, M. A. A. M., & Eliashberg, J. (2011). The antecedents and consequences of restrictive age-based ratings in the global motion picture industry. *International Journal of Research in Marketing, 28*, 367–377.

Luan, Y. J., & Sudhir, K. (2010). Forecasting marketing-mix responsiveness for new products. *Journal of Marketing Research, 47*, 444–457.

Marchand, A., & Hennig-Thurau, T. (2013). Value creation in the video game industry: Industry economics, consumer benefits, and research opportunities. *Journal of Interactive Marketing, 27*, 141–157.

Markusen, A., Wassall, G. H., DeNatale, D., & Cohen, R. (2008). Defining the creative economy: Industry and occupational approaches. *Economic Development Quarterly, 22*, 24–45.

Mazis, M. B., Ahtola, O. T., & Eugene Klippel, R. (1975). A comparison of four multi-attribute models in the prediction of consumer attitudes. *Journal of Consumer Research, 2*, 38–52.

McCarthy, R. J., Coley, S. L., Wagner, M. F., Zengel, B., & Basham, A. (2016). Does playing video games with violent content temporarily increase aggressive inclinations? A pre-registered experimental study. *Journal of Experimental Social Psychology, 67*, 13–19.

Merchant, B. (2013). Studio execs hated the Blade Runner voiceover they forced Harrison Ford to do. *Vice Motherboard*, March 14, https://goo.gl/ijBPNq.

Nelson, P. J. (1970). Information and consumer behavior. *Journal of Political Economy, 78*, 311–329.

Pähler vor der Holte, N., & Hennig-Thurau, T. (2016). Das Phänomen Neue Drama-Serien. Working paper, Department of Marketing and Media Research, Münster University.

Peltoniemi, M. (2015). Cultural industries: Product-market characteristics, management challenges and industry dynamics. *International Journal of Management Reviews, 17,* 41–68.
Petersen, R. A., & Kern, R. M. (1996). Changing highbrow taste: From snob to omnivore. *American Sociological Review, 61*, 900–907.
Prins, M. (1999). Movie review Shakespeare in Love. *Christian Spotlight on Entertainment*, https://goo.gl/gNmHVq.
Qin, A. (2017). At the movies in China, some propaganda with your popcorn. *The New York Times*, July 7, https://goo.gl/yUTsR9.
Ratner, R. K., Kahn, B. E., & Kahnemann, D. (1999). Choosing less-preferred experiences for the sake of variety. *Journal of Consumer Research, 26*, 1–15.
Redden, J. P. (2008). Reducing satiation: The role of categorization level. *Journal of Consumer Research, 34,* 624–634.
Rink, D. R., & Swan, J. E. (1979). Product life cycle research: A literature review. *Journal of Business Research, 7*, 219–242.
Rochet, J., & Tirole, J. (2006). Two-sided markets: A progress report. *The RAND Journal of Economics, 37*, 645–667.
Rössel, J., & Bromberger, K. (2009). Strukturiert kulturelles Kapital auch den Konsum von Populärkultur?. *Zeitschrift für Soziologie, 38*, 494–512.
Runco, M. A., & Charles, R. E. (1993). Judgments of originality and appropriateness as predictors of creativity. *Personality and Individual Differences, 15*, 537–546.
Scaruffi, P. (1999). Paul Simon. https://goo.gl/Y48X98.
Schäfer, T., & Sedlmeier, P. (2009). From the functions of music to music preference. *Psychology of Music, 37*, 279–300.
Schwartz, S. H., & Bilsky, W. (1990). Toward a theory of the universal content and structure of values: Extensions and cross-cultural replications. *Journal of Personality and Social Psychology, 58*, 878–891.
Scrawler, J. (2016). 15 movies that were ruined by just one bad scene, July 28, https://goo.gl/NeFuxj.
Shapiro, C., & Varian, H. R. (1999). *Information rules: A strategic guide to the network economy*. Cambridge, MA: Harvard Business School Press.
Song, R., Moon, S., Chen, H., & Houston, M. B. (2018). When marketing strategy meets culture: The role of culture in product evaluations. *Journal of the Academy of Marketing Science, 46*, 384–402.
Sparviero, S. (2015). Hollywood creative accounting: The success rate of major motion pictures. *Media Industries Journal, 2*, 19–36.
Spielberg (2012). 'Why I'm betting £85 million on one cartoon reporter and his dog': Steven Spielberg brings Tintin to the big screen. *Mail Online*, October 5, https://goo.gl/Edp3Su.
Strahilevitz, M., & Myers, J. G. (1998). Donations to charity as purchase incentives: How well they work may depend on what you are trying to sell. *Journal of Consumer Research, 24*, 434–446.

Supreme Court (2011). Brown, Governor of California, et al. v. Entertainment Merchants Association et al., No. 08–1448, https://goo.gl/BrNQbU.

Szycik, G. R., Mohammadi, B., Hake, M., Kneer, J., Samii, A., Münte, T. F., & te Wildt, B. T. (2016). Excessive users of violent video games do not show emotional desensitization: An fMRI study. *Brain Imaging and Behavior, 10*, 1–8.

Szycik, G. R., Mohammadi, B., Münte, T. F., & te Wildt, B. T. (2017). Lack of evidence that neural empathic responses are blunted in excessive users of violent video games: An fMRI study. *Frontiers in Psychology, 8*, article 174.

Thompson, P., Jones, M., & Warhurst, C. (2007). From conception to consumption: Creativity and the missing managerial link. *Journal of Organizational Behavior, 28*, 625–640.

Troilo (2015). *Marketing in creative industries: Value, experience, creativity*. London: Palgrave.

Unsworth, G., Devilly, G. J., & Ward, T. (2007). The effect of playing violent video games on adolescents: Should parents be quaking in their boots? *Psychology, Crime & Law, 13*, 383–394.

Varian, H. R. (1995). Pricing information goods. Working Paper, University of Michigan.

Varian, H. R. (1998). Markets for information goods. Working Paper, University of California, Berkeley.

Voss, K. E., Spangenberg, E. R., & Grohmann, B. (2003). Measuring the hedonic and utilitarian dimensions of consumer attitude. *Journal of Marketing Research, 40*, 310–320.

Walker, S. (2015). Hollywood's Child 44 pulled in Russia after falling foul of culture ministry. *The Guardian*, April 15, https://goo.gl/1LDG8W.

Walls, W. D. (2010). Superstars and heavy tails in recorded entertainment: Empirical analysis of the market for DVDs. *Journal of Cultural Economics, 34*, 261–279.

Wanderer, J. T. (1970). In defence of popular taste: Film ratings among professionals and lay audiences. *American Journal of Sociology, 76*, 262–272.

Wang, C. (Alex), & Zhang, X. (Michael) (2009). Sampling of information goods. *Decision Support Systems, 48*, 14–22.

Weingarten, G. (2007). Pearls before breakfast: Can one of the nation's great musicians cut through the fog of a D.C. rush hour? Let's find out. *The Washington Post*, April 8, https://goo.gl/zLrEE8.

White, A. (2000). Joy in the struggle: A look at John Milius. *Film Threat*, July 18, https://goo.gl/pFBhoR.

Yaish, M., & Katz-Gerro, T. (2010). Disentangling 'Cultural Capital': The consequences of cultural and economic resources for taste and participation. *European Sociology Review, 28*, 169–185.

Young, S. M., Gong, J. J., & Van der Stede, W. A. (2008). The business of making movies. *Strategic Finance*, 26–32.

Zenatti, A. (1994). Goût musical, émotion esthétique. In A. Zenatti (Ed.), *Psychologie de la Musique* (pp. 177–204). Paris: PUF.

4

Why Entertainment Markets Are Unique: Key Characteristics

The product-related differences we discussed in the previous chapter are not the only reasons why adaptations of the "classical" marketing mix are required for entertainment. Product characteristics also have shaped the development of the markets on which entertainment is offered, in terms of the markets' overall structures, their critical resources, and their dynamics.

Knowing those market-level specifics is essential for understanding what requirements an entertainment company must meet to be successful in the longer perspective. In particular, we name three key characteristics of entertainment markets: a high level of innovation, the existence of substantial entry barriers, and network effects. The latter two characteristics are responsible for a high level of concentration which is typical for entertainment markets. Or at least some parts of them: let us take a general look at entertainment markets and highlight the co-existence of two sub-markets for which somewhat different economic rules apply, before we dive into the details of these key characteristics.

The Big Entertainment Picture: Two Sub-Markets Characterized by High Innovation and Partial Concentration

All entertainment markets consist of two separate, but intertwined *sub-markets* in which the kinds of products that are exchanged overlap only partially (see also Waldfogel 2017). The first sub-market consists of mainly artistic products that are produced for modest budgets by smaller-sized firms. A widely used label for such offerings is "independent," such as in independent

T. Hennig-Thurau and M. B. Houston, *Entertainment Science*,
https://doi.org/10.1007/978-3-319-89292-4_4

films, games, or music. Here, independent can refer to the producer's status and the source of funding (as in produced outside "the industry"). But it can also mean the product's very own character: the product is considered independent because it ignores commercial, "mainstream" requirements and industry "rules." Alternative labels include "art" or "avant-garde," products that stress the less-commercial nature of this sub-market even more strongly. The second sub-market encompasses products with a commercial focus, produced for higher budgets by bigger-sized firms which are an essential part of the global entertainment industry, such as the film and game studios, the major (music) labels, or the "big" publishers.

The two product types (independent and commercial) are targeted at different customer segments. Whereas commercial products by studios target "mainstream" consumers, independent products are aimed at niche or "elite" audiences who focus more heavily on artistic aspects—see our discussion of the corresponding taste differences and segments. As we illustrate in Fig. 4.1 (in which each circle represents a product), entertainment markets usually encompass a very large number of independent products, but only a limited number of studio products. However, industry budgets are allocated quite unevenly; the *sizes* of the circles in the figure symbolize the products' budget sizes. The commercial productions, despite being far fewer in number, capture the clear majority of money that producers spend to create movies, games, books, and music.

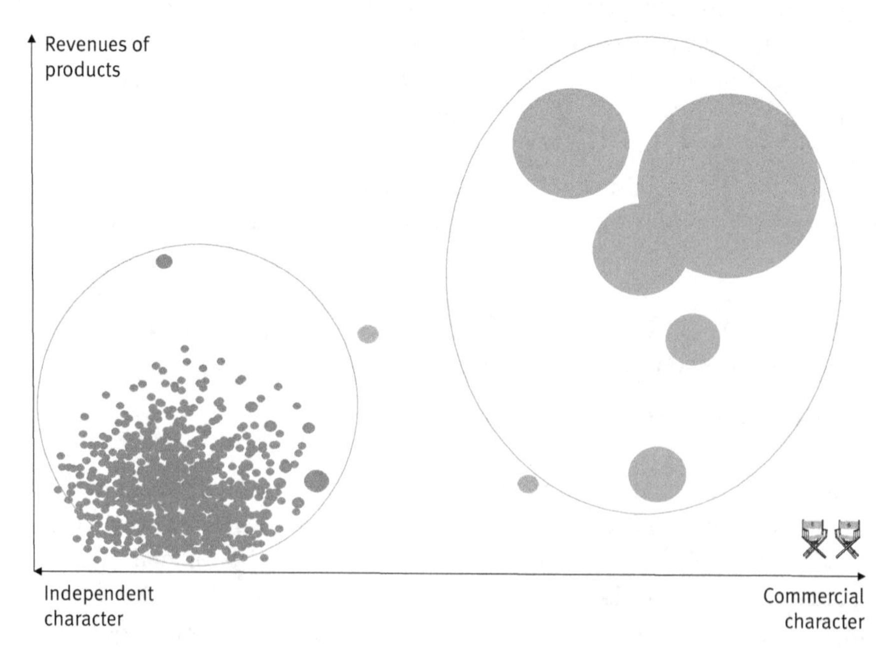

Fig. 4.1 Characteristic structure of entertainment markets
Note: Authors' own illustration.

What are the forces that cause this market structure? The large number of independent products points to a high frequency of innovation for this product type. It is made possible by the "infinite variety" of entertainment, but it is driven by the short life cycle nature of entertainment products and consumers' satiation that feed the need for the continuous creation of new products. Digitalization has further facilitated the creation of independent products, resulting in an additional growth in the number of such titles (see Waldfogel 2017 for details), creating what Anderson (2006) and others have named a "long tail" of entertainment products.[36]

The need for innovation exists also for commercial studio products—short life cycles and satiation also apply for them. Why then the high level of concentration for their sub-market, as indicated by the smaller number of such commercial products in the figure? First, studio products require a set of scarce strategic resources, including access to extensive financial resources. These resources are typically concentrated in a few large firms in each domain of entertainment, a condition that perpetuates the status quo in terms of market structure and that constitutes a massive entry barrier for anyone other than those who dominate the entertainment industry. Please note that the figure also points to the strongly asymmetric allocation of revenues between the two product types—global market success is closely tied to high levels of these strategic resources. A second reason for high concentration levels is that value creation in entertainment markets is often influenced by so-called "network effects" which contribute to a further strengthening of market concentration. These network effects provide advantage to those who have been already successful and a disadvantage to those who have not (including those who are simply new to the game), sometimes furthering a "winner-take-all" pattern.

Statistics show that these forces have shaped the markets for all entertainment products. The six (or five, after the Disney/Fox merger) biggest movie producers regularly assemble a market share of about 80% in theatrical revenues in North America. The three biggest music labels have a combined global market share of about 65% (and above 80% in North America). The ten biggest game companies account for at least 65% of the revenues globally generated by electronic games, and the "Big Five" publishers generate about every second dollar in the market for trade (or "recreational") books.[37]

[36]Please see our discussion of the long tail phenomenon as the underpinning of "niche marketing" in our chapter on integrated entertainment marketing.

[37]Digitalization has somewhat worked against concentration in this market, with the strong help of Amazon. See also our discussion of the book industry in our chapter on entertainment business models.

Networks are also driving concentration in other layers of the value chain for entertainment products, with firms like Netflix and Spotify now dominating certain parts of the chain on a global scale (see our industry overview section for more details).

An important question for entertainment studios is whether, and to what degree, the two sub-markets overlap. How high is the degree of substitution between their own commercial products and independent productions? If there is no substitution between sub-markets, competition would be limited to the relatively few studio productions, and, as a result of dynamic network effects, could be expected to decrease even further in the future. But there is evidence that such cross-market substitution does indeed exist, at least to a certain degree. Movie audiences occasionally prefer a low-budget horror movie, such as THE BLAIR WITCH PROJECT, over expensive studio productions, and some independently produced songs have been hits, as was the case with hip-hop duo Macklemore and Ryan Lewis' song THRIFT SHOP.[38] This is consistent with our discussion of consumer taste, which has shown that no *clear* separation exists between customer segments that value artistic achievements and mainstream audiences that lack such taste—only blurred boundaries. As a consequence, entertainment studios must not only manage their critical resources and deal with competition from other majors and studios, but also fend off competition from independent producers.

In the following, we will explore in some more detail the two forces that mainly shape the entertainment landscape: strategic resources and network effects. But before we begin, let us take a quick look at what we mean when we say there is "a lot" of innovation taking place in entertainment markets. We mean: *A lot.*

High Innovation Frequency in Entertainment Markets

The characteristics of entertainment products require managers to seek continuous innovation if they aim to have a consistently positive cash flow over the years. As a result, a very large number of new entertainment products hits the market every year, from games to movies to music to novels.

[38]We offer some additional perspective on Macklemore and Lewis' hit song and the overlap between independent and commercial products when discussing distribution resources.

What do we mean specifically by "very large"? Let's start with filmed content. There were 718 new feature-length movies released in North American movie theaters in 2016 (almost 14 per week), with just 134 (or one quarter) coming from the six biggest studios. These 702 were a mere subset of the more than 10,000 movies that were produced globally that year, according to the IMDb (not including "adult" films). In addition, 2016 saw the production of 3,761 new TV movies, as well as 172,045 new episodes of 7,027 TV series. In a single month in 2017, almost 13 million hours of new video content were uploaded to YouTube. Despite these "products" certainly not being perfect substitutes for each other, we believe the numbers are indicative of how high the frequency of innovation is for this kind of entertainment.

Numbers are no less impressive for the other forms of entertainment. For music, more than 20,000 new music albums were made available in 2016 in Germany alone, and almost 100,000 songs were newly released in singles format in that year (that is close to 2,000 every *week*), according to Bundesverband Musikindustrie. The International Publishers Association counted the releases of more than 80,000 new book titles and 240,000 re-editions in the U.S. in 2014. And when it comes to games, consumers can choose between close to 800 new console games per year, along with more than 280,000 new game apps that were added to Apple's iTunes store in 2016 alone, as PocketGamer.biz reports (see also Waldfogel 2017 for additional numbers).

For comparison, about 160 new smartphones, 40 new car models, and 10–15 new whiskeys and cognacs are annually released in the U.S. These numbers, both absolute and relative, clarify what we mean when saying that a "very large" number of new entertainment products is created. It's hard to find *any* other industry that matches entertainment when it comes to innovation frequency. Let us now explore the factors behind the concentration levels that exist on the sub-market of commercial entertainment products.

A Tendency Toward Concentration: High Entry Barriers in Entertainment Markets

The resource-based theory of the firm (e.g., Penrose 1959) is quite popular among management scholars to explain why some firms succeed while others fail. In the context of entertainment, the resource-based theory offers sound arguments why it is so difficult for new players to enter the sub-market of commercial products, which is dominated by few studios and labels. At the heart of the resource-based theory is the logic that firms can achieve a sustained competitive advantage in an industry if they own a distinct set of "strategic resources" and implement strategies that make use of these resources.

Via such strategies, strategic resources can create massive entry barriers (or, in the language of the resource-based theory, "resource position barriers;" Wernerfelt 1984). What are those strategic resources, and what makes them powerful? According to resource-based theory, they are firm-controlled assets and capabilities that are rare, valuable, inimitable (i.e., cannot be easily copied by competitors or substituted by other resources), and sustainable (Barney 1991). Such resources create a long-lasting competitive advantage which is responsible for the "strategic" character of strategic resources.

Resource-based theory thus tells us that success in entertainment's commercial sub-market depends on two things: the possession of such strategic resources, along with their transformation into successful market actions via powerful strategies. We discuss those strategies in Part II of our book—the use of marketing instruments for entertainment products. In the remainder of this section, we will take a closer look at the kinds of strategic resources that are essential for success in the market for entertainment, and especially its sub-market for "studio products."

We have identified four types of strategic resources that can help making the difference: (a) financial production and marketing resources, (b) distribution resources, (c) access to or control of creatives and their past works, and (d) technological resources. They stand between the industry incumbents and those who would like to enter the market, embodying quite colossal entry barriers.

Production and Marketing Resources

> "That budget could feed a small country for years!"
>
> —*User comment on Whedonesque.com on the information that* The Avengers *movie will have a production budget of $260 million*

For studio entertainment products to be produced and released, an enormous amount of money is required, most of which is spent up front (the first-copy-cost property of entertainment). The last time the MPAA reported average production costs of Hollywood studio films (in 2007), they exceeded the $70 million mark, and the costs of advertising and distribution that were spent to support the launch in North American theaters averaged

$36 million. In other words, releasing a studio movie required an upfront investment of over $100 million.

Ten years later, the actual capital requirements for releasing a studio film are even higher. McClintock (2014), based on conversations with several leading Hollywood executives, concludes that, *after* paying for production costs, a global marketing campaign for what is called in Hollywood a "tent-pole" movie now requires an investment of up to $200 million—$100 million for North American audiences plus $100 million for what the industry calls "foreign" territories. Consider the James Bond franchise as an example: Sony reportedly invested almost $400 million in the making and global theatrical distribution of the agent series' 2015 entry Spectre (Fleming 2016). This is about 40% more than the total, inflation-adjusted costs for the 1999 James Bond film The World Is Not Enough and a whopping 120% more than for License to Kill in 1989. These enormous amounts are closely tied to the blockbuster marketing concept that now dominates studio productions across the different forms of entertainment; we discuss the concept's economic logic as well as the historic development of financial resources in entertainment in the chapter on integrated entertainment marketing.

It is important to stress that marketing resources are not only critical for studio products, but are now required for *all* entertainment productions that target a broader audience. The main reason is that the costs for advertising an entertainment product are not scalable for products with a smaller production budget; instead, to be heard by potential audiences, any wide-release movie has to spend heavily on advertising. As director-producer Steven Soderbergh words it: "Point of entry for a mainstream, wide-release movie: $30 million. That's where you start. Now you add another 30 for overseas" (quoted by *The Deadline Team* 2013). The low-budget comedy movie The Boss had estimated production costs of $29 million, but still incurred global advertising costs of $70 million (D'Alessandro 2016).

For other forms of entertainment, having access to financial resources is similarly critical, although the absolute amount of required investments differs. The development of a major game costs over $60 million, on average, and can reach up to $200 million, as it was the case with the massively multiplayer online game Star Wars: The Old Republic in 2011 (*Superannuation* 2014). This does not include advertising costs, which are also substantial. Richard Hilleman, as executive of the game company EA, revealed that the firm spends two or three times as much on marketing and advertising than it does on developing games (Takahashi 2009). This might not be to the case for *all* games, but it stresses the need for substantial advertising resources for this kind of entertainment content also.

For music, Alexander (1994) systematically analyzed entry barriers, highlighting the role of production costs. According to his research, they range from $160,000 to more than $800,000 (in 2017 value) for a single record. Alexander also stresses that, as we have shown is the case with movies and games, promotion might be considered an even stronger entry barrier for music because of the existence of a "promotional network" that denies smaller firms radio airplay (and radio is still one of the common ways to make consumers aware of new songs or albums). Since the days of his study, the Internet has not reduced, but increased this critical role of marketing resources for music: whereas digital technology has diminished the barriers for recording music, this has not been the case for promotion. Instead, Jason Flom, former CEO of Capitol Records, states in the documentary ARTIFACT that it has become harder than ever to *make music heard*—because there is so much of it now, and also so much noise.

And what about getting access to financial resources for entertainment? Do digital technologies help in this regard, essentially reducing their strategic role? One approach that digitalization offers is crowdfunding, and a number of entertainment products have indeed been successfully financed this way, including films and games—we have dedicated a section to crowdfunding as a way to reduce the risk of entertainment (see the chapter on entertainment business models).

But crowdfunding projects almost always fall in the independent-products category, with funded production budgets being usually tight and very few resources being available for advertising (versus production). Galuszka and Brzozowska (2016), drawing on 30 in-depth interviews with musicians who crowdfunded their projects, conclude that the approach will not overcome entry barriers in a meaningful way, mainly because of "the difficulties of dealing with promotional activities traditionally conducted by record labels." So financial resources can be expected to remain of strategic importance in entertainment.

Distribution Resources

"It's distribution, stupid..."
—King *(2002, p. 59)*

As a result of the large number of independent entertainment products and the overlap between the sub-markets for independent and commercial studio products, the entertainment industry is characterized by "persistent oversupply" (Peltoniemi 2015, p. 42). This oversupply assigns an important role to a "filtering system" in which gatekeepers make choices that affect the accessibility of products for consumers, ensuring that only a fraction of the many entertainment products are actually distributed to consumers.

In some cases, these choices literally prevent consumers from accessing a certain product (e.g., a movie that is not shown by a theater simply cannot by watched by audiences). But in other cases, the system's choices raise the costs (effort, price, etc.) of access—such as when a song is hidden so deeply in the iTunes database that it can only be found when consumers invest a huge amount of search effort. Without the existence of these filtering systems, entertainment consumers would face an overwhelming overload of choices, which would lead to demotivation and market failure (for more on the consequences of choice overload, see e.g. Iyengar and Lepper 2000).

Distribution resources possessed by major studios, labels, and publishers ensure that a producer gets his or her products past the filter to be distributed to consumers. Netflix, for example, does not possess such resources for theatrical releases, and thus has faced problems findings movie theaters that are willing to show their productions. This lack of consistent access to theaters causes problems for the firm, such as when their products cannot qualify for important industry awards such as the Oscars (which influence movie success in their own right), but also for attracting content and its creators (for whom a theater release offers value on its own).

What exactly do we mean by distribution resources? In short, we have a producer's "assumed capabilities" and "power" as conditions for ensuring consumer access to his or her creations. To define and illustrate these two resources, let us take the case of theatrical movie distribution, where only major film studios are able to motivate a high enough number of theater owners to support a "wide release"—i.e., the simultaneous opening of a new film on several thousand screens.

The first resource of assumed capabilities means that theater owners will only support the distribution of a studio's film if they believe the studio has the capabilities to draw in large numbers of customers with their products. Theater owners only trust the major studios (because of the studios' experience, products, and financial resources) to have this capability, in the same way that radio stations and retailers only trust the major labels and publishers to draw large numbers of people to listen to, or shop for, their products. As Epstein (2010, p. 189) phrases it for movies, "multiplex owners know that the six major studios can [drive] a herd of moviegoers from their homes to the theater on an opening weekend," having what "it takes to fill 2,000 theaters [in North America alone] with popcorn-eating audiences." He notes that this is largely considered "a next to impossible undertaking" for producers outside of the studio system.

"Power," the second distribution resource, refers to producers' ability to leverage their multi-product portfolio to gain distribution for a wide range of their products. Major producers, and only they, can use the power of their future blockbuster projects to ensure distribution of other products which distributors may not rate as highly. According to film producer legend Arnon Milchan, the major studios are the ones "who can tell the theater owner, 'If you are not putting this movie on three screens in your multiplex, you're not getting STAR WARS next month.' So you need the muscle to get in" (quoted in Shanken 2008).

Distribution resources are similarly important for all forms of entertainment, despite differences in distribution mechanisms. For music, Alexander (1994), as a result of his case study, names distribution as a crucial entry barrier that is controlled by major labels and their "distribution arms." He concludes that "[f]ringe firms and new entrants have few alternatives to distribution through a major competitor. [T]he cost of integration into distribution at the national level has been estimated at $100 million." Although digital platforms are reshaping the distribution of music, gaining large actual sales for a song still is the province of the big labels; Macklemore and Lewis' self-produced song THRIFT SHOP did top the Billboard charts in 2013, but it was the first song not released by a major label *in 20 years* to become #1. And even that song was not independent with regard to its distribution, as the artists had hired Warner Music to get continent-wide radio airplay (Chace 2013). As independent producer Benjy Grinsberg states, "can you sell 5 million singles on one song? I doubt it. Radio can take you to that next level, and the majors push singles to radio better than anyone" (quoted in Buerger 2014).

Bottlenecks also exist when it comes to the distribution of games and novels, primarily in the shape of retailers that are dealing with strictly limited shelf space in the physical realm. Publishers compete for that space not only against each other, but also with non-entertainment product categories. For game producers, having access to a distribution platform such as a gaming console or a mobile app market is becoming increasingly important. As both console and app store markets are extremely concentrated, game producers must pay royalties to access the platform's customers. These fees have been reported to reach $80,000 per game and console system.[39]

In addition to financial barriers, awareness barriers can be quite massive for platforms, with search functionalities being limited. Platforms often have

[39]Royalties are substantially lower though for handheld devices (about $100 in 2013) and mobile app stores (between $25 and an annual fee of $100 per game/app for the same time frame) (Marchand and Hennig-Thurau 2013).

a particular interest in highly commercial studio fare, which helps them to sell more platform products such as consoles because of so-called "indirect network effects." It certainly does not help independent game producers that all leading gaming platforms are owned by major game studios Sony, Microsoft, and Nintendo.

With the rise of digital technologies, platforms are also becoming more important for other forms of entertainment—think of Spotify for music, Amazon's Kindle for books, and Netflix for movies. This trend affects the value of established distribution resources (e.g., relations with stationary video rental chains have been discounted), but it does not change the critical role of distribution resources for entertainment success *as such*. It's not by chance that the major labels get the best deals from Spotify et al. (e.g., Lindvall 2011). The new distributors are also eager to get the most popular content, which reasserts the studios' distribution power. As Roy Price, then-Amazon executive, phrased it: "we're increasingly focused on the impact of the biggest shows. It's pretty evident that it takes big shows to move the needle" (quoted in Littleton and Holloway 2017). If the studios want to exploit their distribution power by selling their products to platforms for high fees or by reducing the distributors' attractiveness for consumers by not doing so (but engaging in self-distribution) is a separate issue—one we get back to later.

Access to and Control of Creatives and Their Works as Resources

> "Talent is rare."
> —*Pixar-founder and president* Ed Catmull *(2008, p. 66)*

Talent, defined in entertainment by someone's creative skills, is a scarce resource, as expressed by Pixar's Ed Catmull in this section's introductory quote. Because of this, artists who have the abilities to create entertainment products that resonate deeply with consumers are a strategic resource, as are their past works. In turn, entertainment producers have a competitive advantage when they have access to such creatives or own/control their intellectual creations. The strategic importance of this talent resource is based on both supply-side and demand-side arguments.

Taking a supply-side view, it has been argued that the relative stability of Hollywood as an institution and the long-lasting resistance of its major studios over many decades has been the result of its "relationships with talent" (Friend 2016). With creatives being behind what the industry produces for consumers to enjoy, having better access to these creatives constitutes a major competitive advantage.

Traditionally, such access was secured in entertainment by exclusivity contracts. This type of "star system" still exists today in music and publishing, where artists often sign multi-year contracts. But in film, long-term contractual bonds between creatives and producers have become a rarity since the 1960s; instead, deals are usually made on a per-project basis. Under such conditions, access to creatives depends on "soft" factors—mostly the producer's willingness to provide conditions that enable the creative to realize her or his artistic visions. These desirable conditions certainly include financial leeway, but just as important are respect and artistic freedom.

Interestingly, providing such conditions has been the main gateway for new digital entrants such as Netflix and Amazon. When director David Fincher and star Kevin Spacey were searching for a firm to produce their House of Cards series, they ended up at Netflix because it "was the only company that said, 'We believe in you'" (Spacey 2013). This belief was expressed by promising a full two-season production with "no interference" (Netflix Chief Content Officer Ted Sarandos, quoted in Nocera 2016), whereas all other studios required the filming of a pilot episode first. According to Mr. Spacey (2013), avoiding this request was crucial to the artist team, because "[w]e were creating a sophisticated multi-layer story with complex characters who would reveal themselves over time, and relationships that would need space to play out. And the obligation ... of doing a pilot from the writing perspective is that you have to spend about 45 min establishing all the characters and creating arbitrary cliffhangers and basically generally prove that what you're going to do is going to work."[40]

Netflix has made such a "hands-off policy with the 'talent'" (Nocera 2016) a core element of their market positioning, which has attracted several high-profile creatives, and Amazon has a similar policy in place, also offering uncommon creative freedom (Frank 2014). The relationships of traditional industry leaders with talent are instead often described as taut—a tendency which seems to be only intensifying with the rise of franchise management in entertainment (see Debruge 2017).[41]

---

[40]Please note that Netflix's decision was not an unsubstantiated gamble, but was informed by the firm's analytical insights. As Mr. Spacey recollects, Netflix had run their data, and the data told the firm that its subscribers would watch the series.

[41]The documentary Artifacts offers an instructive documentation of industry practices. It chronicles a $30 million lawsuit by label EMI against the band Thirty Seconds to Mars, that unfolded after the band's decision to exit their contract because they had not received any profits from their albums, despite selling millions of them.

Creative talent is also crucial to entertainment firms for demand-sided reasons. Consumers love entertainment in general, but they in particular love certain stars and the characters and fictitious worlds they bring to life on screens, headphones, and Kindles. Thus, talented creatives and their works can serve as "brand assets" for producers of entertainment, to which consumers can relate and exhibit loyalty. Fans love to hear Lady Gaga sing and dance, they love to watch Will Smith act, and they can't wait for the next Dan Brown novel. Owning the rights for these creatives' next works or the actions of their characters can be an essential competitive advantage—we assume you agree that having Steven Spielberg on one's team and his permission to continue the story of E.T. would be of enormous help. This is what Donna Langley, as chairman of Universal Pictures, has in mind when she argues: "there can be nothing better than a successful source, a brand. Everyone is after such sources these days" (quoted in Beier 2016). We dedicate a full chapter to the critical role of such branded resources and how they should be managed to make the most of them.

Technological Resources

A final resource that can add to an entertainment firm's competitive advantage is exclusive technological expertise. Because entertainment succeeds by providing aesthetic and sensual experiences to consumers, the technology used for creating an entertainment product can have a major influence on customers' reactions. Because of entertainment product's short life cycles, technology is usually not used by producers to advance a single product, but to help the performance of a whole slate of entertainment products, such as a Hollywood studio does apply its capability to produce movies in digital 3D not only to one, but several films. Technology is intended to provide the products with a competitive advantage over other offerings of the same form of entertainment (i.e., 2D movies), but also in a more generic way over other leisure activities.

Nevertheless, the success of some of the biggest commercial winners in entertainment are closely linked to innovative technological milestones. When filming Jurassic Park, its director Steven Spielberg planned to bring the period-defining creatures to life via a blend of animatronics and advanced stop-motion techniques, which had been the industry standard for decades. The movie's commercial watershed moment, however, was when its special effects team substituted the "old" technology with computer-generated images (CGI). As a result, Jurassic Park enabled audiences for the first time to face a "real," *living* brachiosaurus, which millions of consumers did not want to miss (Huls 2013).

In the case of Avatar, director James Cameron delayed the implementation of his cinematic vision for almost a decade because he felt technology was not ripe. After finishing an 80-page treatment in 1996, Cameron worked closely with engineers until 2007 to improve the quality of motion capturing technology to a level which allowed audiences to take his human-like "Na'vi" characters seriously and to empathize with them. Further, he also used the interlude to design the digital 3D "Fusion" camera system that created realistic stereoscopic images far beyond usual blurred 3D effects. When Avatar was eventually released in December 2009, audiences lined up in masses because their ticket entitled them to an unprecedented trip to a different, fantastic world.

Other breakthrough successes that can be linked to technological resources include movies such as Gravity (the "Lightbox" technology), The Matrix (the "bullet time" effect), and the Lord of the Rings trilogy (for new battle scene software), but also games like Myst (for its early 3D designs), L.A. Noire (the "facial capture" technology), and Wii Sports (for its use of the revolutionary motion capture capability of the Wii console) (e.g., Kohler et al. 2013). We find it hard to envision Pokémon Go becoming a big hit with consumers without its innovative Augmented Reality elements.

In all these cases, offering such experiences required extensive technological knowledge, which only major studios are able to accumulate and which sets these companies apart from smaller competitors who cannot master such costly technology without the assistance of a deep-pocketed partner. The legacy of these and many other technological innovations is that even if a producer does not plan to push the technological boundaries further with a new film or game, meeting today's standards is no longer possible without substantial technological resources—something that applies equally to films, series, and games (but less so to music and books).

Technological resources help major studios to differentiate their own products from other products of the same type (people preferred Avatar over other films), but they are also a means to fight more generic competition from *other forms* of entertainment. "I know that Netflix can't do this" said Tom Rothman, as chairman of Sony Pictures, when he was asked why his company made Ang Lee's film Billy Lynn's Long Halftime Walk at 4K resolution and 120 frames per second (quoted in Itzkoff 2016a, b, p. 11). He might also have thought about offering consumers a reason to watch his movie instead of spending their time with Facebook and Instagram—traditional entertainment's new, more generic competitors for consumers' entertainment time budgets.

We have so far only mentioned production technologies, but resources in the field of information technologies can also contribute to competitive advantage. Firms such as Netflix and Amazon have developed proprietary artificial intelligence that they use in automated recommender systems to link products to the preferences of individual consumers. The power of such tools is clearly tied to the smartness of the algorithms, but power also comes from the right databases in order to apply algorithms in a way that creates value for consumers: access to customer data and its organization is a critical technological resource. When we, the authors, are asked by producers how they can make use of "big data" for themselves, access to data is always the key.

Finally, information technology requires powerful and costly hardware. In 2013, Netflix already employed between 10,000 and 20,000 servers to stream its content to mass audiences, which makes it one of the largest users of cloud computing (Brodkin 2016). The master copies of its programs demand more than three petabytes of storage space, which is the equivalent of about 20 billion photos on Facebook's servers. And when devices like a new PlayStation are released, thousands of additional servers adapt to reformat movie files and deal with the new users (Vance 2013). All of these activities require advanced technology that would be quite challenging for potential contenders to imitate.

Even More Concentration: Network Effects

Success in entertainment markets requires strategic resources. But it can also set off certain dynamics known as "network effects" which make the successful even *more* successful and further increase the difficulties for smaller producers to compete with major firms. Network effects essentially foster concentration and sometimes leave room for only a few winners who "take it all." In entertainment markets, two different types of such network effects exist: direct network effects and indirect network effects, which derive from the positive externalities that exist for platforms.

Direct Network Effects

Direct network effects mean that the value of a product for a single consumer increases with the number of other users of the same product, i.e., the size of the "network." The reason that consumers derive utility from the number of users of the product is because the larger the network, the larger the number of potential options and exchange partners. These potential partners provide a consumer with a higher potential of exchange of words, files, or knowledge. As a result, in markets with direct network effects, the number of people connecting to a network depends, to a large degree, on the number of other people who are already connected to it.

Such direct network effects are well known in traditional communication industries: think of telephones, social networks like Facebook, and cloud software such as Dropbox. But also consider software programs such as Microsoft Word. Is Word the best text-processing software? Maybe, but everyone who has ever worked with colleagues or friends on a common document knows that there is hardly any alternative to the program simply because *everyone else* is using it. That combination of compatibility and ubiquity of use is what direct network effects are about.

Similar direct network effects exist in entertainment. They are most obvious for video games, whether massively multiplayer online games (MMOGs) such as World of Warcraft, smartphone quizzes à la Quizduell/QuizClash (which has 29 million players in Germany alone who compete against their Facebook friends to answer questions), or through the online multiplayer mode of console games such as FIFA or Call of Duty. In all cases, users derive benefits from the size of a game's network, as the quality of the consumption experience is clearly influenced by the network size.

Scholarly evidence for the value of such network features is provided by André Marchand (2016) who explores the financial consequences of an "online multiplayer" feature in the context of console video games, where he analyzes almost 2,000 video console games that were released in the U.S. between 2005 and 2014. Using OLS regression with robust standard errors, he finds that an online multiplayer feature exerts no direct effect on game sales. However, a positive interaction effect exists with the age of the console generation: the bigger a console's network, the more beneficial is it for a game to enable consumers to play a game together with others over the Internet.[42] And Liu et al. (2015) show with a regression analysis that for MMOGs from 2003 to 2007 the number of the game's players influences how much the players like a game: adding 1 million users is associated with a 0.17 increase in quality rating (on a 1–10 quality scale).

For other types of entertainment products, direct network effects also exist, but they are of a somewhat different kind. In the case of movies,

[42]For a deeper, multifaceted investigation of the multiplayer function of games and their social benefits, see the articles in Quandt and Kröger (2014).

music, and novels, the value of the network is not primarily a higher level of enjoyment tied to the joint consumption of the product with others; instead, the network increases from the "sign value" of entertainment products and heightens the products' cultural relevance or prominence. The more people who have seen a film, the more people can participate in conversations about it, with these conversations providing psycho-social benefits to consumers and knowledge about the film adding to a consumer's social capital (Bourdieu 2002). We find, in our survey of German consumers, that the ability to engage in a communicative exchange with other consumers is a strong driver of people's watching of a new drama series such as BREAKING BAD; the social capital from knowing these series sets them apart from more traditional formats such as GREY'S ANATOMY (Pähler vor der Holte and Hennig-Thurau 2016).

Similarly, people read novels that are known to have been read by a large network of others, not only because they make inferences about the novels' quality based on network size (see our discussion of "action-based cascades" in our chapter on "earned" entertainment communication), but also because reading the novel provides a consumer with value by allowing him or her to join the discussion. We argue that the same mechanisms are in effect for music. For any type of entertainment, once a critical network size has been reached, its cultural character increases the attractiveness for consumers to join the network of users; in fact, consumers may feel like a "social outcast" if they don't join in.

Such network effects can be argued to also exist for anticipatory processes: if consumers *expect* a product to become popular in the future, a network will form, the virtuous cycle will begin—and consumers' expectations will tend to prove correct in a self-fulfilling prophecy. Finally, automated recommender systems can also provide direct network effects across the different forms of entertainment—the more data a system can employ for generating its recommendations, the better their quality.

Indirect Network Effects and How They Influence the Success of Entertainment Products

Consider a video game such as LEGO STAR WARS and a consumer who plays the game alone on his PlayStation 4 (PS4). This consumer, let's call him

Luke, does not play the game with friends or in multiplayer mode over the Internet, and he doesn't even chat with others about the game. It is obvious that in such a situation, there is no *direct* network effect for the game.

However, there is a different network effect taking place here—one that is of an *indirect* kind and takes place on the platform level. It is called "indirect" because in this case the network benefits to our consumer Luke do not stem directly from the number of *users* of the platform (the PS4 in our example), but from the effect the number of platform users has on the existence of other, complementary products that appeal to Luke.[43] Because the more PS4 consoles are sold, the more attractive it is for game producers to make their game titles available on the platform—which means that more attractive games become available for the PS4 and, eventually, more choice and better games for consumers like Luke. These indirect network effects occur because games and consoles are complementary products, just like butter and bread.

What do such indirect network effects mean for producers of entertainment products which are at the center of this book's attention, such as game makers? A key takeaway is that, in entertainment platform markets, "superstar" titles, such as blockbuster games, are of particular importance for platform producers: they enhance the value of the platform itself more than other titles do, because they not only attract consumers with their game value, but also by exerting an indirect network effect on other game producers and their titles (Binken and Stremersch 2009).[44] As a result, indirect network effects on the platform level further strengthen the already strong market shares of the major studios who are the only ones who can produce such "superstar" titles because of the specific strategic resources this requires.

---

[43]Do direct network effects also exist on a platform level, such as for gaming consoles? It depends—when there are proprietary modes of communication between platform users (such as the Facetime chat on Apple devices), this can be the case, but usually direct network effects are more prominent on the "application" (e.g., game) rather than the platform level. But console operators such as Sony have been making strong efforts to increase the value of their platforms by creating also direct network effects through chat functions etc.

[44]A number of studies have investigated the role of content for hardware/platform success, often in the context of gaming consoles. For those readers who are interested in this perspective, please see, for example, Clements and Ohashi (2005) and Binken and Stremersch (2009), who discuss whether it is the mere size of the content network (the number of titles) or mainly its quality (in terms of "superstar" products) that are responsible for the indirect network effect on the platform. Let us also note that indirect network effects on the platform level are somewhat linked with the existence of the two-sided nature of entertainment products (i.e., the existence of more than one group of customers who influence each other, as discussed later in this chapter). For the platform, the software consumers are one customer group, and the other group consists of the producers of the software which pay royalties to the platform provider. Their willingness to do so depends on the consumers' demand for the software—in other words, positive externalities between customer groups (Marchand and Hennig-Thurau 2013).

Further, the number of adopters of a platform (referred to as the platform's "installed base"), as the result of indirect network effects, influences the demand for entertainment content made available for the platform. Thus, the performance of any product is at least partly determined by this "installed base" and should be taken into consideration by the producers. The success of the PS4 with consumers, including our Luke, has probably been an important consideration for the Lego company and its partners to release LEGO STAR WARS on the console. Scholarly studies confirm this role that the size of a platform's "installed base" plays, but also point to a somewhat more complex relationship.

When Healey and Moe (2016) analyze the weekly sales of 98 console games that were released on all three platforms between 2006 and 2011, they find that two opposing effects are at play: whereas the mere size of the platform's installed base leads to more sales for a game (the "size effect"), the scholars also find a qualitative "aging effect" of hardware networks that implies reduced sales per network member over time.[45] Figure 4.2 illustrates

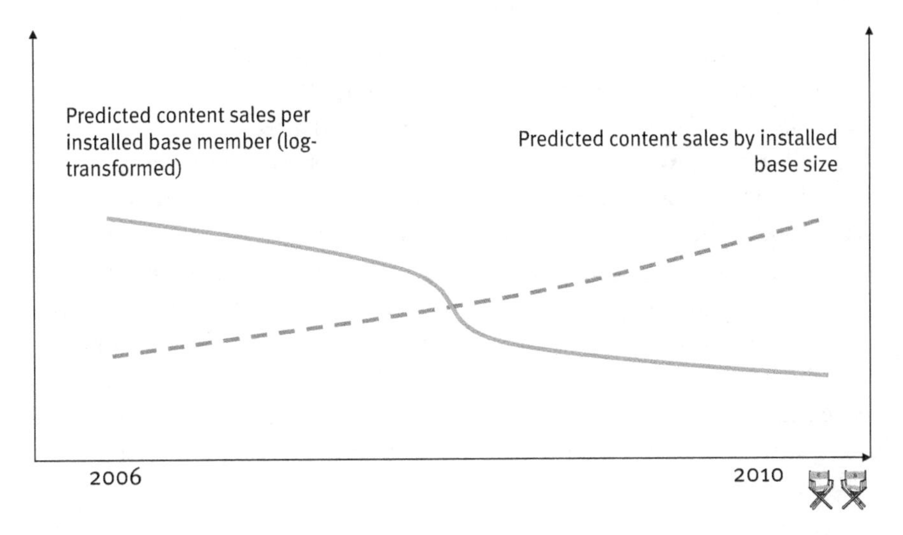

Fig. 4.2 The "size effect" and the "aging effect" of a platform's installed base on entertainment content sales
Notes: Authors' own illustration based on results from Healey and Moe (2016). The solid orange line shows the "aging effect," the dotted blue line shows the "size effect" of platforms on entertainment content sales. Both courses are stylized.

[45]This "aging effect" also explains why Marchand (2016), in his analysis of console games, found a negative trend by console age on game sales while controlling for hardware sales.

these opposing effects. They also identify two factors which make a "good" installed base when it comes to facilitating the demand for games: if the base is characterized by a high level of "innovativeness" (i.e., if a large share of its members adopted the platform early in its life cycle) and if there is a high level of "recency" (i.e., if a large share of the installed base adopted the platform not long ago).[46]

Whereas our discussion here has been dominated by games, platforms exist for other forms of entertainment, and indirect network effects are in play there also. For streaming rights, movie producers can make the choice between platforms such as Netflix and Amazon (or create their own platform, as Disney has decided to do; Castillo 2017). For pay-TV rights, content producers can choose between HBO, Showtime, and Sky, and between networks such as NBC or CBS for TV rights. Music producers can make their songs available on streaming services such as Spotify and/or Apple Music, and book publisher have to decide on which ebook platform (e.g., Amazon's Kindle or Apple's iBooks) they want their products to be available. In every case, indirect network effects exist and the quantity (and quality) of a platform's "installed base" should be considered by content producers.

Why our focus on games then? Because whereas for (console) games the proprietary platform is the primary distribution channel, for movies, music, and books exclusive platforms traditionally played a less focal role. For these forms of entertainment, dominant platform standards exist which are distributed not by a particular platform provider, but across *all* distributors: the printed book, the movie theater, the DVD and the Blu-ray, the CD and the MP3 file. But with the rise of music and film streaming, digital publications, and a trend toward more proprietary platforms, this might not be written in stone—a development that enables, or requires, new distribution models.[47]

We have now reached the end of our discussion of network effects in entertainment. In Fig. 4.3, we summarize the different kinds of such effects (direct and indirect) and provide examples for each type of entertainment product.

[46]Healey and Moe's recency factor is similar to what Marchand, in his study, names the "livelihood" of an "installed base": the console sales that happen in the month in which a particular game is released.

[47]See also our overview of industry developments and our discussion of distribution configurations for entertainment products.

Content type	*Direct network effects –content level*	*Indirect network effects –platform level*
Programmed content	Value added by multiplayer modes (online and offline), communication about games	The number of PS4 owners influences the number and quality of games developed by producers for the PS4, which influences demand for the PS4 (and for the games via "installed base" effects)
Filmed content	Value added by communication about films and series	The number of Netflix subscribers influences the number of films and series made available to Netflix by producers, which influences demand for Netflix (and, subsequently, for the films and series)
Written content	Value added by communication about books	The number of Kindle users influences the number of ebooks offered to Amazon by publishers in its specific format, which influences demand for Kindles (and the ebooks that are accessible through it)
Recorded content	Value added by communication about it	The number of Spotify subscribers influences the number of songs made available by producers via Spotify, which influences demand for Spotify (and for the music).

Fig. 4.3 Direct and indirect network effects in entertainment markets
Notes: Authors' own creation. PS4 means PlayStation 4.

Managerial Consequences of Entertainment Markets' Characteristics

Just as our analysis of product characteristics revealed a number of implications for the management of entertainment products, understanding the characteristics of entertainment *markets* (their innovative nature, the need for strategic resources, and the existence of network effects) offers important insights for entertainment managers.

Let us first note two general, strategic implications. An overarching message is that anyone who wants to produce entertainment content needs to be aware of and carefully consider the existing sub-markets and their dynamics. Strong entry barriers exist for commercial products that nab the lion's share of industry revenues and profits, and competition is enormous, particularly for all others. Given the fundamental differences between commercial and independent products, producers need to judge their resources critically before planning to launch a product into the commercial market dominated by studio fare. They also should accept the realities of the independent market and its limited overlap with the commercial market.

In terms of strategic positioning, independent producers face nearly insurmountable odds if they endeavor to compete with major studios (on the studios' home territory) by trying to produce commercial "superstar" products and employing the blockbuster concept of marketing. Our analysis shows that smaller players simply lack the diverse resources that would be needed to do so satisfactorily. Any attempt to position oneself as a producer of such commercial products like firms such as STX Entertainment and Luc Besson's EuropCorp are pursuing these days in the domain of film, requires a thorough strategy regarding how these entry barriers will be overcome. Otherwise, such efforts are doomed and will end tragically, like they did for the legendary Cannon Group in the 1980s and for Internationalmedia in the 2000s.

In addition, each of the market characteristics that we described has specific strategic implications for managers. The high level of innovation that is typical for entertainment markets sends two important messages to all who hope to be in the business for more than a one-hit wonder. First, innovation needs to be a systematically managed activity. Whereas this seems obvious in most other industry contexts, the "uniqueness" assumption associated with the entertainment industry's "Nobody-Knows-Anything" mantra has often prevented entertainment managers from doing so—why invest in innovation structures, processes, and culture when everything you do is "unique"? Taking an *Entertainment Science* perspective results in a different view—one in which the strategic management of innovation activities is a key element of the entertainment marketing mix. We elaborate this view in our chapter on entertainment product innovation.

Second, high industry-wide innovation means that competition matters, something that is of particular importance for the independent sub-market where entry barriers are less strong. In a strategic perspective, competition affects the choice of product, but given the short life cycles of entertainment products, the issue is most highly relevant for distribution decisions.

The strategic resources that we outline should be used to determine an entertainment firm's market potentials. By identifying its strengths and weaknesses along these heterogeneous resources and by comparing itself with other firms, any provider of commercial products can systematically identify what needs to be done to become competitive (or stay that way) in the industry.

Finally, understanding the network effects which are at work in many areas of entertainment can shape the business model for an entertainment product and particularly for pricing decisions. Network effects provide the logic for so-called "freemium" offers that have become popular in the world of gaming. Such decisions are far from trivial, and even firms as powerful as Lego have failed to adequately value and address the importance of network effects—we discuss the complexities in our chapter on entertainment pricing. Also, what kind of platform is the right one for a specific kind of entertainment content? What kind of content does a platform strive for, and, most importantly, how will the platform develop? Finding good answers to these questions will enable entertainment managers to account for the power of indirect network effects in an effective way.

Concluding Comments

Our examination of entertainment markets reveals that two different sub-markets exist—commercial/mainstream or artistic/independent products are separate, but overlapping offerings that address different consumer segments. We then identified three key characteristics of entertainment markets. The first one is an enormous rate of innovation: we find it difficult to think of another market where more new products are introduced in any given year. This activity stems from the unique characteristics of entertainment, which make constant innovation a requirement for the continued success of the entertainment firm; it requires a systematic approach toward the management of innovations which is, however, the exception rather than the rule among entertainment firms.

The other two key market characteristics contribute to the same effect: a high level of concentration. Strategic resources—financial, distribution, creative, and technical—are essential. A firm's ability to successfully approach the different sub-markets, and particularly the one for mainstream products, is impacted by these resources. In addition, network effects are critical in entertainment markets. Direct network effects exist when the size of the network of product users (such as the players of a game) directly impacts the value that a consumer, as "network member," receives from his or her participation in the network. Indirect effects exist on the "platform level"—when the number of platform users offers value to a user by attracting additional high quality content. Strategic resources and network effects erect barriers to entry for new firms and their products and create concentration effects that cause entertainment markets to tend toward "winners-take-all."

References

Alexander, P. J. (1994). Entry barriers, release behavior, and multi-product firms in the music recording industry. *Review of Industrial Organization, 9*, 85–98.

Anderson, C. (2006). *The long tail: Why the future of business is selling less of more.* New York: Hyperion Books.

Barney, J. (1991). Firm resources and sustained competitive advantage. *Journal of Management, 17*, 99–120.

Beier, L.-O. (2016). Instinkt. Geschmack. Eier. *Der Spiegel*, February 27, https://goo.gl/r5dKPH.

Binken, J. L. G., & Stremersch, S. (2009). The effect of superstar software on hardware sales in system markets. *Journal of Marketing, 73*, 88–104.

Bourdieu, P. (2002). The forms of capital. In N. Woolsey Biggart (Ed.), *Readings in economic sociology* (pp. 280–291). Blackwell: Malden.

Brodkin, J. (2016). Netflix finishes its massive migration to the Amazon cloud, February 11, https://goo.gl/rwos6H.

Buerger, M. (2014). How Macklemore tapped major label muscle to market an indie album. *The Wall Street Journal*, January 28, https://goo.gl/4eHmSQ.

Catmull, Ed. (2008). How Pixar fosters collective creativity. *Harvard Business Review, 86*, 64–72.

Castillo, M. (2017). Disney will pull its movies from Netflix and start its own streaming services. *CNBC*, August 8, https://goo.gl/Egu4mZ.

Chace, Z. (2013). The real story of how Macklemore got 'Thrift Shop' to no. 1. *NPR*, February 8, https://goo.gl/9B6vx9.

Clements, M. T., & Ohashi, H. (2005). Indirect network effects and the product cycle: Video games in the U.S., 1994–2002. *Journal of Industrial Economics, 53*, 515–542.

D'Alessandro, A. (2016). Why 'Boss' is a warning sign for Melissa McCarthy's b.o. machine, even if it taps 'Batman V Superman' on the chin. *Deadline*, April 9, https://goo.gl/U9TF6v.

Debruge, P. (2017). Why movies need directors like Phil Lord and Chris Miller more than ever. *Variety*, June 21, https://goo.gl/udAqYg.

Epstein, E. J. (2010). *The Hollywood economist—The hidden financial reality behind the movies*. Brooklyn: MelvilleHouse.

Fleming Jr., M. (2016). Big Bad Wolves helmers Aharon Keshales & Navot Papushado exit Bruce Willis Death Wish remake. *Deadline*, May 4, https://goo.gl/xUV3uz.

Frank, B. H. (2014). Amazon hopes to attract original video content creators by offering creative freedom. *GeekWire*, August 4, https://goo.gl/wKynDY.

Friend, T. (2016). The mogul of the middle. *The New Yorker*, January 11, https://goo.gl/8hYXxT.

Galuszka, P., & Brzozowska, B. (2016). Crowdfunding and the democratization of the music market. *Media, Culture and Society, 39*, 833–849.

Healey, J., & Moe, W. W. (2016). The effects of installed base innovativeness and recency on content sales in a platform-mediated market. *International Journal of Research in Marketing, 33*, 246–260.

Huls, A. (2013). The Jurassic Park period: How CGI dinosaurs transformed film forever. *The Atlantic*, April 4, https://goo.gl/2TnfcA.

Itzkoff, D. (2016a). How 'Rogue One' brought back familiar faces. *The New York Times*, December 27, https://goo.gl/JyUXBH.

Itzkoff, D. (2016b). The real message in Ang Lee's latest? 'It's Just Good to Look At'. *The New York Times*, October 5, https://goo.gl/k56XAM.

Iyengar, S. S., & Lepper, M. R. (2000). When choice is demotivating: Can one desire too much of a good thing? *Journal of Personality and Social Psychology, 79*, 995–1006.

King, G. (2002). *New Hollywood cinema*. New York: Columbia University Press.

Kohler, C., Groen, A., & Rigney, R. (2013). The most jaw-dropping game graphics of the last 20 years. *Wired*, May 6, https://goo.gl/opNXpT.

Lindvall, H. (2011). Spotify should give indies a fair deal on royalties. *The Guardian*, February 1, https://goo.gl/vSkNHh.

Littleton, C., & Holloway, D. (2017). Jeff Bezos mandates programming shift at Amazon Studios. *Variety*, September 8, https://goo.gl/GFtV37.

Liu, Y., Mai, E. S., & Yang, J. (2015). Network externalities in online video games: An empirical analysis utilizing online product ratings. *Marketing Letters, 26*, 679–690.

Marchand, A. (2016). The power of an installed base to combat lifecycle decline: The case of video games. *International Journal of Research in Marketing, 33*, 140–154.

Marchand, A., & Hennig-Thurau, T. (2013). Value creation in the video game industry: Industry economics, consumer benefits, and research opportunities. *Journal of Interactive Marketing, 27*, 141–157.

McClintock, P. (2014). $200 million and rising: Hollywood struggles with soaring marketing costs. *The Hollywood Reporter*, July 31, https://goo.gl/eoEXYb.

Nocera, J. (2016). Can Netflix survive in the new world it created? *The New York Times Magazine*, June 15, https://goo.gl/e1d2Zu.

Pähler vor der Holte, N., & Hennig-Thurau, T. (2016). Das Phänomen Neue Drama-Serien. Working Paper, Department of Marketing and Media Research, Münster University.

Peltoniemi, M. (2015). Cultural industries: Product-market characteristics, management challenges and industry dynamics. *International Journal of Management Reviews, 17*, 41–68.

Penrose, E. G. (1959). *The theory of the growth of the firm*. New York: John Wiley & Sons.

Quandt, T., & Kröger, S. (Eds.). (2014). *Multiplayer: Social aspects of digital gaming*. Milton Park: Routledge.

Shanken, M. R. (2008). An interview with Arnon Milchan. *cigar aficionado*, September/October, https://goo.gl/7xuGwM.

Spacey, K. (2013). The James MacTaggart memorial lecture in full. *The Telegraph*, August 22, https://goo.gl/Wkwwz1.

Superannuation (2014). How much does it cost to make a big video game? *Kotaku*, January 15, https://goo.gl/MqQVT8.

Takahashi, D. (2009). EA's chief creative officer describes game industry's re-engineering, August 26, https://goo.gl/BzKXf9.
The Deadline Team (2013). Steven Soderbergh's state of cinema talk. *Deadline,* April 30, https://goo.gl/3md7zK.
Vance, A. (2013). Netflix, Reed Hastings survive missteps to join Silicon Valley's elite. *Business Week,* May 10, https://goo.gl/ss45xZ.
Waldfogel, J. (2017). How digitization has created a golden age of music, movies, books, and television. *Journal of Economic Perspectives, 31*, 195–214.
Wernerfelt, B. (1984). A resource-based view of the firm. *Strategic Management Journal, 5*, 171–180.

5

Creating Value, Making Money: Essential Business Models for Entertainment Products

In the previous chapters, we have highlighted the key characteristics of entertainment products and markets. Before studying how marketing decisions should be designed to address these characteristics, we want to shed some light on value creation in entertainment markets. We will examine how economic value is created for producers, but also for other market players who are essential for linking producers and consumers of entertainment content.

We develop what we label the "entertainment value creation framework," a systematization of the key roles and stages that, in its totality, helps firms to understand their own role with regard to the creation of value in entertainment. Our framework identifies core competitors (including those that might not immediately come to mind) and points to potential ways to increase a firm's share of entertainment value by strategic integration and transformation steps. We then use the framework to overview economic developments in entertainment and name some of the industry's key players.

Finally, we also take a look at fundamental business models in entertainment: how can revenues be generated in (two-sided) entertainment markets, and how can risk be systematically managed? All of this information serves as the foundation for the detailed analysis of entertainment marketing instruments and their usage which we conduct in Part II of this book.

T. Hennig-Thurau and M. B. Houston, *Entertainment Science*,
https://doi.org/10.1007/978-3-319-89292-4_5

A Value Creation Framework for Entertainment

"Content is king all over again."
—Rupert Murdoch, *as executive chairman of 21st Century Fox (quoted in* Chmielewski *and* Hayes *2017)*

Early on in this book, we defined entertainment products as those pre-produced market offerings whose main purpose is to offer pleasure to consumers. We explained that such offerings can come in tangible form (e.g., on a DVD or CD), for which we use the term *material* entertainment products, or remain intangible, being transmitted to consumers via technical channels such as the Internet, cable, satellite, and terrestrial TV, and good ol' radio. We label those latter offerings *immaterial* entertainment products.

All economic value originates from consumers' usage of material or immaterial entertainment products—if *consumers* are not interested in an entertainment product, no value is created and no money earned. (Please note that this is even true for other "customers" on two-sided entertainment markets—advertisers and sponsors are essentially interested in an entertainment product's consumers, too.) But between those who create entertainment products, on the one end of the spectrum, and those who consume them, exists a large gap that must be overcome for value creation to take place and a product to become a financial hit. In our value creation framework, we investigate how entertainment content is transferred to consumers. Which steps are necessary, and which parties are involved?

The answers to these questions are provided in Fig. 5.1, which displays the entertainment value creation framework. The figure shows that the road from the producer of entertainment content to the consumer includes two kinds of intermediaries: distribution intermediaries and technical intermediaries. Let us take a closer look at the role of each.

Distribution intermediaries make material and immaterial entertainment products accessible for the consumer, bridging the gap between him or her and the producer of entertainment. They provide the legal (and sometimes not-so-legal) places in which the exchange of entertainment content takes place. We distinguish between three types of distribution intermediaries among which consumers can choose; each type is associated with different conditions and rights. The first type is "retail" distribution—here the entertainment product, such as a song on CD or in digital MP3 format, is transferred to the consumer, who is granted *unlimited* ownership of the copy or

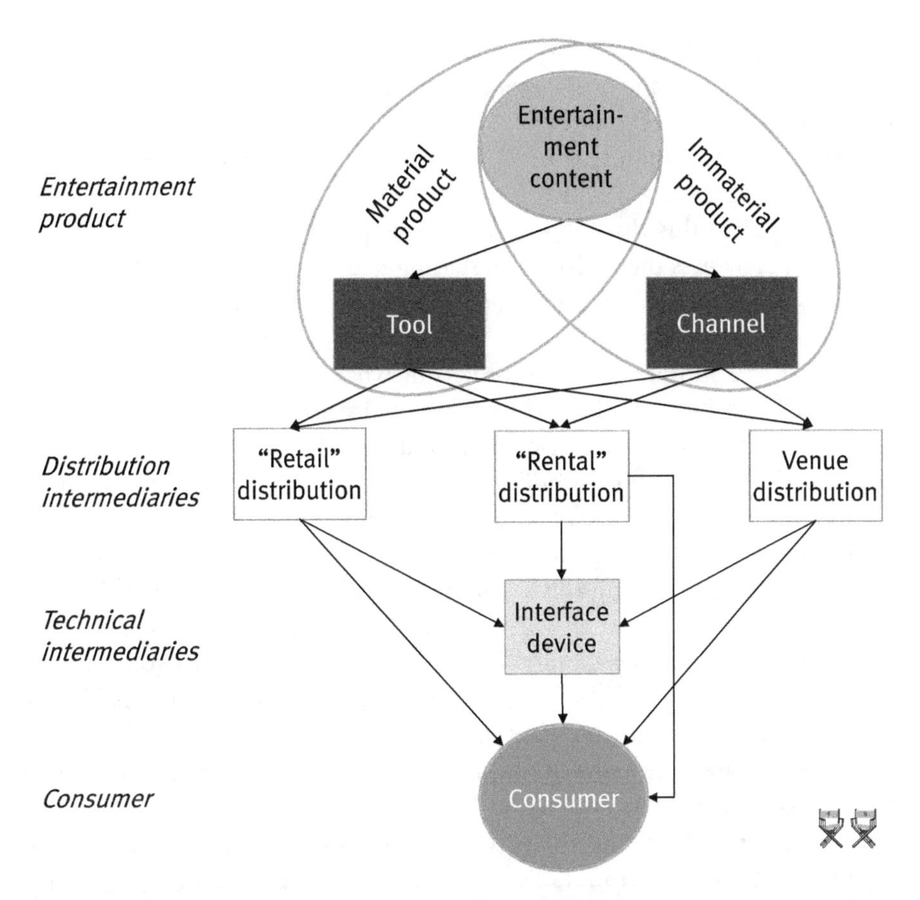

Fig. 5.1 The entertainment value creation framework
Note: Authors' own illustration.

unlimited rights to its usage, as is the case with most ebooks.[48] Although we call this mode "retail," the fee the consumer pays for the content does not necessarily have to differ from zero. For example, Amazon "sells" hundreds of ebooks through its website which are priced at $0, and Sony does the same for several apps in its PlayStation Store. In other words, it is the unlimited nature of the usage rights that define this distribution mode, not the amount of money that is transferred.

[48]For example, Amazon states in the Terms of Use for its Kindle Store (*Amazon* 2016): "the Content Provider grants you a non-exclusive right to view, use, and display such .. Content an unlimited number of times. … Content is licensed, not sold, to you by the Content Provider." Please note that, as with all other distribution modes, the consumer obtains no rights over the content itself (such as a movie or song), but only usage rights for a certain copy of the product.

Popular retail distributors for entertainment include virtual stores such as those by Amazon and Sony, Apple's iTunes, as well as traditional physical retail outlets such as the ones belonging to Barnes and Noble, GameStop, Walmart, and Ceconomy (formerly a part of the Metro Group) across Europe. As we take a practical (versus formal) perspective, illegal providers also populate this distribution mode—despite the fact that they are not proper rights owners themselves. File sharing networks such as The Pirate Bay, as well as the physical *ShanZai* markets for bootleg DVDs in China, provide consumers with products they can keep for life, at least according to their "retailers" (legal authorities might think somewhat differently about this).

"Rental" distribution, the second type of distribution intermediary, provides a *time-limited* transfer of content and/or the rights of its usage to the consumer. As with retail distribution, paying a fee is not mandatory for rental distribution, as the two-sided nature of entertainment enables business models that do not rely on consumer payments (such as the advertising-based version of music streaming service Spotify). Such rental distributors include libraries (either physical, such as public libraries, or virtual, such as Amazon's "Kindle Owners' Lending Library") and rental stores (such as video stores).

Another type of rental distributor is subscription services, such as Netflix (which grants its users the *temporary* right to stream certain films and series—the content remains at the firm's servers and becomes inaccessible if removed by the firm) and Spotify (same for music and audiobook content). Essentially the same is true for Google's YouTube and YouTube Red, only without the subscription element for the former. Illegal rental distributors such as Popcorn Time (a popular app for streaming filmed content) also fall into this category of intermediaries. Finally, we also place TV and radio stations in this category—both pay and "free" stations offer consumers time-limited access to entertainment content, very similar to what the new digital services such as Netflix and Spotify offer, with the main difference being that TV and radio stations "push" the content, whereas streaming content is pulled by the consumer. For "free" TV/radio, YouTube, and similar services, customers "pay" by tolerating advertisements.

The third (and final) distribution intermediary of our framework is venue distribution. Venues such as movie theaters, traditional theaters, concert halls, and discotheques provide consumers access to content as part of a social experience. Such experiences often take place in a physical environment, but they can also happen in digital spaces such as virtual worlds (e.g., SECOND LIFE) or on popular social media platforms such as Facebook and Pinterest. One could argue that these platforms, and particularly those digital ones, also "rent" entertainment content to their users, but we believe that venues differ from rental distributors in an important way: users are attracted by a venue's experiential character, above and beyond the consumption of specific content. It's the unique structure of their offerings that earns venues their place in the framework as a kind of distribution intermediary, separate from retailers and rental agents.[49]

Let us acknowledge that our classification of movie theaters as "distributors" of filmed content causes a language dilemma (maybe even heartburn) for industry veterans. We are well aware that the movie business uses the term "distribution" for the sales of films by producers to movie theaters and thus refers to theater owners not as distributors, but as "exhibitors." From a *market-level* analysis of the industry's value creation process, however, theaters act on the same level as retail stores do with DVDs and CDs, and as streaming providers with music; they all are the final touchpoint, *distributing* content to consumers. Definitions are never "right" or "wrong," but we have made our language choice in a way that ensures that our framework and its application throughout the book is consistent with management theory and also across the different forms of entertainment studied herein.[50] Besides, our market-level vocabulary does not prevent us from including the task of distributing a product to audiences in producers' marketing mix (see the chapter on entertainment distribution decisions)—our thinking is that if a producer plans to give consumers access to a movie via theaters, he or she engages in "theatrical distribution."

Technical intermediaries. In addition to distribution intermediaries, the consumption of entertainment content is often tied to the use of certain interface devices that serve as technical intermediaries. Interface devices are hardware that enables consumers to enjoy material or immaterial products. For movies, they include TVs, computers and laptops, smartphones and tablets, DVD players, satellite and cable receivers, 3D glasses, or an Amazon Fire TV player or stick. For music, the list includes several of those devices, but also contains radios, headphones, and Bluetooth-operated speakers. In their digital format, novels require ebook readers, tablets, or smartphones, and video games are stored on phones, tablets, PCs, or consoles; they are played via controllers and sometimes involve the use of cameras and VR glasses.

[49]One might argue that YouTube, Google's video streaming platform, should also be considered a venue—some consider it a social network, and it offers users room for social articulation. However, we are under the impression that, in contrast to sites like Facebook and Pinterest, people go to YouTube less for the platform itself and its experiential value, but rather for its specific content—and thus consider it predominantly a rental distributor.

[50]Our language choice does not ignore the challenge for producers to manage relationships with distributors. But we consider it as one of business-to-business sales rather than distribution—a task that we do not study in our book that mainly focuses on management activities directed at consumers.

The direct links from the distribution intermediaries to the consumer in Fig. 5.1 mean that some exceptions exist: buying or renting a printed book is a self-sufficient activity, requiring no technical interface device. The same is the case for most kinds of venue distribution when the venue is physical, not virtual; going to a theater to enjoy a film or to a club to listen to dance tracks is usually all that is required from the consumer (except for the occasional 3D glasses). But we have to admit that in our digital times, the role of hardware for entertainment consumption is clearly on the rise.

The framework can (and should) be used to systematically identify the roles that firms and conglomerates play today in entertainment for creating value. It also helps to identify appropriate roles for those firms that might traditionally not consider themselves a part of the entertainment industry (think of cable providers, for example). By doing so, the framework identifies opportunities for a firm to systematically enhance its share of entertainment value creation; we will discuss different paths of integration and transformation later.

Our framework also illustrates that it does not necessarily matter much through which (technical) channel a product is distributed; instead, it points to the critical role of distribution modes. For example, although TV stations air content terrestrially, pay-TV channels use cable to reach consumers, and streaming services send their content via the Internet, our framework points out that they all offer very similar value to consumers: they are all engaged in rental distribution of (filmed) content. Ad hoc evidence for such competition between market offerings—independent of their technical channel—is provided by statistics that show that the enormous growth of streaming services in North America goes along with a decline in TV viewing. Pay-TV channels have lost 10% of the population since 2010, at an continuously increasing rate, and "regular" TV watching was down 11% across all ages in the same period, including down 40% for those between 12 and 24 years of age (*The Economist* 2016, 2017). We study such cannibalization between distribution modes in our chapter dedicated to distribution decisions.

As we will illustrate below, the dynamics of value creation in entertainment have been enormous in recent times, and we expect them not to slow down anytime soon. Our framework points to the close linkages between the different hierarchical value creation levels (production, distribution, hardware) and among the different modes of each level (material versus immaterial products; retail, rental, and venue distribution). That might be the key message of the framework, which also informs our discussion of marketing instruments in the book's Part II: hardly anything happens in isolation when it comes to value creation in entertainment. For producers, this implies the good news that content is crucial for value creation, followed by the no-so-good news that many players who are not traditionally in the business of producing content might see a benefit in controlling it.

Who are Those Who Create Entertainment Value and How They Do It: A Snapshot of Players, Products, and Trends

As shown earlier in this book, filmed, programmed, written, and recorded entertainment content each generate substantial economic value. In the following, we will shed more light onto the products and distribution modes which are most important for each form of entertainment today, along with key players and recent developments. But along with the holistic approach of this book, let us first provide a snapshot of the leading content-producing conglomerates and their business activities, which stretch across the different forms of entertainment.

Major Studios, Labels, and Publishers: The Entertainment Conglomerates[51]

Strategic resources are what distinguish those who are in the market for commercial "studio" products from those who are in the market for independent products. But who are the few big players who own these scarce resources and earn the lion's share of the industry's yearly $750 billion—who are the "studios"? In addition to a few major entertainment companies who specialize in a single type of product, the content-creation side of entertainment is dominated by five conglomerates that each span several forms of entertainment: the Walt Disney Company (including 21st Century Fox), Comcast, National Amusements, Sony, and AT&T/Warner. Despite the fact that they all invest massively in the creation of entertainment content, each

[51]As in other sections of this book, we have worked hard to assure the information presented here to be as comprehensive, topical, and error-free as possible. Please be aware that, despite this effort, things might have changed in the meantime or we might have overlooked or misread information. When writing this in the winter of 2017, at least two major take-overs were in progress: the one of Warner by AT&T and the more recent one of Fox by Disney. Our description of the firms' divisions and assets in this chapter thus have to be tentative; please keep these dynamics in mind. We'll try to keep our readers updated on the book's website http://entertainment-science.com.

conglomerate has a unique structure of assets and subsidiaries; this enables them to play different parts in, and at different layers of, the entertainment value creation framework.[52]

*The Walt Disney Company (including 21*st *Century Fox).* Disney was founded in 1923 by brothers Walt and Roy Disney; it now generates annual revenues of more than $55 billion and an operating profit of about $15 billion, not including the newly acquired businesses of 21st Century Fox, which have accumulated revenues of more than $25 billion and an operating income of nearly $7 billion before the merger.[53]

The company's history and economic backbone is in the creation of filmed entertainment, which these days directly contribute about $9 billion in revenues and $3 billion in profit. Disney's roster includes some of the planet's biggest movie production teams and brands, including Pixar Animation, Lucasfilm, and Marvel. The firm's biggest hits include the latest installments of the STAR WARS franchise, the AVENGER and IRON MAN films (as parts of the "Marvel Cinematic Universe"), and the animated FROZEN, each of which has accumulated more than $1 billion in worldwide revenues during theatrical releases alone. The company has been reported to have generated more than half of the film industry's total operating profits in 2016 (Lieberman 2017).

In addition to this, Disney is now taking ownership of the classic Hollywood studio 20th Century Fox as a result of its acquisition of 21st Century Fox, which traces its roots to the 1930 and has produced several big hits on its own, such as AVATAR (the most successful film in in theater history at the time of writing, with a box office of more than $3 billion, in 2017 value), the first six STAR WARS films (which it made in cooperation with Lucasfilm), and the animated ICE AGE movies (which altogether generated about $4 billion in theaters). Prominent made-for-TV content from Fox includes series such as THE SIMPSONS, X-FILES, and EMPIRE. In 2017, Fox's filmed content generated about $8 billion in revenues and an income of $1 billion. The combination of Disney and Fox is the clear market leader in feature films; its joint theatrical market has been between 25% and almost 40% in the last years.

---

[52]If no other date is mentioned, financial information named in this section refers to 2016.

[53]21st Century Fox itself was the result of a 2013 split of former media and entertainment conglomerate News Corporation initiated by large shareholder and then-CEO Rupert Murdoch. The other resources of the "old" News Corporation formed the "new" News Corp., which, after the acquisition of Fox by Disney, combines various publishing activities, including news services (such as the *Wall Street Journal*) and trade book publisher HarperCollins. HarperCollins and its several imprints have published books by numerous star authors, including Michael Crichton, Clive Barker, and Neal Stephenson. The firm makes about $8 billion in revenues; the trade publishing assets (as the firm's "entertainment branch") accounts for close to $2 billion of them and generates $0.2 billion in income.

Disney is also active in book publishing (with Disney Publishing Worldwide, a part of their Consumer Products division) and in music production (with its Disney Music Group, part of its studio division). It has also produced games (with Disney Interactive producing games such as DISNEY INFINITY), but has stopped publishing console games and now focuses on mobile games only. But all these activities are of a smaller size than the firm's filmed entertainment branch. The firm also owns some of the world' biggest theme parks (such as Disney World) and has an extensive presence in consumer goods (such as toys and clothing, which are also sold through Disney Stores, the firm's own retail outlets).

Both parks and retail are fueled by the firm's film output, as is the firm's powerful distribution arm, which includes a list of powerful TV networks (it owns ABC and sports channel ESPN, among others). The TV segment accounts for more than 40% of Disney (pre-merger) revenues and half of its income. Via Fox, it now also runs specialized networks such as FX, National Geographic, and Sports Media; it also holds a 40% share of the European pay-TV (and streaming) provider Sky. Fox has been particularly active in programming network content for cable providers such as Comcast cable, which it has provided with sports, entertainment, and news content; this business has contributed more than half of Fox' pre-merger revenues and even 80% of its profit.

Disney has announced high-flying plans with regard to streaming its content over the Internet. By the end of 2017, it owns 30% of the SVOD service Hulu directly and another 30% of the service via Fox, making it the service's majority owner. It also operates the multi-content subscription service "DisneyLife" in the UK (which, at the time of writing, offers access to "Movies, TV, Books, and Music"). The firm is expected to transform these services into a global SVOD operation, which shall become the exclusive streaming home of Disney's films and shows (Castillo 2017).

All of this points at one of the most remarkable aspects of the Disney company, which makes it stand out in addition to its size: its high level of systematic integration of the different parts of the company. This integration goes all the way back to the company's origins—the firm's co-founder Walt Disney had a vision of a network of activities which support and build off of each other, with films being at the center of all value creation; this vision still holds true for the company some 70 years later, probably more than ever.[54]

[54]Walt Disney's early vision for links between and synergetic integration across the different areas of entertainment is nicely captured in a map that dates back to 1957 and that can be found at several places on the Internet, such as at https://goo.gl/Acq856.

The Comcast Corporation. Comcast is a media conglomerate that holds assets across different areas and layers of the entertainment industry; it is mainly known for its TV and film business. The firm's origins are in cable hardware, as a means to transport entertainment and information to consumers; Comcast began as a cable operator in Mississippi in the 1960s. The firm today owns NBCUniversal, which assembles Comcast's main assets used for the creation of entertainment content. Among them, the Hollywood-based Universal Studios, which dates back to 1912, plays a pivotal role; they are one of the world's leading producers of theatrical films. In addition, Comcast owns film labels such as DreamWorks Animation (since 2016), Focus Features, and Working Title. In film, the firm's greatest successes include the JURASSIC PARK/WORLD franchise, the FAST AND THE FURIOUS series, and the animated MINIONS movies, which have all brought in more than $1 billion during their respective worldwide theatrical runs.

Additional entertainment content is created by NBCUniversal's TV assets, which feature the major TV network NBC and several other stations that produce shows, sports, and weather, including Telemundo, Syfy, and the Weather Channel. Other than filmed entertainment, Comcast has only limited stakes in the production of entertainment content. Under its NBC Publishing label, it has released a number of ebook versions of company-owned content, usually with multimedia content added (e.g., the ebook version of the novel ROOTS features historical video and audio news footage); its attempt to enter the gaming market with a Universal Games Network in 2012 (created to develop "casual" games for the firm's TV brands) was short-lived. The firm has no music-related activities (anymore); whereas Universal Music carries the same name, it operates fully independently as a result of the acquisition of Universal Studios by General Electric in 2004 (see our section on the music industry).

Almost two-thirds of Comcast's overall revenues and profits (of $80 billion and $17 billion, respectively) come from its cable business; its entertainment content-creation and content-distribution arm NBCUniversal generates annual revenues of $31 billion and a profit of close to $6 billion. Via NBCUniversal, which the firm bought from General Electric and Vivendi in 2011, Comcast also leverages its entertainment content through several theme parks, most of which are associated with the Universal brand. The firm holds a 30% share of the film- and series-streaming service Hulu, and a major stake in Fandango, the leading seller of movie tickets in North America. Through Fandango, Comcast also owns the movie and TV show review aggregation site Rotten Tomatoes (together with Time Warner)—an important platform for "earned" media for filmed entertainment, as we will see later in this book.

National Amusements. National Amusements (hereafter, NA) is not necessarily a household name; it is not even well-known to everyone in the entertainment industry. However, it houses a large repertoire of content creation resources and is also active in the distribution of entertainment. Regarding content, NA owns 80% of the (voting) shares of Viacom, which produces major films via the Hollywood-based movie studio Paramount Pictures. Among the greatest successes in Paramount's 100+ year history are TITANIC (the second most success film of all time on a global scale), the TRANSFORMERS franchise (which has generated close to $5 billion in theaters worldwide, in 2017 value), and the legendary INDIANA JONES movies (through Lucasfilm, which, however, is now owned by Disney).

Additional filmed content is produced by Viacom-owned TV stations, such as MTV and Nickelodeon, but also by the CBS Corporation, of which NA owns an 80% majority share. CBS runs the TV network carrying the same name (as well as the CW network, together with Time Warner). Also in its stable are the pay-TV network Showtime (which has produced series such as HOMELAND, DEXTER, and WEEDS) and radio stations. NA also produces films via CBS Films (the most successful being the comedy LAST VEGAS) and is owner of a number of major publishing houses, with Simon and Schuster being the biggest among them (with rights to numerous star authors such as Ernest Hemingway, John Irving, and Stephen King). Both Viacom and CBS have minor activities in producing music; Viacom hosts niche labels such as Comedy Central Records and Nick Records, while CBS Records (which must not be confused though with the glorious 1970s label of the same name that now belongs to rival Sony) has mainly published soundtracks from the firm's own TV series.

One other part of the entertainment value chain in which NA is active is distribution; in addition to TV stations, it operates some 1,500 movie theaters, mostly in the U.S. Beyond that, the firm owns part of MovieTickets.com, a seller of movie tickets and competitor of Fandango; the movie and games review website Metacritic is part of CBS's interactive group. Total revenues of all operations by NA's majority stakes are not disclosed, but we estimate them to be in the range of $30 billion, made up of revenues of about $13 billion for CBS (and an operating income of $2.6 billion) and similar numbers for Viacom (revenues: $12.5 billion, income: $2.5 billion). For both firms, the clear majority of revenues comes from TV distribution; film contributes close to $3 billion (incurring a rare loss of half a billion in 2016) and publishing around $1 billion (contributing a $100 million profit).

Sony Corporation. Tokyo-headquartered Sony is a leading producer of various forms of entertainment, including film, games, and music. The company's backbone, however, is technology: Sony has a strong history of innovations in home entertainment (such as the Walkman and the Betamax video system) and computers, among others. With regard to content creation, Sony is a leading player in the console game market, producing both hardware (e.g., the PlayStation line) and games which are marketed through Sony Interactive Entertainment (such as hit games THE LAST OF US and the UNCHARTED series).

Sony Music Entertainment, successor to historic labels Columbia/CBS Records and part of Sony since 1987, is among the leading producers of musical content, including the oeuvres of global stars such as Michael Jackson, Beyoncé, and Barbra Streisand. Sony also runs a full-blown studio that produces film and TV content under the title Sony Pictures Entertainment, which the company took over from Coca Cola in 1987. As a film producer, Sony's greatest hits are adaptions of SPIDER-MAN comics (which have generated $5.6 billion in theaters in 2017 value) and Dan Brown's mystery novels (e.g., THE DA VINCI CODE, about $1.7 billion in 2017 value). The only form of (media) entertainment in which Sony is not active is book publishing.

Beyond production, the firm is also active in both distribution and technical intermediaries. It produces most kinds of hardware that consumers use to experience entertainment content (e.g., PlayStation, TVs, CDs/DVDs, and Xperia mobile phones—but no more PCs, which it sold in 2014). Sony has established its PlayStation console as a main hub for distributing content—consumers can purchase or rent different forms of content through the device and its Internet-based PlayStation Store. In contrast to other entertainment conglomerates, Sony has only a niche presence in the TV sector, with its pay-TV Sony Movie Channel. About half of the firm's nearly $70 billion in revenues are affiliated with entertainment; games (hardware and software) contribute the most (close to $14 billion), followed by film and TV content ($8 billion), and music (almost $6 billion). The firm's games and music activities are clearly more profitable these days than the rest (which includes Sony's film and TV division that suffered a loss of almost $1 billion in 2016)—games and music entertainment together provided an operating income of $1.9 billion, or three-quarters of Sony's *total* profits.

AT&T/Warner.[55] The media conglomerate Warner Bros. has a most eventful history. Before being acquired by telco/media giant AT&T, Warner sold its music production arm in 2004 (which continues to operate under the label Warner Music Group, still using Warner's old logo), its cable operations in 2009 (which continue to use the name Time Warner Cable), and also most of its publishing stakes (as part of the split from Time Inc. in 2013). Today, Warner Bros. mostly produces and distributes audio-visual attractions. Its main activity is in theatrical films, which it crafts through Warner Bros. Pictures and other subsidiaries, notably New Line Cinema, Castle Rock Entertainment, and DC Films (which makes films based on comics by DC such as BATMAN). Warner Bros. is also active in TV series and shows (through Warner Bros. Television, but most prominently through its pay-TV asset Home Box Office/HBO), and games (through Warner Interactive). As a consequence of its control over DC, Warner also publishes comics.

The firm's catalogue contains household names in all these areas. In film, its biggest hits include the HARRY POTTER-based movies (almost $10 billion in theaters, in 2017 value), various BATMAN and SUPERMAN films (close to $9 billion), and its cinematic adaptions of the literary works of J. R. R. Tolkien (i.e., the LORD OF THE RINGS and HOBBIT movies; also close to $6 billion). Some of its TV output has drawn a large and loyal global audience, such as GAME OF THRONES, THE SOPRANOS (both produced by HBO), and BIG BANG THEORY (by Warner Bros. TV). Its list of hit video games include the Lego series (such as LEGO STAR WARS) and several BATMAN and MORTAL COMBAT games.

Warner Bros. is also active in the distribution of filmed content via its CNN and TBS TV networks (it also holds a 50% share of The CW) and its leading pay-TV channel HBO (which has 131 million subscribers globally); with HBO Go, it now also runs a streaming service for movie and show content. The film studio division generates about $13 billion in revenues (and close to $2 billion in operating income), and similar revenues come from Warner's TV activities (although these contribute five-times higher profits). However, all these numbers appear small compared to those resulting from the channel-operating activities of now parent AT&T. In addition to its telephone business, AT&T provides, via DirecTV, satellite broadband TV connections for more than 25 million U.S. customers (more than Comcast has cable customers) and mobile Internet access to 50+ million customers. It also

[55]Please see our comment regarding the merger of AT&T and Warner in footnote 51.

extensively distributes entertainment content via its DirecTV and U-Verse streaming services. With these activities, AT&T generates annual revenues of $164 billion and an income of $24 billion, making it one of the planet's largest firms, regardless of industry affiliation.

Figure 5.2 summarizes the business activities of these five leading creators of entertainment content and their respective offerings. The figure demonstrates the major differences that exist between the competitors' portfolios—both with regard to the forms of entertainment provided (e.g., film, games, etc.) and the status of content creation among a firm's overall activities. These differences also affect the firms' valuations, with market capitalizations in 2016 varying from some $40 billion (both National Amusements and Sony) up to about $300 billion (AT&T/Warner), with the others falling in-between (Walt Disney/Fox ~$230 billion, Comcast ~$170 billion). National Amusements and Comcast are controlled by founders and their families; in contrast, Warner, Walt Disney, and Sony do not have such dual share structures. In what follows, we will now look into the specifics of the different forms of entertainment for the creation of value.

The Market for Filmed Content: Movies and TV Productions

> "[A film is] a whole range of elements coming together and making something that didn't exist before. It's telling stories. It's devising a world, an experience, that people cannot have unless they see that film."
> —*Filmmaker* David Lynch *(2007)*

Filmed content is by far the biggest form of entertainment, particularly when you add in TV revenues of circa $400 billion; hardware for the consumption of films, series, and shows would boost its importance even further. With regard to content creation, the studio divisions of the now five conglomerates dominate the production of movies, with an annual cumulative market share of about 80%, and they also provide the lion's share of popular TV shows and series. A major studio spends between $2–3 billion annually for the production and marketing of movies, and between $3–8 billion in total for generating all types of filmed entertainment.

The studios distribute their content via movie theaters, as well as on discs and via broadcast TV stations, pay-TV, digital downloads, and streaming. Theatrical distribution has basically stalled in recent years in North America, but has grown quite strongly in other parts of the planet (by 133% since

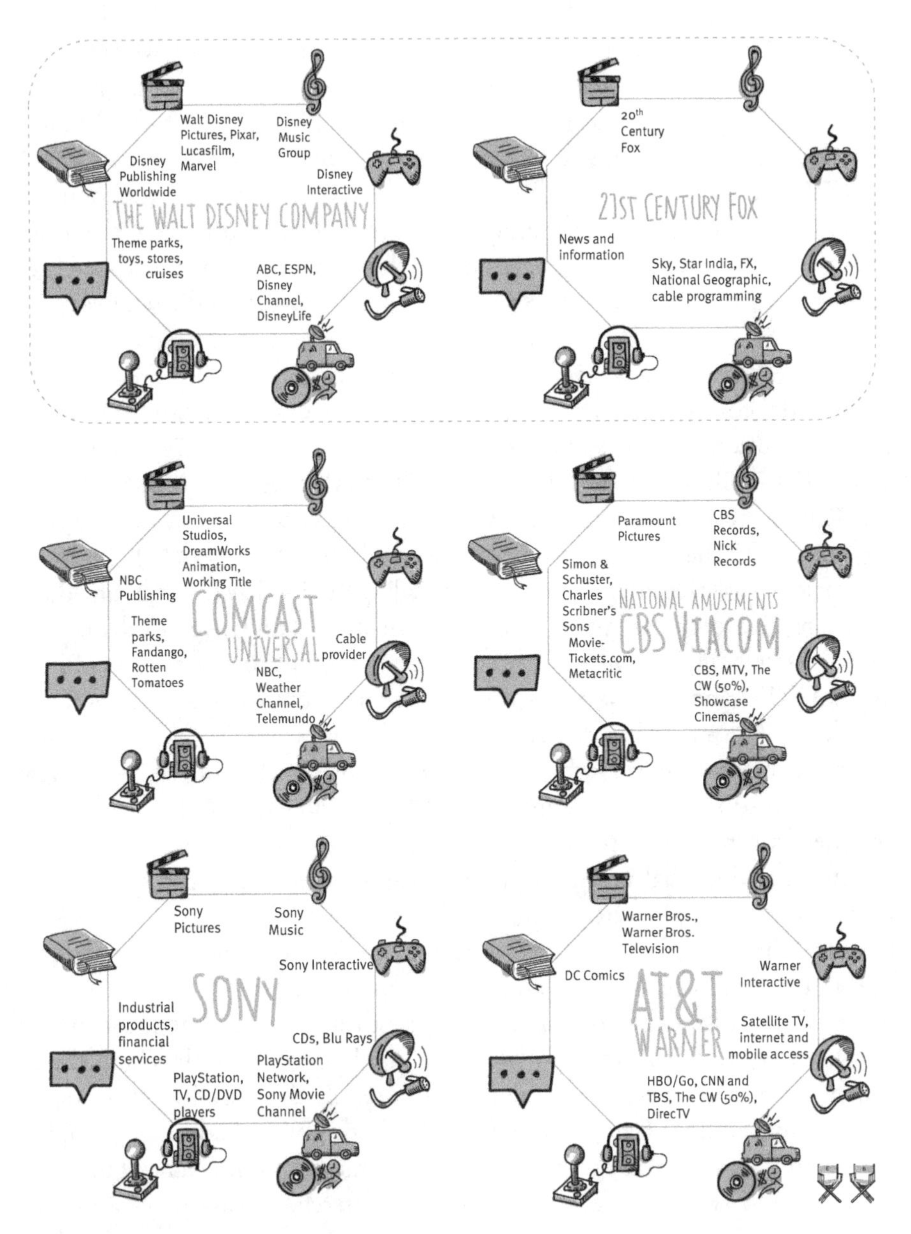

Fig. 5.2 Product portfolios of five leading entertainment conglomerates à la 2017
Notes: Authors' own illustration. Assets listed in the figure are only selections of the companies' activities. AT&T and Warner were still two separate firms at the time of writing, as were Disney and 21st Century Fox. With graphical contributions by Studio Tense.

2001 even when accounting for inflation).[56] Theaters, which have traditionally served as the first step in a sequential chain of distribution channels, are followed by releases of films on DVDs and Blu-Rays (sales and rental) and their digital, immaterial equivalents of "video on demand" or VOD, pay-TV, and finally free-TV. VOD comes in various forms; transactional VOD (via iTunes and other online stores) refers to the purchase ("electronic sell-through") or rental of a single film or show (named "download-to-rent," even though the content is often streamed, not downloaded), whereas subscription VOD ("SVOD") means subscription-based streaming, as popularized by Netflix, Amazon Prime Video, and Hulu.

How do consumers allocate their film-related budgets across the different distribution channels these days? In the U.S. where consumers spent about $10.7 billion for theater tickets in 2016, they expended $18.3 billion (or 70% more) for material and immaterial forms of home entertainment, excluding TV channel subscriptions and cable fees (*DEG* 2017). In 2016, the majority of the latter amount was still spent for DVDs and Blu-Rays, which receive twice as much as their transactional digital equivalents. But SVOD is strongly growing, with consumer spending being almost as high as for DVDs and Blu-Rays. Its disproportional popularity among younger people (in Germany, a 2017 study revealed that while 8% of the 14-and-older consumers watch Netflix at least once a week, 21% of the 14–29 year old segment do so; Kupferschmitt 2017[57]) is an indicator of the distribution channel's future growth potential.

And what does that mean for producers and studios? In Fig. 5.3, we report rough estimates how much each distribution channel returns to those who produce films, based on a sample of recent studio productions. The percentages in the figure are averages, and there is *substantial* variation in the importance of each channel across films—whereas theaters return the most to film producers on average, adding only one standard deviation to the channel percentages can reverse the order. For some films in the sample, transactional home entertainment (sales and rentals from both physical and digital copies) generated the most, and for others, TV was (almost) as important as theaters.

Let us add two comments. First, whereas merchandising plays a limited role for the "average" film, as can be seen in the figure, its contributions can differ quite dramatically. For some popular films, such as Star Wars,

[56]The growth in theatrical revenues outside of North America has recently flattened somewhat, with a 37% increase since 2007 and a 9% increase since 2012, based on data reported by the MPAA.

[57]Kupferschmitt's results are based on a nationwide representative survey of about 2,000 German-speaking consumers. His findings are quite similar for Amazon Prime Video, which is used at least weekly by 18% of consumers between 14 and 29, versus only 12% of the general population.

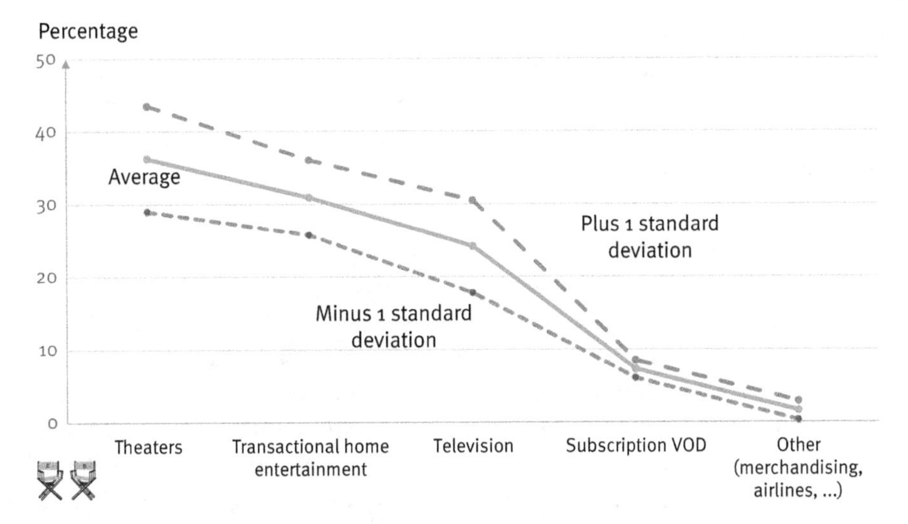

Fig. 5.3 Contributions of different distribution channels to movie producer revenues
Notes: Authors' own illustration based on information from various sources. The data set is a convenience sample of recent studio films that covers different genres and budget sizes. The numbers in the figure are our own calculations and we made several assumptions, so they should be treated as illustrative only.

merchandising revenues have been the major revenue source. And second, don't make immediate judgments about each channel's absolute relevance for a producer based on these numbers: the channels affect each other, with both cannibalistic and complementary effects, as we discuss in our distribution chapter.

Theatrical distribution is certainly among the key channels for most movies. In North America, it is dominated by three firms that own about one-third of all venues: AMC (controlled by China-headquartered Wanda Group), Regal, and Cinemark. The major studios have been cautious about buying into this channel as a result of the "U.S. v. Paramount Pictures" Supreme Court antitrust ruling in 1948, but there are exceptions now. In addition to National Amusements' small theater business, Wanda not only decides which movies are shown in AMC theaters, but also which ones are produced by mid-size studio Legendary Entertainment since taking over the production company in 2016.

Although the rise of streaming services Netflix (the market leader with about 130 million global subscribers as of late 2017) and Amazon Prime Video (about 85 million active users according to Feldman 2017) has substantially increased the revenue potential for the studios, with multi-year fees adding $50+ million to the calculation of a single movie, this growth comes at a cost for the studios. Streaming services now invest large amounts in the creation of their own original filmed content. Netflix spent $1.5 billion for original content in 2017 and plans to increase that number to a studio-sized $2–3 billion for series and about 80 original films in 2018 (Koblin 2017); Amazon is estimated to spend about $1 billion per year (Coen 2017).

Thus, besides competing for consumers' attention, these firms are also competing with the studios (and smaller-sized producers) for rare talent resources. In addition to being the front runners in the use of data analytics for filmed entertainment, they also offer producers unique conditions for marketing a new film or series, through their direct contact with tens of millions of subscribers and also through their affiliations, such as Amazon's IMDb (e.g., Bart 2017). It is thus unlikely that any future studio CEO will compare Netflix with the "Albanian army," as Warner's Jeff Bewkes did a few years ago (Powers 2010). In contrast, the competition by streaming services is widely considered the main reason for Disney's acquisition of rival studio Fox in late 2017—an attempt to beef up the firm's plan to compete with Netflix, Amazon, and maybe other digital giants such as Google, Apple, and Facebook with its own SVOD service, using exclusive access to its productions as a competitive tool.[58]

The gradual shift from material to immaterial entertainment products has also resulted in a much more crucial role for those who provide the broadband infrastructure that is needed to distribute digital filmed content in high resolution—namely cable operators (such as Comcast), providers of satellite TV (such as AT&T), and those who offer wireless high-speed Internet connections (e.g., Verizon). With the pervasive competition created by the digitalization of content, channel providers try to differentiate by packaging their channels with superior content—a driving force behind Comcast's acquisition of NBCUniversal and AT&T's pursuit of Time Warner (*Forbes* 2016). How synergies can be derived from such mergers is not a trivial question, though, as the blockbuster concept used for costly mainstream entertainment requires the widest possible distribution.

Finally, with television sets, gaming consoles, and other hardware intermediaries becoming smarter and digitally connected, hardware manufacturers, such as Samsung, Microsoft, but also Apple are also becoming more interested in movies and series that might set their products apart from the

[58]See also our discussion of the "frenemy" concept later in this chapter.

competition. To summarize, it seems that *everyone* competes for filmed content these days, or is readying to jump into the fray. A lot of money can be earned with great content, but it is in combination with distribution infrastructure and interface hardware where content becomes most valuable.

The Market for Written Content: Recreational Books

> "A reader lives a thousand lives before he dies, the man who never reads lives only one."
> —*Author* George R. R. Martin *(2015) via Twitter*

The size of the market for written content depends on its definition. In its broadest sense, encompassing books, magazines, and newspapers, we calculate that the market generates annual revenues of more than $300 billion, including advertising revenues. Books account for roughly one-third of that number, and the types of books on which we focus, those for entertainment and recreation (versus education), are responsible for two-thirds of global book revenues, or approximately $75 billion per year (e.g., *McKinsey* 2015).

The leading producers of recreational book content are called the "Big Five": Penguin Random House (part of German-headquartered Bertelsmann group)[59] has the largest share, followed by HarperCollins (of the "new" News Corp.), Simon and Schuster (belonging to National Amusements), French-based Hachette (the central division of media conglomerate Lagardère),[60] and Germany-based Macmillan/Holtzbrinck (the only one on the list to focus solely on publishing). For many years, the "Big Five" together represented more than half of recreational book sales; however, their share has recently been falling substantially, to about one-third of total sales in 2015 (Anderson 2016). The beneficiaries of this development are many small publishers as well as self-publishing authors, whose works now combine to account for nearly 50% of the market.

---

[59]In addition to Penguin Random House, Bertelsmann also has stakes in the European TV business (as a "rental distributor," owning Germany's leading free-TV station RTL, but also as a content producer through historic firm UFA), in the music industry (as owner of BMG, which manages the rights of stars such as the Rolling Stones and Janet Jackson), and in magazine-publishing Gruner + Jahr, as well as in education and professional services. The family Mohn-owned conglomerate generates annual revenues of some $17 billion and an operating income of nearly $3 billion.

[60]The Lagardère Group also operates retail outlets at airports and train stations in France and other parts of Europe; it publishes magazines and owns major radio and TV channels in France as well as promotes sport events. Total annual revenues of the group are close to $8 billion, and the operating income is $0.5.

Why does this trend exist that runs counter to the general concentration arguments we presented earlier? The answer has a lot to do with the rise of digital, immaterial products and the fact that financial entry barriers are now somewhat lower for publishing compared to other forms of entertainment—the sub-market of independent products for written content has become a better substitute to studio offerings than is the case with movies. But another huge factor has been Amazon. The firm, which started about 20 years ago as an online bookstore, has set up a fast-growing publishing unit that is estimated to have established a share of about 3% of all book unit sales in the U.S. They feature mostly little-known authors, but also star-writers such as Helen Bryan.

And Amazon's distributor power matters also. The firm's retail and rental offers now largely dominate book distribution in North America and in most other Western countries (nearly every second new book is channeled to the consumer through Amazon today in the U.S.; Mosendz 2014), and Amazon has supported independent content in various ways to gain negotiation power over book prices, etc. Its spurring of ebooks, in particular, has helped to create an immaterial alternative to printed books for both consumers and publishers. Ebooks' much lower marginal costs provide Amazon with room to incentivize independent authors and publishers. Ebooks now earn every fourth dollar American consumers spend for books, a trend that Amazon has facilitated—and benefitted from in terms of market share and power.

This rise of ebooks has also changed the book market in another way: what was once the only form of entertainment media that could be consumed without a technical intermediary has now become "intermediated"—reading ebooks requires a compatible device. Again, Amazon has grown the market for ebook readers with its Kindle devices as part of its interest in selling digital books and, by doing so, has established itself as the clear leader in the ebook reader part of the publishing industry.

In sum, the market for recreational books has developed in a somewhat atypical way, largely influenced by Amazon, which now dominates distribution and is the major player in hardware technology as well. Because Amazon's role in the book market is tied to immaterial versus material content (with the latter still being dominated by the "Big Five" publishers), the firm's future in this market will be partially determined by how strongly consumers will adopt ebooks and devices instead of physical copies.[61]

[61]For some informed speculation about this and potential signs of satiation for ebook demand, see for example Alter (2015).

The Market for Recorded Content: Music

"Music is the only reason. … It'll give you the whole fucking world for free if you just love it and hold back nothing."
—*Character* Grandmaster Flash *in* The Get Down *[Courtesy of Sony Pictures Television]*

Once on par with other forms of entertainment, the music business has suffered during the last two decades. While the radio sector has remained somewhat stable, with global revenues of roughly $50 billion (the clear majority of which comes from advertising; *PwC* 2014, 2015), the sales of recorded music have shrunk by almost 60% since peaking in 1999, dropping from almost $35 billion (in 2017 value) down to below $15 billion (*ifpi* 2017).[62] In some key markets, such as North America, revenues from recorded music have been reduced by two-thirds. Concerts and festivals have gained economic importance (and now generate almost as much revenue as recorded music), but the massive losses have had quite an impact on those who produce and distribute music, altering the way value is generated with (recorded) music. It seems that, as a result of consequential changes, the shrinking has been stopped, and decent growth has been reported recently, both on a global level and in North America (where music sales grew by 11% in 2016; Karp 2017).

Today, the production of recorded musical content is dominated by the "Big Three" firms. Whereas all three carry the names of major entertainment conglomerates, only one of them (Sony Music) actually belongs to one. The other two, Universal Music (the market leader who accounts for close to 30% of recorded music revenues, now part of the Vivendi group[63]) and Warner Music (owned by billionaire Leonard "Len" Blavatnik as part of his diversified holding group Access Industries), have been sold off from their parents, but still carry the names, an eerie reminder of the industry's biggest crisis.

[62]In addition to sales from (digital and physical) music to consumers and a comparably small amount paid by those who use music in movies etc. (less than half a billion), the ifpi report also lists about $2 billion in revenues from radio stations. We exclude the latter from music sales to avoid double counting (it is already included in radio revenues).

[63]In addition to owning Universal Music, Paris-headquartered Vivendi also has assets in TV and film distribution and production (via pay-TV station Canal+ and its subsidiary StudioCanal), gaming (it sold Activision Blizzard, now holds the majority of Gameloft), and live entertainment and ticketing (via "Vivendi Village"). The firm's total revenues were more than $11 billion in 2016, with Universal Music and the Canal+ segment contributing $5.5 billion each. The music division accounted for 80% of the conglomerate's profits of about $1 billion.

The music market's turbulence is closely tied to the digitalization of music, which provided consumers with the opportunity to skip the physical products sold by the major producers through retail stores and, instead, share digital versions of music among each other through file sharing platforms, ignoring artists' and labels' copyrights. Although caught flat-footed, the industry has gradually developed ways to exploit digital distribution itself; within just ten years, the share of global revenues from immaterial music products has grown from 11% to 59% in 2016, with a steep trend further upward.

Figure 5.4 shows the enormous changes in the demand for the main material and immaterial formats among North American consumers. It also illustrates that the distribution intermediaries for music have undergone a disruptive change: in 2016, consumers spent more for subscription-based streaming (about \$2.2 billion) than for digital music and for physical music media (about \$1.8 billion and \$1.7 billion, respectively), which are both in freefall (see also Sisario 2017). Institution-wise, music labels now consider

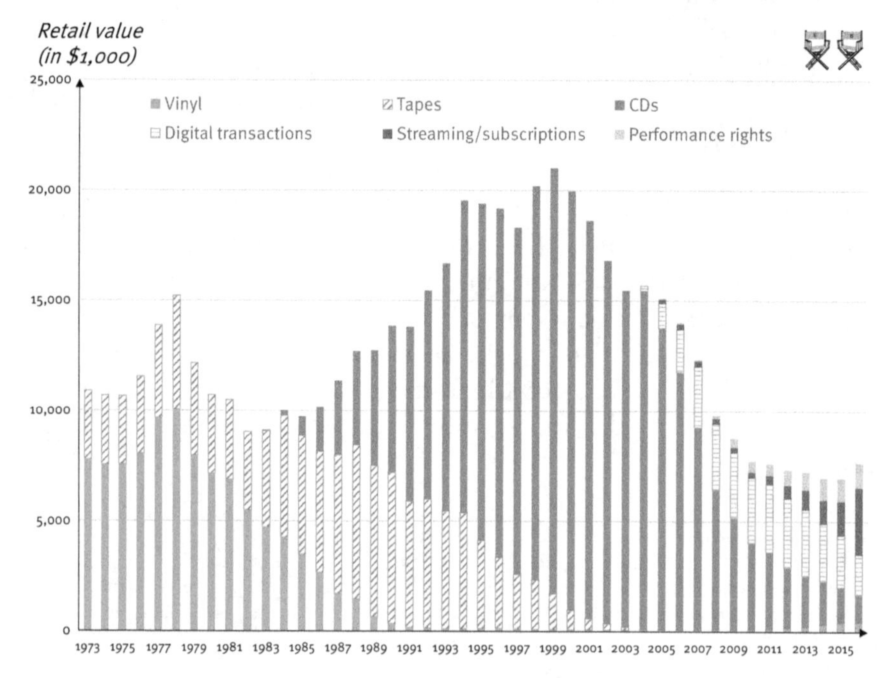

Fig. 5.4 North American music inflation-adjusted revenues for key product types over time

Notes: Authors' own illustration based on data from RIAA (2017). All revenues are the inflation-adjusted North American retail values in thousand 2016 US-\$. Each product type in the figure contains multiple formats, whose revenues we added.

deals with the leading providers of music streaming as essential. Spotify has capitalized on its relatively early market entry and accounts for nearly half of all 140+ million global users (about half of which are paying subscribers, whereas the others actively use its advertising-based offer), followed by Apple Music, which approached 30 million subscribers in late 2017. Apple, who propelled the music business' digital shift through its iTunes store, still dominates the sales of digital music; it accounts for about half of the market, with Amazon and Google (via its Play Store and its own streaming service from 2018 on) being other contenders.

The development toward immaterial music products has also disrupted the hardware intermediary market. Among teenagers, the smartphone has taken over as the most-used device for listening to music, followed by computers; here, it is again Apple, as the producer of the iPhone and its proprietary iOS operating system, along with consumer electronics conglomerates, such as Samsung and LG that make use of Google's Android OS, that help consumers experience digital music. Whereas radio receivers are still in use, the market for CD players is shrinking along with the decreasing demand for material music products (*Statista/Audiencenet* 2015). Smartphones have also mostly retired a generation of devices that were specifically tailored for digital music consumption, such as iPods and others MP3 players.

In summary, the music market's value chain has experienced tectonic shifts as a result of digitalization. Content-wise, it is now dominated by an oligopoly of three mostly music-centered entities who wrestle for power with leading digital distributors Apple and Spotify. Whereas the former generates most of its revenues with its hardware products (with company-owned music services adding value to the iPhone, et al.), Spotify is a music-streaming-distribution-only player who, despite not yet having broken even financially, is shaking the industry with its business model. Let us note that both Spotify and Apple have made initial steps toward entering the music production business to increase their independence, essentially resembling similar steps by filmed-content streamers Netflix and Amazon.[64] Don't be surprised if the firms extend their content production activities in the not-so-distant future.

[64]For example, Spotify has started to produce small-scaled formats of exclusive content, such as "Spotify Sessions" and "Spotify Singles," featuring original recordings in their own recording studio (e.g., Dillet 2016), and Apple signed exclusive deals with a number of musicians (e.g., Frank Ocean) and offers exclusive live events (Sanchez 2017). The industry is, however, sceptical toward such exclusive availability of music content by streaming firms as it believes it triggers music piracy, an issue we discuss in the distribution chapter of our book.

The Market for Programmed Content: Electronic Games

> "We live in a world where a video game can make over $1 billion in less time than it takes most of us to get caught up on laundry."
> —Madigan *(2016, p. XII)*

Among the forms of entertainment featured in this book, electronic games are the fastest growing. Games also contain the highest level of product type heterogeneity, encompassing a wide range of "programmed" products that are consumed via different hardware at different occasions. Each game assigns an active role to the user that goes well beyond the activity level required for reading a book, listening to a song, or watching a film. Estimates from *Newzoo* (2017) note that console games comprise about one-third of the market, generating over $32 billion in annual revenues at a decent growth rate. Not far behind, with revenues of almost $30 billion (but exhibiting a downward trend), PC games include those massively multiplayer online games (MMOGs) played purely on the Internet. The largest part of revenues from electronic games, however, comes from casual mobile games played on the smartphone/tablet—these games today account for almost $40 billion. But after years of enormous growth (climbing a stunning 68% from 2012 to 2015), their growth seem to be slowing down somewhat.

The substantial differences between game types mirror the fact that the leading producers of content vary between game types, with most firms specializing in one type of game. Leading console game producers are Sony and Microsoft (who earned revenues of nearly $10 billion from gaming in 2016, about two-thirds of what Sony made), who both also happen to own the two dominant gaming consoles systems, PlayStation and Xbox (see below). Other major players in the creation of console games are Activision Blizzard, Electronic Arts (EA), Warner, and Nintendo, the third major console owner who accumulates total revenues from games and hardware of some $4 billion. EA generates more than $4 billion revenues with popular sports titles, such as FIFA, and the Battlefield and Need for Speed franchises. Activision generated almost $7 billion in 2016 from titles like the Call of Duty series and long-running MMOG World of Warcraft, but also casual games such as Candy Crush produced by mobile specialists King.com (which Activision bought for almost $6 billion in 2016). Activision is preparing to extend its activities into filmed entertainment in 2018 (Jackson 2017).

Apart from these firms, Asia-based companies have strong stakes in the PC/MMOG segment (with Chinese Internet providers Tencent and NetEase being the largest) and in casual/mobile gaming, where (in addition to King.com) Japan-based DeNA leads the market. This large presence of Asian companies among game producers is not accidental, but reflects the humongous size of the Asian-Pacific gaming market. Nearly half of all global gaming revenues are generated in that part of the world, almost twice as much as in North America.

Although electronic games are digital by their very nature (programming takes place in binary code!), their distribution is only gradually growing beyond cartridges and disks. About 60% of game revenues now come from immaterial, fully digitized game products distributed via the Internet. The lion's share of game revenues come from "retail" distribution channels (rather than from rental and venue distribution), and the consoles themselves have advanced into major retail platforms for console games which can be downloaded over the Internet (and played immediately).

For material game products, Amazon and specialized retailer GameStop dominate the market. GameStop also leads in PC games, whereas for MMOGs the Internet is the sole channel. The distribution of casual mobile games lies almost completely in the hands of Apple and Google via their app stores, where games are by far the most downloaded app format. Amazon operates its own app store, although its attempt to establish a separate mobile OS has failed. Its app store is even more strongly biased toward games, but has about only 10% the number of apps compared to Google, as well as relatively marginal traffic.

Those who control (console) game distribution are also key players when it comes to technical intermediation. For consumers who want to experience console games, there is no way to circumvent the hardware from Sony, Microsoft, and Nintendo. Figure 5.5 shows that Sony dominates the eighth console generation with its PlayStation 4 devices and that Nintendo is only a shadow of its former self, with a market share of less than 10% for its Wii U. For all three firms, both content and hardware decisions influence each other because of indirect network effects. For casual games, consumers need an iOS-powered device or one that runs Android by Google. Finally, the consumption of PC games and MMOGs is tied in a similar way to PCs that employ Microsoft's Windows OS.

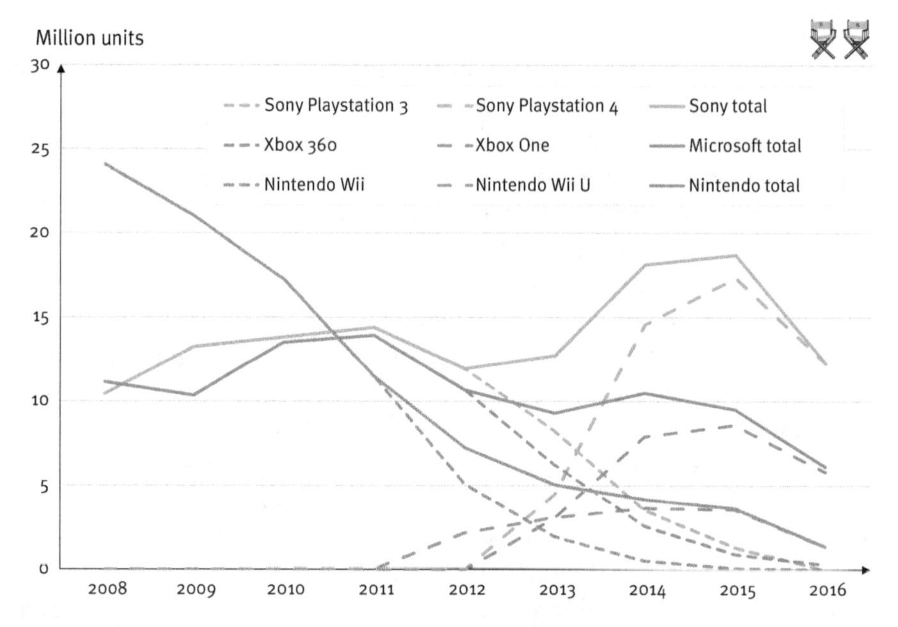

Fig. 5.5 Global market shares of gaming consoles over time
Notes: Authors' own illustration based on data from *VGChartz* (2017). Numbers are globally sold units of a console type per year.

The Dynamics of Entertainment: Some Words on Integration and Transformation Processes

> "Coke is very enthusiastic being in the movie business."
> —*A source close to the negotiations between Coca Cola and Columbia Pictures in 1982 (quoted in* Hayes *1982)*

The history of the entertainment industry has been one of company transformations. Some of those transformations simply involve a company developing new skills or learning to take on new functions in the entertainment value creation framework (the "make" approach). However, others have transformed by merging with other firms or acquiring them (the "buy" approach). Some of the biggest of such moves were when Coca Cola took control of Columbia Pictures in 1982 (and then sold it to Sony in 1989), when the Transamerica Corporation bought the film studio United Artists in 1967 (and sold it in 1981, after the financial unrest caused by the flopping of HEAVEN'S GATE), when Seagram became majority stakeholder of Universal Studios in 1995, and when General Electric acquired NBC TV 1986.

We have also mentioned a number of more recent transformational moves in our industry overview section, including both "make" (Netflix's move into film and series production, Amazon becoming a producer of various forms of entertainment content, Activision Blizzard's extension into filmed entertainment production) and "buy" transformations (e.g., Comcast's takeover of NBCUniversal, AT&T's merger with Time Warner, and Activision Blizzard's acquisition of King.com).

Some of these moves have worked, others not so much. Can scholarly research provide insights into why this has been the case? We believe that a typology of firms' transformational moves, based on insights developed by strategic management scholars, is a sound starting point to do so. Accordingly, transformational moves can be classified into four categories: vertical integration, horizontal integration, concentric integration, and conglomerate transformation (e.g., Walter and Barney 1990).

Vertical integration describes transformations in which a firm extends its business activities from one layer of the value creation framework (such as content production) to other parts of value creation (such as distribution). In vertical integration, a buyer-seller relationship could exist before the transformation, with the transforming firm (whether formerly being the buyer *or* seller) taking over the other's role. Several of the moves named above fall into this category: the Comcast/NBCUniversal merger represented a transformation from infrastructure provider to content production and distribution, just like the AT&T/Time Warner marriage.

Netflix and Amazon each have complemented distribution activities by taking on content production, and Google, Facebook, and Apple are headed the same way (e.g., Otterson 2017; Patel 2017).[65] Adding content production to existing distribution is referred to as "backward" integration because the new activity happens "before" the traditional activity in the value creation framework. In the case of Amazon, a distributer has assumed content production and also hardware intermediation, combining both backward and forward integration, as some of the new activities happen "after" their traditional ones in the value chain. Part of Amazon's backward integration efforts are its massive AWS webserver hosting activities, which help the firm to manage streaming as a popular form of "rental" distribution of entertainment content, and its Prime Video activities in particular. Apple's transfor-

[65]In addition to its rumored steps into music publishing, Apple has announced plans to become more deeply involved in filmed entertainment, both by extending its role as a distributor (e.g., by offering streaming services) and also by becoming a producer of filmed content itself, with an annual budget of about $1 billion (Spangler 2017).

mation could be titled two-stage backward integration—from hardware to distribution (iTunes), and now content production.

What do managers intend to accomplish when engaging in vertical integration? The general logic is to better manage "critical interdependencies" between value creation activities (Walter and Barney 1990). For entertainment, we argue that three specific reasons exist. First, managers may strive to realize synergies between value creation layers to either generate more value (by addressing customer demand better) or to create an advantageous cost structure (e.g., Negro and Sorenson 2006). Incremental customer value can be realized if a firm is able to apply the knowledge it has gathered in one layer of the framework to the new layer. This is a major argument for the Netflix and Amazon transformations, for example—they should be able to produce better content because of the exclusive information about consumers that they collect in the course of their distribution activities.[66] This is a central argument of our book: those who can collect large data and are able to analyze it in smart ways gain a competitive advantage. And one way to realize these advantages is vertical value-creation integration.

Whereas production companies such as Time Warner usually do not have consumer-level behavioral data, mergers could provide them with such information (think of AT&T's DirecTV service). Another example would be Disney's ambition to leverage its brand reputation (earned in production) by exclusively offering its content via its own SVOD channel on a global level (i.e., not licensing it to other SVOD services; Castillo 2017), which could equip the distribution service with a competitive advantage over other distributors and enable Disney to monetize its brand reputation more fully (and also gain more data about customers than it gets when licensing the content). In comparison, cost reduction synergies are somewhat more profane: the basic idea is to increase the profit margin by cutting out the "middleman," i.e., by getting rid of (distribution or technological) intermediaries. For cost-cutting to be successful, the costs of executing a business function must be less than the intermediary's fee, *and* with no loss in quality or effectiveness—which is no small feat, and often failure results from an underestimation of the intermediary's expertise.

Second, entertainment managers may intend to defend a competitive position in one layer of the value chain by taking integrative steps and building a buffer against competitive pressures. Backward integration into con-

[66]Regarding exclusivity, both Netflix and Amazon are said to be notoriously hesitant to share information on how (many) consumers use their services and the products they offer, even with regard to their content partners and the creatives they work with (see Schrodt 2015).

tent creation reduces the dependency of distributors, such as Netflix and Amazon, on content providers. Similar arguments have been named for the AT&T-Time Warner merger, calling it "a hedge against a future where the first point of entry for a media consumer might be Netflix, Facebook, [or] YouTube" (Sharma 2016)—not the content itself. The case also holds for film/music distributor Apple—who at least once also indicated an interest in purchasing content-creator Time Warner (Garrahan and Fontanella-Khan 2016), but now seems to favor the "make" approach for their transformational steps.

Third is a deeply human reason that Mol et al. (2005) have labeled "value chain envy." Here, vertical integration is driven by a manager's observation that profit ratios are higher in other layers of the value creation framework than in the areas in which the manager's firm is active—and envy motivates action. For example, music labels have complained for quite a while that broadcasters fail to "fairly reflect" the "true" value of their creations (e.g., *ifpi* 2017), and many producers feel their products are "exploited" by Google. Whereas Google's mere size prevents any "buy" approach, Mol et al. (2005) show, drawing on interviews with 146 Dutch music managers, that the pattern of vertical integration activities is consistent with the "value-chain envy" concept in the Dutch music industry: in most cases firms expand from low-profit generation activities (the *creation* of music) into higher-profit generation activities (the *publishing* of music).

This perspective might also improve our understanding of why Warner Bros. eagerly expanded its activities from the production of movies to (usually immensely profitable) TV operations, founding The CW and acquiring HBO as part of Time Inc. If no synergies or defensive advantages are realized, the success of envy-driven transformations might be questionable, though, particularly if conflicting corporate cultures hamper the integration effort. For example, a firm with roots in efficiency-driven distribution may struggle to understand the creative element of entertainment production.

For those vertical integration efforts that have failed, it seems difficult to detect either a synergetic or defensive advantage. For example, when Matsushita/Panasonic bought Universal Studios in 1990 (for only five years), the Japanese firm's VHS video format had become the sole home entertainment standard already. Thus, it could benefit little from owning a production company that would compensate for the problems raised by the clash of Hollywood-versus-Japanese culture. Negro and Sorenson (2006) test the power of vertical integration for entertainment firms, although only for a part of the value framework. They analyze survival rates for more than 4,000 film-production companies from 1912 to 1970, studying whether those production firms that also sell their films to theaters directly (versus employing a "sales" intermediary) have a longer life. They provide evidence that firms' vertical integration matters: on average, integrated firms have a 57% lower "rate of exit" than non-integrated producers (i.e., they survive longer).

Overall, vertical integration creates an interesting phenomenon referred to as "frenemies": firms that have been business partners, jointly creating value for consumers by separating and coordinating production and distribution (the "friend" part of the term), become competitors (the "enemy" part) when the distributor competes for production talent or the producer does the same for access to consumers. Such frenemy relations can have far-reaching effects, such as when Disney decided to no longer license its content to Netflix as part of establishing its own SVOD distribution service. We assume that what was partially a step to exploit its own brands better can also be considered a response to the immense growth in market power of its previous customer. This demonstrates that frenemy constellations are notoriously unstable: the success of one party threatens the success of the other. The more success Netflix has as a distributor, the more powerful can it act as a competitor to Disney in producing content. Amazon's backward integration into web services has also created a frenemy relationship with the firm's SVOD rival Netflix, who uses AWS as the hosting platform for its streaming operations (Brodkin 2016).

Horizontal integration means that a firm extends its business activities by engaging in other activities on the same level of the value creation framework, addressing similar customer needs as did their previous activities. Horizontal integration also can take the form of "make" or "buy"—the latter implies that a firm takes over, or joins forces with, an immediate competitor, such as when one major film studio acquires another one (e.g., Walt Disney's acquisition of the Weinstein's Miramax in 1993 and also of 21st Century Fox in 2018),[67] or when two major producers of the same form of content merge (as Activision Blizzard and King.com did in gaming). In addition, the "make" approach of horizontal integration happens when a firm that has focused on a specific distribution channel, or technical infrastructure,

[67]Mergers can also combine elements of vertical *and* horizontal integration. In late 2017, Comcast also reportedly showed interest in acquiring 21st Century Fox (e.g., Chmielewski and Hayes 2017)—which would have implied a merger of two film production studios (i.e., horizontal integration), as Comcast already owned Universal Studios, and also one of a technical infrastructure provider and a producer (i.e, vertcial integration).

extends its business into other distribution channels or technologies (e.g., Apple, already running a retail outlet for music with iTunes, opens a rental service with Apple Music).

Horizontal integration can hamper competition, but it can also intensify it. If the transformation comes in form of a merger (the "buy" approach), horizontal steps are strictly regulated and often interdicted by fair-trade authorities because such mergers reduce the number of players on a market, which can harm consumers. In cases in which the horizontal integration is through the internal transformation of a single firm (i.e., the "make" approach), the situation is quite different: here the efforts often constitute the attempt to use existing technology in novel ways, offering additional consumption options for consumers. Consider the case of Apple Music, mentioned above, which offers consumers an alternative to Spotify for music streaming. Other examples include pay-TV providers, such as HBO or Sky, extending their business horizontally by also offering film streaming subscriptions over the Internet as an alternative to Netflix (i.e., HBO Now in the U.S. and Sky Ticket in Germany) in an effort to stay relevant to consumers.

Concentric integration, a third type of transformation, happens when firms step into parts of the value framework that are on the same layer, but target a different customer need; a company offers new products, but uses existing technology or knowledge to market them. In entertainment, such transformations usually mean that a firm extends its activity from one form of entertainment (e.g., programmed content) to others. When Activision Blizzard continues its game-centered activities, but also begins the production of filmed entertainment, it remains on the same layer of value creation (i.e., production), but offers a new line of products. In these cases, the main driver is usually the managers' interest in exploiting their expertise to gain new customers and expand their business.

In addition, assuming that structural links exist between the different forms of entertainment, a manager may be able to generate synergetic effects by assembling different forms of entertainment in a company—something that we noted in many of the entertainment conglomerates (e.g., Sony, Disney). As Nick Van Dyk, as co-head of Activision Blizzard Studios, argued, "[f]ilm and TV—they are not simply stand-alone, profitable businesses, but they also amplify and extend the tremendous success of our core [gaming] business." Whereas several entertainment brands are now stretched beyond a single entertainment form (or category), defining the synergies between such forms is crucial and certainly not a trivial task—just think of the many failed attempts to transform game brands into successful movies (e.g., Dyce 2015).[68]

Finally, firms can also engage in *conglomerate transformation* (or diversification) by investing in activities that have no structural links at all with their existing products, technologies, or targeted customer needs. The entertainment industry has witnessed several cases of such transformations, including most of the major takeovers we mentioned early in this section: Coca Cola was a beverage producer when they took control of Columbia Pictures, Transamerica was an investment holding company when they acquired United Artists, Seagram was a distiller when it bought Universal Studios, and General Electric's main assets included energy, healthcare, and transportation when it took over NBC TV. None had experience in entertainment.

Research shows that conglomerate transformations often underperform or even fail (Walter and Barney 1990). This is because the main reasons for the "buy" are to exploit new revenue sources and to utilize financial resources, regardless of their product and market specifics. Although synergies are often claimed to exist by the acting firms (for example, Coca Cola managers stressed the opportunity to promote their soft drinks via movies and TV shows), any actual synergies are usually not sustainable and are dominated by the costs of the transformation, which often include cultural conflicts between the acquiring firm and its new entertainment division.

Having studied the value creation processes in entertainment and the repertoire of integration and transformation strategies that a firm can use to improve its position in the value-creation framework, let us move on and take a look at the business models that producers of entertainment content can use to monetize their creations.

Transforming Value into Money: Approaches for Managing Revenues and Risk

To design effective business models in entertainment, a rich understanding of the multiple revenue sources that exist for entertainment products is essential, along with insights into ways to systematically manage the market risk that stems from entertainment products' characteristics.

[68]We discuss the synergic potential of "category extensions" in our chapter on entertainment brands.

We begin with a discussion of revenue sources and strategies for managing them. A market-focused mindset implies that revenues from consumers are essential, so we investigate them first. Then we turn to two other revenue sources with fairly unique applications in entertainment (and related fields): revenues from advertisers (which result from the two-sided character of entertainment) and revenues from "third parties"—mostly subsidies that exist because of entertainment's cultural, aesthetic character. We end this section with an exploration of risk management strategies for entertainment products.

Generating Revenues from Consumers

Our value creation framework highlights the critical role of distribution for linking producers with consumers of entertainment products. Consumers' usage of a product is not only the reason underlying *any* revenue generation, but consumers are also those who contribute a major share of revenues, paying the producer for the right to experience its creation.

In most cases, distribution activities in entertainment are carried out by a different party than the one that produces the content, something that marketing scholars refer to as *indirect* distribution. Examples include movies or musical works that are made available to the consumer through third parties in TV, pay-TV, CD/DVD, or Internet streaming. As we discuss below, a major challenge in such indirect distribution systems is the allocation of revenues between producer and distributor, with the fundamental alternatives being that (1) the distributor pays the content producer a fixed price for the product (and the right to offer it to consumers), and (2) the two parties agree about a percentage allocation and share the price the customer pays accordingly.

However, there are many cases in which the producing firm also distributes its own content to consumers (i.e., *direct* distribution)—think of Netflix or Amazon offering their self-produced shows, or the online stores that several entertainment producers now offer for their books, movies, and TV shows. Although direct distribution was rarely used in the past for entertainment products, the approach has gained traction with the increasing ease of reaching consumers directly via the Internet. Figure 5.6 overviews the basic alternative distribution and revenue allocation approaches for entertainment products—we discuss them in some more detail below.

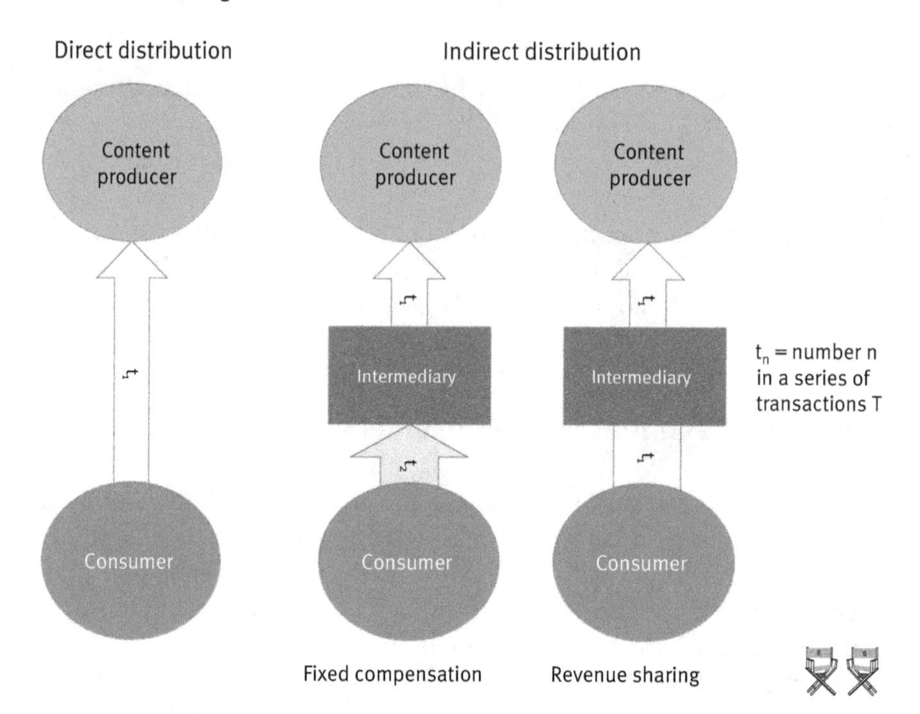

Fig. 5.6 Consumer-related revenue allocation models for entertainment products
Notes: Authors' own illustration. The white-colored arrows represent the first transaction/payment for an entertainment product ("t_1"), the yellow-colored arrow is the subsequent transaction/payment ("t_2").

Direct Distribution of Entertainment

In analogue times, direct content distribution could mainly be found in broadcast TV, where stations self-produced a part of their programming. Examples include legendary shows and series by the British Broadcasting Corporation (BBC), which built a reputation as producer-distributor. Prestigious examples of BBC-produced content include the Doctor Who series (dated back to 1963), Monty Python programs such as their Flying Circus (from 1969 on), and costume dramas based on classical authors such as Jane Austen (e.g., the mini-series Pride and Prejudice from 1995).

Pay-TV channel HBO pushed direct distribution via TV to a new level, making self-produced content a core element of its competitive positioning. HBO's original programming started with sports events shortly after its founding in the early 1970s, then added films in 1983 and high-profile series (such as The Sopranos) in the 1990s. While broadcast TV stations had mostly distributed content produced by others (networks and the Hollywood studios), producing its own content enabled HBO to set itself apart from competition and provide consumers with exclusive films and series that they could not watch anywhere else. Direct distribution also helped HBO to develop a distinct brand image, nicely expressed by their long-time slogan "It's not TV. It's HBO."

HBO's business model was later copied by other pay-TV channels, but HBO's direct distribution model can also be considered the inspiration for other entertainment firms that reach their customers over the Internet. For Netflix and Amazon Prime Video, direct distribution of premium original content (e.g., Netflix—House of Cards; Amazon—The Man in the High Castle) are now a focal part of their respective strategies, as evidenced in their vertical transformations. In combination with the huge volume of usage-related information that distributors can gather via the Internet (when does a consumer watch a movie? At what scenes does the consumer interrupt, or fast-forward, the viewing? And when does he or she stop watching?) and the use of *Entertainment Science*'s data analytics and theory, direct distribution also offers firms the chance to learn—and to create products that more closely meet their customers' needs.

But the adoption of Internet-connected devices by consumers has not only facilitated direct distribution of audiovisual media, it has also spurred direct distribution for other forms of entertainment. In the case of games, even though companies like Sony and Microsoft were already active in other parts of the value creation framework beyond content production (i.e., consoles, hardware), distribution of their game content was traditionally carried out via intermediaries. Since the Internet enabled them to use their hardware as distribution platforms, they have engaged in a multi-distribution mode approach, combining direct distribution of their own-produced games via their consoles (e.g., PlayStation Store) with indirect distribution via other retailers. In addition, Sony and Microsoft also serve as distributor for *other* producers' content.

For books, Amazon employs the Internet for a very similar direct distribution approach, selling and renting its own-produced content via its retail site and its seamlessly integrated devices, including the Kindle ebook reader. And in the case of music, some artists and publisher have offered their content directly via the Internet, but these efforts are often considered mainly as marketing measures to ensure wide attention by media and consumers (see the case of the band Radiohead and their album In Rainbows in our chapter on entertainment pricing).

So, indirect distribution has remained the dominant distribution model for music and prominent for books and also games and filmed entertainment. It warrants a closer look, with a particular focus on how revenues are allocated.

Indirect Distribution of Entertainment

The Fixed Compensation Model

When a producer of entertainment content distributes its product via an intermediary, a fundamental question is how transactions are managed financially. In retail, in general, the dominant model is "fixed compensation"—the producer sells products to the intermediary for a negotiated price, and the intermediary then sells or rents the product to the consumer, earning a profit via the margin between the revenues received from consumers and the price paid to the producer.

This model has been applied in various parts of the entertainment industry as well. Take the example of CDs that are sold via a retail store: Here the retailer pays the producer a fixed fee, either in advance or after having sold the products to consumers, depending on market power. The fee differs between countries, as well as between titles and retailers. In the U.S., this wholesale price (also referred to as "published price to dealer") for a mainstream CD album is between $10 and $12, in Germany it is close to 10 Euro, and in the UK it is £7–8. The retailer than usually adds a margin of 30–40% to determine the price the consumer is asked to pay, plus applicable sales taxes. For all other retail sales, whether vinyl albums, DVDs, Blu-rays, printed books, and games discs, the basic model is the same, although retail margins differ across products, depending mostly on the power of product-specific intermediaries and consumers' willingness to pay for the respective medium.

A similar model also exists for digital products (e.g., MP3 music tracks, movie downloads, or ebook files sold through Amazon or other retailers); online retailers pay a fixed amount to the producer. Because of the immaterial nature of the product and the irrelevance of inventory, no transfer of any kind occurs before the consumer purchases the (digital) product, and the retailer does not pay the producer in advance.[69] For rental of digital products, whether streaming providers like Spotify and Netflix or radio and TV stations, two variants of the fixed compensation model exist. The first is basically identical to the sales model: the producer gets a pre-defined amount for every rental/streaming transaction. For example, Sony's contract with Spotify for 2011 and 2012 required the streaming service to pay Sony

[69]For some constellations in which a large retailer (such as Amazon) has more power than an independent producer, this approach also exists for physical media products where transactions are carried out on a "commission basis" (i.e., the retailer only pays for the copies he or she sells and returns the others).

between $0.00225 and $0.0025 per stream by the service's "free" subscribers (Singleton 2015). The same logic is used for radio stations, which also pay a per-airing fee to the music rights owners.

The second variant for rental is the "flat fee": here, the distributor makes a one-time payment which grants it the right to rent a product to an unlimited number of consumers within a pre-defined time range. For example, Netflix almost always pays a fixed fee to the producer of a film and, in exchange, obtains the right to stream the content an unlimited number of times to its subscribers (e.g., Tostado 2013). The TV model is organized in a very similar way; for content they do not (co-)produce themselves, TV stations and pay-TV firms usually pay an upfront fee to the content's owner for the right to air the content, regardless of the number of people who actually watch it (Hennig-Thurau et al. 2013).[70] In some cases (e.g., Spotify), these fixed models are combined with revenue sharing models (i.e., mixed models), as we describe below.

What are the implications of the fixed-compensation model for producers of entertainment content? The major advantage over other approaches is that the producer receives revenues independently from the actual consumer demand; once the product has been sold, the business risk is solely with the intermediary. The producer is thus shielded from losses if consumers do not adopt the product widely. In such a model, a big focus of marketing efforts is on the producer-intermediary relationship (although many producers create marketing communications in an attempt to build demand by end consumers, which they expect to also influence retailers' inclination to order large numbers of the product).[71] The retailer takes on much responsibility to close the sales with consumers. Because the producer knows the product's true quality better than the intermediary, one might argue that producers will have an informational advantage when they are negotiating volumes and conditions.

However, these advantages of a fixed-compensation model are accompanied by a number of disadvantages for the entertainment producer. Because the parties are separate entities, their interests are not fully aligned (except that each needs the other); thus, synergies between production and distribution of a product will not naturally be realized. Rather than working together

[70]In some cases when the right to air a film is purchased very early in its creation, the fee TV stations pay is tied to its viewer numbers; however, this variable price element refers to the number of viewers of the film in a *different* channel (usually theater attendees), as a measure of its commercial value. Fuchs (2010, pp. 67–95) provides a detailed description of this process.

[71]See the section on "buzz"; we also discuss distributor effects in the context of advertising and the blockbuster concept of marketing.

to provide the consumer with the highest possible value of an entertainment product, there are strong tendencies by each party to maximize their own share of the deal. This self-interest affects pricing, product distribution, consumer-directed communication, and also what products are developed, in general. For example, the intermediary's abilities to pursue different pricing tactics are limited by a fixed transactional fee, as this fee represents a minimum retail price. So, even if an integrated perspective would suggest that the price that would maximize profits for *both* parties is below that fee, there is no way that the intermediary could set such a price without incurring losses—unless the partners cooperate strategically, a rare event.

A second drawback is that, when required in advance, fixed-compensation payments tie up the intermediary's financial resources, limiting its flexibility. Intermediaries will tend to play it safe by avoiding risk, understocking items to avoid costly unsold inventory, and rejecting experiments and innovative approaches that might benefit the sales of "extreme" or "unconventional" titles (out of fear that consumers won't show up). The video rental business provided an illustrative example of such rigidities and their business-hampering implications (Cachon and Lariviere 2005). Until 1998, video rental stores in the U.S. had to purchase video tapes for about $65 per copy from the major studios; they then rented them to consumers for $3, keeping all of the rental fee. Thus, titles became only profitable only when they were rented out at least 22 times. In turn, rental firms clearly had no incentive to stock more copies than what they expected to meet the *long-term* demand for a movie title, leaving a lot of early customer demand unfulfilled (keep the short life cycles of entertainment products in mind). When the Blockbuster company enforced a change from fixed compensation to revenue sharing in 1998, this increased the availability of new titles[72] and grew the firm's revenues by as much as 75% for its outlets. The video rental firm's market share rose from 25 to 31% and its cash flow by 61% in the year following the transition (*Knowledge@Wharton* 2000).

When a joint interest is absent, there is also limited interest by the intermediary to share customer-related information with the producer. As a result, consumer data (how many consume the product? Who? When? Why?) often remain exclusively with the intermediary. Thus, although producers know more about a product's true quality, intermediaries end up

[72]Dana and Spier (2001) analyzed the availability of new video titles in May 2000 in the Chicago area, finding an availability of 86% for Blockbuster, compared to 60% for other nation-wide chains and 48% for independent stores.

knowing more about consumer preferences—information that could be incredibly useful for the new product planning efforts of producers who adopt an *Entertainment Science* approach (but much less so for "Nobody-Knows-Anything" disciples). Yet, information asymmetries are standard in entertainment; producers of books, music, films, and games usually know very little about those consumers who buy their products.

Netflix, who has made data availability a core element of its business strategy, does not reveal even basic viewership statistics with those from whom it purchases its content. When a Netflix manager was asked what he would say if Will Smith asked to know how many people had seen his $90 million movie BRIGHT on the streaming service, the manager's reply was clear: "No" (quoted in Lev-Ram 2016). It is quite obvious that such information asymmetry provides an intermediary that has strong analytical skills with a potential strategic advantage over the producer when it comes to valuing content. As Kevin Tostado, a producer of documentary films states, "if Netflix wants to renew the contract at the end of a year, it will be hard to negotiate a new fee when we won't have the same viewership data they will" (Tostado 2013). We have already mentioned that this asymmetry becomes particularly important when the intermediary extends its operations into production, as Netflix and others have been doing.

But there is a more fundamental drawback of fixed-compensation models, and particularly the lack of data access that often comes with them. It is a cultural consequence that grows over time and, as some argue, has widely pervaded the entertainment conglomerates: producers become increasingly disconnected from the end consumers. Not only can lacking access to information lead to a lack of practical knowledge and understanding, but a sneaky danger is that managers lose interest in the end customer and their preferences, as nicely captured by the adage "out of sight, out of mind."

The Revenue-Sharing Model

The alternative approach to fixed compensation is revenue sharing, which avoids the problems just named—but has some of its own. Revenue sharing means that when a consumer buys a product from an intermediary, the revenues are shared between the intermediary and the producer. In other words, the intermediary sells or rents a product to consumers, retains an agreed-upon percentage from the selling price, and forwards the balance to the producer. Usually no upfront payments will have taken place (e.g., Wang et al. 2004).

Revenue sharing has been used across various forms of entertainment. One of the earliest places it was implemented was in the theatrical (venue) distribution of movies, by which theater owners share ticket revenues with those who provide them with content (the movie producers/studios). But even in this context, revenue sharing should not be taken for granted: Filson et al. (2005) note that theaters and producers used a fixed-compensation model in the early days of theatrical distribution, when films were relatively homogeneous commodities produced for small budgets. The industry switched to revenue sharing in the 1920s when products became more heterogeneous and budgets rose. The differences in budgets would have required producers to charge different prices; simultaneously, theater owners would have needed to predict consumer demand for each film in order to determine how much to pay for a film, despite lacking analytical knowledge. Under these conditions, theaters would have likely had a strong bias against high-budget films, and smaller theater owners would have been overburdened with large upfront payments for the biggest films. Thus, producers had an interest in installing revenue sharing, which quickly became—and remains—the common practice for theatrical movie distribution.

The film business extended the use of revenue sharing to video rental in the hey-day of video rental stores; today it is also the standard for digital sales of movies via online stores such as iTunes and Amazon. For video rental, the industry had begun to experiment with revenue sharing soon after its inauguration in the mid-1980s. But the model became the industry standard only in the late 1990s, when Rentrak, which had fine-tuned the approach, partnered with rising rental firm Blockbuster.[73] Today revenue sharing is also used for digital transactions of books, games (via app stores), and music. In music, it is also used for rental distribution through subscription services such as Spotify; a music producer such as Sony gets a certain percentage of the streaming provider's subscription and advertising revenues, which also reflect the respective studio's share of overall streams by Spotify (Singleton 2015; see also our section on "mixed models" below).

The key question in revenue sharing always is: how should consumer revenues be allocated between parties? In theory, allocations should reflect the relative contributions to value creation by each of the respective partners to

[73]The video rental revenue sharing model was originally developed by manager Ron Berger (under the somewhat misleading name "pay-per-transaction"), who applied the model for his video rental chain National Video as early as 1986. After facing competition from rising chain Blockbuster, Berger sold the 700+ outlets of National Video in 1988 (to West Coast Video), but kept the revenue sharing operation which he offered to all video rental under the Rentrak label.

the sale, recognizing, among other things, each partner's investments and risk. In practice, determining these contributions is quite difficult, so revenue share agreements often reflect the market power of the partners—the more power a partner has, the higher his or her revenue share.

In theatrical distribution, the producer's revenue share rose as production budgets rose: from 20% early on, to 33% in the 1960s, 45% in the 1990s, to 50–60% today in North America (Filson et al. 2005; Guerrasio 2017). This development might be attributed to the rise in production costs over time, and/or to a shift in market power from theaters to producers. But shares also differ strongly between producers, and even titles. Filson and his colleagues report that, in 2005, average shares paid to producers ranged from 42 to 57% for a single theater (with bigger producers getting higher shares), and Disney demanded a then-unique 65% share for its STAR WARS: THE LAST JEDI movie (Guerrasio 2017). The producer's share can be substantially lower in markets outside of North America, below 50% in many countries, and as low as 25% in China (Fritz and Schwartzel 2017).

Particularly for "regular" (i.e., smaller- or medium-sized) films, theatrical revenue allocation also often involves a dynamic component, named a "sliding scale," in which the percentages due back to the producer decrease with the number of weeks a film has been in the market. Filson et al. argue that the logic behind these sliding scales is to provide theaters with an incentive to keep displaying a film for a longer time. Figure 5.7 shows the actual development of revenue shares for a typical North American movie theater for the 2001–2003 period, based on Filson et al. (2007).

In physical video rental, consumer payments were split 45–45 between producer and intermediary (the remaining 10% was kept by Rentrak for its "operational services;" Cachon and Lariviere 2005). For immaterial media products distributed over the Internet, the so-called "70-30" rule has become an overarching heuristic, reflecting the asymmetric allocation of efforts and risks in the digital age (where distributors have only minimal storage costs).[74] According to the rule, the producer gets 70% of the revenues generated from consumers through digital sales, with the intermediary keeping the remaining 30%. This rule is in place for most games sold through the Apple and Google app stores, for independently produced ebooks sold via Amazon, and is the basis, with minor modifications, for

[74]See also our discussion of the "long-tail" phenomenon which is spurred by small storage costs in our chapter on integrated marketing.

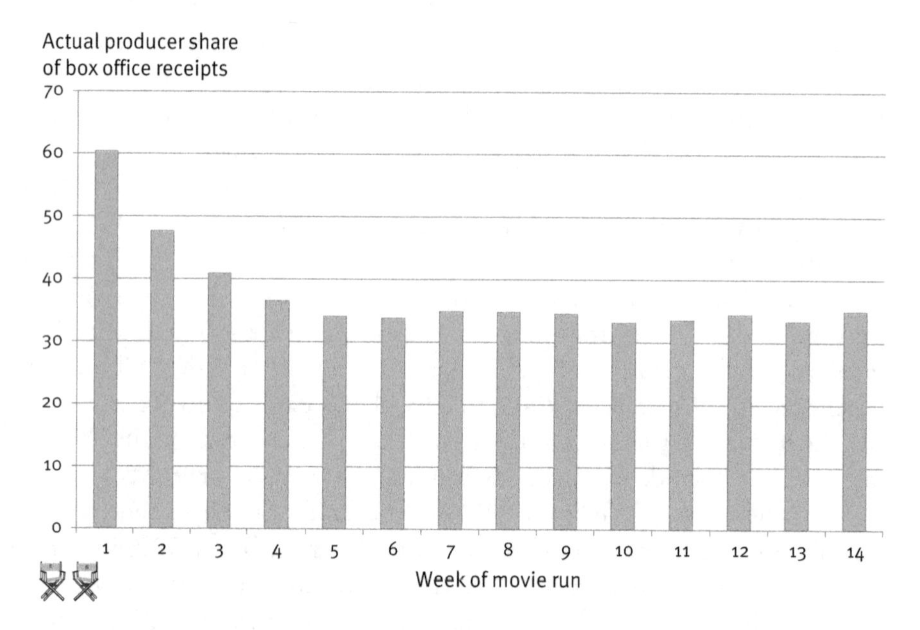

Fig. 5.7 Revenue shares for a "typical" North American movie theater
Notes: Authors' own illustration based on data reported in Filson et al. (2007). During the observation period 2001–2003, the theater had 14 screens and some 2,000 seats.

music sales. For music sold via iTunes and other shops, the retailer keeps the "standard" 30%, but the producer share of 70% is split between the label and the songwriter(s). Contracts between music streaming intermediaries and the music labels also are oriented around the "70-30" rule. When Apple entered the streaming market as a follower in 2015, the firm offered producers a slightly better deal than market leader Spotify, keeping between 27 and 28.5% (Ingham 2015) versus around 33% (Ingham 2016).

The major drawbacks to the revenue-sharing model are cost-related. Research has stressed the critical role of two kinds of costs: operational and monitoring. Operational costs include those related to the processing of sales data, including formatting and transfer of data. Monitoring costs, which are necessary for confidence in any revenue-sharing context, are incurred to verify the sales data of the intermediary. Such costs can be substantial, even prohibitive, depending on the complexities of the partnership and revenue streams. Filson et al. (2005) speculate that concession revenues are excluded from the industry's revenue sharing agreements because monitoring them would be economically infeasible. Specialized parties offer to track revenue streams (like Rentrak for physical video rental), but gains would have to be high enough to overcome their fees to make revenue sharing an attractive approach for both producers and distributors.[75]

Mixed Models

Fixed compensation and revenue sharing are not mutually exclusive ways to divide revenues between producers and distributors. Revenue streams are sometimes allocated in ways that combine elements from both approaches.[76] The most prominent versions of such mixed models are the "basic-fee" model and the "best-of-both-worlds" model.

In the "basic-fee" model, producers and intermediaries share consumer revenues, but, in addition, the intermediary pays the producer a fixed fee. The idea is that the fee covers the producer's variable production costs to acknowledge his/her efforts. When the revenue model for video rentals changed from fixed compensation to revenue sharing, the new model was actually a mixed one, as it included a fixed upfront per-item payment by the rental store. This fee was about $8, a fraction of the previous $65 price and a good approximation of the producer's duplication and shipping costs per video copy.

An alternative way to combine both approaches is the "best-of-both-worlds" model, in which the intermediary also agrees to pay the producer a fixed fee per copy. The difference is that in this model, the fee is not intended to cover a share of the producer's costs, but serves as a "floor" pay-

[75]As we have indicated above, revenue sharing is also used for the allocation of revenues between the producer of an entertainment product and the different creative parties involved in its creation. Such "internal" revenue sharing is standard when it comes to paying music performers and songwriters. It is also applied by movie studios for some of those creative players who have little negotiation power (as a substitute for fixed upfront salaries) and "superstars" (as a complement to fixed fees). "Internal" revenue sharing deals primarily with decisions about intraorganizational processes (i.e., the production of entertainment), rather than with decisions that relate to a product's market and customers, and thus lies beyond the central scope of our book. If you are interested in such internal revenue sharing and contracts between creatives and the producing firm in general, we recommend Caves (2000), who has dedicated several chapters to this issue, starting on p. 19, as well as the work by Darlene Chisholm (e.g., Chisholm 1997, 2004). In addition, very detailed information on how revenues are allocated between artists and labels for different forms of music distribution can be accessed via *Information is beautiful* (2015). Finally, the insightful documentary ARTIFACT sheds some light on the music industry's "360 degree deals," where record labels participate in all kinds of revenue streams of an artist, including touring, merchandise, and endorsement activities.

[76]This is also consistent with Cachon and Lariviere (2005) who provide analytical evidence that fixed compensation and revenue sharing are actually two variations of the same higher-level coordination model between producer and distributors. They show that the two only differ in a "wholesale price" parameter and a "revenue sharing" parameter (which specifies the intermediary's share of revenues); in the case of "pure" fixed compensation the revenue sharing parameter is set to zero, whereas in the case of "pure" revenue sharing the wholesale parameter is set to zero.

ment, or back-up. In the model, the producer and intermediary also include a revenue-share agreement, and the producer receives the higher amount of the two agreements (the "better world"). The fixed fee provides a floor that protects the producer if sales are low (recognizing that pricing authority is in the hands of the intermediary and that the producer had to shoulder the production costs). The revenue share kicks in if revenues are at a level such that the amount due to the producer is higher than the amount that would be produced by the fixed fee per copy.

Thus, the agreement insures that the intermediary will only pay a high price if one is warranted by product sales. A practical example is the deal music labels have worked out with Spotify. The streaming giant must either pay a fixed per-stream fee for its streams by consumers who have signed up for the advertising-based model, or, if it is a higher amount, a pre-defined share of its advertising revenues (Singleton 2015). It is easy to see why music labels, such as Sony, insisted on adding the fixed fee element: the amount of advertising Spotify will generate is difficult to predict and monitoring it is tricky.

Generating Revenues from Advertisers

> "I think you can get confused, you can be advertiser-centric – and what advertisers want, of course, is [consumers] – and so you should be simple-minded about that and you should be focused on [consumers]. If you can focus on [consumers] advertisers will come."
>
> —Jeff Bezos, *as Amazon CEO and Washington Post owner, about the role of advertising revenues for media businesses (quoted in* Rosoff *2017)*[77]

The two-sided nature of entertainment products implies that producers can earn revenues not only from consumers, but also from advertisers. In this section, we discuss how advertising revenues can be generated in entertainment and explore issues that need to be considered when advertising revenues become part of an entertainment producer's business model. The clear majority of advertising money in entertainment is made in the distribution stage, when films or shows are preceded, interrupted, or followed by advertisement when they air on TV stations, and songs receive similar treatments by radio stations. Our focus in this section, however, is once more on the *producer* of entertainment and how he or she can benefit from advertisements.

[77] Mr. Bezos originally speaks of "readers," the news industry's term for consumers. He uses the term "customer" in the sentence that precedes our quote.

Producers can generate revenues from advertisers via two different, but related means: brand placement and "in-product" advertising. The latter is explicit: advertising conveys messages about products via media (of any kind, such as TV, newspapers, or Facebook) by an "identified sponsor" (Kotler et al. 2005, p. 719). Placements, in contrast, are implicit: they also convey messages and are embedded in media, but the sponsor is not explicitly identified and the sponsoring act, as a whole, may not be obvious to the consumer. Instead, the branded product is embedded in the media in a way that can seem "organic."

Examples include an actor driving a car in a key movie scene with the car brand clearly visible or a musician wearing a brand of clothing—with visible brand logos—while performing. We should probably also note the inevitable Apple signs on characters' phones, tablets, and computers (except for Sony-produced films and series, in which the heroes usually use Sony devices—even James Bond, who was reportedly not amused however; Revilla 2015). Even when a film is shown on TV, placements are "presented as an inescapable part" (Marchand et al. 2015, p. 1667) of an entertainment product and thus cannot be skipped by the consumer, as it often the case with traditional forms of advertising.[78]

We begin our discussion with the (dangerous) temptations of brand placement, before we summarize what is known about the factors that make an "in-product" advertisement effective—something that particularly applies to video games.

The Blessings (and Dangers) of Brand Placement

"I see more Benz logos than dinosaurs."
—*An Internet user about the movie Jurassic World*

The Economics of Placements

A producer of entertainment can profit from including brand placements in its creations in different ways. The most intuitive way are direct payments,

[78]Does this mean that placements are more effective than traditional advertising? Not necessarily so. Both communication means have strengths *and* weaknesses. In contrast to advertising, placements offer marketers limited flexibility for the design of placements and usually little control over the final way their brand is presented as part of the content. Whereas advertising enables explicit persuasive messages, placements do not allow that, at least not to the same degree. Placement works best when the goal is to influence a brand's image: when James Bond uses Omega, his image as a premium, daring agent tends to spill over to the watch he wears, reinforcing its image as a premium, daring watch brand.

or "placement fees." Those payments exist and can be quite substantial. For example, Samsung has been reported to have offered Sony a fixed placement fee of $5 million for the inclusion of their phones in their James Bond movie SPECTRE, along with a commitment to spend about $50 million for tie-in advertising, and we assume that a similarly sized part of Heineken's rumored $45 million deal for SKYFALL was due as a direct payment to the producers.[79]

But such direct payments are not the dominant source of advantage and less frequent when it comes to placement remunerations (see also Epstein 2010). Bartering deals between the producer and the sponsor are much more important. The most prominent form of bartering these days is the advertising "tie-in": by featuring the entertainment product in its own advertising, the sponsor can substantially leverage a film's or game's awareness. A prominent example is the commercials that beer company Heineken has produced as part of their placement of Heineken beer in the James Bond movies. In the heavily promoted commercials, Heineken paired its brand with key elements of the sponsored movie, including the series' well-known musical theme and, in the case of CASINO ROYALE, its lead actress, Eva Green. In Heineken spots for subsequent 007 films, even James Bond himself (alias Daniel Craig) appeared.[80]

Via "tie-in" advertising, the producers of JAMES BOND movies have dramatically extended their marketing budgets by drawing on advertising from placement sponsors. Since 1989's LICENSE TO KILL, the producers' advertising budgets for the films have grown substantially (LICENCE's reported $50 million ad spending translates into almost $100 million in 2015 value, and the 2015-released SPECTRE was estimated at $140 million in combined global advertising and distribution costs). But what has changed even more strongly over the quarter of a century are "tie-in" promotions from sponsors. Whereas the approach was essentially non-existent back then, it is now valued clearly higher than the amount of advertising spent by the producers themselves. Heineken alone has been said to have spent about $100 million for its 007-themed advertising campaign for the SPECTRE movie (*Instavest* 2015).

---

[79]As viewers of Spectre will have noticed, Sony eventually decided to let the offer pass and equipped its agent with their own Xperia devices.

[80]At the time of writing, the commercial for CASINO ROYALE could be watched at https://goo.gl/unPRPV. Daniel Craig himself appears in Heineken spots for QUANTUM OF SOLACE (https://goo.gl/9Z-bRcQ) and SKYFALL (https://goo.gl/YDeuQS).

To avoid misunderstandings: "tie-in" advertising is not at all limited to James Bond, but is systematically exploited for many major movies. For example, Universal's BATTLESHIP movie more than doubled its U.S. advertising budget of about $40 million through "promotional partnerships" with consumer brands such as Coke Zero, Cisco, Subway, and firms such as Kraft, Nestlé, and Chevron, who in total spent more than $50 million for TV, print, and online advertising (Finke 2012). Finally, bartering can also take place when sponsors provide costly production elements (e.g., cars) and services (e.g., insurance). Remember the introductory quote to this section? To get its logos in the dinosaur film, Mercedes also provided the costly equipment, which saved the producers substantial amounts of money. And General Motors, whose Cadillac brand was prominently featured in MATRIX RELOADED, gave the producers about 300 cars to film an extensive chase sequence (and destroy all of them; Cowen and Patience 2008).

Overall, global spending for placements reached $10 billion as of 2014, and is still exhibiting strong growth (*PQ Media* 2015). The clear majority of that amount is spent for placements in TV productions; films account for about one-third of placement value. Placement also exists for other entertainment forms such as games and music videos. For major console game releases, "tie-in" advertising is practiced particularly in North America and Asia, such as when UNCHARTED's programmed protagonist Nathan Drake tells TV audiences that Subway restaurants are the place "where winners eat."[81]

But, Two Words of Caution

Placements are not equally suited for every entertainment product. The settings and the content of some products provide more room for placement than others. For example, whereas the release of the fourth film in the action-adventure series MISSION: IMPOSSIBLE was accompanied by extensive TV ads from placement partner BMW, the competing sequel SHERLOCK HOLMES: A GAME OF SHADOWS released around the same time had to do without such support: in its 19th century setting "people were still riding in horse-and-carriages" (Finke 2012). This can have a serious effect on the

[81]The Subway ad can be watched at https://goo.gl/RenYxr.

financial potential of a prospective entertainment product and must be considered accordingly by managers as part of their innovation management. Despite their film being produced for less than its competitor, the SHERLOCK HOLMES producers spent $22 million, or 55%, more for advertising in the U.S.[82]

In addition, entertainment managers better avoid the temptation to give into greed when it comes to placement revenues. Scholars provide evidence that consumers may actually be turned off by the *overuse* of placements. Homer (2009) conducted an experiment, using a 15-minute compilation of four scenes from the actual movie MAC AND ME, and manipulating both the number of placements by the McDonald's brand ("low frequency," i.e., one placement, versus "high frequency," i.e., three placements), as well as the placement's strength (visual or "weak" placement versus verbal or "strong" ones). The scenes were shown to 108 undergraduate students—who reported clearly less positive attitudes toward the film when it contained both strong *and* frequent placements. In this case, attitudes toward the film were about 25% worse than in all other conditions, something the authors empirically linked to high levels of distraction and interference (i.e., the placement "interfered" with the enjoyment of the film). They repeated the experiment with 155 students and excerpts from the MONK TV series, finding similar (although somewhat weaker) patterns.

We also investigated the effects of placements on consumers' perceptions and assessments of entertainment content ourselves (Marchand et al. 2015). We didn't use excerpts from existing films or series, but produced a seven-minute short film ourselves—a professional scene-by-scene remake of AUFGEWACHT, a short film that had been shown successfully at festivals, but was otherwise little known.[83] Creating the stimulus material for our study provided us full liberty to create versions that differed in the "prominence" of the placement—a combination of strength and frequency, similar to Homer (2009). Our results from two studies (with 203 and 312 respondents, respectively) confirm that once a certain threshold level of placement prominence is exceeded, placements hurt consumers' liking of the entertainment content (by 20% and 14% in the first and second study, respectively).

[82]Let us add that MISSION: IMPOSSIBLE had an opening weekend of below $30 million, which was considered a disappointment by many. This raises the question whether its producers should better have used the promotional partnerships to further increase the audiences' anticipation of the film (i.e., influencing the *revenue* side of the profit equation) instead of reducing the *cost* side of the profit equation (i.e., substituting their own advertising with "tie-in" advertising).

[83]A full version of the original film is available at https://goo.gl/YgXp3h.

But we also provide evidence of the psychological mechanism behind this effect: mediated regression analyses show that placements can trigger "reactance" within consumers, as they interpret the brands as threats to their personal freedom, an argument that is in line with psychological reactance theory (Brehm 1966). We show that it is such reactance that is responsible for the worsening of consumers' quality perceptions once placements become overly prominent, fully mediating the link between placement prominence and quality perceptions.

Further insights on the potential destructive effects of heavy placements come from Meyer et al. (2016), who also examine filmed entertainment. They use secondary data for 134 full-length movies released between 2000 and 2007 and listed in "Brand Hype Movie Mapper," a (no longer available) database that provided details of the more than 2,000 placements contained in these films. The results from a generalized method of moments (GMM) regression analysis, which accounts for a potential endogenous role of the number of placements in a movie (brands prefer movies which they believe will be successful), are in line with those derived from the experimental studies. Specifically, Meyer and his colleagues find that consumers' quality perceptions (from IMDb and other sites) are negatively impacted when product placements are used "in excess." As so often, however, context matters: the negative placement effect is stronger for independent versus for "mainstream" films (which consumers expect to be of a commercial nature anyway), and it also takes fewer product placements to hurt independent films (i.e., from 11 placements on) versus mainstream films.

Thus, *Entertainment Science* advises producers of entertainment content to manage a delicate balance, as too many (and too intense) placements can hurt their entertainment assets and counter placement revenues. We also suggest that managers and scholars look beyond the volume of placements, but also study their quality and fit with the host entertainment product. Except for analogies from brand alliance research, no scholarly studies provide evidence so far for this logic. But audiences' reaction to the TRANSFORMERS franchise's massive use of Chinese brands, such as dairy drink Mengniu and the Industrial and Commercial Bank of China, despite being situated in the U.S., offers some ad hoc indication. As a user of the social media site Sina Weibo states, "[e]ven though it is normal to add Chinese elements into the Hollywood blockbusters, it still makes the audience uncomfortable when there are too many *Chinese* brands" (quoted in Rahman 2017).

It remains speculation at this point whether this placement approach is linked with the recent notable revenue decline of the Transformers films. In the U.S., the fifth release returned only about 50% of its predecessor (and less than one-third of the series' best); in China, it made 30% less than the former film. Under any circumstance, entertainment firms should protect their brand assets against short-term placement allures.[84]

How to Design In-Product Advertising

The information-good character of entertainment products enables the modification of a product in a way that embeds advertisements, in addition to brand placements. This is done by distributors for movies (when a TV station interrupts the airing for commercials), books (when Amazon displays ads on its Kindle), music (on the radio and ad-based streaming services), and games (ad overlays on mobile phones). In the case of games, the producers themselves can also make decisions about the integration of advertising into the product—and potentially earn profits from doing so. The common label for this approach is "in-game advertising."[85]

Just like placements, advertising can trigger negative externalities, so that game producers need to keep in mind that it is the consumers who are their essential target group when designing such in-game advertising. This is true even when consumers don't pay at all for their usage directly, and all revenues come from advertisers. Thus, in-game advertisements must be created in a way that interferes minimally with the consumer's enjoyment from playing the game. Verberckmoes et al. (2016) have investigated this issue empirically, employing an experimental approach. They asked 619 fantasy

[84]We recommend Owczarski (2017) for detailed coverage of the role of Chinese placements for the Transformers saga.

[85]Let us note that the managerial challenge can be somewhat similar for other products such as movies and TV shows, when the producer is vertically integrated—and thus has the ability to manage content production and distribution simultaneously. For example, when cable station AMC, (co-)producer of series such as The Walking Dead, interrupted their airing of the series with advertising for Hyundai, using the series settings and atmosphere, this constitutes an attempt to "include" the ad in the product (see https://goo.gl/JoqRrA). More generally, TV stations have to decide on the amount and number of commercial breaks when airing their own content to balance customer satisfaction and advertising revenues. For those who are interested in that challenge, Zhou (2004) offers an analytical econometric model of the optimal number, length, and timing of commercials for a particular piece of content (he does not account though for longer-term effects on the broadcaster's brand or even media channel usage).

Fig. 5.8 Two exemplary in-game adverts
Notes: Reprinted with permission by Elsevier from Verberckmoes et al. (2016, p. 878). The left frame (Panel A) shows the low fit/non-interactive stimulus, the right frame (Panel B) a part of the high fit/interactive stimulus used by the authors.

game-playing consumers to evaluate a number of screenshots ("vignettes") from the MMOG title Lineage 2, which takes place in the medieval ages. All vignettes included an advertisement for a fictitious energy drink. The authors systematically varied the vignettes with regard to (1) the fit of the ad with the game environment (e.g., wooden/historic billboard versus metallic/modern billboard) and (2) the degree to which the advertisement is integrated into the game (i.e., clicking on the ad restored the player's avatar's "energy levels" versus nothing happened). Figure 5.8 shows two exemplary vignettes.

Verberckmoes et al. then ran a series of OLS regressions, in which the participant's intention to play the game in real life served as the dependent variable and the characteristics of the ad as independent, or explanatory, variables. They find that an in-game ad with high fit with the game (versus a low-fit ad) increases the respondent's intention to play; it does so by reducing the perceived intrusiveness of the ad and by heightening the perceived realism of the game. The ad's interactivity leads to a better evaluation of the ad, but does not affect play intentions—a finding that we recommend to treat with care, as it could simply be a result from the static vignette design of the experiment. Nevertheless, just like we have argued for brand placements above, game producers are advised to pay attention to how in-game ads are designed in terms of the ads' fit with the game itself.

Generating "Revenues" from Third Parties: The Case of Subsidies and Other Public Benefits[86]

Because of their cultural nature, entertainment products often qualify for public subsidies by various institutions in various locales, which can provide a substantial amount of revenues. Subsidies are mainly available for filmed entertainment (movies and TV series), but also for literary works, musical productions, and electronic games. In addition to supporting works that are of artistic value, subsidies are often provided for economic reasons: to increase the competitiveness of cultural productions in a region (based on the "societal value" of such creations), or simply to support economic growth in a region or country in which cultural productions are considered an important pillar.

On a global scale, the total dollars awarded via subsidies are vast, with the lion's share allocated to films. To contextualize, our own data shows that out of the 710 movies produced by German companies between 1998 and 2010 that were released in German theaters, almost 93% received some kind of subsidies. The average subsidy per film was 1.5 million Euro (the equivalent of $1.8 million in that time frame); some films such as The Miracle of Bern received up to 8.5 million Euro (or $10 million). On average, each ticket sold for the films was subsidized by 4.5 Euro (or $5.4), or 80% of the total ticket price. The total amount of subsidies given to these films by German public institutions was more than 1 billion Euro, an amount which does not include the country's substantial public spending for international productions.[87] In Europe, more than 20 incentivization models have been introduced for filmed entertainment since 2005, granting more than 1 billion Euro ($1.15 billion in 2017 value) of direct subsidies and more than 400 million Euro ($460 million) in tax incentives per year; similar models exist in Canada and several Asian countries (*Blickpunkt:Film* 2016; Meloni et al. 2015). As of March 2016, 34 of the 50 American states had some kind of film subsidy system in place (Sandberg 2016), collectively spending more than $2 billion (Thom 2016).

---

[86]Economists make a clear distinction between subsidies and tax incentives, stressing that being allowed to keep one's income (as in the case of tax incentives) is different from having it given to you by your competitor (who pays the money through taxes that is then allocated to you). We will discuss them jointly in this section nevertheless, as our focus is on the level of the individual firm (for which both incentives have similar effects), not the economy as a whole.

[87]In 2012 alone, federal and state film subsidies totaled more than 310 million Euro (Posener 2014).

Producers usually welcome subsidies as a way to increase a product's profitability, by reducing the cost or by enhancing the scale of production, marketing, and/or distribution. However, Jourdan and Kivleniece (2017) point to a potential drawback of subsidies for producers. They argue that subsidies have a two-sided effect on the market performance of an entertainment firm that can increase, but also *hamper*, its success. Whereas the budget-enhancing positive effect of subsidies is obvious (in an industry where financial resources are critical), they also carry the risk of corrupting the firm's "incentive system," resulting in production inefficiencies because of the involvement of a political (versus economic) actor.

Specifically, the scholars test the strength of these opposing effects with data from the French film industry, in which subsidies are based on the producer's *past* performance instead of on the commercial prospects of a specific project. Their data set includes 567 film-producing firms and their annual performance from 1998 to 2008. By estimating a fixed-effects regression model, Jourdan and Kivleniece find that subsidies indeed impact firms' return on investment (ROI—which they measured as a firm's total box office divided by total production budget across all films in that year). In their analyses, the inclusion of subsidies explains abouts 4 percentage points of the ROI, increasing its explanation by more than 16%. The effects of subsidies are non-linear; when subsidy levels are low, the positive effects dominate, but things change when subsidies exceed a threshold (360,000 Euro in their study). Beyond that point, inefficiencies caused by subsidies begin to overwhelm the positive effects, hurting market performance (see Fig. 5.9). The results of the research also indicate that these effects differ between firms; they are stronger for producers with a broader product portfolio (i.e., offering products that span a broader range of genres) and for those who work closely with star actors.[88] In contrast, high product quality tends to reduce the effects of subsidies.

Other studies have asked whether subsidies actually improve the quality and market performance of the products that receive them—a fundamental question addressing the adequacy of subsidies in general. Whereas the answer is of less immediate relevance for the individual producer of entertainment, scholars' answers might impact the subsidies system as a whole, so let us take a quick look. Across methods and countries, findings suggest skepticism toward the economic effectiveness of subsidies. For example,

[88]See our discussion of the genre concept and the use of more than one genre in entertainment marketing in our chapter on search qualities for entertainment products and of the contributions stars can offer in the entertainment brand chapter.

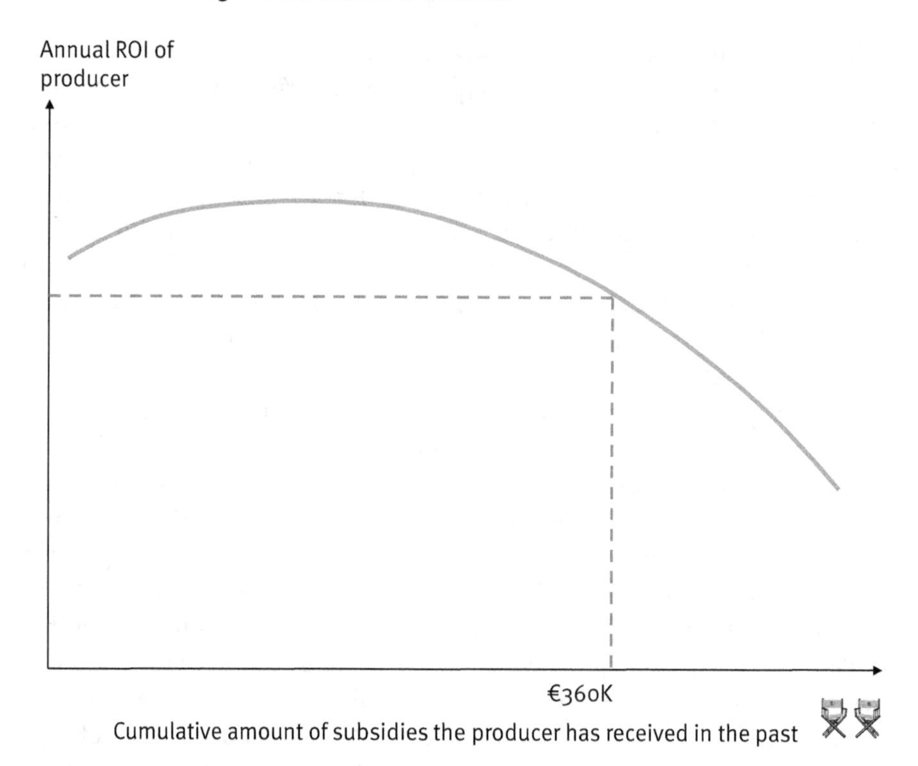

Fig. 5.9 Linking film subsidies with producer performance
Note: Authors' own illustration based on results reported in Jourdan and Kivleniece (2017).

McKenzie and Walls (2013) apply OLS regression to 95 Australian movies released between 1997 and 2007 for which the researchers could locate sufficient data. They find that neither the fact that a movie is granted subsidies by the Australian government nor the amount of subsidies were linked systematically to a film's box office performance. However, results show the budget (which includes subsidies) to be influential, pointing to an potential *indirect* success-enhancing effect of subsidies—one that is independent of the selection of subsidized films in their Australian data set.

Italian film subsidies are granted by the government based predominantly on the "quality" of a film project, with a recent shift toward more commercial success criteria. Bagella and Becchetti (1999) study close to 1,000 movies produced in Italy between 1985 and 1996. Employing a GMM regression (to overcome non-normality issues with the data, not to address a potential endogeneity bias of subsidies), they find that ticket sales are much lower for subsidized versus non-subsidized films. However, they also report that popular stars and directors are much more prominent in non-subsidized films.

When the scholars account for these and other factors, subsidies do not impact movie admissions at all.[89] Jansen (2005) also finds for 120 German films released between 1993 and 1998 that films picked to receive subsidies by committees do not perform differently at the German box office from those films that did not receive subsidies. But his results point to a potential indirect effect of subsidies, independent of a movie's specific characteristics, as the films' budget (which *includes* subsidies) exhibits a positive link to the number of tickets sold.[90]

When reflecting upon these results, let us keep in mind that subsidies are highly heterogeneous, as are the goals for providing them; the data used in these studies is fragmentary and potential selection biases have not been addressed. But one common theme is that committee judgements on how to allocate subsidies do not seem to work effectively across institutions and countries, even when their focus is commercial rather than artistic. In addition to entertainment producers, selection committees might be among those who can benefit from applying *Entertainment Science* knowledge and methods when making decisions. Or they could acknowledge their ineffectiveness and simply allocate budgets on a first-come-first-served basis—an approach practiced by the German Federal Film Fund (DFFF), for example.

Finally, do subsidies benefit those who eventually provide them—that is, the public? Thom (2016) studied the effects that different types of subsidies by U.S. states had on economic parameters, such as movie industry wages, movie industry employment, and state value creation. Examining subsidies provided between 1998 and 2013, he ran a set of cross-sectional fixed-effects regression models to explain changes in these economic parameters (the dependent variables) as a result of different types of subsidies and several economy-level control variables (the independent variables). Effects of subsidies were minor, if they existed at all, and differed between types of subsidy. Tax credits that offered a cash refund created a short-term 5% improvement in wages, whereas tax credits that could be transferred to other projects led to annual employment gains of 0.6 percentage points. Other incentives (e.g., sales and lodging tax waivers) had no measurable effect, and none of the subsidies influenced the gross state product. Thom thus concluded that such

[89]The importance of controlling for alternative success drivers in econometric works can be seen from the study by Meloni et al. (2015) who run a fixed-effects-panel regression with 754 Italian films released between 2002 and 2011. They find a *negative* effect of subsidies on performance—which is most probably a reflection of the lower budgets and commercial appeal of subsidized films, rather than evidence for a causal effect, as the authors do not control for any film characteristics (except genre).

[90]See our chapter on unbranded signals of quality for entertainment products for a closer investigation of the complex link between an entertainment product's budget size and its success.

"incentives are a bad investment," in general (quoted in Gersema 2016). But with the empirical results laid out before you, any reader can make up his or her own mind whether such effects are worth the effort.

After having looked over the different sources of revenues that are available for entertainment producers, let us move on to analyze a second critical element of any business model that requires thorough treatment: the *risk* that entertainment products carry.

Managing the Risk of Entertainment Products

> "[W]hat big corporations want most is risk-averse pictures."
> —Peter Bart, *long-time movie executive and journalist (quoted in Frontline 2001) about the priorities of entertainment conglomerates*

On the Riskiness of Entertainment

We have shown that entertainment products feature a number of specific characteristics. Some of those characteristics make it a far-from-trivial task to forecast how well a new entertainment product will perform. Those include consumers' difficulty in judging entertainment quality ahead of time (making high quality difficult to shine), the aesthetic and symbolic elements (assigning consumers' "taste" a critical role), and the involvement of creatives (complicating the anticipation of what the final product will *actually* look like).

One empirical indicator of the product-level risk that stems from these characteristics is the high variation in financial returns that we observe for entertainment products. A statistical measure of this variation is the "standard deviation" of entertainment products' financial performance. Although standard deviations can be calculated for virtually any set of data points, for the financial returns of entertainment products, they quantify the amount by which returns from the average individual product (say, one movie) differ from the mean value of returns of all products (all movies).[91]

[91]The standard deviation is calculated by (1) subtracting the mean from each data point, (2) squaring, summing across all cases, and then averaging the differences (which produces what is named the "variance"), and finally (3) taking the square root of this variance. Empirically speaking, $\sigma = \sqrt{\frac{\sum_{i=1}^{n}(x_i - \bar{x})^2}{n}}$, where n is the number of data points (or entertainment products to be considered), x_i is the respective value (such as the returns) for a data point (entertainment product) named i, and $\bar{x}$ is the mean value (return) of all data points (entertainment products).

A risk-free product investment would have a standard deviation of zero—the investment's return would be the same as the mean return of all similar investments and will never deviate from this mean. In the real world, an investor's risk is higher the larger the standard deviation for such investments—in other words, the more the returns vary. Ideally, an investor would determine the likelihood of different returns for each specific investment in advance. Finance theory suggests that this can be done—though not with actual future data for the new investment, of course, but by using the returns of *similar* investments in the past and by extrapolating risk from such historical data. The approach is essentially the same as when, to assess the risk and determine the value of shares of a new commercial bank, you would examine a portfolio of existing commercial banks that are similar to the new bank on key characteristics.

Doing so is more challenging for analyzing prospective new entertainment products though—the "infinite variety" characteristic of entertainment and the sheer creativity of artists makes content harder to compare. It is likely that two mid-sized commercial banks are far easier to compare than are two mid-budget drama movies or two mid-budget role-play games. But, as we will show below, this should not prevent entertainment managers from building on learnings and methods from finance theory when deciding the financial prospects of new entertainment products.

But let us first take a general look at the riskiness of entertainment products. In Fig. 5.10, we report the standard deviation of the returns for several thousand movies and games—you find more details about the data set in the note below the figure. The figure shows that for each type of product, the standard deviation is higher than the mean value. Also, the mean is clearly higher than the median (i.e., the return of the product that is right at the 50th percentile). These results demonstrate that returns differ strongly between individual entertainment products of the same types (the high standard deviations) and extreme outliers exist (the means being higher than the medians).

Let us note though that these numbers mark the most extreme end of the "risk spectrum," as they incorporate the full heterogeneity of the film and games business. If we split the data sets into the sub-markets of commercial and independent products and analyze them separately, we find that the standard deviations drop substantially and come back into line with the means. For example, when considering only movies produced by major studios, mean and standard deviation are about the same. And when studying only films that all are in the top 10% in terms of highest budgets (and thus more similar), the mean revenue is now 1.5 times *higher* than its standard

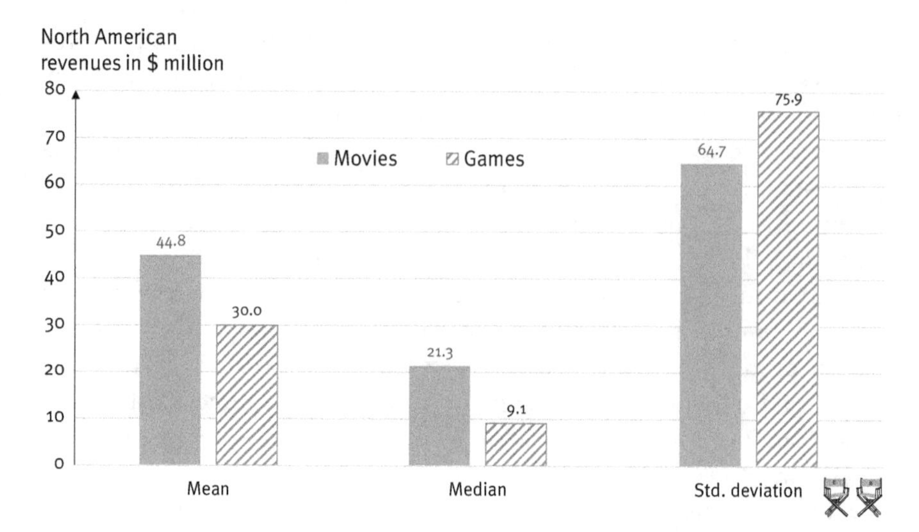

Fig. 5.10 Standard deviations, means, and medians for several thousand movies and games
Notes: Authors' own illustration based on data from The Numbers and VGChartz. Movie statistics reflect the North American box-office revenues for all 3,158 movies released in theaters between 2000 and 2014 (not considering home entertainment revenues); the game statistics reflect the estimated North American retail revenues for all 1,898 console games released between 2005 and 2014. Data is not adjusted for inflation, and we made several assumptions.

deviation (and similar to the median). But this does not change the broader picture: entertainment managers must find ways to systematically manage the risk of the entertainment products for which they are responsible.

Risk management strategies can take place on two basic levels: on the level of the individual product and on the portfolio (or "slate") level. In the latter case, the risk of individual products is reduced by simultaneously managing a larger set of titles (i.e., "spreading the risk"). We now look at both levels of strategies for risk management in entertainment, beginning with slate-level approaches.

Approaches to Manage Risk on the Slate Level

> "If you want to strike it rich in the entertainment business … and you don't want to take huge risks doing so, you are better off investing in a predictable and perhaps humdrum company that distributes a lot of movies rather than in an edgy upstart that hopes to release one or two blockbuster films."
> —*Knowledge@Wharton (2007)*

In the 1990s, Intermedia Films was a well-known entity. Founded and run by a team of German and British managers, the firm's strategy, according to then-CEO Moritz Borman, was to produce "event movies" by hiring star actors and directors, and/or by using prominent "intellectual property," or brands, spending heavily for production and advertising (quoted in *Blickpunkt:Film* 2003). Because resources were limited, the firm produced one film at a time, hoping to create the "next big thing" that would generate the profits that were needed to produce the following film. Intermedia went public in 2000, trading on the Munich Stock Exchange, and used the influx of capital to produce a number of high-budget films. In 2009, however, it had to declare bankruptcy. So, what went wrong?

Their approach worked quite well with several movies. For example, TERMINATOR 3 cost about $220 million to produce and market, but generated enough revenue in theaters alone to nearly cover those costs. But the firm collapsed after just two of its productions failed miserably. The Oliver Stone-directed ALEXANDER cost about $200 million to make and market, but returned less than half that amount to Intermedia, and BASIC INSTINCT 2, which starred Sharon Stone, devoured investments of more than $100 million while returning only about $13 million from theaters worldwide. Intermedia Films was unable to recover from these two flops.

Intermedia's approach was an "all-eggs-in-one-basket" one, where a single failure, let alone *two*, can have fatal financial consequences. But no management paradigm, including *Entertainment Science*, will ever be able to prevent the failure of an individual project in a business world that is defined by probabilities. It is these probabilities which make shifting the focus from individual products to slates of products (or *portfolios*) the most powerful approach for dealing with the inherent risk of entertainment products. The basic idea behind such portfolio-management approaches is that risk is diversified in a manner such that any single flop is absorbed by the overall success of the broad portfolio of products. Probabilistic thinking suggests that, in portfolio management, the prediction error for a single product matters less, and that such errors will be covered by the performance of other products in the portfolio.[92]

However, and this is a key point, this approach only works if the portfolio is large enough and purposefully constructed. Thus, the key question for the management of entertainment portfolios is: how should the portfolio

[92]Empirical evidence is ample for this "portfolio effect": the volatility (the finance term for standard deviation) of portfolios such as the S&P 500 or the Dow Jones is much lower than those of the individual stocks that are included in the portfolio (e.g., Berk and DeMarzo 2014, p. 328).

be constructed? At the practical level this requires deciding which products should be included and which ones should not.

Balancing Diversification and Expertise

Entertainment products, just like any other financial assets, can perform poorly for a large number of reasons. Some of these reasons affect all entertainment products. For example, a recession might cause consumers to drastically reduce spending on any product that is not a necessity. When the choice is between eating food or buying a hard-cover book, survival instincts rule the day. Other reasons for poor performance are specific to a single product. For example, a singer's drunken, race-tinged rant that was captured years earlier on a grainy cell-phone video suddenly is leaked just days before his new album's release. Or a film's sets are destroyed by a natural disaster. Or a lead actor gets ill during filming.

However, still other reasons for poor performance affect multiple products, but only those of a certain type. Consider the case of when another gruesome school shooting shatters the world just a few days before a new first-person shooter game is released. Certainly, societal reaction to that news will impact sales (and perhaps delay the release) of that new product; but it will also impact every game of that genre. Now envision Producer A who has constructed a portfolio that consists only of first-person shooter games. In contrast, Producer B's portfolio consists of a *mix* of first-person shooter games *and* science-fiction movies, balanced between the two product types. Producer A's *entire* portfolio is affected just as strongly as any single shooter game would be—that is because the risk is correlated across all titles within the portfolio. It is called "common," or "systematic," risk. In contrast, producer B's portfolio would be affected less by the shooting than would a single shooter game, because the societal incident that is relevant to shooter games has little influence of the demand for science-fiction movies. Thus, the risk of science-fiction movies and shooter games is not correlated; we refer to this as "independent," or "idiosyncratic," risk.

For entertainment producers, this discussion carries two important lessons. First, portfolios need to consider any relations among the respective risks that are borne by the different assets in the portfolio. If the portfolio elements all share common risk, the risk of the portfolio is not different from that of any single product in it. In turn, the portfolio strategy would not meet its objective of reducing producer risk. Instead, a portfolio needs to be comprised of assets whose risks are independent; if this is the case, the portfolio risk is lower than the risk of its individual assets, because independent risks are averaged out. Constructing a portfolio by combining assets with independent risks is called "financial diversification" (Berk and DeMarzo 2014).

Second, such risk reduction via diversification may come at a price for entertainment producers. Usually, a producer of entertainment, just like any investor, has certain specific area(s) of expertise—not just expertise in entertainment versus other businesses, but areas of expertise *within* specific domains in the heterogeneous entertainment industry. The producer's expertise in other domains will be less pronounced. Because expertise affects the success potential of any product, investments in products that make maximal use of a producer's expertise will, all else equal, provide higher returns than will investments in projects outside of the producer's domain of expertise. So, if the only diversification opportunities require the producer to move into domains of low expertise, the likelihood of generating high returns is reduced, potentially offsetting any benefits to the producer from reducing the riskiness of his or her investments.[93] The negative consequences of diversifying *too* far outside one's expertise has been shown empirically by a number of management scholars, with findings revealing that "too much diversification [both with regard to industry and geography] may actually be detrimental to firm performance" (Pierce and Aguinis 2013, p. 322).

Balancing Risk and (Expected) Returns

Entertainment products differ in both their levels of expected returns and their respective risk: what generates the highest returns on average is not always a safe asset. So, which combination of risk and return is superior? Answering this question should be an important issue for industry executives. However, when it comes to solutions that entertainment firms have actually implemented, it appears that very limited attempts to balance risk and returns have been carried out beyond gut feeling.

Finance scholars offer some promising avenues though—let us translate them into the entertainment context. In Fig. 5.11, we create a two-dimensional graph on which we place several types of movies, based on their

[93]Take the case of a movie company that specializes in producing horror films, but decides to diversify into romantic comedies, because that genre's risks are largely independent from those for horror films. The firm has lower expertise in producing romantic comedies; it has no relationships with top rom-com artists, and also lacks experience in making the final editorial tweaks that often make the difference between a commercial success and a flop. In addition, the decision to diversify across genres also creates major organizational complexities, as the producer is now trying to supervise and control projects that are extremely heterogeneous and that require diverse skills.

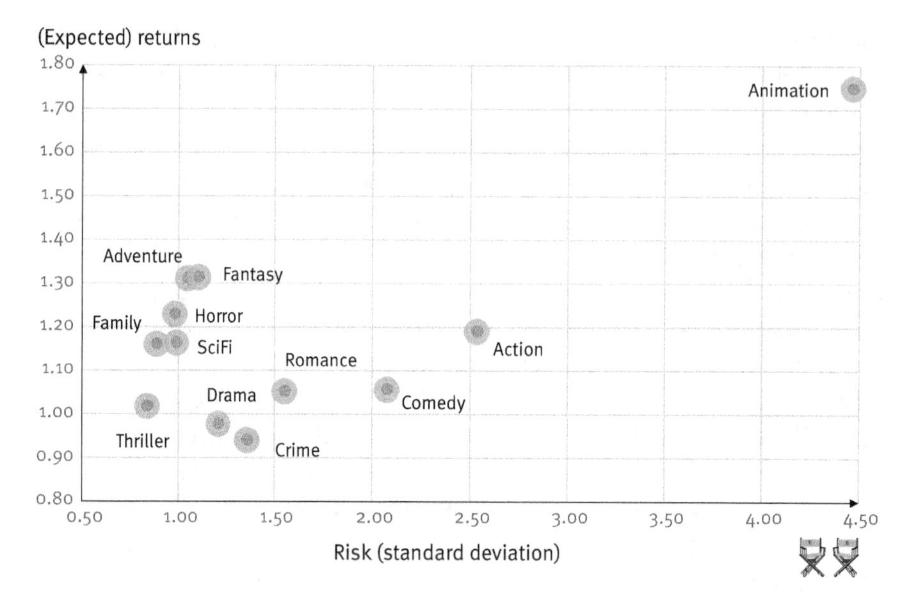

Fig. 5.11 Average empirical returns and risk for different movie types
Notes: Authors' own illustration. Returns are ROIs that were calculated based on (a) actual North American and "foreign" box-office revenues and (b) actual production and advertising costs spent at a movie's North American theatrical release; this information was then combined with industry revenue-split and spending ratios. Data is not adjusted for inflation, and we made several assumptions. Main sources were The Numbers (for box office and most cost information) and Kantar (for advertising).

respective historical return and risk levels (i.e., based on real financial data). We use genres as types, but any other criterion (or combination of criteria) that leads to types which differ in returns and risk would also do.[94] The data is the same that we used earlier in this chapter for calculating overall means and standard deviations of movie returns. For each genre, we calculated the average ROI (our returns measure) and its standard deviation (our risk measure). Across all movies in our database, the average ROI is 1.075 (supporting our earlier argument that movies, on average, turn out a profit) and the average standard deviation of ROI is 1.655 (further evidence that success differs substantially between individual titles).

[94]The nature of this figure is illustrative: we do *not* imply that genres should be used to define movie "types" as the sole criterion. Instead, the decision how to define product types needs to be made by every producer based on his or her own industry expertise and resources. Our discussion of the riskiness of specific kinds of entertainment products such as sequels and remakes in the book's Part II might provide some additional guidance for this difficult task.

The figure illustrates that movie genres differ substantially in the returns that a producer can expect from them, but so do the risk levels. For example, whereas animated films provide producers with exceptional *average* returns, the *variation* across these returns is between 1.7 and 5.3 times higher than for other genres. Although this high-risk/high-return character illustrates the necessity of a trade-off when choosing between animation and other genres, some genres are both above and further to the left than other genres—meaning they are superior in terms of higher return expectations *and* lower risk.

Take the example of horror movies versus comedies. Comedies generate an average profit of 5.6% of their total costs, but these profits have a standard deviation of more than 2. In comparison, horror movies produce an average profit of more than 20% (i.e., they are more lucrative) and their results vary less than half as much as those of comedies (i.e., they are less risky). If the decision of in what kind of movie to invest scarce resources is driven by financial concerns, a movie producer should invest in a horror film rather than a comedy. A producer would have to have very good reasons to do otherwise, such as particular expertise in making comedies (or the lack of such expertise in producing horror).[95]

This information is insightful, but far from sufficient for a producer who is aiming for the "optimal" portfolio. For accomplishing this ambitious task, let us borrow finance theory's "mean-variance portfolio optimization" approach. The approach's idea is to combine the information on each individual product type (treating each movie type/genre as the equivalent to a single "stock") to calculate the expected returns and risk levels for any potential combination (portfolio) of such individual products.

Calculating the expected returns for the portfolio is intuitive—it is the average of the returns of the products in the portfolio, weighted by the products' respective budget share in the portfolio. The standard deviation of the portfolio, as the risk measure, is only a little more complicated to determine: it is the square root of the sum of the squared standard deviations of the products, again weighted by their budget share, and their shared variance ("covariance") (Berk and DeMarzo 2014).

[95]Remember that the data we used to calculate genre averages comes from *multiple* producers who all have somewhat differing levels of capabilities and expertise across genres. Someone skilled in making comedies might outperform the genre's industry averages for revenues and risk, just like Jason Blum's Blumhouse Productions has been outperforming other producers when it comes to making horror films. A producer/studio could conduct this analysis using its own historic data from its own productions to account for its particular situation and expertise.

The following equation describes this in a more formal way, using the example of a portfolio which consists of only two types of products—let's say thrillers (= *THR*) and fantasy movies (= *FAN*):

$$SD(P) = \sqrt{x_{THR}^2 * SD(THR)^2 + x_{FAN}^2 * SD(FAN)^2 + 2 * x_{THR} * x_{FAN} * COV(THR, FAN)}$$

Here, *SD(P)* is the standard deviation of the portfolio we label *P*, *SD(THR)* and *SD(FAN)* are the standard deviations of thrillers *THR* and fantasy movies *FAN*, and x_{THR} and x_{FAN} are the respective weights of the products in the portfolio (i.e., the percentage of the portfolio resources assigned to each of them, with the sum being 1). *COV* is the covariance of the two products, which is the same as:

$$COV = COR(THR, FAN) * SD(THR) * SD(FAN)$$

where *COR* is the statistical correlation between the ROI of thrillers and fantasy movies in the past. The elements of this equation can be linked to our earlier discussion of different kinds of risk: whereas the standard deviations of the two product types reflect their respective idiosyncratic (or independent) risk, their covariance covers the systematic, or common, risk of the portfolio elements.

We can now use these approaches to calculate both the (expected) returns and risk for various portfolios of entertainment products. We have done so in Fig. 5.12—using the real-world ROIs that we calculated for movie genres and their standard deviations. For simplicity, we assume that a producer wants to make ten movies out of a large number of scripts, all of which are all either thrillers or fantasy movies, and that budgets are about the same for all projects; we also assume that the performances of thriller and fantasy movies are uncorrelated.[96] Our data shows that fantasy movies generate higher returns on average than thrillers, but they are also more risky, with their returns varying more than those of thrillers (see Fig. 5.11). Now, which of the portfolios are "better" than others?

As Markowitz (1952) proved analytically in an article that later earned him the Nobel Prize, we can use our figure to separate so-called "efficient" portfolios from "inefficient" portfolios. "Inefficient" portfolios are combinations of risk and return for which other investments exist which are better with regard to *both* criteria. In other words, portfolios are available with

[96]All these assumptions could be released by adding more complexity, but we wanted to keep things simple and relatively straightforward for this illustration.

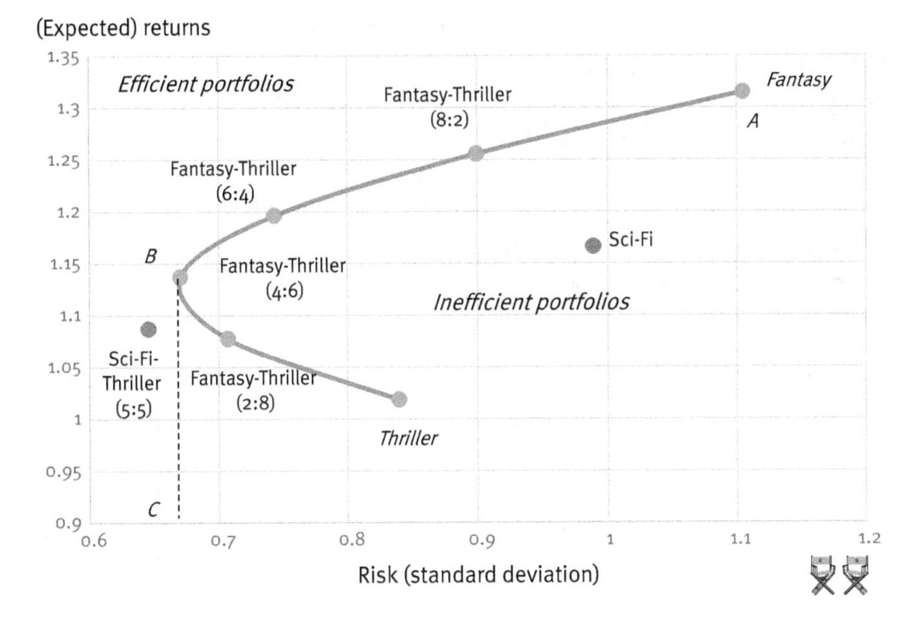

Fig. 5.12 Risk versus expected returns for different movie type portfolios
Notes: Authors' own illustration and calculations. See also our notes to Fig. 5.11.

higher returns for the same level of risk, and portfolios are available with the same returns but with lower levels of risk. In contrast, "efficient" portfolios represent the highest return that can be earned for a given level of risk; to earn higher returns would entail accepting higher risk. Alternatively, for a given level of risk, the efficient portfolio represents the highest return that can be earned; to lower risk would require accepting lower returns.

In Fig. 5.12, every portfolio on the arc from "10-fantasy-movies" to "4-fantasy and 6-thriller-movies" is efficient under the conditions of this example. The arc thus constitutes an "efficiency frontier." In contrast, producing only thrillers or eight thrillers and only two fantasy movies are inefficient alternatives, because higher returns can be expected with different portfolios for comparable levels of risk. More generally, every portfolio below and/or to the right of the line that connects points A, B, and C is inefficient.

Our example is obviously quite restrictive—but we can extend it by adding other portfolios to the figure and judging their attractiveness. For example, producing only science fiction films, instead of thrillers and fantasy, is inefficient, as it is inferior to combinations of thrillers and fantasy (the blue dot for science fiction is in the inefficiency region). However, spending half the budget on science fiction and the other half on thrillers might be an attractive alternative, as this option [the blue "Sci-Fi-Thriller (5:5)" dot] shows less risk for a given level of return and, thus, is outside the inefficiency region. To systematically consider all possibilities that adding the option to also produce a new product type (say, science fiction) brings, the option should be reflected in the generation of a new efficiency frontier that combines all three genres (fantasy, thriller, and science fiction). Let us note again that it is *diversification* that enables such portfolio advantages. As a result, when two product types are perfectly correlated, the arc shown in Fig. 5.12 becomes linear, and every combination of investments along that line is efficient.

Finally, can we also *rank* the efficient portfolios in terms of attractiveness? This can be done quantitatively as long as ranking criteria are clearly defined; however, there is no objectively "right" answer here. In short, the valuation of risk versus return is a tradeoff that lies in the eye of the beholder—namely the respective manager/producer/investor. The ranking criteria of the risk-averse producer would tilt the scales in favor of reduced risk (while accepting lower returns), while the risk-seeker would accept more risk in pursuit of higher returns. Thus, what one can do is to ask an investor to determine the level of risk he or she is willing to take—once this is known, it is possible to focus on efficient portfolios with the highest return for a given maximum level of risk.

And if one could monetize the value of lower risk in terms of ROI, it becomes possible to also determine which efficient portfolio would be the most attractive for the investor. Because risk levels and the value of avoiding risk differ between investors, this "optimal" portfolio would not be same for all investors.

Be Careful, Outsiders: Some Words on Investing in Entertainment Portfolios

> "Studios try to exclude what they think are sure shots and only share risk on things that are not."
>
> —Bruce Berman, *as chairman/CEO of Village Roadshow (quoted in* Fleming *2000)*

If you are an investor from outside the entertainment business who is looking for investment opportunities, let us express a word of caution. Hedge funds have become a part of the entertainment eco-system, but some argue that their financial contributions are systematically disadvantaged by the studios. Epstein (2010), based on interviews with industry executives, argues that studios would have sufficient financial resources to finance their own slates independently. However, they still offer investment opportunities to outsiders when the studio perceives an opportunity to make an "asymmetric deal"—one that allows the studio to generate additional, "abnormal" profits at their financial partners' expense.

Kay Hofmann (2013) provides empirical evidence that is supportive of the existence of asymmetric deals for entertainment. Specifically, he shows that the rise of slate financing in the movie industry from 2006 on has *not* resulted in a higher number of film productions. Instead, Hofmann observes a concomitant reduction in the number of films that the studios financed exclusively, as well as a reduction in projects that the studios co-financed on a per-project basis (see the next section).

Asymmetry in entertainment deals can result from two sources. First, information is asymmetric between studios and external investors: studios have better information than outsiders for judging the financial potential of their projects, and they decide how to allocate resources for marketing these projects (thus constituting some kind of endogenous structure—if one does not allocate sufficient ad spending to a project, it can hardly realize its full potential).[97] For example, external investors are usually not granted participation in a studio's most valuable properties (Owczarski 2012); when Warner Bros. partnered with Legendary Pictures, a film firm backed by $500 million in private equity in 2005, their HARRY POTTER series was excluded from the deal. Also, when both firms ended their financial cooperation some ten years later, it was rumored that keeping Legendary away from Warner's most promising projects, such as the DC Comic movies, was a main factor for the split (Chitwood 2013).

Hofmann (2013) attempts to provide empirical evidence regarding the consequences of such information asymmetry. He uses data for all 235 studio movies released in 2006 and 2007 in North American theaters and compares the financial performance of those 39 films that were part of slate financing deals versus the financial performance of the films that were not. Drawing on rich ROI information (which includes global box-office data, DVD sales, and production and some marketing costs), and controlling for several film elements, Hofmann uses OLS regression to find that the profitability of those slate-funded films is, on average, about 19% *lower* than those of similar films that are produced solely by the studio (or as part of a project-spe-

[97]We discuss the role of advertising for entertainment success in much detail in our chapter on paid entertainment communication.

cific co-financing deal, see next section). Interestingly, he does not find differences in easily observable film characteristics between the slate-funded films and others, which underlines the managerial (and research) challenge of asymmetric information. Although the slate investor might contractually ensure that the slate of projects includes sequels and other promising-looking properties, additional, more subtle facets of the film projects might be only known by the studio.[98]

Second, asymmetry in entertainment often derives from the structure of the deals themselves. Even when external investors get involved in a studio's full slate (and thus cannot be disadvantaged by asymmetric information regarding individual projects), their rate of return is usually not the same as that of the studio. A big reason for these disparate financial results is the industry's special separation of production and "sales."[99] Whereas entertainment studios allow external investors to buy into the production of products, they usually do not do so for the "sales" function. Instead, studios charge a "distribution fee" for their sales efforts, an amount that is usually taken off the revenues that flow back to the producing firms; studios can do this because distribution resources are scarce and valuable. Such fees differ between deals and investors, but in the film industry they rarely fall below 10% (and often climb to 18–20%). For example, when Merrill Lynch invested enormous funds in the production of 26 Paramount studio movies, the studio subtracted a 10% "distribution fee" on Merrill Lynch's share of revenues, making the deal much more lucrative for the studio than for the external investor (Epstein 2010).

In sum, asymmetry exists for those from outside the entertainment industry. Such asymmetry does not mean that such investments do not pay in general, but that they seem to have a higher payoff for players who are *inside* the industry.

Approaches to Manage Risk on the Individual Product Level

In the final section of this chapter on business models for entertainment, we will discuss what producers can do to reduce their risk on the level of an *individual* project (versus by managing slates of products). When shedding

[98]But we also find something positive that slate investors might take from Hofmann's study, particularly those who are more interested in the artistic dimension of the entertainment industry: the slate-funded films in his sample have an above-average chance to be nominated for an Academy Award. This might have to do with those movies' higher risk levels though, as their commercial success requires the hard-to-predict Oscar win.

[99]Remember that entertainment managers, particularly in film, refer to sales as "distribution."

light on this issue, we do *not* discuss which entertainment products may be more risky than others—this topic is something that is tied closely to the specific elements of entertainment products (e.g., being a sequel, having high star power, etc.), and we will investigate it closely in Part II of the book.[100]

Here, we will instead focus on the procedural aspects of producing and financing entertainment—discussing approaches that can help a producer reduce business/financial risk of a new product. Traditional approaches are co-financing and pre-sales; one other, crowdfunding, has garnered attention only recently with an assist from the digital revolution.

Co-Financing of Entertainment

Across entertainment, co-financing is probably the most popular approach for reducing the perceived riskiness of a specific project. Co-financing is an umbrella term that describes various arrangements by which two or more parties share the costs associated with the producing an entertainment product and, in turn, share its revenues.

A number of *Entertainment Science* scholars have empirically investigated the approach's effectiveness, with the clear majority of studies dealing with movies, a form of entertainment for which co-financing (which is sometimes also referred to as "co-ownership" and "equity partnership") has been particularly popular. In their seminal scientific study on entertainment co-financing, Goettler and Leslie (2005) show that Hollywood studios have typically not looked to co-financing as a source of *additional* capital, but mainly as a *substitute* for investing their own money. Among all 1,305 films produced by a major studio and released in North America from 1987 to 2000, about one-third of studio-produced films were co-financed, with the volume trending clearly upward over time. Using probit regression, the scholars' investigation into what types of films are co-financed reveals limited differences: studios partner with another studio when a film's budget is *very* high, but they note no other systematic differences.[101]

For co-financing activities with *external* financers, though, Goettler and Leslie's analyses show that studios prefer certain movie genres (western,

[100]See, for example, our upcoming discussions of "sequel risk" and of "star risk."

[101]The average budget for such inter-studio co-financing is twice that of other studio films! Of course subtle differences might exist for the films which are selected for co-financing (like we have argued in the case of slate financing with external co-financiers), but information asymmetrics will be much harder to establish when the financing partner is another studio who "knows the business."

but not animation and horror), which indicates that the asymmetric information effects we discussed earlier in the context of slate financing also exist here. Palia et al. (2008), for a convenience sample of 275 films by 12 "major companies" (about half of which were co-financed by other studios or external partners), run probit regressions and find that co-financing happens less often for projects which have lower risk (namely PG-rated films and sequels).[102] Their study provides more evidence that film studios reduce their overall risk by co-financing projects that they believe are risky. Hofmann (2013), this time examining all 374 studio movies released from 2003 to 2007, supports these conclusions, also finding that studios tend to finance sequels and PG-rated movies on their own. External partners are invited to invest in dramas instead. And studios seem to be less stingy when it comes to sharing high quality entertainment: as with slate financing deals, Hofmann does not find systematic differences with regard to Oscar nominations and professional reviews.

What do we know about the *performance* of co-financed films? Goettler and Leslie (2005) provide descriptive insights, reporting that whereas revenues of inter-studio co-financed films are, on average, more than two-times those of solo-financed films (which is not a surprise: remember their higher budgets and attractions), the ROI of these films are similar to those of solo-financed films. For their data set, externally co-financed films end up with similar revenues as solo-financed ones, and also a similar ROI. But using a more rigorous statistical approach that filters out alternative influences (namely, OLS regression), Hofmann finds externally co-financed films to be 18% less profitable, on average, than their solo-financed equivalents. This is about the same result he found for slate financing; it seems to be the price that external investors in individual movies pay because of information asymmetry.

Finally, it is important to note that getting a financial partner on board might lower an entertainment producer's financial risk, but can carry problems on its own, such as giving up creative control. Relinquishing creative control is a particular concern for smaller-sized auteurs and "creative" producers in other parts of entertainment than film, where external financing options are much less common. Music "start-up" artists rarely are able to negotiate from a position of power with potential investors such as major labels, but frequently come to regret that they gave up too much of their future revenues, too much of their autonomy, or both, in exchange for

[102]Their risk measures are the standard deviations of the films' ROI.

investments (McDonald 2016). For video games, former Ubisoft producer Suquet (2012) concludes that in the common co-financing model for games, "the balance of power and reward clearly tilts in favor of the business side of the game industry, and not in favor of the creative side" (p. 1).

It is this lack of control that often leads creatives to avoid co-financing with majors. In a study of 349 movies, Fee (2002) finds that "filmmaker-driven" projects (those in which the director also serves as writer and producer) are more often financed independently. Sometimes giving up creative control would be most costly.

(Pre-)Sales Deals

A second means of risk reduction in entertainment is selling certain rights for a product's commercial exploitation. Such rights can be for the coverage of certain regions as well as the usage of certain distributional channels. For example, Hollywood producer Lionsgate sold the rights for its HUNGER GAMES movie's theatrical distribution in Greece to Spentzos Films and for home video releases to Audio Visual Entertainment; in total, the IMDb lists 68 different distributors for the film across channels and regions.

Such deals are made at different points in time, depending on the producers' interests and the project's "pre-sellability." The latter depends on the reputation and track record of the producer and also the characteristics of the project. Projects that feature strong brands have a clear and unique selling proposition (see our discussion of the "high concept" in the integrated marketing strategies chapter), are targeted at a distinct and "receptive" segment of consumers, or are of "high quality" are prime candidates (Follows and Nash 2017). Industry events, such as the American Film Market and the Berlinale's European Film Market for movies and series, are where most deals are made. But with the rise of new distributors such as Netflix and Spotify, their role is challenged, with "interesting IP plus interesting artists going to market every week" (Graham Taylor, co-president of sales agency Endeavor Content, quoted in Goldstein 2017).

Pre-sales deals are those which are made prior to, or during, the production of an entertainment product; they are similar to co-financing agreements. Like co-financing, pre-sales deals reduce the producer's exposure to the volatility of the market performance of a product (and provide the producer with financial leeway for producing other projects), while limiting the producer's earnings (Abrams 2013). The main difference is that buyers of pre-sales rights usually have less influence on the product's creation than do co-financers. Consider the case of TV series Babylon Berlin, which X Filme co-financed with German pay-TV station Sky and broadcaster ARD (who contributed some $10 million and $13 million to the series $45 million budget, respectively). Whereas both co-financing firms executed influence on the final product, this was not the case for buyer Netflix (who secured North American rights early). Pre-sales can be of existential importance: when Luc Besson's Valerian movie flopped heavily, generating global theatrical revenues of only $220 million against production costs of $180 plus advertising and distribution, his EuropaCorp could have been hit just like Intermedia once was hit. But Besson's firm had raised about 90% of the budget via pre-sales, which reduced the losses for the film substantially and let the firm continue its operations, for the time being (Keslassy 2017).

Pre-sales deals are also possible on the slate level: so-called "output deals" are often in place for TV rights of movies when broadcasters purchase the rights for a studio's future productions.[103] A variation of this is when a company obtains the rights to re-sell a producer's titles to theaters in certain territories, just as MGM has recently done for the films produced by Annapurna Pictures' (Wyche 2017).

Crowdfunding Entertainment

A newer approach for sharing the risk of entertainment products is crowdfunding. In crowdfunding, a producer aims to finance his ventures "by drawing on relatively small contributions from a relatively large number of individuals using the [I]nternet" (Mollick 2014, p. 2), without the involvement of standard financial institutions, such as banks or venture capitalists. Platforms such as Kickstarter and Indiegogo provide specialized services for both sides of a crowdfunding arrangement: those interested in getting their products financed and those interested in contributing to the financing of others' products. Incentives for investors are often non-financial, such as privileged access to a product (e.g., receiving a product prior to others, or a customized version, such as a signed DVD) or participation in its creation (e.g., getting to be an extras on a movie set). They can also include a monetary component, though.

---

[103]Such contracts have a long tradition for U.S. pay-TV firms such as HBO and also in many international markets. Although the value of output deals is rarely disclosed, they can be enormous; for example, when German free-TV broadcaster RTL licensed new TV and theatrical productions for 5 years in 2000, the contract was reported to be priced at more than $200 million (Fuchs 2010). And Germany is just *one* country, and free-TV is just *one* channel…

Entertainment is a major arena for crowdfunding. When Mollick (2014) investigated almost 50,000 funding efforts on Kickstarter that were posted in the first three years after the platform's 2009 introduction, he found that almost two-thirds of the projects were films, music, or publishing (games were not listed separately). Examples of crowdfunded entertainment products include film adaptations of TV series (e.g., VERONICA MARS and STROMBERG in Germany), original films (Spike Lee's DA SWEET BLOOD OF JESUS), role-playing games (e.g., SHENMUE III and WASTELAND 2), music projects (e.g., the "final" album by R&B group TLC), and numerous books and comics. Total contributions are usually below $1 million for a single project, but can reach higher in some cases; Spike Lee harvested $1.4 million for his film, and the VERONICA MARS movie attracted more than $5 million from crowdfunding.

The market success of crowdfunded entertainment has varied, as has the list of projects that have used the approach. For example, the VERONICA MARS movie generated a box office of $3.3 million, along with meaningful home entertainment revenues, reportedly exceeding the expectations of Warner Bros. who distributed it. And STROMBERG attracted more than 1.3 million movie-goers in Germany and created a 17% profit for its crowdfunding investors, based on its theatrical performance alone (*Meedia* 2014).

Scientific knowledge on entertainment crowdfunding is, like the approach itself, still in its infancy; most research studies the conditions that contribute to a project's successful funding, rather than the product's subsequent market performance. In his comprehensive analysis of Kickstarter projects, Mollick (2014) used logistic regression to study the determinants of whether projects received their desired amount of funding. He finds that the producer's network size (which he measured as the producer's Facebook friends) is a main predictor, as are quality signals on the site, such as videos, updates, and the lack of spelling errors.

Using a database of 500 randomly selected Kickstarter projects and, again, logistic regression, Marelli and Ordanini's (2016) results confirm the critical role of network size and quality. But they also highlight other success factors: projects are more likely to win investors if a project is initiated by producers with a solid crowdfunding history, if they provide incentives for early investors, if they are to be released in the near (rather than distant) future, and if they are of a predominantly artistic, rather than commercial, nature.

These findings stress the "fan-culture" of crowdfunding: crowdfunding investors seem to be in it for fun and ambition rather than (just) for any financial takeaway. These investors are perhaps more accurately thought of as a social community rather than as an anonymous market. This is where a strong reputation, or brand, also matters: if a project (or its producer) has a powerful background that is consistent with this social and artistic focus, it can help attract the crowd's attention and can signal quality. Consequently, for independent artists with an installed fan base, crowdfunding is of major appeal for financing and risk reduction.

But despite the limited amounts compiled, the approach might still be of use for major labels. As Gamble et al. (2016) reveal through a series of qualitative interviews with music industry managers and artists, crowdfunding might be more valuable for increasing marketing effectiveness by building early interest in the lead-up to the release of a new musical production, to research the market, and as a tool for pre-selling a product, rather than to acquire financial resources and to manage project risk.

Concluding Comments

Digital technologies and new players are changing the business side of entertainment and how entertainment content generates value for those who produce it. Our entertainment-value chain links production with distribution activities and the consumer, defining the "arena" in which all entertainment business decisions (and changes) take place, the main distribution modes and intermediaries, along with an overview of the core strategic transformational moves which entertainment firms can apply. We provided an overview of today's key players in entertainment, their current activities and business portfolios, and the main approaches that they can apply for generating revenues and managing risk.

Despite current dynamics, entertainment firms will continue to look to revenues from consumers and advertisers. Our discussion highlights that, whereas advertising offers enormous financial resources for at least some producers, balancing these sources requires managerial sensitivity to avoid destructive negative feedback effects. When it comes to managing the risk of entertainment products, we showed that at the individual-product level, the array of solutions involves crafting strategies for sharing risks with others. At the slate level, firms can benefit from understanding and applying insights from financial portfolio theory to effectively mitigate risks through diversification based on knowledge on the differences in the riskiness and revenue potential across product types. Our discussion provides detailed guidance and recommendations.

On the following pages, we conclude Part I of this book with an examination of entertainment consumers. Our value-creation framework has made clear that it is those consumers who ultimately determine whether a product will be a hit—or not. An understanding of consumers' feelings and thoughts that lead them to spend time and money with films, games, books, and songs is the final piece of the foundational knowledge that we will then build upon in Part II of the book. There, the task is to learn from scholarly insights which marketing practices work better than others in an entertainment context.

References

Abrams, R. (2013). Even Hollywood studios that churn out hits use pre-sales to minimize downside. *Variety*, August 20, https://goo.gl/WqnEJv.

Amazon (2016). Kindle store terms of use. Retrieved October 19, 2016, from https://goo.gl/76aC83.

Alter, A. (2015). The plot twist: E-book sales slip, and print is far from dead. *The New York Times*, September 23, https://goo.gl/2kpyG1.

Anderson, P. (2016). Glimpses of the US market: Charts from Nielsen's Kempton Mooney. *Publishing Perspectives*, May 20, https://goo.gl/1Kc8YF.

Bagella, M., & Becchetti, L. (1999). The determinants of motion picture box office performance: Evidence from movies produced in Italy. *Journal of Cultural Economics, 23*, 237–256.

Bart, P. (2017). Peter Bart: Amazon raises bet on movie business, but rivals still baffled about long-term strategy. *Deadline*, March 10, https://goo.gl/t46Vvj.

Berk, J., & DeMarzo, P. (2014). *Corporate finance* (3rd ed.). Boston: Pearson.

Blickpunkt:Film (2003). Moritz Borman über die neue Intermedia-Strategie, March 24, https://goo.gl/zQEtRr.

Blickpunkt:Film (2016). Deutschland wird zunehmend abgehängt, November 8, https://goo.gl/cV4KX7.

Brehm, J. W. (1966). *A theory of psychological reactance*. New York: Academic Press.

Brodkin, J. (2016). Netflix finishes its massive migration to the Amazon cloud, February 11, https://goo.gl/rwos6H.

Cachon, G., & Lariviere, M. A. (2005). Supply chain coordination with revenue-sharing contracts: Strengths and limitations. *Management Science, 51*, 30–44.

Caves, R. E. (2000). *Creative industries: Contracts between art and commerce*. Cambridge, MA: Harvard University Press.

Castillo, M. (2017). Disney will pull its movies from Netflix and start its own streaming services. *CNBC*, August 8, https://goo.gl/Egu4mZ.

Chisholm, D. C. (1997). Profit-sharing versus fixed-payment contracts: Evidence from the motion pictures industry. *Journal of Law Economics and Organization, 13*, 169–201.

Chisholm, D. C. (2004). Two-part share contracts, risk, and the life cycle of stars: Some empirical results from motion picture contracts. *Journal of Cultural Economics, 28*, 37–56.

Chitwood, A. (2013). Warner Bros. and Legendary Pictures to part ways. *Collider*, June 25, https://goo.gl/wQSH1S.

Chmielewski, D. C., & Hayes, D. (2017). Comcast Fox deal talks latest entry into media merger mania. *Deadline*, November 16, https://goo.gl/EQCWK6.

Coen, B. (2017). Amazon ramps up content spending in battle with Netflix. *The Street*, July 28, https://goo.gl/ptKaga.

Cowen, N., & Patience, H. (2008). Wheels on film: Matrix Reloaded. *The Telegraph*, August 23, https://goo.gl/scAvZY.

Dana Jr., J. D., & Spier, K. E. (2001). Revenue sharing and vertical control in the video rental industry. *Journal of Industrial Economics, 49*, 223–245.

DEG (2017). 2016 Home entertainment report. *The Digital Entertainment Group*, available via https://goo.gl/KyidSx.

Dillet, R. (2016). Spotify launches new series of original recordings called Spotify Singles. *Techcrunch*, December 1, https://goo.gl/RzMibZ.

Dyce, A. (2015). The 15 worst video game movies. *Screenrant*, March 12, https://goo.gl/844iY4.

Epstein, E. J. (2010). *The Hollywood economist—The hidden financial reality behind the movies*. Brooklyn: MelvilleHouse.

Fee, E. C. (2002). The costs of outside equity control: Evidence from motion picture financing decisions. *The Journal of Business, 75*, 681–711.

Feldman, D. (2017). Netflix remains ahead of Amazon and Hulu with 128 M viewers expected this year. *Forbes*, April 13, https://goo.gl/93xn1n.

Filson, D., Switzer, D., & Besocke, P. (2005). At the movies: The economics of exhibition contracts. *Economic Inquiry, 43*, 354–369.

Filson, D., Besocke, P., & Switzer, D. (2007). Coming soon to a theater near you? A proposal for simplifying movie exhibition contracts. Working Paper, Claremont McKenna College.

Finke, N. (2012). 'Battleship' recruits $50 M in promotional partnerships: Universal's big bet paying off. *Deadline*, February 6, https://goo.gl/WS4QRr.

Fleming, M. (2000). Village people. *Variety*, February 3, https://goo.gl/MwpZQq.

Follows, S., & Nash, B. (2017). Update: What types of low budget films break out? *American Film Market*, https://goo.gl/CsiDQg.

Forbes (2016). Why AT&T is buying Time Warner, October 24, https://goo.gl/FRbCQV.

Fritz, B., & Schwartzel, E. (2017). Hollywood's misses are hits overseas. *The Wall Street Journal*, June 25, https://goo.gl/RFq956.

Frontline (2001). The monster that ate Hollywood. Transcript of PBS program aired November 22, https://goo.gl/QXCufm.
Fuchs, S. (2010). *Spielfilme im Fernsehen*. Lohmar: Eul.
Gamble, J. R., Brennan, M., & McAdam, R. (2016). A rewarding experience? Exploring how crowd funding is affecting music industry business models. *Journal of Business Research, 70*, 25–36.
Garrahan, M., & Fontanella-Khan, J. (2016). Apple executive proposed bid for Time Warner. *Financial Times*, May 26, https://goo.gl/Zj5Rs4.
Gersema, E. (2016). Lights, camera and no action: How state film subsidies fail. *Press Room*, University of Southern California, August 18, https://goo.gl/yWrMNJ.
Goettler, R. L., & Leslie, P. (2005). Cofinancing to manage risk in the motion picture industry. *Journal of Economics & Management Strategy, 14*, 231–261.
Goldstein, G. (2017). AFM: Netflix, Amazon part of forces shaping indie business models. *Variety*, October 31, https://goo.gl/rHukCC.
Guerrasio, J. (2017). Disney's requirements for the new 'Star Wars' movie have angered some movie theaters. *Business Insider*, November 1, https://goo.gl/WuVyF7.
Hayes, T. C. (1982). Coke expected to acquire Columbia pictures. *The New York Times*, January 19, https://goo.gl/n1XGc3.
Hennig-Thurau, T., Hofacker, C. F., Bloching, B. (2013). Marketing the pinball way: Understanding how social media change the generation of value for consumers and companies. *Journal of Interactive Marketing, 27*, 237–241.
Hofmann, K. H. (2013). *Co-financing Hollywood film productions with outside investors: An economic analysis of principal agent relationships in the U.S. motion picture industry*. Wiesbaden: Springer Gabler.
Homer, P. M. (2009). Product placements. *Journal of Advertising, 38*, 21–31.
Ifpi (2017). Global music report 2017. https://goo.gl/WSPh8U.
Information is beautiful (2015). Selling out – How much do music artists earn online? April 10, https://goo.gl/hS36gS.
Ingham, T. (2015). Apple Music is a terrible disaster. Apple Music is a storming success. *Music Business Worldwide*, August 6, https://goo.gl/ivkSmx.
Ingham, T. (2016). Spotify is out of contract with all three major labels—and wants to pay them less. *Music Business Worldwide*, August 22, https://goo.gl/38VjAP.
Instavest (2015). Sponsoring James Bond: 007's branding sweepstakes, November 2, https://goo.gl/tWNfzC.
Jackson, J. (2017). Could the Call of Duty franchise be the next Marvel? *The Guardian*, April 5, https://goo.gl/BkK8Cp.
Jansen, C. (2005). The performance of German motion pictures, profits and subsidies: Some empirical evidence. *Journal of Cultural Economics, 29*, 191–212.
Jourdan, J., & Kivleniece, I. (2017). Too much of a good thing? The dual effect of public sponsorship on organizational performance. *Academy of Management Review*, 60, 55–77.

Karp, H. (2017). In a first, streaming generated the bulk of '16 music sales. *The Wall Street Journal*, March 30, https://goo.gl/Bx5gU8.

Keslassy, E. (2017). Luc Besson, EuropaCorp face day of reckoning with shareholders over flop of 'Valerian'. *Variety*, September 21, https://goo.gl/gbmsJt.

Koblin, J. (2017). Netflix says it will spend up to $8 billion on content next year. *The New York Times*, October 16, https://goo.gl/st7WPp.

Kotler, P., Wong, V., Saunders, J., & Armstrong, G. (2005). *Principles of marketing* (4th European ed.). Harlow: Pearson.

Knowledge@Wharton (2000). Now showing at Blockbuster: How revenue-sharing contracts improve supply chain performance, October 16, https://goo.gl/at3fCt.

Knowledge@Wharton (2007). Investing in the fragmented entertainment industry: Is safe better than sexy? May 2, https://goo.gl/jzddHM.

Kupferschmitt, T. (2017). Onlinevideo: Gesamtreichweite stagniert, aber Streamingdienste punkten mit Fiction bei Jüngeren. *Media Perspektiven, 47*, 447–462.

Lev-Ram, M. (2016). How Netflix became Hollywood's frenemy. *Fortune*, June 7, https://goo.gl/JAFP3P.

Lieberman, D. (2017). Hollywood bear: Why media analyst is gloomy about the movies. *Deadline*, March 17, https://goo.gl/HanNbM.

Lynch, D. (2007). *Catching the big fish: Meditation, consciousness, and creativity*. New York: Jeremy P. Tarcher.

Madigan, J. (2016). *Getting Gamers. The psychology of video games and their impact on the people who play them*. Lanham: Rowman & Littlefield.

Marchand, A., Hennig-Thurau, T., & Best, S. (2015). When James Bond shows off his Omega: Does product placement affect its media host? *European Journal of Marketing, 49*, 1666–1685.

Marelli, A., & Ordanini, A. (2016). What makes crowdfunding projects successful 'before' and 'during' the campaign? In D. Brüntje & O. Gajda (Eds.), *Crowdfunding in Europe* (pp. 175–192). Springer Science + Business Media: Cham.

Markowitz, H. (1952). Portfolio selection. *Journal of Finance, 7*, 77–91.

Martin, G. R. R. [@GeorgeRRMartin_] (2015). A reader lives a thousand lives before he dies, the man who never reads lives only one. #WorldBookDay. https://goo.gl/TJwsuk.

McDonald, H. (2016). Music industry investors. *The Balance*, October 14, https://goo.gl/dL9mnT.

McKenzie, J., & David Walls, W. (2013). Australian films at the Australian box office: Performance, distribution, and subsidies. *Journal of Cultural Economics, 37*, 247–269.

McKinsey & Company (2015). Global media report 2015, July, https://goo.gl/Wd7n5a.

Meedia (2014). Knapp 170.000 Euro Gewinn: So viel Geld bekommen die „Stromberg" Crowdfunder. *Meedia*, October 16, https://goo.gl/ryzp2u.

Meloni, G., Paolini, D., & Pulina, M. (2015). The great beauty: Public subsidies in the Italian movie industry. *Italian Economic Journal, 1*, 445–455.
Meyer, J., Song, R., & Ha, K. (2016). The effect of product placements on the evaluation of movies. *European Journal of Marketing, 50*, 530–549.
Mol, J. M., Wijnberg, N. M., & Carroll, C. (2005). Value chain envy: Explaining new entry and vertical integration in popular music. *Journal of Management Studies, 42*, 251–276.
Mollick, E. (2014). The dynamics of crowdfunding: An exploratory study. *Journal of Business Venturing, 29*, 1–16.
Mosendz, P. (2014). Amazon has basically no competition among online booksellers. *The Atlantic*, May 30, https://goo.gl/uDQjGU.
Negro, G., & Sorenson, O. (2006). The competitive dynamics of vertical integration: Evidence from U.S. motion picture producers, 1912–1970. *Advances in Strategic Management, 23*, 367–403.
Newzoo (2017). The global games market will reach $108.9 billion in 2017 with mobile taking 42%. *Newzoo website*, April 20, https://goo.gl/vBRLMy.
Otterson, J. (2017). 'Karate Kid' sequel series with Ralph Macchio, William Zabka greenlit at YouTube Red. *Variety*, August 4, https://goo.gl/xHJpZN.
Owczarski, K. (2012). Becoming Legendary: Slate financing and Hollywood studio partnership in contemporary filmmaking. *Spectator, 32*, 50-–59.
Owczarski, K. (2017). A very significant Chinese component': Securing the success of Transformers: Age of Extinction in China. *The Journal of Popular Culture, 50*, 490–513.
Palia, D., Abraham Ravid, S., & Reisel, N. (2008). Choosing to cofinance: Analysis of project-specific alliances in the movie industry. *The Review of Financial Studies, 21*, 483–511.
Patel, S. (2017). Inside Facebook's pitch for entertainment content. *Digiday UK*, March 3, https://goo.gl/w4ciRP.
Pierce, J. R., & Aguinis, H. (2013). The too-much-of-a-good-thing effect in management. *Journal of Management, 39*, 313–338.
Posener, A. (2014). Die bittere Bilanz der deutschen Filmförderung. *Die Welt*, November 17, https://goo.gl/p4UNgX.
Powers, L. (2010). Time Warner's Jeff Bewkes: Netflix is no threat to media companies. *Hollywood Reporter*, December 13, https://goo.gl/4LvWYm.
PQ Media (2015). Double-digit surge in product placement spend in 2014 fuels higher global branded entertainment growth as media integrations & consumer events combo for $73.3B, March 13, https://goo.gl/gWhBWj.
PwC (2014). Global radio industry revenue from 2013 to 2018 (in billion U.S. dollars). In Statista - The Statistics Portal, https://goo.gl/a2bfu8.
PwC (2015). Distribution of radio industry revenue worldwide from 2010 to 2019, by source. In Statista - The Statistics Portal, https://goo.gl/5WqBLz.
Rahman, A. (2017). Has pandering to Chinese audiences hurt 'Transformers 5'? *The Hollywood Reporter*, July 2, https://goo.gl/e1pcDY.

Revilla, A. (2015). James Bond turned down Sony's $5 M offer to use an Xperia Z4 because he only uses 'the best'. *Digital Trends*, April 24, https://goo.gl/Pzv5iY.
RIAA (2017). U.S. sales database. https://goo.gl/z9Vf1i.
Rosoff, M. (2017). Jeff Bezos has advice for the news business: 'Ask people to pay. They will pay'. *CNBC*, June 21, https://goo.gl/6osLb1.
Sandberg, B. E. (2016). Film and TV tax incentives: A state-by-state guide. *Hollywood Reporter*, April 21, https://goo.gl/DLwzBU.
Sanchez, D. (2017). Apple music finds a sneaky way around the 'album exclusive'. *Digital Music News*, July 21, https://goo.gl/aASrA2.
Schrodt, P. (2015). Even Hollywood insiders are completely clueless about Netflix's streaming numbers. *Business Insider*, December 2, https://goo.gl/BsrnVX.
Sharma, A. (2016). AT&T-Time Warner deal is mostly about defense. *The Wall Street Journal*, October 23, https://goo.gl/z3cZti.
Singleton, M. (2015). This was Sony Music's contract with Spotify. *The Verge*, May 19, https://goo.gl/jDxm1M.
Sisario, B. (2017). Now on stage: The countdown to a new Taylor Swift album. *The New York Times*, September 3, https://goo.gl/id9SLZ.
Spangler, T. (2017). Apple sets $1 billion budget for original TV shows, movies (Report). *Variety*, August 16, https://goo.gl/YsJiDj.
Statista/Audiencenet (2015). Devices used by teenagers for audio content consumption in the United States as of July 2015. https://goo.gl/w4YJ9P.
Suquet, Y. (2012). Can film-inspired project financing work for games? *Gamasutra.com*, September 20, https://goo.gl/1Gy46w.
The Economist (2016). Cutting the cord, July 16, https://goo.gl/BKU9wp.
The Economist (2017). Traditional TV's surprising staying power, February 9, https://goo.gl/RKcnid.
Thom, M. (2016). Lights, camera, but no action? Tax and economic development lessons from state motion picture incentive programs. *American Review of Public Administration*, 1–23.
Tostado, K. (2013). Answer to thread "How do movie directors/producers/studios get paid for streaming on Netflix?" *Quora.com*, March 3, https://goo.gl/rFMgaF.
Verberckmoes, S., Poels, K, Dens, N., Herrewijn, L., & de Pelsmacker, P. (2016). When and why is perceived congruity important for in-game advertising in fantasy games? *Computers in Human Behavior, 64*, 871–880.
VGChartz (2017). Global unit sales of current generation video game consoles from 2008 to 2016 (in Million Units). *Statista - The Statistics Portal*, https://goo.gl/FL1rbQ.
Walter, G. A., & Barney, J. B. (1990). Management objectives in mergers and acquisitions. *Strategic Management Journal, 11*, 79–86.
Wang, Y., Jiang, L., & Shen, Z. J. (2004). Channel performance under consignment contract with revenue sharing. *Management Science, 50*, 34–47.

Wyche, E. (2017). MGM, Annapurna strike output deal. *Screen Daily*, March 27, https://goo.gl/GZqhMP.

Zhou, W. (2004). The choice of commercial breaks in television programs: The number, length, and timing. *The Journal of Industrial Economics, 52*, 315–326.

6

The Consumption Side of Entertainment

We have stressed early on in this book that the hedonic nature marks a key characteristic of entertainment products. Our discussion of entertainment consumption in this chapter builds on this unique nature which not only describes the specific products that entertain us, but also the fundamental human needs and processes that give them meaning for us as consumers.

On the following pages, we transform the fundamental insights on hedonic consumption and its key facets of emotions and imagery into a more holistic, multi-layered framework of entertainment consumption. The framework follows a "means-end" logic, tracking the link from a product's attractions to the pleasure they provide a consumer. We label it the "sensations-familiarity" framework, as it assigns the sensations that consumers perceive in an entertainment product along with the product's familiarity focal roles for this transformation process.

We will now first overview the "sensations-familiarity" framework, explaining how sensations and familiarity help the entertainment product's "objective" elements to create pleasure, as the "end state" that is usually desired by an entertainment consumer. We then take a deeper look at the emotional and cognitive processes that are triggered by such sensations and familiarity. We end the chapter (and Part I of this book along with it) with an analysis of the *process* of entertainment consumption, disentangling its different stages.

T. Hennig-Thurau and M. B. Houston, *Entertainment Science*,
https://doi.org/10.1007/978-3-319-89292-4_6

Why We Love to Be Entertained: The Sensations-Familiarity Framework of Entertainment Consumption

When we introduced the notion of entertainment products being hedonic, we argued that experiencing a pleasure state is the main aim for consumers spending time, and often money, for entertainment products. We cited scientific research that makes it clear that entertainment consumption can lead to pleasure by activating two different areas of our mind: by triggering emotions, but also by activating cognitive processes, the latter often in the form of what psychologists often call "imagery."

Let us now refine this perspective by adding some more psychological layers. Figure 6.1 extends our previous model of hedonic consumption into a full hierarchical framework of entertainment consumption. With pleasure

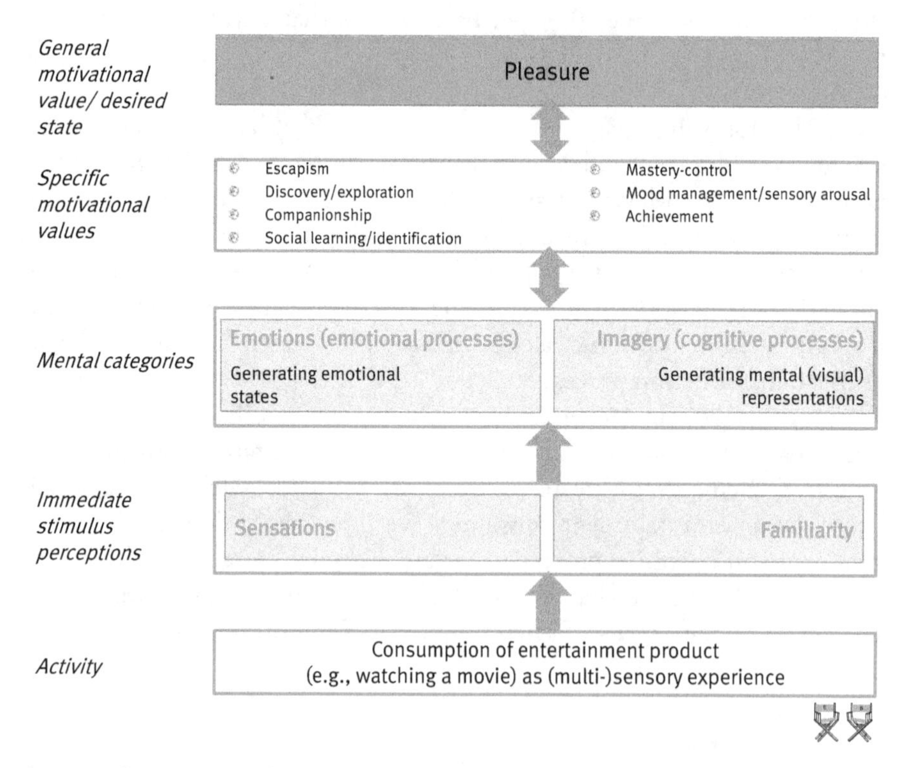

Fig. 6.1 The sensations-familiarity framework of entertainment consumption
Note: Authors' own illustration.

being the desired end state and the ultimate reason for entertainment consumption, let us start from there, the framework's top layer.[104]

A central argument here is that pleasure does not *immediately* result from emotional or cognitive processes. Instead, pleasure is perceived by a consumer when one, or many, consumption motives are fulfilled by the consumption activity. As we discuss below, whereas pleasure is the highest, most general motivational value that consumers seek through entertainment, it is a broad and pretty abstract concept; scholars have identified several more concrete and specific motivational values, or motives, that drive entertainment consumption. One example specific motive is escapism—consumers trying to (temporarily) trade their realities and routines for those of the entertainment experience, such as the galaxies of the Star Wars saga (e.g., Henning and Vorderer 2001). If a consumer strives for such escapism, then experiencing it is the path to experience pleasure, as the ultimately desired state.

How can entertainment products fulfill such specific consumer motivations? This is where emotions and imagery cognitions matter. Both are the key mental consumer categories (Hirschman and Holbrook 1982) that must be activated by the entertainment product if it is to connect to consumers' motivations. All entertainment motivations require a certain amount of *both* emotional and cognitive processing. The relative importance can vary though—whereas one category can be important for certain motives (such as escapism), the other may be more crucial for other motives.

The layer of the framework that precedes emotional and cognitive processes, and that also most immediately follows the consumption activity, is the one after which we name the framework: sensations and familiarity are the central concepts for understanding consumers' reactions to entertainment (Bohnenkamp et al. 2015). Why is this layer so essential? For the

[104]We have to concede that, although pleasure is the desired state behind almost all kinds of entertainment consumption, it can be complemented by another fundamental motivation: a purely social one. Lee and Lee (1995) find that TV viewing can be driven by people's interest in talking with others about the program, and our own results (from Pähler vor der Holte and Hennig-Thurau 2016) show that the ability to chat with others during and after watching a new drama series is a driving force for consumption. Similarly, Schäfer and Sedlmeier (2009) show that music provides "the opportunity to meet other people," and Yee (2006) finds for games (where the product itself can connect consumers) that socializing and being with other MMOG players are influential motivations. In all these cases, the value of consuming an entertainment product is not solely determined by the pleasure the product itself provides, but the product rather serves as a vehicle for experiencing fundamental social needs. It is entertainment's symbolic nature that makes it so well-suited to offer not only pleasure, but also to connect people. Whereas an entertainment producer can certainly gain from providing social benefits, it is not specific for entertainment, and interested readers are recommended to the extensive general literature on social motivation and needs, starting with Maslow (1943).

framework to be useful for entertainment producers, it must shed light on which kinds of product stimuli trigger emotional reactions and the creation of imagery in the mind of the consumer. Simply telling a producer that his or her product must generate a satisfactory level of emotions and imagery to result in consumer pleasure would be too vague of advice to be of any practical value.

Sensations and familiarity, which we define and discuss in detail below, are consumers' immediate perceptions of any *key attractions* provided by an entertainment stimulus. In turn, sensations and familiarity determine, either in isolation or together, how the entertainment product impacts the consumers' mental categories of emotions and imagery. Because we expect producers of entertainment to be able to estimate or predict the degree to which their products will offer consumers sensations and familiarity, including these factors adds usability to the framework.

Let us note that the hierarchical structure of our framework also carries meaning for producers by itself: it informs us that neither familiarity nor sensations are self-sufficing, but only lead to success if they manage to trigger the next layer of the framework, namely emotions and imagery, which requires them to be linked to consumer motives. This is why sensational explosions do not guarantee enjoyment, but can be numbing instead. As Tom Rothman, as Sony executive, noted: "[m]aking the audience care is a lot harder than making things blow up" (quoted in Ford et al. 2017). The hierarchy of the sensations-familiarity framework explains why doing the former is essential for offering pleasure.

In summary, our framework implies that the degree of pleasure a consumer experiences from consuming an entertainment product is the result of a multi-layered process. Pleasure can only result when a product first offers a sufficient level of sensations and/or familiarity ("key attractions"), which are perceived as such by the consumer. Sensations and familiarity, then, can trigger emotional and cognitive processes that are essential for fulfilling the specific motivational values that drive consumption. Pleasure only emerges if those desired specific motivations are fulfilled.[105]

We also want to stress that the links between the different levels of our framework are not one-directional; instead, feedback loops can exist (e.g., the fulfillment of a motivational value can intensify the emotional reactions via cognitive appraisals), and the overall process can be triggered from both

[105]In particular cases in which the consumer aims for mood adjustments rather than pleasure, however, it is possible for the emotional reaction to function as a motivation itself, as we will show below.

the top (the consumer's desire to experience pleasure) and the bottom (by turning on the radio or TV). We also need to keep in mind that whereas the framework describes what happens when we encounter an entertainment product, our decision whether to spend time and/or money for a certain product usually happens before (entertainment products are dominated by experience qualities)—which means that our *anticipation* of what we can expect from a product in terms of sensations and familiarity, of emotions and imagery, of motive fulfillment and pleasure also matters, and does so big time.

In the following, we dive into the key concepts of the framework. We start with an overview of what leads humans to consume entertainment (the specific motivations or motivational values) and what we mean when we describe sensations and familiarity as key entertainment attractions. We then turn our attention to the "heart and mind" of entertainment consumption: the emotions and the imagery triggered by entertainment's sensations and familiarity, which, at the end of the day, determine our reaction to experiencing it.

The Specific Motivational Values that Lead Us to Consume Entertainment

"I sell escapism."
—*Musician* Jimmy Buffett *(quoted in* Leung *2004)*

Trying to understand what inner powers make consumers indulge in entertainment, beyond recognizing just the general motivating desire for pleasure and enjoyment, has kept *Entertainment Science* scholars busy.[106] As is the case for research on consumer motivations in general, structuring the motivations for entertainment consumption is a far-from-trivial task. Whereas boundaries between motivations and other psychological concepts (such as

[106]An important substream of such motivation-focused research carries the label "uses and gratifications." Its roots go back to the early days of radio and television where the uses-and-gratifications approach was developed to understand people's engagement with mass communication; Katz et al. (1973) provide an early overview. Whereas our discussion of entertainment motives in this part of the book encompasses findings from a number of uses-and-gratifications studies, our perspective differs somewhat in that we do not assume entertainment consumers make choices actively to achieve a consciously chosen "goal" (an important tenet of uses-and-gratifications research that has also inspired models of consumer decision making; see Palmgreen and Rayburn II 1982). In contrast, we also allow for subconscious, passive consumer behavior. In a pointed way, the uses-and-gratifications approach is tied to a "rational," heavily cognitive view of consumer behavior, which reflects the approach's historical roots; the approach was developed long before the hedonic consumption models, on which our own perspective of entertainment behavior in this book is based, shifted the scholarly view toward emotions and imagery.

feelings, attitudes, processes, behaviors, or states) are fuzzy and definitions of the motivational concepts are not consistent, a number of key motives have crystallized from the research. Some of them are relevant *across* different forms of entertainment, while others have been more closely tied to specific content.[107]

Escapism. Among the most often-cited reasons for consumers to engage in entertainment is the desire to escape something unpleasant or worrisome, to get away from problems and pressures. This escapism motivation can relate to a person's immediate social environment, his or her general work and life situation (and (dis)satisfaction with the same), or the sense of emptiness that is perceived when there is nothing to do (Henning and Vorderer 2001). Ernest Cline's literary alter ego Wade Watts practices escapism in his Ready Player One dystopic world by logging into the "OASIS," a fictional MMOG, because it allows him "to instantly slip [his worries] away as [his] mind focused itself on the relentless pixelated onslaught on the screen" (Cline 2011, p. 14), and the Saturday audience of Billy Joel's Piano Man enjoys his music because it helps them to not think about their lives for some hours. Escapism has been empirically linked by scholars to various forms of narrative entertainment, including television, film, and novels (e.g., Hirschman 1987 in a study of 364 behavioral science students). It is found to be the best predictor of gaming intensity (in Yee's 2006 survey of 3,000 players of MMOGs).

Discovery and exploration. Consumers also spend time with entertainment to explore and discover "worlds" that differ from their everyday environments. Such exploration is not driven by real-life misery, but by consumers' curiosity to discover something new and inspiring. Writer Almond (2006, p. VII) describes it as opening "the gates to an unknown city" or "the lid of a treasure chest." Empirically, Addis and Holbrook (2010) study all 440 movies nominated for a Best Picture Oscar between 1927 and 2003 and find higher consumer ratings for movies that take place in a setting that consumers have not experienced personally. In his MMOG study, Yee (2006) finds that joy of discovery and desire to role-play are primary gaming motives, above and beyond escapism. The discovery and exploration motive is closely linked to mental states that psychologists refer to as transportation and immersion.

[107]Let us note that our list of entertainment motives, although including what we believe are the focal internal drivers for consumers, is far from comprehensive. Other motives mentioned by researchers are "moral disposition" (i.e., experiencing the good prevailing in the movie, and the bad suffering) and "social comparison" (looking at others, such as characters in a novel, who have it worse off than you do). Bartsch and Viehoff (2010) offer an overview.

Companionship and other relationship functions. People also consume entertainment because entertainment products provide a way to get emotionally involved with the characters, either the ones *in* the product (such as the heroine of a novel) or those *behind* it (such as the lead actor of a movie). Through these "relationships," consumers can experience profound affection and sensitivity to others' feelings (Hirschman 1987). Alan Rubin (1981), based on a survey of 626 consumers, stresses the contributions of entertainment content for companionship, finding that watching movies or listening to songs makes people feel less lonely. And he shows that such companionship has a significant influence on the amount of TV that people watch. Hirschman (1987) extends this finding to movies and novels. The companionship motive also provides the basis for our understanding of entertainment stars as "parasocial relationship partners."

Social learning and self-learning. Only rarely do people consume entertainment products solely for learning "facts." However, particularly for narrative forms of entertainment, consumption can often be driven by peoples' motivation for *social* learning. Social learning is possible because entertainment products permit consumers to self-project into, or to identify with, a particular role or character (Hirschman 1983). The exact nature of the social learning motive spans a broad continuum; it ranges from concrete and pragmatic to abstract and fundamental.

Pragmatic social learning means that consumers observe how others (e.g., movie characters) deal with challenges that the consumer considers to be of potential personal relevance. Think about watching CAST AWAY as a survival guide or SILVER LININGS PLAYBOOK as the parent of a young adult with a mental disorder. More fundamental social learning happens when consumers find role models and heroes who help the consumer visualize an aspirational self—who I really want to be—in entertainment content. The coolness of a James Bond or Eastwood's Man With No Name gives assertiveness to a self-doubting boy; HARRY POTTER's Neville Longbottom offers the ability to stand up to a bully; Katniss Everdeen, Jennifer Lawrence's HUNGER GAMES heroine, inspires an adolescent girl to be brave and daring.[108] Such social learning is not limited to narrative entertainment—Schäfer and Sedlmeier (2009) find, based on a survey of 507 German consumers, that some of the most important motivations for consuming music deal with issues related to the self, with music being an embodiment of one's identity and values.

[108]This motive can be linked to the personal relevance of entertainment.

Mastery-control. Psychologists have long argued that consumers derive value out of the ability to be in control of a situation, as it enables us to make autonomous decisions and manipulate the outcome (e.g., Ryan and Deci 2000). *Entertainment Science* scholars have adopted this logic to explain the use of entertainment products. Hirschman (1987) provides evidence that consumers' mastery-control motivation is correlated with choices of books, television content, and movies—despite the fact that these are *non-interactive* forms of entertainment through which the control motive can only be fulfilled in an imaginary, fantasized manner (Mansell 1980). Ryan et al. (2006) show that (self-)control perceptions are of particular importance for users of *interactive* games in which consumers can actually determine the course of the experience with their actions; their sample consists of 730 members of an online community. In gaming, the fulfilment of the control motivation is closely tied to the consumer's experience of a "flow" state.

Mood management and sensory arousal. Consumers also spend time with entertainment to regulate their moods (Zillmann 1988). According to mood management theory, consumers use entertainment products as a source of external stimulation. By consuming exciting content, consumers can increase their arousal level and escape a "bad" mood state that had been present because of "under-stimulation" (i.e., boredom). Bad mood can also result of "over-stimulation" (or stress), a constellation which "soothing" entertainment content can improve by reducing arousal. But the right entertainment product can also further strengthen an already existing good mood (Bartsch and Viehoff 2010). Empirical evidence for mood management exists for music consumption (Schäfer and Sedlmeier 2009), as well as for consumers' TV viewing patterns (e.g., a survey of almost 2,000 viewers by Lee and Lee 1995, and also Hirschman 1987), and for movie and book preferences (Hirschman 1987).

Achievement. For video games, in which consumers play a very active (versus observational) role, personal achievement has been highlighted as another influential motivation. *Entertainment Science* scholars have collected evidence that gamers are often driven by a strong desire to have a high level of competency (i.e., "be good at") when playing. Achievement can be measured with regards to absolute criteria, such as the advancement in the game, as well as to relative criteria, such as performing better than others (Coursaris et al. 2016, based on survey data for 202 gamers; see also Ryan et al. 2006). The achievement motivation is closely tied to the psychological state of flow, which itself depends on consumer skills.

Our discussion of entertainment motivations so far has shown that motives vary, to a certain extent, between the different forms of entertainment: people might play games for other reasons than they watch TV or listen to music. However, Hirschman's (1987) study puts such variations in a different light: she shows that it is also the *genre* of an entertainment product that determines the level of influence of a particular motive across forms of entertainment. For example, she finds that whereas a mastery-control motivation plays little role in the choice to consume either comedy books or movies, this motive plays a strong role in consumers' preferences for *both* erotic books and erotic movies. And Hirschman shows that entertainment motivations differ substantially based on one additional factor: the consumer's gender. Whereas men enjoy science-fiction movies and history novels for their social learning potential, Hirschman (1987) finds that women are much more likely to consume romantic content for a companionship motive.

Of Sensations and Familiarity

Two factors link entertainment products with consumers' reactions in our framework, constituting the product-consumer interface: the sensations that people experience when consuming the product, and the familiarity it offers them. Both are crucial for triggering the emotions and imagery that address consumption motives and eventually result in pleasure. What exactly are these factors, and what do we know about them?

In consumption, *sensations* are the sensory reactions a consumer experiences as a result of exposure to an external stimulation (Zuckerman 1979). Sensory reactions are bodily, physiological processes and are distinct from cognitive processes such as thinking and interpretation that they can trigger. Sensations can be described as the arousal one feels when nerves are activated and hormones, such as dopamine, are produced. Movies, TV shows, songs, novels, and video games, as the products this book deals with, involve sights, sounds, and tactile sensations that are perceived by consumers' basic senses (via the human "hardware devices," such as ears, eyes, and fingers).

With regard to a consumer's desire to perceive pleasure, not all kinds of sensations are equally well suited. According to Zuckerman's (1979) research, consumers value sensations that are different, new, and rich. Humans have innate preferences for variety (e.g., McAlister and Pessemier 1982), and thus prefer to experience different sensations over time, rather than having the same sensation constantly repeated. In addition to variety over time, humans also have a basic desire for new stimuli (e.g., Hirschman 1980)—we find things stimulating simply because they provide a sense of novelty. Finally, stimuli that are "rich" and multidimensional cause more intense bodily responses than do simple, one-dimensional sensations.

Thus, entertainment products need to be rich, sufficiently innovative, and/or varied enough from previous products to cause sensations in the consumer, thereby avoiding a "same old-same old" feel (Busch and D'Alessandro 2016). Because it is the sensations that consumers are looking for when consuming entertainment (as a vehicle for pleasure), sensations are closely tied to satiation effects; if a product produces only weak sensations, consumers will quickly experience satiation. The potential for sensations differs systematically between types of entertainment products, such as original creations and extensions of existing works (e.g., sequels and remakes), which has major implications for their respective marketing and success potential.[109]

Familiarity, the second factor through which an entertainment product can create pleasure, refers to a consumer perceiving a sense of connection with an entertainment product and/or its elements and characters. This familiarity is based on previous encounters with the product/element or similar others (Bohnenkamp et al. 2015; Green et al. 2004). A consumer's pleasure from a new Nintendo video game featuring Mario comes partly from the new challenges, but also from the familiarity of the beloved character. The character James Bond in the movie SPECTRE will be highly familiar to those who have seen other Bond films; he may even be familiar to others based on his cultural popularity and prominence. People can perceive a new movie starring Daniel Craig as familiar because they know his previous works as an actor. And some might recall memories of other films because of a new film's storyline ("The villain!" "The shootout!") or setting ("I've seen those red sandstone buttes before! They remind me of....") Like the concept of sensations, familiarity is not relevant for narrative forms of entertainment only: songs also strike consumers as sounding more or less familiar (e.g., Ward et al. 2014).

Familiarity is an essential element on the road to entertainment pleasure because it can activate memories and emotions of positive (or negative!) experiences the consumer has had during previous encounters with the familiar product elements and *transfer* them to the new product, thereby sparking positive (or negative...) emotions toward the new entertainment product. On a more fundamental level, familiarity also helps the consumer to cognitively

[109]We discuss the sensations potential for the different product types and also how the richness of sensations can be influenced via technology in earlier chapters.

categorize a new entertainment product, i.e., it helps us to understand and make sense of what the product will be about, or even fantasize about the new product. When consumers can place a new product in their existing "mental maps" of entertainment products they know, they can draw on well-developed cognitive associations. Then cognitive processing is much simpler (a.k.a. of "higher fluency") and takes less effort—a fact that consumers value and which biases them toward familiar choices (Reber et al. 2004).

Now, how influential are sensations and familiarity for consumers' entertainment choices? Schäfer and Sedlmeier (2010), in two experiments with 53 and 210 German students, provide empirical evidence that consumers' musical preferences (i.e., the degree to which they like certain songs) are strongly influenced by music's ability to create *sensations* (namely, to stimulate arousal and activation), as well as to offer *familiar* content and structure. Ward et al. (2014) offer further support for the critical role of familiarity. Also conducting two experiments in which they asked a total of 434 students to choose between pairs of songs (one familiar, the other unfamiliar), they find that the familiarity of a song is closely linked to song choice, even when controlling for consumers' liking of and satiation with a song. In their regression analyses, familiarity is nearly as powerful for explaining song choice as is liking of the songs. And whereas we have said earlier that Askin and Mauskapf (2017) find that being too similar to previous hits can hurt a song's hit potential, this only happens in their study after a critical similarity threshold value is passed. Before this satiation threshold is reached, more similarity with hits *enhances* a song's commercial success—the song is perceived as more familiar by consumers, which the consumer generally considers a good thing (at least until satiation sets in).

For movies, we use more than 6,700 consumer ratings of sensations and familiarity of 648 film trailers to investigate how sensations and familiarity perceptions regarding the trailer relate to the rater's intention to watch the actual movie (Behrens et al. 2017). Using OLS regressions (with watching intention as dependent variable, or DV), we find that the levels of sensations and familiarity that consumers experience when watching a trailer increase their willingness to see the movie. Sensations and familiarity perceptions are also linked with the assumed quality of the film, when we use that variable as our DV. In both cases, although both factors have a strong impact, we find trailer sensations to be even more influential than familiarity.

Our results also offer some richer insights about how sensations and familiarity affect consumers. Figure 6.2 illustrates the courses of the regression functions: whereas trailer-related sensations affect consumers' movie assessments in a linear manner (Panel A), satiation effects seem to exist for familiarity, as the positive impact of familiarity gets smaller as the level of familiarity increases

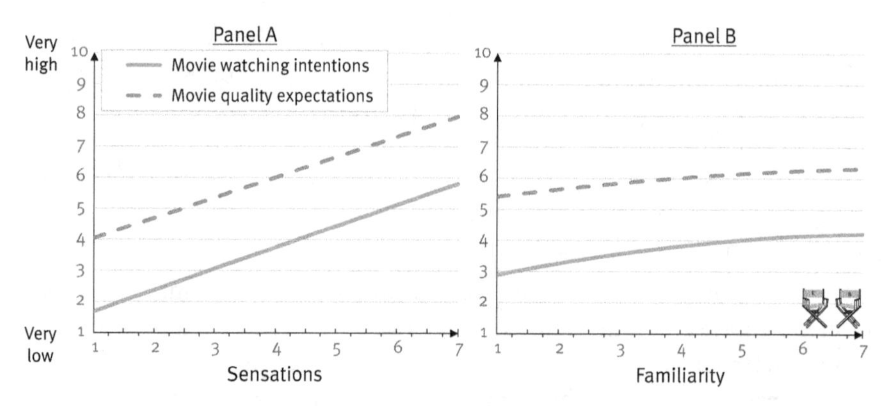

Fig. 6.2 Sensations, familiarity, and consumers' movies assessments
Notes: Authors' own illustration based on results in Behrens et al. (2017). Results show regression functions based on unstandardized parameters of OLS regressions with sensations and familiarity as independent variables. We set the variable that is not shown in the respective figure to its mean to visualize the effects. Movie watching intentions are measured on a 1–7 scale, based on 6,760 viewings of trailers for 648 movies. Quality judgments are movies' mean consumer ratings on Moviepilot.de (scale ranging from 1 to 10).

(Panel B).[110] Interestingly, this is essentially the same as what Askin and Mauskapf found for music, despite the different forms of entertainment and methods.[111] And just in case you wanted to know: sensations and familiarity are significantly correlated, but they do not affect consumers' assessments *jointly*, as adding an interaction term does not affect results.

The Emotional Facet of Entertainment Consumption

> "And when thou in the feeling wholly blessed art.
> Call it, then, what thou wilt, -
> Call it Bliss! Heart! Love! God!
> I have no name to give it!
> Feeling is all in all"
>
> —*The character* Faust *in* Johann Wolfgang von Goethe*'s (1808) novel* FAUST: A TRAGEDY

Emotions are one of the two cornerstone concepts of hedonic entertainment consumption. Our associations with favorite entertainment products

[110]Actually, higher familiarity even leads to *lower* movie assessments after a threshold is reached—but this threshold value lies outside the scale limits.

[111]We will inspect satiation effects more closely as part of our discussion of information strategies for new entertainment product samples (such as movie trailers).

are almost always closely tied to the experiencing of intense emotions. Sometimes entertainment even takes a meta-perspective on the role of emotions, holding up a mirror: such as when actress Rita Wilson's character bursts into tears in the movie SLEEPLESS IN SEATTLE while recounting the plot of the classic drama movie AN AFFAIR TO REMEMBER, just like Tom Hanks' character is touched by the lethal end of war actioner THE DIRTY DOZEN…

Whereas each of us has an intuitive understanding of emotions, the processes underlying the concept are certainly far from trivial. In this section, we first take a look at how emotions work and present a typology of the key emotions at play. We then report empirical findings regarding how, and which, emotions affect consumers' entertainment choices. When doing so, we pay special attention to something that fascinates us, as it has other *Entertainment Science* scholars: why do we, as consumers, pay for entertainment to make us cry or to scare us out of our wits? Why do we enjoy experiencing *negative* emotions?

How Emotions Work

Emotions are studied by various sciences such as psychology, philosophy, neurology, as well as marketing and management; definitions vary substantially across perspectives and disciplines. For the purpose of this book, we take an integrative approach, speaking of emotions as psycho-physiological processes that combine cognitive, physiological, and response-related components (e.g., LeDoux 1996). The cognitive element of emotions refers to the consumer's *perception* of a certain stimulus, such as Alfred Hitchcock's legendary "shower scene" in PSYCHO. The physiological element describes a *hormonal* reaction of the consumer's body, such as the production of adrenaline in situations that are perceived as threatening. This element is central because researchers consider it to be the driver of the "feeling" sensations that are part of emotions. The third element is a response of the body to the two other elements—such as shutting your eyes or screaming in the face of Hitchcock's terror.

Scholars have provided different explanations of how these elements interplay in making up human emotions. "Appraisal" theories have been developed by cognitive psychologists such as Arnold (1960), who argue that cognitive processes and assessments are focal for emotional reactions because they mediate a consumer's sensory perception of a stimulus and his or her experience of feelings, as expressed by a bodily reaction. Neurologists have provided support for such processing, showing that each stage involves a different part of our brain. Accordingly, the initial perception of sensory inputs takes place in the brain's hypothalamus region, whereas conscious cognitive processing happens mainly in the cerebral cortex of the brain—which then frames the eventual emotional response, which itself is orchestrated by the brain's amygdala region.

But appraisal theories are not the only explanation. A different stream of theories, motivated by Zajonc's (1980) work on "subliminal" processing of stimuli,[112] instead stresses the role of unconscious "affective" (i.e., emotional) processes and argues that emotional reactions do not require *any* cognitive processing, beyond perception. And indeed, Posner and Snyder (1975) showed empirically that reaction times to make an affective judgment can be faster than reaction times for the *recognition* of a stimulus. In other words, a person can have a positive emotional response to a photo of a person they love, even before they are able to recognize the person. Neurological findings are also consistent with this "unconscious affect" perspective—they have shown that an animal whose cortex has been removed is still able to exhibit emotional responses (LeDoux 1996).

So appraisal and unconscious affect theories offer conflicting explanations. We argue that this is because both ways of processing exist and it is their *combination* that provides a comprehensive explanation of what happens on the "road to the amygdala." Specifically, every time we encounter a stimulus, the brain's hypothalamus "decides" for us whether to take the "high road" proposed by appraisal theorists, thus involving the cortex in appraising a situation, or to take the "low road" instead (as suggested by unconscious affect theory's proponents), leaving out the complexities of cognitive evaluations, at least for the moment.

The brain will prefer the "low road" in situations in which it judges, in a split second, that there is not enough time for a thorough evaluation of a situation; choosing the "low road" allows a consumer to react immediately without full understanding. But because leaving out the appraisal element

[112]"Subliminal" refers to a kind of processing that takes place when a stimulus is presented to participants for such a short time frame that the participants cannot process it *consciously* and answer corresponding questions reliably.

is, as LeDoux (1996, p. 164) puts it, a "quick and dirty processing pathway," the brain will subsequently re-assess its immediate, unconscious emotional response by taking the "high road." If needed, it will then revise its original interpretation or reaction once it has sufficient time to do so.

Let us illustrate the different emotional "roads" by taking a closer look at what happens when two different people watch Hitchcock's famous horror film.[113] Whereas one of them (we call him "Frederick") is an experienced fan of the genre, the other (let's call her "Claudia") has not seen many horror movies before. When watching PSYCHO, Frederick's brain does not perceive the situation as immediately threatening and, in turn, takes the "high road." There is simply no need for instant reactions for him. Frederick recognizes the combination of dramatic music and on-screen violence as part of a movie-going experience, and his amygdala lets him show dampened surprise when Janet Leigh is slaughtered in the shower with no fear; he grins about the director's virtuosity and indulges in his popcorn.

Claudia, however, does not see room for a closer cognitive inspection of what is happening on the screen—her hypothalamus feels threatened and takes the short cut to avoid negative consequences for her health. As a result, she jumps directly to fear and outright panic, with her eyes wide as she takes in all the action and her body shivering.[114] A moment later, though, she realizes that the artificial character of the situation does not personally threaten her with physical harm, and her neocortex "requests" a re-evaluation of the situation, as a result of which she starts grinning also, and even steals some of Frederick's popcorn.

We have illustrated these basic human emotional reaction patterns to entertainment in Fig. 6.3. Panel A of the figure shows the "high road" (i.e., appraisal) and Panel B the "low road" (i.e., unconscious affect). Panel C, finally, depicts a combination of both processing patterns.

---

[113]In case you have not seen Psycho yet, please do yourself the favor and make up for this omission. If you want to take a shortcut (which we do not recommend for any ambitious entertainment student, scholar, or manager though!), you can still look up the iconic shower scene at several places on the Internet, such as http://goo.gl/XfSvuQ. Enjoy—but take care!

[114]An unconscious, purely behavioral reaction as a response to (scary) entertainment stimuli is also evidenced by neurologists in other areas of the body. For example, Nemeth et al. (2015) noted a significant uptick in blood clotting as a bodily response to watching a horror movie, but not other films. As with most unconscious processes, the explanation to such reactions refers to evolution: in frightening situations, our body prepares itself for the loss of blood, a threat that is countered by more rapid blood clotting.

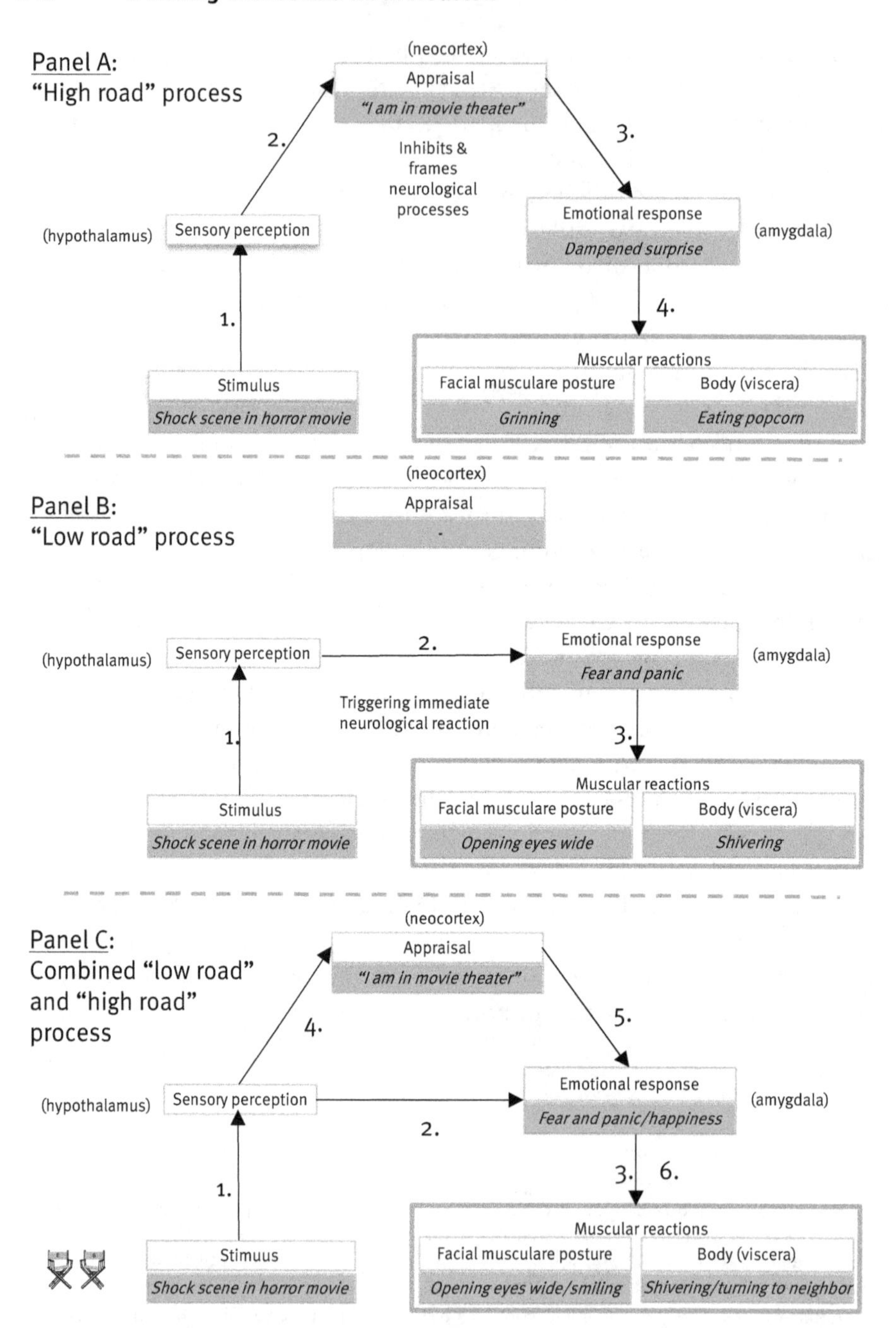

Fig. 6.3 The different roads to emotions when consuming entertainment
Notes: Authors' own illustration based on ideas reported by LeDoux (1996). The numbers in the figure show the order in which the different paths are activated. Terms in parentheses are the parts of the consumer's brain in which an action takes place.

Whether a consumer's brain takes the low or high road when exposed to an entertainment stimulus depends on many factors. The design of the stimulus certainly plays a key role (think about what separates movie scenes that make us cry, laugh, or scream from those that don't provoke such emotions), along with the consumer's idiosyncratic genetic make-up and socialization. For example, some of us have a higher level of empathy than others which makes us respond more strongly to human suffering on the screen. Have you cried when the GLADIATOR is reunited with his family in the afterlife, or when the aged PRIVATE RYAN asks his wife whether he has led a "good life"? You might want to consider this as a litmus test of your empathy repertoire.

But let us not forget the artificial nature of entertainment, which always stages (fakes or simulates) the experiences that it lures us to—there is almost always no actual reason to be afraid or sad when listening to a song, reading a book, or watching a movie. This illusory nature ties the experiencing of emotions closely to the actual situation in which entertainment is consumed: a single misplaced laugh, word, or ring tone from another person in a movie theater or on the sofa at home can prevent us from getting emotionally involved.[115]

Still, why do we react emotionally at all to entertainment stimuli when we know it is all simulated/fake (we pay for the experience, so we should know)? Zacks (2015) compares entertainment stimuli with "supernormal," exaggerated stimuli that have been shown to be highly effective in triggering emotional behaviors. He concludes that it is the "exaggerations of features of emotional expression, dialogue, physical action, setting, color, sound, and so on" that produce our emotional responses to them (Zacks 2015, p. 82). He notes that there are parallels in the animal world, such as when Tinbergen's experiments reveal that a baby gull begs for food more intensely to an exaggeratedly large "parent" gull than to a gull of normal size. Maybe such evolutionary programmed innate responses to the supernormal also offer an explanation of people's fascination with superheroes or with "superproportioned" Disney princesses (Gardner 2013).

What Kinds of Emotions Exist?

A Simple (but Meaningful) Typology of Consumer Emotions

We have mentioned a number of specific emotions in our previous discussion, some of which steer the reactions of the protagonist of Pixar's INSIDE

[115]Please also see our discussion in this chapter of the determinants of being "transported" by entertainment.

Out movie: joy (the "golden" one), sadness (blue, of course!), anger (red), fear (purple), and disgust (green). But a more comprehensive list of consumer emotions, that would ideally would not only name emotions, but also organize them based on their similarities and differences, would certainly be helpful to more fully understand consumers' reactions to entertainment products. Psychology scholars have aimed to create such a typology of human emotions for quite some time; prominent approaches include those by Silvan Tomkins, Robert Plutchik, and Paul Ekman.[116] However, their typologies are essentially one-dimensional and enumerative, which limits their practical usefulness.[117]

Other emotions researchers have tried to overcome this limitation by exploring the fundamental "dimensions" that characterize the various emotions and that explain their differences. None of the resulting typologies is problem-free, but they shed more light on the phenomenon of emotions and help to reduce overlap and redundancies. One particularly powerful approach, named "pleasure-arousal" theory, suggests the existence of two dimensions: a "valence" dimension, which is linked to the positivity (or pleasantness) that characterizes an emotion, and an "energy" dimension that refers to the emotion's level of activation, arousal, or the degree to which it triggers alertness (e.g., Posner et al. 2005).

How are the different emotions positioned in such a model? Whereas most emotions scholars follow Russell (1980) in placing emotions at the outer rim of an emotional circle (or "circumplex"), Reisenzein (1994) took a less restrictive approach: He placed various emotions in a two-dimensional valence-energy space, based on their respective pleasure and arousal levels as rated by 35 psychology students. The resulting positions in Fig. 6.4 show the mean ratings for a number of key emotions.

The figure gives us a deeper understanding of what characterizes each emotion and how they differ one from another. For example, whereas joy is a highly positive, highly aroused emotion, contentment is also experienced as positive (although not as much as joy), but with low arousal. In contrast

[116]Tomkins (1962) suggested eight "basic" emotions (namely anguish/sadness, disgust, fear, joy, interest, rage/anger, shame, and surprise), Plutchik (1980) also eight, with anticipation and acceptance instead of interest and shame, and Ekman (1999) named a total of 15 emotions, with new additions including contentment, excitement, and guilt.

[117]Another question, although one which is mainly of *conceptual* relevance, is whether each concept from one of these lists should be considered an emotion or something else. Take "excitement," for example. The fact that Tomkins and Plutchik do not include it in their list of emotions does not mean that they question whether people get excited, but that they consider it to be an affective state or feeling that is just not complex enough to be considered a unique emotion (which would imply a link to unique bodily responses).

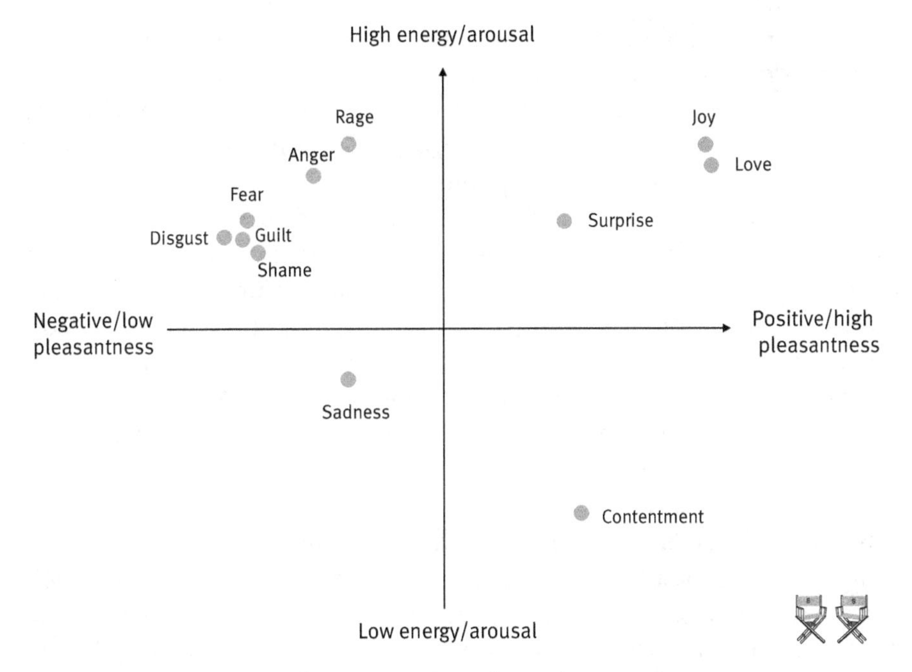

Fig. 6.4 A two-dimensional representation of some key emotions
Notes: Authors' own illustration based on results reported in Reisenzein (1994). The positions of the emotions in the space approximate their respective "typical" intensity levels.

to sadness, which is negative and low arousal, rage is equally negative, but implies a much higher arousal level. The emotions' positions in the figure also show us which emotions share similarities; for example, joy is similar to love, and guilt is similar to shame.

At the same time, questions remain. Some emotions, such as fear and guilt, are in close proximity, but obviously differ in the responses they induce; whereas fear triggers actions, guilt is linked more closely with passiveness. Similar differences exist for joy and love. Such patterns suggest that, although the two dimensions of valence and energy are helpful for understanding consumers' feeling reactions to entertainment, more factors need to be considered for grasping in totality the complex nature of consumer emotions.

Looking Forward: Adding a Time Dimension to Our Understanding of Emotions

Research on consumer emotions usually looks at the emotions that emerge at the very moment a person experiences a product. However, the experience character of entertainment products means that consumers have to make purchase decisions *prior* to actually experiencing these emotions. Does this mean that emotions are irrelevant for our actual choice of entertainment? No—research shows that consumers also produce emotions (and thoughts about emotions) *ahead* of consumption.

Such pre-consumption emotions are not limited to the day of the purchase, but instead can develop days, months, or even years before a product is released. Because of the hedonic and cultural nature of entertainment, there is potential for great anticipation. On the level of the individual consumer, emotions scholars have introduced two kinds of pre-consumption emotions to address this separation of consumers' emotional processing of a product from its actual consumption: anticipat*ory* emotions and the related, but distinct concept of anticip*ated* emotions (Cohen et al. 2006).[118]

The concept of anticipatory emotions describes a situation in which a consumer actually experiences emotions when thinking about the future consumption of a product. For example, a consumer can become excited *today* when he hears about a product he plans to consume at a later point in time. Such anticipatory excitement is clearly evident when Twitter user "Amie" writes about the movie 50 SHADES DARKER, almost half a year before its actual release, gushing that the film's "trailer has got me so excited 😂😂😁😁 I can't wait omg." We note that such anticipatory emotions are closely tied to the creation of (cognitive) imagery (the other key concept of hedonic consumption in addition to emotions). Because the consumer does not experience the entertainment product at that point in time, the emotional reaction depends on how the consumer *envisions* this experiencing. Research has shown that consumers who are good at envisioning rely more on their anticipatory emotions when making decisions (Pham 1998).

Anticipated emotions, on the other hand, are actually not really emotions. Instead, the concept describes the emotions that consumers *expect* to experience when they consume a product in the future. In other words, it is a cognitive forecast of the emotional consequences of engaging in entertainment. Twitter helps us again with a concrete example: when user "Never Say Never" looks forward to the release of a new album by Justin Bieber, he tells his followers that "Friday 13 november going to be a lucky day because justin bieber's new album purpose will make me happy :)." In the next section,

[118]On a collective level, it is this pre-release and pre-consumption anticipation that creates the "buzz" that often accrues before the release of an entertainment product and that can influence the product's eventual success in the market on its own.

we take a look at whether it matters to make such a subtle distinction—and to which concept managers should pay (more) attention.

Which Emotions Affect Entertainment Decisions—and How?

> "If you can make people laugh, cry and feel things with a film you make, you will be successful."
> —*Director and Pixar executive* John Lasseter *(2015)*

Now that we have overviewed the broad repertoire of human emotions, let us take a peek at what *Entertainment Science* research has to say about the role of emotions when experiencing entertainment, as well as for our preceding decisions to do so. We will put special emphasis on the role of negative emotions because the things that make us cry or scare us are often the ones that we enjoy the most.

General Findings on Emotions in Entertainment

But let us begin with more general insights. Aurier and Guintcheva (2014) study how emotions are linked to consumption experiences and consumers' judgments of those experiences. They conduct exit-poll interviews with 400 Parisian moviegoers and link these consumers' emotional states with their satisfaction with the movie they had just seen; the sample includes responses to a heterogeneous set of 28 films. Using a structural equation model in which they control for several aspects of the film (such as the quality of the acting and the script) and an overall "goodness-of-the-film" measure, the researchers find satisfaction-enhancing links for both positive (i.e., joy) as well as negative emotions (i.e., sadness). Interestingly, they also find that higher calmness, a positive, but low-energy emotion, goes along with *lower* satisfaction. In their results, joy has the strongest influence of all factors (even higher than overall "goodness"), followed by sadness.[119]

We also studied the role of emotions in an entertainment context, but focused on the emotions that consumers experience *prior* to consuming an entertainment product (Henning et al. 2012). Specifically, we looked

[119]Aurier and Guintcheva find no significant link between fear and satisfaction. We would assume that this results from the heterogeneous sample of films they use; whereas fear should be a positive state in the context of horror movies, it will probably not affect (or may even obscure) the evaluation of other films. The authors do not report any interactions of emotions with genres or subsample analyses, so their existence remains speculation.

at different kinds of anticipatory emotions and anticipated emotions. In a lab experiment, we offered 308 German college students the opportunity to purchase, among others, a DVD of the movie STAY. We then calculated correlations between the different emotion constructs and (a) the consumers' attitude toward the entertainment product, (b) their purchase intentions, and (c) their actual purchase of the DVD in the experiment. Via a series of regression models, we isolated the role of anticipatory and anticipated emotions from cognitive influences (and also control for consumers' cognitive expectations regarding the product's quality).

So, what did we learn? In comparison to a model that includes only cognitive evaluations (e.g., ratings of key product elements such as movie genre, story, stars, and DVD features), a model that includes consumer emotions explains about one-third more of consumers' attitudes toward the product. Positive/high activation anticipatory emotions (e.g., excitement) explain the most, but positive/low activation (e.g., contentment) also increased the consumers' attitude toward the product. Negative anticipatory emotions, however, significantly worsened it. Whereas anticipated emotions (alias emotional expectations) also correlate with consumers' attitude, their effects are crowded out in the regression analysis by the anticipatory emotions.

Consumers' responses to emotions are quite similar when it comes to respondents' purchase intentions and their actual purchase behaviors. In Fig. 6.5 we show that positive/high activation emotions generally dominate those with low activation, but for negative emotions the pattern reverses: *negative low-activation* emotions tend to explain entertainment decisions more than negative emotions with high activation. Boredom and dullness appear to be worse than fear and sadness when it comes to the emotions that entertainment consumption triggers in advance, something we also address in the following section. Interestingly, for the more cognitive concept of what the consumer *expects* to feel when consuming the product (i.e., anticipated emotions), it is negative feelings with *high activation* that have the stronger impact. We also find that emotions not only influence the consumers' purchase intentions and choices directly, but also indirectly—via their impact on attitudes which, in turn, also affects intentions and choices.

Fowdur et al. (2009) used aggregated data when studying the role of emotions; they linked consumer emotions to the actual box-office success of movies. They infer consumers' emotional reactions to each of the 932 films in their data set[120] from the film's content. Using Latent Semantic Analysis,

[120]The films in their data set were those which received a wide release in North American theaters between 1999 and 2005.

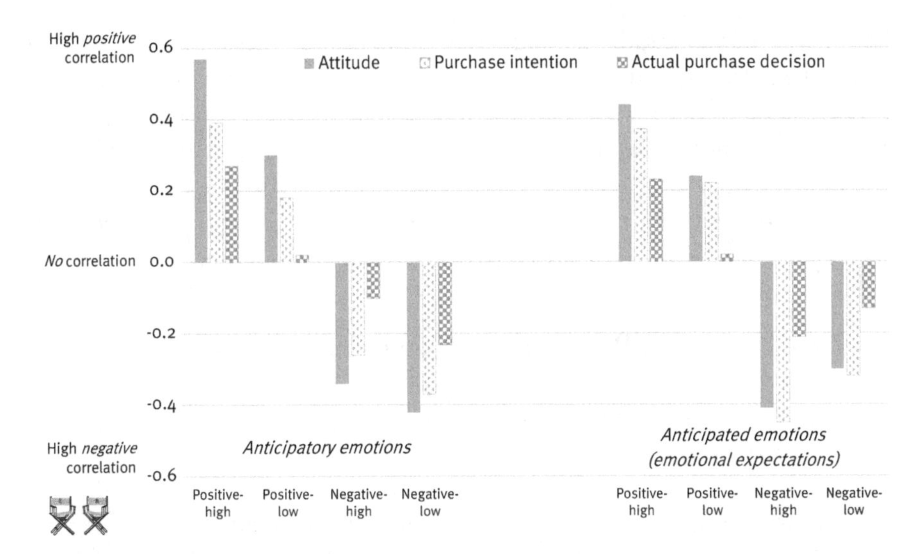

Fig. 6.5 Correlations between different kinds of anticipatory/anticipated emotions and three facets of consumer decision making
Notes: Authors' own illustration based on results reported in Henning et al. (2012). Bars for attitude and purchase intentions are pairwise correlation coefficients; the bars for purchase decision are point-biserial correlations, as the dependent variable here was binary, with 0 = no purchase and 1 = purchase.

a method that determines distances between terms based on a large dictionary archive of text, they create an "emotional profile" for each film based on the "semantic distance" between the film's "plot keywords" (to capture key events and characters of the film) and six key emotions (i.e., the positive emotions of joy, love, and surprise, and the negative emotions of anger, fear, and sadness). The scholars use a Bayesian method to estimate the distances between keywords and emotions and then link the resulting emotional film profiles to the success of each film (i.e., their theatrical "market share" in a specific week). Instead of using the six individual emotions, they employ two composite emotional "factors" that reflect the relations among individual emotions: the movies' emotional complexity (*how many* different emotions are triggered by a movie's elements) and extent of negative emotions.

The researchers' results show that both emotional factors matter. Emotional complexity plays a stronger role for a movie's box office share (with more complexity being linked with higher success). But negative emotions also exert a *positive* effect on film success, independent of the effect by emotional complexity—something we look at in the next section. By decomposing their results, we also learn about the success effects of the different individual emotions: love has the strongest link, but the negative emotions of fear and sadness follow closely. Surprise is also positive, but has the smallest effect on success. As a general takeaway, Fowdur et al.'s findings suggest that stimulating complex emotional reactions by combining positive and negative emotions is a quite powerful approach, at least for movies. Of course, it remains unclear whether a specific movie's content elements actually trigger the emotions to which they are linked on average. That depends at least partly on how (well) the content elements are executed.

The Fascination (and Relevance) of Negative Emotions

> "*Positively* the Most Horrifying Film Ever Made"
> —*Advertisement for the film* MARK OF THE DEVIL

Some of the most successful pieces of entertainment are inseparably tied to deeply negative emotions. Consumers have spent more than $1 billion in theaters alone to be terrified by the violence of the first seven SAW films, and about 1.6 million Xbox One users have paid for the privilege of being slaughtered by zombies while playing the dystopic DEAD RISING 3. Two million readers in the U.S. alone followed Father and Son as they walked a devastated earth in Cormac McCarthy's novel THE ROAD, and is there anyone among our readers who has not enjoyed listening to Eric Clapton suffering about the loss of his boy in TEARS IN HEAVEN?

A considerable amount of research has been conducted to understand why we are so fascinated with entertainment that triggers negative emotions, such as fear or sadness. But despite these efforts, no single, universally accepted explanation has yet emerged (Vorderer 2003). Nevertheless, scholars have proposed a set of explanations, and often supplemented them with at least some empirical support. Some explanations are more general, while others link to specific negative emotions.

A general observation is that entertainment emotions are not the same as "real-life emotions" because of the reappraisal process that takes place during, or after, consumption. In empirical studies, scholars have observed ambivalent emotional reactions to entertainment, with negative emotions such as sadness co-existing with positive ones such as joy; such mixed emotional states are consistent with cognitive reappraisal (Kawakami et al. 2013). Excitation-transfer theory offers a physio-psychological explanation of this reappraisal process (e.g., Zillmann 1971)—it argues that the immediate arousal that is triggered by experiencing a sad or frightening entertainment product lingers on within us until it is cognitively reframed in light of new experiences.

For a movie, these new experiences might be a positive plot twist, a happy ending, or a return to less-miserable real life when the credits are rolling. When Tamborini and Stiff (1987) applied structural equation modeling to survey responses from 155 horror movie goers, they find that consumers' reframing of the experienced cruelty that was enabled by a satisfying resolution was a main driver of liking the movie. It is this reframing that provides the room for euphoria or other positive feelings. A major learning from this finding is that plotting emotional reactions to an entertainment stimulus in the traditional two-dimensional space developed for "real-life" emotions (such as the one in Fig. 6.4 on p. 251) will probably be misleading—if such reframing processes are ignored.

One reason for the occurrence of cognitive reappraisal is that consumers tend to consider entertainment that triggers negative emotions to be "artful" (Kawakami et al. 2013), a characteristic that is highly valued on a societal level. If music, movies, or other entertainment formats are capable of stirring negative emotions in us, we tend to judge these emotional reactions to be the result of artful mastery and virtue. Negative emotions are part of our "darker" side, which humans typically believe to be much more complex and challenging to understand and appeal to than the "bright" side of our identity. The German language gives a nice example by calling classical music "Ernste [serious] Musik," distinguishing it from the more positive, less "valuable" "Unterhaltungs- [entertaining] Musik."

This tie between negative emotions and art is reflected in Fig. 6.6: whereas people are generally more interested in (bright) comedy than (dark) drama movies, this changes in an art context—when people are looking for *artistic* film achievements (as evidenced by adding an award such as the Golden Globe to the search phrase), the interest is reversed, with a higher search volume for dramas than for comedies.

A separate mechanism that has been named as a reason for our enjoyment of negative entertainment experiences focuses on the *simulated character* of entertainment experiences. Simulations provides us with the opportunity to experience what Hirschman (1980) labeled "vicarious consumption": by observing a character engaging in some activity as part of a movie, book, video game, or song, we get to vicariously "live out" that activity (see also Kawakami et al. 2013). The main difference between experiencing a sad or frightening situation in real life and watching a sad or frightening movie is that we are "safe" in the theater—the dangers we confront are only simulated dangers, and neither tragic nor fearful stimuli pose a genuine threat to us. We can simply leave the theater, close the book, put down the game controller, or change the radio station if the negative emotions triggered by entertainment are too much.

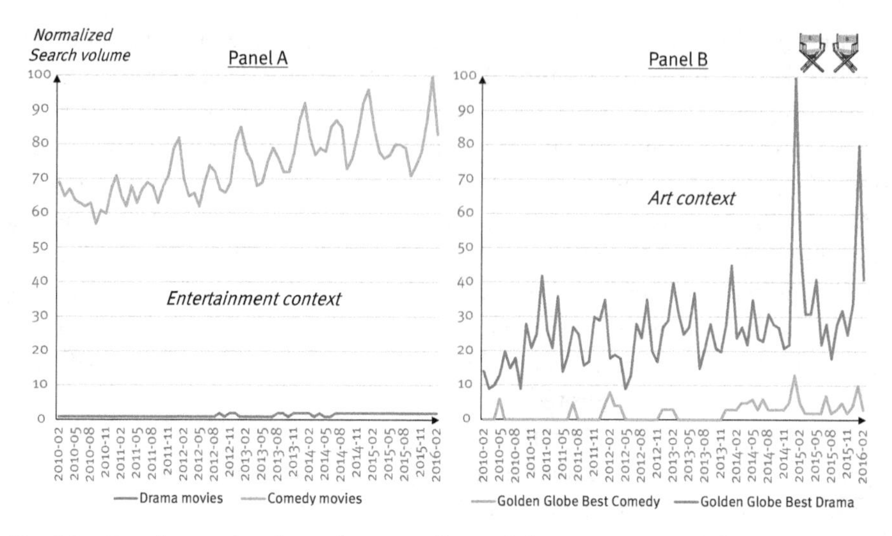

Fig. 6.6 Google search volume for comedies and drama movies in different contexts
Notes: Authors' own illustration based on data from Google Trends. Panel A shows the normalized total search volume over time for comedy and drama movies, and Panel B shows the normalized search volume for comedy and drama movies in a Golden Globe award context. The scale is normalized to values between 0 and 100 as does not account for changes in the total search volume over time.

But why are we tempted to explore such negative experiences in the first place? We assume this desire is linked to the motive of social learning, which makes us explore the deepest, darkest areas of ourselves and the atrocities life can do to humans—essentially the places to which Joseph Conrad sent his narrator Charles Marlow in the novel Heart of Darkness, and the invisible driving force for Martin Sheen's character in Francis Ford Coppola's Apocalypse Now, the book's adaptation. A related reason to such vicarious consumption is "catharsis"—the release of emotions to relieve an underlying state of tension or frustration through entertainment consumption. People play violent games to channel their inner aggressions, or watch a tearjerker to cry their frustration "out of their system" (Vorderer et al. 2004). The catharsis argument can help us understand why Tamborini and Stiff (1987) find the level of destruction shown in a horror movie to have the strongest link to a consumer's liking of the film.

Finally, a complementary explanation for choosing entertainment that triggers negative emotions, and particularly sadness, is emotional support. The concept does not treat entertainment as a way to exploit negative feelings in order to feel "better," but to intentionally *intensify* a negative emotional state. Peter Vorderer (2003) reports that 40% of the consumers he interviewed had selected music for this reason; lists on the Internet, such as

"16 Sad Songs to Listen to When You Need a Good Cry" (Reid 2016), add further ad hoc evidence. Vorderer argues that, in some situations, entertainment might provide a feeling of "togetherness" with the artist, in line with the German proverb "Geteiltes Leid ist halbes Leid," i.e., shared suffering is only half a suffering.[121]

Overall, these insights explain that making people "feel bad" can be a powerful strategy because of the simulated nature of entertainment and the cognitive reappraisal mechanisms that follow consumption. We need to keep in mind, however, that any successful use of negative emotions in entertainment relies on the transfer of negative emotions by consumers and their later reframing—and consumers' willingness and ability to do that. Expectations are crucial, as the reasons for our enjoyment of negative emotion vary in their salience over time,[122] but consumers must also be *able* to do such processing. For dramatic entertainment, a consumer's "empathy potential" is essential for him or her to suffer with the entertainment performers. de Wied et al. (1994) provide evidence that those viewers of the movie Steel Magnolias, a real tearjerker, who have a high level of "empathetic sensitivity" enjoyed the film more than those with lower levels of empathy.

And when it comes to reframing, a separate study of about 100 players of a horror video game points to the role of consumers' individual ability to carry out such reframing as a determinant of enjoyment, in addition to the entertainment product itself (Lin et al. 2017). Whereas highly fearful people, in general, did not enjoy the horror game as much as those who are less fearful, they *did* like it nevertheless—if (and only if) they believed they had an ability to cope with media suspense (i.e., high "horror self-efficacy").

The Imagery Facet of Entertainment Consumption

As we have argued above, great enjoyment from entertainment does not stem only from emotions; it also involves strong cognitive processes. In this section, we introduce the concept of imagery that is at the core of such cognitive processing—the creation of "inner images" which, as we will show, can actually be composed of more than just visual images.

---

[121]See also our discussion of parasocial relationships in our chapter on entertainment brands.

[122]The study by Aurier and Guintcheva (2014) offers an initial exploration of the role of expectations for emotions.

We will then discuss narrative transportation, immersion, and flow—the cognitive states that such inner images enable us to enter. These states are the immediate link between our inner images and the motivational ambitions that spur us to consume entertainment. It is through them we can escape our own realities and explore alternative ones that are inhabited by Jedi knights, hobbits, or mythical men who might have no name, but can make our days.

On Event Models, Images, and Imagery

Some Imagery Basics

When we read a book or watch a movie, our brain uses an approach to cope with the input that is very similar to the approach it uses to deal with the "real world," such as when we go shopping or walk down the street (Zacks 2015). Our mind automatically constructs so-called "event models"—abstract and simplified representations of the things that have happened and we have experienced. These models are not simple one-to-one recordings of the material we saw, heard, or read. Instead, they incorporate insights and logic based on our own knowledge and prior experiences, combining the inputs we now observe with our own logic to create a model that "makes sense" to us. We use such event models when processing entertainment during consumption—and they determine how we remember it afterward. We equip the characters of a novel with a backstory and add motivations to better understand their actions, although these aspects are not necessarily in the book nor have even been envisioned by its author. Event models are idiosyncratic.

But event models go one step further: they include "information about how things look and feel and sound, where objects and people are located, and how you might act" (Zacks 2015). Event models developed from entertainment experiences (but also from the "real world") thus can be envisioned as "inner images." Whereas the term "inner" refers to something that occurs within the mind, let us warn you that the term "image," despite being widely used in this context, is somewhat misleading. The reason is that event model images contain more than just visual impressions, but can instead be multi-sensory, involving smell, taste, and tactile sensations (MacInnis and Price 1987).

When we think of our first romantic love, we might activate pictures of her face, but also the sound of her voice, and maybe even the taste of her skin and the smell of her favorite perfume. In Michael Frayn's novel SPIES, the main character, now long a grown-up man, revisits his boyhood on an imaginary journey through time. He "sees" the "shining" dining room table from decades ago where he and his friend sat, but also recreates in his mind the taste of the chocolate, and once more feels in his fingertips the patterns in the lemon barley tumblers. The visual dimension is often dominant in humans' creation and usage of such models, and also for consumer decisions, which explains why we speak of inner "images," despite their multisensory nature.

Inner images are created, as well as activated, through a process called "imagery." Imagery explains *how* information we perceive through our sensory receptors (the nose, the ears, etc.) ends up in our working memory (MacInnis and Price 1987)—how we create inner images, use them to process stimuli, and also how we store these images.[123] Inner images are created by, and activated by, external or internal stimuli. External stimuli can be visual impressions (a photo, a text, a film, a game screen) or non-visual ones. For example, touching a book you read decades ago brings back images of your youth or listening to a song on the radio evokes the images of a long-ago concert you attended with friends who have become estranged over the years. Or the scent of popcorn in a store creates the image of having a great time in a movie theater—and triggers buying tickets for the new STAR WARS movie via your smartphone.

But sometimes inner images also show up without such an external trigger: they can be caused solely by internal processes, such as when somnolent or daydreaming. Let us add that, although we classify imagery as a *cognitive* process, it mostly happens automatically, with very limited conscious control by the consumer. If you have ever wondered why inner images are often quite difficult for us to verbalize, it's because of their automatic nature.

Types of Entertainment Imagery

Entertainment is a crucial source for the creation and stimulation of imagery. What comes to your mind when you see the numerals *2001*? If you are a dedicated cine-phile, chances are that you think of Stanley Kubrick's classic sci-fi movie of the same name, and you might envision the hypnotic

[123]The concept of imagery has quite a long history in the sciences, appealing to both cognitive psychologists and philosophers since the 1870s. For a classic historic contribution, see Galton (1880); Thomas (2017) provides a comprehensive review of the historical discussion of both theoretical and empirical imagery research, including recent contributions.

red-eyed supercomputer HAL 9000, the bone-turning-into-spaceship, and the Starchild; some of our dedicated readers will even hear Richard Strauss' composition ALSO SPRACH ZARATHUSTRA in their inner ears. Hirschman and Holbrook (1982) label such images as "historic imagery"—images that we have actually experienced in the past. If the underlying experience was positive, such imagery goes hand in hand with positive emotions: stimulating the images rekindles the emotional memories (e.g., MacInnis and Price 1987). However, if a musical piece activates imagery that is tied to negative experiences (a terrible concert performance), those negative emotions will come back too. Such historic imagery, if triggered by a new entertainment product, influences how consumers will think and feel about it.[124]

But Hirschman and Holbrook (1982) list another type of image that might be of even more importance for entertainment producers. They label it as "fantasy imagery"—it describes the construction of images that have no direct connection to a consumer's prior experiences. For such fantasy imagery, the "colors and shapes that are seen, the sounds that are heard, and the touches that are felt have never actually occurred, but are brought together in this particular configuration for the first time and experienced as mental phenomena" (Hirschman and Holbrook 1982, p. 93). These images are particularly powerful for generating entertainment pleasure—they enable the consumer to fulfill the motives that drive entertainment choices and that are directly linked to his or her experiencing of pleasure when experiencing an entertainment product (i.e., the overall goal of hedonic consumption).

Reliving the "historic" screen adventures of Han Solo and Princess Leia in one's mind has great escapism potential, but many people will extend their imagery beyond the specific events in George Lucas' space saga to fantasize about new adventures and new challenges in galaxies far, far away, which offer endless opportunities to escape, explore, and to address other motives, as well. Such fantasy imagery can also be held responsible for the huge success of James Cameron's AVATAR: the film's hyper-realistic imaginary worlds made many of us dream of the things that happen in other parts of the Pandora universe, letting us leave the narrative for a time, during or after watching the film.

As we argue above, inner images are never fully accurate recollections of actual experiences or impressions. Thus, we should not treat historic and fantasy imagery as binary categories—instead they define the end points of a continuum that ranges from purely historic recollections to complete fan-

[124]A concept that is closely related to such historic imagery of entertainment products is the brand image.

tasy. Almost always, the inner images of our event models for entertainment lie somewhere in between these extremes (Hirschman and Holbrook 1982).

The Drivers of Imagery

A key question for entertainment producers is figuring out what causes consumers to produce powerful imagery. Most scholarly research on this topic has been conducted in fields other than entertainment, such as branding and general psychology; we find at least some of the insights to be very relevant for and also transferrable into the context of this book. So what do we know about determinants of imagery?

Research has identified three general factors that explain whether consumers create imagery, and how much: the product (or stimulus), the situation, and the individual consumer. Regarding the product, scholars have often used photos as stimuli, varying their attributes when investigating how the nature of the stimuli impacts the creation of imagery. A key finding is that "vivid" stimuli contribute strongly to the creation of imagery. What is vividness? Although measured in many different ways, vividness is usually associated with a visual (versus textual) character and a high level of concreteness (versus abstractness) (Petrova and Cialdini 2005). Other scholarly results for product determinants point to a close link between the "emotional profile" of the stimulus and imagery creation—photos that are judged by subjects as more arousing trigger substantially more imagery, as do stimuli that are viewed as pleasant (versus unpleasant) (Bywaters et al. 2004).

With regard to the situation, experiments in which consumers were given a task to complete (such as memorizing a list of words) at the same time they were exposed to a stimulus (and ask to "produce" imagery) show that the presence of such "cognitive load" disrupts consumers' production of imagery. The reason is that the task absorbs the cognitive resources that are needed for the production of imagery (Drolet and Luce 2004). For entertainment producers who care about imagery creation, this finding stresses the critical role of the consumption environment, which is part of the distribution mix for entertainment. Do movies that are watched in a theater have a stronger impact on the consumer's creation of imagery than do movies that are watched on Netflix because the theater context captures the consumer's full attention? Or does the presence of others and processing the ambient noise of the crowd siphon away more cognitive resources in the theater? This is an unanswered question which might explain both the consumers' reaction to a specific piece of entertainment, but also those toward another product whose success builds on the existence (or absence) of such imagery.

Finally, consumers have also been found to differ in their individual ability to produce imagery, in general, as a character trait (e.g., Bywaters et al. 2004). Yet research does not tell us very much about the specific characteristics of those consumers who have high/low "imagery ability." Age matters, but in a non-linear way: adults are general superior in creating imagery when compared to children (Kosslyn et al. 1990), but only until a certain age (Craig and Dirkx 1992). Gender effects are occasionally argued to exist, but empirical studies generally find no substantial differences between males and females in imagery generation (Campos 2014).

As a consequence, entertainment producers who aim to heighten enjoyment by making a lasting impression in audiences' minds should carefully craft the product and support its consumption free of disturbances. In comparison, it seems to matter less who the target group is, at least in terms of demographics.

The Power of Imagery: Narrative Transportation, Immersion, and Flow

The creation and activation of strong imagery enables consumers to enter certain unique psycho-physiological states in which entertainment motivations can be fulfilled by the consumer getting "lost" in the alternate world of the entertainment product. Three such states have received particular attention by entertainment scholars: narrative transportation, immersion, and flow. Follow us on a journey that makes a quick stop at each of them (but please avoid to get lost on the way…).

Narrative Transportation

Narrative transportation describes a situation in which a consumer experiences a story and, based on strong imagery, gets lost in it, losing track of the "real world," in a physiological sense, for a while (van Laer et al. 2014). The story element and the existence of characters with whom the consumer can identify are crucial for transportation to occur; thus, it is mostly applicable to narrative forms of entertainment, such as novels, movies, TV series, and certain kinds of games.

As summarized by van Laer et al.'s "extended transportation-imagery model," research has identified a number of factors that determine whether an entertainment product triggers transportation processes. On a basic level, we know that effective transportation depends on both characteristics of the narrative and how it is told (the "story-teller"), along with the person who consumes it (the "story-receiver"). Regarding the narrative itself, research has stressed the need for identifiable characters in a story—if audiences cannot relate to the thoughts and feelings of a novel's hero or heroine, they cannot empathize with these characters (Slater and Rouner 2002). Further, just as storylines can differ in their potential to create imagery depending on their vividness and emotional profile, storylines similarly vary in their "transportation potential." Offering a sequence of events that can stimulate the creation of strong imagery has a better chance of transporting its readers, viewers, or players into the world in which the story takes place.

A third determinant is a story's "verisimilitude," or "fictional realism." This is one that we, the authors, are particularly intrigued by, probably because it reminds us of countless entertainment experiences in which transportation was disrupted by a lack of verisimilitude. What we experienced was unrealistic—but because we are talking about entertainment narratives here, "realism" does not necessarily mean to comply with our "real" world in all ways. Instead, all fictional, fantastic worlds into which we are invited by the creative artists have laws, albeit almost always unwritten ones. When these laws become inconsistent or are broken, the violating action stands out as a disruption of the story and becomes a major distraction—the "oh-come-on-that-is-impossible" moment of entertainment consumption.

Almost all fictional stories change *some* aspects of reality (e.g., what if animals could talk? What if zombies really existed?), while keeping the others intact, which is important to enable the consumer to remain oriented. If a character suddenly develops a superpower for a reason that makes no sense or something else happens outside the laws of the storyline, the fictional world to which we have traveled implodes and we find ourselves back in the real one. If the number of "The-Most-Unrealistic-Movie-Scenes-That-Ruined-the-Entire-Film" lists on the Internet is any indication, we are not the only ones whose transportation has been quashed. Figure 6.7 lists five of our "favorites": movie scenes in which the lack of verisimilitude damages the movie-watching experience.[125]

But transportation not only depends on the product, but also the individual consumer—the "receiver" of the stories. van Laer et al.'s (2014) results show that transportation varies with whether a consumer pays attention and whether

[125]Just in case you want to take a look yourself: all the scenes we list in the figure can be found at several places on the Internet, such as at https://goo.gl/vPFPAs (Catwoman), https://goo.gl/f7HNHn (The Matrix Reloaded), https://goo.gl/MaBWVa (Air Force One), https://goo.gl/TX8KeX (Star Wars: Episode II), and https://goo.gl/xom5wR (Die Another Day).

Film	Scene	Why lacking verisimilitude?
CATWOMAN	Catwoman playing basketball	Instead of showing her skills, it looked "like you showed two people a 15-second clip of a middle school game, and told them to do something vaguely similar and make it all look like a bad '90s music video" [1]
THE MATRIX RELOADED	One Neo fighting numerous Agent Smiths	"[T]he second the fighting begins, both Neo and Agent Smith get replaced with CGI so terrible it makes THE SIMS look like virtual reality." [2]
AIR FORCE ONE	The crash scene	Ruining an action movie "with a rendering that looks like Microsoft Flight Simulator" [3]
STAR WARS: EPISODE II	Anakin and Padme romance scene	The lack of chemistry. The acting. The dialogue. Basically everything [4]
DIE ANOTHER DAY	James Bond surfing a tsunami	A scene that "looks slightly less realistic than playing GOLDENEYE on N64" [5]

Fig. 6.7 Some prominent movie scenes lacking verisimilitude
Notes: Authors' own illustration. The quotes in the figure are from the following sources: (1) https://goo.gl/iy1qhR; (2) https://goo.gl/Am1792; (3) https://goo.gl/NHx868; (4) https://goo.gl/HWnbJ8; (5) https://goo.gl/amw7Qn. Brands are trademarked.

he or she is familiar with a story or the genre in which the story is situated. Familiarity facilitates the understanding of a story and thus transportation, although the link is not necessarily linear—*very* high familiarity levels with a storyline can create perceptions of low levels of novel sensations and could, thus, reduce interest and attention which are needed for being transported.

Some scholars have also argued that consumer transportability is a stable personality trait, i.e., some people are transported more easily than others. Dal Cin et al. (2004), for example, tested a "transportability" scale with four movies and novels, finding that the measure was significantly linked with the extent of transportation. But it remains unclear whether such a "transportation trait" is anything more than a combination of consumers' ability to produce imagery and their empathy skills—at this point, we have to wait for future research to shed more light on this issue.

Transportation has also been shown to vary with consumers' gender. It is stronger for females because they, on average, have a higher empathy potential. No such differences have been found for different age groups (van Laer et al. 2014), despite the fact that younger consumers are less rooted in the "real world" and thus could be expected to have higher imagery potential. It seems that other factors also matter and counter this advantage.

Finally, the circumstances in which we consume narratives might also impact our transportation experience. We assume that such situational factors include whether we are consuming the stories alone or amongst others (who might distract us, but also facilitate the transport) and the devices we use. Is a printed book more suited to transport us than a Kindle, a large TV screen more than a tablet or smartphone? Future technologies, and Virtual Reality in particular, make big bets on their transportation-enhancing powers.[126]

What do we know about the consequences of transportation for consumers' reactions and behaviors? The concept has been linked, theoretically and/or empirically, with several of the key entertainment motives we discussed above. These findings underline the important role that transportation plays in the sensation-familiarity framework, as a mechanism that enables imagery to fulfill entertainment wishes. Green et al. (2004) suggest that transportation is not only a means for escapism (leaving the worries of the "real world" behind), but also for discovery/exploration (by creating an openness to new experiences), companionship (feeling as if one knows the entertainment characters), social learning (by offering simulations of alternate personalities and actions), and mood management (transporting experiences likely being "the most effective at managing moods," p. 319 in Green et al.).

Further support for the concept's relevance also comes from van Laer et al.'s (2014) integrated analysis of existing research findings on the links between transportation and consumers' liking of a storyline and their subsequent "behavioral intentions."[127] Both links are statistically significant and also substantial, with average *r* values of 0.44 and 0.31, respectively. Finally, transportation has also been shown to be highly correlated with the consumers' level of enjoyment, which is the inner driver of all our entertainment activities. In their studies of short stories and novel chapters, Green and her colleagues report correlation coefficients of 0.60 and above between the two concepts, which suggests that the intensity of transportation can, at least in certain entertainment settings, strongly determine the degree to which a story fulfills consumers' desire to be entertained.

Immersion

Immersion is a concept that is closely related to narrative transportation. It describes the consumer's sensory impression of being surrounded by an

[126]See our discussion of technology later in this book for initial empirical findings regarding the use of virtual reality in entertainment.

[127]The data analyzed by van Laer et al. includes heterogeneous settings beyond entertainment in which transportation has been studied by scholars, such as advertisements and website browsing. "Behavioral intentions" thus is a broad concept; example manifestations include a consumer's stated willingness to adopt an advertised product or behavior.

alternate (often virtual) world—this is why some scholars also refer to it as "spatial presence" (Madigan 2010).[128] When fully immersed, the senses of a consumer are tied to the alternate (entertainment) world; the "real world" is screened out and consumers make decisions that only make sense in the context of the imaginary world. A player fully immersed in the western game RED DEAD REDEMPTION will prefer to travel long distances via horse, instead of using "fast traveling" options provided by the game's menu screen. The main difference between immersion and transportation is that the latter is closely tied to the *storyline* of an entertainment product, whereas immersion does not require a narrative at all. Its focus is on an entertainment product's aesthetics and its "physical configuration" (Phillips and McQuarrie 2010, p. 388).

As a result, the concept of immersion is particularly relevant for experiences that are mainly aesthetic, rather than narrative. In the realm of entertainment, this applies to many video games and musical experiences for which narrative transportation is less explanatory. Think of the open worlds of games such as FAR CRY 4, SKYRIM, and MINECRAFT (which is about *creating* an alternate reality), the thrill of participating in a fictitious sports universe (such as in FIFA), or the absorbing experience of listening to a classical piece of music or a soaring movie soundtrack.

So, what are the critical factors that must exist for immersion to happen, and how do they differ from the drivers of transportation? Wirth et al. (2007) have suggested an integrative general model of immersion in which they distinguish two stages of immersion: (1) the cognitive creation of an alternate world and (2) the consumer's acting inside of this world.[129] For immersion to happen throughout these stages, they argue that products must offer "rich" cues, and that these cues need to be consistent among each other. (And yes, as with transformation, individual consumer factors also matter for immersion, but we will get back to this in a moment.)

[128]Let us note that some scholars have tried to set immersion apart from presence by considering immersion as the technological, "objective" element that causes the consumer's psychological perception of presence (e.g., Wirth et al. 2007). Such definition (which restricts the immersion concept to its underlying technical forces), however, conflicts with the common understanding of immersion. Another group, including Bracken (2006), considers immersion as part of a more complex presence concept; these scholars separate immersion from what we consider here as elements of immersion itself (such as the perceived realism of the alternate world), which we do not consider helpful.

[129]We find it an interesting question whether immersion is a binary or continuous concept. Wirth et al. (2007) argue for the former ("you are either in an alternate world or not"), but our own experiences suggest that a continuous interpretation is more appropriate: one's perception of such a world is more or less exclusive, with presence experiences differing in depth and richness. The same question can be asked for narrative transportation.

The richness of cues partially overlaps with the idea of vivid and emotional stimuli that we discussed in the imagery section. But there is something else to it here: the more information the consumer receives regarding the alternate world, the less reason he has to question the world's existence. "Multi-sensory" cues, already highlighted by hedonic consumption pioneers Hirschman and Holbrook (1982), add more "realism" to the consumer's experiences. Seeing a realistically layered horse in a western setting is one thing, but also hearing it nicker or huff is another (Madigan 2010). Such richness makes the blank spots disappear in the consumer's inner image: the less of the alternate world that is left undefined, the easier it is for him or her to accept its existence. Other factors that immersion scholars argue will facilitate the consumer's perception of a rich alternate world include whether the product has a "challenging" character, which absorbs the consumer's mental resources and prevents him from looking at the alternate world with too much scrutiny (Madigan 2010). Also, a strong narrative, although not essential to immersion, *can* help the consumer "stay connected" to the alternate world (Wirth et al. 2007). Whereas the narrative aspect shows closeness to transportation, the challenging character links immersion to flow experiences (which we discuss next).

The consistency aspect of cues that enable immersion resembles the idea of verisimilitude as a transportation determinant. For immersion, scholars argue that cues need to be congruous, both amongst each other and with the rules of the world that the product is trying to establish. For example, if a player enters a tomb that has no burial chamber, sees American police cars in a European setting, or notes misspelled signs, the illusion that the alternate world is "real" is threatened (game designer Toby Gard, quoted in Stuart 2010). In games, congruity is particularly challenging when it comes to the integration and design of menus, heads up displays, tutorial messages, and advertising (Madigan 2010). Consistency is also influenced by the behavior of game characters (can I interact with them? Do they respond in a believable way?) and the technical fluidity of the presentation (e.g., no loading times between scenes).

Turning to characteristics of individual consumers, most arguments we offered in the context of transportation also apply for immersion (familiarity with the genre, etc.). And stable consumer traits have been linked with immersion too: the "immersive tendency" concept by Witmer and Singer (1998), which reminds us of the consumer transportability trait, has been found to explain 13% of the amount of immersion in a role-playing game study with 70 students (Weibel and Wissmath 2011). Wirth et al. (2007) point at two sub-traits: they argue that consumers also differ in their "suspension of disbelief" (some of us pay more attention to "real-word" factors and incongruous cues than others and thus have a better chance to be transported) and in how easily we become fascinated with phenomena that are distinct from our everyday life (see also Wild et al. 1995). Finally, the consumption situation (such as the devices used) will also play a certain role, as it does with transportation, but little research exists to document the exact nature of the impact.

Regarding outcomes, immersion is, like transportation, positively correlated with consumers' enjoyment of entertainment products, although fewer studies have specifically addressed immersion consequences. When Visch et al. (2010) manipulate immersion in a film-viewing experiment, comparing a high immersive condition (i.e., so-called CAVE viewing, where projectors illuminate multiple walls in a cube) with a somewhat less immersive condition (3D viewing), they find enjoyment (measured via a "beautiful" rating) to be more than 40% higher for the former condition.[130] And in their role-playing game study, Weibel and Wissmath measure immersion and enjoyment directly, finding a strong positive correlation of 0.53. However, a path analysis of their data suggests that, in the game context of Weibel and Wissmath's study, the link between immersion and enjoyment is not direct. Instead, they find that it is mediated by consumers' flow state—which we discuss in the next section.

Flow

The concept of flow adds a specific perspective on consumers' cognitive processes. Whereas both transportation and immersion focus on the imaginative aspects of consumption, flow is more interested in the active contributions of the consumer to an experience. Mihály Csíkszentmihályi (1975, p. 43), who introduced and has strongly shaped the concept, describes flow as a state in which consumers "act with total involvement."

A flow state is associated with a holistic energetic feeling; it is characterized by an intense level of immersion, a distorted sense of time and a high level of perceived personal control in the activity. In contrast to transportation and immersion, the flow concept has not been developed with a particular focus on entertainment experiences. Instead, it is a rather general concept that has been be applied to all kinds of hedonic activities, as well as

[130]They also find that immersion goes along with higher levels of consumer emotions (positive or negative), adding further evidence for the coexistence of imagery and emotional processing of entertainment consumption.

other behaviors that are driven by intrinsic motivation, such as the composition of music (e.g., MacDonald et al. 2006) and browsing the Internet (e.g., Hoffman and Novak 1996).

The aspect of flow that sets it apart is "control"—flow necessitates balancing the requirements of the consumer's activity with his or her skills. A flow perspective considers the consumption of an entertainment product as a "task" that a person chooses to accomplish, and flow occurs, and only does so, when the requirements of the task match the skills of that person. The task may be too difficult or too easy for the consumer—in either case, flow will not occur. Skills can be quite heterogeneous, depending on the product and task: cognitive (such as the ability to follow a complex novel plot), aesthetic (such as "seeing" the beauty of an ambitious musical composition), and motor skills (such as swiftly operating the buttons on a PlayStation controller).[131]

Whereas early flow research argued that flow results from *any* match of challenges and skills, Csíkszentmihályi and his colleagues later settled on a "minimum-challenge" condition—a consumer can only enter a flow state when the task provides at least a certain level of opportunities (e.g., Nakamura and Csíkszentmihályi 2002). As we show in Fig. 6.8, a situation in which the consumer faces high challenges, but has low skills (think of a game in which you cannot master a certain level and are killed by the zombies every time you try), will cause anxiety (and probably frustration) instead of flow. On the other hand, if skills are significantly higher than those required to meet the challenges, we experience pure control, relaxation (when watching a soap opera on TV after an exhausting day in the office), or boredom (when the game is too easy for us), but—notably—no flow.[132]

The critical role of the challenges and tasks a customer must tackle, on the one hand, and his or her skills to do so, on the other, affect managers' and artists' production decisions for all forms of entertainment. An intelligent mystery novel can cause frustration instead of flow if the reader lacks the skills to mentally keep up with the complex plot and solve the mystery, whereas an overly simple narration carries the risk of boring readers. Ruth et al. (2016) experimentally manipulate the complexity of music played by a radio program. They find that high musical complexity prevents con-

[131]For readers interested in a discussion of consumer skills required in entertainment, we recommend Sherry (2004).

[132]We find some of the other states in Csíkszentmihályi's current flow model somewhat debatable—given an equal level of challenges, why should higher skills turn boredom into relaxation? Overall however, the model provides us with a sound understanding of what is needed for a consumer to experience flow.

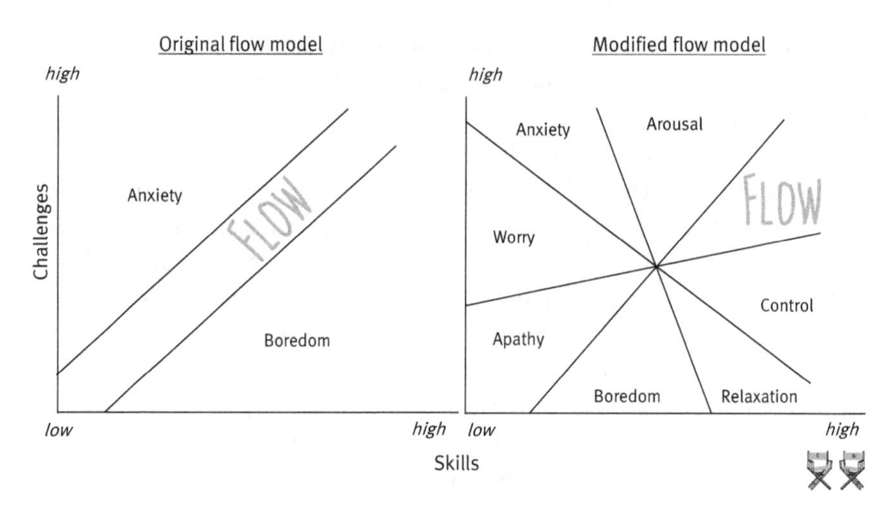

Fig. 6.8 Models of flow
Notes: Author's own illustration based on arguments in Nakamura and Csíkszentmihályi (2002). With graphical contributions by Studio Tense.

sumers with low skills from experiencing flow, whereas for consumers with high skills, complexity actually facilitates flow, in line with our theoretical arguments.

Among entertainment forms, flow is particularly relevant for video games for which the consumption act designates a more active role for the consumer. In this context, flow is common—Hoffman and Nadelson (2010) report that 89% of gamers who played games five hours or more per week experience a distorted sense of time when playing, a key facet of the flow experience. Thus, it should not come as a surprise that most scholarly research on flow in entertainment deals with games.

In addition to control, immersion is a key element of flow, and the close link between the two concepts has been empirically shown. Bachen et al. (2016), applying structural equation modeling to data from players of a role-play game, find that immersion explains 28% of flow, and Weibel and Wissmath (2011) report a correlation of 0.34 (i.e., 12% shared variance) in a similar setting.

As is the case with transportation and immersion, scholars have argued that the level of flow someone experiences is influenced by the consumer's personality. Specifically, Csíkszentmihályi has proposed that people differ in the degree to which they "enjoy life" or in their preference for hedonic activities or primary process thinking in general. In theory, such "autotelic personality" is associated with high levels of curiosity, persistence, and low self-centeredness (e.g., Csíkszentmihályi 1997), but empirical evidence is yet lacking. Others have focused more on consumer abilities than hedonic preferences for explaining different flow levels (Baumann 2012). For example, Sherry (2004) makes an attempt to explain the greater fascination of male versus female consumers with certain kinds of video games with gender-specific abilities such as cerebral 3D rotation (which has been shown to be higher for males). We assume that the consumption context will also influence whether flow is experienced in a given situation, but again, few studies have tackled the issue.

Regarding the outcomes of flow, scholars have accumulated evidence that flow is a major driver of enjoyment and product usage/liking for different forms of entertainment. Because flow builds on intrinsic interest in an activity for which the "end goal" is often simply an excuse for taking part in the activity itself, such a link should not surprise us. In a path analysis for video games, Weibel and Wissmath (2011) find that flow explains 22% of consumers' gaming enjoyment, Choi and Kim (2004) report that consumers' flow perceptions account for even more than two-thirds of their intention to replay a game (based on structural equation modeling of survey data from about 2,000 Korean online gamers), and Smith et al. (2016) show that flow goes along with playing longer in a sample of 422 Australians (the correlation is 0.34). Results are similar for music; Ruth et al. (2016) calculate, in a study of radio music consumption, that flow explains more than 60% of the variation in a participant's liking of the program. However, there is one soft spot in all of these studies. Usually very few (often even no) control variables are included in the empirical models. Thus, the reported effects for flow might actually be caused by transportation or immersion, rather than flow itself.

Finally, there may also be a "dark side" to flow experiences. In a survey of 395 members of virtual communities devoted to Internet games, Chou and Ting (2003) find, via structural equation modeling, that flow is strongly linked to various addictive behaviors, including obsession (i.e., being unable to stop playing) and withdrawal symptoms. Getting people excited with entertainment is certainly a great thing, but producers need to be aware of such unintended outcomes associated with the usage of their products. Ignoring them may eventually harm not only the customers, but also the reputation of a product, company, and industry, as a whole.

Before we move on to the final section of our consumer behavior chapter in which we discuss the process of decision making for entertainment products, Fig. 6.9 summarizes what we have discussed here about the three experiential states of narrative transportation, immersion, and flow, naming their similarities as well as their main conceptual differences.

State/ Concept	*Narrative Transportation*	*Immersion*	*Flow*
Definition	The sensory impression of getting lost in a story, losing track of the "real world" in a physiological sense	The sensory impression of being (spatially) present in an alternate world, with the "real word" being screened out	A state in which consumers act with total involvement
Of main relevance for…	Entertainment for which the narrative is focal (novels, films, TV series)	Entertainment for which aesthetics are focal (video games, music)	Entertainment experiences which assign an active role to consumers (and also other activities in which consumers engage with intrinsic motivation)
Dominant mechanism	Strong imagery and empathy with the story's characters	Strong imagery to create an alternate world and act in it	Immersion and the balancing of challenges of the entertainment product and consumer skills
Product determinants	Identifiable characters; storylines with imagery potential; verisimilitude	Rich depiction of alternate world (multisensory, comprehensive); challenging; strong narrative	See immersion; features that determine the product's challenges for the consumer (e.g., cognitive, aesthetic, motor skills)
Other determinants	Familiarity of consumer with story/genre; transportability trait; demographics; contextual factors (e.g., devices, social constellation)	Familiarity; immersive tendency; suspension of disbelief; absorption trait; contextual factors	"Autotelic" personality and abilities; contextual factors
Empirically suggested outcomes	Linked with entertainment motives; enjoyment; product liking; behavioral intentions	Linked with emotions; enjoyment	Enjoyment; product liking; behavioral intentions; usage time; addictive behaviors

Fig. 6.9 A nutshell comparison of narrative transportation, immersion, and flow
Note: Author's own creation.

A Process Model of Entertainment Consumption

> "We really don't know the decision-making process of moviegoers as well as we should."
> —*Studio marketing executive (quoted in Stradella Road 2010)*

Our sensations-familiarity framework of entertainment consumption links the different psychological responses and states that combine within the consumer to create a desire to watch a movie, read a novel, play a game, or listen to a song. It also highlights those responses and states that people experience while consuming entertainment. But the framework does not tell us the *process* that consumers go through when deciding whether a particular entertainment product, out of the myriad of available options, is well-suited to provide the desired level of pleasure or enjoyment.

Although this question is of obvious relevance for managers, relatively little is actually known about this process, as evidenced by this section's introductory quote. The main reason is that traditional models of "the" consumer decision-making process, which are taught in MBA classrooms and management training sessions around the world, are just too generic, and do not provide sufficient room for the particularities of entertainment, as we discussed them earlier in this book. Models such as the classic "attention-interest-desire-action" chain are constructed at such a high level of abstraction that they enable only limited insights into how consumer confront the entertainment particularities when searching for a product that entertains them.

The model of entertainment decision making we present on the following pages builds on general process models of consumer behavior, but even more so, is it inspired by the work of Hart et al. (2016). To better understand how consumers make decisions about entertainment, these authors applied a qualitative introspective research approach, drawing on the rich personal experiences of just *one* consumer, the smallest of all possible sample sizes. Our own decades-long studying of consumers' entertainment choices suggest that the insights from Hart et al. (2016) align spot-on with reality. Based on their insights, we distinguish between three major stages through which consumers proceed when making decisions regarding an entertainment product: (1) sensemaking, (2) decision making, and (3) the consumption experience itself.

These three stages happen sequentially, but the process is not "linear" —feedback loops between the stages are possible and are the norm rather than the exception. How long does the process take? The decision can be made in what seems like a snap (nearly automatic processing), but it can also be made carefully and slowly over the course of some minutes, hours, or even days (deliberate processing); in either case, we argue that the consumer actually goes through a staged process. Let's take a closer look at what happens in each of the stages and how they are interlinked.[133]

Phase 1: Sensemaking. In the initial stage of the decision-making process, consumers "make sense" of a product to which they have been exposed. Sometimes this exposure happens intentionally when we are purposefully looking for a product to entertain us. But in other cases, we just stumble upon an ad or store display, or a friend or social network contact says "you've got to see/read/hear/play this …." Some of the products to which we are exposed are new to the market (often even yet-to-be-released), whereas others have been out awhile, but we were not previously aware of them (or had ignored them at a prior time).

Regardless of the specifics of the situation, consumers automatically "fit" the product into their very own personal "classification scheme" of entertainment products, based on their sensing of and the processing of the information they receive about the product. This information might be fragmentary (e.g., a first teaser trailer or a friend sharing a vague rumor) or very detailed (e.g., information on every element of the product and its quality). In this sensemaking phase, consumers use their knowledge about, and feelings toward, elements and facets of a new entertainment product (such as the genre of a movie and the actors participating in it) to subconsciously develop imagery and anticipated and anticipatory emotions regarding the product.

Think about your reaction when you first heard about the filming of a BLADE RUNNER sequel. You were trying to "make sense" of it. If you are an ardent fan of the original classic (just like one of this book's authors), the information might have caused skepticism to bubble up, particularly when you hear that the original film's director will only produce (but not direct) this time. You try to figure out if the new director, Denis Villeneuve, is any good by reading reviews about his earlier works, maybe even watching a few of them on Netflix or DVD. Hearing that the writer of the original is crafting the sequel's screenplay excites you, but knowing that his co-writer authored the misguided GREEN LANTERN movie dampens your anticipation. Who will be in it? The fact that Harrison Ford will return and co-star with Ryan Gosling, who was so cool in DRIVE, excites you and creates inner images. The vividness of these inner images is heightened by the first stills shared via the Internet. Watching the teaser trailer then triggers high arousal. You can't wait to see the film anymore; you are filled with desire to do so.

---

[133]Wohlfeil and Whelan (2008) as well as Batat and Wohlfeil (2009) offer additional rich insight into consumers' entertainment consumption process in form of introspection studies.

In other cases, sensemaking will create much less anticipation and desire. A remake of The Magnificent Seven? Why filming again what was perfect the first time? No one can replace Steve McQueen and Yul Brynner. And when you quasi-accidentally stumble over the film's trailer on YouTube and learn that it does not even feature that glorious music theme, your desire cools off even further.

In any case, the outcome of such sensemaking is a certain level of desire for the product that will clearly differ between products; it is, like all consumer judgements about hedonic products, of a holistic type (versus attribute-based) and highly idiosyncratic for each consumer, based on his or her previous knowledge and experiences, preferences and motivations. It is this desire (or the lack of it) which results from sensemaking that mediates all future activities in the process.

Phase 2: Decision making. Based on the desire for a product that is experienced as the outcome of the sensemaking phase, the consumer's brain will, if a critical level of desire is exceeded, produce an intention to experience the product. Then, and only then, the consumer will explore consumption options. If this threshold is not reached, consumption of the product will not take place (at least not until desire changes). In the latter case, the process is interrupted, with the consumer either exploring other entertainment products (of the same form of entertainment—watching another movie—or a different form—e.g., playing a game) or engaging in something completely different (e.g., going to bed, working).

Desire and a resulting inner intention to consume are necessary for consumption, but they are not sufficient. Whether consumption eventually happens depends upon several contextual forces. Such forces include the consumption environment (e.g., is there a movie theater close enough which shows the film? Or can it be downloaded from iTunes? Does the movie's age rating allow the consumer to attend a screening?), the situational environment (e.g., does the consumer have enough time and money? Is the consumer in the right mood?), and the social environment (are friends available or are they insisting on doing something else?). These environmental conditions can also amplify or reduce the level of desire experienced by the consumer (and, subsequently, his or her consumption intention). For example, knowing that you simply do not have the time to watch a movie can subconsciously *suppress the development* of high levels of desire (to avoid disappointment), or can actually *lower an existing desire* when the consumer realizes this time constraint at a later point in the process.

In this phase, the cultural role of the product also matters. Is the product consistent with the subjective social norms of the consumer's local culture or society? For example, is experiencing a critically acclaimed drama movie such as LOVE considered appropriate if the film violates social norms by containing explicit hardcore sex scenes filmed in 3D? Do the film's good expert reviews and "art value" justify the consumer's desire and consumption intention under these circumstances? Like environmental forces, subjective norms can also influence the level of desire experienced by the consumer.

Phase 3: The consumption experience. It is during consumption when the consumer's future behavior regarding the product is determined. The consumption experience can trigger additional search activities. A consumer may look for new information to figure out how producers filmed the uninterrupted six-minute tracking shot in the TV series TRUE DETECTIVE.[134] But the experience of consuming the product can also stimulate the consumption of other (multi-)sensory stimuli, such as watching a movie again, listening to its soundtrack via Spotify, or purchasing its merchandise (think STAR WARS lightsabers). The quality of the experience also determines whether and how consumers communicate about the entertainment product via social media, websites, or personal exchanges with friends.

So, this is the whole picture of how we consume entertainment then? Not yet. Our discussion so far has largely glossed over one important aspect of entertainment products—their *social* dimension. Because entertainment consumption often involves and is influenced by social factors, a deeper look is warranted into the social environment that we have mentioned only cursorily in our discussion of the process model above. Consumers often prefer to enjoy entertainment together, in groups instead of alone. Because entertainment has a vital cultural function, consumers go to the movies together with their friends and spouses, play games with them (and others over the Internet), and listen to music with others at a party or dancing in a club.

Panel A of Fig. 6.10 shows a stylized overview of the process of entertainment consumption, flowing through the three stages of sensemaking, decision making, and the consumption experience, along with naming the different concepts involved in each stage. But it illustrates the full process model for two different consumers: one female who develops a desire for the fantasy classic THE NEVERENDING STORY, and a male consumer who, based on processing information about the film, desires to see the last-days-of-Hitler war drama DOWNFALL. Panel B of the figure then adjusts the process for the case that both want to go *jointly* to the movies. In our example, they end up seeing the raunchy, profanity-rich German hit comedy FACK JU GÖHTE, for which they both feel a desire to watch—together!

[134]In case you want to know: please check out Fukunaga (2014).

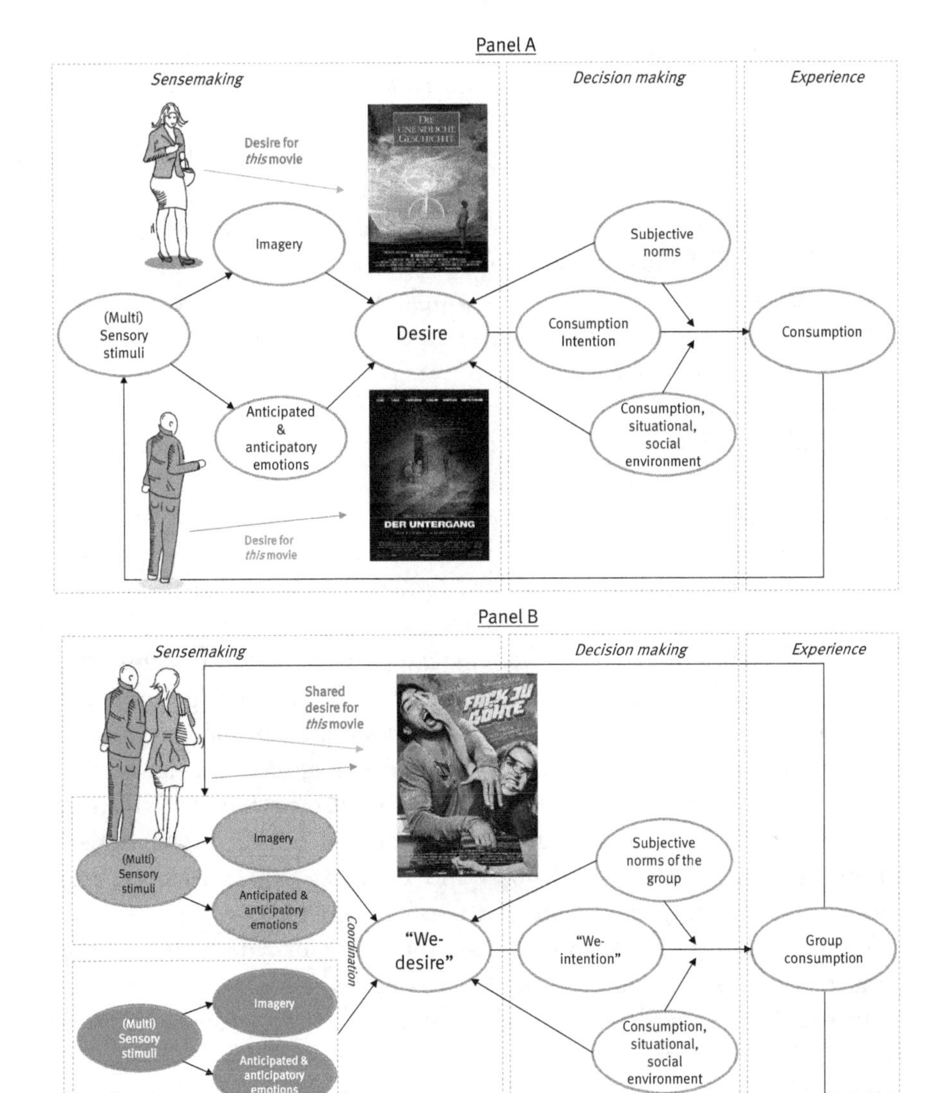

Fig. 6.10 A model of the entertainment decision-making process
Notes: Authors' own illustration. Panel A describes entertainment decision making when consumers act in isolation; Panel B shows the process for groups. Posters for THE NEVERENDING STORY, DOWNFALL, and FACK JU GÖHTE are © Constantin Film Verleih, with permission. With graphical contributions by Studio Tense.

In the "entertainment-consumption-as-a-group" model, we draw from Bagozzi's (2000) theory of "intentional social action." In it, Bagozzi postulates that groups are characterized by a certain "We-ness": the group members share their actions, beliefs, attitudes, and their desires, being aware of this "We-ness." The group's subsequent social action (such as the two consumers watching a movie together in our little example) is based on a joint "We-desire" for a common goal. Such "We-desire" emerges through coordination and interactions between the group members. In the example in the figure, the two consumers skip their original plans because the other group member had discrepant thoughts and feelings about their respective favorite choices. As a group, they develop a "We-desire" for a third movie, FACK JU GÖHTE.

If a critical threshold level of joint desire is passed, a "We-intention" is formed such that the group members become committed to act *as a body* to consume the product together. There is also initial evidence that such social entertainment action can influence the enjoyment derived from the consumption act: Ramanathan and McGill (2007) find, in an experiment with 57 students watching a clip from the TV show SATURDAY NIGHT LIVE, that group consumption results in higher enjoyment.[135] However, it seems safe to say that group cohesion and other factors will moderate such "group effects": going out with a potential love interest who does not reciprocate one's feelings will probably *not* result in a heightened positive evaluation of the dance music.

Finally, Intentional Social Action theory can also help us to explain the activities of groups that have much weaker social ties. Think of brand communities or social networks (Bagozzi and Dholakia 2002), but you could even go more broadly to like-minded fans/consumers, in general. All of us know what can happens when fans share the desire for a new entertainment product: it sparks anticipatory "buzz" behaviors, which are easily observable in our social media feeds, real-life conversations, trending topic lists online, etc. Such buzz expresses the large-scale "We-desire" of fans and consumers who are usually not directly connected, but know each other often only via their buzz behaviors. The "We-desire" of this amorphous group then might send a positive signal to those who have not yet joined the "movement" (and who doesn't want to be a fan of "the next big thing"!), which can initiate a virtuous circle, spreading to even more consumers.[136]

[135]This effect can be subconscious: the consumers attributed this enjoyment to the quality of the show, not to the presence of the other group members.

[136]For a discussion of the buzz concept and its role for product success, please refer our chapter on "earned" entertainment communication; in it, we return to the idea of "We-desire" cascades.

Concluding Comments

Understanding consumers is key for successfully managing entertainment because it is consumers who ultimately adopt (or ignore) new entertainment products. Even in those constellations in which advertisers or subsidy providers are a firm's main direct source of revenue and consumers do not hand over hard-earned dollars to access the products, it is the reactions of consumers that eventually determine an entertainment product's long-term success. In this chapter, we bring together a large number of scholarly studies to present a framework that explains—in a practical way—consumers' cognitive and emotional reactions to entertainment products, and how these reactions result in behaviors that include decisions of whether or not to consume a film, game, book, or song.

In short, firms must ensure that their products provide experiences that create desired levels of sensations for consumers to generate desirable emotions and imagery. But it is not only the sensations that drive enjoyment—consumers also value a new product's familiarity, the return of beloved heroes, places, and tunes which link new experiences with favorite previous ones. Combining sensations and familiarity in the right way attracts consumers and sparks their enjoyment via triggering emotions and cognitive processes that allow consumers to get transported into alternative universes and get "lost" in them. Such transportation then helps to realize key consumer entertainment motives such as escapism and social learning.

Determining the right combination of sensations and familiarity is a huge challenge though, as the links between the framework elements are complex and often subtle—too familiar offerings carry the danger of satiation. Emotions are multi-dimensional and the simulated nature of entertainment implies the distinction between immediate and later reactions. And transportation success is not only affected by the product, but also by consumer and situational factors.

In combination with an understanding of the unique characteristics of entertainment products and markets and the industry's value-creation processes and business models, this understanding of entertainment consumers lays the ground for Part II of this book: the managing and marketing of entertainment.

References

Addis, M., & Holbrook, M. B. (2010). Consumers' identification and beyond: Attraction, reverence, and escapism in the evaluation of films. *Psychology & Marketing, 27*, 821–845.

Almond, D. (2006). Introduction. In D. Hahn, L. Flynn, & S. Reuben (Eds.), *The ultimate teen book guide* (Vols. VII–VIII). London: A & C Black Publishers.

Arnold, M. B. (1960). *Emotion and personality*. New York: Columbia University Press.

Askin, N., & Mauskapf, M. (2017). What makes popular culture popular? Product features and optimal differentiation in music. American Sociological Review, forthcoming.

Aurier, P., & Guintcheva, G. (2014). Using affect–expectations theory to explain the direction of the impacts of experiential emotions on satisfaction. *Psychology & Marketing, 31*, 900–913.

Bachen, C. M., Hernández-Ramos, P., Raphael, C., & Waldron, A. (2016). How do presence, flow, and character identification affect players' empathy and interest in learning from a serious computer game? *Computers in Human Behavior, 64*, 77–87.

Bagozzi, R. P. (2000). On the concept of intentional social action in consumer behavior. *Journal of Consumer Research, 27*, 388–396.

Bagozzi, R. P., & Dholakia, U. M. (2002). Intentional social action in virtual communities. *Journal of Interactive Marketing, 16*, 2–21.

Bartsch, A., & Viehoff, R. (2010). The use of media entertainment and emotional gratification. *Procedia—Social and Behavioral Sciences, 5*, 2247–2255.

Batat, W., & Wohlfeil, M. (2009). Getting lost "Into the Wild": Understanding consumers' movie enjoyment through a narrative transportation approach. In *Proceedings of ACR* (pp. 372–377).

Baumann, N. (2012). Autotelic personality. In S. Engeser (Ed.), *Advances in Flow Research* (pp. 165–186). New York: Springer.

Behrens, R., Kupfer, A., & Hennig-Thurau, T. (2017). Empirical findings on the role of sensations and familiarity for motion picture success. Working Paper, University of Münster.

Bohnenkamp, B., Knapp, A.-K., Hennig-Thurau, T., & Schauerte, R. (2015). When does it make sense to do it again? An empirical investigation of contingency factors of movie remakes. *Journal of Cultural Economics, 39*, 15–31.

Bracken, C. C. (2006). Perceived source credibility of local television news: The impact of television form and presence. *Journal of Broadcasting & Electronic Media, 50*, 723–741.

Busch, A., & D'Alessandro, A. (2016). 'Star Trek Beyond' launches to $59 M; 'Lights Out' electrifies; 'Ice Age' tepid; 'Ghostbusters' no Cinderella story—box office final. *Deadline*, July 25, https://goo.gl/XyFPHm.

Bywaters, M., Andrade, J., & Turpin, G. (2004). Determinants of the vividness of visual imagery: The effects of delayed recall, stimulus affect and individual differences. *Memory, 12*, 479–488.
Campos, A. (2014). Gender differences in imagery. *Personality and Individual Differences, 59*, 107–111.
Choi, D., & Kim, J. (2004). Why people continue to play online games: In search of critical design factors to increase customer loyalty to online contents. *CyberPsychology & Behavior, 7*, 11–24.
Chou, T., & Ting, C. (2003). The role of flow experience in cyber-game addiction. *CyberPsychology & Behavior, 6*, 663–675.
Cline, E. (2011). *Ready Player One*. London: Arrow Books.
Cohen, J. B., Pham, M. T., & Andrade, E. B. (2006). The nature and role of affect in consumer behavior. In C. P. Haugtvedt, H. Paul, & K. Frank (Eds.), *Handbook of consumer psychology*. New Jersey: Lawrence Erlbaum.
Coursaris, C. K., van Osch, W., & Florent, S. (2016). Exploring the empirical link between game features, player motivation, and game behavior. *MCIS 2016 Proceedings*, Paper 53, 1–9.
Craik, F. I. M., & Dirkx, E. (1992). Age-related differences in three tests of visual imagery. *Psychology and Aging, 7*, 661–665.
Csíkszentmihályi, M. (1997). *Finding flow*. New York: Basic.
Csíkszentmihályi, M. (1975). Play and intrinsic rewards. *Journal of Humanistic Psychology, 15*, 135–153.
Dal Cin, S., Zanna, M. P., & Fong, G. T. (2004). Narrative persuasion and overcoming resistance. In E. S. Knowles & J. A. Linn (Eds.), *Resistance and Persuasion* (pp. 175–191). Mahwah, NJ: Lawrence Erlbaum Associates.
De Wied, M., Zillmann, D., & Ordman, V. (1994). The role of empathic distress in the enjoyment of cinematic tragedy. *Poetics, 23*, 91–106.
Drolet, A., & Luce, M. (2004). The rationalizing effects of cognitive load on emotion-based trade-off avoidance. *Journal of Consumer Research, 31*, 63–77.
Ekman, P. (1999). Basic emotions. In T. Dalgleish & M. Power (Eds.), *Handbook of cognition and emotion* (pp. 46–60). Chichester: Wiley.
Ford, R., Kit, B., & Giardina, C. (2017). Hollywood rethinks key movie franchises amid a mixed summer at the box office. *The Hollywood Reporter, 21*, https://goo.gl/YcJgNd.
Fukunaga, C. (2014). How we got the shot. *The Guardian*, March 17, https://goo.gl/qgrA3S.
Fowdur, L., Kadiyali, V., & Narayan, V. (2009). The impact of emotional product attributes on consumer demand: An application to the U.S. motion picture industry. Working Paper, *Johnson School Research Paper Series #22–09*.
Galton, F. (1880). Statistics of mental imagery. *Mind*, 5, 301–318. [Available at https://goo.gl/C39Uuc as part of the "Classics in the History of Psychology" repertoire.].

Gardner, J. (2013). Busting the Disney myth: Artist tears apart the unbelievably perfect anatomies of your favorite characters step-by-step. *Mail Online*, June 4, https://goo.gl/384ana.

Green, M. C., Brock, T. C., & Kaufman, G. F. (2004). Understanding media enjoyment: The role of transportation into narrative worlds. *Communication Theory, 14*, 311–327.

Hart, A., Kerrigan, F., & vom Lehn, D. (2016). Experiencing film: Subjective personal introspection and popular film consumption. *International Journal of Research in Marketing, 33*, 375–391.

Henning, B., & Vorderer, P. (2001). Psychological escapism: Predicting the amount of television viewing by need for cognition. *Journal of Communication, 51*, 100–120.

Henning, V., Hennig-Thurau, T., & Feiereisen, S. (2012). Giving the expectancy-value model a heart. *Psychology & Marketing, 29*, 765–781.

Hirschman, E. C. (1980). Innovativeness, novelty seeking, and consumer creativity. *Journal of Consumer Research, 7*, 283–295.

Hirschman, E. C. (1983). Predictors of self-projection, fantasy fulfillment, and escapism. *Journal of Social Psychology, 120*, 63–76.

Hirschman, E. C. (1987). Consumer preferences in literature, motion pictures, and television programs. *Empirical Studies of the Arts, 5*, 31–46.

Hirschman, E. C., & Holbrook, M. B. (1982). Hedonic consumption: Emerging concepts, methods and propositions. *Journal of Marketing, 46*, 92–101.

Hoffman, D. L., & Novak, T. P. (1996). Marketing in hypermedia computer-mediated environments: Conceptual foundations. *Journal of Marketing, 60*, 50–68.

Hoffman, B., & Nadelson, L. (2010). Motivational engagement and video gaming: A mixed methods study. *Educational Technology Research and Development, 58*, 245–270.

Katz, E., Blumler, J. G., & Gurevitch, M. (1973). Uses and gratifications research. *Public Opinion Quarterly, 37*, 509–523.

Kawakami, A., Furukawa, K., Katahira, K., & Okanoya, K. (2013). Sad music induces pleasant emotions. *Frontiers in Psychology, 4*, 1–15.

Kosslyn, S. M., Margolis, J. A., Barrett, A. M., Goldknopf, E. J., & Daly, P. F. (1990). Age differences in imagery abilities. *Child Development, 61*, 995–1010.

Lasseter, J. (2015). Technology and the evolution of storytelling. *Medium*, June 24, https://goo.gl/dRsCxd.

LeDoux, J. (1996). *The mysterious underpinnings of emotional life*. New York: Simon & Schuster Paperbacks.

Lee, B., & Lee, R. S. (1995). How and why people watch TV: Implications for the future of interactive television. *Journal of Advertising Research, 35*, 9–18.

Leung, R. (2004). Jimmy Buffet rediscovered. *CBS News,* October 4, https://goo.gl/RsBv6o.

Lin, J. T., Wu, D., & Tao, C. (2017). So scary, yet so fun: The role of self-efficacy in enjoyment of a virtual reality horror game. *New Media & Society*, https://doi.org/10.1177/1461444817744850, forthcoming.

MacDonald, R., Byrne, C., & Carlton, L. (2006). Creativity and flow in musical composition: An empirical investigation. *Psychology of Music, 34*, 292–306.

MacInnis, D. J., & Price, L. L. (1987). The role of imagery in information processing: Review and extensions. *Journal of Consumer Research, 13*, 473–491.

Madigan, J. (2010). The psychology of immersion in video games. *The Psychology of Video Games*, July 27, https://goo.gl/znZgNN.

Mansell, M. (1980). Dimensions of play experience. *Communication Education, 29*, 42–53.

Maslow, A. H. (1943). A theory of human motivation. *Psychological Review, 50*, 370–396.

McAlister, L., & Pessemier, E. (1982). Variety seeking behavior: An interdisciplinary review. *Journal of Consumer Research, 9*, 311–322.

Nakamura, J., & Csíkszentmihályi, M. (2002). The concept of flow. In C.R. Snyder & S. J. Lopez (Eds.), *Handbook of positive psychology* (pp. 89–105). New York: Oxford University Press.

Nemeth, B., Scheerens, L. J. J., Lijfering, W. M., & Rosendaal, F. R. (2015). Bloodcurdling movies and measures of coagulation: Fear Factor crossover trial. *BMJ, 351*, 1–7.

Pähler vor der Holte, N., & Hennig-Thurau, T. (2016). Das Phänomen Neue Drama-Serien. Working Paper, Department of Marketing and Media Research, Münster University.

Palmgreen, P., & Rayburn, J. D., II. (1982). Gratifications sought and media exposure: An expectancy value model. *Communication Research, 9*, 561–580.

Petrova, P. K., & Cialdini, R. B. (2005). Fluency of consumption imagery and the backfire effects of imagery appeals. *Journal of Consumer Research, 32*, 442–452.

Pham, M. T. (1998). Representativeness, relevance, and the use of feelings in decision making. *Journal of Consumer Research, 25*, 144–159.

Phillips, B. J., & McQuarrie, E. F. (2010). Narrative and persuasion in fashion advertising. *Journal of Consumer Research, 37*, 368–392.

Plutchik, R. (1980). *A psychoevolutionary synthesis*. New York: Harper & Row.

Posner, M. L., & Snyder, C. R. R. (1975). Facilitation and inhibition in the processing of signals. In P. M. A. Rabbitt & S. Dornič (Eds.), *Attention and performance V* (pp. 669–682). London: Academic Press.

Posner, J., Russell, J. A., & Peterson, B. S. (2005). The circumplex model of affect: An integrative approach to affective neuroscience, cognitive development, and psychopathology. *Development and Psychopathology, 17*, 715–734.

Ramanathan, S., & McGill, A. L. (2007). Consuming with others: Social influences on moment-to-moment and retrospective evaluations of an experience. *Journal of Consumer Research, 34*, 506–524.

Reber, R., Schwarz, N., & Winkielman, P. (2004). Processing fluency and aesthetic pleasure: Is beauty in the perceiver's processing experience? *Personality and Social Psychology Review, 8*, 364–382.

Reid, M. (2016). 16 sad songs to listen to when you need a good cry. *Lifehack*, https://goo.gl/hreJSx.

Reisenzein, R. (1994). Pleasure-arousal theory and the intensity of emotions. *Journal of Personality and Social Psychology, 67*, 525–539.

Rubin, A. M. (1981). An examination of television viewing motivations. *Communication Research, 8*, 141–165.

Russell, J. A. (1980). A circumplex model of affect. *Journal of Personality and Social Psychology, 39*, 1161–1178.

Ruth, N., Spangardt, B., & Schramm, H. (2016). Alternative music playlists on the radio: Flow experience and appraisal during the reception of music radio programs. *Musicae Scientiae, 21*, 75–97.

Ryan, R. M., & Deci, E. L. (2000). Intrinsic and extrinsic motivations: Classic definitions and new directions. *Contemporary Educational Psychology, 25*, 54–67.

Ryan, R. M., Scott Rigby, C., & Przybylski, A. (2006). The motivational pull of video games: A self-determination theory approach. *Motivation and Emotion, 30*, 347–363.

Schäfer, T., & Sedlmeier, P. (2009). From the functions of music to music preference. *Psychology of Music, 37*, 279–300.

Schäfer, T., & Sedlmeier, P. (2010). What makes us like music? Determinants of music preference. *Psychology of Aesthetics, Creativity, and the Arts, 4*, 223–234.

Sherry, J. L. (2004). Flow and media enjoyment. *Communication Theory, 14*, 328–347.

Slater, M. D., & Rouner, D. (2002). Entertainment-education and elaboration likelihood: Understanding the processing of narrative persuasion. *Communication Theory, 12*, 173–191.

Smith, L. J., Gradisar, M., King, D. L., & Short, M. (2016). Intrinsic and extrinsic predictors of video gaming behaviour and adolescent bedtimes. *Sleep Medizine, 30*, 64–70.

Stradella Road (2010). Moviegoers 2010. Company report.

Stuart, K. (2010). What do we mean when we call a game 'immersive'? *The Guardian*, August 11, https://goo.gl/V3wLHA.

Tamborini, R., & Stiff, J. (1987). Predictors of horror film attendance and appeal: An analysis of the audience for frightening films. *Communication Research, 14*, 415–436.

Thomas, N. J. T. (2017). Mental imagery. In E. N. Zalta (Ed.), *Stanford encyclopedia of philosophy*. https://goo.gl/PfTfuC.

Tomkins, S. (1962). *Affect imagery consciousness: Volume I: The positive affects.* New York: Springer.

van Laer, T., de Ruyter, K., Visconti, L. M., & Wetzels, M. (2014). The extended transportation-imagery model: A meta-analysis of the antecedents and conse-

quences of consumers' narrative transportation. *Journal of Consumer Research, 40*, 797–817.
Visch, V. T., Tan, E., & Molenaar, D. (2010). The emotional and cognitive effect of immersion in film viewing. *Cognition and Emotion, 24*, 1439–1445.
von Goethe, J. W. (1808). Faust: A Tragedy. Our English-language cite is from the 1870 edition translated by Bayard Taylor, published by The Riverside Press, Boston.
Vorderer, P. (2003). Entertainment theory. In J. Bryant, D. Roskos-Ewoldsen, & J. Cantor (Eds.), *Communication and emotion: Essays in honor of Dolf Zillmann* (pp. 131–153). Mahwah: Lawrence Erlbaum.
Vorderer, P., Klimmt, C., & Ritterfeld, U. (2004). Enjoyment: At the heart of media entertainment. *Communication Theory, 4*, 388–408.
Ward, M. K., Goodman, J. K., & Irwin, J. R. (2014). The same old song: The power of familiarity in music choice. *Marketing Letters, 25*, 1–11.
Weibel, D., & Wissmath, B. (2011). Immersion in computer games: The role of spatial presence and flow. *International Journal of Computer Games Technology, 2011*, 1–14.
Wild, T. C., Kuiken, D., & Schopflocher, D. (1995). The role of absorption in experiential involvement. *Journal of Personality and Social Psychology, 69*, 569–579.
Wirth, W., Hartmann, T., Böcking, S., Vorderer, P., Klimmt, C., Schramm, H., et al. (2007). A process model of the formation of spatial presence experiences. *Media Psychology, 9*, 493–525.
Witmer, B. G., & Singer, M. J. (1998). Measuring presence in virtual environments: A presence questionnaire. *Presence, 7*, 225–240.
Wohlfeil, M., & Whelan, S. (2008). Confessions of a movie-fan: Introspection into a consumer's experiential consumption of 'Pride and Prejudice'. In *Proceedings of European ACR Conference* (pp. 137–143).
Yee, N. (2006). Motivations for play in online games. *CyberPsychology and Behavior, 9*, 772–775.
Zacks, J. M. (2015). *Flicker: Your brain on movies*. New York: Oxford University Press.
Zajonc, R. B. (1980). Feeling and thinking: Preferences need no inferences. *American Psychologist, 35*, 151–175.
Zillmann, D. (1971). Excitation transfer in communication-mediated aggressive behavior. *Journal of Experimental Social Psychology, 7*, 419–434.
Zillmann, D. (1988). Mood management: Using entertainment to full advantage. In L. Donohew, H. E. Sypher, & E. Tory Higgins (Eds.), *Communication, social cognition, and affect* (pp. 147–171). Hillsdale, NJ: Lawrence Erlbaum.
Zuckerman, M. (1979). *Sensation seeking: Beyond the optimal level of arousal*. Hillsdale, NJ: Erlbaum.

Part II

Managing & Marketing Entertainment—What Makes an Entertainment Product a Hit?

So far, we have focused on the "arena" in which entertainment managers make decisions and the product-sided, market-sided, and consumer-sided conditions under which such decision making takes place. We have shared with you, our reader, what *Entertainment Science* can tell us about the specifics of entertainment products, the markets on which these products are offered, the value-creating strategies, and the consumers which, through their personal time and financial resources, determine the success of any such product.

We now shift our perspective and take a closer look at the drivers of the financial performance of entertainment products. What are the factors that help some movies, games, songs, and novels to become massive hits, while others flop so painfully? Scholars have conducted numerous studies to shed light on this quintessential question, and, in this part of the book, we will build on their findings to contribute to a comprehensive understanding of entertainment product success. Specifically, we will discuss all the factors that make up an "integrated marketing strategy" for entertainment products.

One of marketing theory's strongest achievements is the "Four-Ps" systematization originally introduced by Jerome McCarthy (1960), which we borrow here for our analysis of success drivers of entertainment. Specifically, we distinguish between the four fundamental pillars of such a strategy:

- the design and development of the entertainment product itself ("*product* decisions"),
- the flow of information surrounding the release of the product ("communication decisions" alias "*promotion*"),

- the ways the product is made accessible for consumers ("distribution decisions", or "*place*"), and
- the resources that consumers have to give (or agree to endure) to obtain the product ("*pricing* decisions").

But integrated marketing is not deserving of its name if the whole is not larger than its parts. Thus, we will also discuss two dominant integration strategies through which such synergies can be realized in entertainment markets: the blockbuster strategy and its counterpart, the niche strategy.

A fundamental rule in any market-oriented leadership approach is that understanding customers is the key to success, as it enables producers to link their efforts to consumers' needs and wishes. This must not be confused with the idea of simply asking consumers what they want to watch, play, or read, though; as we noted in our discussion of entertainment product characteristics, consumers have enormous trouble judging entertainment products before experiencing them (and sometimes even afterward), and they are even less able to articulate *future* preferences. Instead, we mean that any producer needs to understand the fundamental motivational forces, drivers, and processes that underlie consumers' entertainment consumption behaviors. Knowing these general behavioral mechanisms helps producers to better judge the power of their own decisions regarding the marketing of entertainment.

Figure II.1 shows how the elements of the entertainment marketing mix, which we will discuss in this second part of the book, are intertwined with consumers' internal processes, from the processing of information about the product itself ("sensemaking"), to information about the conditions under which the product is available ("decision making"), to the eventual consumption action ("experience"). Accordingly, each entertainment product possesses some unobservable qualities that affect the product's appeal to consumers; these qualities result from how the product is made by the producer and his or her team, constituting the outcome of the product strategy.

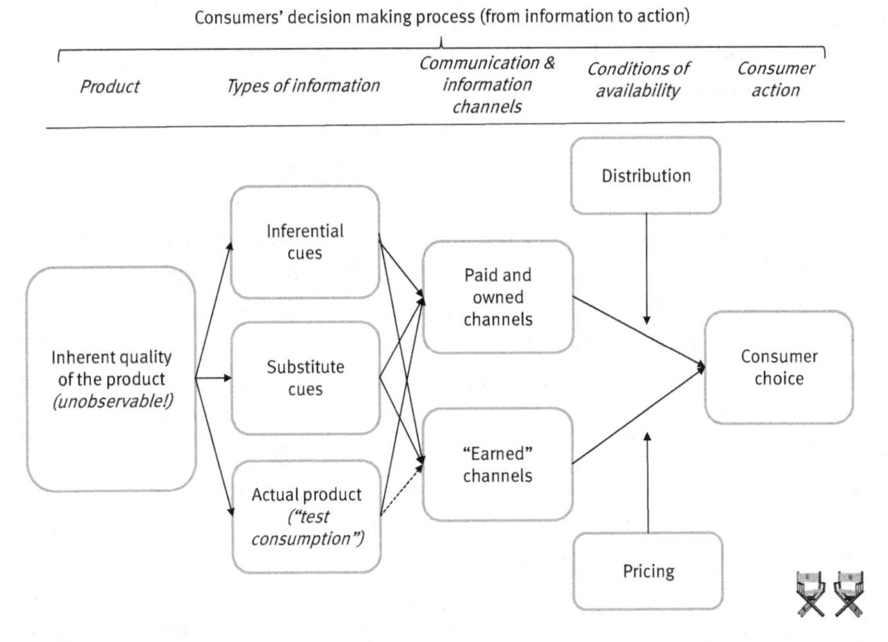

Fig. II.1 Linking consumer decision making and entertainment marketing strategies
Notes: Authors' own illustration based on ideas reported in Hennig-Thurau et al. (2001). The dotted line means that "test consumption" from "earned" channels is a gray area in legal terms, as often consumers do not have official rights to provide others access to the product (or parts of it, such as by sharing a movie trailer on YouTube).

Because consumers are largely unable to inspect these qualities directly (entertainment as experience products), they have to make judgments based on certain types of information that are not part of the product itself. Specifically, three types of information can provide clues to a consumer of an entertainment product's quality. First, "inferential cues" are elements of an entertainment product from which consumers can *infer* how much they will like it. For example, if the director of a movie has won prestigious awards for his earlier works, a consumer might infer that a new movie by this director will also be of an ambitious kind—which s/he might judge as "good" or "not so good," depending on his or her tastes.

Second, "substitute cues" involve information about the quality of the entertainment product that a consumer can use as a *substitute* for his or her own investigation of the product, such as a professional music reviewer's critique of a new album. Finally, consumers may also be able to test at least portions of the actual product. For example, a consumer can read a sample chapter of a book on a Kindle, listen to a song clip on iTunes, or play a trial or "demo" version (with limited capability) of a game.

Next, moving across the figure toward the right, the producer's communication-and-information strategy comes into play. First, marketers can provide the information via channels in which they can "control" the message, including "paid" channels (e.g., any form of paid advertising) and any channels that are "owned" (such as a movie website hosted by the producing studio or a Facebook page for an author's new book). But entertainment producers have to accept our days' reality that such controlled information channels are far from exclusive. Consumers can also access—and are regularly exposed to without actively searching—product-related information from channels in which the producer has little (or no) say.

Both professional reviewers and ordinary consumers now can easily share their thoughts and evaluations of new entertainment products—and they do so frequently and with passion, posting favorable and unfavorable comments (that may or may not be "fair," from the producer's point of view). We use the term "earned" channels for such platforms with tongue-in-cheek because too often entertainment producers get what they consider to be "undeserved" via these channels. But information from these channels is influential in the marketplace; often it helps consumers make up their minds regarding the quality of the entertainment product. Often such "earned" media also provide consumers access to the product itself (or parts of it), such as when consumers post screenshots of a movie, or share its trailer via YouTube or Facebook, or upload a full song or album (which links test consumption with the "sampling" effect of entertainment piracy).

But even if the "earned" information is positive, does this suffice to make a consumer choose the product? Not necessarily, because, as next shown in the figure, the conditions under which the product is available to the consumer also influence the decision—after "sensemaking" comes "decision making," as we discussed in our analysis of the consumer's consumption process. This is where the distribution strategy (e.g., you read about movie, but it is not shown in your area, and is also not available for watching at home) and the pricing strategy (e.g., "The new Mario game sounds fun, but for $65.99? Come on!") come into play.

Let us now dive into Part II of *Entertainment Science* and begin a closer investigation of the wide range of instruments that marketing, in the broad sense we define the concept here, offers entertainment managers. We start by looking at product decisions, and particularly those that deal with the quality of an entertainment product.

References

Hennig-Thurau, T., Walsh, G., & Wruck, O. (2001). An investigation into the factors determining the success of service innovations: The case of motion pictures. *Academy of Marketing Science Review, 1*, 1–23.

McCarthy, E. J. (1960). *Basic marketing: A managerial approach*. Homewood, IL: Richard D. Irwin.

7

Entertainment Product Decisions, Episode 1: The Quality of the Entertainment Experience

"Disney has become the envy of the industry…[by prizing] content over the means to distribute it."
—*The Economist (2015)*

From a marketing perspective, the key instrument for providing enjoyment to consumers is to create a product that delivers the highest levels of enjoyment sought by the consumer. Such great products are the means through which (traditional) producers, in a world characterized by the rise of powerful new entertainment distributors, can keep a strong position. Or, as *The Economist* (2015) named it, the "firms with the most popular stories and characters will have the most bargaining power over whichever distributor." In this chapter, we will explore what exactly makes up a "great" product in the context of entertainment.

Let us keep in mind, though, that the greatness of a product is an *experience quality*—something that consumers can only judge after having consumed the product, but not before. Regarding the financial success of a new entertainment product, this experience quality determines what entertainment managers call its "playability" (e.g., Elberse and Eliashberg 2003). Playability is when customers consume an entertainment product, perceive high experience quality, and are inspired to tell others. Playability produces a reaction among consumers who use their experiences with a product as the basis for triggering informed cascades through word of mouth (Lewis 2003).

We will thus look into what *Entertainment Science* research can tell us about the role of entertainment products' experience quality in driving their market performance. We also mine the existing research to unveil the drivers

T. Hennig-Thurau and M. B. Houston, *Entertainment Science*,
https://doi.org/10.1007/978-3-319-89292-4_7

of playability, paying special attention to the stories that mark the heart of narrative entertainment, not only explaining what "makes" a great story, but also discussing what data analytics can offer for help in understanding and creating them.

This chapter is not the sole one on the product strategy of entertainment; our book offers three more product "episodes" in which we will study the other major determinant of entertainment product success. Two of these other episodes examine products' "marketability," i.e., the ability to attract audiences at product release based on factors other than the product experience itself (Vogel 2015)—"unbranded" and "branded" product characteristics. And the final product chapter studies the innovation process for entertainment success. But first let us look at the (experience) quality of entertainment itself.

Linking (Experience) Quality with Product Success

> "In today's Hollywood, quality doesn't guarantee success."
> —Lang *(2015)*

We have shown that consumers have to make their decisions regarding entertainment products based on quality signals and substitute cues rather than on the product's true (experience) quality. To what degree than does it matter economically how "good" a film, book, or song is, particularly when considering that quality judgments are taste-based and thus subjective (versus objective)?

To take an initial look at this question, we drew again on our sample of 200 movies, but this time added information about each film's North American box office performance. We started by looking at the opening weekend performance because it is a good proxy for a film's marketability; very little experience-based quality information from other consumers is available at this early point in time and consumers have to base their decisions mainly on signals. But we also looked at the films' financial performance in the weeks and months that followed, when more consumers had experienced the product and shared their quality perceptions. At these later times, potential moviegoers can take these judgments by others into consideration—the playability component of success.

Figure 7.1 shows the results of a number of OLS regression analyses we ran to determine the links between consumers' overall quality judgments and two measures of market success: Panel A of the figure lists the results for

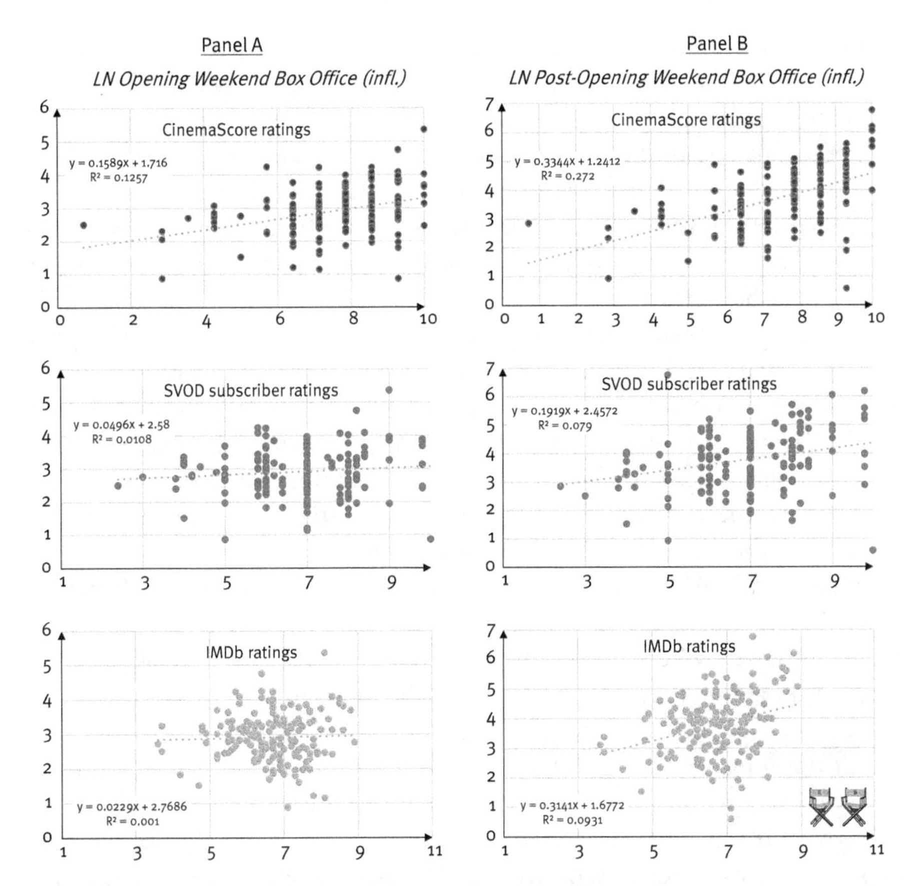

Fig. 7.1 Linking quality ratings of movies and financial success

Notes: Authors' own illustration based on data from various sources (see our notes for Fig. 3.11 on p. 93). Box office information is based on The Numbers; we adjusted all box-office data for inflation and log-transformed it to account for the non-normal distribution of that data. Functions shown are OLS regression estimates without control variables.

the first-weekend box office and Panel B for the box office in the following weeks. We used three different quality assessments: those from moviegoers who saw the film on opening night in theaters (as surveyed by CinemaScore and shown in the upper layer of the figure), from the subscribers of the international SVOD service (medium layer), and from registered IMDb users (bottom layer). As can be seen in the left side of the figure where the results for the opening weekend box-office results are listed, "good" movies, as judged by the expert consumers on IMDb, do not have a performance advantage in their opening weekend (bottom of Panel A)—the correlation is only $r=0.04$ and not statistically significant.

So, does this mean that it does not matter at all whether expert consumers consider a film to be of high quality or not, perhaps because the expert consumers' tastes are too different from the majority of moviegoers? The answer to this question is no—as can be seen in the chart at the bottom layer of Panel B, IMDb ratings correlate positively and clearly more highly with box-office results *after* the opening weekend ($r=0.31$). So, if expert consumers like a movie, on average, the movie's financial performance is better. How much better? The taste rating explains about 10% of the variation in post-opening weekend box office ($R^2=0.31^2=0.096$). So, Lang's quote above is certainly right: the link is *far* from perfect. The same pattern is found for "ordinary" consumers (i.e., SVOD subscribers)—their taste is also independent from early movie success (or the movies' marketability), but correlates with *later* success to a similar degree that we found for IMDb ratings.

However, the analyses show that it does matter for a movie's opening success whether it is liked by the *target* audience (i.e., those who see it on opening night), at least to a certain degree ($r=0.36$, top layer of Panel A). But let's note there might be a selection bias; these quality ratings come from those consumers who have opted to see the movie based on the signals they received, *expecting* they would like it. So the correlation points out that, on average, signals and experiences do somewhat overlap.

But the numbers also point at a different reason why it is important for a movie's success that the target group likes a movie—its taste judgements strongly influence others within the group, and maybe/probably even beyond it. This effect is reflected by the high correlations between opening-night audiences' quality judgments and later movie success ($r=0.51$, i.e., explaining 26% of the success variation across movies; top layer of Panel B). By using the simple models reported in the figure (and ignoring some complexities of the entertainment business for a moment), we can show with the regressions that an "average" movie would make an additional $12.8 million in North American theaters *after* the opening weekend if its core audience rated it as "A-" instead of "B+."

Similar correlations have been reported for other quality measures and also have been verified in other countries. We conducted a structural equation modeling analysis of 331 movies (from 1998 to 2001), in which we measure the effect of a multi-faceted quality variable (comprised of IMDb and CinemaScore ratings, among others) on opening weekend and long-term performance, while controlling for other "success drivers."[137]

[137] Those other success drivers in our study included measures of advertising and distribution.

Results show that the association between this quality measure and films' long-term success is 2.5-times as strong as is the association with initial performance (Hennig-Thurau et al. 2006). And for China, Wu (2015) finds significant and positive correlations of 0.18 and 0.27 between quality ratings on Chinese community websites and Chinese box-office data for 383 Chinese-language movies released in Mainland China in 2010–2013.

In summary, we see that consumers' quality perceptions *do* matter economically. But it depends on whose taste we are talking about, and also what facet of success we have in mind. And even for the target group of an entertainment product, the impact of their quality perceptions must not be over-stated. Empirical data provides evidence that high quality increases the chances for an entertainment product to be successful (particularly in the long run), but findings also conflict with popular industry phrases like "When it's good, in the end it makes money" (PRETTY WOMAN and ONCE UPON A TIME IN AMERICA producer legend Arnon Milchan, quoted in Shanken 2008).

The decent amount of success that quality can explain means that whereas it does usually not hurt to have a great product, quality is neither a "guarantee" for success in entertainment (which, by the way, would run counter to the probabilistic paradigm this book is based on) nor even a necessary condition, particularly when it is the movie experts, not the masses, who are doing the quality judging.[138]

What Makes High-Quality (a.k.a. "Great") Entertainment?

But *what* exactly do people like when it comes to entertainment? In our earlier exploration of consumers' entertainment-related behavior, we concluded that entertainment products must offer both sensations and familiarity to create

[138]Of course, one could always argue that, drawing on Mr. Milchan, if a great entertainment product has not returned its investments, "it's not the end" yet… The problem with this logic is that even if quality finds its way in the long term in some cases, investors usually don't want to wait to get their money back. In addition, ad hoc evidence for our finding that quality is only loosely linked to success is abundant. It comes from those who have produced great works and lost a lot of money (like CITIZEN KANE and ONCE UPON A TIME IN AMERICA, which generate revenues today, but ruined their makers when they came out). It is also true for less well-known films: THE IRON GIANT—IMDb: 8.0 (out of 10), global box office: $23 million at $70 million budget; CHILDREN OF MEN—IMDb: 7.9, global box office: $70 million at $76 million budget; THE INSIDER—IMDb: 7.9, global box office: $60 million at $90 million budget). And evidence also comes from what consumers believe are bad products, but generated a fortune (e.g., ALVIN AND THE CHIPMUNKS: CHIPWRECKED—IMDb: 4.4, global box office: $343 million at $75 million budget; COUPLES RETREAT—IMDb: 5.5, global box office: $172 million at $70 million budget; TRANSFORMERS: AGE OF EXTINCTION—IMDb: 5.7, global box office: $1.1 billion at $210 million budget).

high levels of enjoyment. So the question we want to answer in this section is: what kinds of product elements are linked (according to *empirical* evidence) to the sensations and familiarity that can lead to the desired consumer reactions, i.e., perceptions of being "great entertainment"? Keep in mind that, as we just argued, these might not be the same as the factors that distinguish commercially successful from less successful entertainment products.

Scholars have shed some light on this issue by searching for factors that empirically correlate with consumers' quality perceptions of entertainment products. Probably the most revealing study on the topic was conducted by Morris Holbrook (1999), who assembled a rich data set of 1,000 movies that were produced prior to 1986, including several critically acclaimed and commercially successful films. Through an OLS regression in which he used the movie quality ratings of a representative sample of viewers of the pay-TV station HBO as dependent variable, he finds that consumers' quality perceptions are statistically associated with certain genres (more positive for family fare, more negative for dramas), with a film's country of origin (higher for films made in the U.S.), English-language dialogue (versus other languages), and the participation of star actors and prominent directors. Holbrook's results suggest that consumers also like certain technical aspects (color and a longer runtime), which can serve as search-quality indicators that are so rare in entertainment.[139] He also finds that consumers appreciate "up-to-dateness"—the older a movie is, the more it is discounted (and at an increasing rate).

Among the most intriguing insights of Holbrook's study is that, on average, "offensive" violent content and "exploitative" sexual material lead to a perception of *lower* quality by the audiences he surveyed. This finding might be attributed to the "mainstream" character of the surveyed consumers; we assume it differs between consumer segments. Holbrook's findings that the quality judgements of "experts" (movie guide book author Leonard Maltin in his study) are not impacted by displays of violence, and that sexual material has a *positive*, quality-enhancing effect, can be considered an empirical indicator of heterogeneity of tastes.[140]

---

[139]See our earlier section on the dominance of experience qualities for entertainment. We will discuss the role of technological attributes as entertainment search qualities in the next chapter.

[140]Holbrook (1999) also finds other discrepancies between the tastes of HBO audiences and his "experts." Experts' judgments are negatively influenced by recency (older films are rated systematically *higher* by them!), whereas a foreign language soundtrack and a non-U.S. origin are associated with more *positive* quality perceptions. In another study of expert quality criteria for movies, Wallentin (2016) explains professional critics' judgments of almost 2,000 movies shown in Sweden from 1999 to 2011 using regression analysis. Like Holbrook, he finds that experts see a higher quality in non-U.S. films. Focusing on genre effects for experts, he finds a positive effect for dramas and documentaries, whereas action, family, comedy, romance, and horror have negative effects on the experts' quality perceptions.

Overall, the factors in Holbrook's study explain a remarkable 38% of consumer-perceived quality judgments. In addition to giving us initial ideas about what defines high quality in consumers' minds, some of the findings help us to better understand the potential relative impacts of these factors on a product's *commercial* success (such as for genres and stars).

Moon et al. (2010) build on Holbrook's seminal work and examine 246 movies released in North American theaters from 2003 to 2005, using average consumer ratings from Yahoo Movies. With a series of stepwise regressions, they find that consumers incorporate two other factors into their quality judgments, both of which are now much easier to access for consumers than in the pre-digital era of Holbrook's sample: the commercial success of a movie, up to that point in time, and how well the movie has been rated by other consumers. The effects of both factors are positive, which also aligns with our discussion of success-breeds-success effects (see our chapter on "earned" entertainment communication); market success in a given period exerts a feedback effect on success in future periods, causing even greater success.

To what extent do these findings hold for other entertainment contexts? Using data for 383 Chinese films from 2010 to 2013, and using the ratings of the quality of films by consumers on two Chinese community websites as dependent variables, Wu (2015) also runs OLS regressions. He finds some of the same quality determinants from Holbrook's study to be relevant in his context. As for their American counterparts, Chinese audiences' quality perceptions are positively linked to the participation of star directors and actors, as well as certain genres. Interestingly, he finds a different quality patterns for genres, probably because of cultural differences in entertainment preferences: thrillers are seen as being of lower quality, crime movies as having higher quality.[141] In total, the variables in his analysis explain between 24 and 31% of consumers' quality perceptions.

Much less is known about quality determinants for other entertainment products. In our own investigation of new drama TV series, we asked 1,800+ German respondents to rate a selection of the series they had watched on 16 attributes (Pähler vor der Holte and Hennig-Thurau 2016). We then used this information in a regression analysis to explain consumers' quality judgments of those series. Our findings shed detailed light onto what consumers consider a "great" new drama series. Accordingly, the strongest impacts on consumers' quality ratings result from a series' particular atmosphere and its production value. These variables are followed by "smart" dialogues, deep, original characters that are focal to the series and show a dynamic development, and surprising plot turns.

[141]We discuss international differences in genre preferences in more detail in the following chapter.

Getting Closer to the Product's Core: What Makes a "Great" Storyline?

> "All screenwriters think their babies are beautiful. I'm here to tell it like it is: Some babies are ugly."
>
> —Vinny Bruzzese, *as CEO of Worldwide Motion Picture Group (quoted in* Barnes *2013)*

The studies we cited in the previous section give us some ideas of what consumers have in mind when they think of (or feel) "great" entertainment. Beyond genre conventions, however, a product's plot, which is the product element that gives the concept of narrative entertainment its name, has been largely left out in these analyses. The reason is quite obvious: the complexities of a plot are notoriously difficult to measure empirically.

Nevertheless, some *Entertainment Science* scholars have made attempts to explore what sets "great" stories apart. We extract their key insights in this section, looking at two interrelated issues: the use of data analytics and theories to *understand* what characterizes a great story, and the use of these two key elements of *Entertainment Science* to *create* great stories, taking the concept of artificial intelligence to its most extreme.

Using Analytics and Theory to Understand "Great" Storytelling

Crafting a good story, and understanding what makes it good, is quite challenging because of the "multiplicative production function" of entertainment: each element has to work well with every other element for the end product to be of high quality. Just like multiplying a large number by zero results in a zero, having cool characters but combining them with a weak plot results in a poor story. And a story is very complex, consisting of myriad elements on its own (characters, dialogue, settings, plot, etc.); because of their multiplicative relationship, a single weak element can have a massive negative impact on the quality of the overall result. In other words, a weak ending can kill the whole movie, as a tsunami-surfing super-agent can end our trip into his alternative universe.

In the entertainment industry, crafting a good story has always been "guru" territory. That is, an elite set of people have claimed that they (and they alone) *know* what it takes to craft a successful storyline (whether in a novel or movie, or in a narrative game or song). One of them, Robert

McKee, names himself the "most sought after screenwriting lecturer," quoting director Peter Jackson for calling him "The Guru of Gurus" (McKee 2016). Hollywood's ADAPTATION movie is essentially all about him and his approach to storytelling.

Now what has McKee's work to do with *Entertainment Science*? The common element is that McKee's arguments, like those of most other gurus, are neither the result of "genius" or "inspiration," but are instead interestingly similar to some of the oldest *theories* on entertainment. They trace back to ideas from Greek philosophers such as Aristotle, Roman poets including Horace, and German novelist Freytag, all of whom have proposed that great storytelling would require a certain number of parts (often referred to as "acts"), with the recommended number of parts varying between three (Aristotle) and five (Horace and Freytag)—very similar to what today's "gurus" recommend.

Further, entertainment scholars have adapted these historical ideas for specific forms of entertainment and also used empirical data to refine our understanding of stories. For movies, the most renowned classification is the "four-acts" theory by Thompson (1999), which distinguishes between the "setup" act (during which the audience is introduced to key characters, their motivations, and the environments), a "complicating action" act (in which difficulties are introduced), a "development" act (in which the story broadens and the characters struggle), and finally a "climax" act (in which the action concludes and conflicts are resolved). Based on her manual coding of 73 movies, Thompson (1999) finds that the industry standard is an almost equal allocation of runtime across the four acts.

Cutting (2016) builds on her findings, making a major step toward integrating "acts theory" into a more coherent narrative theory of entertainment. He decomposes 150 popular movies (i.e., commercially successful and/or highly rated) from between 1935 and 2010, using objective measures such as the average shot duration, the amount of motion and brightness levels, use of music, when in the story new characters and locations were introduced, and shot types. His findings confirm the four acts identified by Thompson, but also add two optional subunits (which he names prolog and epilog) and a few turning points and plot points. By using detailed measures, he also links each act with certain specific styles and elements. Such structural insights are certainly very interesting on a descriptive level, but we have to admit that their prescriptive value is limited—neither Thompson nor Cutting link narratives to consumers' quality perceptions or commercial performance, beyond that such narratives are found in "good" films. But do they discriminate those from others?

In a study that aims to combine descriptive and prescriptive insights via the use of data analytics, Reagan et al. (2016) compute a book's "emotional arc"—an illustration of how happiness develops as the story unfolds. They do so by applying automated sentiment analysis (which calculates the "happiness level" of the words used by a book's author) to 10,000 excerpts of 1,327 English-language books (mostly fiction) from the Project Gutenberg website at multiple points of the story. They then identify (through a factor-analytic approach) three main emotional arcs that, together with their inversions, capture 85% of the books in their sample. The arcs, which we display in Fig. 7.2, are all representations of classic story patterns: (1) a "rags to riches" arc which is characterized by a continuous rise in happiness (top-left graph); (2) a "tragedy" arc (or "riches to rags") as an inversion of arc 1; bottom left) which implies a continuous fall in happiness; (3) a "man in a hole" arc (with happiness first falling, then rising; top middle); (4) an "Icarus" arc as an inversion of arc 3 (happiness first rising, then falling; bottom middle); (5) a "Cinderella" arc (happiness rising-falling-rising; top right); and (6), as the inversion of arc 5, an "Oedipus" arc (happiness falling-rising-falling; bottom right).

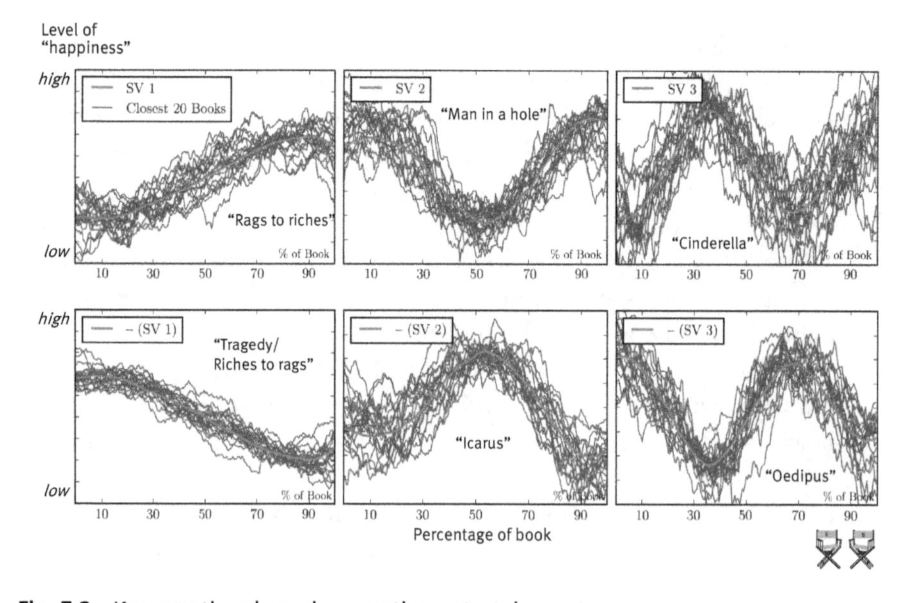

Fig. 7.2 Key emotional arcs in narrative entertainment

Notes: Reprinted with minor modifications from Reagan et al. (2016, p. 7), which is distributed under the terms of the Creative Commons Attribution 4.0 International License (https://goo.gl/n6FkkT). "SV" is the respective mode/factor identified from matrix decomposition. The orange line shows the average "happiness" level for a mode over the course of a book; each grey lines is for one of the 20 books in the data set that most closely follow it.

The prescriptive contribution of Reagan et al.'s investigation is an analysis of the respective download numbers from the Project Gutenberg site for books that follow a particular arc. The authors find that book *downloads* are highest for "Oedipus" books and those with a "rags to riches" arc. Interestingly, those are not the patterns which most books in their data set follow—these are "tragedy," "Icarus," and "man in a hole" stories instead. Let us add that when the authors account for outlier books when assessing downloads (by using the median instead of the mean value), the "Oedipus" arc ranks still highest, but "Icarus" and "Cinderella" take second and third rank.

We find it interesting that downloads are highest for emotional arcs that contain a larger number of emotional ups and downs over the course of the book—maybe this suggests that people often consume entertainment for quick emotional stimulation, rather than appreciating the long emotional developments. However, success-related implications should be interpreted with care; the database used by Reagan and colleagues does not necessarily reflect the taste of "mass consumers." The statistics also tell us to be cautious: the variance in downloads is *very* high for all arcs (several thousand times the mean).[142]

Prescriptive insights also come from studies that do not look at the narrative as a whole, but that instead focus on certain aspects of a story and the links between these story aspects and product success. Just like narrative structures and emotional arcs, such aspects can affect success by heightening (or lowering) the quality of the entertainment experience instead of functioning as a quality signal (in most cases, the audiences just don't know the aspects prior to consumption).

What specifically can research tell us then about such story aspects? Eliashberg et al. (2007) derive 22 aspects of screenplays from drama theory and screenwriting practice and test their roles in enabling the prediction of a movie's performance (return-on-investment, or ROI).[143] They use human coders to determine whether the aspects are present in short versions of the early scripts of 200 movies from 2001 to 2004 and then apply

[142]We assume that, as a result of this high variation, the reported differences in downloads are probably not significant (the authors do not report any statistical tests).

[143]Specifically, these script aspects encompass: a clear premise, a familiar setting, an early exposition, the avoidance of coincidences, interconnectedness, surprises, anticipation of what happens next, no flashbacks, a linear timeline, clear motivation of characters, a multidimensional hero, a strong nemesis, a sympathetic hero, believable characters, hero character growth, important conflict, multidimensional conflict, conflict build-up, conflict lock-in, an unambiguous resolution, a logical ending, and a surprising ending. Let us note that Eliashberg et al. also include some other interesting language characteristics, such as the use of passive sentences and the average word length of sentences. Because they use fan-created texts instead of original treatments in this study, the results for these variables are probably of limited generalizability though.

a decision-tree approach to determine the links between each story aspect and movies' ROI. Jehoshua "Josh" Eliashberg, a legend of the *Entertainment Science* field, and his colleagues find that the presence of a clear premise is of prime importance for a story to influence a film's commercial performance. They also find that believable characters, an early exposition (in which the characters are introduced early on in the film), and a plausible story that does not rely on coincidences help a film to be successful. Based on such script-related information, Eliashberg et al. apply their model to a set of 81 additional movies whose short scripts they coded accordingly and, with 62% accuracy, are able to predict whether a movie has an above- or below-average ROI. This rate might be far from perfection—but what we find more relevant is that it is clearly superior to models that lack such storyline information.

In a follow-up project, Eliashberg et al. (2014) test the relevance of the same set of script aspects with real, full-length screenplays instead of short scripts. Again they let humans code the presence of various aspects, in this case for 300 movies. Then, using a kernel-based regression approach, Eliashberg et al. link each script aspect to movies' North American box office performance. As in their earlier study, results show that several aspects predict a film's success, but some difference in importance can be seen. For the full screenplays, the most important aspects are, in descending order, an early exposition, a strong nemesis ("every character gets what s/he deserves in the end"), a continued anticipation of what will happen next, a setting that is familiar to audiences,[144] an unambiguous resolution, and growth in the hero's character. Their analysis also includes measures of the language used in the screenplays, which they code via machine learning techniques; Eliashberg et al. find hints that some "bag-of-words" predict movie success better than others, but no definite insights can be derived from their results here. We have to wait for more studies to better understand the role of language in stories.

Can Computers Craft "Great" Stories?

As the use of data analytics continues to rise, the limits of what the approach can accomplish continue to be pushed back. In entertainment, this progress raises a particularly controversial question: can data analytics be employed to "program" a great entertainment product? In other words, can entertainment managers use artificial intelligence and big data not only to understand the

[144]An aspect that is tied closely to the concept of familiarity, by the way.

dramaturgic patterns that trigger certain consumer reactions (which is what we have discussed in the preceding section), but also to *autonomously* create new work that translates such learnings into new stories, songs, or games? To do so, an algorithm would essentially need to be capable of creativity. Keep in mind that to generate pleasure, entertainment must offer consumers familiarity, but it must also create new and stimulating sensations.[145]

So, are computers capable of doing so—engaging in creativity? Certainly the possibility can be envisioned. Fascination with the mere idea of using algorithms to produce creative work is at the heart of several influential works by novelist Philip K. Dick (e.g., Do Androids Dream of Electric Sheep?), whose robots are more human than humans. But numerous filmmakers have also mined this territory, being more (Ridley Scott's Blade Runner films) or less (e.g., James Cameron's Terminator, Spike Jonze's Her, and Alex Garland's Ex Machina) closely inspired by Dick's ideas, as have top-shelf TV visionaries, such as Jonathan Nolan (Westworld). But besides such visions, what do we *know* about the issue from research? What is science fiction—and what is science fact?

David Cope, a composer and computer scientist, is leading a stream of scholars who have proposed computational processes which produce musical works.[146] His approach indeed combines data analytics with theoretical considerations, *Entertainment Science* style. He deconstructs classical compositions by human composers such as Bach and Chopin (the "data analytics" part), identifies common elements in them (the first "theory" part), and then uses an emergent understanding of "musical rules" (the second "theory" part) to "recombine" song elements into new works (see Cope 1996, 2005). As a consequence of this recombination process, Cope often names his algorithmic compositions after the composer whose work they are based on, such as with Sonata Movement (After Beethoven). A related attempt at "algorithm-generated" entertainment (in the context of narrative entertainment) is the theater play Beyond the Fence, which had a two-week experimental run in London. A team of computer scientists employed algorithms to develop the play's central premise, the plot structure, the "core narrative arc," and the music material. The computer scientists did not reveal, however, details on how this was done, and which parts of the result were actually computer-generated (Jordanous 2016).

[145]Please refer back to our discussion of the fundamental role of creativity in entertainment and the sensations-familiarity framework in prior chapters.

[146]For a broad review of the state of algorithmic music composition, see the articles in McLean and Dean (2018).

Listening to Cope's works (which you can do for free at his website at the time of writing)[147] and watching the London play raises the question of what actually constitutes creativity—and whether it isn't something more than the mere recombination of product elements in a new way. Innovation scholars have used the concept of "innovativeness" to deal with related questions; the concept distinguishes offerings that provide only incremental changes to existing products from those that are "radically" new (e.g., Garcia and Calantone 2002).[148] In essence, whether something is radically new eventually lies in the eye of the beholder (that is, the entertainment consumer), but new products are usually only considered to be radically new if they tap uncharted territory, not just offer a rearrangement of existing products. In entertainment, the fundamental concept of sensations is essential for judging a new product's innovativeness. New entertainment products are of high innovativeness when they stun us with sensations that we have never before experienced. In contrast, low innovative products lack such sensations, but offer us variations of familiarity.

Accepting this logic tells us something about the powers and limitations of computer-generated entertainment: the "newness" of entertainment products built by artificial intelligence comes solely from their reconfiguration of things that already exist. As a result, whereas their familiarity will tend to be quite high (we have actually experienced all ingredients before!), algorithms' potential to create sensation is inherently restricted, and so seems to be their level of innovativeness. Cope's algorithmic creations illustrate that data analytics, in combination with smart theory, can indeed craft new creative products; however, these products will sound (or look, or read, or play) quite like the ones we have experienced before. Other scholars have reported a massive 63% overlap between an original Chopin composition and one which Cope has claimed to be generated by algorithms, suggesting that the recombination step of his approach is in large parts a *reassembling* task (Collins and Laney 2017).[149]

Consistent with conclusions regarding music, the critical responses to the algorithm-authored London play named it "formulaic" and "pattern-driven,"

---

[147]You can find Cope's website and his musical compositions at https://goo.gl/1W3pnm.

[148]We discuss the concept of innovativeness as part of our innovation management chapter.

[149]This finding, along with the fact that Cope has never released the algorithms underlying his creations in full, has led some to question the credibility of his work; for example, Collins and Laney (2017) name him a "somewhat controversial figure," something that does not necessarily conflict with his role as a pioneer in the field of computer-generated music and entertainment.

criticizing that "nothing felt fresh."[150] In addition to the lack of fresh sensations, reviewers of the play complained about a lack of awareness of context and of longer-term structures. These weaknesses point to another, more technical challenge for the use of artificial intelligence in entertainment (particularly for narrative products such as novels, films, or games, more than for individual songs): it is nearly impossible for a simulation to account for the overwhelming level of complexity of narrative entertainment. This limitation is acknowledged by one of the music programmers for Beyond the Fence: "Even if they give you a stroke of genius, they can never follow that up… every thought is a new thought" (Benjamin Till, quoted in Jordanous 2016).

Be aware though that all of this evidence does not provide a straightforward answer to the question whether computer-generated entertainment products will be able to succeed *commercially*. To do so, one has to determine the role of radical sensations for consumers' liking of new entertainment—which is definitely not a trivial thing to do, considering that sensations are only *one* element of the sensations-familiarity framework and that algorithms appear better equipped to provide familiarity, as Cope's work indicates. The current state of mainstream entertainment suggests that, at least in the short-run, people are fine with offers characterized by a dominance of familiarity but comparability low levels of new (radical) sensations. Some have described recent successful blockbuster products as having an "artificial-intelligence" quality, arguing their formulaic cleanness would "appear to be designed by computers who have studied human taste" (The R-gument(or) 2016).

But there seem to be limitations. Empirically, Collins and Laney (2017) found (for a similar algorithm to Cope's) that even when musical experts could not distinguish the algorithmic from the human creation, they liked the algorithm's work much less than the original compositions.[151] But it is unclear to what extent it would apply also to the mass market. Our original analysis of consumer desires makes us believe that an overdose of familiarity will not work in the longer term, as people will eventually become satiated by familiarity. But it is not clear what "the longer term" means—weeks? Months? Years? Or even decades?[152]

[150]Let us note that one might question the low-innovativeness character of reconfigurations—think of the first iPhone as a reconfiguration of existing products (Steve Jobs, in his original announcement speech in 2007, referred to it as a combination of an iPod, a mobile phone, and an Internet device; quoted in Wright 2015). But whereas high innovativeness products might build on existing technology and offers, they take large liberties at integrating them—which again would require high levels of creativity.

[151]See also the reactions from a small sample of student listeners who Simoni (2018) repeatedly exposed to different pieces of algorithmic music.

[152]See also our analysis of the state of the industry as part of the integrated entertainment marketing chapter for a discussion of the prospects of familiarity-dominated entertainment offers.

Let us end by conceding that not everyone agrees with our argument that limitations of artificial intelligence in entertainment are systematic and structural. Google's "Brain team" follows Cope's approach, with their "Magenta" project being driven by the mission to develop intelligent machines that "can learn how to generate art and music, potentially creating compelling and artistic content on their own" (Eck 2016). Sony also has ambitious goals with its Computer Science Laboratory and has already released a number of pop songs that Sony claims were "composed" solely by algorithms, again mimicking the works of certain artists like The Beatles or Irving Berlin (Freitag 2016; see also IBM's attempt to create a movie trailer based on artificial intelligence which we discuss in the context of paid entertainment communication).

These people believe that computers will surprise us one day, just as android David surprises his predecessor Walter by composing his own tunes in the movie ALIEN: COVENANT. But this remains science fiction (the ALIEN tunes were, in real life, composed by a human being, Harry Gregson-Williams), and we all will have to wait for the next editions of this book to find out whether these efforts will disrupt entertainment or just disappear, like so many other "moonshots" by the technology giants of our digital age.

Concluding Comments

To what degree does it influence the success of an entertainment product how good it actually is? Research provides evidence that such "experience" quality does indeed matter economically. The patterns differ depending upon whether you look at the quality judgments of experts versus "ordinary" consumers, but the end result is that higher quality, on average, draws higher sales—at least over the long haul and, even then, only to a degree that is able to explain just a limited share of sales.

Our analysis has shown that the question of "what constitutes experience quality?" is a more difficult one to answer. We show that there are some differences between genres and other factors, but the most insightful findings deal with the storylines that underlie narrative forms of entertainment. Scholars have fused drama theory and data analytics to empirically measure types of "emotional arcs" in entertainment stories; their studies shed some initial light on how these arcs in are linked with commercial success.

We wrapped up the chapter by discussing whether artificial intelligence is the future creator of "great" entertainment. Story and song writers might breathe a sigh of relief: early efforts to use artificial intelligence to automate entertainment creations show that the creation of fresh sensations is a domain where human creativity and art beats computers, and we name strong arguments that this is not going to change soon. Thus, we recommend that entertainment producers focus their quality-related efforts on judging how strongly human creations address the theoretical requirements of "great" entertainment, rather than letting computers write the next song or screenplay.

But the experience character of entertainment products limits the role of consumption experiences and assigns an important role to those factors that influence consumers' decision to spend time and money for a certain offering—search qualities and particularly "signals" of quality. This is what the next chapter is about.

References

Barnes, B. (2013). Solving equation of a hit film script, with data. *The New York Times*, May 5, https://goo.gl/S5RvQt.

Collins, T., & Laney, R. (2017). Computer-generated stylistic compositions with long term repetitive and phrasal structure. *Journal of Creative Music Systems, 1*.

Cope, D. (1996). *Experiments in musical intelligence*. Madison: A-R Editions.

Cope, D. (2005). *Computer models of musical creativity*. Cambridge, MA: MIT Press.

Cutting, J. E. (2016). Narrative theory and the dynamics of popular movies. *Psychonomic Bulletin & Review, 23*, 1713–1743.

Eck, D. (2016). Welcome to Magenta! *Magenta*, June 1, https://goo.gl/gWNt8B.

Elberse, A., & Eliashberg, J. (2003). Demand and supply dynamics for sequentially released products in international markets: The case of motion pictures. *Marketing Science, 22*, 329–354.

Eliashberg, J., Hui, S. K., & Zhang, J. Z. (2007). From story line to box office: A new approach for green-lighting movie scripts. *Management Science, 53*, 881–893.

Eliashberg, J., Hui, S. K., & John Zhang, Z. (2014). Assessing box office performance using movie scripts: A Kernel-based approach. *IEEE Transactions on Knowledge and Data Engineering, 26*, 2639–2648.

Freitag, C. (2016). Listen to 'Daddy's Car,' a track composed by artificial intelligence. *Pigeons & Planes*, September 22, https://goo.gl/9vvqJ2.

Garcia, R., & Calantone, R. (2002). A critical look at technological innovation typology and innovativeness terminology: A literature review. *Journal of Product Innovation Management, 19*, 110–132.

Hennig-Thurau, T., Houston, M. B., & Sridhar, S. (2006). Can good marketing carry a bad product? Evidence from the motion picture industry. *Marketing Letters, 17*, 205–219.

Hennig-Thurau, T., Walsh, G., & Wruck, O. (2001). An investigation into the factors determining the success of service innovations: The case of motion pictures. *Academy of Marketing Science Review, 1*, 1–23.

Holbrook, M. B. (1999). Popular appeal versus expert judgments of motion pictures. *Journal of Consumer Research, 26*, 144–155.

Jordanous, A. (2016). Has computational creativity successfully made it 'Beyond the Fence' in musical theatre?, Conference Paper, University of Kent.

Langfitt, F. (2015). How China's censors influence Hollywood. *NPR*, May 18, https://goo.gl/fNtSM9.

Lewis, J. (2003). Following the money in America's sunniest company town: Some notes on the political economy of the Hollywood blockbuster. In J. Stringer (Ed.), *Movie blockbusters* (pp. 61–71). London: Routledge.

McCarthy, E. J. (1960). *Basic marketing: A managerial approach*. Homewood, IL: Richard D. Irwin.

McKee, R. (2016). The Aristotle of our time. https://goo.gl/ApH3B9.

McLean, A., & Dean, R. T. (Eds.). (2018). *The Oxford handbook of algorithmic music*. New York, NY: Oxford University Press.

Moon, S., Bergey, P. K., & Iacobucci, D. (2010). Dynamic effects among movie ratings, movie revenues, and viewer satisfaction. *Journal of Marketing, 74*, 108–121.

Pähler vor der Holte, N., & Hennig-Thurau, T. (2016). Das Phänomen Neue Drama-Serien. Working Paper, Department of Marketing and Media Research, Münster University.

Reagan, A. J., Mitchell, L., Kiley, D., Danforth, C. M., & Dodds, P. S. (2016). The emotional arcs of stories are dominated by six basic shapes. *EPJ Data Science, 5*, 1–12.

Shanken, M. R. (2008). An interview with Arnon Milchan. *Cigar Aficionado*, September/October, https://goo.gl/7xuGwM.

Simoni, M. (2018). The audience reception of algorithmic music. In A. McLean & R. T. Dean (Eds.), *The Oxford handbook of algorithmic music* (pp. 531–556). New York, NY: Oxford University Press.

The Economist (2015). The force is strong in this firm, December 19, https://goo.gl/E7vbr4.

The R-gument(or) (2016). Why Hollywood might want to scale back (a little): A rant against modern tentpole filmmaking, August 26, https://goo.gl/vGeiys.

Thompson, K. (1999). *Storytelling in the new Hollywood*. Cambridge, MA: Harvard University Press.

Vogel, H. L. (2015). *Entertainment industry economics: A guide for financial analysis* (9th ed.). Cambridge: Cambridge University Press.

Wallentin, E. (2016). Demand for cinema and diverging tastes of critics and audiences. *Journal of Retailing and Consumer Services, 33*, 72–81.

Wright, M. (2015). The original iPhone announcement annotated: Steve Jobs' genius meets genius. TNW, September 9, https://goo.gl/fxpx5t.

Wu, S. (2015). An empirical observation of Chinese film performance drivers. *Empirical Studies of the Arts, 33*, 192–206.

8

Entertainment Product Decisions, Episode 2: Search Qualities and Unbranded Signals

As we have argued above, the quality of the entertainment experience is important because it determines the "playability" of a product. However, it is only one piece of a more complex commercial-success puzzle. Playability is complemented by the "marketability" of entertainment, which is not determined by experiential product quality (at least not solely, and not even directly)—because the consumer has yet to experience that quality first-hand when making a decision about its adoption. Instead, marketability is driven by the information about a product that is available to consumers when making that adoption decision.

In addition to holistic "substitute cues," such as word of mouth and expert reviews, consumers can glean information from some observable elements of the product itself to make such consumption-related decisions. Among the very few "true" search quality attributes that entertainment products possess, we will study the role of technological attributes (such as 3D and virtual reality features) in this chapter.[153] But there are other

[153]One other product element that constitutes a search attribute for some kinds of entertainment deserves a special mention here: the *packaging*. An entertainment product's package can add value on its own, such as the utility and coolness of a special DVD box set (Plumb 2015 lists some impressive examples). Packaging may also be a hidden force that contributes to the current revival of vinyl albums—one of this book's authors (guess who!) has a history of collecting vinyl soundtrack albums for his most beloved films. The topic of packaging in this section overlaps with our discussion of digital technology when we describe the value of the physical package for haptic qualities that digital versions lack. But the main commercial relevance of packaging in entertainment comes from its informative and communicative capabilities, which we discuss more thoroughly later in the context of communication.

T. Hennig-Thurau and M. B. Houston, *Entertainment Science*,
https://doi.org/10.1007/978-3-319-89292-4_8

informative elements beyond such search qualities that are much more interesting to us here and, hopefully, also to managers: *signals* of an entertainment product's quality that scholars often refer to as "quasi-search attributes." These signals are not part of a product's actual, "objective" quality, but rather serve as inferential cues for consumers about this quality.

What exactly are such signals, or quasi-search attributes? The present chapter focuses on categorical attributes of entertainment products, such as a product's genre and country of origin—attributes for which consumers can hold strong cognitive and emotional associations and which differ between alternative offerings. Whereas these categorical attributes provide information and orientation, they are, unlike specific brands, rather broad and abstract quality "categories," which somewhat limits their influence on consumers, as we will see. In addition, their categorical nature prevents managers from shaping an attribute like this as is possible for brands, but their inclusion can still take an essential part in an entertainment marketing strategy. In the chapter after this one, we analyze what we consider "branded" signals of entertainment quality, including sequels and stars.

But first, in this chapter, let's now take a look at technology and its role as a rare "true" search quality, followed by a discussion of a number of key "unbranded" signals of entertainment.

Technology as a Search Attribute in Entertainment

> "It's not the technology that entertains people, it's what you do with the technology."
>
> —*Director and Pixar executive* John Lasseter *(2015)*

One notable element of entertainment products that is observable before consumption and that conveys useful information to consumers (and, therefore, can function as a search attribute) is the use of technology. The histories of entertainment and technology are tied together closely; each of the entertainment products we discuss in this book has undergone fundamental technological changes since its invention. Our goal here is not to provide a historical overview of such changes, but instead to highlight *Entertainment Science* research that helps to explain more recent technological developments and help managers to assess the relevance of such technologies and their potential impacts on the commercial success of entertainment products.

So, what factors influence whether a newly created technology will pay off for a producer of entertainment? Let us first clarify that in entertainment almost any new technology (at least any successful one) diffuses across products and companies, becoming generic and industry-wide. For example, 3D technology is not exclusive to a single movie but utilized by many of them. The Fusion 3D Camera System developed by James Cameron in cooperation with his business partner Vince Pace and with help from Fujifilm and Sony has been used in more than 20 major movies (*Wikipedia* 2016). The reason is the short life cycle of entertainment products which prevents product-specific technology from being lucrative; this is a major difference that contrasts with industries with longer life cycle products, such as in health care. But nevertheless, technology is almost always first developed with a specific product in mind, just like Cameron created the Fusion cameras for his AVATAR film. And quite often, this first product's success determines not only the career paths of its makers, but also the technological innovation's licensing potential.

We identify three key factors that influence whether a new technology as part of an entertainment product will facilitate that product's success. Because our perspective in this book is one that puts customers at the core of all things, the first factor that determines whether an entertainment product can profit from a new technology is the *customer benefits* it offers. Does the new technology offer consumers meaningful increases in the level of pleasure they derive from experiencing the entertainment product, compared to alternative products that use the existing technological solutions instead? Technologies (in entertainment and elsewhere) are developed by engineers who tend to be biased toward optimizing the "objective" quality of products, but it is the "subjective" character of quality—as perceived by consumers—that determines whether a technology can influence a product's success.

Technological innovations are effective if they heighten the level of sensory stimulation and trigger sensations (Holbrook and Hirschman 1982), making entertainment experiences richer and more engaging (Netflix's Gomez-Uribe and Hunt 2015). Technologies can offer consumers pleasure by intensifying already existing sensory stimuli (e.g., better sound quality) or by adding new sensory stimuli (e.g., adding the sound element to silent movies).

As large parts of the entertainment industry lack a customer focus, we consider this factor the primary reason why so many seemingly promising technological developments have failed to change the game—they have been embraced by entertainment managers for their "coolness" or engineering quality, but not for their potential to make movies, games, or novels more fun.

Hollywood legend George Lucas appears to agree when he says that a "lot of the hype on new technology is overhyped" (quoted in Patton 2015).

A second important factor is the technology's *costs*: are the incremental revenue potentials associated with the new technology (i.e., higher sales volumes and/or higher prices) sufficient to cover the costs for its development and implementation, not to mention to compensate for risk? Depending on the creator's time frame, costs can be considered for a single product, a slate of products, or (if the technology is to be licensed to other producers) an industry-wide rollout of a new technology—as it eventually happened for Cameron and Pace's camera technology.

Third, the *infrastructure required* for the benefit-generating use of a new technology needs to be considered. Even if consumers tend to like a new technology, in general, the technology's effective usage can suffer from insufficiently developed framing factors. A new CGI-producing software might be limited by the capabilities of the hardware it requires ("production infrastructure"—think of the early low-quality computer-generated special effects), the network of consumers that own the required hardware ("network/platform infrastructure"—see our discussion of indirect network effects in our chapter on entertainment market characteristics), and the "supply/distribution infrastructure" (i.e., is the number of movie theaters equipped with 3D projectors sufficient to exhibit my 3D movie? Are consumers able to stream movies in HD over the Internet?).

Director James Cameron delayed his AVATAR film for several years, advancing the motion capture technology so essential for his film. In contrast, the producers of THE POLAR EXPRESS, while being able to use a well-develop animation method, had to face severe supply limitations—very few movie theaters were able to show the digital 3D version of the film, which was the first of its kind back in 2004, so that the film's rollercoaster attractions appeared pretty flat for the majority of audiences.

In what follows, we discuss a number of key findings from *Entertainment Science* scholars that shed light on the role and contributions of new technologies for the different forms of entertainment that we feature in this book.

Technology and the Quality of Games

No entertainment product category is more closely tied to technology than games. Technology is the very essence of video games; their inherent nature is embedded in technology and they cannot be consumed without a proper device at hand. Technological innovations began with vector graphics (versus pixels) in LUNAR LANDER and then proceeded to range from the use of a three-dimensional perspective (first employed in 3D MONSTER MAZE) to the motion-capture capability used for WII SPORTS. Today, crucial technologies include Augmented Reality (AR) and Virtual Reality (VR). Whereas AR uses technological means to add to and/or modify consumers' perceptions of the real world, VR replaces the real world with a fictitious one into which the consumer is transported/immersed.

Regarding VR effects, research evidence demonstrates that virtual reality hardware devices, such as head-mounted displays, can increase consumer immersion when used for games. For example, Nichols et al. (2000) compared the immersion perceptions of 24 students who played a "duck shoot" game in a virtual reality setting (using a head-mounted display) versus those who played the game in a desktop setting. The scholars found reflex responses to be higher and background awareness lower for the VR condition; VR players also rated several measures of immersion more highly (differences ranged from around 15% to over 25%).

But there's a catch: other studies of VR have observed nauseogenic or disturbing reactions by players, particularly after consumers have been exposed to the VR condition for more than 45 min. For example, Cobb et al. (1999) report "serious" sickness symptoms for seven out of 148 participants across a number of VR experiments, and "minor" and "short-lived" symptoms for the majority for the rest of the participants. Nichols and her colleagues also ran a study with a 20-minute "virtual house" simulation, noting substantial feelings of nausea and disorientation among participants. Finally, Lin et al. (2002) report similar findings from a study using a driving simulator. Note that these findings tend to be rather old, so that the observed negative reactions might be triggered at least in part by lower-resolution and -frame rate devices which have been replaced today.

So, how will VR impact consumer enjoyment, and for which consumers—what is the net result of VR's presence-enhancing capability and potential negative effects? A recent experiment by Shelstad et al. (2017) offers some insights: when the scholars let 40 undergraduate students play in sequence a VR-enhanced and a standard version of the commercial "tower-defense strategy game," DEFENSE GRID 2, using a state-of-the-art Oculus Rift headset and a 24-in. monitor, respectively, they find that the overall satisfaction with the experience and the participants' enjoyment are significantly higher for the VR-enhanced game version (by 6 and 10% on average). It is unclear though how closely these results are tied to the specific game and hardware used in the experiment and the people that participated in it.

Producers will have to continue to improve the technology, reducing potential sickness effects (which Shelstad et al. did not measure separately) with even higher-resolution tools, while increasing the fun of using VR. But higher levels of realism are not necessary a good thing in this context: in Lin et al.'s study, enjoyment does not increase with higher realism, while sickness increases. A challenge for the technology might be the way it was presented in extant works of entertainment: we doubt that consumers' expectations toward the performance of VR, shaped by uber-impressive science-fiction versions (think: Star Trek's holodeck, the Matrix's *deja vues*, but also the limitless virtual "OASIS" simulation in the novel Ready Player One and its big-screen adaptation by Steven Spielberg), will be met by real life products anytime soon. Thus, satisfaction and the benefits consumers' perceive might be limited by those unrealistic expectations for quite a while (*The Economist* 2017a).

The commercial potential of AR technology is even more difficult to isolate empirically. Designing an authoritative, generalizable study is a challenge because of the endless range of possible applications of the technology (which part of the real world should be enhanced? How?) and the difficulty of choosing a meaningful alternative condition (i.e., non-AR). But the release of the Pokémon Go app game on Android and Apple smartphones in July 2016 indicates that AR can offer attractive benefits for consumers: in the week after its release, the game, which lets consumers search their "real-world" neighborhoods to track down virtual creatures, was installed on more than 10% of all North American Android smartphones and played by more than half of the users on a daily level (Perez 2016).

In an analysis of the Pokémon case, Tang (2017) attributed the game's success to the fact that the AR technology enabled users to "fulfill their [childhood] dreams [of becoming Pokémon trainers] in reality"—in other words, combining the AR sensations with the attractions of high familiarity. In a rare scholarly experiment, Avery et al. (2016) tested consumer reactions to a self-developed outdoor AR game ("Sky Invaders 3D") and compared them to consumer reactions to a desktop version of the game. Their results, based on 44 student participants, showed a higher level of enjoyment for the AR version, but the difference was not significant; however, replay intentions were significantly higher for the AR version.

Overall, whereas AR's immersive potential might be limited compared to VR, its adoption does not appear to be accompanied by the negative side effects that scholars have observed for VR, a difference that might be important for a broad acceptance of the technology. In predicting that AR "is going to become really big. VR, I think, is not gonna be that big, compared to AR," Apple's CEO Tim Cook is not alone (quoted in Strange 2016; see also *The Economist* 2017b). And the analysis by Ailie Tang which suggests that partnering AR with strong, emotion-laden brands can offer consumers strong benefits, may guide the way for future adaptation of the technology in entertainment. The rather short-lived nature of the Pokémon Go hype might raise some questions, though; in November 2017, the search volume for the game was only about 2% of what it was four weeks after its release 16 months earlier.

Finally, can the enjoyment of games also be enriched by olfactory stimuli (i.e., smell), maybe in combination with VR? Howell et al. (2016) ran a very small-scale within-subjects experiment with six consumers, who participated in a VR simulation (using the Oculus Rift head-mounted display) showing a bowl of oranges. The researchers found that adding the scent of oranges to the environment resulted in only small increases in immersion for four participants, with no increase for the other two. Future studies might trigger different results, but these very preliminary findings do not instill much hope that entertainment's somewhat less-than-impressive history with olfactory stimuli will change anytime soon.

Technology and the Quality of Movies

Technological advances have also played a big role in movie producers' (and distributors') efforts to increase the sensory stimulation (and, eventually, consumer pleasure) from the movie-watching experience. Some of those attempts have been highly successful: consider the addition of sound ("talkies") to what was before a visual-only experience, then the subsequent progression through mono soundtracks, stereo, and surround sound (Block and Wilson 2010) The addition of color to a medium that had consisted of black-and-white images was also a game-changer.

The success of many action, science-fiction, and fantasy films has been tied to innovative special effects—which enable consumers to be transported into mythical worlds (the dinosaurs in Jurassic Park, the movie), but can also serve as attractions on their own (the same could be said for games, by the way). Innovative special effects can also facilitate the transfer of heroes and visions from other entertainment categories, as has been experienced with the superhero genre since the late 2000s: "All that changed was visual effects. When Iron Man came out, visual effects had caught up so that going to see a superhero movie was worth it to see for the spectacle, and not [only] worth to see it because you were a pre-existing fan" (movie director James Gunn, quoted in D'Alessandro 2017a).[154]

[154]Please also see our discussion of the role of technological resources for entertainment firms in the market characteristics chapter.

Several other so-called "advances" have turned out to be short-lived, however. Remember "Sensurround" sound? It was an attempt to add a "physical" element to the viewing experience, as a movie's soundtrack was played through specially developed, low-frequency bass speakers so the sound could be "felt," not only heard, by the audience. It ended up being used only for a handful of action films, such as EARTHQUAKE, because it caused damage in some theaters and it disturbed audiences of films shown in adjacent screening rooms; Fuchs 2014). But we assume that the main reason was that the technology simply did not offer substantive benefits to consumers, particularly taking the quite substantial implementation costs into account.

Another failed attempt to use technology to enhance the audio-visual experience introduced a smell element. In the 1960s, "Smell-O-Vision" blew thirty different odors, synchronized with the action of the film SCENT OF MYSTERY, into specifically prepared theaters, involving non-trivial costs. In contrast, the "Odorama" approach required much less preparation: cards were handed out to audience members, who were then asked to scratch a certain spot on the card when a corresponding number was shown on the screen; the card then released a specific smell (including one of a *fart*; Nowotny 2011). Again, we doubt that the audience saw a major benefit in those approaches.

Nevertheless, some companies believe that digital technologies can enhance the benefits for moviegoers provided by physical and smell elements, developing approaches that turn movie going into a truly multi-sensory experience. By the end of 2017, Seoul-based CJ E&M ran about 400 theaters, mostly in Asia and the U.S., which combine physical sensations (moving chairs and exposure to "wind," "rain," and "mist") with smell in an attempt to "draw you into the movie as if you're living inside its world" (*CJ* 2017). The technology, which tracks the on-screen movement scene-by-scene and thus requires involvement of the producers, has been compared with theme-park rides, which might point at its future role as a niche attraction for *some* kinds of effects-heavy film content. It is unclear though how sustainable its attractions are (how often are we in the mood for roller coasters?) and how appealing the experience is considered by consumers in general. As one wrote, "on the whole, being shaken isn't very fun" (Grierson 2014).

Other current key technologies for filmed entertainment include 3D and higher frame rates.[155] 3D, a stereoscopic presentation technology, has received uneven reactions from consumers over several decades, but is

[155]Some movie executives have also articulated interest in the use of VR as a means to enhance the movie-watching experience, such as by using VR headsets as "virtual movie theaters" (Busch 2017). In addition to enormous (and costly) technological requirements, the consumer value of such applications appears questionable at best, however.

enjoying more lasting success since the early 2000s when digital presentation technology became available.[156] Whereas film makers and marketers have focused on the immersion-enhancing power of (now digital) 3D, and studios and theaters have invested enormous amounts in it, the long-term perspectives for the technology are subject to a controversial debate (see, for example, the letter from film editor Walter Murch in Ebert 2011).

With 3D being an expensive technology for those who produce films, the core issue once more is whether 3D offers consumers benefits that outweigh the higher costs. When marketing research firm YouGov asked 3,000 consumers in the UK whether a 3D screening "makes the cinema experience better" for them, 22% agreed. But more reported no difference (28%), and nearly the same number (19%) felt that watching a movie in 3D even *worsened* the experience (Follows 2017). These being self-reported judgments, however, what can scholars tell us about consumer reactions? Rooney and Hennessy (2013) conducted a survey of 225 consumers who had just seen Marvel's THOR in cinemas either in 2D or in 3D. The scholars find that those who watched the movie in 3D reported a greater degree of perceived realism and self-reported attention (during the film), but no more emotional arousal or satisfaction than 2D viewers. Results by Ji and Lee (2014), who showed 102 consumers a 15-minute segment of a Hollywood movie on a large TV screen, also found no difference in terms of enjoyment between 2D and 3D audiences; in their case 3D did not even imply higher levels of immersion and flow.

But Ji and Lee's results are interesting for another reason: they point to a potential moderating role of movie genre for these effects, as the levels of immersion and flow tend to differ between film genres.[157] Cho et al. (2014) also stress the role of potential moderating forces. When they showed a 15-minute self-produced 3D film (for a $100,000 budget!) to 188 participants in a theater,[158] the results indicate that 3D effects vary with consumers' seat locations: consumers who were seated in the front or the back of the theater were less satisfied with the film than those in the middle.

But Cho et al.'s study also points to another pattern that we consider to be critically important for film producers: consumers' previous experiences with 3D screenings were accompanied by reduced arousal and immersion, an insight

[156]Previous historical periods in which 3D films bloomed were the early 1950s (with films including Alfred Hitchcock's DIAL M FOR MURDER from 1954) and the early 1980s (e.g., JAWS 3-D).

[157]This result needs to be treated with care though, as the authors do not report a formal moderation test.

[158]The $100,000-budget the authors used for producing the film was remarkable and indicates a high level of professionalism.

which indicates a "wear-out" or satiation effect of 3D.[159] Understanding such satiation was focal for our own study in which we examined the economic consequences of the 3D presentation format based on actual market data (Knapp and Hennig-Thurau 2014). Specifically, we collected box-office results for all 73 digital 3D movies that were widely released in North American theaters from 2004 to 2011, starting with THE POLAR EXPRESS. We compared the box office of each of these films not with the box office of all other films, but with those of what we call a "statistical twin"—a 2D film that is similar across other key movie characteristics that drive a movie's success, as we argue at various points in this book.[160] Such a selection process is necessary to avoid comparing apples with oranges, or what econometricians call a "treatment bias" that results from an independent variable being not truly independent (or exogenous), but endogenous.[161] 3D movies are often treated differently (better!) than other films made in 2D. An econometric technique named "propensity score matching" helped us identify those movie twins.

Then we compared the performance of the 3D movies with those of their twins. The results were somewhat sobering: across all 73 3D movies in our data set, the 3D movies did, on average, not generate more revenues than their 2D lookalikes. Even worse, they attracted *fewer* attendees than did their 2D twins, and they were significantly *less* profitable (they had a lower average ROI). Follow-up analyses revealed a more fine-grained picture though, with time being the essence: while 3D movies blew up the box office in the early years of digital 3D, they no longer provide an advantage.

This is the main message of Panel A of Fig. 8.1, which shows the development of the average financial effects of 3D on the box office over time, per our estimations.[162] Please note that this refers to box-office results: the

[159]For more on satiation in entertainment, please refer to our discussion in the chapter on entertainment product characteristics.

[160]These other determinants of movie success included genre, production budget, advertising spending, number of opening screens, participation of stars, and being a sequel. In essence, these other determinants of movie success helped us to rule out the possibility that the 3D movies in our data set differed systematically from their 2D movie "twins" with regard to any criterion that could cause a potential difference in success. Not controlling for the relevant determinants could have resulted in wrongly attributing a difference in success between our 3D and 2D movies to their 3D versus 2D nature. Let us add one methodological note: the twin identified by the approach we used here is not a "real" movie—instead, it is a "hybrid" movie that represents a weighted combination of all 1,082 2D movies in our database that were released in the same time frame as the 3D movies (from 2004 to 2011).

[161]We explain the general problem of such endogeneity in regression models in our introductory chapter. "Treatment biases" in entertainment are by far not limited to the use of 3D, but hamper our understanding of how several other product characteristics, such a product being a sequel, a remake, or featuring a star, influence product success. We will get back to this issue in the respective chapters and sections of our book.

[162]The estimates are the result of a polynomial weighted least squares (WLS) regression model in which we included the linear and squared interaction terms of the 3D variable and a film's production year.

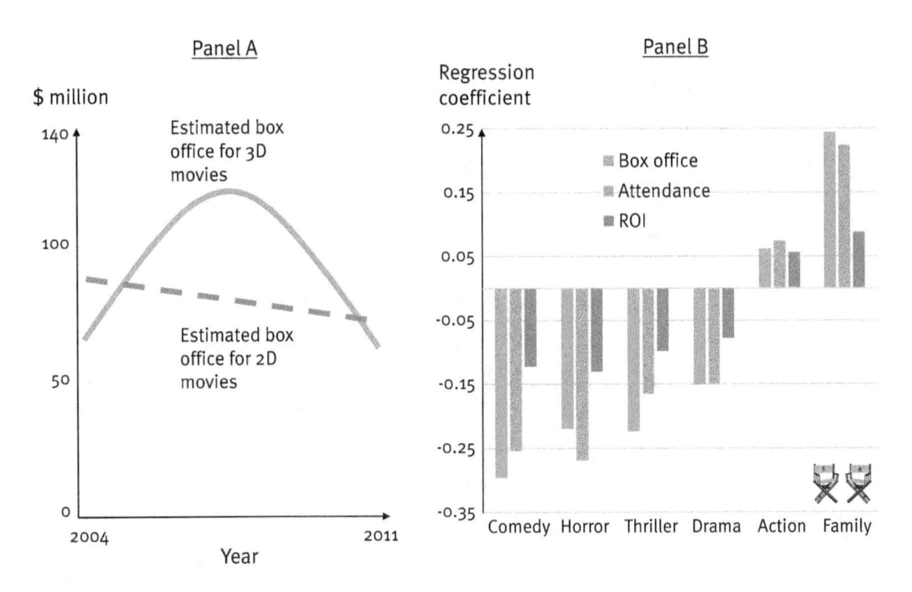

Fig. 8.1 The impact of 3D on movie success

Notes: Authors' own illustration based on results reported in Knapp and Hennig-Thurau (2014). The parameters in Panel B are unstandardized regression coefficients for a 3D × genre interaction from a WLS regression with North American box office as dependent variable and the 3D variable, the respective genre, and controls as independent variables. Weights are taken from a matching procedure that accounts for the different treatment 3D films receive from producers.

estimates suggest that the higher revenues per ticket are neutralized by the fewer tickets sold for a 3D movie, and this does not take the extra costs for producing a film in 3D into account.

However, the figure's Panel B show that the effect of 3D differs between movie genres: the regression coefficients for a 3D film are positive for action movies and much more so for family films, informing us that movies of these genres tend to perform better in 3D, on average. The box office performances of thrillers, drama, horror, and comedies instead tend to suffer from the 3D format—a finding that is well hidden under the higher budgets and heightened attention these 3D movies usually get from their producers. Our study is restricted to data up until 2011, but the pattern we found seems to not have changed more recently. Follows (2017) reports that the North American box office share of 3D movies has declined from more than 20% in 2011 to just 7% in 2016, despite the number of released 3D movies being at an all-time high.

Compared to 3D technology, we know much less about the financial impacts of higher frame rates and screen resolutions. Audiences' evaluations of movies that are filmed with a higher-than-usual frame rate (i.e., a film recorded at a higher number of frames per second—up to 60 instead of the traditional 24) have been linked to the psychological theory of the "uncanny valley" (Yamato 2012). This theory, which goes back to robotics scholar Masahiro Mori's work in 1970, suggests that consumers react much more negatively to aesthetic stimuli that look *almost* real than they do to stimuli that are either very real or are very unrealistic (see Mori 2012).

Mori's logic, which he developed with robots in mind, has been studied by a number of scholars, but little is known about it in entertainment. Empirical findings tend to support his argument that the depiction of a face generally increases likability the higher its "humanness" (and the less "mechanical" it is); however, likability falls to low levels if humanness is presented in "close-to-realistic," but just not *fully* realistic ways. Figure 8.2 shows this pattern as theoretically proposed by Mori (Panel A) and as empirically found (Panel B) in a study of 80 faces whose humanness varied from "fully mechanical" to "fully human" by Mathur and Reichling (2016). It has been argued that "close-to-realistic" depictions seem eerie to consumers and make it difficult for them to empathize with characters—reactions that we assume will hamper immersion.

According to such "uncanny-valley" logic, a higher frame rate can let fantasy movies appear *almost* real for audiences, but not quite; such "failed attempt at reality" would be disliked by most consumers. This argument is consistent with Michelle et al.'s (2017) finding of a "hyperreality paradox" for audiences of Peter Jackson's HOBBIT. In a qualitative study, they observed that some consumers found the movie's visuals to be spectacular and immersive, but at the same time "experienced this same visual aesthetic as unconvincing and distracting and as undermining suspension of disbelief." They perceived the effects to look "fake." As with other technologies, entertainment producers must be aware of these potential negative effects and learn more about them, hopefully in fruitful collaboration with *Entertainment Science* scholars.[163]

[163]Let us add that the "uncanny valley" theory might be of value beyond the understanding of higher frame rates and even beyond filmed entertainment. By building on the immersion-related argument above, the theory could help to better understand the impact that CGI elements in visual entertainment have on audiences. For example, Itzkoff (2016) discusses it in conjunction with audiences' reactions to the digital revitalization of dead actors, such as Peter Cushing as evil Grand Moff Tarkin in the STAR WARS film ROGUE ONE. But given the huge commercial success of clearly imperfect animation tricks in commercial super hits such as the initial STAR WARS movie (and also those recent entries which use digital revitalization), the link between realism and immersion/success is certainly not a trivial one and requires a thorough extension of Mori's original thinking.

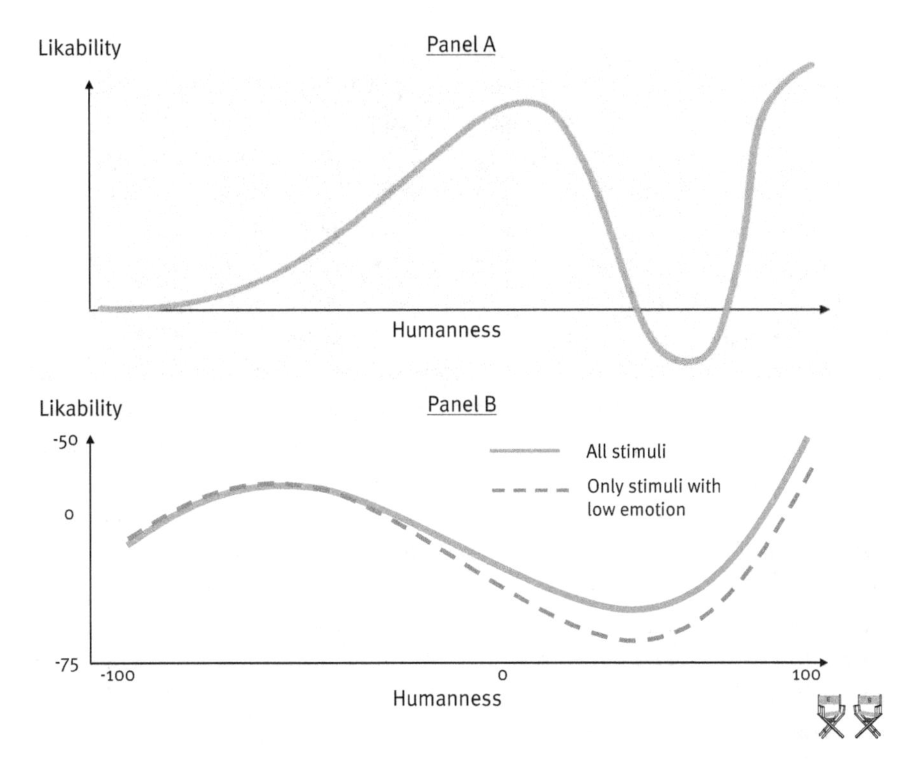

Fig. 8.2 The "uncanny valley" as theorized and empirically observed
Notes: Authors' own illustration based on Mori (2012; Panel A) and Mathur and Reichling (2016; Panel B). The results from Mathur and Reichling are based on a coding of the "humanness" of 80 faces by 66 consumers; the average scores of this exercise were used as the "humanness" ratings. Likability of the different faces was collected by asking 342 consumers to rate a random subset; overall, each likability rating is the average of 64 individual ratings. The course of the function is a third-degree polynomial which had a better fit than other functions (R^2 of 0.29 for all stimuli; not reported for only low-emotion stimuli). Graphical contributions by Studio Tense.

Technology and the Quality of Books

Books have been a low-tech entertainment medium for most of their 500+ years existence; with the exception of some special, niche-targeted high-quality print editions, the content dominated its presentation mode. Digitalization and the rise of e-readers, with Amazon's Kindle eco-system having led the way, are in the process of changing that, offering readers the choice between a printed book and a digital version. Digital versus print constitutes a search quality for readers. In contrast to the technologies we discussed for games and filmed entertainment (e.g., VR and 3D), the creation of the digital ebook version of a novel entails very limited marginal production costs, so that today almost all novels are available in both formats.

Nevertheless, we find it informative to examine the potential differences in impact that the technological format might have on consumers and their reactions to novels. Is the consumer's reading experience different if the text is on paper versus on an electronic screen, which lacks the haptic and olfactory sensations of printed books? Anne Mangen and her colleagues conducted a series of experiments in which they address this question by comparing reader reactions to text on paper versus reactions to text on iPads. Specifically, for a sample of 145 readers of the dramatic fiction short story MURDER IN THE MALL, Mangen and Kuiken (2014) find no difference in terms of narrative transportation and empathy levels between the two forms of reading. Quality ratings were consistently higher for the digital device, even though the authors did not control for participants' previous digital experience.

In another experiment with 72 high-school students, Mangen et al. (2013) compared reading comprehension between a paper version of a (non-fiction) text and a version shown on a computer monitor with "HD ready" resolution. The results of a regression analysis, in which the authors control for gender (but not e-reading/computer experience), show comprehension to be higher for the paper condition, pointing at possible differences in the cognitive processing of texts by consumers.

But iPads and computer monitors are certainly not the best-suited and most-used devices on which to read novels digitally these days. It thus remains unclear to what extent these insights can be generalized to specialized e-readers such as Kindles or Kobos. Let us also keep in mind that these analyses are based on the assumption that the medium differs, but the context does not. Findings might thus change if this assumption is challenged by new literary forms, such as hypertext novels.

Technology and the Quality of Music

For music, similar to books, it traditionally has been the content that mattered, much more so than the technological production format. Except for the occasional format switch (from shellac to vinyl in the 1950s, then from vinyl to CD in the 1990s after some flirtations with cassettes), the music business was dominated by one single production standard at a time, not counting technological extravaganzas such as Katy Perry's TEENAGE DREAM album, which was scented to smell like cotton candy (Bauer 2010).

However, when music moved to the Internet, different formats began to co-exist, and consumers were enabled to choose among them. These formats differed in the degree to which they provided haptic sensory features, but also in their compression level, which is tied to sound quality—factors which presumably are responsible for the recent revival of vinyl records.[164] As with books, the role of format for consumers' choices and consumption behavior has not received systematic attention from researchers or managers (in contrast to distribution-related issues which we discuss in that later chapter), probably because almost all titles are made available both in digital formats (where copies are produced for virtually zero marginal costs) *and* physical forms (which still provide substantial revenues).

Nevertheless, and particularly because of the decreasing demand for expensive-to-produce physical versions, better understanding the roles of format and production technology for consumer choices seems promising. Music research has found that compression affects quality judgments of consumers only after a threshold is exceeded (e.g., Croghan et al. 2012). But what is this threshold, and is it within a practically relevant spectrum? Existing studies are usually of limited explanatory power for assessing commercial impact because participants are highly trained music experts, rather than ordinary music consumers. For example, Pras et al. (2009) let 13 "trained listeners" (namely, musicians and sound engineers with 15 years of experience) rate pairs of the same music that varied systematically across seven sound criteria. Results showed significant differences between CD/WAV quality and lower bitrate formats in particular—listener ratings are higher for CDs versus 96 and 128 kb/s compressed formats and marginally higher ratings for CDs versus 192 kb/s from 96 to 320 kb/s. The scholars do *not* find differences between CDs and higher bit rates, though.

We don't know yet whether such differences would also be recognized by ordinary consumers and how such perception differences would impact enjoyment levels. Spotify users can choose between 96 kb/s and 320 kb/s, with the "normal" mobile usage being at the lower end of the spectrum; Internet chatter as well as sales trends do not suggest that this is a major hindrance for mass audiences to prefer the service. Our skepticism is somewhat fueled by a finding from Plowman and Goode (2009), who, based on a survey of 206 students and Spearman correlations, report that expectations regarding the sound quality of the music do not influence consumers' usage of file sharing as an alternative to purchasing music. So, are we consumers

[164]For details, see the development of different music formats in Fig. 5.4 in our chapter on entertainment business models.

just musical ignorami? Remember that sound quality is determined by many factors in addition to the song material, such as hardware (playing or listening devices), as well as user and situational characteristics, so that recording (and compression) technology will hardly hamper the consumers' listening experience, at least not for higher bitrate formats, and at least not in an *objective* way.[165]

But this does not mean sound technology has no influence on some consumers' enjoyment of music—the streaming service Tidal, for example, offers "lossless" music for twice the price as "compressed" versions. And several vinyl fans adore the vinyl format not for its perfection, but for its *lack* of it—the pops and noise, the smooth crackles, the tighter bass (which is basically an engineering artifact; Richardson 2013), all of which are considered as parts of an active, intense, aesthetic "high-quality" experience (for a list of subjective examples, see *TRCG* 2015).

Signals of Quality for Entertainment Products

We have shown that the *experience* quality of entertainment has little ability to drive the early sales of a new product's release, and dedicated search qualities are generally rare in entertainment: the previous section has illustrated that managers should not place too much hope in the power of superior technologies as a search quality because they usually provide only limited *enduring* competitive advantage (and are quite often double-edged swords in terms of customer value). But the early adoption decisions of consumers account for a non-trivial share of entertainment success, particularly for those products marketed in line with the "blockbuster concept,"[166] and their role is further intensified by cascading effects in which a product's initial success sends a distinct quality signal to consumers regarding whether or not they should adopt it.

Under these conditions, "pseudo-search" qualities that serve as signals from which consumers can infer a product's quality (that's also why we also call them "inferential cues") play a powerful role. In this section, we discuss

[165]In case you want to test your own ability in distinguishing between different compression formats, we recommend the little test that NPR has put together at https://goo.gl/wXJmHg: for six songs from different genres, it asks us to judge three versions that differ only in compression levels. [At least one of this book's authors didn't recognize *any* differences with his Sennheiser PC headset.]

[166]For a detailed discussion of the blockbuster concept as the dominant integrated marketing strategy for entertainment products, see our chapter on integrated entertainment marketing.

a number of such pseudo-search entertainment qualities and their respective link to (early) product success, including genres and themes, age ratings and the (controversial) content on which they are based, a product's country of origin, and the production budget (which, as we will see, sends not one, but various signals regarding a product's standard of execution). Our discussion of these factors will be followed in the next chapter by another category of specific pseudo-search qualities: those signals that carry a "brand label," such as stars or sequels.

When we dive into the realm of the different signals, keep the core learning from our study of consumer behavior in mind: that the effectiveness of *all* pseudo-search qualities, branded and unbranded, eventually depends on the degree to which they convince potential customers that the product will offer powerful aesthetic sensations and familiarity, the core drivers of consumers' enjoyment of (and demand for) entertainment.

Entertainment Genres and Themes

> "Genre is one way movies have been pre-sold throughout the history of Hollywood."
> —King *(2002, p. 119)*

The genre is often the first thing we hear about a new entertainment product. Let us inquire into the concept itself, before examining what *Entertainment Science* scholars can tell us about how genres impact product success. We then look into whether it is more promising from an economic stance to tie a product to a single genre or to combine elements from different genres. And finally, we also take a quick glance at international differences in genre effects.

What, Exactly, is a Genre?

A genre, taken from the French term "type" or "kind" (King 2002), is an abstract concept that describes a certain category of entertainment or art. Any genre contains links with a semantic network of associations that is activated in the consumer's mind once they hear that an entertainment product belongs to that genre, i.e., a movie is a thriller or a western, a song is pop or jazz, or a game is a shooter or a role-playing game (e.g., Cutting 2016). A consumer will roughly classify an entertainment product in their very own hyperdimensional cognitive network based on information about its genre.

Each consumer's network is unique, and it is these idiosyncratic differences between networks that explain why some of us become excited when we hear about a new horror movie that is coming out, others react with indifference, and still others are annoyed by the mere thought of another horror flick.

So, what kind of cognitive associations are we talking about? Genre associations are about basic types of characters, dramaturgic routines, and aesthetic patterns. In Fig. 8.3, we show a stylized semantic network for three major movie genres (westerns, thriller, and romance movies), based on the most frequently assigned key words for each genre by IMDb users. The thickness of a line shows how strongly a concept is linked with a genre. So, when we hear that a movie is a western, most people think of cowboys and sheriffs and outlaws as characters who will populate the film; they think about gunfights and fistfights as part of what will happen (i.e., the dramaturgic routine), and they think about rural Arizona or Texas landscapes as where the action will take place, framing the consumers' aesthetic expectations.

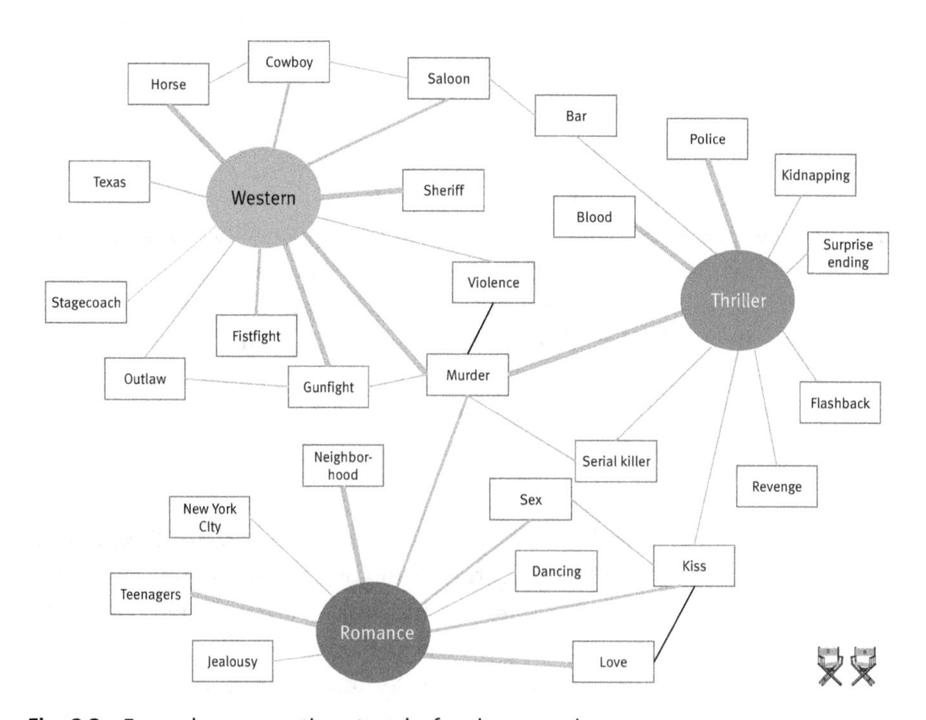

Fig. 8.3 Exemplary semantic networks for three movie genres
Notes: Authors' own illustration based on publically available data from IMDb. The associations represent a selection of the most-named key words on IMDb for each of the genres in the figure. The links between associations are purely hypothetical and for illustrative purposes only.

The figure shows that the genres are linked with many unique associations, but also that the associations sometimes overlap. For example, murder is a prominent theme in thrillers *and* westerns; even romance movies are not free from murder, although topics such as love dominate. Thrillers and westerns have more in common than do westerns and romances. Sometimes genres' aesthetics involve a specific narrative style, such as thrillers often telling parts of the story through flashback scenes. The same logic applies to other entertainment products; when we hear that someone has recorded a bluegrass song, banjos and acoustic guitars will come to our minds, spurring aesthetic expectations in the form of acoustic imagery, and often also visual imagery (such as the rural south-east of the U.S.).

The assumed importance of genres for product success is mainly based on the fact that genres provide consumers with a first reference point for judging an entertainment product. The associations we hold help us to make quick judgments regarding what to expect, and whether or not we might like it (Zhao et al. 2013). As film scholar Geoff King (2002, p. 120) words it, a genre label "is an implicit promise" to the consumer. The genre concept is tied closely to the benefits that come from familiarity in entertainment; it is the promises that provide us with familiarity regarding something we otherwise do not know, allowing us to draw a mental picture of the unknown product. Genres give us "enough familiarity to generate a sense of comfort and orientation" (King 2002, p. 120).

But even beyond that consumer-related role, genres are important because they contribute to the organization of entertainment industries on many levels. Think of companies who define themselves as producers of "rock music" or offer a movie channel specialized in science fiction (such as the Comcast-owned "Syfy Channel"), teams which assemble around a genre (e.g., when musicians agree to form a "blues band"), and news media who structure their coverage of entertainment around genres, such as *Billboard*'s genre-specific charts, like "Hot Country Songs" or "Hot Rock Songs" (Silver et al. 2016).

Are Some Genres More Successful Than Others?

> "To pigeonhole a genre as being successful or unsuccessful is weird."
> —*Musician* Chester Bennington *(2016)*

When it comes to comparing genre performances, let us clarify one thing upfront: given the subjective nature of genres, there is no definitive genre typology for any of the products we study in this book. Genre definitions are blurred and overlap; whereas Boxofficemojo considers PASSENGERS to be "Science Fiction," The Numbers classifies it as "Thriller/Suspense," and IMDb codes the movie as "Adventure, Drama, Romance." Genres are also multilayered; they blend with sub-genres and themes with increasing levels of concreteness. So a comedy (the main genre) can be a romantic comedy or a slapstick comedy (the sub-genre), and a main theme in a romantic comedy could be "unrequited love," a situation in which one character loves the other without his love being returned (e.g., Sarantinos 2012). Genres tend to be more consistently defined for narrative forms of entertainment, whereas genre typologies in music (beyond the broadest categories) are less clean and also less-consistently used.[167]

Determining empirically how genres affect success across products is challenging because the concept's basic nature means that genres differ massively—which is the logic behind our introductory quote in this section. Not only do genres differ with regard to the other quasi-search attributes we discuss in this book (such as casting a major star in a movie), but also with regard to further marketing variables, such as advertising budgets—action movies might have higher box office returns than dramas, but producers also spend more money to make and promote them. Thus, to determine a genre's "true" commercial appeal, it is important to not only look at average success numbers, but also to examine the nature of a genre in terms of many individual attributes and variables. And of course, because costs differ between genres, revenues cannot be equated with profit/ROI.

When it comes to movies, a large number of studies have empirically linked genres with success metrics, mostly box-office results. For our own data set of more than 3,000 movies released in North American theaters between 2000 and 2014,[168] we calculate the mean North American box-office revenues for 13 key genres, as well as a proxy of each genre's average ROI (using the same formula as for our risk analysis in the chapter on

[167]Some scholars have tried to use empirical data and statistical techniques for developing music genre typologies by investigating common elements between music pieces. Schäfer and Sedlmeier (2009) use consumers' preferences toward 25 popular music genres and condense them via factor analysis to six musical genres (i.e., sophisticated, electronic, rock, rap, pop, and beat, folk, and country music). Silver et al. (2016) employ a network analysis approach to discover patterns in how 3 million musicians presented themselves and their work on the social media site MySpace.com in 2007. They find three musical genre "complexes"—a *Rock* complex (encompassing what the authors refer to as "Countercultural," "Mainstream," and "Punk Offshoots" subgenres), a *Hip-Hop* complex (dominated by Rap, Hip-Hop, and R&B), and a *Niche* complex (which covers several less popular musical styles, such as Electronic, "Dark/Extreme" Metal, and World Music). Whereas these attempts can be applauded, the biggest problem with the empirical determination of music genres is about *selling*: gaining industry and consumer acceptance for such typologies is tough, but indispensable for having a "real-world" impact.

[168]For more information about the data set, please see our earlier Fig. 5.10 in our entertainment consumption chapter.

entertainment business models). When doing so, we adapt the IMDb's genre coding for our analysis because it does not restrict a movie to a single genre but allows movies to be assigned to multiple genres—an approach that better fits the realities of today's hybrid entertainment world (and avoids arbitrary genre assignments).

Panel A of Fig. 8.4 shows that movie genres differ strongly in revenues and ROI. On average, fantasy, animated, and thriller movies generate the highest revenues, about three times higher than those of drama and romance movies—and almost ten times higher than those of documentaries. The same basic order holds for our measure of movie profitability, with the exceptions that documentaries (because of their *lower* costs) are as profitable as more popular genres, and crime movies (because of their *higher* costs) have lower ROI than dramas.

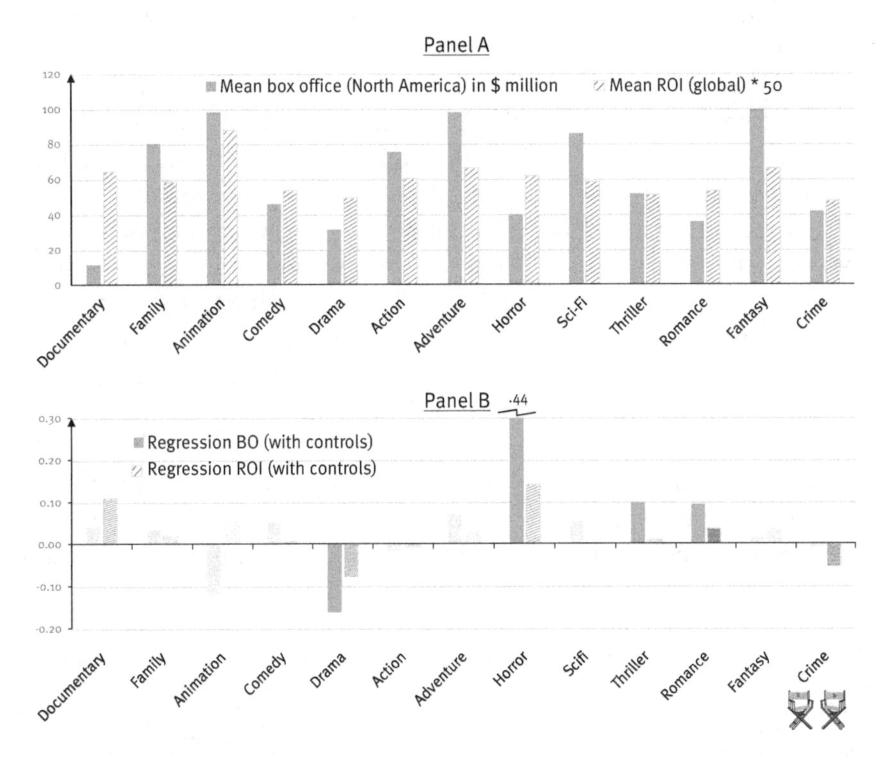

Fig. 8.4 Revenues, ROI, and regression parameters for key movie genres

Notes: Authors' own illustration based on data from various sources, including The Numbers, Kantar Media, and IMDb. Numbers in Panel B are unstandardized regression parameters from OLS regressions, with movie box office and ROI as dependent variable, respectively. In the analyses, we log-transformed the dependent variables to approximate a normal distribution. The true regression value for horror is actually as high as 0.44; it is capped in the figure. In Panel B, shaded bars indicate that a parameter was not significant (at $p<0.05$).

In Panel B, we show the effect of each genre on a movie's box office and ROI when controlling for the existence of several other "success drivers."[169] These results paint a different picture: only three genres lead to a higher-than-average box office (horror, thriller, and romance) and drama is the only genre that results in a lower-than-average box office. The two genres that generate the highest average revenues (fantasy and animation) do not exert any significant impact on movie box office. With regard to ROI, horror, romance, and documentaries are also effective, while crime films and dramas tend to lower a film's profitability. Here, none of the three genres that we found to generate the highest average ROI are significant, not even thriller. Because the dependent variable is a log-transformed measure, the coefficients tell us that, everything else equal, the average increase in box office for horror movies against other films is an impressive 55% ($=e^{0.44}$) and +15% in ROI, whereas dramas underperform at the box office by about 15% (and have an 8% smaller ROI). You can do the math for the other genres yourself.

Why do we see these differences between the descriptive analyses (Panel A) and the predictive analyses (Panel B)? According to the analysis, it is not the fantasy genre's own attractions (like orcs and elves and magicians) that draw people into see movies of this genre. Instead, people are drawn in by other characteristics, such as the popular novels on which fantasies such as LORD OF THE RINGS are based, along with production and advertising budgets that promise a visual spectacle. Our regression results suggest that thriller, horror, and romance attractions stand out for luring audiences to the theater among the films in our data set, and the latter two genres create these attractions in a way in which the incremental revenues exceed the costs of producing them. And why are documentaries profitable? Probably because of a "selection effect" that takes place outside of our data set, on the "supply side" of the film industry. Out of the numerous documentaries that are produced, only a few outstanding ones end up being shown in a movie theater, and we find these few to perform quite profitably.[170]

[169]Specifically, we control in the analyses for whether a movie was distributed by a major studio, the advertising budget, the production budget, whether a film featured a star (based on the annual "Quigley" star ranking—see footnote 232 on p. 417), was a sequel, a remake, or a version of a previous movie, was based on a novel, book, or bestseller, was produced in the U.S., and ratings of the film's quality by critics and IMDb users.

[170]Our data set is limited to those films that made at least $1 million in North American theaters—whereas this barrier will hardly matter for films of other genres, it might contribute to the high ROI of documentaries which are often produced for a small budget. Separately, researchers have pointed to the role of distribution as a *mediator* of the effect of genres on success. According to such logic, genres not only impact consumers directly, but also via an influence on movie theater owners and their screen-allocation decisions for a movie (e.g., Clement et al. 2014).

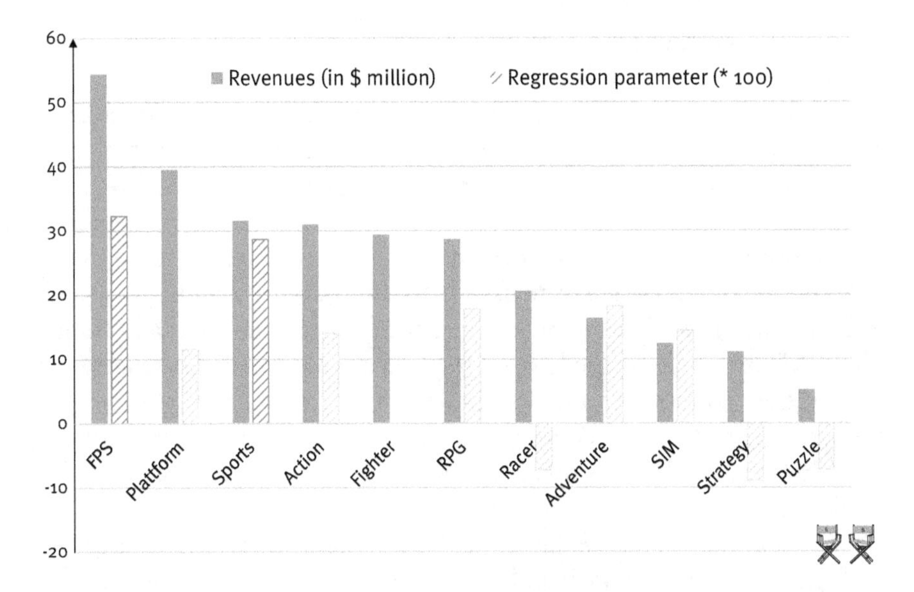

Fig. 8.5 Revenues and regression parameters for key game genres
Notes: Authors' own illustration based on results reported by Marchand (2016) and data from VGChartz. FPS means first-person shooter. The regression parameter for fighter games is missing because that genre was left out of the regression analysis for methodological reasons. Bars in light blue are not statistically significant at $p<0.05$. The dependent variable was log-transformed.

For video games, we find a similar pattern: average sales differ substantially between genres. Figure 8.5 shows the average revenues for 11 video game genres (the orange bars in the figure) across the seventh generation consoles Microsoft Xbox, Sony PlayStation, and Nintendo Wii, based on a data set of 1,898 games released between 2005 and 2014. First-person shooters are, on average, the most successful (a finding that does not change even if one uses the median instead of the mean to account for "outlier" titles, such as the Call of Duty games), with average revenues of $55 million in North America. Next come platform games (such as Little Big Planet) and sports games (e.g. Wii Sports or Madden NFL).

And again, when we account for the fact that genres differ in other product characteristics (which also impact success), we see that few genres exert a direct effect, above and beyond those other characteristics. Specifically, in a regression with robust standard errors that includes not only the genres, but also a game's advertising budget, price, and several other factors, only first-person shooter (FPS) and sports game genres explain a unique share of a game's sales (Marchand 2016)—the blue-striped bars in the figure show the respective coefficients.[171] By transforming the coefficients, we see that FPS have 38% and sports games 33% higher sales than an "average" other game, respectively. But let's keep in mind that these numbers are averages across the different consoles, and that the impacts differ between them quite strongly (see Marchand 2017 for additional insights).

There is less empirical evidence for the economic effects of genres for books and music. In an exploratory analysis of the decision-making process of 50 Dutch book buyers, Leemans and Stokmans (1991) find that a book's genre and its theme are among the critical influencers of book choices. Genre and themes are named by *all* respondents as a selection criterion for reducing the choice set of alternatives, and they are noted more often than all other criteria for finally choosing one book out of the consideration set of titles. The scholars do not provide insights into the relative attractiveness of specific genres and themes, however.

Also examining books, Schmidt-Stölting et al. (2011) use a large data set of fiction books to understand the factors that impact sales of hardcover and paperback titles in the German market. Out of almost 38,000 releases between 2003 and 2006, the scholars study the market performance of 1,206 books (603 hardcover novels and their paperback versions). Regarding genres, they distinguish between "novels" (i.e., drama titles), thrillers, fantasies, (fictional) biographies, and a catch-all category of "other" genres. They analyze hardcover editions separately from paperbacks, while accounting for their related nature by estimating models for the two formats using a seemingly-unrelated-equations regression (SUR) approach, controlling for the impact of many other "success factors."[172]

Schmidt-Stölting et al.'s results show that genres matter in both formats, above and beyond the other book characteristics; interestingly, the attractiveness of a genre varies with a book's format. For hardcover editions, biographies exert a stronger positive impact on consumers than do drama novels, whereas neither fantasy, thriller books, nor "other" genres make any difference in sales. For paperbacks though, biographies have *lower* sales than drama novels (as do fantasy books), whereas thrillers, on average, are more attractive to consumers. In other words, (German) consumers snap biographies up at

[171]The other factors included in the analysis are: the platforms on which a game was released, price, number of previous versions, published by a major studio, advertising budget of the game and its competitors, consumer and expert evaluations of its quality, hardware variables (the installed based and console age), and existence of a multiplayer feature. For details, see Marchand (2016).

[172]Specifically, they also include measures for the stardom of the author, whether the book is a sequel, publisher status, and the books' price (which is set for each book by the publisher in Germany and must be respected by *all* retailers).

their initial releases, despite the usually higher price for hardcover versions, whereas consumers' demand for thrillers is biased toward the cheaper paperback format—an effect that the authors attribute to the genre's lesser symbolic value.

Finally for music, Lee et al. (2003) analyze the North American success drivers of 245 music albums, using weekly sales from SoundScan as dependent variable. Their prediction model, based on ambitious Bayesian statistics, suggests that positive genre effects exist for the rhythm and blues (R&B) genre, even after controlling for the artist's previous sales, the record's quality, and the advertising budget. In the model, sales for R&B are 35% higher relative to the other music types in their data set (country, pop, rap, alternative, rock, and hard rock).

We assume you agree with us that these are exciting initial insights, but we would love to learn much more about the role of different genres for the success of entertainment products, particularly for books and music.

The More Genres the Merrier?!

Most of us think of Star Wars as a science fiction film. But others have argued that it is actually a collage of multiple genres, combining elements of fantasy (the Jedis' mystical powers), western (did Han Solo shoot first?), war (the final battle), samurai movies (the lightsaber battles), and romance (who gets the girl?); it also offers connections to many other genres (the Mos Eisley Cantina scene made many people think of *Rick's Café Américain* from Casablanca). Let's turn this observation into a question of managerial relevance: does it help or hurt the success of an entertainment product when it contains elements of *more than one* genre?

Using a data set that encompassed almost 3,000 movies released on North American screens between 1982 and 2007, Zhao et al. (2013) found that the number of genres in a film (as measured by movie sites such as the IMDb) has a negative impact on opening box-office results. The scholars argue that this is because high "genre spanning" makes it troublesome for consumers to cognitively categorize the entertainment product—and so we simply don't know what to expect (and thus don't watch/read/play). In other words, the potentially success-enhancing effect of genres seems to be countered if multiple genre labels are attached to a product, at least in the first weeks of availability.[173]

[173]Let us mention one statistical caveat: the authors do not include the genres themselves in the equations. The same limitation applies for Hsu's (2006) study we discuss below.

But do such negative effect also apply for the *total* success of a movie over its full theatrical run? Using a data set of 949 U.S. movies released between 2000 and 2003, Greta Hsu (2006) also found a negative parameter for a "number of genres" variable, but in her case, the link did not reach statistical significance. This provides at least suggestive evidence that genre-spanning effects might fade over a movie's run. But take note: she also reported *other* negative consequences of multiple genres. Both audiences (on IMDb) and professional critics rate a film less positively as the number of genres increases.

In a follow-up study, this time looking at firm survival, Hsu et al. (2012) report evidence that a higher share of "hybrid" movies (i.e., those that are attached to more than one genre) increases the likelihood that the producing company will go out of business. In other words, producing multi-genre films impairs economic survival. But before you drop your plans to produce a genre-spanning piece of entertainment: the scholars also point to at least one potentially interesting *upside* of genre spanning. They found that doing so tends to increase the probability of producing the *most successful* film of the year—probably by attracting fans of all the different genres of the "hybrid" film.

Hsu et al.'s data in this study are quite historical, covering the years 1914–1948, but our introductory anecdote of the Star Wars movies might suggest that this finding still holds today. Nevertheless, we would be interested in empirical testing with more recent data sets and ideally also for other entertainment products besides movies. Such studies might also reveal whether the genre-spanning effect is indeed linear (as existing studies imply). Or is there is an optimal number of genres that a movie should span, and what is that number?

International Differences: Not Everyone Loves Baseball

> "You don't say foreign, anymore. It's 'International.'"
> —*Actor/director* Mel Gibson *(quoted in* Fleming Jr *2016)*

Genres are deeply buried within a culture's fabric of values, attitudes, and rituals. Take the western, which mirrors the foundational mythos of America, the conquering of wild and unexplored landscapes against all odds and by heroic loners. In contrast, a samurai film celebrates Japan's iconic noble warriors. As a consequence of this cultural embeddedness, the attractiveness of a genre often differs between countries because of the (mis)match of the values celebrated by the genre and the values of the culture itself.[174]

[174]Regarding the specific values of a culture, please also note our discussion of country-of-origin signals later in this chapter.

Sometimes the desire to avoid a cultural mismatch has an immediate impact on how entertainment products are made. For instance, the sub-genre of invasion fantasies, in which brave resistant fighters defend their homeland against the country's arch enemies, is tied closely to the cultural value of American patriotism, but is hardly compatible with how Russian and Chinese audiences see the world. To avoid such a culture clash, the producers of the HOMEFRONT games and of the 2012 movie remake RED DAWN gave the invaders a North Korean background, instead of a Russian nor Chinese one, as originally planned (Totilo 2011).

How strong overall are such differences in genre preferences? In a descriptive analysis of the market shares of movie genres in different parts of the world, Follows (2016) demonstrates that they are quite substantive. For example, in his data set of 3,000 films released from 2012 to 2016, action films have 50% higher-than-average market shares in large Asian countries (e.g., China and Japan), but underperform in Italy. Comedies are twice as popular in Italy than in other countries, whereas South Korean and Japanese consumers show less appetite for humor (the market share of comedies is 65 and 45%, respectively, below the global average). Drama is also popular in Italy and in South Korea, but Japanese and Chinese audiences see only half as many dramas at the cinema than does the rest of the world.

Such insights are instructive, but one needs to be careful about interpreting them in a causal way. Consider the case of horror films, which Follows finds are not popular at all with Chinese audiences (with a genre market share that is 90% lower than the global average). Because horror films are heavily censored in China, their lower share is probably due to supply-sided factors instead of reflecting low demand. Clearly, there are other factors that will also influence the local genre market share; understanding them is crucial for making inferences about the popularity of a genre among *consumers* in a given country.

Accordingly, some *Entertainment Science* scholars have attempted to shed additional light on the international genre preferences of consumers by using more rigorous statistical methods and including more factors. Among them are Akdeniz and Talay (2013), who explore a data set of 1,116 U.S.-produced movies that were released in 27 countries between 2007 and 2011. They use a hierarchical linear regression approach in which they control for several other film elements, such as budget, participation of (American) stars, and professional reviews; they do not control for "local" variables (such as distribution in a given country), however.

The regression parameters for five main genres in 14 countries suggest that romance and action each exert a positive box office effect in

Scandinavia, Israel, and the Netherlands, above and beyond every other "success driver" the authors consider. Thrillers are only effective for Dutch audiences, and dramas are mostly negative across cultures (but mostly insignificant). The authors also find a negative effect of (American) comedies for South Korea (consistent with Follows' market share analysis), as well as for Israel, Austria, and Germany.[175]

We have argued that the genre concept is multilayered, and scholars have gone beyond the "main-effect" level to explore cultural differences in the appeal of more fine-grained themes. Specifically, Moon and Song (2015) compare the North American and "foreign" box office performance of 240 Hollywood movies from 2003 to 2005, distinguishing between movies that have what the authors name an "American theme" (such as dealing with American football; 120 such themes are identified and used in their study) and those who have a "non-American theme" (e.g., the Samurai culture).

The authors then determine for each movie the degree to which it features American and/or non-American themes; they do this by applying machine-based text categorization to more than 100,000 consumer movie reviews. Using OLS regressions, they find no influence of the presence of American and non-American themes on the North American box office of movies. But the films' performance outside of North America profits from a non-American theme and also suffers from an American theme. In other words, an American theme shifts the foreign-to-domestic ratio of movie revenues toward the domestic component.

We also looked into the economic effects of cultural themes in movies, focusing on three specific American themes, namely sports, military, and African-American themes (Hennig-Thurau et al. 2003). Drawing on a sample of 231 U.S.-produced films, we first determined the "expected" German performance for each film (via regression analysis with its North American box office as predictor) and then investigated whether a film's under- or overperformance could be explained by the three cultural themes.

We find that six out of the 20 most *under*performing Hollywood films in Germany contain an African-American theme; among them are MALCOLM X (which made only 16% of what could be expected based on its North

[175]For action and drama films, Akdeniz and Talay also find negative effects for South Korea, which conflicts with these genres' higher-than-average market shares as reported by Follows—something that points at the existence of "hidden" factors for these genres that are not accounted for in mean comparisons. Separately, please keep in mind that their results only reflect a culture's reception of *American* genre films. So, whereas the underperformance of comedies in Germany seems to confirm the country's reputation of being "not funny" (Evans 2011), such an interpretation would ignore the enormous successes of native comedies such as DER SCHUH DES MANITU (12 million attendants), OTTO—DER FILM (9 million) and the FACK JU GÖHTE trilogy (which attracted more than 20 million moviegoers in total).

American performance) and BOYZ IN THE HOOD (only 19%). For sports movies, we find a similar, although somewhat less pronounced bias, as four of the 20 biggest underperformers deal with sports that are not as popular in Germany as they are in the U.S. Three of the films dealt with American football (e.g., JERRY MAGUIRE, which generated 26% of its North American equivalent) and one with baseball (BULL DURHAM, which returned just 1%); these films failed at the German box office despite the participation of popular stars such as Tom Cruise and Kevin Costner. The effect was not as obvious for military themes as for the other themes, but at least two military films are among the 20 that underperformed most strongly at the German box office (e.g., MEN OF HONOR, 36%).

In Panel A of Fig. 8.6, we illustrate this "theme-bias" for baseball movies, comparing the North American and German performances for the five best-rated baseball films on IMDb. For comparison sake, we do the same for the five best-rated action films in Panel B.

In sum, culture-dependent genre effects should not prevent a U.S. producer from turning an exciting story with a specific cultural genre or theme into an entertainment product. However, the producer should realize the potential commercial limitations that such a product can expect to face in markets outside North America. The producer can make changes to the project itself (such as limiting the budget or tweaking it in a way that is less

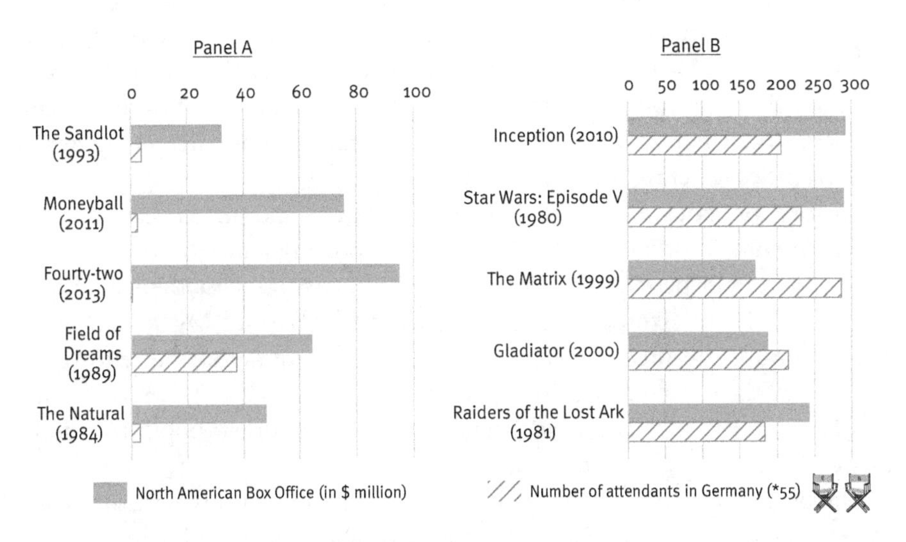

Fig. 8.6 Success of baseball and action films in North America and Germany
Notes: Authors' own illustration based on data reported by The Numbers, Insidekino, and IMDb. Our film selection is based on key words (for baseball) and IMDb genre (for action).

culture-dependent) or alter the communication strategy. Sony followed the latter strategy with its baseball movie MONEYBALL, which was a hit in North America (with a box office of $75 million): the firm's managers de-emphasized the sports element and instead stressed the film's "man-fights-against-the-system" element. But making effective product or communication changes is a far-from-trivial challenge, as the figure shows: MONEYBALL attracted fewer than 50,000 moviegoers in Germany and also flopped in other parts of the world, despite the studio's repositioning efforts.

Let us end our discussion by saying that a cultural bias will also almost certainly exist for other themes in movies and other forms of entertainment than the ones we mentioned in this section; for example, Hollywood producer Victor Loewy claimed that "[faith-based] films don't travel well" (quoted in Cieply 2014). Knowing those themes would help producers make more accurate estimates, so further *Entertainment Science* work on the topic is desirable. It could contribute to an even more-nuanced understanding for other themes, entertainment forms, and other regions. The question whether such cultural preferences are stable over time is another related question—one we address next.

Culture is a Dynamic Phenomenon: The *Zeitgeist* Factor

People's interest in cultural themes can be stable over a long period of time, as is the case with sports. Baseball, for example, has been "America's game" for more than 150 years, and Germans' lack of fascination with the game has been similarly constant. However, some other themes that occupy a culture are much less stable; these dynamic themes might offer even greater opportunities for entertainment products.

History provides examples that novels, films, and music can become embodiments of a certain cultural *zeitgeist*, capturing and reflecting the lifestyle of a certain period. The Beatles and Rolling Stones set the soundtrack for the rebellious sixties, Simon and Garfunkel's BRIDGE OVER TROUBLED WATER album captured the early seventies' disillusionment, the self-expressiveness of the disco era was spurred by the Bee Gees STAYIN' ALIVE tunes, and Michael Jackson's and Madonna's rhythms then turned pop into a dominant (sub)cultural element.

The degree to which an entertainment product captures the attitudes and values that are particularly salient for a culture at a given point in time can heavily influence its commercial success. All these artists and their songs became highly successful because of their *zeitgeist* fit, an explanation that has also been named as the reason for the unexpected enormous success of several recent films. Rambo offered a very simple (and very right-wing) resolution to the nation's traumatic Vietnam war (Nathan 2006). The patriotic and pro-military themes of American Sniper resonated with American audiences in 2015 at a time when the complexities of globalization and frustrating war experiences contested the nation's global leadership rule (Barnes 2015).

But let us note that *zeitgeist* is not always a good thing for makers of entertainment; it can also hurt an entertainment product's commercial performance when *zeitgeist* is missed. To understand the hateful reactions to Michael Cimino's epic western movie Heaven's Gate (which even liberal reviewer legend Roger Ebert called the "most scandalous cinematic waste" he had ever seen; Ebert 1981) seems impossible without considering the *zeitgeist* of a torn, post-Vietnam war America at the film's release. After just having made a fragile peace with this war trauma, it seems that America was not ready for one of its foundations, the frontier myth, to be critiqued by an entertainer. The film's affront to the guiding *zeitgeist* of the time kept consumers from watching the film and destroyed United Artists, its legendary producing studio.

You will have noted that our arguments here are largely anecdotal—this is because no empirical research that has yet addressed such dynamic effects of cultural themes in entertainment. That needs to change, as *zeitgeist* certainly deserves a more prominent place in *Entertainment Science* theory. But for now, we will leave the topic and move on to a related quasi-search quality of entertainment: the content out of which entertainment is made and from which, in its aggregation, genres and themes are formed.

Entertainment Ratings and the Controversial Content on Which They Are Based

One consequence of entertainment's cultural character is the existence of content-rating institutions that decide whether a product is suitable for a country's population, and for which parts of it. These institutions' recommendations play a dual role in the success of an entertainment product.

The first role of ratings is the focus of this chapter—they send a signal to consumers of the "radicalness" of the experience that awaits them. Ratings deal mainly with three facets of aesthetical radicalness: the degree to which a product contains violence, profanity (in the form of offensive language), and nudity or sexual content. How do consumers value these signals? This is far from a trivial question: in some cases, more consumers may be attracted by killing, cursing, and/or simulated intercourse in a movie than may be turned off by these "qualities."

Ratings' second role is that they limit the number of potential customers for a product by restricting it from some portion of the population. For example, a movie rated "16" in Germany prevents everyone under the age of 16 from seeing it—this group accounts for 13% of all theater visits, and a much higher percentage for some kinds of films (*FFA* 2016). In the following, we will first look at the aggregation of these two roles of ratings. We will then attempt to separate the two effects, looking first at the effects of radical content and then the "restriction effect" for entertainment products, in general and for different genres.

Linking Entertainment Ratings with Product Success

Most empirical research on ratings has been conducted in the contexts of movies and games, the two areas for which access restrictions are most prominently applied. The majority of studies includes age-rating categories in their regression analyses that aim at explaining product success, in addition to other "success drivers," such as genres and advertising.

In the context of movies, researchers have either included the actual rating categories (as categorical variables) or a scale that measures a rating's restrictiveness (i.e., the more restrictive the rating, the higher the score). Some studies report more restrictive ratings to be a commercial disadvantage. For example, when S. Abraham "Avri" Ravid (1999) applies OLS regression to a data set of 175 films from 1991 to 1993, he finds positive effects on revenues and ROI for G- and PG-rated films. And De Vany and Walls (1999) find that R-rated films are outperformed by other ratings in terms of both revenues and profits when fitting a non-normal Pareto distribution to a large data set of 2,000 films from 1985 to 1996.

Such results are consistent with the traditional tendency of movie studios to avoid restrictive ratings, particularly for action and drama films, because of their "restriction effect;" it hurts producers to see consumers complain of being unable to spend money for their products, as did eight-year old Deadpool-fan Matthew (Derisz 2016). When director Ridley Scott was asked about the PG-13 rating for the film Prometheus, the prequel to his science-fiction classic Alien which holds a reputation for its strong R-rated violence, his reply reveals the studio's way of thinking: "The question is, do you go for the PG-13, *[which] financially makes quite a difference*, or do you go for what it should be, which is R?" (quoted in De Semlyen 2012, with italics added by us).[176]

But the matter is somewhat more complicated. The studies cited above use somewhat older data; they are also limited in their use of controls. The results of more recent studies show no clear pattern regarding the economic effect of ratings. Some note a restriction effect, but only for a limited time period. For example, Leenders and Eliashberg (2011), applying a hierarchical regression to pooled data from nine countries, conclude that a more restrictive rating negatively influences the *opening weekend* of a movie, but not the *total* box office. Other recent studies find no systematic differences for the performance of rating categories (e.g., Clement et al. 2014, who analyze the North American and German box office of about 2,000 movies for the 2000–2010 period using 3SLS regressions).

And some studies, such as our own, even report a *positive* effect of more restrictive ratings (e.g., Hennig-Thurau et al. 2009, for an OLS regression of 202 movies from 1998 to 2006, half of which are sequels). For video games, Cox (2013) finds, for a data set of 1,770 games released for the seventh platform generation, that "mature"-rated titles (which he measures with a dummy variable) have 12% higher sales in North America than do games available for "everyone." And Dogruel and Joeckel (2013) report that M-rated games, which represent a share of just 8% of seventh generation games, accounted for 26% of "best-selling" games between 2008 and 2010.

So, why the conflicting results? One explanation is that the effect of the different rating categories is not linear, but that some categories have a distinct effect; thus, continuous measures of rating restrictiveness will lead to systematically different results than will binary ones.[177] But, as we will show below, it is the two distinct roles of ratings that are critical for understanding rating effects: the appeal of radical content and the restriction of audiences. When we take a closer look, we will also examine the context in which radical content is offered, such as an entertainment product's genre.

---

[176]In case you're interested in the studios' handling of ratings, we also highly recommend the documentary THIS FILM IS NOT YET RATED. Prepare yourself for some radical content, though.

[177]Marchand (2016), who uses basically the same data set for the same console generation as Cox, finds *no* effect of a *continuous* measure of rating restrictiveness. When we reanalyze the same data and substitute this measure with a binary one ("mature" rating or not), we find the same sales-enhancing effect that Cox reports.

Disentangling the "Appeal Effect" of Radical Content and the "Restriction Effect"

> "Rated 'G' is nobody gets the girl. 'PG' is the good guy gets the girl. 'R' is the bad guy gets the girl. 'NC-17' is everybody gets the girl."
> —*Meme on the Internet*

Information about the radicalness of an entertainment product will spread via ratings, but also through the information provided in trailers, posters, and other material. How do consumers react to radical content?

In Holbrook's (1999) study of HBO viewers, he finds that radical content *negatively* influences mainstream consumers' liking of movies. But Lang and Switzer (2008) go beyond studying effects on liking to connect content ratings to movies' commercial success; they use violence, sex, and profanity codings of 1,160 movies from 1993 to 2004 made by the family-recommendation service kids-in-mind.com. In an OLS regression, they include all three dimensions of radicalness as predictors, along with dummy variables for G, PG, and PG-13 ratings. Whereas profanity turns audiences away, and sex shows no effect, the researchers find that, across all films, a higher level of violence indeed attracts more consumers. It is this "consumer appeal" that attracts some of us to restrictively rated content, just as it did former Warner Bros. executive Lorenzo di Bonaventura: "[w]hen I was in my late teens, I wanted to see R-rated movies" (quoted in D'Alessandro 2017b).[178]

Do we find the same patterns in more current data? We investigated this question ourselves for a data set of 1,309 movies from 2005 to 2013 that were rated either PG-13 or R, as the two critical rating categories when it comes to "appeal" effects. Also measuring radicalness via kids-in-mind.com codings and controlling for the rating categories and several other success factors (such as advertising spending), we find that a one-point increase in violence (on the 10-point scale) leads to a 3% increase in box office. Profanity and sex both have negative parameters, but they are small and insignificant.[179]

[178]As Lang and Switzer's analysis controls for the ratings categories (as well as critics' judgement and distribution intensity/screens), with the coefficients of the ratings variables indicating how ratings themselves influence movie success, separate from the content they signal (the "restriction effect"); the coefficients for the radicalness dimensions reflect the dimensions' average "consumer appeal."

[179]In another study, using a data set of 2,000 films from 1992 to 2012, Barranco et al. (2015) code radicalness based on reasons given by the MPAA and arrive at similar insights regarding the appeal of average content. In an OLS regression across all age ratings (for which they do *not* control in the analysis, though), they replicate the success-enhancing effect of violence. They also obtain a negative sign for profane language, but it does not reach significance.

The restriction effect from the rating categories is much stronger in our data; we learn that a movie can, on average and controlling for radicalness and other characteristics, expect to lose 33% of its revenues with an R rating instead on a PG-13 rating.[180] In other words, higher violence can indeed attract additional audiences, but producers have to pay a high price to win them over. And there might be one additional hidden cost of restrictive ratings which our data does not capture: if ratings hurt potential merchandise revenues associated with an entertainment product.[181] This is what movie director James Mangold has in mind when he argues that for R-rated films "the studio has to adjust to the reality that there will be no Happy Meals. There will be no action figures. The entire merchandising, cross-pollinating side of selling the movie to children is dead before you even start" (quoted in Hayes 2017).

But we have to keep in mind that these are *average* effects across all kinds of movies. So, do findings vary by rating category or by product type and context? To answer this question, let us take a deeper look at the appeal of radical content for each rating category and also study whether differences in the appeal and restriction effects exist between genres.

The Appeal of Radical Content, Contextualized. Or: Nobody Wants to See Sex (in a Galaxy Far, Far Away)

Let's look at rating categories first and the differences in impact of the appeal of radical content of each. When Lang and Switzer run rating category-specific analyses, they find that higher levels of violence increase the attractiveness of R-rated and PG-13-rated films (but not of G/PG-rated films). Profanity hurts only R-rated films, and high levels of sexual content hurt

[180]The strength of this restriction effect assigns meaning to the processes through which such ratings are determined. Leenders and Eliashberg (2011) conduct an empirical investigation into the determinants of ratings for movies across nine countries, finding that not only the movies' ingredients are impactful (e.g., violence etc.), but also are the characteristics of the rating board (e.g., membership structure, size, and a country's culture). Related, Waguespack and Sorenson (2011) investigate potential biases in the assignment process of ratings. Analyzing 2,408 films that have been released in North American theaters between 1992 and 2006 and using content classifications by both kids-in-mind.com and IMDb, they show through linear models that distribution via an MPAA member firm reduces the chances of receiving an R-rating, as does the previous experience of the distributor. In addition, they find that it helps to have directors and producers that are well-connected within the film industry. In contrast, using a director who has a reputation for R-rated films reduces the chance of being rated less restrictive than R.

[181]See our chapter on entertainment branding for a more detailed discussion of the revenue streams of entertainment brands and franchises.

only PG-13-rated movies. In our analyses with more recent data, we find the "violence-is-good-for-business" effect to be only present for R-rated movies, and although the parameters for sex and profanity are mostly negative in our re-analyses, they are not statistically significant for *any* rating category. Be aware that all these results of secondary market data are only meaningful for how Hollywood has been using radicalness in the past, namely limiting more radical elements to the less restrictive ratings. Thus, our findings do not allow any generalizations regarding non-existent scenarios (e.g., nudity in a G-rated film).

Genre-specific effects of radical content are next. We study them by running analyses on a genre-by-genre basis, only considering the films of a certain genre at a time (while leaving all other films out of the analysis). In our data set of 1,309 movies rated either PG-13 or R, both the restriction effect of rating categories and the appeal effect of radical content vary between genres. The negative restriction effect of an R-rating ranges from −8% for horror movies (where it is statistically insignificant) to around −54% for romantic films; it is also insignificant for science-fiction and fantasy movies. Such differences will probably also exist for merchandising revenues—not every PG-13 movie (or T game) is equally headed for a release at Burger King, so that restricting merchandising potentials matter little for them (while they certainly matter much for others). For some entertainment products, selling expensive "Collectors' sets" merchandise to older, well-heeled consumers might be the real deal—one that is will not be affected by age restrictions, but based on the appeal of (radical) content only.

Figure 8.7 show how this *appeal* effect of radical content influences not merchandising, but box office revenues for selected genres. The figure reveals that whereas a higher level of violence hurts the commercial performance of PG-13-rated action movies, it helps the performance of R-rated thrillers. Higher levels of sexual content turn audiences away from R-rated adventure and science-fiction movies (telling us that consumers have not been very excited by the idea of seeing simulated sex in space on the big screen), whereas it does not influence action movies and thrillers, regardless of their ratings.

Whereas profanity has no significant impact on most genre-rating constellations, it helps R-rated adventure movies find their audiences, at least in our data set. The latter finding is consistent with the doubling of industry expectations by Deadpool, the foul-mouthed Marvel character from 2016, when the movie generated opening weekend revenues of $135 million in

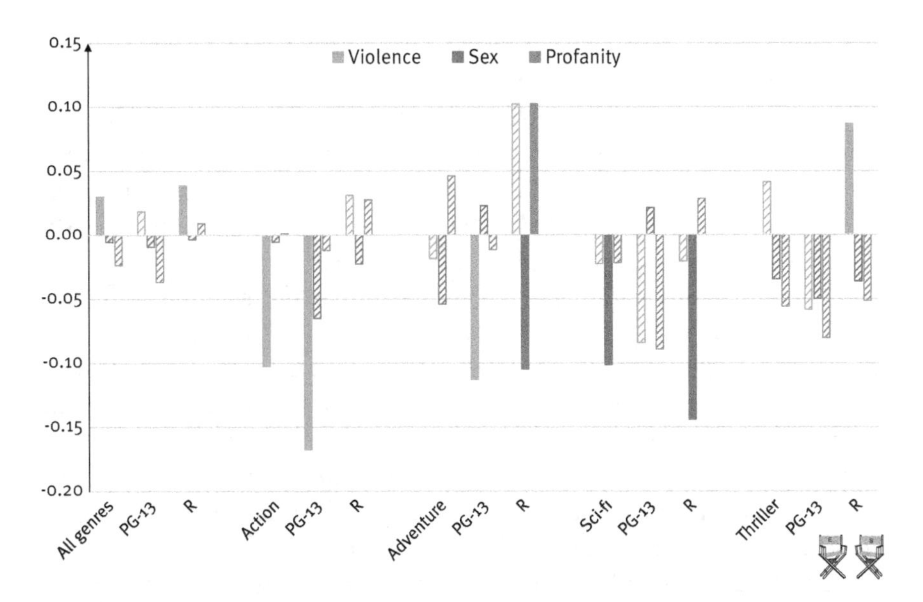

Fig. 8.7 The appeal of radical content for different movie genres
Notes: Authors' own illustration based on publically available data from kids-in-mind.com, MPAA, and The Numbers. All numbers are unstandardized regression parameters from OLS regressions, which also included as predictors whether a movie was a sequel, the movie's quality rating by critics, the number of opening-weekend theaters, the advertising budget, and major studio distribution. Shaded bars show parameters that are not significant at $p < 0.10$.

North America (Fritz 2016). The film's performance provides ad hoc support that restriction effects are only *one* side of a two-sided coin—and that situations exist when "[w]hatever they lose in teenage audiences who can't see the film without a parent, … they can more than make up for with people 17 and older who are more attracted by the bloodier or funnier material an R-rating allows" (Fritz 2017).

In addition to genres, the value of radical content also varies with the *context* in which entertainment products are consumed. Movie research suggests that theater visits and home entertainment follow a separate logic when it comes to rating effects. When we applied a partial least squares estimation to a data set of 331 films from 1999 to 2001, we found that movies' restrictiveness *positively* influenced the video rentals of films (Hennig-Thurau et al. 2006). This result is consistent with Jozefowicz et al.'s (2008) result that, for their data set of movies that were highly successful in theaters, PG-13 and R-rated movies performed better in rental markets than films that were accessible for all ages (VHS rentals were +20%, DVD rentals +50%).

For video games, it is the console that defines the context more than anything else. When we crunched the data set of games also used by Marchand (2016), we learned that although a more restrictive rating adds to the attractiveness of an average Xbox or PS3 game, restrictiveness *hurts* game sales when the console is Nintendo's Wii, similar to "average" movies. Violence, as the force behind most restrictive ratings in video games, tends to lure Xbox and PS3 players, but keeps Wii players at a distance.

Some other questions remain unanswered. What is the "optimal" degree of radicalness for different forms of entertainment? Our own analyses, like most others, imply a linear effect of radical content. But it seems plausible that the appeal of violence, sex, and profanity is non-linear, such that a moderate level of radicalness may be preferred to both low or (very) high levels—although the success of ultra-violent fare such as the SAW movies make such an effect far from obvious, at least for the horror genre. *Entertainment Science* scholars will hopefully continue to explore the matter, shedding even more light on what some might consider a "dark side" of entertainment.

Finally, a Few Words on Risk and Radicalness

If restrictive ratings can hurt the success potential of films, why then do so many of them feature radical content and carry a restrictive rating? Scholars have named this the "R-rating puzzle" of the movie industry (Ravid and Basuroy 2004), referring the issue to the importance of risk for entertainment producers.

One potential explanation scholars have pointed to is that although films that contain radical content elements do not produce higher revenues, they might involve *less risk*. For a sample of 175 films from 1991 to 1993, Ravid and Basuroy (2004) use of the MPAA's explanation of their ratings to compare the standard deviations of the ROI of films that contain different levels of violent content and sexual content. The scholars find that the standard deviation of the ROI of films which are *very* violent is lower than for the average film in the sample. The same is true for those films that contain both violence *and* sex. Based on these findings, Ravid and Basuroy suggest that radical content may serve as a means to hedge the risk of film productions.

But why the lower risk for radical content? We speculate that it could be because radical content appeals to people's most base needs. These needs, although suppressed by more civilized processes, have at least *some* influence on most of our behaviors when consuming entertainment (as in other parts of life). Although pleasing such needs is not a *sufficient* reason to let entertainment enthuse us, there is always a market for "cheap thrills" and other stimuli that address such fundamental needs, regardless of other quality criteria, just as there is always demand for adult entertainment, which might explain the existence of films that would otherwise attract only very little interest.[182] But this logic remains, at least at this point, speculation.

Entertainment's Country of Origin

"Made in Hollywood" as a Quality Signal

In a globalized world, the country of origin is a straightforward quality signal. We know it because of the many products we have purchased that express their origin via stickers and labels, such as "Made in the USA." The power of country-of-origin cues differs between product categories: Germany has a reputation for high-quality cars, Italy for trend-setting fashion, and Columbia for premium coffee. Extensive research by marketing and management scholars has compiled evidence that country of origin influences customers' quality perceptions and purchase intentions for utilitarian consumer products and industrial products, across a large variety of conditions (e.g., Peterson and Jolibert 1995).

For many people, such country-related quality associations also apply to entertainment products. What kind of associations are triggered when we hear a film is a "Hollywood movie"? Filmmaker Robert Altman has offered a pointed description of the Hollywood stereotype in his film THE PLAYER, when he lets studio executive Griffin Mill tell his assistant that a story just pitched to him lacked certain elements which are needed to market a film in a success way. When asked what elements he was thinking of in detail, Mr. Mill's answer is a succession of nouns, including suspense, violence, hope, heart, laughter, sex, and happy endings. Mr. Mill then specifies that happy endings are the main concern, not reality.

Although offered in a tongue-in-cheek manner, several of the elements in Mr. Altman's characterization actually overlap with entertainment qualities

[182]We have mentioned the Cannon Group's approach to making movies earlier in this book and will return to it in more detail in our discussion of entertainment innovation. At this point it is informative that most of the firm's works have been labeled "exploitation" films, as they almost always featured "cheap attractions"—high levels of exploitative (i.e., dramatically unmotivated) violence, sex, and profanity. Despite this fact, or, following our argument here, because of it, they have developed a devoted fan base which still celebrates Cannon's creations some 25 years after the firm went out of business, on Facebook (e.g., "Cannon Films Appreciation Society") and elsewhere.

we have discussed in previous chapters of this book.[183] It is their *accumulation* that links to the narrative conventions that are often associated with Hollywood films. Other widely shared Hollywood associations are American values (i.e., the individual hero or the importance of "achievements" that are reflected in "rags-to-riches" movies, such as Rocky and The Pursuit of Happyness), a distinctive aesthetic style, as well as the use of a high budget and stars (Hennig-Thurau et al. 2001)—elements we will discuss on the following pages.

Very different associations are triggered by entertainment products from other countries. French movies are believed to be "art house" rather than entertaining, focused on non-conformist characters instead of special effects, the disappointments of life instead of happy endings, and complex and ambitious stories and styles (e.g., Porter 2010). Indian films from Bollywood, in contrast, are known for their "hybrid" nature; most plots involve a love story, but also singing and dancing.[184] They also combine heavy melodrama with slapstick humor. And Russian works are often associated with deep philosophical questions and a pessimistic outlook rather than a straightforward optimistic narrative.

But the country-of-origin concept is not limited to films. It also applies to pop songs (we expect different tunes and sounds from British bands and French singers), video games (doesn't an American game sound cutting-edge?), and novels (aren't British writers masters of dark humor and Russian novelists, like the country's filmmakers, obsessed with reflection?).

In addition to those aesthetic differences, there is also a much more practical one: language. If entertainment products from a country use a different language than that practiced by the consumer, novels have to be translated, films and games have to be dubbed or subtitled, and music lyrics become sound elements instead of conveyors of meaning. But be aware that the country-of-origin concept must not be reduced to language. Whereas British and most Canadian film makers use the same language as their colleagues from California, audiences' associations won't be the same, and their products will thus not benefit from being perceived as "Hollywood films."

[183]See in particular our discussion of "great" storylines in the chapter on entertainment product quality.

[184]Or, as an Internet user suggested humorously, there's always a boy, a girl, and a tree in Bollywood movies—the boy falls for the girl, the girl, after some hindrances are overcome, falls for the boy, then they (and others) sing and dance around the tree in various locations (Valan 2010).

Empirical Findings on How Entertainment's Country of Origin Influences Success

The impact of language, as the pragmatic layer of country of origin, is obvious for countries in which dubbing is disliked by consumers (as is the case in the U.S.). With the exceptions of sub-titled Taiwanese action film CROUCHING TIGER, HIDDEN DRAGON and Mel Gibson's PASSION OF THE CHRIST (which featured dialogue purely in ancient biblical tongues), no foreign-language film has ever generated more than $100 million in North American theaters; only ten have made at least $20 million. The number of non-English language pop songs that became hits in the U.S. is also quite small, with few eccentric exceptions such as Austrian singer Falco's ROCK ME AMADEUS which in 1986 made it to the top of the Billboard charts despite its mainly German lyrics. From a statistical perspective, those exceptions are "outliers" or "artifacts," which cannot be replicated and thus provide no basis for learning.

Instead of trying to imitate such rare occurrences, entertainment producers have developed strategies to address this country-of-origin malice. Musicians sometimes record "localized" versions of their hit songs.[185] And American film makers regularly produce English-language remakes of foreign-language films for their home market (e.g., Gore Verbinski's THE RING was a remake of the Japanese horror film RING), which, in addition to overcoming the language gap, also allows them to get rid of other disadvantages that are tied to the importing of a product with a different country of origin, crafting an "Americanized" version of the original.[186]

A language bias is not exclusive for American consumers, but exists for all cultures. Schmidt-Stölting et al. (2011) show that German readers have a bias against translated hardcover books. They find that, compared to books originally written in German, sales of translated books are 16% lower, using a sophisticated SUR analysis that accounts for several other book characteristics. But their results also point to the role of context: findings show no impact of language for paperback books.

---

[185]For example. German singer Nena released an English-language version of her 1983 song 99 LUFTBALLONS under the title 99 RED BALLOONS. Whereas the German version climbed up to #2 in the U.S., the English-language version indeed became #1 in the UK, Ireland, and Canada (but interestingly had no success in the U.S.). In the pre-globalized world of the 1950s, 1960s, and 1970s, recording songs in other languages was done by many international stars, such as The Beatles (SHE LOVES YOU—SIE LIEBT DICH in German), and The Beach Boys (IN MY ROOM—GANZ ALLEIN, also in German). As late as in the 1980s, some stars still recorded versions of their songs in other languages, such as Michael Jackson did with a Spanish version of I JUST CAN'T STOP LOVING YOU in 1988.

[186]But the challenges that exist for every remake of an existing entertainment product also apply here.

But country-of-origin effects are far bigger than language. Using a conjoint analysis approach to determine the importance of certain movie characteristics for New Zealand moviegoers, Gazley et al. (2011) find clear preferences for movies within certain countries of origin. For their sample of 225 consumers, the "Made in Hollywood" label is as important as a movie's genres—and even more important than the presence of a favorite star. A New Zealand country of origin, in contrast, strongly *reduces* a film's attractiveness.

We have employed secondary data for 231 movies released in both North America and Germany from 1998 to 2001 to study the impact of "Made-in-Hollywood" associations (Hennig-Thurau et al. 2003). Using a measure of films' "Hollywood style" from (now defunct) Reel.com, we find a significant positive link between the degree of a film's "Hollywood style" and its box office success in both regions. The variable explains about 13% of success in North America, and it explains 8% of the German box office performance.

Other researchers have included one or more country-of-origin variables in their models when explaining the success of entertainment products. The consistent finding for movies is that an American origin provides movies with a competitive advantage among *American* audiences. For example, Litman and Kohl (1989) conduct an OLS regression with 697 movies from 1981 to 1986, estimating that between $5.6-$8.5 million in theatrical rentals can be attributed to a film's North American origin, which equal about twice the effect in total box office. Wallace et al. (1993) obtain similar results from a stepwise regression using 1,687 movies from 1956 to 1988; they estimate an average rental effect of $5.6 million. Consistent with such a bias, we learn when examining 575 movies from 1998 to 2002 with partial least squares (and controlling for an extensive list of other factors) that both a European origin and a "neither-North-American-nor-European" origin are disadvantages in North American theaters (Hennig-Thurau et al. 2013).

Additional findings point to the heterogeneity that exists among country-of-origin associations for entertainment on a *global* scale, that is, outside of North America. Look at Fig. 8.8 which lists the market share of U.S.-produced films in the 15 largest movie-going countries in 2007/2008. It illustrates that Hollywood productions capture about nine out of ten tickets sold in Mexico and Canada, and about two-thirds of tickets sold in Spain, Germany, and Italy. But U.S. films take only about one-third of ticket sales in Japan and China, and even lower in India (Epstein 2011). In addition to supply-side issues, those massive discrepancies point at a varying appeal of the "Made-in-Hollywood" image.

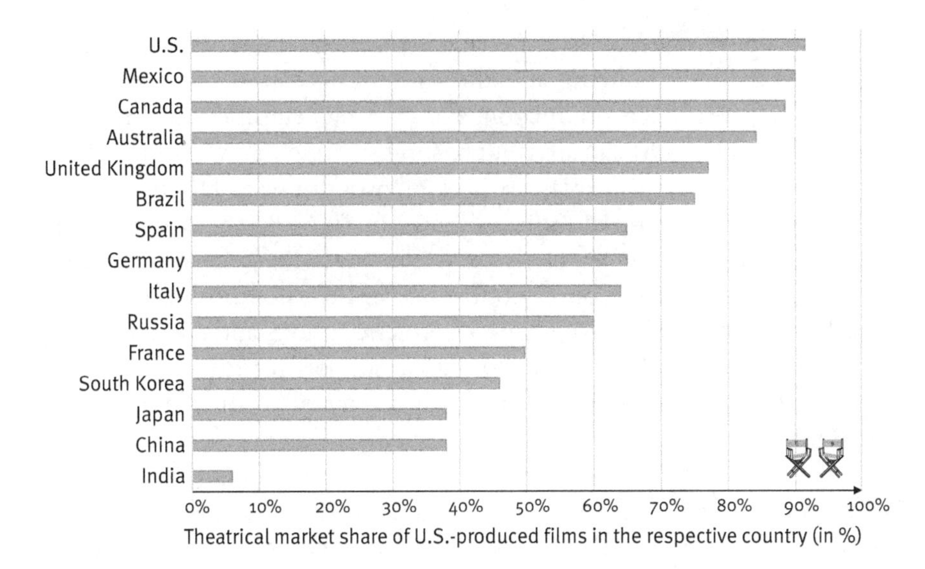

Fig. 8.8 The market share of U.S.-produced films in different countries
Notes: Authors' own illustration based on calculations by Epstein (2011). Data refer to 2007/2008.

So, although the image of America, in general, and Hollywood, in particular, can help U.S. entertainment products enormously in several parts of the world, there is clearly more to global success. How can we explain such enormous differences? Drawing on entertainment's cultural nature, scholars have argued that the "cultural distance" between the producing country and the consuming country is crucial for explaining entertainment-related country-of-origin associations and their impact on success. Let us take a look into this issue.

Both Sides Matter: Cultural Discount

The basic idea here is that we, as consumers, "discount" a cultural product based on the distance between the culture in which we live and the culture in which the product is manufactured. Hoskins and Mirus (1988) introduced this idea in an attempt to explain the dominance of American TV productions in several, but not all, parts of the world. They explain that such cultural discount reflects the "diminished appeal" of content produced in a different cultural context than a consumer's own. Such discount exists because people "find it difficult to identify with the style, values, beliefs, institutions and behavioral patterns of the material in question" (p. 500).

The concept of cultural discount suggests that the larger the cultural distance between the producing and the consuming country, the lesser consumers value the foreign production. For example, cultural discount would argue that because U.S. culture is more similar to Germany's culture than it is to Indian culture, this explains why Hollywood movies fare much better in Germany than they do in India, as shown in Fig. 8.8. And it suggests that American moviegoers prefer Hollywood films over those from other countries (as we have reported above) not because of their patriotism, but because of the higher cultural similarity between American films and audiences.

Can *Entertainment Science* offer empirical support for this logic? A number of studies have empirically tested the impact of cultural discount and related factors on entertainment product success. Craig et al. (2005) analyze the performance of close to 300 American-produced films (the U.S. "Top 50" from 1997 to 2002) in eight foreign countries. The dependent variable in their regression analysis is the "per-capita box office" of a film, i.e., the revenues a movie made in a country (expressed on a per-citizen basis to control for the population size). They then examine whether this measure of film success can be explained by the cultural distance between the U.S. and the respective country, which Craig and his colleagues calculate based on Hofstede's culture dimensions (i.e., individualism, power distance, uncertainty avoidance, and masculinity; Hofstede 1991). Controlling for a film's genre (but no other movie elements), they confirm that higher cultural distance diminishes a film's box office: a one-point increase in distance corresponds to a 15.7% decrease in per-capita revenues. The findings also reveal a positive association between a country's "Americanization" level (which they measured by the number of McDonald's restaurants in a country): the more Americanized a country, the better American films perform there.

Other scholars corroborate these findings and extend our understanding. Moon et al. (2016) employ Hofstede's culture dimensions when analyzing the performance of 846 American movies (from 2008 to 2015) in 48 countries using 2SLS. They also find that cultural distance is negatively related to film performance in a country. But they also show that the impact of cultural distance is non-linear: it hurts box office most strongly when distance is increased from low to medium, but much less so as it is increased to higher distance levels. And there's one more interesting insight we learn from the work by Sangkil Moon and his colleagues. Using a text mining approach in which they look for cultural terms in movie reviews, they determine each movie's cultural "compatibility" with the country to which it is exported. Their results show that higher compatibility helps a movie's box office in a country, above and beyond the cultural distance between the countries of production and consumption.

In another study, Hanson and Xiang (2009) study the performance of 284 American films (from 1995 to 2006) in 46 countries, using language dissimilarity and geographic distance as proxies of cultural distance between countries. Running OLS regressions on the country level, they find that language dissimilarity explains more than 20% of the average performance of American films relative to the performance of local movies; larger geographic distances also are negatively related to the (relative) success of American films in a country.

Finally, cultural distance also works when using an "import" perspective, i.e., explaining why movies from one country succeed and those from others fail when released in a specific culture. Fu and Lee (2008) apply this logic to Singapore (as the importing country), empirically explaining the performance of 441 films from 2002 to 2004 with the cultural distance (measured again with Hofstede's dimensions) between Singapore and the 22 countries in which the released movies were produced. OLS regressions on the country level show that a higher cultural distance has two consequences: fewer films are imported (the "supply effect" of cultural distance) and lower demand for those that are imported (the "demand effect").[187]

The Production Budget

Before we move on to branded signals, let us look at one last factor that is also often considered a quality signal in empirical studies: an entertainment product's budget. The logic is that consumers treat information about a product's budget as an indicator of the talent involved in the making of a product (which should be reflected, for movies, in great acting, dialogue, and special effects), *or* as an indicator of the popular appeal of the product (which also implies "high quality"). After all, what kind of producer would spend $200 million for a movie if the project is anything less than exciting for a lot of people (Hennig-Thurau et al. 2001)?

Several scholars, including us, have included the production costs in their regression models to help explain product success. The majority of these studies deals with movies; budgets are much less often disclosed for games, musical productions, or novels. The existing results provide strong evidence that the budget size indeed correlates positively with entertainment products' revenues. For example, Ravid (1999) reports elasticities of larger than

[187]Park (2015) reports similar findings for both supply and demand of 222 movies in Australia. Her findings have to be taken with some care, though, as she conducts the analysis on the movie level, not country level, and does not account for the hierarchical nature of the data.

+1 for revenues earned via several distribution channels (North American and international theaters, home video rental). Other studies that show similar effects include those by Litman and Kohl (1989), De Vany and Walls (1999; higher budgets are associated with higher "hit probabilities"), and Lampel and Shamsie (2000).

However, although we agree with the idea that the budget of an entertainment product can indeed serve as a quality signal for consumers, we are skeptical that these findings should be interpreted as *causal* effects. Such logic would imply that mainstream consumers are aware of a specific product's production budget. But we see little evidence that this is the case; even though the media occasionally reports when films have extreme budgets or have exceeded their initial budgets, consumers rarely mention budget as a reason for selecting a certain entertainment product. On the Internet, we find some enthusiasts discussing record-breaking budgets on fan sites (e.g., *Whedonesque* 2011), but we do not find budget to be a regular topic of consumers' social media chatter. This kind of chatter would be essential for budget information to spread via word of mouth from enthusiasts to "normal" consumers.[188] In sum, we don't think that a substantial share of consumers knows a film's budget or would have high interest in it, at least when other information is available, as is usually the case when consumers make entertainment choices. And if someone is unaware of something, then she or he simply won't use it as part of the decision-making process.

Instead of being treated as a causal effect, it seems much more plausible to interpret the correlations between budgets and success to be the result of a complex, multifaceted relationship. Managers of entertainment products make up their minds early in the production process regarding the commercial potential of a project. Among the first decisions affected by their judgment is the production budget, which influences the way the product is made (how much money is spent for the writing, the acquisition of brands, for stars). But the judgment is also mirrored in later decisions regarding key elements of the marketing mix, such as the amount of money spent for advertising and distribution. As we discuss in this book, all these follow-up decisions then influence the consumer's perception of, and anticipation for, the product in a causal way.

If these "other" variables are missing from a statistical model, the budget variable will absorb the variation in success that is actually due to these other variables and we will mistakenly conclude that budget has a

[188]See our chapter on "earned" communication for a discussion of the role of word of mouth in entertainment.

more important direct effect as a signal for consumers than it truly does. This situation is a classic case of a spurious correlation caused by an omitted variable bias.[189] Empirical support for this logic comes from studies in which scholars control for other factors and which consistently show that, the more comprehensive a statistical model, the smaller the effect size of the production budget; some studies even report it to be insignificant (e.g., Elberse and Eliashberg 2003; Hennig-Thurau et al. 2007; Liu 2006). And even if the budget indeed remains significant in such models, we suspect it is because the budget stands in for further factors that are not included, such as high-quality special effects or exotic locations, not because consumers buy an entertainment product directly because of its budget.

So, what then should be the role of the production budget in empirical models of entertainment success? Because it is determined so early, the budget is a good proxy for managers' expectations. Interpreted as such, the budget can be a useful instrument for other factors such as advertising or the use of stars (which are both also affected by managerial success expectations, rather than chosen independently, as a standard regression model would assume), an approach that would enable an unbiased measuring of these other factors. For example, Basuroy et al. (2006) use budget information as an instrument of advertising spending. Luan and Sudhir (2010) have done the same, and they also employed it to explain other DVD characteristics, such as the price. Ekaterina "Kate" Karniouchina (2011) used the budget as an instrument of a movie's "buzz" in her analyses, and we used it to create an unbiased measure of the role of movie stars for film success (Hofmann et al. 2016). Another way to make use of the budget is to explain *distributors'* decisions regarding new products, such as in how many theaters a new movie opens (see Elberse and Eliashberg 2003, Clement et al. 2014, and De Vany and Walls 1999). These industry experts have, compared to consumers, a much better knowledge of the ingredients of new entertainment products.

Does all this mean the budget is not a relevant factor for the success of an entertainment product? Not at all—as we have laid out before, financial resources are a key source for competitive advantage in entertainment. But it means that increasing a product's production budget does not necessarily lead directly to higher consumer demand, because its effect on consumers is of an indirect and complex kind. The budget is only impactful when it influences one or more of the more direct drivers of entertainment success.

---

[189]We discuss the matter of spurious correlations and warn you, our readers, about them in this book's inaugural chapter.

Concluding Comments

With the "true" quality of any entertainment product being hidden inside the consumption experience, what can we, as consumers of entertainment, do to pick the "right" product out of the myriad of offerings—the one that we will enjoy? In this chapter, we first looked at search qualities, which are notoriously limited in number and also relevance for entertainment. Technology, such as stunning special effects, but also virtual reality and 3D presentations can ensure consumers of certain benefits, and we discussed the key achievements and developments in this chapter for the different forms of entertainment. Our discussion demonstrated that no technology can guarantee success; what matters is whether meaningful consumer benefits are created by the technology that exceed costs—for those who offer the technology and those who have to pay for consuming it.

In the rest of the chapter, we examined the research concerning signals of product quality, or pseudo-search qualities: factors that are provided by a producer to lead consumers to infer from them that a new entertainment product is one that we will enjoy, based on previous experiences with those factors. You have enjoyed romantic comedies in the past? Here's a new one for you! Knowing the genre of a movie, video game, novel, or song can narrow the field for consumers because we have learned what to expect from genres and have formed genre-based preferences, loving some while despising others. We reported that spanning multiple genres can be dangerous, and that the fit of product's genre with broader cultural preferences in a market matters: a winning genre in one country may be a big loser in another market. Genre preferences also evolve over time and with societal shifts.

Age ratings and the controversial content they regulate are a two-sided coin in that the restriction on market size (because some portion of consumers are restricted from consuming the product) may be offset by the product becoming more appealing to certain target audiences, because of the "edgy" content that inspired the rating to begin with. We showed that, on average, restrictions dominate such appeal, but that a more detailed look at genres is recommended, as effects vary strongly between them. An entertainment product's country of origin is another pseudo-search quality: consumers hold strong expectations and preferences for entertainment that comes from different countries of origin (think Hollywood versus Bollywood). We argued that the production budget of a product, although often treated as just another signal, plays a more complex role and should be treated as such when it comes to configure an entertainment product and to predict its success.

For the producer of entertainment products, thinking through the impact of these various signals enables one to make more informed product decisions. Is the goal to win a narrow niche or market or to play broadly across customer segments and markets? Signals will open some options and close off others. This is particularly true for the type of signals we will discuss in the next chapter—signals that use brands of various kinds and origins to enable consumers to infer the quality of the new entertainment product to which they are attached.

References

Akdeniz, B. M., & Talay, M. B. (2013). Cultural variations in the use of marketing signals: A multilevel analysis of the motion picture industry. *Journal of the Academy of Marketing Science, 41*, 601–624.

Avery, B., Pickarski, W., Warren, J., & Thomas, B. H. (2006). Evaluation of user satisfaction and learnability for outdoor augmented reality gaming. *Proceedings of the 7th Australasian User Interface Conference, 50*, 17–24.

Barranco, R. E., Rader, N. E., & Smith, A. (2015). Violence at the box office: Considering ratings, ticket sales, and content of movies. *Communication Research, 44*, 1–19.

Barnes, B. (2015). 'Sniper' rules weekend box office. *The New York Times*, January 18, https://goo.gl/8yKNZd.

Basuroy, S., Kaushik Desai, K., & Talukdar, D. (2006). An empirical investigation of signaling in the motion picture industry. *Journal of Marketing Research, 43*, 287–295.

Bauer, C. (2010). The making of Katy Perry's cotton candy scented packaging. *Unified Manufacturing*, September 3, https://goo.gl/mwbtET.

Bennington, C. (2016). Chester Bennington: Quotes. *IMDb*, January 20, https://goo.gl/bM3M5s.

Block, A. B., & Wilson, L. A. (2010). *George Lucas's blockbusting: A decade-by-decade survey of timeless movies including untold secrets of their financial and cultural success*. New York: Harper Collins.

Busch, A. (2017). Christopher Nolan shows off 'Dunkirk,' says "The Only Way To Carry You Through" the film is at a theater—CinemaCon. *Deadline*, March 29, https://goo.gl/L6QhGz.

Cho, E. J., Lee, K. M., Cho, S. M., & Choi, Y. H. (2014). Effects of stereoscopic movies: The positions of stereoscopic objects and the viewing conditions. *Displays, 35*, 59–65.

Cieply, M. (2014). Hollywood works to maintain its world dominance. *The New York Times*, November 3, https://goo.gl/3adCoQ.

CJ (2017). Website of CJ 4DX. Accessed December 12, https://goo.gl/W8whqk.

Cobb, S. V. G., Nichols, S., Ramsey, A., & Wilson, J. R. (1999). Virtual reality-induced symptoms and effects (VRISE). *Presence, 8*, 169–186.

Cox, J. (2013). What makes a blockbuster video game? An empirical analysis of US sales data. *Managerial And Decision Economics, 35*, 189–198.

Clement, M., Wu, S., & Fischer, M. (2014). Empirical generalization of demand and supply dynamics for movies. *International Journal of Research in Marketing, 31*, 207–223.

Craig, S. C., Greene, W. H., & Douglas, S. P. (2005). Culture matters: Consumer acceptance of U.S. films in foreign markets. *Journal of International Marketing, 13*, 80–103.

Croghan, N. B. H., Arehart, K. H., & Kates, J. M. (2012). Quality and loudness judgments for music subjected to compression limiting. *The Journal of the Acoustical Society of America, 132*, 1177–1188.

Cutting, J. E. (2016). Narrative theory and the dynamics of popular movies. *Psychonomic Bulletin & Review, 23*, 1713–1743.

D'Alessandro, A. (2017a). As exhibitors fret over studios' push to crush windows, here's the sobering reality about PVOD—CinemaCon. *Deadline*, March 27, https://goo.gl/iWRTjd.

D'Alessandro, A. (2017b). Lorenzo di Bonaventura on moviegoing in a streaming world: "We Still Have The Advantage Of Spectacle". *Deadline*, October 19, https://goo.gl/MW5Ye2.

De Semlyen, P. (2012). Exclusive: Ridley Scott on Prometheus. *Empire*, March 28, https://goo.gl/M5XzL9.

De Vany, A., & Walls, W. D. (1999). Uncertainty in the movie industry: Does star power reduce the terror of the box office? *Journal of Cultural Economics, 23*, 285–318.

Derisz, R. (2016). This child's adorable letter stating why he wants to see 'Deadpool' has launched a PG-13 petition. *Movie Pilot*, January 20, https://goo.gl/uwz4j8.

Dogruel, L., & Joeckel, S. (2013). Video game rating systems in the US and Europe: Comparing their outcomes. *International Communication Gazette, 75*, 672–692.

Ebert, R. (1981). Heaven's Gate movie review and film summary. *RogerEbert.com*, January 1, https://goo.gl/qrFHSY.

Ebert, R. (2011). Why 3D doesn't work and never will. Case closed. *Roger Ebert's Journal*, January 23, https://goo.gl/uFNbsT.

Elberse, A., & Eliashberg, J. (2003). Demand and supply dynamics for sequentially released products in international markets: The case of motion pictures. *Marketing Science, 22*, 329–354.

Epstein, J. (2011). World domination by box office cinema admissions. *GreenAsh*, July 18, https://goo.gl/DcMi4L.

Evans, M. (2011). Germany officially the world's least funny country. *Telegraph*, June 7, https://goo.gl/c3gY1B.

FFA Filmförderungsanstalt (2016). Kinobesucher 2015 – Strukturen und Entwicklungen auf Basis des GfK-Panels, April, https://goo.gl/LuouGA.

Follows, S. (2016). The relative popularity of genres around the world, September 19, https://goo.gl/ipQNCq.

Follows, S. (2017). Are audiences tiring of 3D movies? November 20, https://goo.gl/YvCQyJ.

Fleming Jr, M. (2016). Mel Gibson on his Venice festival comeback picture 'Hacksaw Ridge'—Q&A. *Deadline*, September 6, https://goo.gl/xKRp9R.

Fritz, B. (2016). Hollywood now worries about viewer scores, not reviews. *The Wall Street Journal*, July 20, https://goo.gl/K8CR95.

Fritz, B. (2017). Why more movies will be R rated this summer. *Wall Street Journal*, April 26, https://goo.gl/geD4sn.

Fu, W. W., & Lee, T. K. (2008). Economic and cultural influences on the theatrical consumption of foreign films in Singapore. *Journal of Media Economics, 21*, 1–27.

Fuchs, A. (2014). Earthshattering: FJI salutes the 40th anniversary of Sensurround's quakes and battles. *Film Journal*, August 15, https://goo.gl/EpXFwM.

Gazley, A., Clark, G., & Sinha, A. (2011). Understanding preferences for motion pictures. *Journal of Business Research, 64*, 854–861.

Gomez-Uribe, C. A., & Hunt, N. (2015). The Netflix recommender system: Algorithms, business value, and innovation. *ACM Transactions on Management Information Systems, 6*, 13–19.

Grierson, T. (2014). 8 things you need to know about the 4DX theater experience. *Rolling Stone*, May 19, https://goo.gl/gQ66Nh.

Hanson, G. H., & Xiang, C. (2009). Trade barriers and trade flows with product heterogeneity: An application to US motion picture exports. *Journal of International Economics, 83*, 14–26.

Hayes, D. (2017). 'Logan' director James Mangold: If Fox film fades out post-merger, "That Would Be Sad To Me". *Deadline*, December 11, https://goo.gl/bHZ5Pb.

Hennig-Thurau, T., Walsh, G., & Wruck, O. (2001). An investigation into the factors determining the success of service innovations: The case of motion pictures. *Academy of Marketing Science Review, 1*, 1–23.

Hennig-Thurau, T., Walsh, G., & Bode, M. (2003). Exporting media products: Understanding the success and failure of Hollywood movies in Germany. Working Paper, Bauhaus-University of Weimar.

Hennig-Thurau, T., Houston, M. B., & Walsh, G. (2006). The differing roles of success drivers across sequential channels: An application to the motion picture industry. *Journal of the Academy of Marketing Science, 34*, 559–575.

Hennig-Thurau, T., Houston, M. B., & Walsh, G. (2007). Determinants of motion picture box office and profitability: An interrelationship approach. *Review of Managerial Science, 1*, 65–92.

Hennig-Thurau, T., Houston, M. B., & Heitjans, T. (2009). Conceptualizing and measuring the monetary value of brand extensions: The case of motion pictures. *Journal of Marketing, 73*, 167–183.

Hennig-Thurau, T., Fuchs, S., & Houston, M. B. (2013). What's a movie worth? Determining the monetary value of motion pictures' TV rights. *International Journal of Arts Management, 15*, 4–20.

Hofmann, J., Clement, M., Völckner, F., & Hennig-Thurau, T. (2016). Empirical generalizations on the impact of stars on the economic success of movies. *International Journal of Research in Marketing, 34*, 442–461.

Hofstede, G. (1991). *Cultures and organizations: Software of the mind*. London: McGraw-Hill.

Holbrook, M. B. (1999). Popular appeal versus expert judgments of motion pictures. *Journal of Consumer Research, 26*, 144–155.

Holbrook, M. B., & Hirschman, E. C. (1982). The experiential aspects of consumption: Consumer fantasies, feelings, and fun. *Journal of Consumer Research, 9*, 132–140.

Hoskins, C., & Mirus, R. (1988). Reasons for the US dominance of the international trade in television programmes. *Media, Culture and Society, 10*, 499–515.

Howell, M. J., Herrera, N. S., Moore, A. G., & Mcmahan, R. P. (2016). A reproducible olfactory display for exploring olfaction in immersive media experiences. *Multimedia Tools and Applications, 75*, 12311–12330.

Hsu, G. (2006). Jacks of all trades and masters of none: Audiences' reactions to feature film production. *Administrative Science Quarterly, 51*, 420–450.

Hsu, G., Negro, G., & Perretti, F. (2012). Hybrids in hollywood: A study of the production and performance of genre-spanning films. *Industrial & Corporate Change, 21*, 1427–1450.

Itzkoff, D. (2016). The real message in Ang Lee's latest? 'It's Just Good to Look at'. *The New York Times*, October 5, https://goo.gl/k56XAM.

Ji, Q., & Lee, Y. S. (2014). Genre matters: A comparative study on the entertainment effects of 3D in cinematic contexts. *3D Research, 5*, 5–15.

Jozefowicz, J., Kelley, J., & Brewer, S. (2008). New release: An empirical analysis of VHS/DVD rental success. *Atlantic Economic Journal, 36*, 139–151.

Karniouchina, E. V. (2011). Impact of star and movie buzz on motion picture distribution and box office revenue. *International Journal of Research in Marketing, 28*, 62–74.

King, G. (2002). *New Hollywood cinema*. New York: Columbia University Press.

Knapp, A.-K., & Hennig-Thurau, T. (2014). Does 3D make sense for Hollywood? The economic implications of adding a third dimension to hedonic media products. *Journal of Media Economics, 28*, 100–118.

Lampel, J., & Shamsie, J. (2000). Critical push: Strategies for creating momentum in the motion picture industry. *Journal of Management, 26*, 233–257.

Lang, D. M., & Switzer, D. M. (2008). Does sex sell? A look at the effects of sex and violence on motion picture revenues. Working Paper, California State University and St. Cloud State University.

Lasseter, J. (2015). Technology and the evolution of storytelling. *Medium*, June 24, https://goo.gl/dRsCxd.
Lee, J., Boatwright, P., & Kamakura, W. A. (2003). A Bayesian model for prelaunch sales forecasting of recorded music. *Management Science, 49*, 179–196.
Leemans, H., & Stokmans, M. (1991). Attributes used in choosing books. *Poetics, 20*, 487–505.
Leenders, M. A. A. M., & Eliashberg, J. (2011). The antecedents and consequences of restrictive age-based ratings in the global motion picture industry. *International Journal of Research in Marketing, 28*, 367–377.
Lin, J. J., Duh, H. B. L., Parker, D. E., Abi-Rached, H., & Furness, T. A. (2002). Effects of field of view on presence, enjoyment, memory, and simulator sickness in a virtual environment. In *Proceedings of the IEEE Virtual Reality 2002* (pp. 164–171). Los Alamitos: IEEE Computer Society.
Litman, B. R., & Kohl, L. S. (1989). Predicting financial success of motion pictures: The 80s experience. *Journal of Media Economics, 2*, 35–50.
Liu, Y. (2006). Word of mouth for movies: Its dynamics and impact on box office revenue. *Journal of Marketing, 70*, 74–89.
Luan, Y. J., & Sudhir, K. (2010). Forecasting marketing-mix responsiveness for new products. *Journal of Marketing Research, 47*, 444–457.
Mangen, A., & Kuiken, D. (2014). Lost in an iPad: Narrative engagement on paper and tablet. *Scientific Study of Literature, 4*, 150–177.
Mangen, A., Walgermo, B. R., & Brønnick, K. (2013). Reading linear texts on paper versus computer screen: Effects on Reading comprehension. *International Journal of Educational Research, 58*, 61–68.
Marchand, A. (2016). The power of an installed base to combat lifecycle decline: The case of video games. *International Journal of Research in Marketing, 33*, 140–154.
Marchand, A. (2017). Multiplayer features and game success. In R. Kowert & T. Quandt (Eds.), *New perspectives on the social aspects of digital gaming: Multiplayer* (2nd ed., pp. 97–111). New York: Routledge.
Mathur, M. B., & Reichling, D. B. (2016). Navigating a social world with robot partners: A quantitative cartography of the uncanny valley. *Cognition, 146*, 22–32.
Michelle, C., Davis, C. H., Hight, C., & Hardy, A. L. (2017). The Hobbit hyperreality paradox: Polarization among audiences for a 3D high frame rate film. *Convergence, 23*, 229–250.
Moon, S., & Song, R. (2015). The roles of cultural elements in international retailing of cultural products: An application to the motion picture industry. *Journal of Retailing, 91*, 154–170.
Moon, S., Mishra, A., & Mishra, H., & Young Kang, M. (2016). Cultural and economic impacts on global cultural products: Evidence from U.S. Movies. *Journal of International Marketing, 24*, 78–97.
Mori, M. (2012). The uncanny valley. *IEEE Robotics and Automation Magazine, 19*, 98–100.

Nathan, I. (2006). Rambo: First Blood Part II review. *Empire*, July 31, https://goo.gl/hpcQaY.

Nichols, S., Haldane, C., & Wilson, J. R. (2000). Measurement of presence and its consequences in virtual environments. *International Journal of Human-Computer Studies, 52*, 471–491.

Nowotny, B. (2011). Aroma-Scope? A history of 4-D Film sensations. *Movie Smackdown*, August 14, https://goo.gl/wgYKTF.

Park, S. (2015). Changing patterns of foreign movie imports, tastes, and consumption in Australia. *Journal of Cultural Economics, 39*, 85–98.

Patton, D. (2015). George Lucas slams Hollywood & 'Circus Movies' at Sundance panel. *Deadline*, January 29, https://goo.gl/4K4S5S.

Perez, S. (2016). Pokémon Go tops Twitter's daily users, sees more engagement than Facebook. *Techcrunch*, July 13, https://goo.gl/zjfc1Q.

Peterson, R. A., & Jolibert, A. J. P. (1995). A meta-analysis of country-of-origin effects. *Journal of International Business Studies, 26*, 883–900.

Plowman, S., & Goode, S. (2009). Factors affecting the intention to download music: Quality perceptions and downloading intensity. *Journal of Computer Information Systems, 49*, 84–97.

Porter, H. (2010). French films glow with confidence and culture. Ours should do the same. *The Guardian*, August 8, https://goo.gl/RnXWMn.

Pras, A., Zimmerman, R., Levitin, D., & Guastavino, C. (2009). Subjective evaluation of MP3 compression for different musical genres. *Audio Engineering Society Convention Paper 127*.

Plumb, A. (2015). The most ludicrous DVD/Blu-ray box sets ever. *Empire*, October 9, https://goo.gl/5n9w1g.

Ravid, S. A. (1999). Information, blockbusters, and stars: A study of the film industry. *The Journal of Business, 72*, 463–492.

Ravid, S. A., & Basuroy, S. (2004). Managerial objectives, the R-rating puzzle, and the production of violent films. *Journal of Business, 77*, 155–192.

Richardson, M. (2013). Does vinyl really sound better? *Pitchfork*, July 29, https://goo.gl/KoTdrK.

Rooney, B., & Hennessy, E. (2013). Actually in the cinema: A field study comparing real 3D and 2D movie patrons' attention, emotion, and film satisfaction. *Media Psychology, 16*, 441–460.

Sarantinos, J. G. (2012). Types of romantic comedies. *Script Firm*, June 13, https://goo.gl/s62gY5.

Schäfer, T., & Sedlmeier, P. (2009). From the functions of music to music preference. *Psychology of Music, 37*, 279–300.

Schmidt-Stölting, C., Blömeke, E., & Clement, M. (2011). Success drivers of fiction books: An empirical analysis of hardcover and paperback editions in Germany. *Journal of Media Economics, 24*, 24–47.

Shelstad, W. J., Smith, D. C., & Chaparro, B. S. (2017). Gaming on the rift: How virtual reality affects game user satisfaction. In *Proceedings of the Human Factors and Ergonomics Society 2017 Annual Meeting* (pp. 2072–2076).
Silver, D., Lee, M., & Clayton Childress, C. (2016). Genre complexes in popular music. *PLOS ONE, 11*, 1–23.
Strange, A. (2016). Apple's Tim Cook says augmented reality, not VR, is the future. *Mashable*, October 4, https://goo.gl/W8yAR8.
Tang, A. K. Y. (2017). Key factors in the triumph of Pokémon Go. *Business Horizons, 60*, 725–728.
The Economist (2017a). Alternative realities still suffer from technical constraints, February 11, https://goo.gl/e79idH.
The Economist (2017b). Better than real. *The Economist*, February 4, 67–69.
Totilo, S. (2011). China is both too scary and not scary enough to be video game villains. *Kotaku*, January 13, https://goo.gl/KaYyfb.
TRCG (2015). Why we love (& listen to) vinyl records. *The Record Collectors Guild*, July 2, https://goo.gl/t72zEF.
Valan, G. (2010). Answer to thread "What Are Defining Characteristics of a Bollywood Movie?". *Quora.com*, May 14, https://goo.gl/R7mqsb.
Waguespack, D. M., & Sorenson, O. (2011). The ratings game: Asymmetry in classification. *Organization Science, 22*, 541–553.
Wallace, W. T., Seigerman, A., & Holbrook, M. B. (1993). The role of actors and actresses in the success of films: How much is a movie star worth? *Journal of Cultural Economics, 17*, 17–27.
Wikipedia (2016). Fusion camera system. https://goo.gl/urQ63C.
Yamato, J. (2012). The science of high frame rates, or: Why 'The Hobbit' looks bad at 48 FPS. *Movieline*, December 14, https://goo.gl/5UZbUP.
Zhao, E. Y., Ishihara, M., & Loundsbury, M. (2013). Overcoming the illegitimacy discount: Cultural entrepreneurship in the US feature film industry. *Organization Studies, 34*, 1747–1776.

9

Entertainment Product Decisions, Episode 3: Brands as Quality Signals

"Unlike consumer package goods, the movie product has no established brand leaders, no brand loyalty among consumers, and, indeed, no actual brands."
—Austin *(1989, p. 8)*

"Everyone in Hollywood knows how important it is that a film is a brand before it hit theaters. If a brand has been around, HARRY POTTER for example, or SPIDER-MAN, you are light years ahead."
—*Director and producer* James Cameron *(quoted in* Oehmke *and* Beier *2011)*

Until a few years ago, entertainment managers hardly ever used the word "brand," a term that refers to a core marketing asset and a widely applied business concept. Most likely, the term just too openly conflicted with the "Nobody-Knows-Anything" mantra of the industry, with its self-conception as art rather than mere commerce, and the self-declared uniqueness of entertainment, per se (versus other "conventional" businesses). Rejecting the mere possibility of brands in entertainment was just too tempting, and the industry's managers were so convincing in their negation of the concept that even some prominent scholars of the industry agreed (see the introductory quote by Bruce Austin).

More recently, the entertainment industry created its own vocabulary to describe brand-like phenomena, with the term "franchise" gathering the most prominence among entertainment managers. In recent times, however, the industry's terminology has changed, and the term "brand" has now become quite prevalent. We tie this development closely to the enormous success that Marvel and its parent Disney have experienced with their building of a

T. Hennig-Thurau and M. B. Houston, *Entertainment Science*,
https://doi.org/10.1007/978-3-319-89292-4_9

"Cinematic Universe" since its introduction in 2006—a case study in brand management excellence that has been widely noticed inside the film and entertainment industry, but also outside of it. The second introductory quote to this chapter, by Hollywood luminary James Cameron, gives evidence of this change.

You, the reader, might wonder if it matters whether the term "brand" is used or some other phrase. Doesn't William Shakespeare answer that question when lovely Juliet stated that "A rose by any other name would smell as sweet"? Although Juliet's perspective turned out tragically (as we all know, it actually made *quite* a difference to change her name from Capulet to Montague), we are less concerned about the term "brand"—but much more so about taking advantage of a vast body of knowledge that has been developed by business and marketing scholars with regard to managing the immense power of brands. These scholars, over the course of half a century, have devoted enormous resources toward understanding and documenting how brands affect consumers and firm success. And although they rarely had entertainment products in mind when developing their theories, we will show that a lot can be learned from these theories by transferring them into the realm of entertainment. In other words, we do not adopt the term "brand" mainly as a linguistic device, but to open the gates to a large body of theory on how intellectual property, or "I.P.," can be effectively managed. As we will show, brand management concepts enhance our understanding of franchises, explain how and why sequels follow a different logic than remakes, and also help us understand why star actors and bestselling authors make such major contributions to the success of their entertainment products.

In what follows, we will take you, the reader, on a short trip to the core elements of brand theory and apply them to the context of entertainment. We make clear what we consider an entertainment brand and uncover the strategic options that exist for managing such brands. Our trip will stop at the creation of new brands, before we explore how brand assets that already exist can be exploited strategically by managers. We will discuss line extensions (such as sequels and remakes) and category extensions (such as the transfer of a bestselling book into a successful movie), but also *human* brands (such as actors) that serve as "ingredients" of entertainment products.

Although our discussion will focus on the decisions involved in the creation of individual products, we will also take a more holistic perspective and see what branding theory can teach us regarding the long-term management of brands. And those in the entertainment industry who own I.P. might be eager to learn about the well-developed econometric approaches that can help them to measure their financial value.

The Fundamentals of Entertainment Branding

What's in an Entertainment Brand?

About 10 years ago, Sood and Drèze (2006, p. 352) noted that "Hollywood has begun branding movies in a way similar to that in which consumer-packaged-goods manufacturers brand their products." Their statement foreshadowed a central change in the entertainment industry; large parts of the industry have now re-organized themselves around brands. Former Disney executive Jay Rasulo noted: "everything we do is about brands and franchises, and that wasn't true 10 years ago" (quoted in *seekingalpha.com* 2014).

So, what exactly defines a brand? As with most abstract concepts, there is not one single definition that all scholars agree upon. Whereas earlier definitions of brands focused on features that set a product or a service apart from its competitors (such as a logo or a name), more modern definitions put consumers in the spotlight and broaden the understanding of what can be a brand. For the purpose of this book, we follow this logic and consider entertainment brands as anything (1) for which cognitive associations are held by consumers or other relevant stakeholders (such as those who invest in entertainment), and (2) that can be managed professionally (e.g., Thomson 2006).

The "anything" element of the definition includes products and services, but also people. The "associations" element means that the essence of a brand is the semantic network that people hold in their minds for the "anything." To be a useful brand, such a cognitive network must exist. When we think of Star Wars, we think of heroes, themes, and space ships; James Bond reminds us of drinks that are shaken (not stirred!), high-tech gadgets, and charismatic villains; Lady Gaga conjures a plethora of associations such as pop-music, fancy dresses, and startling hairstyles.

The final element of the definition is the need to be professionally manageable—something that applies to products, services, and people (such as stars, who are managed by themselves or by talent agencies). But it does *not* apply to the concepts that we discussed in the previous chapter. Think of a genre like the western, a theme like baseball, or a movie's country of origin. Whereas we also hold associations about them (which are at the core of their signaling role), none of these can be managed professionally, at least not by the manager of a single entertainment firm.[190]

Why, then, are brands valuable? Brands have two key functions (Keller 1993). The first is an "awareness function": a brand can generate immediate attention. Whereas the name of something for which consumers have no associations will mostly pass unnoticed, brands catch consumers' attention. Because of this, they can help companies build consumers' awareness of new products by attaching the known brand name when the products are launched. Given the short-lived character of entertainment and the enormous costs of advertising, such "built-in" awareness can mark a particularly crucial competitive advantage.

The second key function of brands is their "image function." Inside the consumer's mind, the semantic network that surrounds a brand stores meanings which provide the basis by which consumers' identify with a brand and differentiate the brand from other products. Such meaning is particularly crucial for entertainment because of the problems consumers have in being unable to evaluate the quality of entertainment prior to experiencing it. A strong brand image can promote trust which helps to overcome such evaluation hurdles.

Take the example of bestselling author James Patterson, who argues that his own brand assures readers that they will be unable to put his books down before the end. He described his brand image as follows: "There will be tension. And pace. And some kind of human identification, not just with the heroes but also with the villains. Above all my brand stands for story. I became successful when I stopped writing sentences and started writing stories" (quoted in Streib 2009). Hachette rewarded Patterson's brand with $150 million. For filmed entertainment, Marvel has developed a strong brand image as a provider of high-quality entertainment with a constant stream of content that is consistent in subject and tone, but also professionalism: "whenever [Marvel] takes risks on an unknown [intellectual property], audiences believe it's going to be phenomenal" (D'Alessandro 2017).

The awareness and image functions of brands illustrate that the crucial question in branding is not whether something should or should not be called a brand; instead, it is how *strong* a brand is. We, as consumers, have associations with many things, but it is when we hold strong, positive

---

[190]In the case of a country, some actually argue that it can be considered a brand (see for example Kotler and Gertner 2002). But in that case, the brand managers are the country's politicians or dedicated country marketers, far beyond the realm of a producer of entertainment products. A similar logic applies to country-like labels such as "Made in Hollywood" (see p. 349).

and unique associations toward something what makes a difference (Keller 1993). Some entertainment products resonate with a large part of the population, whereas others leave hardly a mark. When we asked about 750 German consumers which movies they had heard about out of a list of 39 equally successful titles (all were the tenth most successful in Germany in their release year), the proportion of consumers aware of the films ranged from below 5% to above 80% (Hennig-Thurau et al. 2009).

Similarly, the richness and positivity of consumers' associations differ massively between entertainment products, further adding to the separation of strong and weak brands. Whereas time certainly plays a major role in brand awareness in entertainment,[191] other factors that are largely determined by a brand's manager also matter strongly, both prior to the release of a new product and over its life cycle—we will get to them later in this chapter.

Strategic Options of Entertainment Branding

Entertainment brands can—and should—be managed strategically, which requires a thorough understanding of the repertoire of brands entertainment managers have at hand and also the types of strategies that can be applied. We first offer the "Entertainment Brandscape" as a typology of brands in entertainment. And then, we develop a procedural framework that guides producers of entertainment in their brand strategy choices for new products.

A Typology of Brands: The "Entertainment Brandscape"

Entertainment producers can chose among a variety of brands when designing their entertainment products. We distinguish between four fundamental types of entertainment brands that co-exist and can be combined in highly creative ways by producers: product brands, character brands, company brands, and human brands.[192]

Product brands. Product brands are straightforward: whenever consumers maintain cognitive associations for an existing product, this product constitutes a brand. Accordingly, existing movies, books and comics, games, and musical compositions *are* product brands, regardless of whether they are

[191] When running a simple OLS regression with the awareness of the films among consumers as dependent variable, the release year explained more than 60% of the variation in awareness.

[192] The first two types in this list are what entertainment executives often refer to as "I.P."

treated as such by their owners or managers. The product brand is not the title or lead character, but a product's very own "universe" of elements and characteristics. For example, for an existing movie, the consumer can hold distinct associations regarding elements like the characters, places, and the fundamental narrative configurations that are combined in a holistic way by the product brand. As we argued earlier, awareness levels and image profiles differ heavily between existing products, and so does their financial brand value.

Character brands. Sometimes characters grow larger than the original products in which they were introduced and the characters take on a life all their own. Consider Spider-Man, Indiana Jones, and Lara Croft: they are all well-known by audiences, and audience members often have strong associations connected to them. It is when a character is taken out of his or her respective "universe" that the character brand idea becomes important. When Spider-Man appears in Marvel's CAPTAIN AMERICA: CIVIL WAR movie, it is a big deal: the product brand (i.e., SPIDER-MAN movies) and the character brand (the Spider-Man character) are separated from each other here. Marvel/Disney rented the character from competing studio Sony in order to add Spider-Man to its squad of super heroes, The Avengers.

To be able to do so, Sony requested (and was granted) extensive rights by Marvel: they could veto each scene in which their character brand was involved in order to protect the brand image of the character (Gonzales 2015). In addition to character combinations like The Avengers (which have a long tradition in comics), many "spin-offs" were inspired by the idea of granting a character its own host product. For example, THE PINK PANTHER's side character, inept Inspector Clouseau, became the star of its own film A SHOT IN THE DARK, which then set the tone (and narrative) for many more PINK PANTHER films. SHREK's Puss in Boots became the lead character in a film carrying the same name, and STAR WARS' Han Solo experiences his very own adventures, outside the main narrative, in SOLO: A STAR WARS STORY.

Company brands. In many industries, the name of the production company is used for the branding of most of its products (just think of "Heinz Tomato Ketchup" or "Chevrolet Cruze"). Such company brands have been used less often in entertainment. There are some prominent exceptions though. Disney is known for premium family entertainment, its Pixar subsidiary for modern animation, and Marvel (also owned by Disney) for its superhero tales. Consumers have strong associations for HBO and now also Netflix as a producers of "quality" filmed entertainment; and for music labels such as Motown which created its own "Motown Sound" of soul (Landau 1971).

The challenge for most entertainment firms in building a company brand is that their product portfolios are so heterogeneous, in terms of content and quality, that consumers may be aware of the producing firm, but will be unable to develop consistent brand associations about it. To become a strong company brand, intentional strategic choices are essential, such as a careful selection process and the rejection of projects which might be successful on their own, but would conflict with the company brand's intended brand image. As Disney's Sean Bailey phrases it, the key question resulting from a company brand orientation for any project then becomes "should it be Disney" (quoted in Fleming 2017a), rather than "will it be a hit." It is certainly not by accident that Marvel movies always involve "heart and comedy" (D'Alessandro 2017) in combination with a lot of mayhem and spectacle. Instead, it is the fact that audiences know what to expect from a Marvel movie which makes it a strong brand. On the company brand level, such image focus can get in the way of ambitious projects: whereas the Disney brand certainly helps to establish a new SVOD service, it conflicts with offering a vast variety of titles such as the ones now obtained via the take-over of Fox. When you subscribe to something like Disney Life, you don't expect your children to be scared away by xenomorphs and predators.

Human brands. Finally, there are also the human participants in entertainment who are known and valued by consumers. Such human brands can be visible or audible in an entertainment product, such as star actors (e.g., Harrison Ford) or musicians (e.g. Lady Gaga, or soundtrack composers such as John Williams). Alternatively, they can be involved in the product behind the line of visibility, such as movie directors (e.g., Steven Spielberg) or producers (e.g., Jerry Bruckheimer), writers (e.g., James Patterson, Stephen King), music producers (e.g., Mike Will), or in other roles.

As we are talking about creative products which depend on the contributions of artists and are released to fickle consumer tastes, steering any entertainment company brand consistently over time is certainly a challenging task. But the huge potential of a powerful brand has been empirically demonstrated by entertainment managers. Let us see in the following sections how producers can employ these types of brands when developing—and marketing—new products. Branding strategies are complex because they require the manager to balance the advantages of entertainment branding with its intricacies.

Entertainment Branding Strategies

When releasing a new product, an entertainment producer always has to make decisions regarding two strategic branding dimensions (Sattler and Völckner 2013). The first is whether or not to integrate an existing brand name into a new product (and if yes, which one). We call this decision the "brand integration strategy." In essence, the strategic alternatives are to either build a new brand from scratch or to extend an existing one—which entails a plethora of options for how to implement the extension.

The second strategic dimension of branding is whether the new product should function as a stand-alone brand or be combined with existing brands. Such partnering approaches, which we refer to as "brand alliance strategy," can take the form of co-branding or ingredient branding. In the remainder of this section, we will take an introductory look at brand integration and brand alliance strategies and their business potentials (and limitations). We then discuss the details of each strategy along with other facets of entertainment branding.

Brand Integration Strategies

Figure 9.1 illustrates the strategic brand integration options available to an entertainment producer for a new product. Because the success of all entertainment products is closely tied to consumers identifying and positively anticipating those products, the question whether or not one wants the new product to be a brand is not really a question. With rare exceptions, there simply is no success in entertainment unless the product takes on the character of a strong brand. It deserves much more thought to decide whether you want the new product to become an original brand (i.e., applying an "original brand strategy") or you prefer to make use of an existing brand—a "brand extension strategy." In the latter strategy, the new entertainment product extends a so-called "parent" brand that consumers already know well because of familiarity with the parent brand's prior use.

When designing a brand extension strategy, two crucial decisions have to be made: what kind of parent to use and which product category to target. Let's take a closer look at two options for parent brand choice. One type of brand extensions is "creator branding"—in this case, the brand that is extended is the one of the creator of the product. Creators can be companies that have commissioned the new product's production (such as Disney, Pixar, or now Netflix and Amazon), as well as the human brands that produce, direct, or

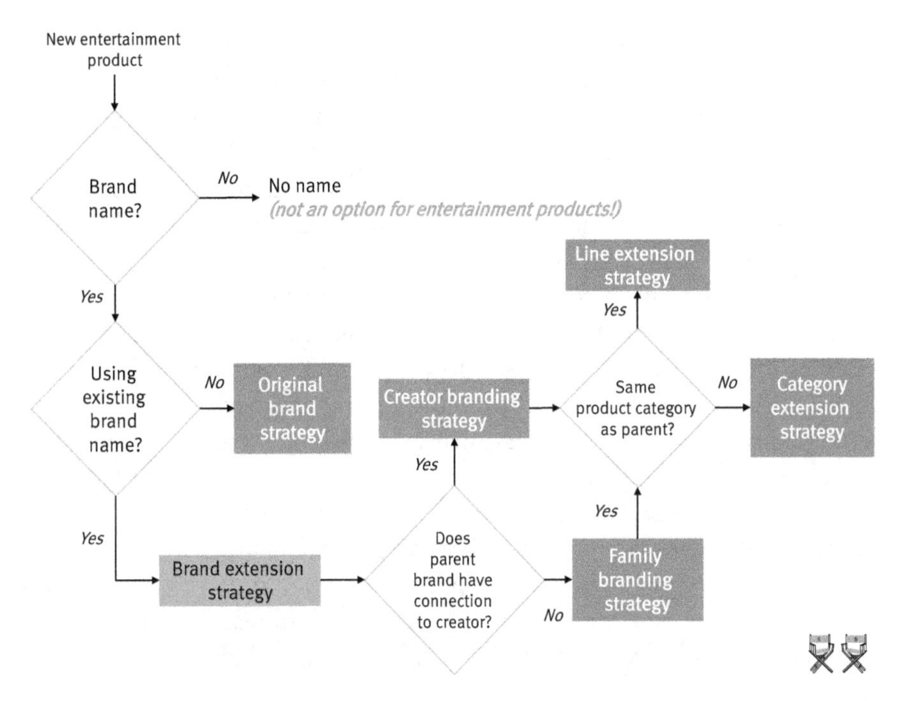

Fig. 9.1 A step-by-step framework of brand integration strategies
Note: Authors' own illustration based on ideas from Hansen et al. (2001) and Sattler (1999).

otherwise leave their mark on the product, Consider self-titled albums such as BEYONCÉ by Beyoncé[193] or filmed entertainment that contains a creative person's name in its title, such as ALFRED HITCHCOCK PRESENTS, TIM BURTON'S THE NIGHTMARE BEFORE CHRISTMAS, or STEPHEN KING'S IT.[194]

The other type of brand extension is "family branding." Here, the new product makes use of an existing brand from another entertainment product or market offering. For example, a new movie can take on the brand of a previous piece of entertainment media, such as a movie like TOY STORY (the existing brand used for TOY STORY 2) or a book like THE HUNGER GAMES (as in THE HUNGER GAMES, the movie). But the parent brand can also be adopted from other fields beyond media entertainment; think of board games-turned-movies (such as CLUE or BATTLESHIP) and toys-turned-movies (the TRANSFORMERS, MASTERS OF THE UNIVERSE, or THE LEGO MOVIE).

---

[193]The essay by Demarais (2009) has some very interesting (though certainly speculative) thoughts on when such self-titling works, and when it does not.

[194]Peden (1993) names several additional examples and also gives background information about the practice of using "human brands" as part of movie titles.

The brand extension strategy is also shaped by the choice of the target product category for the extension product. When the target category is the same as the category for which the parent brand is best known (as is the case with sequels and remakes), we call the new product a "line extension." In contrast, when the target category is different from the parent's category (for example, when a book or game is turned into a film, or vice versa), branding theory usually refers to this as a "category extension."

Every brand extension requires a decision about both the parent and the target category, and all combinations are possible, as we illustrate with the examples in Fig. 9.2. Extensions can even combine a real-world parent or a fictitious one as part of a multi-level branding approach. Take the example of games maker EA, who uses such a multi-level strategy for marketing its extensions to sports simulation games. Whereas EA SPORTS FIFA 17 is a family-branded line extension of both the FIFA brand and the EA Sports brand, it also makes use of creator branding by including the brand of the production company EA in the EA Sports family brand. In case you want to know what works best for a particular new entertainment product: we will get to the pros and cons of these strategies in the following sections.

Brand Alliance Strategies

This second branding strategy dimension might appear to offer producers fewer options at first glance, but a closer look reveals that the strategy can be used to create equally complex branding configurations as brand integration. Essentially, two basic kinds of brand alliances are possible: "co-branding" and "ingredient branding."

In co-branding, two (or more) brands mark a joint product. In entertainment, co-branding can take place on the creator level (such as in "Disney & Pixar present…"), but also when two or more well-known people cooperate to bring a product to life. As an example of such "human creator co-branding" in entertainment, you can think of every musical duet between otherwise autonomous artists, such as when Diana Ross and Lionel Richie promised each other ENDLESS LOVE (and led the Billboard charts for nine weeks) or when Michael Jackson and Paul McCartney collaborated on the hit single SAY SAY SAY.[195] Such versions of co-branding also happen in other areas of entertainment than music, such as when THE ADVENTURES OF

[195]Caulfield et al. (2011) have compiled a wonderful overview of the most remarkable, and most successful, duets in music history. Enjoy reading (and listening to) it!

		Parent brand	
		Creator branding	*Family branding*
Target product category	*Line extensions*	DISNEY'S FROZEN (extension of the Disney brand which is closely tied to the movie category)	BACK TO THE FUTURE PART II (extension of the first BACK TO THE FUTURE movie)
	Category extension	DISNEY INFINITY (extension of the Disney brand, which is closely tied to the movie category, to the video games category); Disneyland (similar logic, but to the amusement park category)	GAME OF THRONES TV series (extension of the novel series A SONG OF ICE AND FIRE)

Fig. 9.2 Exemplary combinations of brand extension strategies
Notes: Authors' own illustration. Brands are trademarked.

TINTIN, the movie adaptation of Hergé's comics and thus a category extension itself, was branded as "Steven Spielberg and Peter Jackson present …."

Co-branding is also used in entertainment when *characters* serve as family brands in what is sometimes called a "crossover," i.e., a product "in which characters or concepts from two or more discrete texts or series of texts meet" (Nevins 2011). Take the example of the movie BATMAN V SUPERMAN: DAWN OF JUSTICE. Here the main attraction was, as stressed in the film's marketing campaign, the joint appearance of the Superman and Batman characters, which otherwise (in movies) belong to separate "universes." Similarly, consider the joining (or clashing) of forces of the species from the ALIEN and PREDATOR franchises in the ALIEN VS. PREDATOR comics, games, and movies as an example for such "character co-branding."[196]

Ingredient branding, the second kind of brand alliance strategy, was made most famous by the "Intel Inside" ingredient-branding campaign of the Santa Clara-headquartered processor manufacturer. In this approach, the overall product, regardless of its brand, attempts to leverage the brand strength of one (or more) of its components or ingredients. The overall product is also named the "host," and ingredients differ from other products in that they usually are not or cannot be purchased in isolation—a processor does not offer much value to the consumer on its own.

[196]Crossovers have a long history in popular culture, with Nevins (2011) tracing them back to the Greek myths, when several mythical heroes, including Castor, Pollux and Heracles, join Jason in his search for the Golden Fleece. Other historic cases of crossovers include the teaming of legendary detective characters in Carolyn Wells' novel PURSUIT OF THE HOUSEBOAT (published in 1905) and its sequels, DC Comic's myriad superheroes forming The Justice League of America, and team-ups of Universal Studio's horror characters such as Dracula, Frankenstein, and The Wolf Man in the 1930s and 1940s (e.g., HOUSE OF FRANKENSTEIN from 1944, featuring all three "monsters"). Such crossovers sometimes serve as the basis for the development of their own family brands and also for today's meta-franchises.

The most popular ingredient brands in entertainment are stars (or "human brands"), such as when Arnold Schwarzenegger participates in a Terminator movie or Game of Thrones star Peter Dinklage was prominently featured as the voice of an artificial intelligence in the ultra-expensive Destiny game.[197] But fictitious characters can also be considered a branded ingredient in cases in which they are simply an addition to a story, such as in the case of Spider-Man's guest appearance in Captain America: Civil War.

Which Branding Strategy Has the Most Potential?

So, which of the different strategies is the most promising for branding entertainment? As we will discuss in the following sections, each carries its unique set of strengths, but also brings particular difficulties and limitations. In general, the biggest advantage of any of these strategies is that a brand provides built-in familiarity, one of the key drivers of consumers' choice of entertainment products. Just think of the collective sentimentality that pervaded the Internet when, in a late trailer of Star Wars: Episode VII, Harrison Ford, reprising his role as space cowboy Han Solo, intoned, "Chewie, we're home." Although Mr. Solo was speaking to his Wookie companion Chewbacca, his message to legions of Star Wars fans was unmistakable. As Harrison Ford words it: "[f]amiliarity was unlocked at that moment" (quoted in *The Hollywood Reporter* 2016).

Although brands can be powerful because consumers recall familiar past experiences with the brand, our sensations-familiarity framework also clarifies that more than familiarity is needed to entertain consumers. A general challenge for entertainment brands is to provide enough novelty to create a sufficient level of sensations. As Megan Colligan phrased it as marketing president at Paramount: "People are looking for something that really reimagines what they're expecting and yet they want the things that are very familiar to them" (quoted in Fritz and Schwartzel 2017). So, whereas a branded entertainment product has the *potential* for a competitive advantage over unbranded alternatives, its success depends on how the brand is used. We will explore the context factors of brand usage on the following pages.

Overall, however, decision makers in entertainment must see some value in branding because they have strongly embraced the use of brands for their

[197]As an aside, Mr. Dinklage's performance was not particularly appreciated by parts of the games community—and later re-dubbed by another actor (though the producer gave other reasons for this move; Philips 2015).

products, often as a focal part of the blockbuster concept of entertainment marketing.[198] Consumers have followed suit; for example, in 2011 eight of the ten top-grossing films in North America were sequels (and the other two were comic adaptions, a kind of category extension). Thirty years before, only two top-ten films were sequels—and all others were "originals" in the sense of brand integration (Allen 2012). Figure 9.3 provides a more comprehensive look at the current prominence of branded products across the different forms of entertainment—in each field, a substantial share of the hits are products that use one or more of the branding strategies we highlight.

The figure shows that even if we neglect the creator-brand role of James Cameron for both AVATAR and TITANIC, the most successful 15 films include 13 that carry an existing brand. With not even a single exception, all

Rank	Movies	Console Games	Music	Novels
1	Avatar	Call of Duty: Black Ops	Michael Jackson: Thriller	The Da Vinci Code
2	Titanic	Grand Theft Auto V	Eagles: Their Greatest Hits	Harry Potter and the Deathly Hallows
3	Star Wars: The Force Awakens	Call of Duty: Modern Warfare 3	Billy Joel: Greatest Hits I & II	Harry Potter and the Philosopher's Stone
4	Jurassic World	Call of Duty: Modern Warfare 2	Led Zeppelin: IV	Harry Potter and the Order of the Phoenix
5	Marvel's The Avengers	Call of Duty: Black Ops II	Pink Floyd: The Wall	Fifty Shades of Grey
6	Furious 7	Grand Theft Auto IV	AC/DC: Back in Black	Harry Potter and the Goblet of Fire
7	Avengers: Age of Ultron	Call of Duty: Ghosts	Garth Brooks: Double Live	Harry Potter and the Chamber of Secrets
8	Harry Potter and the Deathly Hallows Part 2	Call of Duty 4: Modern Warfare	Fleetwood Mac: Rumors	Harry Potter and the Prisoner of Azkaban
9	Disney's Frozen	Call of Duty: World at War	Shania Twain: Come on Over	Angels and Demons
10	Iron Man 3	Guitar Hero III: Legends of Rock	The Beatles: The White Album	Harry Potter and the Half-Blood Prince
11	Minions	Just Dance 3	Guns N' Roses: Appetite for Destruction	Fifty Shades Darker
12	Captain America: Civil War	Battlefield 3	Boston: Boston	Twilight
13	Transformers: Dark of the Moon	EA Sports' Madden NFL 10	Elton John: Greatest Hits	The Girl with the Dragon Tattoo
14	The Lord of the Rings: The Return of the King	Lego Star Wars: The Complete Saga	Garth Brooks: No Fences	Fifty Shades Freed
15	James Bond 007: Skyfall	Guitar Hero: World Tour	The Beatles: The Beatles 1967-1970	The Lost Symbol

Original | Line extension | Category extension | Creator branding | Ingredient branding | Co-branding

Fig. 9.3 Branding strategies among entertainment successes
Notes: Authors' own illustration based on various sources of information. Movie ranks reflect global theatrical revenues (based on The Numbers). Console games ranks reflect global unit sales for all games released at two or more 7th generation consoles (based on VGChartz). Music ranks reflect U.S. "certified" unit sales (based on RIAA/Business Insider). Novels ranks reflect UK sales between 1998 and 2012 (based on Nielsen BookScan/The Guardian).

[198]We provide an in-depth discussion of the blockbuster concept in our chapter on integrated entertainment marketing.

bestselling games are branded, and only one among the top 15 albums does not feature a brand (i.e., the band Boston's debut recording). Novels have the highest share of original hit products, with only ten out of the top 15 being branded. The figure also illustrates that all of the branding strategies are represented to a certain degree, but that their frequency varies between products. Whereas line extensions are prominent for all forms of entertainment, they are *particularly* dominant for games. Many hit films are (also) adaptations of brands from other categories outside of movies (comics, novels)—a strategy we have named category extensions. Hit novels usually have a star author as creator brand, and music hits are closely tied to their star performers, or human-ingredient brands.

But one has to be careful before attributing such success solely to the branded components that are obviously involved. Instead, we have to account for how entertainment producers handle such branded products, compared to how they handle original products. Because entertainment producers invest heavily in brands, perhaps it is simply the higher spending that creates the competitive advantage, not the brand itself—which would make the underlying correlations spurious. Thus, we need to take a closer look at the different branding strategies in entertainment, isolating the share of their financial performance that can be attributed directly (and solely) to the brand.

In addition, and of equal importance, whereas Fig. 9.3 shows that some branded products perform well, our readers will have little problems coming up with examples for branded entertainment that have not worked so well. We thus need to learn about the reasons for such differences, identifying contingency factors that determine the success of each branding strategy. But before we do so, let us take a short look at how every brand is born—that is, as an "original" product. At this beginning point, can the name make a difference?

Brand Elements: What is a Good Brand Name (and Does it Matter Financially)?

"What's in a name? That which we call a rose
By any other name would smell as sweet."
—*The character* Juliet *in* William Shakespeare's *play* *Romeo and Juliet*

At the beginning of all brands is—a name. In entertainment, just as in many other industries, coming up with a great brand name for a product is often considered crucial for future market success. Entertainment firms pay a lot of attention (and money) for title specialists to come up with great titles, but very much like the "Nobody-Knows-Anything" mantra as a whole, mystique beats analytics when it comes to defining what makes a title great.

And it is quite a fertile area for myth-building because hindsight is 20–20 when it comes to judging titles: didn't the wonderful movie THE SHAWSHANK REDEMPTION fail only because of its awkward title? (Even Tim Robbins can't memorize it in the film's making-of!) And didn't WHEN HARRY MET SALLY and SLEEPLESS IN SEATTLE soar because of their ingenious titles? At least this is what industry gurus such as Matthew Cohen (who make a living out of crafting brand names for entertainment products) argue, and what journalists support ("A great title matters;" Patterson 2008).

Brand-name gurus often refer to scientific arguments when describing the quality of a title, but the evidence they cite is anecdotal and counter-examples abound. For example, Mr. Cohen explains that WHEN HARRY MET SALLY is a great title because "it has a rhyme," and rhymes are better remembered by humans (quoted in Sugarman 2011). Although, interestingly, the German title, HARRY UND SALLY, lacked any rhyme, but didn't stop the film from becoming the ninth most successful film in Germany in 1989, compared to ranking #12 that year in North America. SLEEPLESS IN SEATTLE was great because of its "alliterative 'S's" (which have a "tendency to stick to your brain"). But why then did GRACE IS GONE, a good movie with stars and a soundtrack by Clint Eastwood, flop dramatically, despite being equipped with a similar linguistic feature?

Maybe empirical research can shed some light onto what makes a great entertainment brand name and how the name impacts product success. Outside of entertainment, research has studied several linguistic aspects of brand names and found that names can indeed influence consumers' reactions to brands. Most of these studies are experiments conducted in tightly controlled laboratory settings (for an overview, see Lowrey et al. 2003); aspects of brand names that are found to affect how consumers respond include the name's phonetics (certain *sounds* of the name trigger reactions, such as via the use of vowels and consonants), its orthography (such as unconventional spelling), morphology (it helps if words are formed in special ways, such as by adding a prefix or combining words), and semantics (it's good if a name "means" something, such as via metaphors). These aspects support consumers' drawing of cognitive inferences about the brand and trigger imagery processes, which might then influence consumers' brand-related attitudes and behaviors.

But considering the particularities of entertainment products (such as their holistic evaluation and short life cycles) and markets (with high innovation levels and a large number of products from which consumers have to choose), we should not take it for granted that such findings can be transferred to entertainment. And indeed, empirical results indicate that skepticism regarding brand name effects might be appropriate. In the film context, Zhao et al. (2013) code consumers' familiarity with the titles of close to 3,000 movies and link this "title familiarity" to movie success. When two independent coders judged a movie's title to be similar to an earlier movie (but were not part of the same series or franchise), familiarity was set to 1 (and to 0 in all other cases). The scholars then link this familiarity to the films' North American opening weekend box office via OLS and 2SLS regressions. They find no significant effect of title familiarity; even the isolated correlation of familiarity with box office is close to zero.

Schmidt-Stölting et al. (2011) include a similar investigation in their study of book success. They ask four coders to rate the "appealingness" of the title of the more than 1,000 hardcover and paperback books in their data set on a 5-point scale (which ranges from "not appealing at all" to "very appealing"). When they then link this appealingness measure with book success as part of their research model (a SUR approach in which they control for several other drivers of book success), the scholars also find no significant effect of the title's appealingness on sales of either hardcover books or paperbacks.

Whereas these studies rely on secondary data, we tried to shed additional light on the matter by combining historical information for 300 German movies[199] with survey data from 1,063 German consumers (Pähler vor der Holte et al. 2016). The survey approach enabled us to understand the effect that names have on consumers' cognitions. We find that title-based cognitive inferences ("will the movie be of a genre I like," "contain a star I like," etc.) and the imagery potential of a title (such as the perceived vividness of the title) explain a good share of the amount of information search consumers intend to do about a movie, as well as their intention to watch the movie. But when we then link the average inferences and imagery potential of the movie titles with the films' actual opening weekend box office,

[199]Those 300 films were all German co-/productions that were theatrically released between 2000 and 2008, with the exception of the most successful films and sequels, which we left out because of their popularity. Respondents only rated movies they had not heard of before the survey.

they do not explain success over and above other "success drivers" (e.g., star power, genre, advertising). This suggests that whereas titles can indeed play an important role in early stages of a new entertainment product's release (such as green-lighting) and might also be helpful to stimulate early buzz and interest by consumers (i.e., prior to the ramp-up of glitzy advertising campaigns), our results suggest that any impact of the title is crowded out by other cues, such as stars and trailers, as the release date approaches.

Finally, where we focused on original brand names, Sood and Drèze (2006) looked only at sequels. Their research question was: can a movie title create satiation in consumers because of the underlying family brand?[200] In an analysis of 317 movie sequels from 1957 to 2005, Sood and Drèze show that titles do indeed matter in this specific context. They find that sequels with named titles (such as INDIANA JONES AND THE TEMPLE OF DOOM) received higher ratings by consumers on the IMDb than those with numbered titles (think of ROCKY 2). By reducing familiarity, the naming (versus numbering) of a brand extension appears to reduce consumers' satiation with a movie brand (and consequently increase their liking of it).

So we see that a title *can* matter to a certain extent (for building a successful entertainment brand and also when extending it later), but its role differs with time and context. What we also don't know yet, at least not for new brands, is what *kind* of name is superior to others—which of the many linguistic variations name researchers have identified are most effective for which kind of entertainment in its early stages? Future research will hopefully provide more answers. And let us add a disclaimer: studies that use secondary data might be biased by a "selection effect"—because they only compare titles that have been turned into a product, but not those alternatives that have been rejected. So one might argue that all titles in the historical databases are great, and differences would only point to those which stand out among this elite crowd.

We now move on to look at brand extensions more closely—as illustrated above, they are the one branding strategy that stands behind most major entertainment successes. We begin by exploring the economic logic of two major types of line extensions (namely sequels and remakes) and then investigate category extensions.

[200]See also our discussion of the satiation phenomenon in our chapter on entertainment product characteristics.

Entertainment Line Extensions: The Case of Sequels and Remakes

> "51% [of the moviegoers] said they attended STAR TREK BEYOND because it's part of a franchise they love."
> —Busch *and* D'Alessandro *(2016)*

What Sequels and Remakes Have in Common—and What Sets Them Apart

The term "sequel" describes, in both the public and the scholarly discourse, an entertainment product that is a line extension which continues a previous product in the same family brand (Basuroy and Chatterjee 2008). Classifying them as line extensions means that sequels belong to the same product category as their predecessors.

Movie and novel sequels continue the narrative of a previous movie or novel, respectively. TV series and comics, often referred to as "serials," follow the same logic of continued storytelling—one season of a TV series builds on the previous season(s), just as one comic issue follows the last one, with production decisions being made on a per season/per issue basis (so the new product would be the season or issue, respectively). Sometimes the sequel is more loosely tied to its predecessor in that the continuity is less in the narrative, but more in other elements, such as the setting, the characters, and the general "concept." Think of Bart Simpson, who does not grow over the years, but continues his permanent struggle with his father, his family, and other members of THE SIMPSONS' "Springfield universe."

Game sequels often also pursue the "concept" of a previous game, such as GRAND THEFT AUTO V follows previous entries of the GTA series by focusing on a character's rise up through the criminal ranks, with settings, characters, and/or storylines varying. And music knows sequels too; such music sequels follow a previous album's sound and musical style (as in the case of the band Chicago, which even connoted sequels by numbering their releases). Occasionally a music sequel also continues an earlier song's narration expressed in lyrics.[201]

[201]Examples for such song sequels are David Bowie's revisiting of his Major Tom character from the song SPACE ODDITY in ASHES to ASHES 11 years later, Eminem's BAD GUY single which introduces Matthew as the brother of his previous record character STAN, and Austrian singer Falco's COMING HOME (JEANNY PART II, ONE YEAR LATER), in which he traces the titular character of his scandal hit JEANNY (various more sequels of the song were released after the singer's death).

The concept of sequels originated in the movie context. In the 1920s and 1930s, Hollywood studios had "B units" which produced what was then called "series"—loosely connected episodes of a popular character's adventures, such as Charlie Chan, Mr. Moto, or Cisco Kid. These serials were manufactured for a low budget and with limited artistic ambitions. BENEATH THE PLANET OF THE APES, made in 1970 by 20th Century Fox, is often considered the first "modern" sequel—the continuation of the studio's previous box office hit PLANET OF THE APES, itself the screen adaption of Pierre Boulle's classic novel. Rather than being the result of strategic managerial thinking, it was the studio's then-troubled financial state that made its managers look for ways to exploit its existing properties. At a time when sequels were basically considered as both artistically inferior and commercially unattractive, Fox made what has developed over the following decades into a family brand with numerous products attached to it.[202]

Remakes are a related, but distinct, kind of line extension. The concept describes a *new version*, or "re-representation," of a previous entertainment product, again in the same product category (Horton and McDougal 1998). Like sequels, remakes can also be found in most forms of entertainment—prominent examples are movies such as PSYCHO (1998), Gus van Sant's re-filming of Hitchcock's classic thriller, the 2013 version of the game TOMB RAIDER (a remake of the original title from 1996), and new arrangements of musical hits (such as the recent re-recording by rock band Disturbed of Simon and Garfunkel's THE SOUND OF SILENCE). Remakes are less frequent for novels (where the single-sensual character provides limited room for fresh interpretations), but occasionally happen—an example for the latter is the "Hogarth Shakespeare project," in which modern authors such as Anne Tyler and Margaret Atwood rewrite classic novels by Shakespeare (e.g. Smiley 2016). Alternative names for remakes are "reboots" (often used when a film or game brand has been extended by several sequels and the brand's origins are reimagined) and "cover versions" for music remakes.

Is it important, beyond rhetoric, to distinguish between sequels and remakes? We argue that it indeed is—both on a theoretical and also on a very practical/empirical level, as we will show later. First, let us stress that both sequels and remakes contain a potentially important advantage over

[202]The highly recommended film documentary FROM ALPHA to OMEGA: BUILDING A SEQUEL is a rich source for *Entertainment Science* scholars and fans on its own, reminding us how strongly today's version of the industry is shaped by managers, rather than by "natural" forces. We elaborate on this later in this book when we discuss the development of the blockbuster concept as an integrated marketing strategy.

original products, as they both are affiliated with an existing brand and thus can offer consumers familiarity (which, as we have laid out earlier, can trigger emotions and imagery, and, eventually, pleasure). But as part of our discussion of the benefits of familiar stimuli, we have also stressed that familiarity is not the only factor that matters for entertaining consumers: successful entertainment also requires sensations. And sequels and remakes differ systematically with regard to their potential to offer consumers such fresh sensations.

Sequels can provide sensations by exploring new facets of characters, tap new territories, or take surprising narrative turns in their continuation of previous stories. The potential of remakes, in contrast, to offer new sensations is systematically limited as they, by definition, tell an existing story again, and do so in the same modality (Bohnenkamp et al. 2015). As Mendelson (2013) phrases it, the "very thing that causes such a remake to get made (brand awareness) … cripples the ability to differentiate itself. Tell the same story, and you've rendered yourself useless. Go in remarkably different directions, and the hardcore fans will condemn your existence on one hand while condemning your lack of source fidelity on the other."

What are the financial consequences of these two central forms of entertainment line extensions? Do sequels and remakes perform better than original entertainment products on average? Do they differ from one another? After we explore the financial consequences of these line extensions, we will take a closer look at what distinguishes the successful from the unsuccessful of such extensions—with a particular interest in how closely extensions should resemble their predecessors.

"Average" Return and Risk Effects for Entertainment Line Extensions

We begin our investigation of the financial performance of line extensions by taking a look at "average" effects. Of course no such thing exists, but we find it still informative to learn whether line extensions, in general, provide producers of entertainment with economic benefits. (Just like we want to know whether education, in general, leads to higher income. Or using digital media makes us smarter and/or happier.) Such a generalized perspective leaves room for differentiation, which we then fill with more refined insights in the second half of this section. We begin our analysis with sequels and then compare their performance with those of remakes.

What Can Entertainment Producers Gain from an "Average" Sequel?

Some Basic Insights on Sequel Value

Most of the empirical work on sequels has been conducted in the context of movies. Let us begin by comparing sequels with their predecessors. When Basuroy and Chatterjee (2008) analyze the weekly performance of 11 film sequels released between 1991 and 1993 with those of their predecessors using a Generalized Estimating Equations approach, they find that the sequels, on average, do not match the revenues of the parent films they are continuing.

Dhar et al. (2012) confirm this finding with a much larger data set of 2,000 movies that were widely released in North American theaters from 1983 and 2008;[203] they report that whereas sequels attract, on average, an audience of 20.5 million North American moviegoers over their full theatrical life cycle, their predecessors attract an average of 24 million. However, sequels do better than their parent films in the *first* week of release, with 8.2 million versus 6.6 million attendees on average (which they achieve with higher distribution efforts—opening on 2,700 versus 2,100 theaters—and higher production costs of $32 million versus $19 million, on average).

But as the entertainment products that are turned into sequels tend to be the more successful ones, a more powerful comparison is between the performance of sequels and other products that are similar to them in all ways—except for the fact that they are not sequels. A number of studies have aimed to make this comparison by adding a "sequel" variable to their econometric models of product success. Early studies did not control for other product elements and marketing actions; thus, all of the effects of these variables were absorbed by the sequel variable, and these papers often report quite enormous effect sizes.

But more recent studies use a larger set of such control variables and produce more realistic estimates. Generally, these studies find that sequel movies generate between 20% and 30% more revenues than nonsequels. Basuroy et al. (2006), who re-analyze Ravid's (1999) data set of 175 films with a simultaneous equation approach (i.e., 3SLS), find that sequels generate 24% higher revenues than nonsequels in the opening week. Clement et al. (2014), applying the same method to a much larger, and more recent, sample of

[203]They do not name the exact number of sequels in their data set, but it appears to be more than 100.

films, estimate a 20% premium on North American opening weekend revenues for movie sequels. And Akdeniz and Talay (2013), in a joint analysis of 14 countries, report an average increase of 31% in the opening box office if a movie is a sequel.

The findings by Akdeniz and Talay and by Clement et al. point to an interesting insight: in contrast to what we have seen for genres and other variables, the sequel effect seems to be of a global nature. But the strength of the sequel effect differs across countries—Akdeniz and Talay find it ranging from 10% (in Japan) to 90% (in Chile). On the country level, the scholars find empirical hints that such differences are related to a country's "uncertainty avoidance" tendency—when people "feel threatened by ambiguous situations" (Hofstede et al. 2010, p. 191), they seem to particularly appreciate the comfort provided by the sequel character of an entertainment product.

Whereas these studies look mostly at the opening success of movies, there is also some evidence that sequels lose their power afterward, as more and other quality cues become available. Basuroy and Chatterjee (2008) show that the sequel effect is highest immediately at the release and shrinks to about half that size during the first nine weeks of a film's run (few movies are shown longer in theaters). Dhar et al. find in the analyses of their 25-year data set that the sequel effect is one-third weaker for total theatrical revenues then for the opening week (which is included in the total revenues). Their study also offers two additional insights. First, it is not only audiences that are affected by sequels, but also distributors, who tend to assign sequels to a larger number of theaters than other movies. And second, the scholars find that the sequel effect has increased over the years from 1983 to 2008 in terms of first week attendance, but with no change in movies' total theatrical revenues.

Average sequel effects have also been estimated for other forms of entertainment than movies, and the general takeaway is that sequels also offer an advantage, but the level differs between the type of product. For books, Schmidt-Stölting et al. report that the sequel effect can be quite strong, but that is depends on the book format. A book sequel (versus an original) increases paperback sales by a massive 46%, but has no effect in the hardcover market.[204] For video games, Cox (2013) finds 6% higher revenues for game sequels, and Marchand (2016), who controls for additional factors in his analysis of about 2,000 video games, estimates a sequel elasticity of 0.07

[204]The authors use book ranks at Amazon.de as a proxy of sales.

on total game sales, which translates into about 5% higher sequel sales compared to an original.

We have seen earlier that most game super hits *are* sequels (see Fig. 9.3 on p. 381), so why this comparably small "sequel premium" for games? One explanation is that, as this is an average effect size, "bad" sequels dilute the sequel effect of strong sequel games. In addition, one has to keep in mind that these results are derived from multivariate analyses which control for the success impact of other factors that will also differ between sequels and originals, providing an additional explanation for the strong performance of several big-budgeted game sequels.

But the latter argument also points to a related, but even more delicate econometric matter: if sequels are systematically treated differently (better!) by entertainment producers than originals, this might establish an econometric artifact known as "treatment bias," which can cause us to overestimate the sequel effect—just like a statistical comparison of 3D movies with all 2D movies would have misled us. Let's take a closer look at this bias—and some remedies that *Entertainment Science* scholars have applied to produce unbiased estimates of the financial impact of a sequel.

The "Treatment Bias" Problem—and a Solution

Because entertainment managers believe in the power of sequels, it seems kind of obvious that they treat them systematically "better" than original products. They equip them with higher production budgets, higher advertising, and give them wider distribution than they do for the average unbranded product. Dhar, Sun, and Weinberg's extensive data set gives us an idea of the size of this preferred treatment of sequels: sequel movies, on average, have about 50% higher (inflation-adjusted) budgets ($32 million versus $21 million) and open on 35% more screens (2,700 versus 2,000) than all other movies.

While there is certainly nothing wrong with managers giving such preferred treatment to products that they believe carry a higher success potential, doing so introduces a statistical problem in predictive analyses: basic regression methods erroneously assign a part of the success effects of the treatment variables (such as higher budgets, higher advertising, and wider distribution for sequels) to the fact that a product is a sequel, thereby causing a tendency to exaggerate brand effects. Whereas the inclusion of realistic control variables can mitigate this problem, it unfortunately does not fully eliminate it.

So, what to do? One approach to solve this problem is "statistical matching." In statistical matching, those products which presumably have received a better treatment than others by managers (i.e., the sequels in our case) are not compared with *all* other products, but only with those that have received similar treatment in all other regards that matter for their financial performance. These products have received a similar treatment as sequels—despite not being sequels themselves.

We applied this approach to movie sequels by finding statistical matches for all 101 movie sequels (only those which were the *first* sequels in a series to avoid any complications) that were released in North American theaters from 1998 to 2006 (Hennig-Thurau et al. 2009). To locate such twins, we drew from all 1,536 theatrically released movies that were not sequels from the same time period, using a multivariate procedure.[205] Before the matching, the sequels in our data differed quite strongly from the nonsequels (for example, budgets and opening theaters were 37 and 35% higher, respectively, similar to what Dhar et al. reported)—the matching process removed these differences almost completely, allowing us to compare apples with apples.

So, what remains of the sequel effect when accounting for the treatment bias? Panel A of Fig. 9.4 compares the average total revenues of the sequels and matched nonsequels in our sample; whereas we find that sequels are significantly more successful on average, the 27% difference is clearly smaller than the 144% reported by Dhar et al. who do not apply any kind of matching. When we run a regression for the data set of sequels and matched nonsequel movies in which we also add numerous controls and filter out their direct success effects, we find in this most restrictive analysis that being a sequel results in a 14% increase in a movie's North American total revenues (from theaters, home-video retail, and home-video rental, all inflation-adjusted).

Panel B of the figure points at a second facet of sequels that is of economic relevance for managers, comparing the financial risk of sequels versus that of matched nonsequels.[206] Specifically, we ran separate regression

[205]Specifically, we calculated the "distances" between each of the 101 sequels and all nonsequel movies based on a set of key success variables: namely a movie's production budget, its distribution intensity (in terms of the number of opening theaters), the age-rating, the star power, the existence of a family brand from another product category, and several genres. Then we picked the three nearest "neighbors" for each sequel in the sample. For details about this procedure, see Hennig-Thurau et al. (2009).

[206]See our discussion of the important role of risk in entertainment and the need for systematically managing it in our chapter on business models for entertainment.

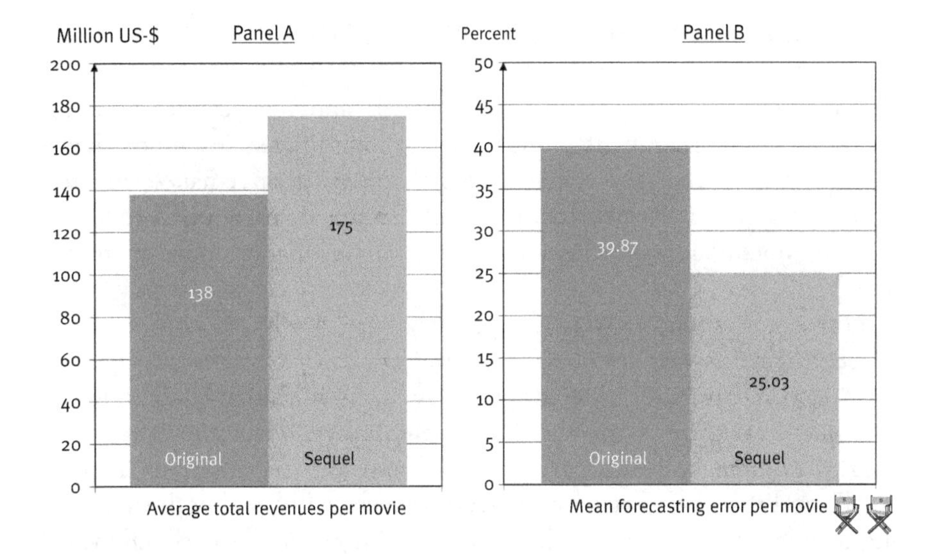

Fig. 9.4 The average returns and risk of movie sequels
Notes: Authors' own illustration based on results reported in Hennig-Thurau et al. (2009). The sequel results are based on 101 sequels and the nonsequel results on 301 matched original movies. In Panel B, the mean forecasting error is the Mean Average Percentage Error (MAPE) of two separate regression analyses (one for sequels, one for original movies). The MAPE values are weighted with each film's actual total revenues.

analyses for the sequels and nonsequels in our data set and then used the resulting equations to predict the total revenues for each movie in our data set. Our measure of financial risk is the average percentage error of these predictions compared to the films' actual performance. The results show that producing a sequel involves a prediction error that is 15 percentage points (or 37%) lower than producing a movie that is not a sequel—a major advantage.

Our results, based on the matched sequel data set, imply that we neutralize cost factors in our analyses. As a consequence, they not only demonstrate that sequels, on average, generate more revenues and are less risky, but they are also more *profitable*. This interpretation is consistent with other scholars' findings of the link between sequels and profitability (such as Hofmann 2013 and also Gong et al. 2011, who both report higher ROI effects of sequels, but do not control for treatment bias).[207]

[207]Gong et al. actually make an attempt to control for such bias, but use some of the sequels' predecessors' attributes (star power, rating, season of release, and year) as basis for their matching.

And What Can be Gained from an "Average" Remake?

We have argued above that although sequels and remakes are both line extensions of entertainment brands, they follow different economic logics. So, what do existing studies tell us about the value of remakes? Are they equally attractive as sequels, commercially speaking, or does their limited sensation potential make them a less attractive entertainment marketing strategy?

Those scholars who have studied remakes are clearly far from euphoric. Although remakes were not at the center of our study of differences between movie success drivers for different channels (Hennig-Thurau et al. 2006), we nevertheless calculated correlations between a movie being a remake (versus being a "nonremake") and both its initial and long-term success in theaters. We find that the correlation between remake status and both short- and long-term success variables is literally zero—in contrast to sequels, which share above 20% of the films' variation in movie success.

In a study we exclusively dedicated to movie remakes and their economic impact, we compare the North American box office of all 207 remakes released in North American theaters from 1999 to 2011 to a sample of almost 2,000 other movies, or "nonremakes" (Bohnenkamp et al. 2015). Here we control for the success effects of several other movie variables, but also use an elaborated matching approach ("propensity score matching"—the same we used in the case of 3D) to prevent a treatment bias from distorting the results. With this approach, we find (and remove) treatment biases for the production budget and advertising/distribution (which are generally higher for remakes), other kinds of brand extensions such as sequels (which remakes are less often), being a comedy (less often the case for remakes), and being a horror or thriller movie (which are more often remade than other films in our data set).[208]

How do remakes pay? Although the average remake's box office is about $8 million (or 30%) higher than for all nonremakes ($34.8 versus $26.7), this difference disappears almost completely when removing the treatment bias—matched nonremakes have an average box office of $33.1. This is also what we find when we include a remake variable in a weighted least squares regression;[209] the variable's parameter is very small (0.013) and not significant.

[208]In our matching estimation, we also test, but do not find for a potential bias by a movie's star power, the judgment of professional critics, age ratings, and the action, drama, and science-fiction genres.

[209]The matching weights served as regression weights.

In other words, movie remakes, in contrast to sequels, do not enjoy an advantage in terms of box-office revenues. At least not *on average*—but we'll take to a more refined look in the next section below.

But there is one other financial reason to prefer a remake over original films. Just as we found with sequels, remakes are *less risky* than original films. The standard deviations of revenues and ROIs, as risk measures, are clearly lower for remakes (by 14% and 41%, respectively) than for the matched nonremakes. In other words, you might not generate much extra revenues by producing a remake instead of an original film, but your chance of losing your money is reduced.

A Closer Look at the Factors that Make a Successful Entertainment Line Extension

Whereas our previous discussion addresses "average" sequels and remakes, let us now dive into those factors that will make line extensions perform "above average" in terms of success. *Entertainment Science* scholars have empirically unearthed three categories of contingency factors that matter: characteristics of the family or parent brand, characteristics of the "fit" between the family/parent and the extension, and characteristics of the extension. Figure 9.5 shows these basic categories of contingency factors and also lists two prominent example variables for each. We now discuss what we can learn about these contingency factors from *Entertainment Science* research.

Characteristics of the Family (or Parent) Brand

Research on the role of brand elements of the family brand on a new product's success has focused on two concepts: consumers' level of awareness of the family brand and that brand's image. Very often scholars have drawn on a specific previous parent product (the original Rocky would be the parent of Rocky II, for example), instead of the more abstract concept of the brand "family."

Let us talk about brand awareness first. In our own study of (initial) movie sequels, we measure the awareness of a sequel's parent brand as a combination of the parent movie's total box office in North American theaters and its opening number of theaters, adjusting this awareness metric for the loss of awareness that happens over time with a "forgetting-curve" function

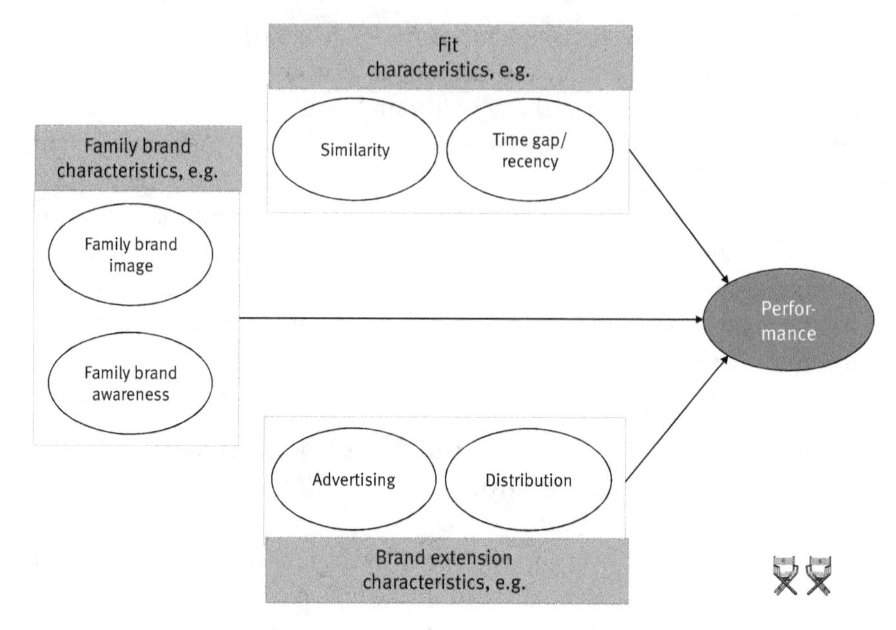

Fig. 9.5 Contingency factors of entertainment line extension success
Note: Authors' own illustration based on ideas reported in Hennig-Thurau et al. (2009).

(Hennig-Thurau et al. 2009).[210] When we use an OLS regression analysis that also contains the other variables shown in Fig. 9.5, we find that brand awareness has the by-far greatest impact on sequel success. Specifically, awareness' standardized coefficient is more than *twice* as large as the second most impactful variable.[211]

In the movie remake context, we operationalized parent brand awareness by making use of the parent film's IMDb's MovieMeter search rank at the point in time when the movie remake was announced (Bohnenkamp et al. 2015). We then conducted regression analyses for different subsets of the data: each subset contains data for a certain type of remake and the respective matched nonremakes. In these finer-grained analyses, we find that when it comes to remakes, maximum awareness is not the best precondition for success: instead, we learn that remakes of parents with *medium*

[210]The logic behind this is that a sequel to a film that had a $300 million box office has a higher awareness than one to a film which made only $100 million, but that this might reverse if the former parent movie was released in 1985 and the latter one in 2015.

[211]Our forgetting-adjusted measure of brand awareness alone captures more than *one-third* of the sum of all coefficients. Elasticities are not available for this analysis because we used a standardized version of the variable.

brand awareness perform best, providing a 6% increase over nonremakes. They are followed by parent films with low awareness, and the parameter for high awareness parents is even negative (though not statistically significant), which suggests that remakes of films that have high awareness among today's moviegoers may tend to *lose* money compared to other films.

Film maker Steven Soderbergh seemed to have this finding in mind when he stated that the studios "get simple things wrong sometimes, like remakes. I mean, why are you always remaking the famous movies?" (quoted in *The Deadline Team* 2013). Martin Rackin addressed the problem when producing a remake of John Ford's classic STAGECOACH: he made sure that the original film was withdrawn from public distribution for several years before his own remake was released (Pfeiffer 2015, p. 33)—an effort to reduce the target audiences' awareness of the parent brand in our scholarly perspective.

Mr. Rackin's efforts hardly paid though: his remake just covered its production costs. But that result might have to do even more so with the second major facet of the family/parent brand, a film's parent brand image—we will get back to his STAGECOACH movie in a moment. In our study of movie sequels, we measure films' brand image as a combination of quality judgments of the parent by experts, consumers, and industry members. We find the variable to be influential in our regression analyses: it has the third-highest standardized regression parameter of all variables in the estimation. This insight somewhat corresponds with industry wisdom. A rival studio manager articulated his surprise about Disney's decision to make a sequel to its live-action ALICE IN WONDERLAND, based on the film's commercial success (or brand awareness) alone—despite the fact that "[t]he first movie wasn't that good" (quoted in D'Alessandro 2016).[212] ALICE THROUGH THE LOOKING GLASS ended as a major financial disappointment, returning less than $50 million to its producer from North American theaters (with a $170 million production budget).

The important role of the parent brand image also carries valuable lessons for producers for managing franchises.[213] Hollywood analyst Doug Creutz was right to look beyond the box-office results when discussing the success of STAR WARS: THE FORCE AWAKENS, which Disney considered a focal element in their plans for future STAR WARS movies. He stressed the relevance of image-related responses when saying "[n]o matter how much [THE FORCE AWAKENS] does this weekend…we think the strong critical reviews

[212]The film received only 52% positive reviews at Rotten Tomatoes, for example.

[213]See also our more detailed discussion of the franchise concept later in this chapter.

have more significant implications for the long-term health of the franchise" (quoted in Lieberman 2015).

For remakes, however, the effect of parent image is, once again, more complicated. We measured the image via consumers' ratings of the parent film on the IMDb in Bohnenkamp et al. (2015), finding that remakes of parent movies which have a "good" image made on average almost 9% *less* than nonremakes. In contrast, the remakes of movies with a medium or bad image were significantly more successful: they generated 7% more revenues than nonremakes.

And producing a successful remake is even more difficult if the parent brand is tied to an acclaimed artist, such as a director ("an Alfred Hitchcock movie!") or a star ("Sylvester Stallone *is* Rocky!"). We find that remakes of such "signature" brands were a whopping 38% less successful than nonremake films; in comparison, remakes without such signature elements make, on average, 5% more than nonremakes (Bohnenkamp et al. 2015). This is what might have most troubled Mr. Rankin's remake of STAGECOACH—the parent was not only a very well-received movie that is widely considered one of the best westerns ever put on film, but is also as a key element in the oeuvre of its famous director.[214]

"Fit" Characteristics

Research on entertainment line extensions highlights the contributions of two facets of the "fit" between the extension product (i.e., the remake or sequel) and the parent/family brand: the similarity between them (a content-related facet of fit) and the time gap between the extension and previous installment (i.e., extension recency, a time-related facet).

Regarding "similarity fit," we tested the relevance of 11 different fit variables in our study of movie sequels, each of which covered one particular aspect of how similar the parent and the extension product are (e.g., in terms of stars, release date, budget, and title). A stepwise regression procedure showed that having the same lead actor can have a huge positive impact on sequel success. And there is corroborating ad hoc evidence for this insight: out of the first six sequels to the original FAST AND THE FURIOUS movie, those four which starred the original movie's lead actors Vin Diesel and Paul Walker generated more revenues in North America (between

[214]As just one example, the American Film Institute in 2008 ranked the film #9 among all westerns (*AFI* 2008).

$155–$353 million) than did the original ($144 million). In contrast, the one that starred only Mr. Walker made slightly less ($127 million), and the sequel that had none of its original stars on board returned less than half of what the original made ($63 million).[215]

In addition, similarity in terms of having the sequel distributed by the same firms, a similar poster design and title, the same producers, and being released at a similar time of the year are, though not included in the final regression, positively correlated with sequel performance (listed in descending order of impact). And the regression results also revealed that similarity fit can also impact the size of the main effect of the parent's image on the success of the extension: having the same age rating and belonging to the same genre as the parent facilitates the transfer of a positive parent brand image on sequel success.

Again, remakes behave differently with regard to similarity. Remakes that resemble the original to only a "limited" degree (we measured this in Bohnenkamp et al. via a "cumulative" score that considers genre, characters, narrative, and settings) exerted a positive effect on performance (+7% versus nonremakes), while remakes that are highly similar to the original attracted about 10% *fewer* viewers than average nonremake movies. One might tie these results to our earlier discussion of satiation effects: whereas sequels can avoid the "satiation trap" by combining familiar brand elements (such as characters) with new sensations (such as new dramatic turns), so that high similarity can be considered an economic virtue in general,[216] combining sensations with familiarity is much more challenging in the case of remakes, where high similarity easily causes satiation in consumers' minds ("I have seen that before, so why should I see that again?").

The time gap between the parent and the extension is the other fit facet. For sequels, Basuroy and Chatterjee (2008) find that movie sequels tend to

---

[215]The most recent installment, The FATE OF THE FURIOUS, is a peculiarity in the series—being released after Mr. Walker's death, only Mr. Diesel was able to star in it, but the film was a success nevertheless. Let us add that the commercial power of such lead-actor continuity can benefit not only film producers, but also star actors: when Arnold Schwarzenegger negotiated an unprecedented $29.25 million "pay-or-play" deal for his participation in TERMINATOR 3, plus 20% of the film's global gross receipts (after breaking even) and also a "pre-approval" clause with regard to the director and other key positions, this was because the film's major financiers had made his participation a condition for becoming involved in the project. In other words, no Schwarzenegger—no TERMINATOR sequel... (for more details, see Epstein 2010).

[216]Even for sequels, high similarity can become problematic *over time*. We will get back to this when discussing the dynamics of line extensions.

perform better when they are released sooner (rather than later) after their parents, consistent with our "forgettingness" logic discussed above in the context of brand awareness. In contrast, for remakes, the original must not be too old (the "forgetting effect" also applies), but also not too new: if the parent is too "top-of-mind," a remake does not offer sufficient sensations for consumers. Instead, remaking a film of "medium" age is most promising: in our sample, we find that remakes are most successful (compared to nonremakes) when they are released after a period of 11–30 years (Bohnenkamp et al. 2015). Within this time gap, the producer is rewarded with revenues that are about 8% higher than could be expected for a similar nonremake movie.

Characteristics of the Line Extension

In addition to having a powerful parent brand to use and fitting the extension with it, producers of extensions also must get the basics of *Entertainment Science* right—the rules for product, communication, distribution, and pricing decisions we describe in the other chapters of this book also apply to entertainment brand extensions.

We find in our sequel study that the sequel's distribution intensity (i.e., number of theaters at release) has the second strongest association with sequel performance of all factors, accounting for 18% of cumulative importance (Hennig-Thurau et al. 2009). The production budget is also correlated, but to a somewhat lesser degree (accounting for 5% of importance).[217]

Whereas most scholars implicitly assume in their analysis that marketing variables are similarly influential for line extensions and original products, Basuroy et al. (2006) show that their importance can instead vary. They provide evidence that advertising spending is more effective for sequels—which implies that the potential advantage that a sequel's brand awareness and image offer is larger the more a producer spends on advertising. We suspect that more interactions of this type might exist, but they have not been systematically studied.

[217]Before treating this budget result in a causal way, please see also our discussion of the production budget as a quasi-search quality in entertainment in the previous chapter.

Using Contingency Information to Develop a Return-Risk Portfolio

On the previous pages, we compiled scholarly findings on how the performance of entertainment brand extensions differs with parent, fit, and extension characteristics. Let us now illustrate how entertainment managers can combine these various pieces of information to develop tailored marketing strategies for extensions. We do so for movie remakes, combining the insights on return effects we reported earlier with similarly fine-grained findings on the risk of different kinds of remakes (Bohnenkamp et al. 2015).

Figure 9.6 shows a return-risk portfolio for remakes, with risk on the horizontal axis and returns on the vertical axis. The figure provides straightforward guidance regarding which kinds of remakes are most promising in economic terms: those that are placed in the upper right quadrant offer not only above-average revenues, but they are also less risky than the average nonremake movie in the matched data set. This quadrant includes remakes of horror films, those of brands with a medium level of recency, a medium awareness, and a medium image. In contrast, those remakes that land in the lower left quadrant are less attractive from a financial standpoint, as they combine a tendency for low returns with relatively high levels of risk—they include science fiction remakes, remakes of low recency brands, and those of high awareness brands.

Of course each entertainment producer/investor has to weigh the importance of the two financial criteria based on his or her personal preferences and make decisions accordingly. If you are highly risk averse, you might pay more attention to the figure's horizontal axis, for example. Producers might also consider developing similar portfolios for other types of extensions, namely sequels. There is also room to use other measures for returns (such as ROI instead of revenues) and for risk, building on the insights derived by *Entertainment Science* scholars we have reported in this section (and also in other parts of the book). It is also important to stress that in reality, each remake is a combination of the different extension facets we discussed above, despite them being listed separately in the figure, and the financial appeal of a specific remake project has to be judged based on the gestalt of *all* its facets.

Take the example of Disney's current approach to remaking their classic brands, such as CINDERELLA, THE JUNGLE BOOK, and BEAUTY AND THE BEAST. The insights we have reported above suggest that the original brands'

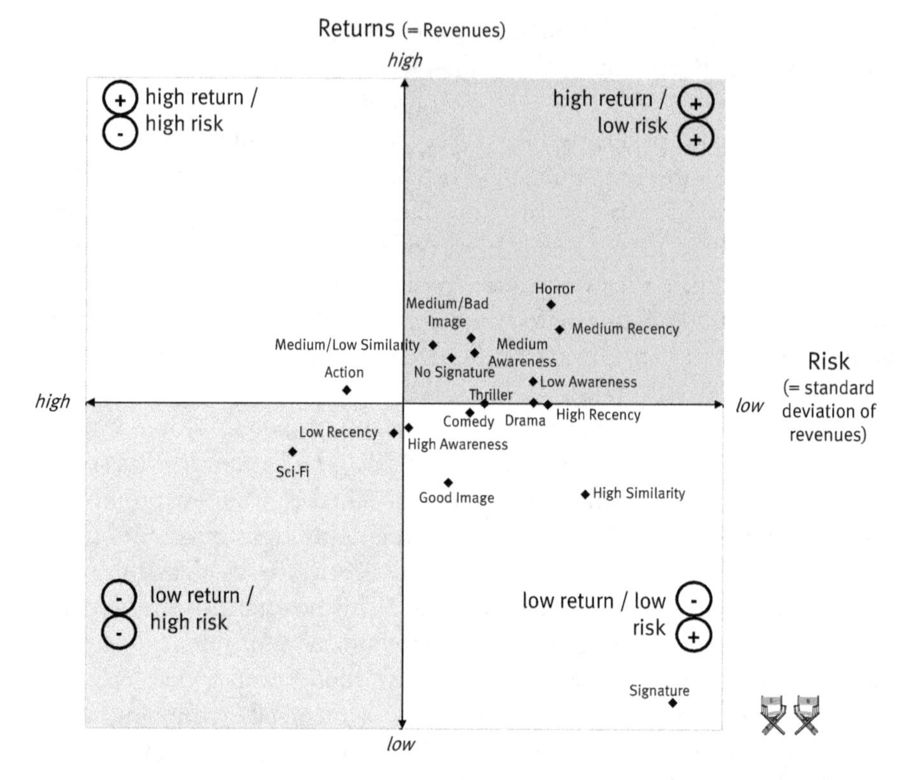

Fig. 9.6 A return-risk portfolio of different kinds of movie remakes
Notes: Reprinted with minor modifications from Bohnenkamp et al. (2015) with permission by the Journal of Cultural Economics. The figure includes only a selection of the kinds of remakes studied in the article. The returns in the figure are the regression parameters for a remake dummy variable in a subsample regression, and the risk parameters are the standard deviations of the revenues of a certain kind of remake.

stellar images and high awareness present a challenge for the remakes, because these factors imply the risk of limited new sensations, triggering satiation. But Disney approaches their remakes in a creative way, producing low-to-medium similarity versions by moving from animation to a blend of life-action and state-of-the-art digital rendering technology created with enormous budgets (Barnes 2017).

Thus, these remakes promise "fresh takes" to those who know the original, while also aiming at a new target group that only has limited awareness/knowledge of the original brands. New audiences appear to be motivated to see the remakes of something their parents love so much. The financial results so far suggest that the approach works well, with the new technological mode offering sufficient sensations to fight the satiation threat and to attract large numbers of consumers.[218]

The Dynamics of Line Extension Similarity: Of Thresholds that Vary with Time

Our findings above regarding the similarity facet of line extension fit suggest that, although too much similarity can hurt a remake, higher similarity is generally better for sequels. However, there are reasons to assume that there is probably also a threshold in the level of similarity for sequels—once it is passed, higher similarity may hurt sequel success because it limits the room for fresh sensations.

Let us first take a look at the work of Sood and Drèze (2006) and their study of the quality ratings of 317 sequel movies. They gathered genre classifications from the IMDb, which usually assigns a movie to three different genres—for example, PSYCHO is a thriller, but also horror and mystery. Sood and Drèze's analyses compared sequels for which genre classifications were completely identical to their predecessors and those sequels which differed in at least one genre classification. They find that consumers rated a sequel with the *exact* same genres as its parent worse than a sequel whose parent differed in at least one of the genres, supporting the idea of a limit for sequel similarity.

In a separate study, we show with a series of experiments and movie data that encompasses 341 films from 92 movie brands (essentially all major franchises of a 50-year time frame) that the threshold for "too-much" similarity is a moving target that changes over the life cycle of a brand (Heath et al. 2015). The results of a GMM regression tell us that for sequels early in a franchise, a high level of similarity produces higher revenues and ROI, but for later sequels less similarity is advantageous. These results also provide an explanation why we find that higher similarity is better in our earlier study (Hennig-Thurau et al. 2009): we used only "first" sequels, for which high similarity is the recommended approach.

Simulations based on the regression results demonstrate that the effect of similarity, and its change over time, are both quite substantial. Specifically,

---

[218]The CINDERELLA (2015) remake made more than $500 million at the global box office, The JUNGLE BOOK (2016) almost $1 billion, and BEAUTY AND THE BEAST (2017) $500 million in its first *six* days alone, easily covering their production costs of $95 million, $175 million, and $160 million, respectively.

whereas an "average" *initial* sequel earns about $25 million at the North American box office, a low-similarity version of the same film would make only $3 million when seven of its elements (such as stars, genre, age rating etc.) are changed. However, the performance of the average *third* sequel in a series is hardly influenced by its level of similarity. But it is for fifth sequels when too much similarity can hurt success: for sequels that come so late in a series, we estimate that if no changes are introduced ("maximum similarity") the average box office is only $2 million, which is much less than the $26 million in revenues we predict for the same late sequel with seven changes. Obviously, audiences' valuations of seeing the same attractions once again lessen over time, an eventuality that requires the producer to fight audience satiation by varying a brand's recipe more extensively.

Sony thus did the right thing when their SKYFALL entry to the James Bond series (the 23rd "official" film, made 50 years after the first) differed notably from its predecessors, obviously aiming for a fresh start and a new exploration. The film's huge success (it generated more than $300 million in North American theaters alone, more than any other Bond film) confirmed this logic based on the sensations-familiarity framework.[219] In essence, letting "final cut auteur" Quentin Tarantino direct the *first* sequel to your beloved hit movie might not be the greatest idea, but having him add an episode to an aging franchise could offer fresh sensations and fight brand satiation. *Entertainment Science* thus approves Paramount's idea to add a new movie to the decade-old STAR TREK brand (Fleming 2017b).

Despite the many things we know about the role of similarity for line extension success, there is a lot that still needs to be learned. Among the open questions is the need to localize the exact threshold point for similarity: how much similarity is too much, and how should we best measure such sequel/remake similarity in "objective" and generalizable ways, beyond comparing a number of product elements?

Entertainment Category Extensions

Whereas line extensions exist in the same category as their parent, category extensions of brands stretch a familiar brand name beyond the product category for which it is best known. Such an approach is widely used in entertainment. Think of films which are made based on books (e.g., THE

[219]See our in-depth discussion of the sensations-familiarity framework in the chapter on entertainment consumption.

Hunger Games), comics (e.g., the Marvel/DC superhero movies), games (e.g., Tomb Raider), TV series (e.g., Star Trek, Mission: Impossible), toys (The Lego Movie and Transformers), and even music (think Pink Floyd's The Wall, or Convoy, based on the C. W. McCall song).

But category extensions go far beyond movies. Games adapt films (e.g., Alien: Isolation), books (e.g., Destination: Treasure Island), and TV content (e.g., Law & Order: Dead on the Money and Dancing With the Stars). And many major films get a "novelized" book adaptation, spark a book series (the numerous Star Wars novels), and sometimes also inspire a comic series (again: Star Wars). This phenomenon now is even beginning to emerge for TV series (e.g., the novel Bratva as an extension of the series Sons of Anarchy) (Alter 2015).

We will now examine how such category extensions work and show their potential benefits—and limitations. As with line extensions, we will also summarize the insights that *Entertainment Science* scholars have compiled regarding the financial outcomes that can be realized by producing category extensions.

Why Do a Category Extension?

> "[I]f you're selling books, you're selling movie tickets."
> —Joe Drake, *as co-COO at Lionsgate, responsible for the* Hunger Games *movies (quoted in* Orden *and* Kung *2012)*

Entertainment producers adopt brands from other product categories for two primary reasons. The first is the same rationale that motivates a firm to do line extensions: an existing brand helps make audiences aware of a new product, as well as helping that audience to build strong positive associations toward the product. The target is usually the consumer who is already a fan of the brand. As TV producer Michelle Lovretta words it, "[a]daptations can offer decision-makers the security of a presumed built-in audience" (quoted in Liptak 2017). We label this the "brand effect" of category extensions.

The second reason for adapting products from other categories is their demonstrated quality, which is an important determinant for any new product's playability. This "quality effect" implies that category extensions are based on products with proven quality in their original categories. This existing sense of quality is of importance in an industry context where product quality is notoriously difficult to judge—and particularly so for those products whose success is closely tied to high quality judgments from consumers.[220]

However, transferring a brand into a new category is not a sure thing in terms of producing good financial results. Interpreting the category extension strategy via the sensations-familiarity framework tells us that the new extension product has a familiarity advantage over original new products. But, for this advantage to positively influence consumers, the extension has to master the "category gap." Whereas consumers might recognize the brand in the new product, positive emotions and imagery will only emerge when the brand's quality associations are also relevant for consumers in the new category. For a wide array of product categories, the lack of such "category fit" is often cited as a main reason for a brand extension's flop (think of perfumes as extensions of the BIC and Zippo brands!), and it seems intuitive to also blame a lack of category fit for the commercial failure of entertainment products such as BATTLESHIP, the board game-turned-movie. Each entertainment product category has its own characteristics (e.g., board games are interactive, requiring active consumer involvement), and entertainment producers must find ways to transfer a product's main attractions into a category that may lack these characteristics.

And as with line extensions, there are also systematic limitations regarding the ability of a category extension to offer new sensations to consumers. Conceptually, category extensions are more akin to remakes than to sequels because they usually tell the same story that was previously told in the other category, suggesting satiation. But the new medium might provide room for compensation, as when the book character Harry Potter now arises in color and with a voice of his own in the movie version of the novel, or can be guided by the consumers through the halls of Hogwarts in the video game version. Similar to what we have seen to be a crucial factor for Disney's movie remakes, here the new category setting carries the potential to offer sensations on its own.

How Category Extensions Affect Revenues and Risk: Averages and Contingencies

Just like for line extensions, the existing empirical work on category extensions has also focused on movies. Our study on distribution channels offered

[220]Please see our discussion of the difficulty of quality judgments as a core characteristic of entertainment in the chapter on product characteristics.

a first glance into category-extension effects (Hennig-Thurau et al. 2006). We found no correlations between book adaptions (which includes both bestsellers and non-bestsellers) and movie success, as measured via the North American box office and video rentals. But we did find a significant correlation between the fact that a movie was a TV series adaptation and its theatrical opening ($r = 0.23$).

But a much more systematic in-depth exploration was Joshi and Mao's (2012) study. In it, they distinguished general book effects on movie success from those of bestsellers, and also took a closer look at different facets of a book's bestseller status. The distinction between books in general and bestsellers is crucial: books that were not bestsellers reveal the "quality effect" of category extensions, while bestsellers approximate the strategy's "brand effect." Joshi and Mao use a sample of 482 book-based movies that received a wide release in North America between 1973 and 2007, along with a convenience sample of 242 original films from the late 1990s.

In an OLS regression of both book adaptations and original films (in which the scholars control for other movie variables such as budget, distribution, and genres), they find that when a film is based on a book, it helps the film's opening weekend box office, but has no impact on the later box office. In absolute terms, the effect is rather marginal, though; book adaptations provide, on average, additional revenues of just \$230,000, or 1.8% of the sample's mean opening box office.[221] As the book variable includes bestsellers, as well as books that did not reach that status, this result might be treated as an indicator that the quality effect of books should not be overrated.

For the "brand effect," the scholars find clearly stronger linkages. In a separate regression in which Joshi and Mao include only book adaptations, they find that a film based on a book that made it to the top of the USA Today, New York Times, or Amazon charts would, on average, generate an \$8 million higher box office on its opening weekend compared to a film based on a book that never hit the charts; an additional \$4 million would be generated in the weeks that follow. It also helps if the bestseller is turned into a movie quickly to counter forgetting: an immediate release in their data set generates box-office revenues that are almost \$11 million higher than an adaptation of a book that was a bestseller a decade ago. Similarity, measured by Joshi and Mao as the inclusion of the book's author in the making of the

[221]Joshi and Mao do not log-transform the box office variables in their study, so that the regression parameters constitute absolute values, not percentage effects.

movie, also helps; such "high similarity" adaptations make, on average, half a million dollars more at their opening than others.

In a follow-up study we also investigated the brand effect of movie adaptions of bestsellers, this time using accumulated "bestseller points" from the USA Today charts (Knapp et al. 2014).[222] We applied OLS regression to a somewhat more comprehensive data set, which contained all 446 film adaptations of books from 1998 to 2006, and added the films' advertising spending to the list of control variables. Just like Joshi and Mao, we find that bestsellers, on average, are more successful at the box office (we use the *total* box-office revenues as our dependent variable). But we also learn that this advantage might be elusive: when the movies' ad budget is included in the analysis, the bestseller effect fades into insignificance.

Does this mean bestsellers offer no financial advantage for movie producers? No—instead, our results show that bestsellers still make a difference, but *only* when they are recent. An interaction between bestseller and a recency variable (that measured whether a book was a bestseller in the year before the movie's advertising campaign started) yielded an elasticity of 0.125—meaning that if a film is based on a recent book hit, a 10% higher "bestseller score" is associated with 1.2% higher box-office revenues. A score that is twice as high (i.e., plus 100%) is linked to an average box office increase of 9%.

The competitive advantage for movies offered by book adaptations has also been argued to exist for TV series, and some adaptation have indeed been huge successes (think of HBO's GAME OF THRONES, based on George R. R. Martin's book series, and Amazon's THE MAN IN THE HIGH CASTLE, based on Philip K. Dick's alternate reality novel). But as with all kinds of category extensions, there might also be "category fit" challenges caused by creative limitations for this kind of category extension. For example, the finite narrative of a novel may hinder, or complicate, a TV producer's ability to develop an ongoing open screen experience.

Hunter III et al. (2016) provide initial empirical insights on the role of adaptations from books and other product categories on TV series' viewership. Their data encompass 1,441 episodes of 98 new dramatic TV series introduced in the 2010–2014 period. They report a negative correlation between whether a series was adapted from another source and its viewer

---

[222]Our study's focus was on "feedback effects" that describe the impact of the film adaptation on the book—we discuss those in a later section.

numbers ($r = -0.21$),[223] and the variable's parameter in a generalized least squares regression explaining viewership is also negative (the regression does not include many series-specific controls, though). According to their results, adapted series attract, on average, 10% fewer viewers than do original ones. This effect tends to get stronger as more episodes have been aired (it is −14% for episodes 11–15).

It is not clear whether the lack of category fit or some other factor is to be blamed for these results. But drawing once more on the sensations-familiarity framework, it might also be that in a TV series context, where familiarity with characters and setting is inherent (it results from the multi-episode character of such programs), the sensations that can be derived from original characters and stories are valued more highly by consumers than is the additional familiarity that comes from the category brand extension. The fact that in Hunter III et al.'s analyses the parameter increases with the length of the run of a series (when people have become more and more accustomed to the series' setting and characters) would be consistent with this argument. We will observe whether any of this will get in the way of Amazon's ambitious (and enormously costly) plans to turn J. R. R. Tolkien's LORD OF THE RINGS novel into a multi-season series (Andreeva 2017). Based on what *Entertainment Science* can tell us, it is a wise decision to not re-tell the novel (and its movie adaptations), but to instead develop a new storyline set in the RINGS universe.

There is certainly much more to be learned about brand extensions and brand integration strategies in general. But we will nevertheless now shift our attention to brand alliance strategies for entertainment products, and, more specifically, the role of stars as human brands.

Stars as Human Entertainment Brands

> "The actors in the earliest films hadn't been credited, but audiences nonetheless came to have favorites. … [E]xecutives recognized the potential for establishing brand names, and they signed the crowd-pleasers. … The star system was born."
> —Brands *(2016, p. 23)*

Stars are an essential element of the entertainment eco-system. They personify the industry's creations, breathing life into movies, music, and novels that would otherwise be only material or digital products. Stars catch our eyes on

223Hunter III et al. do not study the various sources separately, but aggregate them into a "prior source material" variable, so no source-specific results are reported.

posters, win the hearts of professional reviewers, and communicate with us via tweets or Facebook posts. We are all, to differing degrees, fascinated with stars. Not even entertainment executives are always immune to this fascination; it can interfere with their economic decision making (as in the case of the Golan-Globus cousins and the downfall of their Cannon company).[224]

In this section, we will explore the power of stars in their role as human brands. In our *Entertainment Science* way-of-thinking, stars *are* brands because they have the core elements that define brands: consumers hold associations regarding them, and they can be professionally managed (see also Luo et al. 2010). This perspective is also shared by entertainment stars themselves. Actor Kevin Hart made that quite clear in a message to his fans via the social media site Instagram: "I look at myself as a brand… I OWN MY BRAND… I MAKE SMART DECISIONS FOR MY BRAND…I PROTECT MY BRAND…" (quoted in Stedman 2014).[225]

We will first discuss the psychological mechanisms through which stars, as human brands, can influence consumers' reactions to an entertainment product—the "roads" to their success-enhancing role. Then we will share what scholars have found empirically regarding star effects, both on average and when looking more closely at contingency factors. Finally, in case you are prepared to take over the star role yourself, we also try to provide an econometric answer to the question what is needed to become a star: is it talent—or mere luck?

How Do Stars Generate Value for Consumers?

What roles do stars play in influencing consumers' entertainment choices? We will first discuss stars' role as quality signals, or cues, which enable consumers to make inferences about an entertainment product's quality. This cognitive explanation views stars in the role of human ingredient brands. A second explanation puts the (para-)social relationships we, as entertainment consumers, maintain with stars in the spotlight, highlighting the emotional and social factors that make us (at least *some* of us) watch a movie in which Ryan Gosling or Jennifer Lawrence plays the lead role.

---

[224]See our later discussion on the rise and fall of Cannon in the context of innovation management decisions, which had a lot to do with the temptations that stars can offer.

[225]Mr. Hart sent his message in response to being called a "whore" by a Sony executive in a leaked email because he had requested additional payments for promoting one of his films to his fans via social media (Stedman 2014). Please see also our discussion of social media communication by stars as part of our chapter devoted to owned entertainment communication.

The "Cognitive Route": Stars are Ingredient Brands

According to this explanation, consumers are interested in human brands and stars because they signal a product's high quality. When a star actor participates in a movie, the two (the film and the actor) form a brand alliance in which the star takes the role of a branded ingredient. Thus, ingredient branding theory can tell us more about the value of stars—we illustrate its logic in Fig. 9.7 (see also Hennig-Thurau and Dallwitz-Wegner 2004).

Accordingly, when a star participates as an ingredient brand in a movie (i.e., the "product"), building a brand alliance together with the movie's own brand (which we call the "host brand"), this is intended by the movie's producer to have an *awareness* effect. Consumers who are aware of the star should also become aware of the brand alliance product of which that star is a part.

In addition, the star's image is hoped to "spill over" to the alliance product, adding his or her own allure to the attractiveness of the total package. There are two aspects to the power of this spillover from the star to the entertainment product, according to Albert (1998): the ingredient's "drawing power" and its "marking power." The drawing power refers to the *valence*

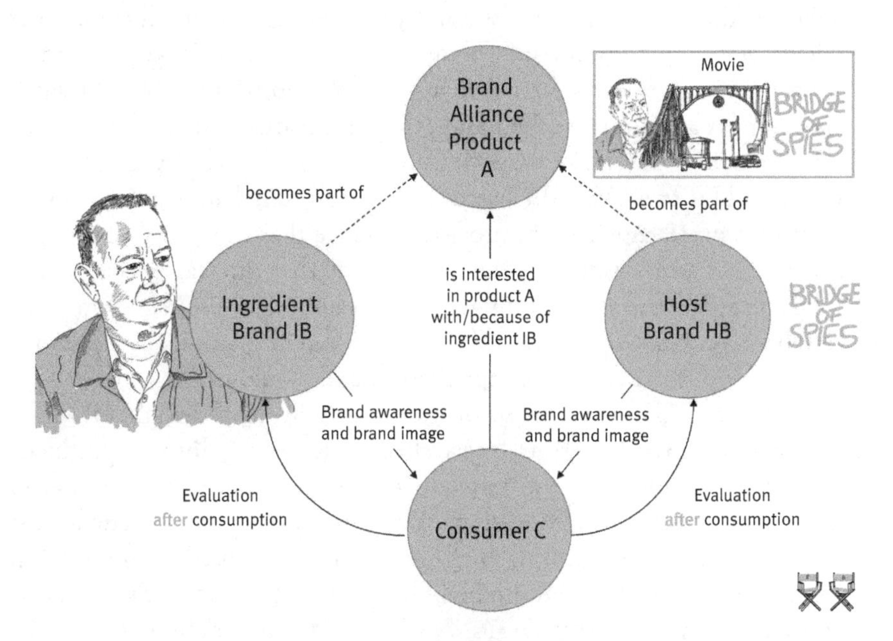

Fig. 9.7 How stars function as ingredient brands
Notes: Authors' own illustration based on ideas in Hennig-Thurau and Dallwitz-Wegner (2004). With graphical contributions by Studio Tense. Film title is trademarked.

of the star's brand image—people who are fans of the star will also tend to think more positively of the film in which the star appears ("I love Tom Hanks movies!"). Thus, the star "provides the movie with some immediate consumer fan base (Luo et al. 2010, p. 1115).

In contrast, the marking power is less about how good or bad the new product will be, but rather about what *kind* of product one can expect. Each star has a certain image profile that will spill over to the alliance product. For instance, the Arnold Schwarzenegger brand is tightly associated with action and science-fiction films based on hits such as The Terminator. In contrast, Tom Hanks might primarily activate consumer associations with dramatic and more ambitious forms of entertainment, based on his most successful films (such as Forrest Gump and Cast Away).

Because of the ingredient brand's awareness and image, consumers will have higher interest in the alliance product, which can translate into higher sales. In the figure, we illustrate this with the link from the consumer to the alliance product, the drama movie Bridge of Spies. The fact that Tom Hanks played the lead role will have helped certain audiences get an idea of the movie's nature and ambitions, based on Mr. Hanks' prior awareness and image.

The arrows that point from the consumer to the two components of the alliance (the host brand and the star/branded ingredient) illustrate that ingredient branding does not stop there. After having experienced the product, consumers will reassess their evaluation of the host/family brand (which is important for additional sequels and other extensions), but will also adjust their perception of the star's image, both in terms of favorability and content associations. Luo et al. (2010), based on a longitudinal industry survey of image ratings of 48 movie stars, provide evidence that the image of a star is highly dynamic and varies with the host brands that the star selects as alliance partners. The scholars' findings from a method-of-simulated-moment regression show that after a series of flops in different genres, a star brand is diluted, just as any other type of ingredient brand would be.

Ingredient branding theory also provides more in-depth insights into the conditions under which stars are most effective for entertainment products. Research suggests that using a branded ingredient is particularly powerful for host brands under four conditions. First, ingredient brands benefit host brands that do not have existing high quality associations. For example, in movies, less is known about an original film, such as Bridge of Spies, versus a sequel such as Angels and Demons, which has a high awareness and distinct brand image on its own as a result of being part of the Robert Langdon film series.

Second, ingredient brands are powerful for host brands for which relatively few other quality signals are available. For example, does a film involve any other known ingredient brands beyond the star? Bridge of Spies also featured director legend Steven Spielberg. Third, ingredient brands are more effective when the ingredient has a great (versus bad or mediocre) image. The fourth condition is when there is a strong fit between the ingredient's image and the host's image (as was the case in Bridge of Spies, but probably less so in Arnold Schwarzenegger's father-daughter drama Maggie).[226]

The "Emotional Route": Stars are Parasocial Relational Partners

But stars have more to offer than only sending quality signals. An alternative, though complementary, explanation of the value of stars for consumers is that they help people to fulfill *relational* motives. We, as humans, are innately relational beings, and this tendency extends even to the formation of relationships with fictional characters. We can link ourselves on a personal level with singers and other entertainment stars and the characters they bring to life in movies, books, and games. Our relationships with them transcend the professional performances of the stars.

Such relationships with human brands can take on various types, including "friendship," idolatry, fandom, and even celebrity worship.[227] A common element of these connections is that consumers feel a significant level of *attachment*, i.e., an emotional bond toward the relationship partner. The value that such bonds can offer consumers results from the deep human need for relatedness: feeling close to others, being connected with them, and cared for by them. In a seminal study on human brands, marketing scholar Thomson (2006) conducted an experiment with 164 students. Using structural equation modeling, his results demonstrate that a human brand's capacity for relatedness is the main driver of the strength of attachment that consumers feel toward the brand.

Communication scholars call these connections "parasocial relationships," a label that recognizes their mediated, unreal nature. The term was coined

[226]We discuss such star-product fit in more detail later in this chapter.

[227]Wohlfeil and Whelan (2012) offer unique, rich insights into the nature of such fan relationships via an introspective exploration of the first author's adoration of actress Jenna Malone over several years. And in their complementary work in Wohlfeil and Whelan (2008), they offer qualitative insights into how this fan relationship can affect interest in a movie in which the star appears.

by Horton and Wohl (1956) based on their observation that audience members of the (then-new) mass media of TV, radio, and movies often developed "relationship-like" feelings toward the media personalities, emotional connections that were similar to those with "true" social friends.[228] Horton and Wohl describe parasocial relationships as "illusion of intimacy" in which consumers feel they know and understand the (media) stars, just like they know and understand people in real life. Consumers even empathize with the parasocial relational partner when he or she makes a mistake. And one must not think that such illusions are simply the result of the consumer living an otherwise-isolated life style: Rubin et al. (1985) tested and empirically rejected a "social deprivation" or "compensation" hypothesis.

Research on parasocial relations suggests, and provides empirical evidence, that such relations affect consumers' entertainment choices, as consumers look forward to spending time (e.g., seeing, interacting with, or listening to) with their entertainment "friends." Summarizing a total of 15 studies on parasocial relations via a meta-analysis approach, Schiappa et al. (2007) find a mean correlation of 0.22 between consumers' level of parasocial relationships with television personalities and their overall TV consumption. On the level of an individual program (instead of "general" consumption behaviors), Rubin and Step's (2000) OLS regression analysis of data from 235 listeners of talk radio shows demonstrates that a consumer's parasocial relationship with the show host was a (much) stronger determinant of listening behavior than other motivations, such as "passing time" or being "entertained."

Consumers not only watch more content that features their parasocial partners, but also end up liking it more. Addis and Holbrook (2010) show that IMDb users rate those movies higher (from a data set of 440 Oscar winners and nominees) in which the leading star (a) is of the opposite gender than the consumer and (b) is younger (or of the same age) as the rating consumer. Addis and Holbrook argue that it is these conditions that movies provide room for (romantic) parasocial relationships with their lead actors.

Let us add that with today's prevalence of digital and social media, parasocial interactions may have reached a level of realism and socialness that is unprecedented. Stars are now able to directly respond to their fans' comments and love letters. This implies a blurring of the distinction between parasocial relationships and real-life ones, which offers implications for the marketing of entertainment products, and particularly the communication

[228]For those readers who want to know more: Giles (2002) offers a more recent review of the literature on parasocial relationships.

element of it. "Relationships" with consumers can now be, to some degree, actively managed by the stars and/or the producers who hire stars for their entertainment products. We discuss such opportunities and their implications in our chapter on owned entertainment communication.

The Financial Impact of Stars on Product Success

> "But the fact that THINNER did 28,000 copies when Bachman was the author and 280,000 copies when Steve King became the author, might tell you something, huh?"
>
> —*Novelist* Stephen King, *quoted in* Levin *et al. (1997, p. 179)*

Now that we have looked at the theoretical mechanisms through which stars affect consumers' perceptions, choices, and evaluations of entertainment products, let us see whether hiring a star is actually a lucrative strategy—and under which conditions. Whereas this might sound like a taken-for-granted business strategy, we argue that it is not so much for at least two reasons.

First, stars are usually well aware that their brand status makes them a rare strategic resource for entertainment producers, so they tend to charge high fees for their participation. The world's 20 highest-paid actors earned a cumulative $700 million in 2016 (*Forbes* 2016a), and singer Taylor Swift alone received a stunning $170 million for her records, tours, and endorsements (*Forbes* 2016b). Second, some have argued that stardom is an outdated concept, at least in film (Bernardin 2016), and what happened when star Peter Dinklage was hired for the massive DESTINY game might be considered as ad hoc evidence for this argument.

So, what can *Entertainment Science* tell us about the contributions that stars make? Are they worth the financial investments they demand, and under what particular conditions do such investments pay off?

"Average" Star Power Effects

The fascination of stars is not limited to audiences. Scholars also find it hard to resist—the role of stars in movies and other forms of entertainment has stimulated more research than most other *Entertainment Science* topics. In what follows, we first summarize what scholars have found regarding the "average" impact of stars on entertainment success. We then take a look at contingency effects and the treatment bias problem that also hamper the valuation of star brands in entertainment.

Basic Insights on Star Value

Because the role of the star differs quite substantially between the various forms of entertainment and there has been so much attention paid by *Entertainment Science* scholars to star effects, let us present average effects on a per-product basis. We start with movies, before moving on to music and novels.

Movies. We counted more than 60 studies that have empirically investigated the impact of a star in a movie on the film's success. The normal approach is to add some kind of "star variable" to the list of factors that explain movie performance in a regression-type analysis, an approach that reveals the "average" advantage the star generates over films which feature no such star (e.g., Basuroy et al. 2003; Wallace et al. 1993). Whereas most of these studies find that the participation of a star helps the success of a film, the size of the impact differs quite enormously between studies.

To understand these variations and to determine the average of these star effects, we applied a meta-analysis approach to the existing studies, taking into account, among other things, the different ways star value was measured across the studies (Hofmann et al. 2016). The meta-analysis showed that the average correlation between the participation of a star and movie revenues was significant, but pretty weak ($r = 0.10$). But when looking more closely, we found that the size of the star effect differed systematically by the type of star power. The star effect was clearly higher ($r = 0.16$) for commercial star power (measured as the commercial performance of a star's previous movies) and lower for artistic star power (as measured by the awards a star had won before, for which the correlation was only 0.07).

Also, we learned that the star effect is higher, and likely exaggerated, when other important movie variables, such as advertising and distribution, where left out of the analysis. The star effect also varied with the data set; when only successful movies are considered, there is less variation in success and the star effect tends to be lower. Star power effects do not differ for movies, though, between the opening weekend and the weeks that follow. This is probably because the two star mechanisms we described before balance each other over time: whereas the ingredient branding mechanism might be stronger early on in a movie's life cycle, the relational partner role of stars might be most influential at a later point.

And then there is a country effect: we find that star effects are systematically higher when North American data is used versus data from other countries. Does that mean that stars matter less outside of North America? Not so fast: because most empirical studies have focused on *Hollywood* stars only, the result means that *their* influence can differ regionally. But other countries have other stars, who will themselves have less commercial appeal in North America. Consider Til Schweiger, who, with lead roles in three of the most successful German movies of all times, is considered a superstar among German audiences, but is nowhere near a household name for American moviegoers.

Hollywood producers aim to address such regional stardom by casting local stars for key regions, such as Chinese actress Yang Yin (a.k.a. "Angelababy") in INDEPENDENCE DAY: RESURGENCE, and sometimes by creating a "local version" of a film with added scenes that feature local stars (Langfitt 2015).[229] No empirical evidence exists yet regarding the effectiveness of such approaches, but some have argued that such strategic casting of local stars as "flower vases" might fall flat and even carry the risk of patronizing local audiences (Schwartzel 2016). A related question for producers is how audiences in other markets than the home market of the local star will react to strategic casting approaches.

Overall, context factors of star power (such as those we have named above: how star power was measured, whether other success variables were also included, whether North American or non-North American stars were studied) explained a substantial 42% of the variation in star power effects between the studies. Using a regression approach, we estimate an average correlation of 0.29 for a commercial Hollywood star at the North American box office, when control variables are included and the data set is not restricted to successful movies.[230] Some scholars have also pointed out that star effects are not limited to audience reactions; stars also impact distributor decisions, or the "supply side" of the movie business. Specifically, Clement et al. (2014) find that a 10% increase of a star's IMDb MovieMeter ranking (their measure of star power) corresponds roughly with a 1% increase in the number of screens allocated to a film in North America; they report a similarly sized effect for German theaters.

Music. Scholars have dedicated somewhat less attention to star power effects for other forms of entertainment, but existing studies show that stars matter for music and books also. For music, singers and bands function as stars, and results tell us that star products consistently sell better and stay longer on the charts. For example, in a Bayesian analysis of music album

[229]IRON MAN 3 provides an example for the latter approach: here Fan Bingbing and other Chinese stars perform surgery on the superhero in an infamous scene that was only included in the Chinese version of the film.

[230]We determined this average correlation by inserting the respective values in the regression equation that is reported in Table 3 of Hofmann et al. (2016).

sales, Lee et al. (2003) find that every additional "Platinum album" previously earned by a musician increases the market potential of his or her new album by about 8%. And Gopal et al. (2006), who apply OLS regression to a data set of 314 albums from 1995 to test whether artists' past reputation affects the performance of new albums, find that a musician's reputation explains 8% of success in the first week and 10% in the weeks that follow.[231]

Other authors have shown that star power also enhances charts survival. Asai (2008) finds that new singles and albums stayed longer on the Japanese charts if the music artist was a star, as measured by exceeding a threshold in sales in the previous year. She used data from two different years (1990 and 2004) and measured the impact of the star variable on the number of weeks a product remained on the charts; her method was an exponential regression in which she also controlled for label and genre factors. Star power had a stronger impact on albums (which "survive" more than twice as long on average if the artist is a star) than singles (survival time is 45% longer in this case). Bhattacharjee et al. (2007) studies album survival using a less-restrictive star variable (everyone is a star who already had an album in the charts) and a larger set of controls. For their data set of nearly 1,500 albums that appeared in the (American) Billboard charts from 1995 to 1997 or from 2000 to 2002, they apply an "accelerated failure time" survival model and find that albums from stars stay about 35% longer in the Top 100.

Novels. In the context of books where authors take the star role, Schmidt-Stölting et al. (2011) are the only researchers that use actual success data to study star effects. They link an author's star power with book sales in their data set of more than 1,000 books released for sale in Germany. Schmidt-Stölting et al. study two facets of such star power: "fame" (a score based on the author's previous bestseller list placements) and "celebrity" (if an author has a reputation outside of publishing). They find in their SUR analysis that fame has a strong effect for both hardcover and paperback books, but celebrity status only helps hardcover books.

Other scholars provide additional evidence for the link between star authors and book success via experiments and surveys. When Levin et al. (1997), as part of an experimental design, asked 138 marketing students how much they might like a new book, they learned that a book by a "star" author such as Michael Crichton received (clearly) higher ratings, particularly when critical reviews are negative. And Kamphuis (1991) found that, out of 218 Dutch book customers, 55% self-reported the author as the main

[231]They measure artists' past reputation via the amount of time the artist was on the Billboard Top 200 charts from 1991 to 1994. Album performance of artists' new albums is an album's entry rank in the Billboard charts and also its charts positions in the following weeks and months.

factor driving purchase. Similarly, when Leemans and Stokmans (1991) studied the decision-making process of 50 Dutch book buyers, "knowing the author" was the most often-mentioned reason in the early stages of the consumer-choice process, and the third most often-named reason when overall comparisons are made among all considered alternatives.

The "Treatment-Bias" Problem Once More

As with parent brands that are chosen for sequels, stars are also selected for a reason by entertainment producers; their "assignment" is not random. Thus, empirical results that link stars' participation with product success are also potentially affected by the "treatment bias"—because producers expect stars to be particularly lucrative for a project, they equip them with better material and stronger marketing (see also Liu et al. 2014). As we have noted before, this better treatment might make good business sense, but also elevates the difficulty in disentangling how much of the star-product's success is due to the presence of the star versus how much is due to the better treatment.

How strong is this bias in the case of star brands? We once again put to work a statistical matching approach to calculate unbiased estimates of star power (Hofmann et al. 2016). We used a set of 1,545 movies, all of which were released in North America between 1998 and 2006, made a minimum of $1 million, and were neither sequels nor animated films. Following the same approach we used for examining remakes, we created a "hybrid" twin movie for each star movie via the propensity score matching technique. We did this two times here: one time for each of the 361 movies that featured a "commercial" star[232] and the other time for all 334 movies with an Oscar-winning "artistic" star.

Figure 9.8 compares the average box office of films with and without stars. We find that commercial stars add, on average, close to $13 million (or 25%) to a film's box office success. Artistic stars, in contrast, "only" increase movie revenues by $4 million (or 9%), but this star power effects is, like the one by commercial stars, still significantly greater than zero. The results also stress how important it is to address the treatment bias when measuring star value: the uncorrected (i.e., apples-to-oranges) comparisons show much higher average star values of $32 (versus $13) million and $14 (versus $4) million.

[232]More specifically, we defined "commercial" stars as those actors or actresses who had been recognized prior to appearing in a film for their previous works in the so-called "Quigley star power list." The list was published annually since 1932 by Quigley Publishing; it is based on a survey of theater owners and film buyers which are asked to name "the ten stars that they believe generated the most box-office revenue for their theatres during the year" (*QPMedia* 2013). We believe the last year the survey was conducted was 2013; historical lists are available at https://goo.gl/U9ube2.

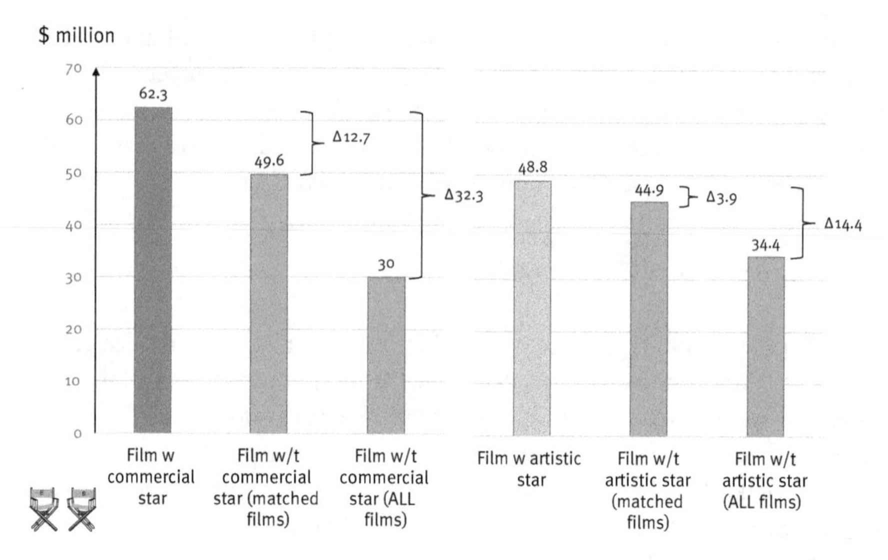

Fig. 9.8 The (average) financial value of movie stars
Note: Authors' own illustration based on results reported in Hofmann et al. (2016).

In a follow-up analysis, we also tested what remains of these incremental financial contributions of movie stars when one controls not only for the treatment bias, but also for various other aspects of the movie marketing mix (basically everything we discuss in the second part of this book). When we ran a weighted least squares regression (in which the matching scores are used as "weights" for each film in the data set), we saw that the participation of a commercial star still increases the North American box office of a film by almost 13%, while having an Oscar-winning star on board adds 8% more revenues.

Contingency Factors for (Movie) Star Power

So, does it pay for a producer to cast a star in his or her next film? Whereas the previous pages give us a general idea of how valuable stars can be, they cannot fully answer this question. Because entertainment products are so dissimilar in terms of budgets and many other factors, a contingency approach is needed that explores how the value contribution of stars affects different kinds of products.

We have conducted such an approach for movies (Hennig-Thurau et al. 2014), drawing on the same data set we use in Hofmann et al. (2016) and again controlling for the treatment bias of movie stars. We looked at three groups of factors that might influence how valuable a star is for a movie: (a) characteristics of the star himself (or herself—gender is a contingency variable on its own!), (b) characteristics of the movie, and (c) characteristics of the fit between the star and the movie in which he appears. For each characteristic, we split the data set accordingly into subsets, ran separate regressions for each subset, and then compared the size of the star effect between the different subsets. Let's summarize the learnings.

Star Characteristics

When it comes to a star's impact on the success of movies, it matters a great deal what "kind" of star we are talking about. Our analyses show that when the star is a "recent" one, which we defined as having been included in the Quigley star power list within the last three years prior to the release of the new movie, North American revenues are increased by 33% on average. When the star was only on the list four or more years ago, his or her participation did not boost the box office in any significant way. We find similar differences for leading versus supporting roles; a star who is top-billed adds 24% to revenues on average, whereas the same star makes no significant difference when not listed first.

We also find that audiences favor younger (24–41 years) and "medium"-age (43–53) stars (who both boost box office by about 20%) over older stars (who do not offer a statistical advantage). At least in our data, audiences do not discriminate in their interest between male and female stars, everything else equal, a finding that contrasts with the assumption that male stars are more valuable and which is often named as a justification of the "Hollywood pay gap" (i.e., females get paid less; Berg 2015). Lindner et al. (2015) report a similar result, finding that once you control for budget sizes and other film factors, moviegoers do *not* penalize films that have a female star. Finally, results show that having one star in a movie is nice, but having more than one is *much* nicer—a second star multiplies the star effect not by 2, but by 3.7!

Movie Characteristics

The value that stars provide also varies strongly between movies of different characteristics. When a film has a below-median production budget (< $37 million), a below-median advertising budget (< $18 million), or opens in a below-median number of theaters (< 2,500), having a star does not make a meaningful difference. In contrast, for *above*-median budgeted films, a star adds an impressive 40% to box-office returns, on average. And for both widely distributed films (additional revenues of 49%) and those with above-median advertising spending (+68%), the star effect is even stronger!

Whereas these product and marketing characteristics go along with a higher impact of stars, having a parent brand in addition to the human star brand reduces the star's incremental contribution—the star effect is only half the size for brand/line extensions (i.e., sequels) compared to other films. Obviously, the different brands (parent movie and star) cannibalize their respective effects; something we pointed to when discussing the role of stars as ingredient brands. In terms of genre, we find that the average family movie benefits most strongly from having a star on board, followed by thrillers, and action and adventure films. In contrast, stars add the least to comedy and romance films.

"Fit" Characteristics

Fit also matters when it comes to casting stars. If the overlap between a star's "genre image" and the genre of the movie in which the star is cast is low, we find that the star adds no value at all. When the fit between the star and film fit was "medium," the star increased the film's box office by an average of 21%, a value that increased to 29% when there is a *high* fit between the star's genre image and the film's genre.

This is consistent with an experimental analysis of film-star effects we conducted (Hennig-Thurau and Dallwitz-Wegner 2004): when we added comedic actor Jim Carrey to an otherwise "unbranded" movie project, the respondents' attitude toward the movie and their viewing intentions increased for comedies. However, if the film was an action thriller, the lack of fit between Mr. Carrey's image and the film's genre instead caused a *reduction* in the respondents' attitudes and also their viewing intentions.[233]

In our subset regression analyses, we also learn that it makes a difference if a star and the film's director have had a hit film together before. In such a case, the revenues were more than 50% higher than for a comparable film without a star. And finally, if more than one star participates in a film and these stars have had a joint hit film before, the producer can even expect a stunning 72% increase in box-office revenue (compared to a similar film without any star).

Figure 9.9 highlights the key findings of our contingency analysis of star power effects. Let us note that other contingencies likely exist as well. For example, Akdeniz and Talay (2013) find, in their large data set

[233]This is also what happened at the box office. When Mr. Carrey starred in the horror-thriller THE NUMBER 23 a few years after our experimental study, the film underperformed quite drastically, and is one of the actor's biggest flops.

Fig. 9.9 Contingency factors of movie star value

Notes: Authors' own illustration based on results reported in Hennig-Thurau et al. (2014). All numbers are percentage change estimates in a movie's North American box office when a certain condition is met (versus no star participates in the otherwise identical movie). All estimates for star and fit characteristics are based on OLS regressions that combined the "no-star" films with the subset of films that meet a certain condition; estimates for film characteristics are based on sample splits.

of movies, that the effect of star power varies systematically with a country's culture and its tendency to avoid uncertainty. Further, in countries characterized by high uncertainty avoidance, moviegoers rely more on the presence of stars.

The Effect of Stars on Financial Risk

But the advantage of stars is not limited to only bringing in higher revenues. Just as is the case with family/parent brands, human star brands can also *reduce the risk* of entertainment products. Amit Joshi (2015) provides evidence for such a risk-reducing role of film stars when he investigates a data set of 41 stars and their participation in 467 movies that were released over a 26-year period, namely between 1981 and 2007. Joshi studies the volatility of weekly movie revenues (which he divides by production costs to make star movies and those without stars more comparable) over a five-week time frame for each movie. His results show that the revenues of star movies indeed vary less than other films in the same genre, with the average revenue variation being 34% lower. On the level of individual stars, he finds such low variation pattern for 32 (or 78%) of the 41 stars in his data set. In case you are curious: George Clooney, Bruce Willis, and Russell Crowe reduced their films' respective levels of risk most strongly in Joshi's study!

We also investigated risk in our analysis of movie star effects (Hennig-Thurau et al. 2014). We compared the prediction accuracy of a regression analysis of the 363 movies with "commercial" stars in our data set to a regression analysis with their statistical "twins." Using the films' North American box office as dependent variable, we find that the standard error of the regression estimate for the star-movie regression is 24% smaller than the standard error of the no-star regression, and that this difference is significant. The pattern is quite similar for a regression with a measure of "global revenues" as dependent variable; here, the prediction accuracy of the star regression is 29% better. With predictability being a solid proxy for financial risk, these results provide evidence that, in general, star movies are less risky to produce than those that do not feature a star.

When You Wish to be a Star: The Controversial Roles of Talent and Serendipity

In addition to studying the value that stars provide for entertainment consumers (and producers!), scholars have also tried to demystify the star concept itself. So, what is it that makes one person an entertainment star, while another works in relative obscurity? Two alternative arguments have been offered: the first stresses the role of talent, whereas the second argument puts "chance" in the spotlight. And then there are those who let the data speak to find out which side is favored by empirical evidence. Be aware that the proponents of both arguments are economists by training, which quite clearly has shaped their ways of thinking about consumers.

Rosen's "Talent Argument"

In Sherwin Rosen's (1981) "theory of superstars," it is a person's talent that turns him or her into a star and is the force behind the person's ability to demand enormous amounts of money for his or her contributions. Rosen argues that small differences in ability are transformed into large differences in "success" through two kinds of mechanisms: demand-side and supply-side processes.

Regarding the demand side, Rosen argues that "in certain kinds of economic activity" (p. 845) in which stars can be found, the quality of a person's performance is a function of his (or her) talent. It is this quality that consumers demand. But consumers do not judge the quality of the most talented person independently—instead, according to Rosen, they always put it in relation to the quality that can be provided by others. And they consider the (lower) quality offered by one person as a weak, "imperfect" substitute for the (higher) quality of others' performances, i.e., lesser talent is a poor substitute for greater talent. As a consequence, if one artist is only 10% better than another, consumers are willing to pay far more than just an additional 10% for the "better" artist. And assuming that prices would be similar, then the demand for talent is not a linear function, but increases exponentially with the person's talent level: a 10% more-talented person will command much more than a 10% higher market share than his competitor. We all prefer a single exceptional performance over several "solid" ones, don't we?

On the supply side, Rosen recognizes the copy-cost characteristics of entertainment products.[234] Because of entertainment's first-copy cost characteristic, i.e., the vast majority of costs go into creating the first copy, artists can reproduce their performances for low marginal costs. This makes it easy for a large number of consumers to inexpensively experience the performance of the best artist, bypassing the second-best offers.

It is the combination of demand- and supply-side effects which, according to Rosen's logic, explains why people who have superior talent can satisfy the

[234]Rosen's theory is not limited to entertainment, but applies also to other aspects of life where humans offer creative deeds, such as in health care and education.

disproportionately large demand for their performances. Because this will often happen at a price that is higher than the offerings of less talented performers, talent should have a multiplicative effect on profits for the star.

Adler's "Chance Argument"

Moshe Adler (1985) challenged the critical role of talent for superstardom. His logic is based on the assumption that to be able to enjoy a certain kind of cultural products (such as classical or pop music), having knowledge about these kind of products is essential.[235] This assumption itself implies a concentration process: in order to become knowledgeable about products, consumers have an incentive to "patronize" the same artists as other consumers do. The reason is that information and knowledge about the most popular artist is always much easier to access than knowledge about less popular alternatives. Why might this be the case?

If you are looking for great pop music and want to learn about it, it makes sense to start with Beyoncé rather than some unknown local pop musician, simply because the search costs are much smaller, both among your friends (from who you can learn by discussing things) and via media outlets such as Rolling Stone magazine. And if other artists are neither dramatically better nor dramatically cheaper (by enough to compensate for the lower costs of knowledge for the most popular artist), sticking with the most popular artist is, economically speaking, the logical choice for consumers.

Adler's line of thinking carries an important implication: stars become stars *not* because they have more talent than other performers, but because they are more popular at a given point in time than others who have a similar level of talent, so that consumers will prefer them over other artists for the lower search costs. And Adler argues that among those with similar talent, it is mere luck (his terms for "everything else besides talent") that determines stardom. Whoever is the lucky one in the first place should then, in Adler's terms, just "snowball into the star" over time.[236]

---

[235]As a side note, this argument bears some resemblance to Bourdieu's idea that "cultural capital" is required to enjoy cultural products.

[236]Let us note that this logic is very similar to the phenomenon of "success-breeds-success" cascades that we discuss later in this book. But whereas Adler discusses consumer choices between stars, success-breeds-success deals with choices between entertainment products.

What Data Analytics Can Tell Us About the Talent Versus Chance Controversy

Can data analytics tell us which of the conflicting superstar theories is "right"? Scholars have tried to shed light on this issue by using measures of a star's talent, or "luck," or both.

For a data set of the cumulative sales of 107 popular singers from 1955 to 1987, William Hamlen (1991) uses the "average harmonic amplitude" to measure each singer's "voice quality," an attempt to capture the singer's talent. Regressing sales on voice quality and other variables (such as "years in business" and the singer's gender), Hamlen finds talent to be positively related to artists' success. However, the elasticity indicates that a 10% higher talent corresponds only with a 1.4% increase in cumulative sales, which does not support Rosen's idea of increasing returns to talent.[237] Franck and Nüesch (2012) use a similar approach in the related field of professional soccer, measuring a player's talent with a set of 20 objective performance indicators. They also find that some of the indicators are positively linked to the player's future monetary market value; but again, the link is far from perfect.

Wirtz et al. (2016) circumvent the problems associated with such talent measures by using a survey approach to determine the link between talent and success. For 554 German movie actors, they find substantive correlations from 0.53 to 0.37 between various aspects of talent, such as an actor's repertoire size, physical appearance, and language skills, and the actors' artistic and commercial success. Whereas these results are again in line with Rosen's arguments (although the shared variance is once more far from perfect), some methodological limitations apply.[238]

Others have searched for empirical support for Adler's chance logic. Chung and Cox (1994) focused on "luck" by using a purely stochastic distribution (known as the "Yule-Simon distribution") to explain the number of

[237]But the low elasticity Hamlen finds in his study might also be the result of measurement error: whereas it seems intuitive that Barbra ("the voice") Streisand leads the talent ranking, it feels rather counterintuitive that Whitney Houston, often considered the "greatest voice of her generation" (Gill 2012), receives a *much* lower talent score. And isn't music talent about more than a singer's voice anyway?

[238]The self-reported character of both talent and success in Wirtz et al.'s study carries the danger of inflating the contribution of talent, because of a "same-source bias." This bias results from the statistical rule that two subjective judgments made by one person are systematically correlated beyond substantive reasons.

Gold Records earned by a singer (a measure of sales). They find that the stochastic model, which does not include any talent factor, explains about 94% of the actual distribution of Gold Records among artists. But these analyses look only at success patterns *across* artists in general, and do not look at developments over time. And, for a larger data set, Giles (2006) shows that the same stochastic model is *not* well suited to explain how many No. 1 hits singers achieve during their careers.

Another key element of Adler's logic is that popularity makes one artist even more successful (because it lowers the search costs for consumers). Some researchers have compiled evidence for this claim: Luo et al. (2010) show, in their longitudinal survey-based analysis of movie-star value, that the volume of media coverage about a star positively affects his or her brand value. And Franck and Nüesch find that coverage in newspapers and magazines about soccer players that is *not* related to their performance on the field increases their market value, over and above talent factors.

So, it appears that there is some truth to both theories: star power, in reality, requires a combination of talent *and* luck. But even when considered in combination, these factors do not seem to offer a comprehensive explanation of the star phenomenon. Instead, further factors have also been shown to influence star power. These include a person's career decisions: Luo et al. show that star value is influenced by how well an actor's film choices are received by consumers, critics, and industry peers.[239] Similarly, Mathys et al. (2015), in an analysis of consumers' interest in 161 actors from 2004 to 2010, report that the commercial success of the films the actors appear in and the films' fit with the actors' images drive interest in an actor. These results are also consistent with Wirtz et al.'s finding that an actor's ability to effectively manage relationships with casting directors is highly correlated with the actor's success.

We will now close the debate of the specifics of stars and the other brands and branding strategies that entertainment managers can use to generate awareness among consumers for a new project and send quality signals. In entertainment, brands are today not only employed to increase the success potential of single products, but are also approached from a more abstract, holistic perspective: as multi-product, multi-category franchises.

[239]One learning for actors from Luo et al.'s study is that risky decisions do not pay off in this context, as the authors find that *bad* choices leave a stronger mark than do good ones.

Franchise Management: A Holistic Look on Entertainment Brands

> "Every single first meeting I have on a movie in the past two years is not about the movie itself, but about the franchise it would be starting."
> —Shawn Levy, *director of the three-film* NIGHT AT THE MUSEUM *series (quoted in* Suderman *2016)*

Most of what we have discussed on the management of entertainment brands has taken a "product-level" perspective: we have discussed the potentials, and problems, that producers face when they apply a branding strategy for their next entertainment product instead of developing an "original" product. Such a perspective is certainly valuable and important, as it highlights how brands affect consumer decision making in an entertainment context.

However, in addition to this product-level perspective, an important second perspective considers individual branded products to be part of a bigger concept that is often labeled a *franchise* by entertainment managers and scholars. The question here no longer concerns the advantage of brands over unbranded alternatives, but how such a franchise should be managed to generate maximum "franchise value."

In the following, we first discuss how such a perspective shift changes the economic logic of managing entertainment, increasing the effectiveness of certain decisions while letting others become suboptimal. We then turn to reciprocal effects—how do extension products in a franchise affect the value of their parent brand? We end the section by analyzing the historical development of the franchise concept, making the connection from the first STAR WARS movie to current "mega-franchises," as so successfully illustrated by Disney and its "Marvel Cinematic Universe."

How Thinking in Franchises Shapes the Economic Logic

Figure 9.10 illustrates the economic logic of a franchise approach to entertainment. In a franchise, a producer creates a branded entertainment product with the idea to extend that brand into additional products, letting it become a family/parent brand to them. The resulting set of products, in its entirety, then constitutes the "franchise." It doesn't matter if the initial product is an original product (i.e., a new brand), like the first HUNGER GAMES novel in the figure, or an extension of an existing brand (such as the first

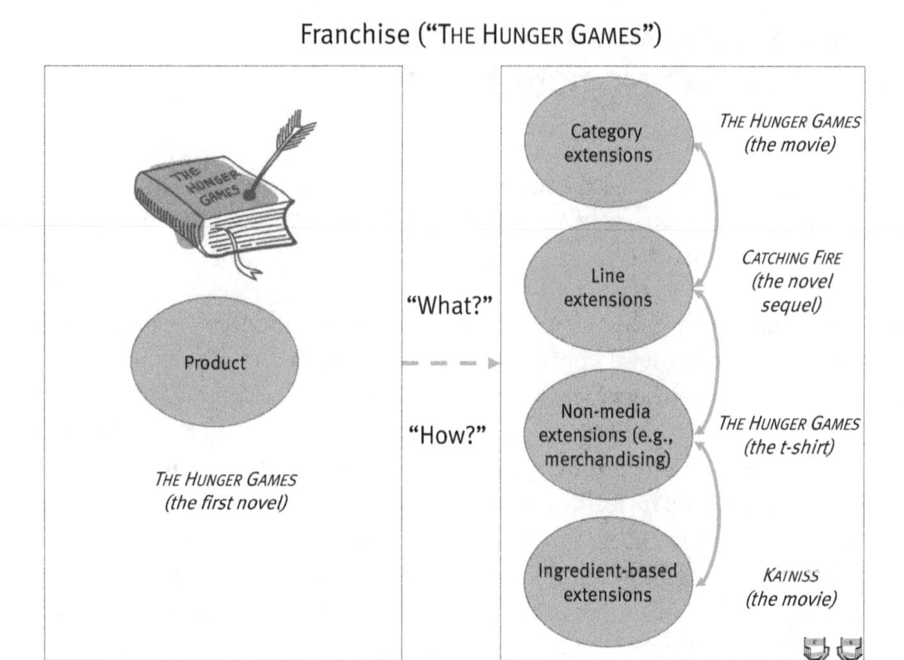

Fig. 9.10 The franchise logic
Notes: Authors' own illustration. Extension products named in the figure are partly hypothetical (such as a movie spin-off using the lead character "Katniss"). Brands named in the figure are trademarked. With graphical contributions by Studio Tense.

movie of the HUNGER GAMES series, adapted from the novel). The firm that builds and manages the franchise might only hold the rights to exploit the brand in certain forms of entertainment or other product categories (e.g., whereas Lionsgate operates the HUNGER GAMES films, Scholastic has kept the book rights). What is crucial for the franchise logic is that success is not judged solely on the success of just the initial product, but on the success of the franchise—*as a whole.*

The franchise logic implies that a producer treats the rights to extend an initial product at a later time as an investment option (Gong et al. 2011). Such extension options can refer to producing, at a future time, line extensions (i.e., sequels, remakes) and category extensions. Options also include extending the brand into non-media categories, such as theme park attractions (e.g., the BACK TO THE FUTURE ride at Universal Studios) or whole theme parks (e.g., as Lionsgate plans to open one for the HUNGER GAMES in South Korea; Swertlow 2017).[240] A regular extension category for entertainment products is merchandising, which ranges from putting a brand on T-shirts and light saber toys to action figures and costumes. Such merchandise today can "get up to half of [a Hollywood blockbuster's] budget back" (Follows 2016). When Lionsgate released its initial HUNGER GAMES movie in 2012, it also licensed the brand for use on more than 160 products, including a replica bow, lunchboxes, action figures, and all kinds of memorabilia (Orden and Kung 2012).

The main characteristic of an option is that its value changes over time, depending on other events. As we have shown that the success of a movie sequel depends on the parent brand's awareness and image, among other factors, the reception of the initial movie impacts the option value of a sequel right, as well as other extension rights (Van der Stede 2015). An option perspective enables producers to invest higher amounts when an option value is estimated to be substantially higher than zero. It might make sense to spend more on an initial film's production budget and hire stars if a studio believes this can lead to an increase of the film's marketability and playability, because such increases would also spill over to the brand's option value. Following the same logic, it could make sense to spend more on the initial film's distribution and advertising in order to increase brand awareness, as this could then pay off for the sequel(s) and category extensions.

Hollywood studios have, at least implicitly, started to act upon such option logic. Didier Lupfer, as StudioCanal's CEO, noted that the firm "invest[ed] a lot of money on the first film … to win more on the second one, the sequel, the spinoff and so on" (quoted in Jaafar 2016). Such option logic has also guided Lionsgate to shepherd the development of the first HUNGER GAMES movie with enormous scrutiny, including sponsoring an online giveaway of the source novel's first two chapters to further broaden brand awareness (Orden and Kung 2012). The company considered the first HUNGER GAMES movie's success to be crucial for making three sequels (and also selling large volumes of brand merchandise). Similar behaviors have been observed for other forms of entertainment, such as when game studio Ubisoft develops games with upfront plans to extend them into other fields of entertainment (Graser 2013).

[240]Like other category extensions, it is not relevant for the franchise logic whether a franchise owner operates the extension product him- or herself, or licenses its right to someone else. Whereas Universal operates the BACK TO THE FUTURE theme park attraction on its own, Warner has licensed the firm the rights for also operating a theme park attraction around its HARRY POTTER film-based brand.

But a franchise perspective also changes what products are considered valuable. A franchise approach implies a longer-term perspective, taking into account not only the success potential of a single product, but also its "extension potential." This extension potential needs to be weighed with the uncertainty surrounding this potential and discounted for the time that will pass before extension revenues can be expected to pour in.

For example, Disney now devotes at least 80% of its production budgets to films that it considers to have strong extension potential for sequels and merchandising. As the firm's CEO, Robert Iger has conceded that the decision to make a sequel to the Pixar film CARS was "very much an extension of the franchise discussion" (quoted in Smith 2011). The company considered the combination of toy vehicles and beloved characters in a movie as ideal for driving sales in multiple product categories. And sales figures support that decision: when the sequel was released in 2011, merchandise for the brand had already accumulated total revenues of about $10 billion, of which $1.2 billion had flowed back to studio (Smith 2011). This number was a multiple of the roughly $230 million the studio earned from the film's theatrical release.

But not everything sells like toy cars, and applying a franchise logic also implies that some products which should appeal to audiences will *not* be made nevertheless—if they are believed to have only limited extension potential.[241] With regard to the branding strategies for single entertainment products we discussed on previous pages, some have a systematically higher franchise potential than others. Specifically, the familiarity-sensations framework informs us that remakes and category extensions, such as best-seller adaptions, are limited in their potential to be extended. Their problem is that a sequel to a remake or adaptation (without being based on a book sequel itself) offers lower familiarity than the remake or adaptation it is extending—the popular characters of the original book or movie no longer serve as the direct reference point for consumers.

So, unless the remake/adaptation has managed to establish itself as a strong brand (beyond attracting audiences who came for the curiosity of seeing the "old" in new form), such sequels will not benefit from brand familiarity, but lack the "tantalizing and paradoxical familiar-newness"

[241]Making the franchise logic the *de facto* standard of the business also influences the expectations of investors and business partners, making it harder to justify the production of products with limited franchise potential. For example, Disney had to face criticism from financial analysts and retailers when it released Pixar's UP movie—which not only lacks cars, but any kind of "merchandizable" characters (see Barnes 2009).

(Bramesco 2016) of their predecessors. This was likely a reason for the failure of Disney's Alice Through the Looking Glass—the sequel to what was both an adaptation (of a classic novel) and a remake (of a beloved movie classic). The studio seems to have learned about the franchise limitations of remakes since then. Despite the enormous successes of their real-life remakes of animated classics such as Beauty and the Beast, Disney has announced no plans to produce sequels for them at the time of this writing (Fleming 2017a).

The course of a story also determines the franchise potential of an entertainment product. If the hero dies at the end of the story, a large part of the franchise option value usually dies with it—remember what *Entertainment Science* tells us about the value of star continuity for sequel success. For directors of franchise films this means, quoting Logan's James Mangold: "You can't kill the characters because they're worth so much effing money" (quoted in Hayes 2017). But entertainment products are never of a binary kind, so that modifications to their narratives can influence franchise potential. When Sylvester Stallone, who was also turning his Rocky character into a successful franchise, was filming an adaptation of the novel First Blood, he faced a dilemma. In the novel, the main character dies in the end. Mr. Stallone decided to not follow the literary source at the last minute. Despite having already filmed the death scene,[242] he insisted on attaching a different ending to the film in which his character stays alive. In terms of franchise development, this turned out to be a wise decision, as it paved the way for several additional—and hugely successful—films featuring Vietnam veteran John Rambo, as well as an array of related "Rambo" products.

Let us note that such changes are delicate, because they can violate the integrity of the artistic work that essentially constitutes an entertainment product. Violations can leak through and turn critics and audiences against a product, as happened when the producers of Blade Runner added a blue-sky happy ending to an otherwise distinctly dystopian film. And some of entertainment's greatest hits take a decidedly negative path that is closely tied to their success: it's hard to think of Titanic with Leo and Kate living happy ever after, or of Love Story with Ali MacGraw's Jenny being cured from cancer in a last-minute turn. Entertainment managers must accept the fact that some stories are just better told in a single-product way, instead of fiddling with them to make them "franchiseable."

[242]You can find the original ending of the film on some DVD/Blu-ray versions and also on YouTube (e.g., https://goo.gl/Yym7YV at the time of writing).

Finally, a franchise approach implies that managers should adopt a holistic perspective, because the different products that are part of a franchise affect each other in more ways than one. Because it serves as the "power source" for all future franchise revenues, the image of the parent/family brand is pivotal for all extension activities. This affects extension decisions, because extensions usually also impact their parents' value, something that branding and *Entertainment Science* scholars refer to as "reciprocal spillover effects." A coordinated and integrated planning of extension activities is needed instead of an ad hoc, per-product approach to deciding about future extensions.

Let us discuss these reciprocal spillover effects, which are so essential for successful franchise management, in some more detail, before taking a look at "universes" as extra-large versions of entertainment franchises.

When the Extension Affects the Original Brand: Reciprocal Spillover Effects

Reciprocal spillover effects describe the change in revenues of the parent brand that can be attributed to an extension (e.g., Balachander and Ghose 2003). Franchise managers want such reciprocal effects to be positive—the release of a sequel movie should increase the home entertainment sales of its predecessor, as consumers are motivated by the new entry to re-live the original or explore it for the first time. Similarly, the adaptation of a novel for the big screen is expected to trigger not only sales of movie tickets, but also book copies. In both cases, the extension and the parent are complementary products.

But reciprocal effects can also be negative. In entertainment, we envision two ways this can happen. First, an extension product might serve as a *substitute* for the parent. When consumers talk about SCARFACE today, they will usually have Brian de Palma's film adaptation from 1983 in mind, whose cultural presence largely suppresses any thoughts about the 1932 movie it remade. We think of Whitney Houston's voice when the song I WILL ALWAYS LOVE YOU is mentioned, whereas the original song by Dolly Parton has been largely forgotten. Of course, substitutive effects would imply that the original versions would be remembered more if they had not been remade, something which is challenging to show empirically.

Second, an extension's lack of quality or lack of consistency with the parent might drive consumers away from the parent. Consider LOST: the widespread disappointment over the series' dramatic resolution has probably hurt the reputation of the series as a whole. We assume that the problems of the second season of True Detective might have hurt interest in the show's first season. And the Matrix sequels might have triggered consumer demand for the original Matrix movie when they were announced and released, but their uneven quality threatens the film's standing among movie fans in the longer run. Again, no empirical evidence exists for this effect, but branding theory offers strong arguments. The problem is also acknowledged by at least some executives: when Disney exploited its core brands with low-budget direct-to-DVD sequels (such as Lion King 1 ½), this approach was harshly criticized by then-Disney board member Steve Jobs (who called the extensions "embarrassing") for the sequels' diluting effects on the original films' reputations (Spence 2007).

We will now take a look at empirical studies that have empirically tested and quantified such reciprocal spillover effects in the context of movies and books. In addition to "average" effects, the results also shed some light into contingencies. Which factors multiply positive reciprocal spillover, and which factors hurt it instead?

Reciprocal Spillover Effects of Line Extensions: The Case of Movie Sequels

Given the short life cycles of entertainment products, how can a sequel affect the success of its parent? In the case of movies, reciprocal spillover effects can influence the home entertainment performance of the original brand. In our study of movie sequel effects (Hennig-Thurau et al. 2009), we also looked at the impact of the sequels' releases on their respective parent's DVD sales. For 76 initial sequels for which parent DVD sales data was available, we used actual sales data from Nielsen (which represents about 65% of total sales) to conduct an "event study"—an approach developed by finance scholars to determine "abnormal changes" in a stock's price resulting from specific events.

In our case, the event was the release of the sequel film in North America. In essence, we first estimated for each parent movie how many of its DVDs would have been sold if there had been no sequel release, using weekly cumulative DVD sales data. Then we subtracted the "normal" sales from the actual sales that occurred to estimate the "abnormal" sales caused by the sequel release. Figure 9.11 illustrates that approach for two example movies—Rush Hour and Underworld.

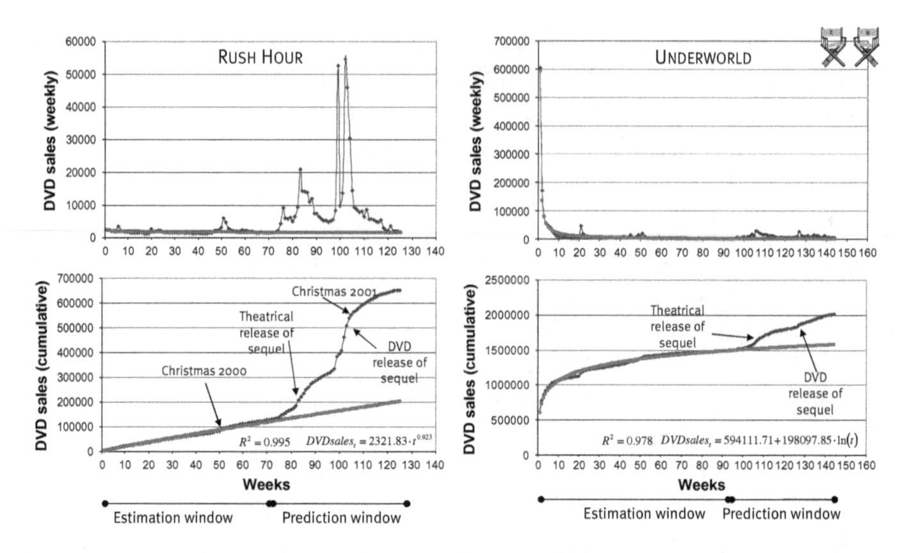

Fig. 9.11 Determining "abnormal" DVD sales caused by two sequels
Notes: Reprinted with minor adjustments with permission from Journal of Marketing, published by the American Marketing Association, Hennig-Thurau et al. (2009). Conceptualizing and measuring the monetary value of brand extensions: the case of motion pictures, November 2009, Vol. 73, No. 6, pp. 167–183. Brands are trademarked.

The results show that all 76 parent brands benefitted from the release of their sequels, with an average bump in sales of DVD units of 217,000, which translated into additional sales for the studio of about $4 million.[243] But our results also show that the reciprocal effects varied quite massively between films. Whereas State Property sold just 219 additional DVDs, Shrek sold an impressive 1.4 million additional copies solely because of the sequel release. So, why these differences?

To uncover what factors are responsible for the differences in abnormal DVD sales, we ran a stepwise cross-sectional OLS regression, using the extra sales as dependent variable. As explanatory variables, we used those which also determine the success of a sequel—characteristics of the parent and the sequel, their fit, and also a number of DVD-specific variables (such as time lag, number of DVD versions available, whether there was a joint promotion campaign, etc.).

The results explain about 70% of the variation in abnormal DVD sales between the movies. The most influential factor is whether the parent brand

[243]The $4 million is the result of the following equation: 217.000 units × $20 (average retail price per unit) × 0.6 (average studio share of retail price)/0.65 (to correct for the incomplete data we got from Nielsen, which misses, for example, the sales from Walmart stores).

is widely known itself (which helps the parent to benefit from an extension—it accounts for about 25% of the variation). The second most-impactful factor then is the success of the extension in theaters (higher success triggers higher reciprocal effects, contributing about 20% of additional parent sales). The third factor that matters is the image of the parent brand—much-liked parent brands can clearly expect to benefit more from extension products (also accounting for about 20%). And finally, the *combination* of a positive image and a high awareness also triggers additional sales of the parent. At least in our data set, no other variable exerts reciprocal effects.

These results are for line extensions—movies followed by movies. Let us now investigate whether things look different for category extensions—when books are adapted for the "big screen."

Reciprocal Spillover Effects of Category Extensions: The Case of Book Adaptations

> "Nothing sells books more than a movie."
> —Minzesheimer *(2004)*

For book authors and publishers, having their products extended into the movie category is widely considered a lucrative venture. But exactly how strong are these reciprocal spillover effects when the extension is outside of the parent's own product category, and which factors determine their size?

Let us take a look at all 446 novels that were turned into movies and released in North America between 1998 and 2006 (Knapp et al. 2014). Using the books' ranking in the USA Today Top 150 bestseller list *after* the film adaptation was released as dependent variable in a OLS regression (and controlling for their ranking prior the adaptation's release),[244] we focused on the role of extension factors (namely, the success and the advertising of the movie adaptation) and the parent-extension fit, as well as the parent book's sales prior to the movie's release. We also wanted to find out whether an

[244]More specifically, we transformed each weekly position of a book on the list into a "point score," and then added up the points for each book over time. We calculated one score for the period *after* the film adaptation was released and another for the period *before* the film was released (in fact, before the advertising for the movie started). When doing this, we accounted for the non-linear distribution of attention that the ranks of such bestseller lists get from consumers by using an exponential transformation approach which ensures that the highest ranks get disproportionally higher point scores than do lower ranks.

integrated approach helps: does it matter whether the book, once the film is released, actively promotes its "parental" role, either by mentioning the movie on the cover ("Now a major motion picture!") or by replacing its cover with the movie poster—and how much?

Together, these factors explain about 63% of the "extra" book success after the movie release. The reciprocal spillover is strongly enhanced by advertising efforts for the extension (i.e., the movie adaptation)—any increase in extension advertising leads to an almost *equally sized* increase in parent success. The parent also profits strongly from the success of the extension, with a 10% higher movie success being linked to 4% higher book sales. Whereas fit does not exert a direct effect on additional book sales, high fit heightens both the impact of advertising and extension success on book sales in a substantial way (by 75% and 50%, respectively). The results also show that integrative efforts pay off well for managers of the parent brand: when the book's cover points at the movie release, extra book sales in response to the movie release are about 50% higher, on average. Such integrative measures further increase the spillover caused by the success and advertising of the extension product.

In a separate analysis, we also learned that the reciprocal spillover effect is not limited to the book on which the movie is based: other books from the same franchise/series can also benefit. The reciprocal effects for these other books are less strong than those for the adapted ones, but still substantial (elasticities of about half the size as for the adapted book). Producers of more complex franchises should thus invest even more managerial attention in extensions, as they can spill over to a broader product range, not only to a single parent product.

From Star Wars to Marvel: The Rise of Entertainment Universes

Let us round out our investigation of entertainment franchises by looking at how the concept has evolved over time, reviewing now-established facets of the concept and highlighting new developments. We start with George Lucas of Star Wars and end with Kevin Feige, the managerial brain behind Marvel's "Cinematic Universe." We speculate that it is not accidental that both Star Wars and Marvel Studios are today part of one and the same entertainment studio—Disney makes use of the franchise concept in a way so radical that other entertainment firms have trouble keeping pace. It says a lot that the company's consumer products division in 2015 generated almost the same revenues as entertainment did—at an even higher profit margin (39% versus 27%).

"A Long Time Ago": Star Wars as the First Entertainment Franchise

George Lucas is not only a famous film writer, director, and producer, but also, and maybe even more so, a highly innovative business person. Similar to the massive impact that his Star Wars movies have had on pop culture, his stewardship of the Star Wars brand has changed the film and entertainment industry in drastic ways. Whereas (spoiler alert!) Darth Vader is Luke Skywalker's father, George Lucas' Star Wars is the father of the franchise concept.

When Mr. Lucas, who was at that time best known as the director of the 1950s-themed hit movie American Graffiti, discussed the terms of his first Star Wars movie with the studio 20th Century Fox, he traded away a higher salary for the sequel rights to the movie. Back in 1976 when sequels were rare exceptions, Fox did not anticipate making a "multi-part saga" or even a single sequel and thus valued the sequel rights for Star Wars at little more than zero. But Lucas must have envisioned exactly that—a saga consisting of several connected movies. After the first movie's huge success, he then extended his franchise ambitions by obtaining all the merchandising rights for the brand from Fox, offering the studio the distribution rights of the first sequel for a period of seven years (Fleming 2015).

These deals, as Squire (2006) phrases it, "rewrote the economics of the movie business" (p. 7) because they enabled George Lucas to systematically develop and exploit his Star Wars movie into the first true franchise. In addition to planning sequels and prequels, Lucas systematically extended the brand into several product categories besides film, including numerous books and comics. The first book alone, a novelization with Lucas credited as author, has generated sales of $200 million; in total, about 170 official novels have been published (*Yodasdatapad* 2016) and nearly 2,000 Star Wars comics have been written, as of 2017 (*Wookieepedia* 2017). Over the years, games have become another essential element of the franchise; more than 140 different games have been released for various platforms.

Merchandising was hardly recognized as a revenue source by entertainment producers until Mr. Lucas turned it into one—otherwise, he likely could not have secured the rights from Fox in the first place. In Mr. Lucas' franchise plan, merchandising played a central role from the very beginning, despite retailers still being hesitant even when the second film was released. But obviously, he couldn't have been more right. Over time, the Star Wars brand has generated more than $32 billion in retail merchandising sales (Taylor 2015), which, assuming an average royalty fee of about

12% for the brand, suggests that Mr. Lucas and Disney, after buying the franchise rights in 2006, had earned close to $4 billion from merchandising by 2015. Within the 12 months following the release of 2015's STAR WARS: THE FORCE AWAKENS, an additional $5 billion in merchandising retail were expected (Rohbemed 2015). In comparison, the global theatrical box-office returns for the first nine STAR WARS films themselves (seven official "Episodes," one spin-off, and one animated CLONE WARS movie) have been roughly $7 billion, plus a similar amount from various home entertainment releases.

With total revenues of more than $40 billion, not accounting for inflation, the STAR WARS franchise has become the most successful in entertainment history, being almost twice as big as the HARRY POTTER franchise. The only true contender today is Marvel's Cinematic Universe—we will get to that in a moment. The key to the success of all these efforts was George Lucas' visionary franchise approach to entertainment. In addition, his ongoing handling of the franchise also teaches a lot. Mr. Lucas, and later Disney, paid close attention to insure that *all* extension products were carefully placed within the STAR WARS franchise so that they could add value to the brand (or at the minimum, would not hurt it). And in case something turned out badly (such as the now infamous TV show STAR WARS HOLIDAY SPECIAL; Conterio 2015), they made sure that it was quickly made unavailable to the public.

From Franchises to Mega-Franchises a.k.a. Universes

When the Disney company bought Lucasfilm in 2012 for $4 billion, the options to extend the STAR WARS franchise were a crucial argument. And since then, Disney has systematically transformed these options into extension products, with new movies, games, and merchandise at the center of their actions. But it is with its Marvel subsidiary that Disney has taken the franchise logic to the ubernext level—by developing a hypercomplex mega-franchise consisting of multiple interconnected characters and brands known as the "Marvel Cinematic Universe."

After the birth of STAR WARS, a number of other multi-product franchises had been developed. One example is the ALIEN franchise: originating from Ridley Scott's space-horror thriller from 1979, various movies, games, and comics have been produced around the titular xenomorph monster. Some of these products link the film's monster with other original characters and brands, most prominently the Predator species in a number of ALIEN VS. PREDATOR games and movies. But the franchise also encompasses the alien encountering comic heroes, such as Superman, Batman, and Green Lantern. Among the wildest of such crossovers is certainly the comic ALIENS VS. PREDATOR VS. THE TERMINATOR, released in 2000. So why is this chapter not featuring the ALIEN franchise and its management? Because the rights owners, along with the partnering film studio (again: Fox), never strategically explored its brand's potentials. Instead, the franchise evolved more in an ad hoc release-by-release way, with each new product planned one-at-a-time. As a result, the franchise's elements lack connectedness, missing both a consistent narrative and aesthetic vision.[245]

The idea of the Marvel Cinematic Universe is quite different and shares many more similarities with the STAR WARS franchise than the evolvement of the ALIEN brand. As with ALIEN, crossovers play an important role, but they are explored much more systematically by Marvel. Here, the "universe" consist of several interwoven superheroes and plots that are made available across different entertainment products (or "platforms"). Figure 9.12 illustrates how strategically Marvel has developed the various character and product brands in their films, over time, from its beginning in 2008 until the end of 2017. Parallel to treating each brand as a franchise, the brands are also treated as integrated elements of a joint "meta-franchise." By the end of 2017, the first 16 movies of this meta-franchise had generated global revenues of almost $13 billion in theaters alone, not including the close-to-$1-billion income by Sony's tangential effort SPIDER-MAN: HOMECOMING.

The history of this franchise traces back to Marvel's insolvency in the 1990s, when Marvel had licensed its main character brands to several other studios: Spider-Man to Sony, X-Men, Elektra, and Ghost Rider to Fox, the Hulk to Universal, the Punisher to Lionsgate, and Blade to New Line. The resulting film adaptations were not coordinated in any way, with highly

[245]An example for this lack of a long-term vision is that whereas the PREDATOR 2 movie indeed contains a reference to the ALIEN brand (an ALIEN skull is present in the trophy case on board the predator ship), that scene was *not* strategically chosen. Instead, it was the result of an inside joke and reference to the original ALIENS VS. PREDATOR comic (which had been released the year before) by two special effects artists who had also worked on the first ALIEN movie (*Xenopedia* 2017). Similarly, the most recent addition to the franchise, a new film trilogy by Ridley Scott, the creator of the initial ALIEN film, resulted not from any studio plan, but from Mr. Scott's vision, and there were even parallel plans for other ALIEN movies circulating.

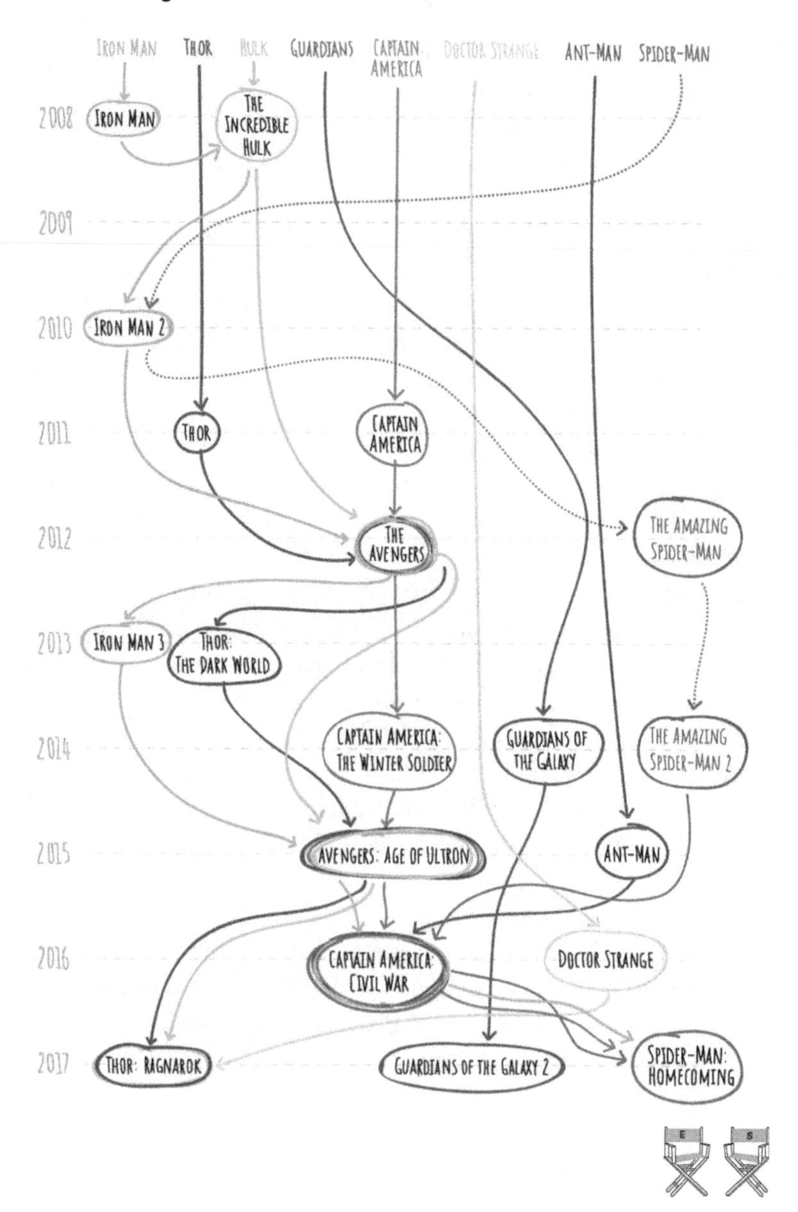

Fig. 9.12 The Marvel Cinematic Universe as a meta-franchise à la 2017
Notes: Authors' own illustration, design by Studio Tense. Characters and brands are trademarked. The line between the Spider-Man character and the film IRON MAN 2 is dotted, because the link was revealed retroactively by Marvel (Peter Parker, before he became Spider-Man, appeared as a child). The 2012 and 2014 Spider-Man films are grey-shaded, because they were produced by Sony and are not considered as parts of the Marvel Cinematic Universe, unlike the Sony/Marvel cooperation SPIDER-MAN: HOMECOMING.

incongruent stories, aesthetics, and production budgets (e.g., some adaptations like THE PUNISHER were made for as little as $33 million, whereas SPIDER-MAN 3 cost nearly eight times that much). Further, several of the films failed creatively or commercially (or both),[246] resulting in a further downgrading of Marvel's assets.

To counter this, Marvel managers David Maisel and Kevin Feige in 2005 crafted the idea of the cinematic universe—a series of self-produced connected and interlinked mainstream films that featured the most popular (and several lesser-known) characters for which Marvel had kept the rights (or bought them back, such as retrieving Iron Man from New Line). This vision was backed by a credit line of half a billion dollars from Merrill Lynch.[247] Each film had to work on its own, but also as a part of the larger joint framework; the stories and characters were developed accordingly. For example, the initial THOR movie was not only about the Nordic god character, but also was a key building block for THE AVENGERS, a film in which Thor would meet with other iconic heroes, such as Iron Man and Hulk (whose characters had been developed in other separate screen ventures).

By featuring the ensemble of heroes, THE AVENGERS was the initial culmination of the cinematic universe concept—a "six-in-one superhero supermovie" (Stork 2014, p. 79). Marvel had carefully laid the groundwork for this joining event. One approach was inserting "mysterious" guest appearances into each individual super hero's films in a way that created a frenzy of consumer speculation and anticipation. For example, Tony "Iron Man" Stark was featured in a post-credits scene in HULK. And S.H.I.E.L.D. agent Nick Fury fanned the flames in a similar scene in IRON MAN by stating that he had come to talk to Tony Stark about the "Avengers Initiative." Further,

---

[246]SPIDER-MAN 3 and GHOST RIDER were among those screen adaptations that were relatively successful in theaters, but received little love from audiences and critics; the films have Metascores of 59 (out of 100) and 35, respectively, and IMDb user ratings of 6.2 (out of 10) and 5.2. Among the films based on Marvel characters that were both disliked *and* considered commercial failures were the FANTASTIC FOUR sequel RISE OF THE SILVER SURFER (with a Metascore of 45, an IMDb score of 5.6, and global revenues of just $289 million, at a $130 million production budget), ELEKTRA (Metascore = 34, IMDb = 4.8, global revenues = $56 million), and THE PUNISHER (Metascore = 33, IMDb = 6.5, box office = $54 million).

[247]Our coverage here is based on Stork's (2014) in-depth analysis of the cinematic universe. As an interesting aside, whereas Merrill Lynch left Marvel quite extensive creative freedom regarding how to spend the money, they had one condition: the movies all had to be rated PG-13, not R (Masters 2016).

Marvel created special featurettes such as the "One-Shot" series that were released as "extras" on DVDs and the Internet.

Since the first collaboration of the heroes in The Avengers, the characters have continued to deal with their own individual challenges. But they have also remained in close contact through a second Avengers movie and more frequent and extensive guest appearances in seemingly stand-alone character films. For example, Iron Man and Black Widow, among others, had significant roles in Captain America: Civil War.[248] Similar to the way that audiences' awareness of and interest in the individual heroes has spilled over to their collaborative efforts in The Avengers (which became the second most successful film of all time globally), the subsequent separate ventures have benefitted strongly from the attention that has been gathered by the ensemble, as a whole. For example, Iron Man 3, released one year after The Avengers, doubled the box-office revenues of its predecessors.

The continued combination (and occasional confrontations) of various heroes played by the same set of actors, together with the use of different directors, enabled a combination of variety (which ensures new sensations) and similarity (which ensures familiarity). This is a tricky balancing act as familiarity also requires being "true" to the source comic book material held in reverence by the most devoted fans. In the case of Marvel's Cinematic Universe, the coherence of this highly complex venture has been tied to the involvement of Kevin Feige, a Marvel superfan himself. Mr. Feige has served as producer for all universe entries and supervised all activities, from scripts to marketing, ensuring that the underlying comics are taken seriously (similar to "sacred texts"; Mr. Feige quoted in Fleming 2016) and preventing "auteurist eccentricity" (Stork 2014, p. 89).

The uber-producer's strategic franchise vision dominates the creative ambitions of his directors for their individual films; he also assures that they have the franchise's key atmospheric ingredients, such as "heart" and "comedy," in addition to "action" (D'Alessandro 2017). Some have thus compared the development of new entries to the Marvel Cinematic Universe to the creation of new meals—with Feige always crafting the recipe, "which is handed to a director, then the parts are churned through a machine of

[248]Even Marvel favorite Spider-Man made his first appearance in the Marvel Cinematic Universe in this film, having been rescued from an exile imposed by licensing deals that started in 1985. The rights for the Spider-Man character are still owned by Sony, but Sony and Marvel figured out mutually beneficial ways to cooperate. For details of the deal between the two firms, see Chitwood (2017).

studio notes, fattened up with threads that connect each film in the universe to each other, and sanded down to ensure global appeal" (Kelley 2017). Let us note that this unique division of labor carries the potential for conflict given the nature of creative talents, as evidenced by the departure of AVENGERS director Joss Whedon from the franchise (Vary 2015).[249] Mr. Feige and his team have made the films the sole core of the universe's narrative; the universe contains other products, like TV series and games, but what happens in those products does not flow back into the films. Taking the "cinematic" in the universe's name by its meaning has prevented the mega-franchise's complexity from becoming unmanageable.

The enormity of Marvel's brand management achievement in building and managing its cinematic universe is made even clearer when one observes other studios' efforts to establish similar meta-franchises. Whereas the management rules for universes are the same as for "ordinary" brand franchises, their multi-brand nature implies an exponentially higher level of coordination and integration. Any individual problem has the potential to escalate and threaten the whole meta-franchise. This became obvious when Warner released its movie BATMAN V SUPERMAN: DAWN OF JUSTICE to jump-start a competing DC superhero universe in March 2016: although the studio spent about $400 million making and promoting the film, the concept lacked a carefully developed array of interconnected stories and films. Instead, BATMAN V SUPERMAN was based on a single prior SUPERMAN film that had itself received only a lukewarm reception and an artistically distinct film series that featured a different Batman actor.

The film also lacked what we have highlighted as a key requirement of any brand extension and franchise: a high level of quality. It was criticized by many, as reflected by a Metascore of only 44 and an IMDb rating of 6.6 (out of 10). At a result of all these factors, Warner's first AVENGERS-like ensemble film JUSTICE LEAGUE (starring Batman, Superman, Wonder Woman, the Flash, Aquaman, and others) then failed to connect strongly with its audiences. Despite at least one more beloved entry to the "universe" (WONDER WOMAN) and despite spendings of roughly $400 million again, the ensemble film opened at under $100 million in North America, which is less than half of what the first AVENGERS movie made in its opening week-

[249]For a more general discussion of the potential conflict of managing franchises and relationships with creative talent, see the section on the state of the entertainment industry in the integrated marketing chapter.

end. JUSTICE LEAGUE was also criticized for its own quality problems (its Metascore was 46). We will see whether the studio can eventually overcome such fundamental teething problems by following the rules of *Entertainment Science* more closely.

What's an Entertainment Brand Worth? Using Econometric Approaches for Measuring Brand Equity in Entertainment

Going Beyond Averages and Subsets: On the Valuation of Individual Entertainment Brands

In the previous sections, we have discussed the results of scholars' exploration of the value of various entertainment brands, ranging from line extensions to category extensions to human brands. These findings, achieved through rigorous statistical approaches, should be insightful for entertainment managers, as they illustrate the financial contributions such brands can provide when applied in entertainment.

A limitation of these findings is that they are all *averages*, either calculated across all variations of a certain kind of brand (such as sequels) or across a more refined subset of this kind of brand through a contingency approach. Although averages are helpful for comparing a strategy (such as producing a sequel) with its strategic alternatives (such as producing an original new product), they are less powerful when a manager has to determine the value of one or more specific brands. For example, a manager may face a decision about whether to obtain the extension rights for a certain book or a particular movie brand. He or she then has to determine for what price, and decide which actor or actress would be best suited for a new movie (along with figuring out how much to pay for his/her talent).

But there is some good news on this front. The investigations into brand value that *Entertainment Science* scholars have undertaken enable entertainment managers to make forward-looking estimates for these finer-grained purposes. Whereas the specifics of scholarly approaches toward the valuation of individual projects differ, the general logic is the same in all cases: the specific parameters of a specific entertainment project are inserted in the general success equations that were determined for a certain type of brand. Doing so then allows the comparison of the financial success of the project to a project with alternative specifications, such as using an alternative parent brand or no brand at all. On the following pages, we will illustrate this logic for the case of movie sequels based on our work in Hennig-Thurau et al. (2009).[250]

We concede that market power, mutual agreement about the "rules of the game," and simple pragmatism ("it works") can be solid reasons for employing established heuristics when it comes to valuing a star, a sequel, or a movie as a whole (such as when selling the TV rights for it). But we are convinced that these reasons should not prevent managers from learning more about the economics of such deals. Such a better understanding might enable a manager to make decisions that his less-informed competitors would not make when rules of thumb made the decisions look unattractive or overly risky, or avoid projects that heuristics suggest to be unprofitable. That's when *Entertainment Science* provides a competitive advantage.

Valuing the Next Spider-Man Sequel

The dominant logic of valuing brand equity is one of differences (e.g., Simon and Sullivan 1993). Consider the case of author J. K. Rowling: whereas the crime novel The Cuckoo's Calling was originally a "slow seller" according to retailers, this changed overnight when the Harry Potter author revealed that she had written the book under pseudonym. When this information broke, the novel immediately skyrocketed to bestseller lists' No. 1 spots (Trachtenberg 2013).[251]

According to brand equity logic, the difference in sales for the "unbranded" book version and the version which was connected to the author's name provides a proxy for the equity of Mrs. Rowling's brand. To determine the full "Rowling brand equity," it would still need to be multiplied by the number of books that the author would write in the coming years and take into account potential contingency effects. One of those contingencies we discussed earlier in this book is the author-image "fit." Whereas Mrs. Rowling is best known for the Harry Potter children-targeted

[250]Readers with a particular interest in this topic might also enjoy learning about a similar approach for the valuation of individual book adaptations we developed in Knapp et al. (2014), our work about the monetary valuation of individual stars for specific movie projects (see Hennig-Thurau et al. 2014), and about the monetary valuation of the international TV rights for individual movies, as illustrated in Hennig-Thurau et al. (2013).

[251]See also our introductory quote in the book's section on the financial impact of stars which describes a similar occurence for Stephen King's book Thinner.

fantasy series, The Cuckoo's Calling was an adult-targeted thriller, which might have limited the effect of her brand in this particular case for some fans. But such "natural experiments" are, of course, quite rare for authors and for other kinds of entertainment brands. So a different approach has to be used to determine brand equity econometrically.

Before we lay out one method that does not require experiments to be run, but is based on historic data, let us take a look whether determining the value of a brand is a relevant matter at all. Maybe fees are trivial and intuitive agreement exists about what would be the "right" price? Ad hoc evidence suggests quite the contrary. Whereas producers Mario Kassar and Andrew Vajna bought the rights to make a third Terminator movie for $14.5 million in the early 2000s (Epstein 2005), a hedge fund paid $29.5 million when the brand was auctioned ten years and two films later. Another year later, the brand was sold to producers Megan and David Ellison for $20 million (Fleming 2011). For the right to turn The Lord of the Rings into a series, Amazon reportedly paid the Tolkien estate $200–250 million—"*just* for the rights, before any costs for development, talent, and production" (Andreeva 2017). We assume that you agree that the mere sizes of these amounts are as impressive as the differences in valuation. Besides, even if a rights holder does not plan to sell its existing intellectual property to others, he or she should know what its equity is, and make extension and franchise decisions accordingly. Knowing the value of such equity could also be an important element in the market valuation of the studio itself.

So, what *is* the "right" price for the Terminator or Lord of the Rings brand then? Based on what we have discussed so far in this chapter on entertainment brands, we argue that for any entertainment brand, its equity (or "extension value") has to account for, first, the success an extension can be expected to achieve, second, the success that could be expected to be realized by a "similar" product that does not contain the brand, third, the risk of both the brand extension and its similar unbranded alternative, and fourth, potential changes in the success of the parent brand, i.e., reciprocal spillover effects.

Figure 9.13 illustrates the roles that these different elements play for the monetary value of an entertainment brand, and how they need to be combined. This combination process requires managers to complete three steps: (1) estimating the revenue differences between the extension and its unbranded alternative (the orange boxes in the figure), (2) adjusting these results for potential differences in risk (the blue boxes), and (3) estimating any reciprocal spillover effects (the grey box). In the following, we discuss what entertainment managers must pay attention to in each of the three steps.

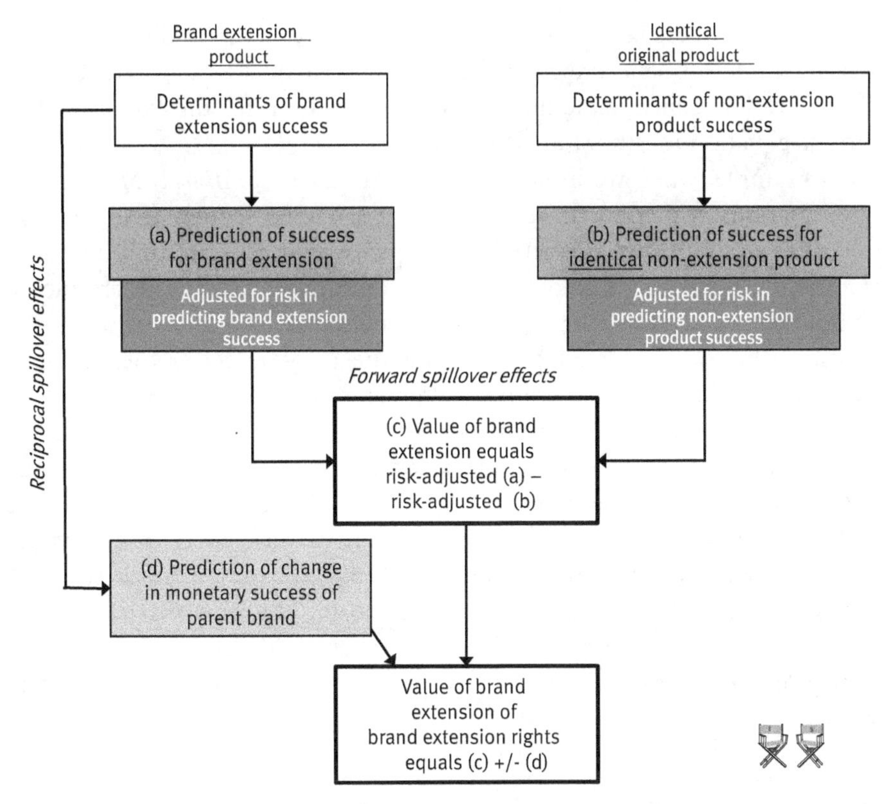

Fig. 9.13 How to determine an entertainment brand's extension value
Source: Reprinted with minor adjustments with permission from Journal of Marketing, published by the American Marketing Association, Hennig-Thurau et al. (2009) Conceptualizing and measuring the monetary value of brand extensions: The case of motion pictures, November 2009, Vol. 73, No. 6, pp. 167–183.

Step 1: Estimating Forward Spillover Revenue Effects

In the first step, separate prediction models (which include the factors that drive a product's performance) have to be developed for both brand extensions and their unbranded alternatives. For unbranded products, the relevant "success drivers" are the factors that we are discussing at length throughout this book, ranging from product elements (such as a product's genre, budget, and star power), to communication (advertising spending), distribution intensity, and price (if relevant). For brand extensions, information on all these factors is critical again, but additional extension-specific factors which we have shown to influence extension success are also available (namely parent characteristics and parent-extension fit). After a producer calculates how these factors influence revenues by estimating different equations for extensions and unbranded alternatives, he or she can insert the specific values for each variable into the success equation for the extension and into the equation for unbranded alternatives.

For example, when we inserted the parameters for SPIDER-MAN 2 in the two equations which we have calibrated for movie sequels (see Hennig-Thurau et al. 2009), we predicted this sequel would generate total revenues of $763 million, $373 million of which would flow back to the producing studio (the rest remains with the theaters). For an unbranded, but otherwise identical film we calculated by inserting the parameters for SPIDER-MAN 2 into the "unbranded" equation that the studio would earn $320 million (from $655 million of total revenues).

The difference between those two values (in this case, $373 million minus $320 million = $53 million) is the "forward spillover revenue" component of the total extension value. It's a raw number, because we do not consider any differences in risk between the sequel and an unbranded alternative. Let us note that these numbers result from using the actual characteristics of SPIDER-MAN 2—all other kinds of variations could also be used (such as using a different star, opening the film in fewer theaters, or having a different age rating).[252]

Step 2: Adjusting Results for Risk Effects

But as we have shown before, the attractiveness of brand extensions in entertainment is not limited to higher average revenues—they are also less risky. So the value we determined in Step 1 should be adjusted for such differences in risk. To do this, we recommend a technique developed by finance scholars—the so-called "Value at Risk" approach. It corrects the expected

[252]For example, when we calculate the value of a SPIDER-MAN sequel without the participation of the original film's star Tobey Maguire, we arrive at 50% lower revenues and, consequently, a negative brand extension value—in other words, making an otherwise similar film without the Spider-Man brand would have made more sense economically. When Sony made the first SPIDER-MAN sequel without Mr. Maguire (i.e., THE AMAZING SPIDER-MAN), the film sold about 34 million tickets in North America, compared to about 60 million for SPIDER-MAN 2—despite having consumed $30 million higher production costs. When asked by the team of The Guardian to demonstrate the approach's validity, we also used our equations to predict the success of NEW MOON, a sequel to TWILIGHT, prior to the film's release. The model suggested North American revenues of $267 million, or just 10% less than what the film actually generated—and $69 more than the model predicted for an unbranded "twin" movie (see Allen 2009).

revenues by using the statistical variation of these revenues, also taking into account a manager's personal risk orientation. The logic here is that for highly risk averse managers, risk correction plays a bigger role than for those who are willing to take higher risks.

We use the standard error of the regression estimate as our measure of revenue variation. When we correct the revenues for risk, the expected revenues drop accordingly to lower levels for both extensions and unbranded films. However, the *difference* between them *becomes larger* because of the higher riskiness of unbranded films (originals) versus branded films (sequels). For the SPIDER-MAN sequel, we calculated that in this case the value of producing an extension increases by between 7% (for a relatively risk-averse manager) and 34% (for someone who is highly risk averse). In monetary terms, for the highly risk-averse manager the total brand extension equity is $71 million.

Step 3: Estimating Reciprocal Spillover Effects

For a fully holistic estimation of an entertainment brand's equity, one also has to acknowledge the existence of reciprocal spillover effects, i.e., how the extension affects its parent. How can this be accounted for econometrically? For this purpose, one can use the regression equation we had calculated earlier in this chapter with the purpose of predicting the abnormal sales for the parent's DVDs that resulted from the extension's release. Like the "forward" equation of Step 1, one can insert the values for *any* movie (actual or hypothetical) into this equation and, by doing so, calculate the abnormal sales for that particular movie.

In the case of the SPIDER-MAN sequel, we found that sales of the parent film's DVDs in North America were about 1.25 million units higher than they would have been without the sequel (as actually produced). We estimate that these additional DVDs represent a financial value of about $15 million for Sony. Again, these revenues can be adjusted for risk using the "Value at Risk" method. Doing so reveals that a highly risk averse producer could expect only about $11 million higher revenues because of abnormal DVD sales of the original SPIDER-MAN.

So in total, our approach calculates the financial value of the SPIDER-MAN sequel rights at ($53 million + $15 million =) $68 when we do not consider potential differences in risk between the sequel and a similar "unbranded" film (i.e., taking a "risk-free" perspective). When adding risk effects and taking a highly risk-averse stance, the financial value of the brand rises to $82 million (=$71 million + $11 million). Let us keep in mind that all these calculations are for a single sequel only, not taking the full franchise value of the brand, and do also not include additional revenue channels, such as DVD rentals and the sale of SVOD and TV rights. But those could be easily added to the calculation using the same approach, given the availability of historic data for the respective channels.

We want to end our discussion of brand equity estimation by clarifying that the addition of the reciprocal effects to the forward spillover value is not a trivial matter. If the producer of the brand extension does not also own the rights to the parent brand, he or she could argue that the reciprocal value should even be *deducted* from, instead of added to, the forward spillover-based brand equity. The argument could be that the more effort he or she puts into the extension product, the more the owner of the parent will benefit from that. For the same reason, some publishers do not sell the extension rights for their novels to those who offer the most, but prefer those instead who can be expected to produce the most promising adaptations—a strategy that pays financially when novels are turned into movies (as we demonstrate empirically in Knapp et al. 2014).

Concluding Comments

Only recently entertainment firms have begun to adopt the brand concept. In this chapter we show that *Entertainment Science* scholars, however, have compiled massive evidence of brands' power when applied to an entertainment context, along with rich insights regarding how the concept can be used most effectively by entertainment managers. Brand logic works for entertainment because consumers' minds do include strongly-held cognitive associations—the hallmark of a brand—regarding entertainment entities.

We overview the two key options of strategic entertainment branding, namely brand integration and brand alliances, and provide evidence regarding their commercial power. Brand integration encompasses different types of brand extensions—extending the brand to new products in the same line (e.g., movie and game sequels, book series, etc.) or into new product categories (e.g., merchandise, theme parks, etc.). Brand alliances deal with the combination of multiple brands, such as the systematic use of stars for movies or other entertainment offers. For each strategy, we spell out what is known regarding the conditions under which such strategies make sense. Considering these conditions can trigger incremental revenues, but also reduce product risk by having more predictable performances. Not all brand strategies are equal when it comes to their financial power; remakes face particular limitations.

We discuss brand franchises, which imply a holistic look on entertainment brands, and their specific and complex requirements that Lucasfilm and Marvel address so impressively while others struggle with them. Managers must shift their focus to the performance of the brand franchise as a whole, versus simply maximizing the short-term payoff of a single product. We end the chapter by introducing our readers to scholarly ways to measure the financial value of an entertainment brand, acknowledging both revenue and risk effects and also the forward and backward spillover effects through which brands create value.

With our look at entertainment product quality and signals of quality (unbranded and branded) now complete, our next chapter dives into one final product-related challenge: how can entertainment managers enable the *continuous* development of high-quality new products which is so essential because of entertainment's short life cycles? Let us see what *Entertainment Science* can teach us in this regard.

References

Addis, M., & Holbrook, M. B. (2010). Consumers' identification and beyond: Attraction, reverence, and escapism in the evaluation of films. *Psychology & Marketing, 27,* 821–845.

Adler, M. (1985). Stardom and talent. *American Economic Review, 75,* 208–212.

AFI (2008). AFI'S 10 Top 10. *American Film Institute,* https://goo.gl/C6t7Zw.

Akdeniz, B. M., & Talay, M. B. (2013). Cultural variations in the use of marketing signals: A multilevel analysis of the motion picture industry. *Journal of the Academy of Marketing Science, 41,* 601–624.

Albert, S. (1998). Movie stars and the distribution of financially successful films in the motion picture industry. *Journal of Cultural Economics, 22,* 249–270.

Allen, A. S. (2012). Has Hollywood lost its way? *Short of the Week,* January 5, https://goo.gl/6fnnj3.

Allen, K. (2009). Mathematicians find the formula for a hit film sequel. *The Guardian,* November 8, https://goo.gl/sM9npB.

Alter, A. (2015). Popular TV series and movies maintain relevance as novels. *The New York Times,* January 4, https://goo.gl/GKjZaj.

Andreeva, N. (2017). 'Lord of the Rings' TV series shopped with huge rights payment attached, November 3, https://goo.gl/jY1T8h.

Asai, S. (2008). Factors affecting hits in Japanese popular music. *Journal of Media Economics, 21,* 97–113.

Austin, B. A. (1989). *Immediate seating: A look at movie audiences.* California: Wadsworth Pub. Co.

Balachander, S., & Ghose, S. (2003). Reciprocal spillover effects: A strategic benefit of brand extensions. *Journal of Marketing, 67,* 4–13.

Barnes, B. (2009). Pixar's art leaves profit watchers edgy. *The New York Times*, April 5, https://goo.gl/jqLKSb.

Barnes, B. (2017). 'Beauty and the Beast': Disney's $300 million gamble. *The New York Times*, March 8, https://goo.gl/LnhPYz.

Basuroy, S., & Chatterjee, S. (2008). Fast and frequent: Investigating box office revenues of motion picture sequels. *Journal of Business Research, 61,* 798–803.

Basuroy, S., Chatterjee, S., & Abraham Ravid, S. (2003). How critical are critical reviews? The box office effects of film critics, star power, and budgets. *Journal of Marketing, 67,* 103–117.

Basuroy, S., Desai, K. K., & Talukdar, D. (2006). An empirical investigation of signaling in the motion picture industry. *Journal of Marketing Research, 43,* 287–295.

Berg, M. (2015). Everything you need to know about the Hollywood pay gap. *Forbes*, November 12, https://goo.gl/5hnqC1.

Bernardin, M. (2016). Marvel, 'Star Wars,' 'Harry Potter' and more: Why the movie star no longer shines as bright as the franchise. *Los Angeles Times*, June 17, https://goo.gl/q63rWC.

Bhattacharjee, S., Gopal, R. D., Lertwachara, K., Marsden, J. R., & Telang, R. (2007). The effect of digital sharing technologies on music markets: A survival analysis of albums on ranking charts. *Management Science, 53,* 1359–1374.

Bohnenkamp, B., Knapp, A.-K., Hennig-Thurau, T., & Schauerte, R. (2015). When does it make sense to do it again? An empirical investigation of contingency factors of movie remakes. *Journal of Cultural Economics, 39,* 15–31.

Bramesco, C. (2016). Where did 'Alice Through the Looking Glass' go wrong? *Rolling Stone*, May 31, https://goo.gl/fCT4uu.

Brands, H. W. (2016). *Reagan: The life.* New York: Knopf Doubleday Publishing Group.

Busch, A., & D'Alessandro, A. (2016). 'Star Trek Beyond' launches to $59M; 'Lights Out' electrifies; 'Ice Age' tepid; 'Ghostbusters' no Cinderella story—box office final. *Deadline*, July 25, https://goo.gl/XyFPHm.

Caulfield, K., Comer, M. T., Concepcion, M., Letkemann, J., Lipshutz, J., Mapes, J., & Trust, G. (2011). The 40 biggest duets of all time. *Billboard*, February 14, https://goo.gl/BgpRtg.

Chitwood, A. (2017). Marvel and Sony 'Spider-Man' rights explained: What's MCU and what's not? *Collider*, July 3, https://goo.gl/v6ngWh.

Chung, K. H., & Cox, R. A. K. (1994). A stochastic model of superstardom: An application of the Yule distribution. *Review of Economics and Statistics, 76,* 771–775.

Clement, M., Steven, W., & Fischer, M. (2014). Empirical generalization of demand and supply dynamics for movies. *International Journal of Research in Marketing, 31,* 207–223.

Conterio, M. (2015). May the farce be with you: The Star Wars Holiday Special they want us to forget. *The Guardian*, December 1, https://goo.gl/XtgJzf.

Cox, J. (2013). What makes a blockbuster video game? An empirical analysis of US sales data. *Managerial And Decision Economics, 35,* 189–198.

D'Alessandro, A. (2016). 'Apocalypse' & 'Alice' take a dive on saturday as memorial day b.o. bloodbath continues—late night update. *Deadline*, May 31, https://goo.gl/6qHmhj.

D'Alessandro, A. (2017). What's the secret to Marvel's b.o. superpowers? A look inside the 'Guardians Vol. 2' superhero hit house. *Deadline*, May 7, https://goo.gl/8FqcJQ.

Demarais, K. (2009). The do's and don'ts of self-titled record albums. *Monkey Googles*, https://goo.gl/ep7pe9.

Dhar, T., Sun, G., & Weinberg, C. (2012). The long-term box office performance of sequel movies. *Marketing Letters, 23,* 13–29.

Epstein, E. J. (2005). Concessions are for girlie men. *Slate*, May 9, https://goo.gl/UCe3JG.

Epstein, E. J. (2010). *The Hollywood economist—The hidden financial reality behind the movies*. Brooklyn: MelvilleHouse.

Fleming Jr., M. (2011). Cannes: Megan Ellison wins 'Terminator' rights auction. *Deadline*, May 13, https://goo.gl/nbA7Jq.

Fleming Jr., M. (2015). 'Star Wars' Legacy II: An architect of Hollywood's greatest deal recalls how George Lucas won sequel rights. *Deadline*, December 18, https://goo.gl/s92zqd.

Fleming Jr., M. (2016). Kevin Feige on 'Captain America: Civil War' and all things Marvel—Deadline q&a. *Deadline*, May 6, https://goo.gl/pJo7BX.

Fleming Jr., M. (2017a). Sean Bailey on how Disney's live-action division found its 'Beauty and the Beast' Mojo. *Deadline Hollywood*, March 21, https://goo.gl/4aLMxM.

Fleming Jr., M. (2017b). Quentin Tarantino hatches 'Star Trek' movie idea; Paramount, JJ Abrams to assemble writers room. *Deadline*, December 4, https://goo.gl/3StjEK.

Follows, S. (2016). How movies make money: $100 m+ Hollywood blockbusters, July 10, https://goo.gl/uYwnJe.

Forbes (2016a). The world's highest-paid actors 2016. https://goo.gl/jMjswJ.

Forbes (2016b). The world's highest-paid celebrities. https://goo.gl/9mVrKi.

Franck, E., & Nüesch, S. (2012). Talent and/or popularity: What does it take to be a superstar? *Economic Inquiry, 50,* 202–216.

Fritz, B., & Schwartzel, E. (2017). Hollywood's misses are hits overseas. *The Wall Street Journal*, June 25, https://goo.gl/RFq956.

Giles, D. C. (2002). Parasocial interaction: A review of the literature and a model for future research. *Media Psychology, 4*, 279–305.

Giles, D. E. (2006). Superstardom in the US popular music industry revisited. *Economics Letters, 92*, 68–74.

Gill, A. (2012). Whitney Houston, the greatest voice of her generation. *The Independent*, February 16, https://goo.gl/gwPTXY.

Gong, J. J., Van der Stede, W. A., & Young, M. S. (2011). Real options in the motion picture industry: Evidence from film marketing and sequels. *Contemporary Accounting Research, 28*, 1438–1466.

Gonzales, U. (2015). 'Captain America: Civil War' screened for Sony execs for Spider-Man clearance. *Heroic Hollywood*, December 9, https://goo.gl/G7Wm6t.

Gopal, R. D., Bhattacharjee, S., & Lawrence Sanders, G. (2006). Do artists benefit from online music sharing? *Journal of Business, 79*, 1503–1533.

Graser, M. (2013). Ubisoft to make movies based on 'Watch Dogs,' 'Far Cry,' 'Rabbids' (exclusive). *Variety*, June 12, https://goo.gl/Ver9K7.

Hamlen Jr., W. A. (1991). Superstardom in popular music: Empirical evidence. *Review of Economics and Statistics, 73*, 729–733.

Hansen, U., Hennig-Thurau, T., & Schrader, U. (2001). *Produktpolitik* (3rd ed.). Stuttgart: Schäffer-Poeschel.

Hayes, D. (2017). 'Logan' director James Mangold: If Fox film fades out post-merger, "that would be sad to me". *Deadline*, December 11, https://goo.gl/bHZ5Pb.

Heath, T. B., Chatterjee, S., Basuroy, S., Hennig-Thurau, T., & Kocher, B. (2015). Innovation sequences over iterated offerings: A relative innovation, comfort, and stimulation framework of consumer responses. *Journal of Marketing, 79*, 71–93.

Hennig-Thurau, T., & Dallwitz-Wegner, D. (2004). Zum Einfluss von Filmstars auf den Erfolg von Spielfilmen. *MedienWirtschaft, 1*, 157–170.

Hennig-Thurau, T., Houston, M. B., & Walsh, G. (2006). The differing roles of success drivers across sequential channels: An application to the motion picture industry. *Journal of the Academy of Marketing Science, 34*, 559–575.

Hennig-Thurau, T., Houston, M. B., & Heitjans, T. (2009). Conceptualizing and measuring the monetary value of brand extensions: The case of motion pictures. *Journal of Marketing, 73*, 167–183.

Hennig-Thurau, T., Fuchs, S., & Houston, M. B. (2013). What's a movie worth? Determining the monetary value of motion pictures' TV rights. *International Journal of Arts Management, 15*, 4–20.

Hennig-Thurau, T., Völckner, F., Clement, M., & Hofmann, J. (2014). An ingredient branding approach to determine the financial value of stars: The case of motion pictures. Working Paper, SSRN.

Hofmann, K. H. (2013). *Co-financing Hollywood film productions with outside investors: An economic analysis of principal agent relationships in the U.S. motion pictureindustry*. Wiesbaden: Springer Gabler.

Hofmann, J., Clement, M., Völckner, F., & Hennig-Thurau, T. (2016). Empirical generalizations on the impact of stars on the economic success of movies. *International Journal of Research in Marketing, 34*, 442–461.

Hofstede, G., Hofstede, G. J., & Minkov, M. (2010). *Cultures and organizations: Software for the mind* (3rd ed.). New York: McGraw-Hill.

Horton, A., & McDougal, S. Y. (1998). *Play it again, Sam: Retakes on remakes.* Berkeley: University of California Press.

Horton, D., & Wohl, R. (1956). Mass communication and parasocial interaction: Observations on intimacy at a distance. *Psychiatry, 19*, 215–229.

Hunter III, S. D., Chinta, R., Smith, S., Shamim, A., & Bawazir, A. (2016). Moneyball for TV: A model for forecasting the audience of new dramatic television series. *Studies in Media and Communication, 4*, 13–22.

Jaafar, A. (2016). Studio Canal Chief Didier Lupfer lays out challenges & opportunities ahead for Europe's major player—Cannes. *Deadline Hollywood*, May 23, https://goo.gl/q1htyu.

Joshi, A. (2015). Movie stars and the volatility of movie revenues. *Journal of Media Economics, 28*, 246–267.

Joshi, A., & Mao, H. (2012). Adapting to succeed? Leveraging the brand equity of best sellers to succeed at the box office. *Journal of the Academy of Marketing Science, 40*, 558–571.

Kamphuis, J. (1991). Satisfaction with books: Some empirical findings. *Poetics, 20*, 471–485.

Keller, K. L. (1993). Conceptualizing, measuring, and managing customer-based brand equity. *Journal of Marketing, 57*, 1–22.

Kelley, S. (2017). 'Guardians 2': Why James Gunn is now the Marvel Cinematic Universe's biggest winner. *Variety*, May 7, https://goo.gl/Cyx8Av.

Knapp, A.-K., & Hennig-Thurau, T. (2014). Does 3D make sense for Hollywood? The economic implications of adding a third dimension to hedonic media products. *Journal of Media Economics, 28*, 100–118.

Knapp, A.-K., Hennig-Thurau, T., & Mathys, J. (2014). The importance of reciprocal spillover effects for the valuation of bestseller brands: Introducing and testing a contingency model. *Journal of the Academy of Marketing Science, 42*, 205–221.

Kotler, P., & Gertner, D. (2002). Country as brand, product, and beyond: A place marketing and brand management perspective. *Journal of Brand Management, 9*, 249–261.

Landau, J. (1971). The Motown story. *Rolling Stone*, May 13, https://goo.gl/WpQ8cf.

Langfitt, F. (2015). How China's censors influence Hollywood. *NPR*, May 18, https://goo.gl/fNtSM9.

Lee, J., Boatwright, P., & Kamakura, W. A. (2003). A Bayesian model for prelaunch sales forecasting of recorded music. *Management Science, 49*, 179–196.

Leemans, H., & Stokmans, M. (1991). Attributes used in choosing books. *Poetics, 20*, 487–505.

Levin, A. M., Levin, I. P., & Edward Heath, C. (1997). Movie stars and authors as brand names: measuring brand equity in experiential products. *Advances in Consumer Research, 24,* 175–181.

Lieberman, D. (2015). Why is Disney's stock price falling as 'Star Wars' breaks box office records? *Deadline*, December 18, https://goo.gl/R9X7Vn.

Lindner, A. M., Lindquist, M., & Arnold, J. (2015). Million Dollar Maybe? The effect of female presence in movies on box office returns. *Sociological Inquiry, 85*, 1–22.

Liptak, A. (2017). Why Hollywood is turning to books for its biggest productions. *The Verge*, January 26, https://goo.gl/fBdN6b.

Liu, A., Mazumdar, T., & Li, B. (2014). Counterfactual decomposition of movie star effects. *Management Science, 61*, 1704–1721.

Lowrey, T. M., Shrum, L. J., & Dubitsky, T. M. (2003). The relation between brand-name linguistic characteristics and brand-name memory. *Journal of Advertising, 32*, 7–17.

Luo, L., Chen, X. (Jack), Han, J., & Whan Park, C. (2010). Dilution and enhancement of celebrity brands through sequential movie releases. *Journal of Marketing Research, 47*, 1114–1128.

Marchand, A. (2016). The power of an installed base to combat lifecycle decline: The case of video games. *International Journal of Research in Marketing, 33*, 140–154.

Masters, K. (2016). Marvel Studios' origin secrets revealed by mysterious founder: History was "Rewritten". *The Hollywood Reporter*, May 5, https://goo.gl/MT816d.

Mathys, J., Burmester, A. B., & Clement, M. (2015). What drives the market popularity of celebrities? A longitudinal analysis of consumer interest in film stars. *International Journal of Research in Marketing, 33*, 428–448.

Mendelson, S. (2013). Trailer talk: 'Robocop' and the problem with remakes. *Forbes*, November 7, https://goo.gl/M96n9z.

Minzesheimer, B. (2004). 10 years of best sellers: How the landscape has changed. *USA Today*, March 10, https://goo.gl/1L7UMz.

Nevins, J. (2011). A brief history of the crossover. *Gizmodo*, August 23, https://goo.gl/4gzzeB.

Oehmke, P., & Beier, L. (2011). All der Mist passiert wirklich. *Der Spiegel*, January 3, https://goo.gl/HtLkQv.

Orden, E., & Kung, M. (2012). Lions Gate hungers for a franchise. *The Wall Street Journal*, February 21, https://goo.gl/nhguVP.

Pähler vor der Holte, N., Gless, F., Knapp, A., Riehl, U., & Hennig-Thurau, T. (2016). What's in a name? Analyzing the influence of brand names on entertainment product success. In *Proceedings of the 44th Academy of Marketing Science Annual Conference* (Vol. 1441). Orlando: Academy of Marketing Science.

Patterson, J. (2008). It's in the name. *The Guardian*, April 5, https://goo.gl/N2tdNL.

Peden, L. D. (1993). The possessive is nine-tenths of the title. *The New York Times*, November 28, https://goo.gl/Lvgk2C.

Pfeiffer, L. (2015). Movie classics: The American westerns of Clint Eastwood. A Cinema Retro special edition magazine. *Cinema Retro*, 2015.

Philips, T. (2015). Why Destiny ditched Peter Dinklage. *Eurogamer.net*, November 8, https://goo.gl/RsRBab.

QPMedia (2013). QP money making stars all years. https://goo.gl/U9ube2.

Ravid, S. A. (1999). Information, blockbusters, and stars: A study of the film industry. *The Journal of Business, 72*, 463–492.

Rohbemed, N. (2015). For Disney, biggest payday on Star Wars won't be at the box office. *Forbes*, December 16, https://goo.gl/j4gJYc.

Rosen, S. (1981). The economics of superstars. *American Economic Review, 71*, 845–858.

Rubin, A. M., & Step, M. M. (2000). Impact of motivation, attraction, and parasocial interaction on talk radio listening. *Journal of Broadcasting & Electronic Media, 44*, 635–654.

Rubin, A. M., Perse, E. M., & Powell, R. A. (1985). Loneliness, parasocial interaction, and local television news viewing. *Human Communication Research, 12*, 155–180. Winter.

Sattler, H. (1999). Markenstrategien für neue Produkte. In F.-R. Esch (Ed.), *Moderne Markenführung* (pp. 337–355). Wiesbaden: Gabler.

Sattler, H., & Völckner, F. (2013). *Markenpolitik* (3rd ed.). Stuttgart: W. Kohlhammer.

Schiappa, E., Allen, M., & Gregg, P. B. (2007). Parasocial relationships and television: A meta-analysis of the effects. In R. W. Preiss, B. M. Gayle, N. Bureell, M. Allen, & J. Bryant (Eds.), *Mass media effects research: Advances through meta-analysis* (pp. 301–314). Mahwah: Lawrence Erlbaum.

Schmidt-Stölting, C., Blömeke, E., & Clement, M. (2011). Success drivers of fiction books: An empirical analysis of hardcover and paperback editions in Germany. *Journal of Media Economics, 24*, 24–47.

Schwartzel, E. (2016). Hollywood under pressure to put more chinese actors in the spotlight. *The Wall Street Journal*, September 19, https://goo.gl/BvqSWp.

seekingalpha.com (2014). The Walt Disney company's management presents at Goldman Sachs Communacopia Conference—transcript, September 10, https://goo.gl/HWBLUx.

Simon, C. J., & Sullivan, M. W. (1993). The measurement and determinants of brand equity: A financial approach. *Marketing Science, 12*, 28–52.

Smiley, J. (2016). Touch up your Shakespeare: Anne Tyler recasts 'The Taming of the Shrew' for our time. *The New York Times*, July 6, https://goo.gl/akj9km.

Smith, E. (2011). Disney's 'Cars 2' a hit already—in stores. *The Wall Street Journal*, June 20, https://goo.gl/ag374V.

Sood, S., & Dréze, X. (2006). Brand extensions of experiential goods: Movie sequel evaluations. *Journal of Consumer Research, 33*, 352–360.

Spence, N. (2007). Steve Jobs directs Disney. *Macworld*, June 25, https://goo.gl/3RZetW.

Squire, J. E. (2006). Introduction. In J. E. Squire (Ed.), *The movie business book* (International 3rd ed., pp. 1–12). Maidenhead: Open University Press.

Stedman, A. (2014). Kevin Hart responds to Sony 'Whore' comment: 'I Protect My Brand'. *Variety*, December 11, https://goo.gl/3bTLDv.

Stork, M. (2014). Assembling the Avengers: Reframing the superhero movie through Marvel's Cinematic Universe. In J. N. Gilmore & M. Stork (Eds.), *Superhero synergies: Comic book characters go digital* (pp. 77–96). Lanham: Rowman & Littlefield.

Streib, L. (2009). Why James Patterson is worth $150 million. *Forbes*, September 9, https://goo.gl/n4Fzfq.

Suderman, P. (2016). Hollywood is stuck in a bubble of expanded movie universes. It's time for it to pop. *Vox*, January 27, https://goo.gl/u8zAap.

Sugarman, J. (2011). The right and wrong ways to name a movie. *Salon*, February 11, https://goo.gl/7Bo7kb.

Swertlow, M. (2017). You're reading this correctly: A Hunger Games theme park is coming. *E! Online*, August 15, https://goo.gl/PfTT56.

Taylor, C. (2015). 'Look at the size of that thing!': How Star Wars makes its billions. *The Telegraph*, May 4, https://goo.gl/Ghu8mu.

The Deadline Team (2013). Steven Soderbergh's state of cinema talk. *Deadline*, April 30, https://goo.gl/3md7zK.

The Hollywood Reporter (2016). Hollywood's 100 favorite movie quotes, February 24, https://goo.gl/aj1XuV.

Thomson, M. (2006). Human brands: Investigating antecedents to consumers' strong attachments to celebrities. *Journal of Marketing, 70*, 104–119.

Trachtenberg, J. A. (2013). Rowling's second adult novel flies off shelves. *The Wall Street Journal*, July 14, https://goo.gl/37iQcT.

Van der Stede, W. A. (2015). Hollywood studios appear to plan sequels before they produce the original movie, September 10, https://goo.gl/f6jCKa.

Vary, A. B. (2015). Joss Whedon's astonishing, spine-tingling, soul-crushing Marvel adventure! *Buzzfeed*, April 21, https://goo.gl/Rma2u7.

Wallace, W. T., Seigerman, A., & Holbrook, M. B. (1993). The role of actors and actresses in the success of films: How much is a movie star worth? *Journal of Cultural Economics, 17*, 17–27.

Wirtz, B. W., Mermann, M., & Daier, P. (2016). Success factors of motion picture actors—An empirical analysis. *Creative Industries Journal, 9*, 162–180.

Wohlfeil, M., & Whelan, S. (2008). Confessions of a movie-fan: Introspection into a consumer's experiential consumption of 'Pride and Prejudice'. *Proceedings of European ACR Conference*, 137–143.

Wohlfeil, M., & Whelan, S. (2012). 'Saved!' by Jena Malone: An introspective study of a consumer's fan relationship with a film actress. *Journal of Business Research, 65*, 511–519.

Wookieepedia (2017). List of comics. https://goo.gl/Ts5vZw.

Xenopedia (2017). Predator 2. https://goo.gl/atfCNR.

Yodasdatapad (2016). Star Wars book list, December 16, https://goo.gl/q8A6Ex.

Zhao, E. Y., Ishihara, M., & Loundsbury, M. (2013). Overcoming the illegitimacy discount: Cultural entrepreneurship in the US feature film industry. *Organization Studies, 34*, 1747–1776.

10

Entertainment Product Decisions, Episode 4: How to Develop New Successful Entertainment Products

"Innovation (n). The introduction of something new. A new idea, method, or device."
—Merriam-webster.com/dictionary/innovation

Our discussion in Part I of this book about the short life cycles of entertainment products made a strong case that continuous innovation is critical for firms in this industry. At the same time, the unique characteristics of entertainment products demand approaches to innovation that address these particularities, such as consumers' holistic judgment of (hedonic) entertainment products and the critical role of artists and creatives for the development of new entertainment content. These characteristics are behind the industry's traditional skepticism regarding systematic approaches toward the management of innovation in entertainment, an attitude well in line with the industry's "Nobody-Knows-Anything" mantra.

In contrast to such industry skepticism, we are convinced that entertainment innovations can indeed be managed systematically and quite powerfully. Just look at the track record of animation producer Pixar, whose films have been a remarkable string of successes. The firm's first 18 full-length releases have generated more than $13.2 billion globally in theaters alone (in 2017 value), with an average of more than $700 million per film, and not a single one of their films has made less than $300 million just in ticket sales.

Ronny Behrens co-authored this chapter with us.

T. Hennig-Thurau and M. B. Houston, *Entertainment Science*,
https://doi.org/10.1007/978-3-319-89292-4_10

We find it hard to believe that such stream of successes could be attributed to mere luck—particularly when considering that 12 of the 18 films were *not* based on an existing brand, beyond the firm itself. And we are not the only ones who think so; Pixar has become a role model for innovation management within and even beyond film and entertainment (e.g., Catmull 2008). The firm's innovation skills were a main reason why Disney paid more than $7 billion for Pixar in 2006—and indeed, through sharing personnel and creative resources, Disney has been able to revitalize its own animation division, which has crafted a string of recent hits on their own, including as FROZEN and ZOOTOPIA (Lussier 2016). Remember that *Entertainment Science* is all about managing the probability of success, and the way Pixar handles innovations obviously influences this probability.

For innovation to occur, there is always a continuum between how much the innovating firm relies on collaboration among a team and, at the other extreme, on individual creatives. The relative importance of collaboration and individuals varies over the innovation process, but also across the different forms of entertainment. Whereas movies and games put more weight on collaboration because of the mere scale and scope of the projects (i.e., bringing a full-blown video game or blockbuster film to market), the individual takes on a somewhat greater weight in the creation of books and music. But even for the latter forms of entertainment, creating a new product is almost always a long process along which those in charge must get numerous ideas right, and collaboration plays an important role.

In any case, we agree with Pixar's Ed Catmull (2008, p. 4) that for creating great entertainment, creativity "must be present at every level of every artistic and technical part of the organization." In what follows, we will take a look at issues at different levels to examine how entertainment firms can be successful through innovation. At the strategic level, we ask what the right environment looks like, what level of innovativeness works best, and whether firms should engage in innovation themselves ("doing it in-house") or partner with others. At the cultural level, we look for values that support effective innovation processes. And at the organizational level, we investigate the people and the structures that are required to make innovation happen.

We complement this firm-level discussion with a detailed look at what *Entertainment Science* can tell us about product-level innovation decisions, with a particular focus on managers' ability to accurately forecast the success of new entertainment projects at different points of the innovation process, a challenge to which *Entertainment Science* scholars have dedicated substantial effort. We will show that systematic forecasting approaches can be powerful alternatives to "Nobody-Knows-Anything" thinking when testing early product concepts and also when optimizing those concepts in later production stages.

The Strategic Dimension of Entertainment Innovations

> "I don't make pictures just to make money. I make money to make more pictures."
> —Walt Disney *(quoted in IMDb 2017)*

If innovation is the creation of something new, and entertainment product firms need a continuous supply of new products, then managers of entertainment firms need a systematic innovation management process that coordinates the planning, implementation and control of innovation-related actions—to ensure that meaningful innovations continue in a sustainable flow. Even among non-entertainment firms, there is huge variation in the ways innovation activities are handled.

A strategic approach toward new products builds on the distinction between creativity and innovation—whereas the former describes coming up with novel ideas, the latter means the process of actually *implementing* a creative idea and turning it into something marketable (see Amabile 1996). Unless supported by a wealthy benefactor (or the ultra-rare venture capitalist who expects no return on investment), entertainment firms, like any other businesses, are required to generate net positive revenues. Being creative is nice, but the firm will not make money without turning that creativity into commercially successful innovations.

Regarding the strategic dimension of innovation management, an entertainment firm must make decisions regarding three fundamental issues: (1) its innovation goals, (2) the aspired degree of innovativeness of the new products, and (3) its own role in the innovation process (versus the role of other companies). Let's take a look at the options entertainment managers can chose from for each of these issues.

Artistic Versus Economic Innovation Goals

For sustainable innovation management, managers always have to find a balance between the goals they are trying to achieve with new products (usually a combination of economic and artistic goals), the cost of innovation associ-

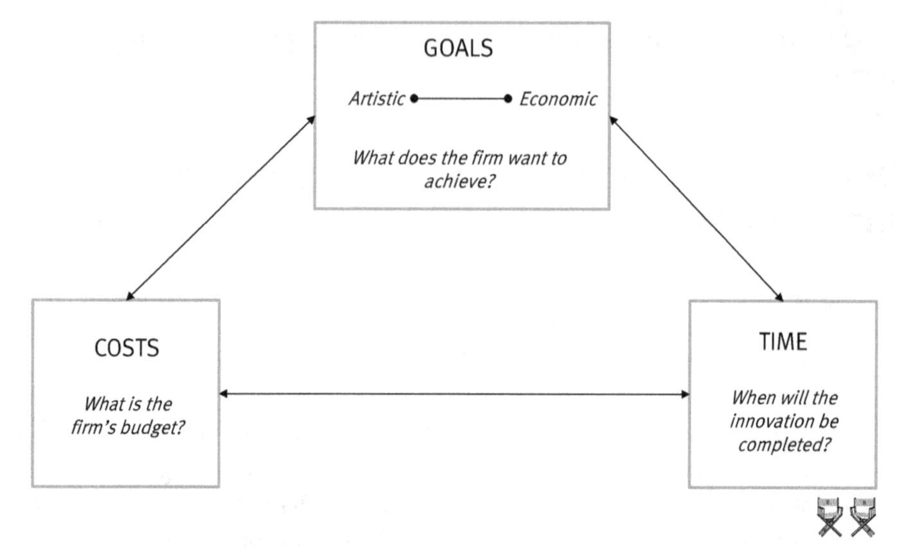

Fig. 10.1 Strategic innovation management requires balancing between three strategic considerations
Note: Authors' own illustration.

ated with reaching those goals, and the time it requires to develop the innovation (see Fig. 10.1).

Although cost and time are widely applicable to entertainment,[253] the desired goals are where we confront the intricacies of entertainment. Economic goals and artistic goals are usually both aimed for by entertainment companies (although not necessarily by the same people in the firm), but they oftentimes clash—something that can be linked to our earlier discussion of taste differences between mass consumers and cultural experts. So before focusing on the "nuts-and-bolts" of how to manage entertainment innovations successfully, entertainment managers first have to determine how to weigh economic and artistic goals.

The tension between artistic and economic goals is inherent for creative products, and it is not difficult to note examples of this clash in firms that produce movies, books, music, and video games. Some entertainment producers aim to avoid this tension by clearly favoring one goal over the other. For example, Israeli cousins Menahem Golan and Yoram Globus, when establishing their Cannon Films in the late 1970s, clearly emphasized economic

[253]Let us note that with regard to the timing dimension of innovation, some particularities also exist for entertainment—we discuss them in our chapter on distribution decisions.

goals. Their business model was to produce low-budget exploitation films for which quality did not matter, or was even a hindrance, according to their argument that "[i]f you make an American film with a beginning, a middle and an end, with a budget of less than five million dollars [the equivalent of about $15 million today], you must be an idiot to lose money" (quoted in Slifkin 2014). But in entertainment, such an approach carries some problems, which eventually led to Cannon's demise—we return to their sad, but instructive story a little later.

Other firms have instead emphasized the artistic aspects. For example, the founding of United Artists was rooted in the belief that innovation decisions in entertainment should be made by the creative artists themselves (see Kehr 2008, Thomson 2008). The firm was founded by four Hollywood heavyweights in 1919: director D.W. Griffith and actors Charles "Charlie" Chaplin, Douglas Fairbanks, and Mary Pickford, who built a distribution company for their own productions, independent of other commercial interests. However, although the creatives had ambitious artistic goals and also experienced strong initial success, their company faced harsh economic realities and volatile profitability when radical artistic visions produced at high costs flopped heavily.

A key event was the collapse of the film Heaven's Gate by high-profile director Michael Cimino. It suffered from the combination of escalated costs ($44 million in production costs alone), terrible reviews, and a bad reception by the viewing public (just $3.5 million in domestic box office). Heaven's Gate ended up as a main contributor to the demise of the United Artists studio (Barber 2015)—it did not help that the film is today held in high esteem by many experts and audiences, including one of this book's authors. Further evidence that having a creative in the driver's seat of an entertainment company is far from a guarantee for long-term success is the story of DreamWorks. The movie studio, co-founded in 1994 and led by acclaimed director Steven Spielberg to enable talented filmmakers to pursue "their more personal projects" (Russell 2004, p. 233), "struggled for financing—and for hits—for much of its existence" (Masters 2016).

Between the extremes of being fully driven by either economic goals or artistic goals, a continuum exists. Our review suggests that some of the most successful entertainment companies have fleshed out an approach near the center of the continuum, fusing artistic ambitions with economic considerations. The reason is that although artistic goals are crucial for creating powerful entertainment products, the opportunities for the artist to *continue* creating new products will be limited without business discipline. Similarly, a focus on business without honoring the relevance of artistic creativity is also doomed in a field where consumers are, at least to a certain extent, driven by creative inspirations. Thus, the uneasy symbiosis between the two kinds of goals should be accepted and managed.

Brad Bird, a director *and* producer at highly innovative *and* successful Pixar, praises this coalescence of commercial and artistic ambitions. He paraphrases Walt Disney's statement we quoted at the beginning of this section, stating that he wants his films "to make money, but money is just fuel for the rocket. What I really want to do is to go somewhere. I don't want to just collect more fuel." Because most, if not all artists will share Mr. Bird's motivation, "making money can't be the focus … for imagination-based companies to succeed in the long run" (quoted in Rao et al. 2008).

Firms that effectively balance artistic and economic goals employ approaches that usually entail the close cooperation of creatives and business managers (e.g., directors and producers), rather than a hierarchy in which one "class" rules over the other. Ed Catmull, Pixar's president, argued that without an understanding of economic realities, resource demands by creatives can be almost unlimited as people can always justify spending more money and time in order to "make a better movie" (Catmull 2008). And while creatives generally understand that money and time are not unlimited, they often do not fully appreciate the costs of various processes. Pixar addresses this via a "Popsicle" approach: starting with a number of Popsicle sticks that represents their total capacity (one stick is a "person-week"—what a typical creative could normally accomplish within a week), creatives and managers work together to allocate the sticks across the elements of the production process (e.g., a certain character in a film). Then, when creatives later ask for additional time and/or more money to devote to one element, the creatives and managers work together to identify other elements from which an equivalent number of sticks can be taken away—in order to stay on time and on budget.

When used in such a way, Catmull and Wallace (2014) argue that budget constraints can be actually facilitating creative processes, as they make people think differently and come up with creative solutions.[254] But they clarify that they can also hurt the work of creatives massively if used differently: when managers of the "Disney Oversight Group," in the mid-2000s, gave the economic goals stark priority and micromanaged even the smallest

[254]Think of the wonderful animated travel map in Raiders of the Lost Ark—which the film's director Steven Spielberg invented "[t]o save money" (*Total Film* 2006).

aspects of production, they impeded the innovation process by robbing creatives of their required freedom.

In sum, the entertainment industry provides examples that both the economic approach and the artistic approach can work for a certain period if done right, but such unbalanced approaches are unstable and tend to collapse at a certain point. In contrast, a purposefully chosen balance between artistic freedom and the careful management of resource constraints (among managers *and* creatives) can be a key to success.

The "Right" Degree of Innovativeness

A second strategic question is: how innovative should an entertainment firm's new products be? According to the sensations-familiarity framework, consumers love familiarity, but can also become satiated by it; at the same time, they love sensations, but only when those sensations trigger the "right" emotions and images. This demand perspective can be matched with a "supply perspective": whereas artists love to indulge their creativity, their radical creations can create new sensations that, even if great pieces of art, are too demanding and radical for consumers to fully appreciate. Meanwhile, managers may crimp the style of artists by preferring to invest in less radical (and more familiar) projects. Thus, the "how innovative?" question is somewhat linked to the prior strategic question about artistic and economic goals.

Innovation research has framed the "degree of innovativeness" challenge as one of *exploitation* versus *exploration*—the former concept describes making use of existing assets when creating new products, whereas the latter refers to the pursuing of new ideas and intellectual properties (March 1991). We have previously shown that the management of entertainment brands, as a case of exploitation, at least at certain points in their life cycle, requires the infusion of new creations, a kind of exploration. But a bigger question is how to optimally manage the brands a producer owns: it deals with the allocation of an entertainment firm's resources *in general*.

So, what does the theory of exploration-exploitation tell us about the "ideal" level of innovativeness? We can learn from it that over-reliance on exploitation can reduce the firm's ability to discover entirely new opportunities, leaving it trapped at a "suboptimal equilibria" (March 1991, p. 71), and can make the firm vulnerable if the market environment changes in ways such that the existing assets (which have been effective for earlier products) become obsolete or less attractive (Greve 2007). Exploitation in entertainment essentially means to use existing brands and products and to focus on extensions within existing categories and with existing brands.

Whereas our discussion above has shown that this can mitigate a variety of risks and be a viable strategy for entertainment companies on a product level, a starkly exploitative strategy, if practiced on the company level, can be a dangerous strategy for an entertainment firm in the longer run for two reasons. First, in the words of John Lasseter, "[s]equels are financially less risky. But if that's all we did, we would become creatively bankrupt" (as Chief Creative Officer for Pixar's and Disney's animation studios; quoted in Franklin-Wallis 2015). What he means is that an over-reliance on entertainment product extensions (not necessarily only sequels, but also other kinds of low-innovative new products) would have an adverse effect on a firm's creatives, who suffer mightily from producing too much of the same content. The departure of the best talent would set off a negative creativity spiral, further reducing a firm's ability to offer sensations to consumers.

The second argument against an "exploitation-only" approach in entertainment is linked to consumers' reactions: because of the satiation embedded in entertainment products and brands, the firms' assets will eventually lose value over time. We have discussed ways to mitigate such value reduction in our brand management chapter, but in the longer run, even the best brand management skills cannot fully counter the satiation inherent in all entertainment and the negative effects it has on revenues. Consumers will eventually grow tired of consuming the "same old, same old." Besides, this is what we observe these days at the entertainment-industry level where exploitation has become the norm—a tendency we consider a threat to the industry as a whole.

In contrast, exploration-based innovations are more likely to be radical, using technologies or pursuing markets that are new to the firm. Innovations of these types are high-risk/high-return: they can produce spectacular market breakthroughs, but because of the high levels of uncertainty involved in exploration-based innovation (de Ven et al. 1999), firms with too heavy reliance on exploration run the risk of incurring the high costs of constant exploration without gaining sufficient net benefits. What United Artists and also DreamWorks did might fall into this category—pointing at the systematic nature of the firms' economic evolution.

Organizations scholar James March (1991) and others conclude that it is wise to try for a delicate balance between the two types of innovation in order to reap the benefits of each while avoiding the traps. Consistent with such scholarly insights, Mr. Catmull has decided that his firm will produce both original films and sequels. "[Originals] are high-risk ideas. So in order to take the high risks, which is very important to us, then we do things which are lower risk. We have to make sure we're also smart as a business."

Let us note, however, that balancing is not trivial from an organizational perspective. The problem with doing so is that the two types of innovation are not easily compatible, because "the mindsets and organizational routines needed for exploration are radically different from those needed for exploitation" (Gupta et al. 2006, p. 695). Exploitation requires tight control around current strategy and current organizational practices—whereas exploration instead eschews tight control and cannot emerge without greater emphasis on discovery (Dougherty and Heller 1994). So, how can the unbalanceable be balanced nevertheless?

Innovation scholars have pointed to two potential paths forward to address this challenge—we illustrate them in Fig. 10.2. One option (shown in Panel A of the figure) is to do both exploration and exploitation at the same time. Doing so has been labeled "ambidexterity," drawing on the analogy of a person who can perform tasks with either hand (Gupta et al. 2006). As successful innovations emerge from the frequent experiments in the firm's exploration unit, those products can be passed along to the exploitation unit for ongoing success. The exploitation unit has a useful supply of new ideas without being distracted by constant experimentation and failures.

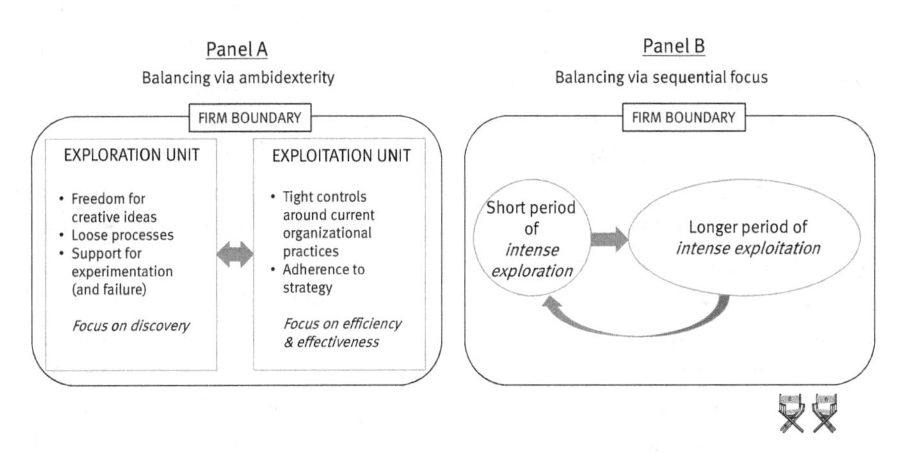

Fig. 10.2 Balancing exploration and exploitation via ambidexterity and sequential focus
Note: Authors' own illustration.

For this approach to work, it is important that the tightly controlled processes that are necessary in the exploitation unit not be allowed to infect the culture of discovery in the exploration unit (Benner and Tushman 2003). Ambidexterity tends to work best when the work of the two cooperating units can be done independently of *other* units (meaning that the firm is large enough to support multiple autonomous units). Disney has acted in line with the ambidexterity approach to innovation on a company level by having kept the originals-focused Pixar a distinct division after purchasing it in 2006, instead of merging it with either Disney Animation (which are put in charge for exploitations of Pixar brands beyond sequels, such as PLANES to CARS) or Disney Studios (which are focusing on exploitation by systematically remaking Disney's original brands such as THE JUNGLE BOOK).[255]

The other option (which we illustrate in Panel B of the figure) is for a firm to switch its attention from one focus to the other at different points in time (i.e., decoupling across time or sequential focus, Gupta et al. 2006). Under this model a firm engages in short periods of intense exploration to create a slate of new innovations. At some point, the firm then shifts to a longer period of exploitation in which those new innovations are fully refined and leveraged in the marketplace. The firm can repeatedly iterate through these stages, over time. If the firm in question is essentially one large system that cannot be easily subdivided into autonomous units, then "sequential attention to exploitation and exploration" is required (Gupta et al. 2006, p. 698).

This is what happens at Disney at the Pixar level: the firm aims for this balance by targeting about two original films for every sequel they produce. With risk-and-return expectations being different between exploitations (i.e., sequels) and explorations (i.e., originals), this balancing approach also resembles ideas we presented in our discussion of managing risk at the slate level, where we concluded that it is important for a firm to balance the risk and returns to their offerings.

In sum, innovation leaders must manage high-wire acts, balancing familiarity against sensations for the consumer's experience and, along with that, balancing risk and freedom for the firm and its artistic creatives. They have to decide about the appropriate ratio of exploration and exploitation to employ in their production slate—and how to accomplish that balance. Our arguments, inspired by innovation theory, have shown that the decision is consequential, as it impacts the attitude of creatives, the responses of consumers, and the risk profile of a firm's slate of innovations.

[255]Other reasons, such as culture, also contributed to this decision—we will get back to them.

Make, Cooperate, or Buy?

A third strategic issue that entertainment firms face when developing innovations is deciding upon the degree to which these innovations are built solely with in-house capabilities—or with assets or capabilities purchased from other firms in the market. This issue is the entertainment version of the classic "make-or-buy" dilemma that economists have wrestled with since the seminal works of Nobel Prize recipients Coase (1937), Arrow (1962), and Williamson (1985). Their foundational works take a "transaction costs" approach that advocates for governing an activity in a way that most efficiently deals with uncertainty and the risk incurred because of that uncertainty (e.g., Walker and Weber 1984). A main insight of the theory is that uncertainty and risks of cooperation require firms to safeguard their interests, which implies the need to evaluate the "relative … merits and demerits of make-or-buy options" (Kurokawa 1997, p. 124). When transaction costs are too high, it is better to "make" the output yourself, but when these costs are lower, a "buy" approach (i.e., cooperation with a partner) would be advantageous.

Early research on transaction costs economics focused on more traditional manufacturing firms, but since then extensive work has found support for the basic tenets of this theory in information technology (e.g., Poppo and Zenger 1998) and high-technology R&D applications (e.g., Kurokawa 1997)—both applications that are somewhat more akin to entertainment innovation. In practice, entertainment innovations are produced across the entire gamut of "make-or-buy" options—let's take a quick look at each option in theory and practice, applying *Entertainment Science* thinking to identify their pros and cons for entertainment product innovations.

The "make" option. The first alternative, and the one often preferred by artists, is to innovate internally, doing everything—from raw idea to finished product—using internal resources. The advantages of such a "making" approach include having complete creative and managerial control over the innovation. The firm's innovation capabilities are improved as each innovation fosters learning, and the firm retains full control of any new intellectual property created for further exploitation. (Be reminded that learning advantages can only be realized in a culture that sees value in such learning—but not in one that honors the "Nobody-Knows-Anything" mantra.)[256] However, there are downsides too—they include "owning" all technological and market

[256]Please see our link between learning and the Goldman mantra in our introduction to this book.

risks in production, while requiring significant investments of time and managerial bandwidth when the innovation is created.

Pixar has always been a distinct proponent of the maker approach, having never bought scripts or ideas from others. "All of our stories, worlds, and characters were created internally by our community of artists" (Catmull 2008). For the firm, the learning that comes from exploring new things on your own has been a main reason: "in making these films, we have continued to push the technological boundaries of computer animation, securing dozens of patents in the process." In the world of video games, Andy Gavin and Jason Rubin, the founders of games producer Naughty Dog, have worked together creating original material since they were 12-year olds. Now leading a team of strong creatives and technical whizzes, Naughty Dog, being part of Sony Interactive Entertainment since 2001, has internally created very successful franchises for the PlayStation console, including THE LAST OF US and UNCHARTED (Moriarty 2013).

The "buy" option. The second alternative is to buy innovations that have already been produced by others. For example, Amazon, Netflix and other distributors, in addition to their own creative endeavors and licensing activities, often buy property rights to promising movies at (or even before) film festivals and markets. For example, Amazon paid $10 million for MANCHESTER BY THE SEA at the 2016 Sundance (Ingram 2017). When Vivendi-owned Studiocanal bought Michael and Rainer Kölmel's Kinowelt, a major reason was Kinowelt's back catalogue of several thousand film titles (Kirschbaum and Hopewell 2008). And in the video game world, Microsoft acquired MINECRAFT, one of the best-selling games to date with over 120 million copies sold as of 2017, when taking over Mojang, the Swedish indie-game developer that created it (Jones 2014).

An upside of acquisitions is that the quality of a finished product tends to be easier to evaluate and, thus, customer appeal and commercial success can be forecasted more reliably.[257] And there is a potential advantage regarding costs, too—by buying a completed product, the firm foregoes the uncertainty and investments of time and managerial effort that would be required to make innovations (for example, HEAVEN'S GATE eventually consumed three-times its scheduled production costs of around $10 million in 1980—a major contributor to United Artists' troubles). However, a major disadvantage of the "buy" option is that other potential suitors can spot high quality innovations, too, and the increased competition

[257]We discuss the role of timing for success predictions later in this chapter.

can drive costs up considerably, while reducing the likelihood of winning a desired property—something that can be observed today at all major festivals.

Further, depending upon the deal, the seller may impose restrictions on further exploitation of the property by the buyer. And, if the movie or game is acquired as an essentially "finished" product, the buyer may not have the ability to make desired tweaks to content or appearance, along with his or her own company brand image. Finally, an over-reliance on buying finished products may cause the buyer's own in-house innovation capabilities to stagnate, as creatives spend their time polishing up someone else's artistic creation (while potentially browsing for other, more stimulating, employment opportunities).

And finally: *the "cooperate" option*. Innovations usually do not come solely from "100% make" or "100% buy"; firms can make some innovations (or elements of an innovation) and buy others. Further, just as manufacturing firms can enter joint ventures and other alliances, entertainment firms can co-develop innovations with outsiders. This is the traditional Hollywood approach in which a deal on an idea is made before it has gone into production. The studio then partners with the producer (and often also various distributors) to make and market a new film. And the new players in entertainment also act this way. For example, Netflix bought Martin Scorsese's IRISH MAN, complete with a deal to star Robert De Niro, before it went into production (McNary 2017). The pros and cons of this cooperating option are, just like the strategy itself, a blend of those for make and buy. The buyer gives up some creative and managerial control and accepts more dependence on (the ideally skillful) outsiders. The buying firm might also have to share revenues in the end, depending on the kind of cooperation deal, but it shares risks and, more importantly, gains access to creative content, unique capabilities, and talents that it otherwise would not have.

In sum, it is readily apparent that these three options are not "all-or-nothing" alternatives. Larger firms can combine make with buy actions in their portfolio of projects; whereas Microsoft's use of the MINECRAFT game was a clear case of "buy," it also purchased the right to "make" extensions of the game on its own, which it has done since the acquisition. In a similar way, film studios often blend their own projects with movies made by others that they buy at festivals and markets. It is critical to examine the relative tradeoffs that come with each solution in light of the entertainment firm's goals, from both a transactional as well as a portfolio perspective.

Some Threats to Systematic Innovation Management in Entertainment

Although some successful entertainment companies have adopted a strategic approach to innovation management, there are threats that can impede the long-term implementation of strategic innovation in entertainment. Figure 10.3 names four such threats—pieces of cheese that can bait entertainment managers into the "innovation mousetrap," if you will. We discuss them next.

The "Artistic Temptation" Threat

A temptation entertainment managers are confronted with is a growing desire to be affiliated with, and respected by, shining artists and stars. When managerial vanity gets paired with a craving for prestige, a manager's economic goals can become overwritten by the artistic value system, reconfiguring the firm's strategic goals. The firm begins to strive for the Oscars (or for just being part of the "star system"), regardless of financial implications.

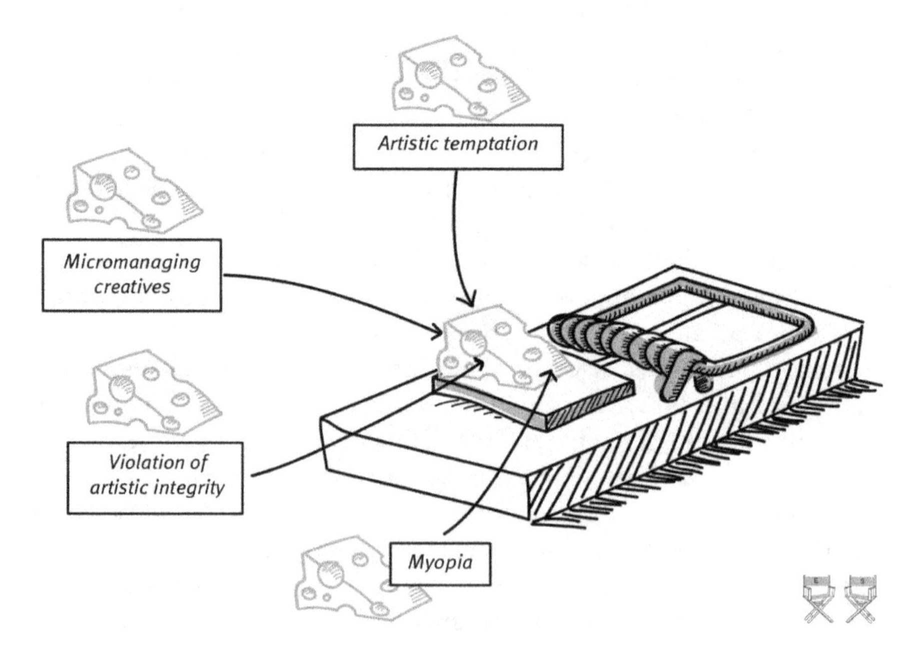

Fig. 10.3 Four threats to systematic innovation in entertainment
Note: Authors' own illustration, design by Studio Tense.

This temptation mostly plays out in smaller ways, but history shows that it can be big enough to bring down a whole firm. Let's take a closer look at the downfall of Cannon Films. As we mentioned earlier, the firm's original business model was to produce low-cost, high-action films with low-to-zero artistic ambitions. However, after having experienced enormous success with that model for a number of years with films such as MISSING IN ACTION, HERCULES, and AMERICAN NINJA (and their numerous sequels), the firm's founders became increasingly "eager for prestige" by the mid-1980s (Howe 2013). They no longer wanted to sit at the side table, but to be invited to the big parties, together with the cultural icons of the time.

To achieve this, they dumped their "$5-million-max!" business model for pursuing deals with prominent directors (e.g., Martin Scorsese) and big-name actors such as Dustin Hoffman and Al Pacino, offering some of the biggest paychecks ever seen at that point in time; they also purchased iconic character brands such as Superman and Spider-Man (Pond 1986).

When Cannon paid Sylvester Stallone a then-unprecedented $12 million for his Rocky-like role in the arm-wrestling drama OVER THE TOP ("This time he's fighting with his bare hands," the trailer touted), Mr. Golan and Mr. Globus' craving for *artistic* appreciation was clearly a major motive, and it was not by accident that Mr. Golan put himself in the director's chair for the film. In their quest for awards and respect, the studio also produced radical projects by cultural elitists Jean-Luc Godard, John Cassavetes, and Norman Mailer. The studio lost a lot of money on both their high-budgeted projects and on their "artistic" projects. Figure 10.4 contrasts some of Cannon's initial films with productions after its executives had begun to go for artistic recognition instead on money.

In the end, it was the artistic temptation which led the Israeli cousins astray from their formula—to lose everything (Howe 2013). But the fate of Cannon is certainly not the only case where the artistic dimension of entertainment has influenced business decisions. For example, Comcast CEO Brian Roberts only recently stated that working with director Steven Spielberg has more to offer for him than movies: "It's such an interesting life he lives. How can you not want to be in business with him?" (Masters 2016).

As one film producer told us, the involvement of artists and creatives, along with the products' popularity among consumers and cultural experts, makes the entertainment industry a "vanity business, an emotional business." For effectively managing innovations in such an environment, resisting the artistic temptation threat is a "must" for entertainment managers.

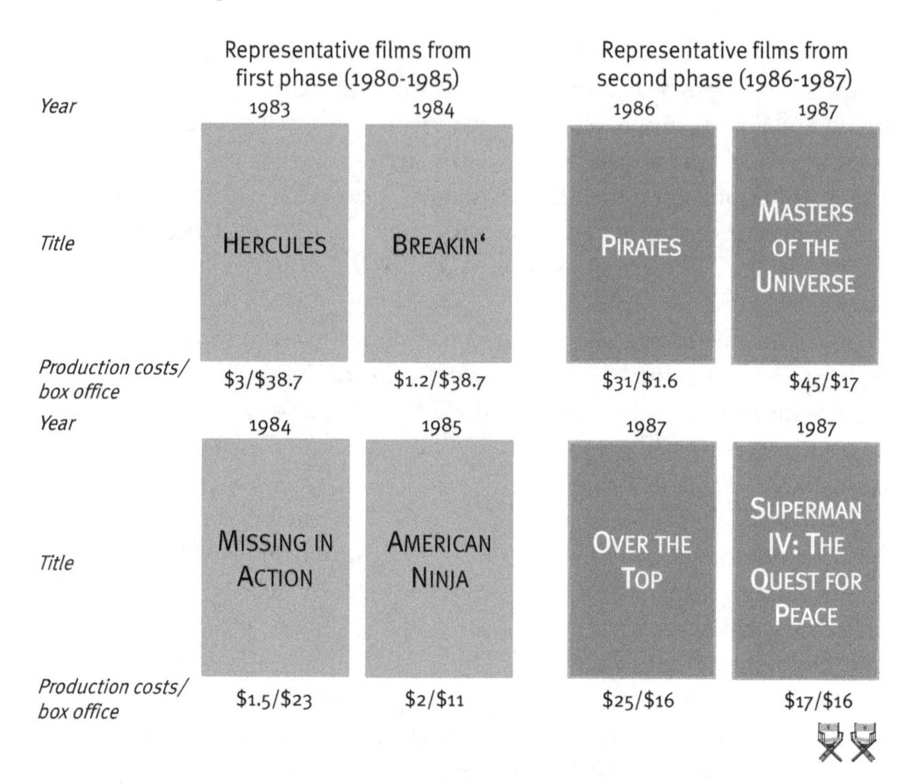

Fig. 10.4 The transformation of Cannon Films
Notes: Authors' own illustration. Box office and production budget data are based on The Numbers and other sources; all numbers are estimates only and not adjusted for inflation. Titles are trademarked.

The "Micromanaging of Creatives" Threat

The "micromanaging" threat is nearly the mirror image of what we just discussed. Whereas artistic temptation can make managers lose sound business judgment, it can also become a serious problem when an entertainment manager treats the creative organization just like any other firm, failing to recognize the unique challenges that are associated with managing creatives in pursuit of business goals.[258] Although we understand the temptation to tightly control the behaviors of artists, whose value systems differ from those of the manager him- or herself, translating business goals into a tight micromanagement of

[258]We outlined these challenges when discussing the "art-for-art's-sake" property of entertainment earlier in the book.

creatives can have the exact opposite effect on the efficiency and effectiveness of the artists' work. In fact, micromanaging creatives can pose a threat not only for art, but also for business.

For movies, we have already mentioned the example of the well-intentioned "Disney Oversight Group." Designed to encourage effective and efficient use of resources, this group actually halted progress on key projects as part of questioning "everything." The results ended up being detrimental to the quality of products and tended to drive away good creative people (Catmull and Wallace 2014). This observation is in line with single-case evidence from the video game industry compiled by scholars Hotho and Champion (2011). Using a case study approach, they collected data over a period of eight months from a small computer games development studio (around 20 artists, developers, and coders) during a time the studio was strategically shifting from doing work-for-hire (small and fast projects for others) to the creation of their own intellectual properties (i.e., two games that were self-funded).

In interviews, the scholars learned that early after the transition, all parties seemed to understand that innovation and creativity were quintessential for the new endeavor, and everyone worked diligently to establish a corresponding goal and value system. However, after mixed initial results (in terms of quality and timeliness), the climate changed—now the artistic "vision" became controlled from the top, and risk-taking was neither encouraged nor present, but replaced by strict plans that led to a downfall in artist autonomy. As motivation and commitment of the creatives waned, creativity evaporated as well. In other words, the attempt to stimulate creativity by controlling it actually led to the opposite, with some employees desiring to return to work-for-hire. In the end, the company announced a reduction in workforce and less emphasis on new intellectual property work.

The "(Perceived) Violation of Artistic Integrity" Threat

But threats to systematic innovation success do not come only from managers (who can either fall in love with artistic goals or launch ill-fated attempts to micromanage artists), but also from artists. All artists value their artistic integrity, but some tend to overemphasize it to a degree that makes it difficult (or even impossible) to fit into a system that simultaneously pursues legitimate business goals.

A well-established explanation of human motivation is self-determination theory (e.g., Deci and Ryan 2000). It states that "intrinsic" motivation—a person's inner drive to perform, a type of motivation that is closely linked to creative performance—is determined by three primary factors: personal autonomy, control, and relatedness. Whereas the relative importance of these drivers varies between people, artists usually value their autonomy most highly. Thus, if an artist feels his or her autonomy is threatened, whether through restrictions on the artist's behavior (or work habits) or by giving final decision making to a manager (and, thus, threatening the "integrity of the art"), his or her motivation can be damaged.

This is crucial, because the performance of creatives and artists is driven by such intrinsic motivation much more so than for other people. Artists typically do things because of an inner impetus rather than for extrinsic stimuli, such as money (which serves more as "validation for creative skills"; Peltoniemi 2015, p. 48). Thus, violating artistic integrity can threaten an otherwise-promising innovation project. Let us stress that it doesn't really matter whether the artist's feeling that his or her autonomy is under siege is objectively true—the mere *impression* that his or her autonomy will be crimped is sufficient for producing these detrimental effects on motivation.

Take the example of young directors Aharon Keshales and Navot Papushado, who had gotten their "dream job" and a "legendary salary" when they were hired for a remake of the 1974 hit movie DEATH WISH (Fleming 2016). The artists went through stressful interviews with the presidents of MGM and Paramount, as well as being personally approved by star Bruce Willis. A big part of their excitement came from their artistic vision to revive the spirit of the source novel, moving away from the vigilantism-celebrating previous film adaptation; they imagined a thriller in the range of "TAXI DRIVER, FALLING DOWN…with a bloodcurdling finale like SICARIO." However, Keshales and Paushado felt that they could not realize what they had envisioned due to a tight time table and "creative differences." They left the project, despite the fact that this film would have been their entry ticket to Hollywood.

Other artists have shown a similar unwillingness to compromise, perceiving their artistic integrity to be threatened by *any* deviation from their vision. For example, rapper Kanye West stated in an interview: "I do not negotiate. I can collaborate. But I'm an artist, so as soon as you negotiate, you're being compromised" (Bailey 2016). Managers can influence the artist's autonomy perception, but in such radical cases, it might be impossible to integrate an artist into a system in which artistic freedom is balanced with economic considerations.

The "Myopia" Threat

Finally, success itself can be dangerous, threatening systematic innovation. Anyone who is successful with a certain business model or activity is tempted to keep doing what he or she is good at and focusing on staying ahead of the current competition. Successful people become myopic, ignoring the broader framing in which they operate. The danger is that continuity can only work as long as the environment is stable—but once the environment changes, myopia-caused continuity will only lead to failure.

Catmull and Wallace (2014) describe early tech companies in Silicon Valley that got successful fast, attracted lots of smart people, but then failed due to clearly observable and avoidable error. They were myopic—they focused on continuing what made them successful in the first place, paying external attention only to what their direct competition was doing, but without any self-reflection or broader view.

Entertainment firms have fallen prey to a myopia threat in the past. When significant social, political, economic, and technical upheavals were changing society in the late 1960s and early 1970s, Hollywood did not link them to their business model, believing that the changes would not affect movie going. The studios made the same movies they had made for decades, but audiences did not want to see them anymore—myopia brought the "old Hollywood" to the brink of bankruptcy. Finally the financial duress forced the studios and their managers to look outward, resulting in a period of radical transformation that has been called the Hollywood Renaissance, or "New Hollywood" (Kokonis 2009).

More recently, Faughnder (2017) overviewed the high level of turnover among executives within entertainment firms, arguing that the "legacy movie business is under siege" from emerging digital platforms and mobile technologies that have emerged from outside of entertainment firms.[259] As these technologies, which have also disrupted book publishing, the music industry, and reshaped video gaming, gradually alter consumers' entertainment consumption habits, firms must not continue to do what they currently do and only do it "better"; fundamental environmental changes require firms to change their fundamental business models.[260]

[259]See also our analysis of the entertainment industry's value chain today in our chapter on entertainment business models.

[260]See also Smith and Telang's (2016) detailed analysis of how digitalization is transforming the entertainment business.

So, how can the myopia threat be countered? The key to avoiding such myopia is *self-awareness*—knowing the skills you have, but also what is happening around you, and being open to adapt your business accordingly. These things are closely tied to an entertainment company's culture, which also affects the other threats to innovation we mentioned in this section. So let us take a closer look at the role that culture plays for managing innovation successfully.

The Cultural Dimension of Entertainment Innovation

Strategy is important for powerful innovation, but won't live up to its potential if a firm does not have an innovation-friendly culture. A regular series of breakthroughs will happen only when strategic considerations take place in the "right" culture (and in the "right" organization, but that comes later). For understanding the role of culture for innovation success, a good place to start is to actually clarify what "culture" is. Deshpande et al.'s (1993, p. 24) definition of culture as "the pattern of shared values and beliefs that help individuals understand organizational functioning and thus provide them with the norms for behavior in the organization" is not only consistent with the colloquial meaning of culture as "the way things are around here," but it also stresses the essential role of (shared) values for culture. And it explains why values are so important—because it is those values that drive the behaviors that we observe of a culture.

Organizations vary in the values, beliefs, and norms that comprise their cultures. Likewise, cultures vary in the degree to which they help organizations accomplish their goals—or hinder them (Cameron and Freeman 1991). Whereas a mechanistic culture that features competitiveness, top-down control, and regulations might help to accomplish a clearly defined sales goal, such a culture will likely *not* create the context for the most creative innovations. As Archer and Walcyzk (2006, p. 16) phrase it: "the mechanism that inspires creative people to come up with the perfect design is hardly the same as the one that inspires a salesperson to make a big sale." Research has stressed that in high-creativity environments such as entertainment, effective organizational cultures are those which help to unleash the (typically high) intrinsic motivation of employees, whereas cultures that celebrate the tight management of employees do not work.

More specifically, we have identified four common cultural elements, or "themes," whose presence can contribute to entertainment innovation: (1) the combination of high autonomy and responsibility, (2) the adherence to a shared core goal, (3) an entrepreneurial orientation, and (4) a peer culture built on candor and trust. Their absence, in turn, can inhibit creativity. Our arguments combine scholarly findings with insights from Netflix and Pixar—two firms that have established themselves as highly innovative (and successful) performers in the entertainment industry by stressing the role of a strong culture that facilitates innovation. Let's take a look at the four themes one by one.

Theme 1: Autonomy *and* Responsibility

We have already highlighted that the intrinsic motivation that drives creatives implies a strong emphasis on their ability to act autonomously, thus cultures that celebrate such autonomy should in general facilitate entertainment innovation. But things are somewhat more complicated. Netflix's Reed Hastings (2009) notes something important when he stresses that providing room for autonomy is not the same as the absence of accountability. Instead, he argues that autonomy in entertainment only works for those who also accept the responsibility it carries for the organization. He argues that autonomy and responsibility must be considered as two sides of a coin to stimulate powerful innovation: "Responsible people thrive on freedom and are worthy of freedom."

Figure 10.5 shows combinations of autonomy and the acceptance of personal responsibility as part of a culture. A mechanistic culture that solely focuses on responsibility while not offering freedom to act will cripple creativity, draining the life (and intrinsic motivation) from creatives (lower-right box of the figure). Skipping the responsibility does not improve things; it just enhances passivity and stagnation (lower-left box). What about a culture that stresses autonomy, but does not require accepting responsibility for one's actions? It might have a tendency toward anarchy and chaos (upper-left box). But when both values co-exist, innovation will tend to thrive (upper-right box). So control is not a bad thing per se, particularly when it refers to the outcome of an activity (rather than the process). Taken at face value, to be supportive of innovation, the people who are responsible for quality (i.e., those doing the innovation work) must be empowered to make decisions without having to wait for approval.

An entertainment firm that marries autonomy with responsibility is Guerrilla Games, an Amsterdam-based first-party video games developer

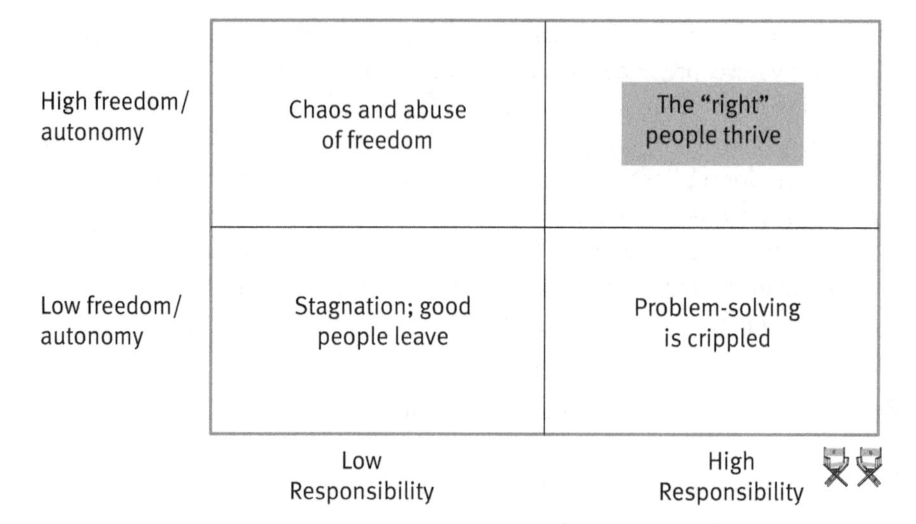

Fig. 10.5 Autonomy must be accompanied by responsibility
Note: Authors' own illustration.

that is owned by Sony Interactive Entertainment since 2005. The firm has successfully launched some of the biggest game innovations for the PlayStation console, such as HORIZON ZERO DAWN. Managing director Hermen Hulst describes the enormous freedom given to their firm by Sony and to creative employees within Guerilla quite bluntly: "We can do what we want." However, this freedom comes with conditions: "So far all games at Guerrilla have been profitable. As long as we hold that, we keep our creative freedom" (*AT5* 2017).

A similar balance between freedom and responsibility has been reported for Amazon's film division. According to Christoph Schneider, president of the firm's German business, Amazon considers "creative/artistic freedom .. a fundamental condition for the creation of extraordinary and unique series" (quoted in Kloo 2016). "We at Amazon believe in an idea and a team, which we let work freely. We'll get the cook in the kitchen and just let him cook.... [W]e won't tell him: Hey, you shouldn't use sweet potatoes. As the expression goes, too many cooks spoil the broth" (quoted in Schillat 2017). But he adds that the firm also has performance expectations for new content, which are, like everything else at Amazon, measured via success metrics such as the number of people who have watched the series, how many episodes have been streamed, and if the series is able to attract new customers to Amazon. So the creatives have artistic freedom in the development of a series without interference from management, but the creatives also have to accept responsibility for the market performance of their series.

Theme 2: Adherence to a Shared Core Goal

Scholarly research has compiled strong evidence about the importance of shared goals within an innovation culture (e.g., Gilson and Shalley 2004). For Pixar, the one shared core goal is "excellence." The firm is very reluctant to compromise on quality, prioritizing a great story over anything else (including technology). For example, what we know today as the beautiful Toy Story 2 film was initially planned as a much less ambitious direct-to-DVD release, following demands from then-Pixar partner Disney. But Pixar's executives noted that the lower quality and compromises were destroying morale among employees and decided to aim for the highest ambitions possible instead, creating a theater-worthy premium product.

By doing so, Pixar not only solidified the Toy Story franchise, but learned an important lesson about the motivational benefits of having a relentless focus on quality. According to Pixar's Brad Bird, "If you have low morale, for every $1 you spend, you get about 25 cents of value. If you have high morale, for every $1 you spend, you get about $3 of value. Companies should pay much more attention to morale" (Rao et al. 2008).

Please note that we are not arguing that "highest quality" must be the shared goal of every entertainment company. Instead, our point is that *some* important goal consistent with the firm's values and resource constraints must be shared among the team members to energize coordinated efforts throughout the firm. There can be different shared goals in different entertainment firms, consistent with our discussion in the prior section on the strategic dimension of entertainment innovation. Remember that Cannon was very successful as long as they stuck to their shared goal of making highly profitable exploitation films. A lot of artistic *and* financial failures in entertainment result from the lack of a shared vision among those involved. Take the fifth Dirty Harry film, The Dead Pool, for example: whereas it was certainly not director-star Clint Eastwood's interest to ruin the franchise, the film suffered from the lack of shared goals among those involved. Mr. Eastwood himself later acknowledged that he simply did not know where to take the character anymore when making the film (Brunsdale 2010, p. 362). He agreed to make it in exchange for Warner's support of his own personal ventures (such as the jazz movie Bird).

So, whatever the self-chosen core goal, adherence to and impassioned pursuit of it should be cultivated as a central part of the entertainment firm's culture.

Theme 3: Entrepreneurial Orientation

Innovative cultures benefit from what has been called by scholars an "entrepreneurial orientation": the welcoming of proactivity, a desire for learning (including from failure), an openness to new ideas and experimentation, and a high tolerance for risk (e.g., Lumpkin and Dess 1996). Such an orientation has been found to have a particularly high positive impact on firms' performance in contexts where dynamic market conditions and rapid changes are common. With this being an apt description of entertainment, with its constant need of new ideas and creations and now also the challenging impact of digital technologies, we expect entrepreneurial orientation to immensely benefit entertainment firms also.

The role of failure and how to treat it in a firm's set of shared values deserves special attention. We find it important to note that embracing failure as part of acting entrepreneurial must not be equated with a passive acceptance of it. Only when a person or a firm *learns from failure* does failure become valuable, providing progress towards the accomplishment of a goal. Thus, an entrepreneurial orientation implies that failure must be, especially during the ideation process at the beginning of an entertainment project, regarded as a learning process, and when this the case, people should be encouraged to fail fast and often. However, Pixar's Ed Catmull argues that failure must not be accepted, with people moving on without reflection: "The better, more subtle interpretation is that failure is a manifestation of learning and exploration. If you aren't experiencing failure, then you are making a far worse mistake: You are being drive[n] by the desire to avoid it" (quoted in Clarkson 2016).

Related is the question of what to do with someone who fails. Part of the entertainment manager's job is to make risk-taking safer. But it is not safety per se that works, but the way it is provided. In discussing the notion of psychological safety, Edmonson (1999, p. 350) does not advocate "… a careless sense of permissiveness, nor an unrelentingly positive affect but, rather, a sense of confidence that the team will not *embarrass, reject, or punish* someone" for taking risks.

In sum, successful companies consider continuous changes, the taking of risks, and the challenging of routine approaches as integral parts of their culture. An entrepreneurial orientation makes a company more agile and ready to respond to changing market demands and new technological opportunities. And such an orientation needs to be part of a firm's DNA—it can't be imported from outsiders and consultants. An entrepreneurial orientation works best with the "right" people, which is why firms such as Netflix and Pixar put substantial effort into recruiting the best talent. And, as any cultural cornerstone, it should not be limited to a few, but be a pervasive element of culture that is shared by *all* members, from janitor to CEO. Look at Pixar, where all employees are encouraged to solicit suggestions, pose aggregated discussion topics, and work to implement solutions (Catmull and Wallace 2014).

Theme 4: Peer Culture Built on Candor and Trust

Open and candid communication among all organizational members, regardless of hierarchical level, is also a vital condition for creative innovation. The infinite possibilities that come with creating entertainment require the weighing of alternative options, and having multiple constructive perspectives available helps with that task. Each of us benefits from critical feedback—just like we rewrote each sentence of this book multiple times, it is rare (if not impossible) to find a screenplay whose initial draft was "the best."

Research provides strong evidence that such open communication only takes place when a culture is characterized by trust-based cooperative relations that exhibit mutual respect (e.g., Jucevicius 2010). Only in such a culture will team members encounter others in pursuit of solutions: to reveal problems, have candid, free, and open exchanges with people across levels of the firm, and to expose ideas to criticism (i.e., trying to get the right answer, not to win arguments).

The notion of psychological safety, mentioned earlier as an important condition for an entrepreneurial orientation, can only exist if people perceive mutual respect and trust among colleagues (Edmondson 1999). How can entertainment managers signal and support such respect and trust? Learning from the experiences of Pixar and Netflix, the key is to create a safe platform for offering ideas and thoughts, one which meets a number of criteria (see Hastings 2009 and Catmull and Wallace 2014). One criterion is that factors that inhibit candor must be carefully monitored and removed.

A second one refers to techniques and processes that can foster candor—they need to be established. But what are those "techniques"?

Pixar names "postmortems," or post hoc analyses of the reasons why a product has been a success or a failure, as an example. After a project has been completed, the company asks the team involved to name five things they would do again next time, along with five things they won't repeat. This is complemented with a detailed analysis of data about the various process steps—very much in line with the idea of *Entertainment Science* (see Catmull and Wallace 2014). Usually entertainment firms, driven by the "Nobody-Knows-Anything" mantra, do not see much value in such approaches: nobody likes being criticized, and there is nothing that can be learned from the last entertainment product anyway, is there? But powerful innovation depends on becoming used to open and critical debates on a daily basis. Let us note that no finite set of candor-ensuring techniques exists, and *cannot* exist: as soon as we, as employees, know a technique, we might find a routine to avoid its uncomfortable implications.

Finally, architecture also plays a role. Simply creating an environment in which people come into unstructured contact with others of differing backgrounds and views can lead to the serendipitous exchange of "seemingly unrelated bits of information" that can be the trigger for innovation. At Pixar, the Steve Jobs Building is constructed in a way that leads people from across function and hierarchy level to bump into each other and have conversations (Smith and Paquette 2010). Here and elsewhere, employees will take note and "get the message" that their bosses want them to exchange ideas.

Strategy and culture need a third force to enable powerful systematic innovation: the organization's foundational processes and structures, including the people who embody them. We will take a look at them now.

The Organizational Dimension of Entertainment Innovation

An entertainment firm's organization essentially consists of two key and interlinked aspects: *who* is part of the organization, and *how* do those members coordinate their efforts to create value? Let's take a look at them and at the ways they affect a firm's innovation performance.

The "Who" Question: The Importance of Human Resources

> "In procedural work, the best [people] are 2-times better than the average. In creative/inventive work, the best are 10-times better than the average."
> —*Netflix founder and CEO* Reed Hastings *(2009, p. 36)*

We have mentioned that highly talented people are fundamental to innovative entertainment enterprises. Such highly talented people are rare, as Mr. Hastings argues in this section's introductory quote above. So what do we know about the role of human resources management for successful entertainment innovation? Because innovations can be made, bought, or created in cooperation with others, we will address the management of both internal and external human resources in our discussion.

Internal Human Resources

Pixar has outperformed other entertainment firms with an approach that builds on the "in-house" development of all its projects. A pivotal aspect of the firm's approach to innovation has been the prioritization of talent over ideas. Pixar lives by the belief that if you give a mediocre team a brilliant idea, the transformation of the idea into a product will not live up to the promise of the idea; however, if you give an idea, even a mediocre one, to a brilliant team, the team will make it better, and often great. Thus, they focus on building a team comprised of the "right" people. The example of Toy Story 2 shows how the same story line can result in outcomes that differ dramatically in quality when executed by teams of differing abilities (Catmull and Wallace 2014; p. 379).

So a focus on attracting and retaining top-flight talent in the internal entertainment innovation operation is recommended. But what characterizes such talent? Netflix, who puts a similarly strong emphasis on internal personnel, prioritizes mature people, acting under the assumption that brilliant "rookies" are more costly for the firm than what they are worth in comparison (Hastings 2009). Further, they regularly employ the "Keepers Test," a mental exercise that asks a manager to consider how hard they would fight to keep an employee if that employee announced an intention to leave. If you wouldn't fight hard to keep a person, perhaps it is best to cut them loose and hire someone you would fight to keep.

But the management of human resources for innovation is not limited to hiring decisions. Another way to boost performance is to provide opportunities for development and support to key creative people, recognizing that what these people might be able to do tomorrow is more important than what they can do today. Employees with high innovation capabilities and high intrinsic motivation will exhibit a willingness to learn new things and grow their own creative innovation capabilities—and the firm's (Savitskaya and Järvi 2012).

The individuals and their respective skills are important, but there is more that needs to be considered. We have already stated that across entertainment it is usually the coordinated efforts of a *team* that results in new entertainment products that win in the marketplace, which is why Pixar's focus is on brilliant *team* performances rather than on brilliant individuals. Teams are more than the accumulation of great skills—their output requires ongoing coordination, which depends on social harmony. Narayan and Kadiyali (2016) provide empirical evidence for the role of a creative team's composition with dynamic panel data estimations of interactions between members of production teams of 1,123 movies. Studying weekly North American box-office revenues with a GMM regression, they find that the shared previous experiences between a film's producer and other team members are most impactful for a film's success. Interestingly, it is not the *success* of earlier collaborations that matters, but the mere existence of a social connection between them that results from having collaborated in the past.

A related question is whether creating an internal team that is highly diverse in background and orientation is beneficial for innovation. Evidence is fairly anecdotal so far and exists for both perspectives—diversity has been argued to bring benefits of serendipity (Smith and Paquette 2010), whereas homogeneity has been associated with creative synergy (Harvey 2014). We argue, based on our discussion of the critical role of culture and the different goals associated with the creation of entertainment, that the benefits of diversity end when diversity exists regarding fundamental values or goals (see Gilson 2015 for a discussion).

External Human Resources

For creating innovations, it can also be beneficial to maintain a network of cooperation with resources *outside* the firm. Packard et al. (2015) study empirically two aspects of such cooperation: the degree to which a creative is tied to prominent others in the industry (which might help the reputation of the projects he or she is working on among consumers) and the degree to which the creative maintains connections with other "sub-communities" in the industry (which might provide access to unique skills and resources). In an analysis of 15,000 movie professionals and the performance of their films with regression analysis (in which they control statistically for several other "success drivers"), the scholars find that both kinds of networks can help—but for different groups of creatives. Specifically, whereas ties with prominent others ("high positional embeddedness") exert a positive influence on a movie's box office performance for creatives who are part of the onscreen cast (i.e., actors), maintaining bridges and thus access to other segments of the industry (or "high junctional embeddedness") pays off for members of the behind-the-screen crew (including directors and producers).

Academia plays a particular role for such outside networks—one that is neglected by many. For example, Pixar values close connections to the outside world, sharing their early work on computer animation with the scholarly world (e.g., through publications and conference presentations), instead of cloaking it in secrecy like other companies. This participation in the intellectual discourse has also taken place in other academic fields such as business, and our numerous references to Pixar and its approaches in this book give evidence of it. Though the exact impact of this sharing is difficult to pinpoint, connections with information technology scholars were formed over time and have been named as valuable for fueling innovation and the firm's understanding of creativity (Catmull and Wallace 2014).

Netflix, too, practices active ties to the academic community. We give evidence of it at various points of this book, such as when discussing their $1 million recommender algorithm competition known as the Netflix Prize, as well as their contributions to the academic discussion on recommenders (e.g., Gomez-Uribe and Hunt 2015) and movie trailers (Liu et al. 2018). The company also worked together with cultural anthropologist Grant McCracken to gain deeper insights into consumers and the "spoiler" phenomenon (Steel 2014). Particularly in times characterized by high dynamics, such connections can allow a firm to overcome traditional restrictions and problem-solving patterns by thinking "out of the box." As we have noted earlier in this book, we have come under the impression that such openness to outsiders and scholars, in particular, is not strongly developed among many traditional producers of entertainment.

At the end of the day, an entertainment manager's job must be to bring the *right people* together, combining internal and external talent, to construct a team that complements each other and works collaboratively toward the pursuit of shared goals.

The "How" Question: Creativity Needs Freedom

> "I am a firm believer of the chaotic nature of the creative process needing to be chaotic. If we put too much structure on it, we kill it. So there's a fine balance between providing some structure and safety—financial and emotional—but also letting it get messy and stay messy for a while."
>
> —*Pixar-Executive* Ed Catmull *(quoted in* Catmull *and* Wallace *2014, p. 142)*

Assuming you have assembled the right team of creatives, what kind of organizational structure will allow them to flourish and create high-quality entertainment innovations? In the introductory quote above, Mr. Catmull stresses that creativity is inherently chaotic, which implies that the structure of innovation processes must be designed in a way that accounts for such chaos, leaving room for it while, at the same time, preventing escalation. This is also in line with the "micromanagement threat" for innovation we noted earlier.

Based on our review and analysis, we argue that a good organizational design for supporting innovation must meet the following three conditions: (1) it requires a relatively "flat" organizational structure to facilitate communication and decision making, (2) it must leave room for failure to happen in the *early* phases of the innovation process to allow quick adaptations in case of failure and minimize the economic consequences, and (3) it must avoid over-structuring, in the case a company grows, to not harm innovation processes. Let's see what kinds of organizational approaches are helpful for meeting these requirements.

Condition 1: (Relatively) Flat Hierarchical Structures

To enable free-flowing communication and quick, effective decision making, there is evidence for the advantageous nature of flat organization structures for innovation activities. A flat structure supports communication and decision making in several ways:

- it reduces bureaucracy levels (e.g., Archer and Walczyk 2006);
- it avoids a restrictive vertical chain of command (with "top-down" dictates and no exchange of thoughts and ideas between members of the firm) and restrictive roadblocks to change (e.g., Kanter 1997);
- it avoids rigid routines and supports flexibility in terms of when or where to work (e.g., Savitskaya and Järvi 2012); and

- it offers more decision-making options as a reward for good performance (e.g., Archer and Walczyk 2006).

When the "right" people are given big jobs along with the authority and responsibility to work together to be innovative, good results normally happen when there is no excessive hierarchy to keep them from doing so. One powerful example for a flat hierarchical structure for innovation activities is Google, where communication is seen as peer-to-peer rather than manager-to-minion (Hamel 2006).

To enable the best ideas to bubble up regardless of where they occur within the firm, job titles and hierarchy are relatively meaningless at Pixar, with unhindered communication being a core structuring principle (Rao et al. 2008). One specific mechanism to insure this open communication across a flat structure at Pixar is named "Braintrust"—an ad hoc, problem-solving unit formed to analyze the emotional aspects of a story without getting emotional (Catmull and Wallace 2014). Braintrust transitioned from a fixed group of senior people to a larger group of people that assemble as needed to solve concrete problems—the best ideas get challenged and tested.[261] An important credo within the Braintrust is that the person who proposes an idea is separated from that idea during the debate, which supports candor and honesty in the spirit of constructive criticism. The approach does not magically solve all problems, but it can help recognize when something is off.

In addition, there are no mandatory notes handed down from executives to directors at Pixar, as is common in most traditional entertainment firms. Instead, it is the responsibility of a director to make a good movie and he/she is trusted with figuring out how to accomplish the goal. However, this does not mean that the whole system is *laissez-faire* until the end—if a team is truly stuck, the rules then require changing the team. Which turns out to be not such a rare occurrence after all at Pixar—the initial director was not only substituted in Toy Story 2, but in five out of the firm's first 18 films, or 28% of the projects (Spiegel 2013).[262]

But how do you steer people in entertainment while employing a "hands-off" approach implied by flat hierarchies? With Netflix in mind, Hastings (2009) argues that management should be done through context, not

[261]Any scholar will be reminded of the ideal version of the peer-review process in academia.

[262]Flat structures and open communication should not be confused with eternal harmony: at least four of the five substituted director were no longer with Pixar in 2013 (Spiegel 2013).

control, putting emphasis on some of those aspects of innovation management we have mentioned earlier in this chapter, such as strategy (including accurate assumptions and objectives), culture, success metrics, and people. This view is consistent with general scholarly research on creativity and innovation that finds that performance is enhanced when goals are provided to talented people with creative thinking skills, but they are given as much autonomy as possible concerning *how* to achieve the goal (e.g., Walesh 2012).

Condition 2: Leave Room for Failure in the Early Phase of the Innovation Process (Only)

Careful planning is difficult in the realm of creativity with the limitless number of potential combinations of elements in an entertainment product innovation. Errors are somewhat unavoidable for the firm that is pushing forward and trying to accomplish something great. Managers cannot avoid missteps; however, they can design the innovation process in a way that biases failures toward the *early* stages of the process.

Doing so is crucial from an economic perspective, because the costs associated with wrong-doing are not equally distributed over the course of the process. Instead, costs escalate in later stages, increasing exponentially from the creation of ideas, through the formulation of the product concept, to the production of the actual product. Underlying this exponential cost development is an inverse, non-linear reduction in the firm's flexibility to make changes to the project over time, as we display in Fig. 10.6.

And what kind of structures help to shift errors toward the front-end of the process? One potential response is the so-called "loose-tight" concept, which suggests that the organization of the innovation process should be designed loosely in the early stages, leaving large room for creatives to experiment (e.g., Albers and Eggers 1991). In the later stages, however, the concept suggests a change toward a more restrictive form of process management, in which creatives are provided much less room for changing the product idea and concept. In other words, efficiency and project implementation are prioritized over creativity in the later stage. We have to note though that the loose-tight concept has so far received only partial empirical support and has not been tested in an entertainment context.

A different approach that has been suggested to minimize the impact of failure is to focus on the speed of making changes and adjustments to the product concept during the process by creating a "high-velocity environment."

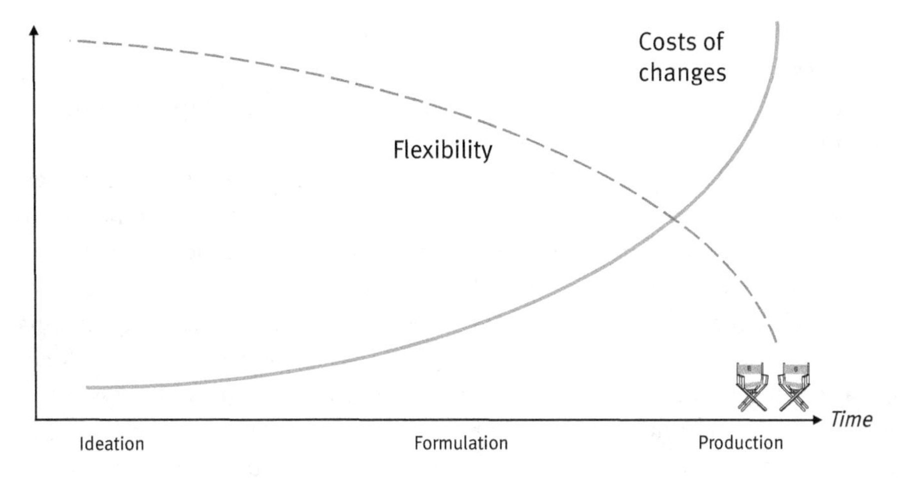

Fig. 10.6 Escalating costs of failure during the innovation process
Note: Authors' own illustration.

Netflix, for example, operates under a "rapid recovery" model that is based on the assumption that it is better to learn and fix errors than to prevent them from happening in the first place. Keep in mind, though, that such a model conflicts with the cost development we describe in the figure above and that is so typical for innovation processes; its use might thus fit better with some entertainment innovations (e.g., the adding of features to online games) than others (e.g., the development of a big-budget movie or series). Finally, *Entertainment Science* scholars have accumulated evidence that thorough testing in the early stages of the innovation can be a powerful way to meet this condition in a satisfying way, despite the industry's traditional skepticism toward it.

Condition 3: Avoid Over-Structuring When Growing

Success makes an organization grow, which is inherently a good thing. For innovation, however, growth introduces additional challenges. Growth brings an increased complexity because of the higher number of individual workers whose efforts must now be coordinated, and because a growing firm usually manages a higher number of projects and/or product lines simultaneously. Look at Pixar, which has moved from a biannual release schedule to now releasing two new films each year. As firms grow, a tendency exists to institutionalize vertical chains of command, put in place formal (and often rigid) policies and procedures, control employees communications and access to resources, and create approval processes that may improve efficiency (e.g., Kanter 1997)—essentially all those things we have argued in this chapter should be avoided to the degree possible in entertainment innovation.

Netflix's Reed Hastings (2009) argues that hiring additional high-performance employees provides the solution to this organizational challenge—they are motivated, and able, to maneuver the innovation process themselves, without the need for additional structures and bureaucracy. To assist such people in operating a growing organization informally, Netflix works to minimize structures whenever possible. For example, they focus on a few big products instead of many small ones. The firm also systematically works to eliminate distracting complexity that comes in the form of process-focused policies that might have worthy goals, but saddle employees with onerous compliance burdens or restrict their agility.

Overall, our conclusion, based on scholarly research and anecdotal evidence from growing entertainment firms, is that there are certain strategic, cultural, and structural parameters that determine creativity and innovation. The complete process—from initial idea to finished product—is inherently chaotic and involves at least a few errors. Attempts to formalize the process can quickly devolve into micromanagement and risk avoidance, approaches that are, overall, bad for the performance of an entertainment firm. Improving processes must not be the main goal; the main goal is making the product great or achieving whatever goal a firm has set.

At the same time, although freedom is foundational for creativity, freedom cannot be unlimited. Some rules are necessary, such as those to prevent irrevocable disaster and those that relate to moral, ethical and legal issues. We like Mr. Hastings' (2009) distinction between "good" processes (those which empower talented people to get more done and help them avoid serious missteps) and "bad" processes (those that incur high costs of time and/or frustration to prevent mistakes that are, in reality, easy to rectify if they occur). We believe that most readers of this book, regardless of their background, will be able think of examples where budget management policies were "good" (e.g., spend a proportion of budget without asking for permission) and where they were "bad" (e.g., requiring multiple layers of approval for routine repurchases). The goal must be to create operational policies that are highly aligned with goals and enable flexibility and creativity for those working within it.

Let us now move away from the innovation-related matters that take place at the company level and take a look at actions that deal with a specific new product idea: how the use of smart forecasting of an innovation's performance can help firms turn an idea into a successful new entertainment product.

The Product Level: How to Forecast the Success of New Entertainment Products

> "If my fanny squirms, it's bad. If my fanny doesn't squirm, it's good. It's as simple as that."
>
> —Harry Cohn, *founder and president of Columbia Pictures from 1919–1958, describing how he predicted the success of a new movie (quoted in* Austin *1989, p. 1f.)*

For any new product, whether internally developed, co-produced, or externally acquired, a need exists to forecast the success potential of the respective idea, concept, or product. Success predictions are needed for green-lighting a concept, to guide substantive pre-release changes to a product, and to allocate budgets to an innovation that are in line with its commercial prospects, an essential facet to the long-term health of any business.

We are convinced that managerial intuition (in the way used by the legendary Mr. Cohn in the quote above, or somewhat more subtly), can help with such predictions. However, a central theme of this book is that a manager's experience-based judgment can quite often be supplemented with *Entertainment Science*, i.e., the thorough combined use of theory and data, to the manager's advantage. We argue this to particularly be the case for the prediction of the success of innovations. In this section, we describe how such support can come in the form of econometric prediction models. Like any part of *Entertainment Science*, innovation prediction models should be seen as complementary to, not as substitutes for, decisions based on hard-earned managerial intuition.

Entertainment Science scholars have developed various approaches for using existing data for predicting the success of new entertainment products. These approaches differ in several ways—most importantly with regard to the econometric method used and the data employed. Data availability differs between the stages of innovation processes, and so do prediction models that are applied in the early conceptual phase of entertainment products differ from others which are used in the crucial pre-release phase shortly before release, and also from early post-release models that incorporate initial sales data.

After a short primer on the essentials of success prediction, we will overview the key prediction methods, namely feature-based and diffusion-based models, before we illustrate the use of predictions at different points of the innovation process. We will restrict our discussion strictly to the methods and fit of predictions, but refer to the other, substantive chapters of this book for a discussion of the predictor variables themselves. As in other areas of *Entertainment Science*, most research we report uses movies, but insight should be usable with some modifications for any entertainment product for which similar data is available. And what about predictions offered by consultants and made by entertainment firms? For the most part, we will leave them aside, because they usually lack information about the method and their goodness of fit, things that we consider essential for any kind of useful prediction.

Some Words on the Essentials of Success Prediction

Regardless of the specific analysis method used, success prediction studies in entertainment rely on a pattern that comprises four steps: (1) choosing the predictors, (2) choosing the prediction objects and the data sources, (3) running the analysis using the holdout-sample approach, and finally (4) assessing the quality of the forecasts. Let us take a closer look, step by step.

Step 1: Choosing the predictors. The first step in any prediction analysis is to determine the variables that will be used for the forecasts. A key question here is whether to restrict the set of predictors to those variables for which causality has been demonstrated (or can at least be argued to exist), or to allow *any* kind of variable to serve as a predictor if it helps the prediction to be more "accurate" (a *terminus technicus*; see below what it means in a prediction context). For most of the marketing variables we have presented in the various chapters of this book's Part II, from quality to different forms of pricing, we argue that they indeed have a causal relationship with success.[263]

But in these times of abundant data availability, there is a lot of additional information out there that could also be used in prediction tasks, and several prediction analysis conducted by scholars do indeed include them. We have one very clear recommendation for every entertainment manager though: stay away from predictors for which you cannot claim a causal effect. Using them might work in the short-term, but it can lead those who apply them into serious trouble once the measurement, the meaning, the usage, or the context of that predictor changes. In the face of environmental change,

[263]But this causal nature should not be taken lightly.

including variables without a clear causal explanation can result in substantial under- or overpredictions—and corresponding misallocations of marketing budgets, or plain wrong decisions. Further, we argue that the same care is suitable for methods that are "black boxes" with regard the *course* by which the predictors and success are linked. Unknown or arbitrary functions can lead to serious mispredictions if your predictor takes new values, outside the original spectrum or even within it. Do not trade in causation (or at least its justified assumption) for some supposed formal increase in accuracy; in the long run, chances are that it will turn out to be an artifact and lead you toward bad decisions.

If you follow our recommendation, this book provides a sourcebook for selecting powerful predictors. Their availability will vary between prediction tasks, and not all will be equally relevant—there is a tradeoff between the increase in prediction accuracy another predictor adds and the costs for collecting the respective information. But keep in mind the fundamental logic of this book, and of *Entertainment Science* in general, when selecting prediction variables: the better the model reflects what is actually going on in the market and the consumer, the more powerful the prediction will be.

Step 2: Choosing the prediction objects and collecting the data. Before data collection can commence, the researcher has to answer the question: "For what object(s) do we want to predict success?" The answer can range from trying to predict the success of a single product (e.g., as Eliashberg et al. 2000 predict, and measure, the performance of the movie SHADOW CONSPIRACY in Rotterdam theaters) to trying to establish a model applicable to *any* product in an entertainment category (e.g., any movie). There's no right or wrong—the choice of the object(s) is determined by the strategic decisions that need to be made based on the prediction. But the choice of the object influences the adequacy of the prediction approach.

Regarding the data itself, predictions can be made based on secondary data (which exist separately from the prediction analysis) and primary data (which is collected specifically for the analysis); these types of data can be used separately or in combination. Secondary data is widely available in entertainment and can be either accessed freely for many usages (via websites such as IMDb, Google Trends, or Facebook's API) or purchased (from services such as Kantar, Rentrak, or Linkfluence); its advantage is that it is readily available for collection and for analysis. But secondary data often has limits, as it often contains "missings," and the quality of the available data is not easily verifiable (such as for production budgets). Also, keep in mind that secondary data is always historic data (about something that has already happened, either years or seconds ago), so its usefulness for predictions of things *to* happen in the future depends on whether the context from which it stems will still be valid. The alternative is primary data, whose collection can consume enormous resources (such as when done via surveys or panels) and can suffer from quality problems itself (do respondents tell the truth?). However, primary data can be exclusive when collected by a company itself and its collection can be designed specifically to meet the prediction task's demands. But depending on resources, exclusivity can also exist for secondary data—think of Netflix's viewer statistics and individual usage patterns. Some of the best prediction models combine both kinds of data.

Step 3: Running the analysis using the holdout-sample approach. A key question for making predictions is to pick the "right" econometric method; we name some key alternatives that can be used in the context of entertainment below. One important component of predictions is the "holdout" approach. The approach's idea is intuitive: when the analyst does not want to wait for predicted events that lie in the future to happen, he or she selects a subset of the objects from historic data and puts this "holdout sample" aside until later in the analysis. The remaining objects are used to "train" the prediction model, and its goodness of fit is judged against how well it "predicts" the objects in the holdout sample. If all available objects are used for training and evaluation, the prediction results are biased toward the positive and are thus only meaningful for the objects in the data set that was used for calibrating the model. Selecting a holdout sample is far from a trivial task: the way it is done affects the fit of the prediction model.

Step 4: Assessing the quality of the forecasts. To assess the "goodness of fit" of a prediction, researchers rely on several accuracy metrics. There is not a single "best" one; instead each of them brings its own limitations of which the user should be aware. Among the most common metrics are the Mean Absolute Error (MAE), the Root/Mean Squared Error (R/MSE), and the Mean Absolute Percentage Error (MAPE). Whereas both the MAE and the R/MSE compare *absolute* differences between predicted and actual values, the MAPE compares *relative* (percentage) differences.

The MAE describes the average absolute difference between a predicted value and the corresponding actual value for the success metric (such as a film's box office). As an absolute value, it depends heavily upon the magnitude of the chosen success metric. If you have one AVATAR in your data whose performance your model cannot predict correctly, the mean error will be quite high, even if the model works fine for all other films in the data set. In contrast, it doesn't matter much if the smaller products in a data set are predicted wrongly, even if these errors are high percentage-wise. The R/MSE builds on the logic of the MAE; it is the square root of the sum of the squared deviations of all predicted and actual values in the data set. As the measure penalizes prediction outliers by squaring, it is of most use when large prediction errors for single cases are particularly undesirable (and a higher, but more equally distributed prediction error is preferred over drastic fails for certain cases).

MAPE is the average prediction error measured in percent; it is intuitive, but can be influenced strongly by the error of smaller absolute values, and it penalizes extreme upward deviations more strongly than for lower deviations (the former can be—much—higher than 100%, the latter cannot). Whether to use an absolute or relative criterion depends on the user's preference—if the objects in the data are of similar importance despite their differences in absolute success, then a relative criterion makes more sense. If the bigger hits are more important for the firm, then it should be an absolute criterion instead. Having a Blair Witch Project in the data set, which generates 100-times the expected returns, drastically inflates relative accuracy metrics such as the MAPE, in particular.

Prediction Methods: Feature-Based Versus Diffusion-Based Success Prediction

At the heart of any prediction task is the statistical algorithm that transforms the values of the predictor variables into forecasts. Entertainment scholars (and managers) can chose between a plethora of specific prediction methods, but the underlying more fundamental decision is whether to pursue a "feature-based" or a "diffusion-based" approach. Feature-based predictions combine knowledge about the predictors of entertainment success, namely the different "success drivers" we discuss in this book (such as product characteristics like story elements or quality or ad spending), and generate predictions based on linear or non-linear transformation functions.

Diffusion-based predictions, in contrast, use theoretical models of how entertainment products diffuse among consumers over time. Neither type of approach is generally superior; their effectiveness depends on how well the user understands the linkage between features and success or products' diffusion patterns, respectively. We will now overview the most prominent techniques for both approaches, starting with feature-based predictions.

Feature-Based Approaches of Success Prediction

The most-employed feature-based prediction method is regression analysis, along with its many extensions. A second important stream consists of methods powered by machine learning, such as neural networks and decision trees.

The majority of the causal insights on how marketing variables drive entertainment success that we report throughout this book stems from regression-type analyses, which is why we discussed the foundations of the method in the book's introductory section. In a nutshell, regression analysis assumes and estimates a systematic (often linear) relationship between one or more independent variables and a dependent (or outcome) variable. Regression uses an optimization process that minimizes a measure of how much deviation there is between the observed data points and those that are calculated by the estimated regression function.

But in addition to explaining relationships, regression also provides everything the user needs for a *prediction* task: a regression function allows the user to predict the success of an entertainment product by entering specific values in the function. Explanation and prediction with regression analysis are methodically the same; they differ with regard to the intentions of the user and if the causality assumption is waived when making predictions. Scholars occasionally blur the distinction, speaking of predicting when the actual goal is to gain causal understanding (e.g., Litman 1983; Simonoff and Sparrow 2000; Chang and Ki 2005).

An alternative approach for making predictions is to use machine learning. Machine learning has become something of a buzz word (Gartner 2016), but the concept is not quite as well-understood by the public and industry as the term is familiar to them. Broadly speaking, machine learning encompasses algorithms that can learn from data and, through optimization routines and multiple iterations, find the "best" analytical solution for a given question (e.g., Brownlee et al. 2013; Kohavi and Provost 1998). Predictions are just one application for machine learning, a family of approaches that sprouted from the artificial intelligence community.

Models that are used in success predictions tend to belong to the category of *supervised* machine learning—because the predictors and success variables are pre-specified (e.g., Kelleher et al. 2015). Machine learning can offer stunning fit values, which, however, come at a price for the entertainment manager: the approach tends to treat the world as a "black box," with little emphasis on reliable explanation of causes and effects (see Breiman 2001 on statistical modeling cultures). We have already mentioned the problems that can result from such an approach—and kindly ask our reader to keep them in mind.

One stream of machine learning for which these concerns are particularly valid uses artificial neural networks, methods that attempt to represent mathematically the same type of information processing that occurs in our biological brain system.[264] Neural networks consist of artificial neurons that are arranged in layers and interconnected. Given a certain set of input variables, corresponding nodes get activated in the following layer, which then activate certain nodes in the next layer (and so on, depending on the number of layers). The unique set of activated neurons in the last layer provide a solution.

Sharda and Delen (2006) are among those who applied the method to predicting entertainment product success. Their model is based on 834 movies released in North America between 1998 and 2002 and data for several marketing variables we discuss in this book (e.g., MPAA rating, competition, star power). They report that their model correctly assigns about three out of four films into one out of nine success categories (or to one of two categories that are directly contiguous to the correct category). However, sensitivity analyses show that variables like competition, MPAA ratings, and most genres, whose influential role we have explained and empirically demonstrated in this book, play no role in their estimation. This result should raise plausibility concerns—and underlines our caution regarding prediction methods that model the complex links between predictors and entertainment success in a solely data-driven way, using black boxes instead of theoretical considerations.[265]

A second stream of machine learning techniques are called "decision trees." They work through a sequence of questions that have categorical answers (e.g., yes/no, high/medium/low, values above/below a certain value) to arrive at a mathematically optimal solution; continuous variables have to be categorized to become part of the process (see Kelleher et al. 2015 for an overview). A first categorical variable considered in a decision tree is called the "root," or "starting node." The decision tree then asks which potential value the variable has taken on, and for each possible answer or variable manifestation the node splits to consider further variable questions. These

[264]Please see Algobeans (2016) for a more accessible description of artificial neural networks and Bishop (2006) for a more technical overview.

[265]We have similar issues with other studies that predict box office using neural networks, such as Sharda and Delen (2010), Ghiassi et al. (2014), and Zhou et al. (2017).

next layers are called interior nodes. Some variables contain more information than others and can help the algorithm progress more quickly. After working through all interior nodes, the model will eventually reach a conclusion, called the "leaf" (or "terminating node").

In entertainment, decision trees have been used to predict the commercial success of films (e.g., Eliashberg et al. 2007; Parimi and Caragea 2013),[266] as well as for how much consumers will like a film (Asad et al. 2012). In addition to the black box problem for the links, decision trees tend to have stability issues, with the choice of the root impacting the solution. Interpretation and stability are particular issues when either many variable categories or continuous variables are used, which are clearly the norms when predicting the success of entertainment products, as this book demonstrates.

Diffusion-Based Approaches of Success Prediction

Based on the notion that the patterns of revenues generated over the lifetime of entertainment products "display remarkable empirical regularity" (Sawhney and Eliashberg 1996, p. 113), scholars have drawn on the diffusion literature to explain, in a theoretically meaningful way, the empirical diffusion patterns of entertainment products and, ultimately, to forecast their future success. Let us explain the basic logic of diffusion-based predictions by drawing on the fundamental diffusion model by marketing scholar Frank Bass (1969) before highlighting key learnings from the empirical application of diffusion models to entertainment products.

The "Bass Model" of Diffusion

Frank Bass' basic idea was that there are two primary segments of customers in any market that drive the diffusion of a new product: Innovators and Imitators. According to his model, Innovators are those consumers who are drawn to new innovations, willing to take risks and deal with uncertainties. Most importantly, these consumers adopt a new product independent of other people's actions, such as word of mouth. Most of these Innovators gain access to a new product early on; the number of Innovators still available to

[266]Josh Eliashberg and his colleagues apply the method in their analysis of movie script features, whereas Parimi and Caragea use it for a general prediction exercise.

adopt the new product for the first time decreases quickly for a successful product.

Imitators, in contrast, are more risk-averse and will not tend to adopt new products until some of the uncertainties surrounding technology issues and market acceptance are reduced; thus, they tend to observe Innovators' behaviors and learn from those behaviors. Imitators make decisions based on the behaviors of—and word of mouth from—the part of the population that has already adopted the product. They do not adopt the product at its launch, but only begin to adopt the new product as they eventually copy or learn from the behaviors of the Innovators and other Imitators who have already experienced the product.

The Bass model features parameters for the adoption behavior of each of these two consumer groups and describes the two groups' respective contributions to a product's diffusion over time, as well as the links between them. The innovation parameter, or α, represents the reactions of Innovators when a new product becomes available and how they generate revenues over the diffusion process, and the imitation or word-of-mouth parameter β captures the behavior of the Imitators. According to Bass, the number x of a new product sold in a time period t follows the following pattern:

$$X_t = \alpha \times \left(\bar{Y} - Y_{t-1}\right) + \beta \times \frac{Y_{t-1}}{\bar{Y}} \times \left(\bar{Y} - Y_{t-1}\right)$$

where X_t is the total number of units of the new product that are sold in period t, $\bar{Y}$ is the number of potential buyers of the new product (or the "market size"), and Y_{t-1} is the number of buyers who have already purchased the new product by the end of the *previous* time period. The equation shows that the segments of Innovators and Imitators affect adoption in a largely additive way; however, as their respective sizes influence the number of adopters up to a given point in time Y_{t-1}, a higher β reduces the number of products that will be sold to Innovators in the next period. The effect of a higher α on the number of products sold to Imitators is less straightforward—on the one hand, a higher α reduces the number of products that are left for Imitators, but on the other it increases the chance that Imitators are influenced by word of mouth from those who have already adopted the product.

At its core, the Bass diffusion model (like all other diffusion models) uses statistical distribution functions; the response of Innovators follows an exponential distribution, whereas the adoption by imitators describes a logistic distribution. By combining these distributions, the model offers substantial flexibility; it provides room for a number of different patterns, depending on the value of the specific parameters for the innovation parameter α, the imitation parameter β, and also Y, as the number of total buyers. In Fig. 10.7 we illustrate this flexibility by showing the diffusion patterns the model calculates for four different combinations of α and β.

Whereas the upper left pattern (Panel A) exemplifies a new product with a relatively low innovation parameter and a moderate imitation parameter, the upper right pattern (Panel B) shows how diffusion changes when a product appeals more strongly to innovators; that trend is illustrated with an even more radical example in the lower left part of the figure (Panel C). Finally, the lower right pattern (Panel D) has the same innovation parameter as the one above it, but shows the impact of a higher imitation level.

The figure also illustrates that the two diffusion parameters of the Bass model are closely tied to the concepts of marketability and playability that

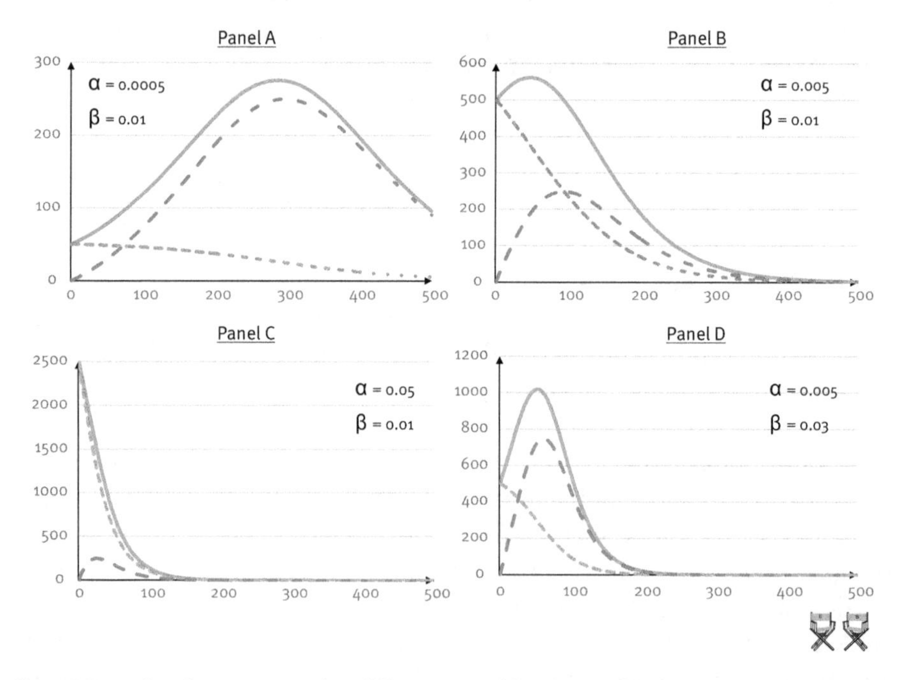

Fig. 10.7 Adoption patterns for different combinations of innovation and imitation parameters

Notes: Authors' own illustration based on the diffusion model by Bass (1969). All patterns in the figure assume a market potential of 100,000 units for the new product and study sales during the products initial 500 days. The straight orange line indicates the total adoption of the new product; the dashed blue line shows the adoption by Imitators only and the dotted grey line are Innovator adopters only. All data is hypothetical.

we discussed earlier. Specifically, as can be seen in the figure by comparing the upper and the lower left pattern, a higher innovation parameter leads the function to peak earlier, because a large number of consumers adopt the product independent of quality-related recommendations from friends or others—which is essentially the same as high *marketability*. In contrast, a higher imitation parameter reduces the time needed for word of mouth about a new product to spread among consumers, because a large number of consumers choose the product based on quality recommendations from friends and others—the equivalent to high *playability* of an entertainment product. Statistically, if α is larger than β, the total sales function (the orange line in the figure) follows an exponential distribution (which is typical for entertainment blockbusters, see this book's chapter on integrated marketing strategies), whereas if β is larger than α, the function follows a logistic distribution (which is typical for "niche" products).

Both the innovation and imitation parameters of the Bass model, as well as the total sales volume, can be estimated using either early historical sales data from similar products or from early sales data of the new product and then applying regression analysis to them. If sufficient parameters of the model are already known, the equation can be used to solve for the missing one. Clearly, the quality of the forecasted adoption pattern depends on the quality of these inputs. If there are no reasonable "comparables" for which data is available, then the quality of the output will be lower. The version of the Bass model we have described here does not account for marketing strategies and interventions that may alter market acceptance of the new product. However, Frank Bass himself and also other scholars have advanced the model in several ways, including the provision of room for marketing actions (e.g., Muller et al. 2009 provide an overview).

Applying Diffusion Models to Entertainment Products

Whereas the Bass model was originally developed with consumer durables (such as TVs and lawn mowers) in mind, *Entertainment Science* scholars have developed diffusion models which are attuned to the peculiarities of entertainment products. A prominent example is BOXMOD, a diffusion model that Sawhney and Eliashberg (1996) developed with movies in mind. At the heart of BOXMOD is the assumption that an individual consumer's time to adopt a movie is the sum of (a) the time he or she needs to *decide* to see the movie (based on available information) and (b) the time he or she needs to *act* on that decision (i.e. to actually go out and see it).

Depending on these parameters, the observed diffusion follows different patterns again, which in this case include an exponential distribution and a Generalized Gamma distribution. Combining the two parameters with the estimated market potential of a movie, Sawhney and Eliashberg developed a diffusion model which requires three weeks of success data to generate forecasts, as well as an extended version which also incorporates the "supply side" (i.e., the number of theaters in which a film was shown in a given week).[267] To be of use *prior* to a film's release, firms would need to anticipate the two key parameters of the model, relying on experience and/or the historical results of comparable movies.

Combining their diffusion-based prediction approach with the feature-based approach by linking their diffusion parameters with certain movie characteristics (such as star power, MPAA rating, sequel, professional reviews, and genres), Sawhney and Eliashberg explain about 12% of the variation of their "time-to-decide" parameter and about 22% of their "time-to-act" parameter with regression analysis (and 42% of their total sales variable).

In a separate study, Ainslie et al. (2005) suggest an alternative approach to BOXMOD. Their diffusion model for movies assumes a Gamma distribution and uses three parameters, namely (a) the expected attractiveness of a film in its opening week, (b) the point in time when the film's distribution peaks, and (c) a "speed" parameter that reflects the speed with which a movie's attractiveness builds and decays. Like Sawhney and Eliashberg, Ainslie et al. also integrate the feature-based approach by linking their parameters with movie characteristics, for example star power and professional reviews. They further make an attempt to integrate "market" forces such as distribution and timing effects,[268] as well as competition, into their forecasts and estimate it with an approach known as a "Markov Chain Monte Carlo," or MCMC, algorithm (which essentially estimates switching probabilities from one phase to another by drawing samples of data).

When Ainslie et al. applied their model to predict the *total* box office of 404 movies released in North American theaters from 1995 to 1998, they calculate a MAPE of just 6% when using only demand factors, and of below 4% when adding in information on "market" forces. Their estimations for

---

[267]A separate approach to include distribution into a model of movie diffusion is the approach by Jones (1991), who essentially suggests a modification of the Bass model.

[268]A study by Radas and Shugan (1998) focuses on how seasonal variations of demand for entertainment could be embedded in diffusion models. Please see our chapter on entertainment distribution decisions for a discussion of the impact of products' release timing on success.

the basic Bass model and the BOXMOD model are comparable. Predictions become clearly less accurate, though, if used to predict the films' *opening weekend* box office results. Here, the Ainslie et al. model has a MAPE of 33% (remember that it also uses product features), whereas the MAPEs for BOXMOD and Bass are four- and five times higher, respectively. When performing a kind of out-of-sample analysis (using only data for the weeks up until the week prior to the prediction), the MAPE of their model worsens to 74%, as do those for Bass and BOXMOD. The importance of what kind of information is used to calibrate the prediction model is also visible when Sawhney and Eliashberg estimate their model for 111 movies from 1992 (and using a holdout set of ten movies). Whereas using movie characteristics only shows a decent MAPE of 71%, including one week of sales reduces this error to 52%, and including three weeks of sales data result in a MAPE of only 7.2%.

You, our reader, might wonder: are these levels of error too high for the models to be of *any* use at all? Whereas more accurate models are clearly desirable, their value is not really gauged by an absolute error number; instead, the true comparison is to having to take action without *any* such information. These MAPEs indicate that the best prediction models are far better than chance or than having *no* information. As is commonly said, a good forecasting model is like using your headlights when driving a car at night; your vision of the road ahead is not perfect, but it is far better than it would be without headlights.

With the available information obviously being a crucial factor for the effective use of prediction models, let us now take a look into the critical points in time for a manager to make success predictions—and the respective requirements for these models.

The "When" of Success Predictions: Early Versus Later Approaches

Early-Stage Predictions for Entertainment Products

An initial key point in time for making success predictions is when an entertainment product is still in its idea or concept phase—relatively little money has been spent so far for its development, and valid information on the product's success potential can either enable the firm to greenlight a huge hit or save a lot of money from being wasted. At this point, predictions can be made based on a product idea's genre and content elements (such as the script of a movie), and its brand characteristics, among others.

For each of these variables, our book describes studies that have empirically explored their role for product success—see, as just one example, the studies of Eliashberg and his colleagues on the role of movie scripts. Netflix used econometric analyses to predict the appeal of the HOUSE OF CARDS to its subscribers at an early stage based on the combination of genre and stars; the results encouraged the firm to greenlight the series without requiring the makers to develop a pilot episode, which was a critical artistic criterion for the makers (Nocera 2016).

In addition to such secondary data-based approaches, concept testing can be employed. The history of concept testing in entertainment is a pretty troubled one, as Austin (1989) recalls. But we argue that the reputation of concept testing has been damaged mostly by the insensitive way it has been handled by entertainment managers, who often do not address the inherent tension between audiences' reactions to a new product and the involved artists' vision for it. Gitlin (1983) points out that estimates based on concept testing rely heavily on a mix of expertise and data. This mix requires responsible handling—we have noted simply too many anecdotes in which executives instead misused the approach to support previously held preferences (i.e., to support the launch of a favored film or the "killing" of another).

Our book gives evidence of the many product characteristics that exert a causal influence on entertainment product success, and concept tests should find ways to anticipate these factors' impacts on target audiences before those audiences have actually experienced the final product. We have stressed that entertainment success contains a marketability and a playability element, and concept testing should be particularly effective for the marketability element of success that so strongly shapes product diffusion patterns.

We concede, however, that running predictions at such an early stage faces major challenges. Probably the most serious one is that not much information is yet available about the product, and entertainment products evolve dramatically, over time, as they are developed. At this stage, predictions also require a particularly careful framing of and sensitivity regarding what to do with the prediction results. We have shown that early predictions, although being far from arbitrary, are error-prone, and decisions must acknowledge this high uncertainty. Using them in the wrong (i.e., deterministic) way can threaten creativity, turning artists against the firm and its managers. At the same time, when used in a sensitive way, early predictions can help managers to position a project better, anticipating audience reactions and allocating adequate funds. This is when everyone involved still has the chance to fundamentally modify the product and rework the strategy—including its artistic vision, as our previous discussion of the strategies of successful entertainment innovators such as Pixar has demonstrated.

Later-Stage Predictions for Entertainment Products

When the development of the product advances, additional information can be added to prediction models, and new prediction methods become available. When the product moves from concept to reality, prediction models can be used (1) to help a firm with a decision of whether or not to buy a product being made by others, or (2) with the aim of fine-tuning the final formulation of a product being developed in-house in order to optimize its eventual market performance.

Except for the product's actual performance, almost all factors we discuss in this book can be included in later-stage predictions, ranging from product factors, such as content and age rating, to advertising spending. Also available are distribution decisions and consumers' reactions to the product so far, as expressed in pre-release buzz and pre-orders.[269] Regarding the latter, Moe and Fader (2002) show for 66 albums that were released in 1997–1998 that such weekly pre-release ordering data can help to make more accurate predictions of post-release sales.

We have shown that the accuracy of such later models can be substantially higher than those which are run early; however, this gain in accuracy comes at a price. We have demonstrated that creative decisions usually have path dependencies—they cannot be easily altered in later stages of the development process. Also, contracts with distributors and decisions within the producing firm may prevent the making of major changes at this point. And the vast amount of data available at this point in time increases the tendency of basing decisions on the results in a superficial way, purely optimizing fit measures and treating correlations as causal effects.

Again, primary data from product tests can be collected at this stage and used for predictive purposes, such as by fitting audience reactions to a diffusion pattern. In such product tests, consumers are usually provided with the (nearly) completed product or parts of it, with the idea being that their reactions can help the team make final editing and positioning tweaks (DeVault

[269]Please see our discussion of the crucial role of buzz in today's entertainment marketplace in our chapter on earned entertainment communication.

2016; Marich 2013). For example, Avirgan (2015) describes how test marketing with a representative sample of potential viewers predicted the box office problems of the remake of the Marvel Entertainment film, FANTASTIC FOUR. But it also shows that such later tests might lack the ability to address the true causes of the problem; it was just too late in development to make the needed cast changes, so promotional activities were altered to cut the studio's losses.

One promising way to improve tests is to employ methods that do not rely solely on asking questions of test audiences, but instead infer audience members' reactions to concepts or product elements from measures of their bodily responses. Marketing scholars have been experimenting with biometric markers via techniques such as fMRI (functional magnetic resonance imaging) brain scans, eye-tracking, pulse rate measures, etc. In a specific approach, researchers from Disney have used infrared cameras and motion-capture technologies to measure the facial/body language of movie audiences to determine moviegoers' emotional reactions to specific scenes (i.e., smiling, laughing, or neither of the two) for nine Disney movies from 2015 to 2016 (Deng et al. 2017). By combining primary observations with technology and insights derived from large databases, such approaches might breathe new life into the under-used method of product testing for entertainment products.

An example for the power of combining multiple data sources for later-stage entertainment predictions is Eliashberg et al.'s (2000) MOVIEMOD approach. MOVIEMOD employs a diffusion model that is developed before a movie's release to forecast consumer awareness and adoption intentions for the movie and, subsequently, its success. MOVIEMOD uses a large set of the success drivers we discuss in this book ("features")[270] and derives audience reactions from a three-hour consumer "clinic," in which consumers are asked to fill out questionnaires, while some actually watch the movie in question. Based on this information, the authors classify consumers into certain stages (from "undecided" to "negative spreader") and statistically model consumers' transitions between stages, based on the number of people that are already in a certain stage. When the scholars implemented the approach in a real-word setting (for the film SHADOW CONSPIRACY), they report that their predictions had an impressive prediction

[270]In the case of MOVIEMOD, the variables range from product variables (such as the quality, theme, story, and cast) to advertising and distribution.

error of only 4%, which also outperformed other prediction approaches they used as comparison standards.

Our previous discussion has shown that prediction models can strongly benefit from the inclusion of a product's initial sales; the combination of diffusion-based models and actual sales makes a great team. This is particularly relevant for products whose financial viability is less determined by short-term results, but their long-term performance, such as those for which the "niche concept" of marketing is used. One way to make full use of the post-release period is to measure the actual word of mouth for a product by those consumers who have already experienced it.[271] In line with our later analysis of word of mouth as a substantial influencer of entertainment product success, Dellarocas et al. (2007) prediction study shows that a diffusion model with word-of-mouth information forecasts the success of 80 movies (from 2002) better than one without.[272]

Multi-Stage Prediction Models

Finally, some *Entertainment Science* scholars have made the differing availability of information over the innovation process and its life cycle the very topic of their efforts. They have developed multi-stage models which enable managers to predict the performance of an entertainment product at different points in time.

One of these achievements is the model by Neelamegham and Chintagunta (1999), which predicts opening-week movie performance in the U.S. and in international markets at different stages, ranging from the concept stage, to the pre-release stage, to an international release stage (in which the film's domestic box office is included as predictor variable).[273] Using an MCMC estimation procedure, the scholars calibrated their model for 25 movies that had been released in the U.S. and one or more of 13 other countries in 1994–1996; ten others were used as holdouts.

Their results demonstrate that the prediction error decreases as more information becomes available; for example, the RMSE for the U.S. opening

[271]Please note that we make a clear distinction between experience-based word of mouth and other kinds of (speculative) consumer articulations.

[272]Let us note that Dellarocas et al.'s study measures only the *amount* of word of mouth and only compares the word-of-mouth model with one that does include neither word of mouth nor sales data.

[273]At the time Neelamegham and Chintagunta developed their model, sequential international releases were dominant in film. Please see also our discussion of such "intermarket success-breed-success effects" in the context of entertainment distribution decisions.

decreases by 24% when moving from concept to pre-release stage. And for the international opening performance, the RMSE declined across all 13 countries and the ten holdout films by 14% from the concept stage to the post-domestic stage (i.e., when information about a movie's U.S. performance is used). The improvement reached 20% when information on the film's local distribution is also included in the model. This shows the benefit of waiting to determine the value of the rights of a movie until such information becomes available.

In our own study of TV rights of films, we also applied a multi-stage prediction approach, but took the perspective of an international TV broadcaster (Hennig-Thurau et al. 2013). Using partial least squares for making rating predictions, we estimated a separate set of equations for a total of five stages, each time incorporating only the information available at that point in time for a broadcaster to use to predict the future TV ratings for a film. From a baseline model (which did not include *any* movie-specific information) to a model that includes a film's success in the foreign country's theaters, the prediction error (measured here as the RMSE[274]) shrank by 31%, while the explained variance increased by 34%. Figure 10.8 shows the changes in prediction error and variance explanation across the different stages.

In addition to RMSE and explained variance, the figure also illustrates the impact that such improvement in predictive power can have for the estimated monetary value of an example film for the TV station. We used Sony's SPIDER-MAN movie and calculated the value of the film for the TV station based on the actual advertising fees per million viewers around the time we conducted the study. As can be seen in the figure, the value does not necessarily always increase—the change in valuation depends on the added information in each phase, and in this case the German box office did not meet the lofty expectations raised by the film's immense North American box office performance.

So, Better to Wait for the Fanny to Squirm or Use Prediction Models?

In this final section of our chapter on entertainment innovation, we have provided you with an overview of prediction models, illustrating the value of several specific models that have shown promise in aiding entertainment

[274]In the article, we refer to the standard error of the (regression) estimate, or SEE, which is mathematically the same as the RMSE.

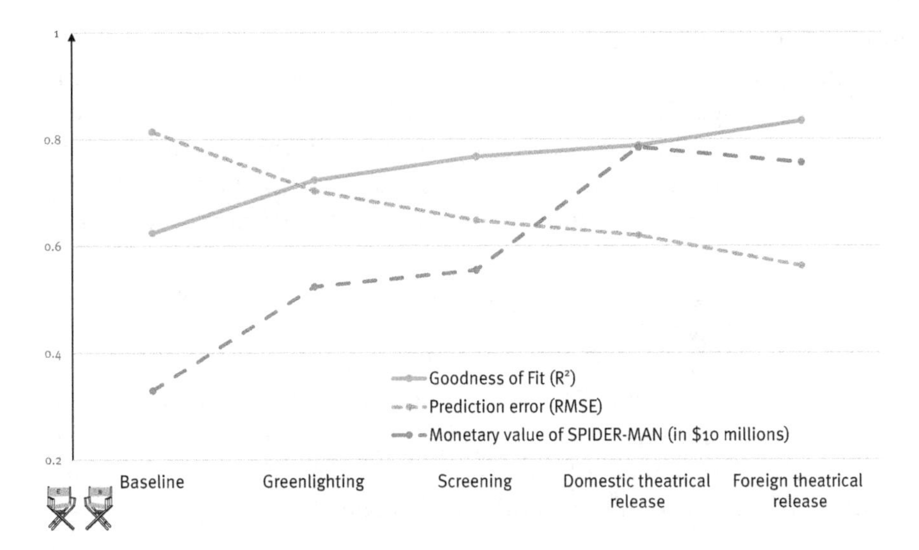

Fig. 10.8 Development of prediction accuracy over different innovation stages
Notes: Authors' own illustration based on results reported in Hennig-Thurau et al. (2013). The baseline model describes a model in which no film-specific variables are considered in the prediction. Film title is trademarked.

managers making tough innovation decisions. Despite the imperfections of any given model, the predictions provided by the modeling of data will improve the quality of decision making. Managers should not rely blindly on model-based predictions. But the other extreme—making "100 million-dollar decisions" based on "gut" (STX Entertainment's CEO Fogelson, quoted in Friend 2016)—is a ditch that needs to be avoided.

The insights that emerge from analytics can complement a manager's personal experience and increase his or her odds of making the right decisions at each stage of the product development process. The quality of prediction is connected to the types and amounts of data that are used, the statistical methods employed, and, last but not least, the theoretical model that underlies the estimations. But the most challenging part is something else: it's the development of a mind-set that integrates prediction results into decision making. The first half of this chapter provides guidelines on how to master that challenge.

And which of the several prediction models we have covered is best-suited? Certainly the specific conditions of the situation, such as the intended contributions of using prediction models, matter for this decision, but so do the resources a firm has at hand and the economic importance of the innovation(s) whose performance is to be predicted. Please keep in mind that all the models we cited are based on what was known about entertainment success drivers at the time of their creation. Our book offers an updated perspective of what influences product success in entertainment, and we recommend that entertainment managers include as many relevant features as possible and link them with product success in a way that is in line with state-of-the-art knowledge about the theoretical mechanisms and linkages.

There is one more thing we would like to ask you for. Please be careful if someone boasts that they can make predictions for you which are just "too good" to be true. Such approaches should be examined *very* carefully, paying attention to the conceptual foundation on which they are based, whether they purport to explain or to only predict, and the quality of the procedures used to select the sample of focal cases and to collect data. If such "behind-the-curtain" information is not offered, it might be better to let the offer pass.

Concluding Comments

In this chapter, we worked through state-of-the-art findings from organizational science and prediction research and applied them to the management of innovation in entertainment firms. By considering the unique characteristics of entertainment products and markets, we believe these are useful insights for entertainment managers, addressing the pressing challenge to innovate new entertainment products on a continuous basis.

In entertainment, those with artistic visions often consider bowing to commercial considerations as sullying—or even defiling—the actual work of art. Finding a way to respectfully balance artistic and economic goals is the foundation for the entire chapter. Our analysis showed that this can only be achieved by creating a culture that combines autonomy and responsibility, along with an organizational structure that attracts people with the right skills and values and equips and enables them to be creative, but with discipline.

We complemented this firm-level analysis of factors that contribute to prolific innovation activities with a product-level analysis of approaches that can be used to improve managers' understanding of a new product's commercial potential. We reviewed the different econometric prediction methods that are available for such a purpose and discuss concrete scientific models that have been developed for predicting new product success at different stages of the innovation process. No approach is perfect, but used thoughtfully, predictions can be generated that will provide valuable input into key managerial decisions.

Whereas the decisions we discussed in our four product episodes are essential for entertainment managers, having the right product is only one part of a complex equation for success. We now move on to the second "P" of the entertainment marketing mix: promotion. Specifically, we examine the rich repertoire of tools that firms can use for communicating with consumers (or to get consumers to communicate with each other) regarding an entertainment product, along with other, less controllable information sources such as success cascades, word of mouth, and expert reviews.

References

Albers, S., & Eggers. S. (1991). Organisatorische Gestaltung von Produktinnovations-Prozessen – Führt der Wechsel des Organisationsgrades zu Innovationserfolg? *Zeitschrift für betriebswirtschaftliche Forschung, 43*, 44–46.

Algobeans (2016). Artificial neural networks (ANN) introduction, March 13, https://goo.gl/hprXNQ.

Ainslie, A., Drèze, X., & Zufryden, F. (2005). Modeling movie life cycles and market share. *Marketing Science, 24*, 508–517.

Amabile, T. M. (1996). *Creativity and innovation in organizations*. Boston: Harvard Business School Background Note.

Archer, A., & Walcyzk, D. (2006). Driving creativity and innovation through culture. *Design Management Review, 17*, 15–20.

Arrow, K. (1962). Economic welfare and the allocation of resources for invention. In *The rate and direction of inventive activity: Economic and social factors* (pp. 609–626). Princeton University Press.

Asad, K. I., Ahmed, T., & Rahman, S. (2012). Movie popularity classification based on inherent movie attributes using C4.5, PART and correlation coefficient. In *Proceedings of ICIEV* (pp. 747–752).

AT5 (2017). Herman Hulst (Guerrilla Games) We can make what we want of Playstation. *AT5*, March 16, https://goo.gl/AJkgG1.

Austin, B. A. (1989). *Immediate seating: A look at movie audiences*. California: Wadsworth Pub. Co.

Avirgan, J. (2015). Podcast: The guy who predicts whether a movie will bomb, months before it's made. *FiveThirtyEight*, September 7, https://goo.gl/eNTnSC.

Bailey, S. (2016). Kanye West: Free form. *Surface*, November 18, https://goo.gl/P5e8HD.

Barber, N. (2015). Heaven's Gate: From Hollywood disaster to masterpiece. *BBC*, December 4, https://goo.gl/TvSji6.

Bass, F. M. (1969). A new product growth model for consumer durables. *Management Science, 15*, 215–227.

Benner, M. J., & Tushman, M. L. (2003). Exploitation, exploration, and process management: The productivity dilemma revisited. *The Academy of Management Review, 28*, 238–256.

Bishop, C. M. (2006). *Pattern recognition and machine learning*. Cambridge: Springer.

Breiman, L. (2001). Statistical modeling: The two cultures. *Statistical Science, 16*, 199–231.

Brownlee, Alexander E. I., Regnier-Coudert, O., McCall, J. A. W., Massie, S., & Stulajter, S. (2013). An application of a GA with Markov network surrogate to feature selection. *International Journal of Systems Science, 44*, 2039–2056.

Brunsdale, M. M. (2010). *Icons of crime and detection*. Santa Barbara: Greenwood.

Cameron, K. S., & Freeman, S. J. (1991). Cultural congruence, strength, and type: Relationships to effectiveness. *Research in Organizational Change and Development, 5*, 23–58.

Catmull, Ed. (2008). How Pixar fosters collective creativity. *Harvard Business Review, 86*, 64–72.

Catmull, Ed., & Wallace, A. (2014). *Creativity, Inc*. New York: Random House LLC.

Chang, B.-H., & Ki, E.-J. (2005). Devising a practical model for predicting theatrical movie success: Focusing on the experience good property. *Journal of Media Economics, 18*, 247–269.

Clarkson, N. (2016). Why failure is a key part of Pixar's culture. *Virgin*, January 8, https://goo.gl/LfWY1t.

Coase, R. H. (1937). The nature of the firm. *Economica, 4*, 386–405.

Deci, E. L., & Ryan, R. M. (1985). The general causality orientations scale: Self-determination in personality. *Journal of Research in Personality, 19*, 109–134.

Deci, E. L., & Ryan, R. M. (2000). Self-determination theory and the facilitation of intrinsic motivation, social development, and well-being. *American Psychologist, 55*, 68–78.

Dellarocas, C., Zhang, X., & Awad, N. F. (2007). Exploring the value of online product reviews in forecasting sales: The case of motion pictures. *Journal of Interactive Marketing, 21*, 23–45.

Deng, Z., Navarthna, R., Carr, P., Mandt, S., Yue, Y., Matthews, I., & Mori, G. (2017). Factorized variational autoencoders for modeling audience reactions to movies. In *IEEE Conference on Computer Vision and Pattern Recognition*.

Deshpande, R., Farley, J. U., & Webster, F. E., Jr. (1993). Corporate culture, customer orientation, and innovativeness in Japanese Firms: A quadrad analysis. *Journal of Marketing, 57*, 23–37.

DeVault, G. (2016). The market research behind Hollywood movies. *The Balance*, September 14, https://goo.gl/KjfQtb.

Dougherty, D., & Heller, T. (1994). The illegitimacy of successful product innovation in established firms. *Organization Science, 5*, 200–218.

Disney, W. (2017) Walt Disney: Quotes. *IMDb*, no date, https://goo.gl/5FaCr3.

Edmonson, A. (1999). Psychological safety and learning behavior in work teams. *Administrative Science Quarterly, 44*, 350–383.

Eliashberg, J., Jonker, J., Sawhney, M., & Wierenga, B. (2000). MOVIEMOD: An implementable decision-support system for prerelease market evaluation of motion pictures. *Marketing Science, 19*, 226–243.

Eliashberg, J., Hui, S. K., & Zhang, J. Z. (2007). From story line to box office: A new approach for green-lighting movie scripts. *Management Science, 53*, 881–893.

Faughnder, R. (2017). The reason Hollywood's studio leadership is in flux: The business model is changing. *Los Angeles Times*, March 26, https://goo.gl/KWkK2D.

Fleming Jr., M. (2016). 'Big Bad Wolves' helmers Aharon Keshales & Navot Papushado exit Bruce Willis Death Wish remake. *Deadline*, May 4, https://goo.gl/xUV3uz.

Franklin-Wallis, O. (2015). How Pixar embraces a crisis. *Wired*, November 17, https://goo.gl/dfVMm8.

Friend, T. (2016). The mogul of the middle. *The New Yorker*, January 11, https://goo.gl/8hYXxT.

Gartner (2016). Gartner's 2016 hype cycle for emerging technologies identifies three key trends that organizations must track to gain competitive advantage. *Gartner*, August 16, https://goo.gl/pmmHPs.

Ghiassi, M., Lio, D., & Moon, B. (2014). Pre-production forecasting of movie revenues with a dynamic artificial neural network. *Expert Systems with Applications, 42*, 3176–3193.

Gilson, L. L. (2015). Creativity in teams: Processes and outcomes in creative industries. In C. Jones, M. Lorenzen, & J. Sapsed (Eds.), *The Oxford Handbook of Creative Industries* (pp. 50–74). Oxford: Oxford University Press.

Gilson, L. L., & Shalley, C. E. (2004). A little creativity goes a long way: An examination of teams' engagement in creative processes. *Journal of Management, 30*, 453–470.

Gitlin, T. (1983). *Inside prime time*. New York: Pantheon Books.

Gomez-Uribe, C. A., & Hunt, N. (2015). The Netflix recommender system: Algorithms, business value, and innovation. *ACM Transactions on Management Information Systems, 6*, 13–19.

Greve, H. R. (2007). Exploration and exploitation in product innovation. *Industrial and Corporate Change, 16*, 945–975.

Gupta, A. K., Smith, K. G., & Shalley, C. E. (2006). The interplay between exploration and exploitation. *Academy of Management Journal, 49*, 693–706.

Hamel, G. (2006). Management à la Google. *The Wall Street Journal*, April 26, https://goo.gl/REAaRj.

Harvey, S. (2014). Creative synthesis: Exploring the process of extraordinary group creativity. *Academy of Management Review, 39*, 324–343.

Hastings, R. (2009). Netflix culture: Freedom & responsibility. *SlideShare*, August 1, https://goo.gl/QHvJAP.

Hennig-Thurau, T., Fuchs, S., & Houston, M. B. (2013). What's a movie worth? Determining the monetary value of motion pictures' TV rights. *International Journal of Arts Management, 15*, 4–20.

Hotho, S., & Champion, K. (2011). Small businesses in the new creative industries: Innovation as a people management challenge. *Management Decision, 49*, 29–54.

Howe, S. (2013). How Martin Scorsese's Elmore Leonard movie LaBrava is one that got away. *Vulture*, August 23, https://goo.gl/YLrYnq.

Ingram, M. (2017). Here's what Netflix and Amazon spent millions on at Sundance. *Fortune*, January 30, https://goo.gl/thLE4t.

Jones, R. (1991). Incorporating distribution into new product diffusion models. *International Journal of Research in Marketing, 8*, 91–112.

Jones, O. (2014). Yes, we're being bought by Microsoft. *Mojang*, September 15, https://goo.gl/psZGmM.

Jucevicius, G. (2010). Culture vs. cultures of innovation: Conceptual framework and parameters for assessment. In *Proceedings of the International Conference on Intellectual Capital* (pp. 236–244).

Kanter, R. M. (1997). Strategies for success in the new global economy: An interview with Rosabeth Moss Kant. *Strategy & Leadership, 25*, 20–26.

Kelleher, J. D., Namee, B. M., & D'Arcy, A. (2015). *Machine learning for predictive data analytics*. London: The MIT Press.

Kirschbaum, E., & Hopewell, J. (2008). StudioCanal buys Kinowelt. *Variety*, January 17, https://goo.gl/oNCSEn.

Kloo, A. (2016). Amazon-Video-Chef Schneider: Kreative Freiheit ist Grundbedingung für einzigartige Serien. *Blickpunkt: Film*, October 18, https://goo.gl/w2gjLS.

Kohavi, R., & Provost, F. (1998). Glossary of terms. *Journal of Machine Learning, 30*, 271–274.

Kokonis, M. (2009). Hollywood's major crisis and the American film 'Renaissance'. In R. B. Ray (Ed.), *A certain tendency of the Hollywood cinema, 1930–1980* (pp. 169–206). Princeton: Princeton University Press.

Kurokawa, S. (1997). Make-or-buy decisions in R&D: Small technology based firms in the United States and Japan. *IEEE Transactions of Engineering Management, 44*, 124–134.

Kehr, D. (2008). United Artists—90th anniversary. *New York Times*, March 27, https://goo.gl/LgxWam.

Litman, B. R. (1983). Predicting success of theatrical movies: An empirical study. *Journal of Popular Culture, 16*, 159–175.

Liu, X., Shi, S., Teixera, T., & Wedel, M. (2018). Video content marketing: The making of clips. *Journal of Marketing, 82*, 86–101.

Lumpkin, G. T., & Dess, G. G. (1996). Clarifying the entrepreneurial orientation construct and linking it to performance. *Academy of Management Review, 21*, 135–172.

Lussier, G. (2016). Walt Disney Animation is officially as good as Pixar now. *Gizmodo UK,* February 18, https://goo.gl/JyLJ1s.
March, J. G. (1991). Exploration and exploitation in organizational learning. *Organization Science, 2*, 71–87.
Marich, R. (2013). *Marketing to moviegoers: A handbook of strategies and tactics.* United States: Library of Congress Cataloging-in-Publication Data.
Masters, K. (2016). Steven Spielberg on Dreamworks' past, Amblin's present and his own future. *The Hollywood Reporter*, June 15, https://goo.gl/LqFjKR.
McNary, D. (2017). Netflix buys Martin Scorsese's 'The Irishman' starring Robert De Niro. *Variety*, February 21, https://goo.gl/pcLFjC.
Moe, W. W., & Fader, P. S. (2002). Using advance purchase orders to forecast new product sales. *Marketing Science, 21*, 347–364.
Moriarty, C. (2013). Rising to greatness: The history of Naughty Dog. *IGN*, October 4, https://goo.gl/bXQVyS.
Muller, E., Peres, R., & Mahajan, V. (2009). *Innovation diffusion and new product growth*. Cambridge: Marketing Science Institute.
Narayan, V., & Kadiyali, V. (2016). Repeated interactions and improved outcomes: An empirical analysis of movie production in the United States. *Management Science, 62*, 591–607.
Neelamegham, R., & Chintagunta, P. (1999). A Bayesian model to forecast new product performance in domestic and international markets. *Marketing Science, 18*, 115–136.
Nocera, J. (2016). Can Netflix survive in the new world it created? *The New York Times Magazine*, June 15, https://goo.gl/e1d2Zu.
Packard, G., Aribarg, A., Eliashberg, J., & Foutz, N. Z. (2015). The role of network embeddedness in film success. *International Journal of Research in Marketing, 33*, 328–342.
Parimi, R., & Caragea, D. (2013). Pre-release box-office success prediction for motion pictures. In *Proceedings of International Workshop on Machine Learning and Data Mining in Pattern Recognition* (pp. 571–585). Springer, Berlin, Heidelberg.
Peltoniemi, M. (2015). Cultural industries: Product-market characteristics, management challenges and industry dynamics. *International Journal of Management Reviews, 17*, 41–68.
Pond, S. (1986). Dateline Hollywood. *The Washington Post*, August 28, https://goo.gl/7o2UFN.
Poppo, L., & Zenger, T. (1998). Testing alternative theories of the firm: Transaction cost, knowledge-based, and measurement explanations for make-or-buy decisions in information services. *Strategic Management Journal, 19*, 853–877.
Radas, S., & Shugan, S. M. (1998). Seasonal marketing and timing new product introductions. *Journal of Marketing Research, 35*, 296–315.
Rao, H., Sutton, R., & Webb, A. P. (2008). Innovation lessons from Pixar: An interview with Oscar-winning director Brad Bird. *McKinsey Quarterly*, https://goo.gl/rHqcDv.

Russell, J. (2004). Foundation myths. *New Review of Film and Television Studies, 2*, 233–255.

Savitskaya, I., & Järvi, K. (2012). Culture for innovation: Case Finnish game development. *Proceedings of ISPIM Conference, 23*, 1–15.

Sawhney, M. S., & Eliashberg, J. (1996). A parsimonious model for forecasting gross box-office revenues of motion pictures. *Marketing Science, 15*, 113–131.

Schillat, F. (2017). Amazon-Prime-Video-Chef im Interview: 'Wir haben kein Interesse daran, dass unsere Kunden nicht mehr fernsehen'. *Meedia*, March 15, https://goo.gl/zWmUzg.

Sharda, R., & Delen, D. (2006). Predicting box-office success of motion pictures with neural networks. *Expert Systems with Applications, 30*, 243–254.

Sharda, R., & Delen, D. (2010). Predicting the financial success of Hollywood movies using an information fusion approach. *Industrial Engineering Journal, 21*, 30–37.

Simonoff, J. S., & Sparrow, I. R. (2000). Predicting movie grosses: Winners and losers, blockbusters and sleepers. *Chance, 13*, 15–24.

Smith, S., & Paquette, S. (2010). Creativity, chaos and knowledge management. *Business Information Review, 272*, 118–123.

Smith, M. D., & Telang, R. (2016). *Streaming, sharing, stealing—Big data and the future of entertainment*. Cambridge: The MIT Press.

Spiegel, J. (2013). The Pixar perspective on replacing directors. *The Pixar Times*, September 3, https://goo.gl/PhX4oH.

Steel, E. (2014). Those dreaded spoilers that can torpedo dramatic plot take on a new meaning. *New York Times*, September 21, https://goo.gl/tFapLw.

Slifkin, I. (2014). Cannon fodder: In praise of Golan and Globus. *MovieFanFare*, June 13, https://goo.gl/SZkN2y.

Thomson, D. (2008). The history of United Artists. *The Guardian*, February 23, https://goo.gl/AqaTym.

Total Film (2006). The story behind Raiders of the Lost Ark. *Gamesradar+*, August 24, https://goo.gl/h5eMrT.

de Ven, V., Andrew, D. P., Garud, R., & Venkataraman, S. (1999). *The innovation journey*. New York: Oxford University Press.

Walesh, S. G. (2012). Staging a creative culture. *Leadership and Management in Engineering, 12*, 338–340.

Walker, G., & Weber, D. (1984). A transaction cost approach to make-or-buy decisions. *Administrative Science Quarterly, 29*, 373–391.

Williamson, O. E. (1985). *The economic institutions of capitalism. Firms, markets, relational contracting*. New York: The Free Press.

Zhou, Y., Zhang, L., & Yi, Z. (2017). Predicting movie box-office revenues using deep neural networks. *Neutral Computing and Applications*, Online Publication, 1–11.

11

Entertainment Communication Decisions, Episode 1: Paid and Owned Channels

To Control or Not to Control: Some Words on the Three Basic Communication Categories

The myth that "quality finds its way" is just that—a myth. A great product (one that meets the requirements mentioned in the previous chapters of this book) is certainly helpful, but is never solely sufficient to warrant the success of a new entertainment product. Instead, communication with consumers about the product is also crucial, as it levers the potentials embedded in the product, may they be experience, search, or quasi-search qualities. Often, product and communication strategies are tied closely together, as the success potential of various product characteristics relies on specific communication approaches. The importance of communication as part of the entertainment marketing mix is illustrated by the enormous budgets that are dedicated to communication efforts; these investments constitute strategic resources on their own.

In this book, we refer to communication simply as the informational flow regarding an (entertainment) product. Today, entertainment producers face a multitude of different communication channels. We classify them into three general categories: paid channels (such as TV and print advertising), owned channels (including a film's social media domain on Facebook, as well as the product's packaging), and "earned" channels. Whereas the first two categories should be more or less intuitive, the latter one might require some explanation. It contains different kinds of communication by consumers and other "independent" market actors: it captures the communication

T. Hennig-Thurau and M. B. Houston, *Entertainment Science*,
https://doi.org/10.1007/978-3-319-89292-4_11

about a product that is (1) articulated and shared by consumers in the form of word of mouth (when communication is evaluative) and pre-release buzz (when it is anticipatory), (2) information that reflects consumers' reactions to a product (e.g., when charts and bestseller lists signal quality), as well as (3) communication by experts, such as professional critics (when they review a product) and members of the entertainment industry (when they give awards to a product). Note that we often put "earned" in quotation marks; although what is communicated through this channel is often positive, it certainly isn't always so. Consumers' word of mouth can be devastating, as can be professional reviews (just think of the movie GIGLI, which experienced such hateful backlash from consumers and critics that it threatened the careers of Ben Affleck and then-fiancé Jennifer Lopez).

The main difference between the three categories is the degree to which they can be controlled by entertainment managers. Controllability ranges from very high (i.e., you largely get what you pay for in paid communication channels) to somewhat high (for owned channels which usually require active contributions by consumers) to low (for "earned" media, where anything goes—fair or unfair). A big challenge for managing entertainment communication is that controllability does not correlate very highly with effectiveness. For starters, Fig. 11.1 illustrates what consumers consider as the sources to become aware of a new movie they end up seeing in a theater. The numbers are based on the German Federal Film Board's annual panel of 20,000 consumers that is representative of the German market.

We see that highly controllable paid media is still named by many consumers as their main source of movie awareness, although the proportion of consumers who name offline paid media has declined from a decade ago.[275] But word of mouth, which is *much* more difficult to control for managers, follows closely behind. And owned media (e.g., online movie trailers) shows the highest growth rate; it is now a more-frequent awareness source than TV and print media content, according to these consumer reports, and almost as frequently used as word of mouth. We also note strong differences between consumer segments: TV is (still) king among kids and teens (who hardly even notice editorial—offline—content); they rely much more on friends and (company-)owned online sources. Older consumers, in contrast, say that their movie-going inspirations mostly come from the same

[275]Most of the decline of the offline advertising effect in the figure comes from outdoor, not TV and print. But editorial content of both TV and print has lost large parts of its awareness-related power.

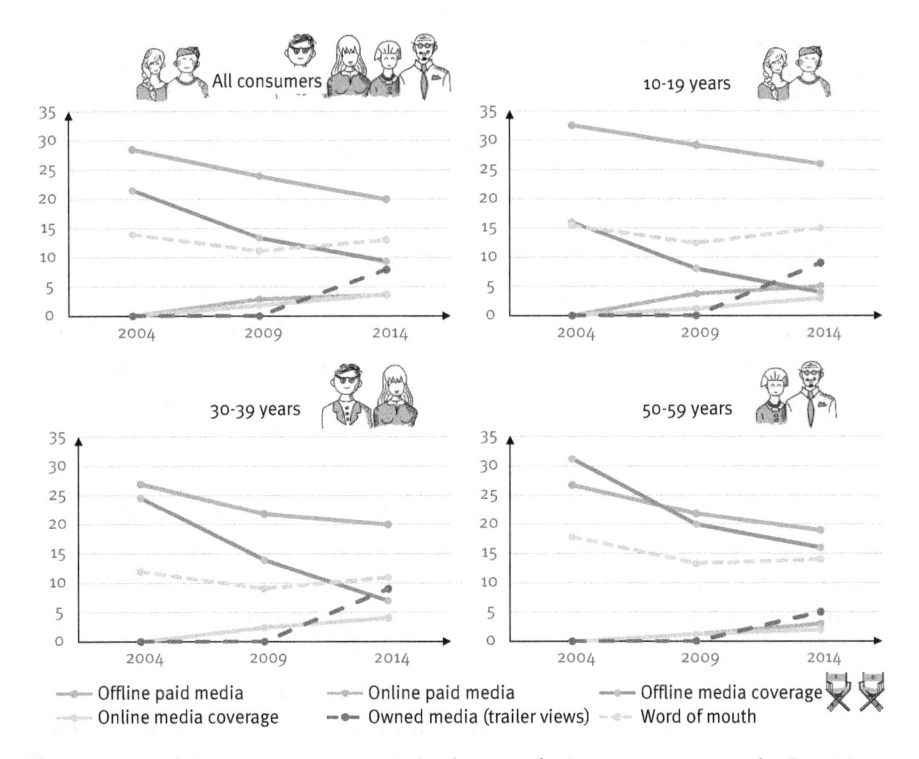

Fig. 11.1 Movie awareness sources over times and consumer segments in Germany
Notes: Authors' own illustration based on data reported in FFA (2005, 2010, 2015). Numbers are percentages of consumers who reported becoming aware of a movie mainly through a particular information source in a given year. "Offline paid media" encompasses TV, newspaper, and outdoor advertising, and "Online paid media" is Internet advertising. "Owned media" are views of trailers hosted on Internet platforms such as YouTube. "Offline media coverage" contains TV and newspaper reports, while "Online media coverage" is non-advertising information on the Internet, including social media. With graphical elements by Studio Tense.

editorial offline content that teens ignore; these older consumers are much less impacted by digital sources. So, movie producers had better know their target group.

But as valuable as such self-reported survey results are for an initial look, they are also fraught with problems. In short, we consumers are not very good at remembering how we made decisions. Our memory systems are biased toward retaining active processes, such as watching trailers in theaters or the Internet, and less likely to recall passive ones (such as seeing an ad on TV or on a website). There is a social desirability issue too; whereas being influenced by TV advertising is nothing we are proud of, reading a professional movie review is something most of us are more prone to report.

Entertainment Science research tries to overcome such biases by using more rigorous methods. But rigor comes at a price: most academic studies focus on only one communication category at a time, trading detail and depth off against broadness. At the time of writing, we noted only two studies that included all three communication categories in an entertainment context: Chen et al. (2015) look at music artists, whereas Lovett and Staelin (2016) analyze viewership of six episodes of a TV series. Neither study is *fully* comprehensive in their coverage of communication channels, but both consider data from all categories simultaneously. We will pay particular attention to their findings when we discuss the different communication categories in the following. Our use of "normed" metrics, such as elasticities, should also help in comparing the findings across studies.

In this chapter, we will explore the first two categories, paid and owned communication, and their respective impacts on product success. With managerial controllability serving as the structuring element of the discussion, we will begin with paid channels before entering the less-controllable terrain of the owned channels of social media, a setting in which we compare entertainment communication to the act of playing pinball. In the next chapter, we will then study the least-controllable "earned" channels.

Our discussion of paid and owned communication in this chapter circles mostly around the *logistics* of communication—the "how-much-to-spend" and "when-to-communicate" questions. But *Entertainment Science* scholars have also shed some light on the *content* of communication and its design—the "what-to-communicate" question, if you will, which is of complementary and fundamental concern. This is where we will begin.

What to Communicate: Designing the Content that Fuels Entertainment Communication Channels

> "[The trailer] is the single most important piece of advertising… There's nothing else that comes close."
>
> —*CBS Films president* Terry Press *(quoted in* LaFrance *2014)*

Most, if not all, entertainment communication aims to make consumers aware of a new product's existence (that is, building brand awareness), and/or to provide consumers with information about the product's potential to provide desirable familiarity and sensations (i.e., establishing a compelling brand image). Which criteria must entertainment communication meet to achieve these aims?

Our discussion begins with what can be considered the backbone of all entertainment communication: the trailer. As "previews of coming attractions" (Kernan 2004, p. 1), trailers are the focal communication format for movies, TV content, and video games (Grainge and Johnson 2015). Trailers are even sometimes used by publishers to inform consumers about new novels (just search for "book trailers" on YouTube). According to the German Federal Film Board's panel, about 22% of consumers say that they become aware of a new film from its trailer; for some age groups the format's reach is even higher. It has been estimated that there were more than a billion movie trailer views in 2013 on YouTube alone (Kehe and Palmer 2013), and we assume that this number has not shrunk since then.

Some of our readers might remember Amanda from the movie The Holiday, in which she is paid "big bucks" by Hollywood studios for her trailer-editing skills. We will look what *Entertainment Science* can tell us about what kind of skills those might be—skills that set "good" trailers apart from "not-so-good" ones. We will then do the same for posters, which are a less dynamic tool of entertainment communications. And after having shed light on the essentials of trailer (and poster) design, we will address a key question that affects the design of *any* kind of entertainment communication: how *much* information about a new product should be provided by its producer? There is a delicate balance between "not enough" and "too much"!

What Makes a Powerful Trailer?

The Roles of Trailers Then and Now

Trailers are a highly complex marketing tool because they combine all three kinds of quality-related information we discussed earlier in the book. First, trailers tell consumers which unbranded and branded attractions are involved in a product, such as which stars will participate ("inferential cues"). Second, they often mention awards or critical acclaim that signals the product's overall quality ("substitute cues"). Third, trailers also provide consumers with excerpts of the actual product, enabling them to sample it. The sampling aspect is what sets trailers apart from other forms of advertising and communication—they are a hybrid format, mixing communication elements with elements of the product itself (e.g., Grainge and Johnson 2015).

Whereas consumers usually acknowledge that a trailer is biased (showing only the "best" elements of the product), they also appreciate them as "valuable short-form content" (Grainge and Johnson 2015, p. 149), as reflected by the many views that trailers attract on digital video platforms.

The digital age and its platforms have not only increased the accessibility of trailers for consumers, but have also shaped the way trailers are designed. When posted on the Internet, a trailer has to *attract* an audience, instead of only being shown to a "captive" audience that is locked in the theater and has no choice but to watch. And like digital samples of other forms of entertainment, movie trailers can spread like wildfire, being shared and re-shared virally among members of large consumer networks.[276]

Trailers historically "hypersold" their films (in the 1940s) and applied advertising techniques (in the 1950s and 1960s), such as speaking directly to the audience (e.g., google how actor James Stewart does so in the trailer for Hitchcock's REAR WINDOW).[277] However, by the 1980s, studios developed more subtle selling strategies, as audiences became largely desensitized to hard-sell advertisements (LaFrance 2014).[278] But to stimulate viral sharing and to make them the talk of the (digital) town, producers now present trailers as kinds of complementary "standalone products."

Elements of the final product are complemented with additional, original footage, sometimes created just for the trailer. For example, the first teasers for THE HUNGER GAMES: MOCKINGJAY PART 1 were fictitious television addresses from evil President Snow. Also, the fact that trailers are now analyzed on a frame-by-frame basis by fans and media (e.g., Plumb 2015) enables triggering questions and embedding hidden messages. When the trailer for STAR WARS EPISODE VII did not show legendary Luke Skywalker, millions joined the conversation on the Internet. But trailer's standalone character might carry its own problems—one executive told us he was concerned that modern trailers might be treated as (gratuitous) products on their own by at least some consumers, satisfying their entertainment needs instead of triggering them. But with no empirical evidence, such cannibalization effects remain speculation at this point.

---

[276]See the section about the pinball character of entertainment communication in the digital age.

[277]Or just go to https://goo.gl/rMU426.

[278]If you are interested in a comprehensive historical review of trailers, please do not miss Kernan (2004).

Three (or More) Principle Appeals of Trailers

Media theorists have tried to identify the main appeals of trailers for audiences by content analyzing individual trailers. As a result of such efforts, Lisa Kernan (2004) finds five key aspects of trailers that can be managed by their producers. Three are content-related appeals: the film's genre, its story, and its stars. The other two, "spectacle" and "realism" (another term for verisimilitude—see our discussion of the latter concept in our consumer behavior chapter), are transformational factors that describe how the content elements are brought to life in the film. As Staiger (1990) shows, these appeals have a long history, having been used as selling points as early as 1915 by producers when selling their films to theater owners.

Kernan links these appeals to the basic concepts of sensations and familiarity. She argues that the promotional appeal of genre in a trailer "rests heavily on familiarity" (p. 45). In contrast, the more specific story-related information is essentially about presenting "new" events that producers hope will create sensations that consumers will find exiting. The content-related appeal of stars is the least-specific. Stars bring the associations we hold from their former films and public life (i.e., familiarity attractions), but also kindle our hope for new adventures.[279] We have referred to the familiarity appeal of Harrison Ford's line "Chewie, we're home" in the trailer for STAR WARS EPISODE VII elsewhere in this book, but we did not sense familiarity alone. Wasn't there also the proclamation of new adventures and sensations resonating in his aged voice? Spectacle, by definition, is a promise about the sensations that a film will offer, and realism/verisimilitude moderates how much we will enjoy the content appeals and spectacle.

Some scholars have made initial steps to empirically determine the relative roles each of these appeals play for consumers' liking of a trailer, along with their eventual adoption decision regarding the product. Finsterwalder et al. (2012) focused on how film trailers influence the expectations of audiences, distinguishing between "quality-" and "content-related" expectations. Based on qualitative interviews with 12 consumers in New Zealand who had to watch several trailers of then-forthcoming films, the authors suggest that actors are the greatest influencers of film *quality* expectations. Consumers' expectations regarding a film's *content*, in contrast, are most strongly shaped by the genre-related information in the trailer. Finsterwalder et al. further name the "style" of the trailer, its music, and the story as determinants of consumers' expectations toward a film.

[279]See also our discussions why we value entertainment stars in general in our chapter on entertainment brands.

A rare quantitative investigation of trailer appeals is by Karray and Debernitz (2015). They code the trailers of 140 movies that were wide-released in North American theaters (in 2010–2011) with regard to (1) what a trailer reveals about the film's story, using some of the story criteria that have been found to positively influence a film's success, (2) content elements, such as humor and violence (a combination of genre and content variables), and (3) technical aspects, such as the number of scenes in the trailer and its release time. The scholars then use an event-study approach to see how these variables influence the "success" of the corresponding film. Because trailers are usually released before a film opens, they use the movie's "stock price" on the *Hollywood Stock Exchange* (hsx.com), a virtual stock market, as a proxy of its commercial performance.[280] With an OLS regression, in which the trailer variables and a number of movie controls serve as explanatory factors, and the movies' "abnormal returns" (i.e., the change in a movie's "stock price" due to a trailer's release) as the dependent variable, Karray and Debernitz find that story factors have the strongest impact on moviegoers.

Specifically, adding one "successful" story element (e.g., a happy ending) to a trailer resulted in an expected box-office increase of $0.6 million, and adding all ten elements that the scholars studied corresponded with $6 million higher (expected) revenues. Showing violence and humor also bumped commercial expectations; one additional scene adds value of about $300,000. However, the total number of scenes exerts a negative impact—too many cuts appeared to confuse viewers and limit the trailer's emotional appeal. Please keep in mind that all these results are only approximations of the true value of trailer elements, as they reflect the collective wisdom of the trader crowd instead of measuring actual effects.

In a recent study in which *Entertainment Science* scholars cooperated with Netflix, Liu et al. (2018) showed 100 consumers trailers for comedy movies and coded their facial expressions with software. Their "frame-by-frame" MCMC analysis of the lab experiment shows that participants' happiness is impacted by several trailer features, with happiness being linked with higher movie-watching intentions. The number of trailer scenes has a negative effect, whereas longer scenes placed late in the trailer increase viewer happiness (and also movie watching intention directly!); the trailer music's volume and trend also matter.

So, can't we just let data analytics create trailers then? In 2016, the studio Fox hired IBM to do exactly that for their film Morgan, a horror thriller

[280]At the HSX consumers trade virtual stocks of upcoming films; the "stock prices" reflect the expectations of the game's "investors" (i.e., players) regarding a film's financial performance.

dealing with, you might have guessed it, artificial intelligence (Smith 2016). IBM trained their system on the trailers of 100 horror movies, decomposing each trailer into what they called "moments," including the visuals, the audio, and the composition of each scene. They then fed the system with the full-length MORGAN film and let it search for moments that resembled those featured in previous trailers. From this set, a (still human) filmmaker edited ten moments into a trailer.

In line with our logic of *Entertainment Science* that a thorough understanding of what constitutes "effective moments" and the use of human-exclusive creativity would be essential, the resulting trailer was bloodless and derivative. Still, the case might point at how the industry could make use of data analytics—not as a tool to craft great trailers (in an attempt to address consumers' desires better), but to save time and reduce costs instead. "Reducing the time of a process from weeks to hours—that is the true power of AI" (Smith 2016).

What Makes a Powerful Poster Ad?

Trailers are the dominant, but not the only, content format for entertainment advertising. Whereas *Entertainment Science* scholars have dedicated less research time to more static communication formats, such as posters, in general, one noteworthy study is the work by Rao et al. (2017), who analyzed the contents of print advertisements for movies.

The scholars focus on the elements of posters that provide *information* to consumers, leaving out aesthetic aspects—something we return to when discussing the integrated blockbuster concept of entertainment marketing. They distinguish between two fundamental kinds of entertainment information we covered earlier: inferential and substitutive cues. For inferential cues (which signal to the consumer that a product will be of "high quality"), the researchers note whether a poster's tag line stressed the movie's star(s), director, or content. With regard to substitutive cues (i.e., judgments of a product's quality by those who have already experienced it, cues that Rao et al. label "external validation variables"), the scholars measure whether a movie poster quoted critical reviews (and how many of them), whether a "top reviewer" was named (from the New York Times, Los Angeles Times, or Time magazine), and any mention of awards. In addition to these two basic information categories, the scholars also look at the size of the ad as a technical criterion.

Rao et al. empirically test the role of these factors for a data set of major print advertisements for 206 movies from 2003 to 2004, all of which were published in the New York Times around a film's opening day. After coding each poster, they ran random effects panel regressions in which they linked the different content elements, as well as a number of other "success drivers," or controls, to the movies' weekly box office numbers. In addition, they also conducted OLS regressions with a movie's opening weekend performance and its total box office. The results are straightforward and consistent across all analyses: the one poster element that they find makes a difference is including an endorsement from a "top reviewer" (a kind of substitute cue). The other elements don't have much of an effect; neither the number of reviews nor any of the inferential cues are associated with above-average box-office results.

So when designing effective posters, the challenge for the producer is to create enough interest by leading critics so that they write about the film, while also assuring that the quality of the product is high enough so that the review will be positive. In other words, the effectiveness of communication links back to the quality of the product itself. The reward for succeeding in this challenge appears to be substantial: Rao et al. estimate an average box office increase from a positive review by a top critic of $8 million for the opening week and of $16 million in total. We suspect that these results might also offer insights for another related aspect of entertainment communication—the design of the product's packaging, such as a book's or Blu-ray's cover. We get back to this issue at the end of this chapter.

The "How-Much" Question: Can There Be Too Much of a Good Thing?

> "[P]eople really want to know exactly every thing that they are going to see before they go see the movie. … What I relate it to is McDonald's. The reason McDonald's is a tremendous success is that you don't have any surprises. You know exactly what it is going to taste like."
>
> —*Film director* Robert Zemeckis *(quoted in* Ebert *2000)*

> "50 Trailers That Ruined The Movie."
>
> —*Website title (*Kinnear *2012)*

In their analysis of trailer elements, Karray and Debernitz (2015) have found one additional factor that we have not mentioned so far to be influential: if a trailer *leaves out* key plot twists, the (expected) box office increases by half a million dollars in their study. This leads us to the question of "how much" should be revealed to consumers about an entertainment product in the communication campaign (via trailers but other means of communication such as poster advertisements). We will show that we are dealing with a delicate balancing act: whetting consumers' appetite without spoiling their supper.

The Pros and Cons of Spoilers

Entertainment communication must address the experience character of entertainment products by providing information that helps a potential customer judge the product's quality. But remember that entertainment products are also subject to satiation effects, so too much information about a product may hurt its success instead of helping it.

The findings by Karray and Debernitz (2015) which we have mentioned above point to the practical relevance of such a "satiation" threat. And there is also quite a bit of anecdotal evidence for it: when the airing of David Lynch's TV series Twin Peaks by broadcaster RTL faced low ratings in 1991, several blamed competitor SAT.1—that station used its teletext service to reveal the murder of Laura Palmer before RTL could air the series (*Das Fernsehlexikon* 2008).[281] Also, Internet sites (such as the one we quote at the beginning of this section) list "spoilers" which they claim have hampered the success of entertainment products, and individual tweets cite "too much" information as a reason for skipping a product.[282]

Whereas some firms thus are hesitant to give away any "surprises" (such as Sony, for the James Bond movie Skyfall, did not reveal the true meaning of the film's title or its connection to its lead character), others doubt the existence of any "spoiler effect"—the statement by Mr. Zemeckis at the beginning of this section gives evidence of such thinking. Consistent with the renowned director's argument that people "want to know," Fritz (2015) reports that Hollywood studios consider "revealing plot points and showing the most exciting action scenes... [as] the most effective way to draw big audiences." So, who is right? How much information *should* actually be spoiled for a new entertainment product to maximize its success potential?

In Fig. 11.2, we describe the different routes through which information about a new entertainment product influences consumers, whether provided by trailers or other forms of communication. The upper route links information to anticipatory processes in the minds of consumers and their decisions to consume a new product. In contrast, the lower route describes the role that information plays in a consumer's evaluation of the product after having experienced it. In what follows, we analyze these two routes in more detail.

---

[281]This spoiler was heavily criticized by many at that time, including a court which named it "immoral" and forbade SAT.1 from revealing it—again...

[282]For example, Twitter user "luckymojo" told his followers that he has "no interest in seeing [the film] Life As We Know It, especially since they tell you the entire plot/outcome of the movie in the trailer..."

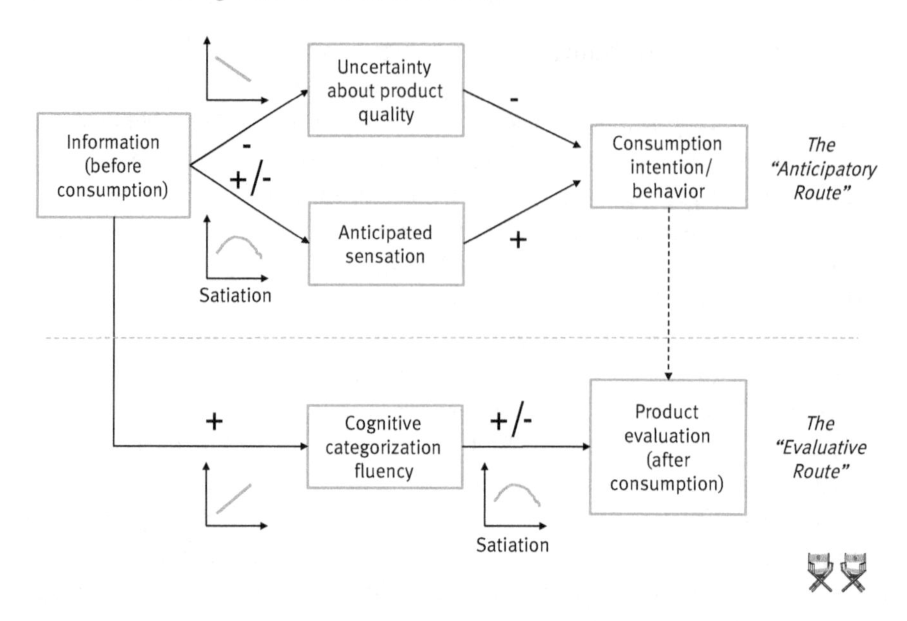

Fig. 11.2 The spoiler effect
Note: Authors' own illustration.

The "Anticipatory Route": How Spoilers Influence Consumption Intentions

In the upper route, we distinguish between two effects. Information regarding a an upcoming entertainment product impacts the consumer's consumption decision by influencing his or her perception of uncertainty. But such information also impacts the extent to which the consumer anticipates that the product will provide new sensations. As entertainment products are inherently risky for consumers because of their experience character, any ex-ante information about them may reduce the uncertainty of the consumer regarding the "things-to-come." Thus, as consumers generally prefer less consumption risk, more information will, all else equal, increase the consumer's intention to see a film (or play a game, etc.)—an argument that captures the essence of Mr. Zemeckis' logic.

But we have made clear early on in this book that entertainment products differ from hamburgers in some important ways, with consumers' desire to experience new hedonic sensations being among them. Here, the role of pre-consumption information for our anticipation of expected sensations is not so clear cut. Whereas some new information will foreshadow the sensation potential of a product and thus increase a consumer's interest in experiencing it, *too much* pre-consumption information (e.g., learning the outcome of a thriller before seeing it) might serve as a *substitute* for the actual entertainment experience. As such, the information might reduce the consumer's curiosity and interest—a satiation effect.

It is this satiation effect that explains what Yan and Tsang (2015) find when they asked 180 Hong Kong undergraduates to forecast the enjoyment they would get from watching the short film TICKER starring Clive Owen, a thriller with a twist ending. Their results show that those who were given a "low-intensity" spoiler that provided information about the story, but did not reveal the ending, reported nearly the same level of anticipated enjoyment as those who received only basic information ("unspoiled"). In contrast, a "high-intensity" spoiler that gave away not only the story, but also the film's twist, *reduced* anticipation enjoyment by 25%.[283]

The "Evaluative Route": How Spoilers Influence Enjoyment

Let's take a look at the lower route now which deals with the *post*-consumption processes that result from information given to a person prior to consuming an entertainment product. It argues that spoilers can also impact how a consumer evaluates his or her actual consumption experience—which can then impact product success via word-of-mouth processes.

The logic here is that pre-consumption information influences our cognitive processing of the consumption experience. Knowledge helps us to assign our experiences into "cognitive categories" more easily, facilitating our understanding of stories that are told to us, and making our thoughts more "fluent" (Leavitt and Christenfeld 2013; see also Hennig-Thurau et al. 2006). Because consumers, in general, prefer less effort to more effort, a basic psychological tendency of humans is that we like such fluency. When Leavitt and Christenfeld (2011) had 800 students read short stories, they found that those students who received a spoiler that discussed the story and also mentioned the outcome ("in a way that seemed inadvertent"), rated the stories more highly than those who received no information up front. The

[283]Yan and Tsang reported similar patterns in other constellations: when they asked 92 consumers to watch a recorded 8-minutes clip from an NBA finals game, those consumers who were told which team won the game had a 12.5% lower anticipated enjoyment than others. And for a *fictitious* thriller movie, spoiling the identity of the murderer reduced watching intentions by almost 20%.

result was consistent across different types of stories; ratings were higher by 10% for mysteries, 9% for "evocative literary stories," and 7% for "ironic-twist stories." Increased cognitive fluency might also explain why in Yan and Tsang's football watching experiment, enjoyment was 13% higher for those who were told the winner prior to watching the match.[284]

But although higher fluency tends to make entertainment consumption less effortful, this does not mean that enjoyment *must* be higher. This results from the upper, "anticipatory" route: if we know what will happen the level of experienced sensations might be lower, which, as we argue in the sensations-familiarity framework, hurts the consumer's enjoyment. Consistent with this, when scholars have studied *high-intensity* spoilers, the results point to lower levels of enjoyment. This is the case for one twist-ending story in Leavitt and Christenfeld's (2011) study, the thriller short film used by Yan and Tsang (2015),[285] and also when Johnson and Rosenbaum (2015) analyzed the reactions of 412 undergraduate students to a short story. They found that an "ending spoiler," compared to a "medium-intensity" spoiler, reduces the sensations perceived ("experiencing suspense" was 9% lower) as well as consumers' enjoyment in terms of being "fun" (6% lower).

Although, with the exception of Johnson and Rosenbaum's study, the findings reported here do not reach statistical significance, the pattern is clear and consistent: a *high intensity* spoiler goes along with lower enjoyment.

Beware of Who You Spoil—and for Which Product You Do so

Research also points out that the specific patterns and relative strengths of these routes will differ between consumers and between products. Yan and Tsang (2015) find people with higher imagery potential to be more negatively affected by pre-consumption information—they tend to enjoy creating their own visions, and spoilers hinder them from doing so. But this also means that if people lack imagery potential, spoilers won't hurt them (as much).

---

[284]That difference was not statistically significant, though.

[285]The difference is only significant though for the *low* intensity spoiler (which corresponded with worse film evaluations than the high intensity spoiler here), which indicates that the two spoilers variants differed also with regard to other, more qualitative criteria.

Relatedly, Rosenbaum and Johnson (2016) find, based on a short story-reading experiment with 368 undergraduate students, that those consumers who seek emotions (i.e., have a high "need for affect," Appel and Richter 2010) enjoy stories more without a spoiler. A similar tendency is found for people's enjoyment of thinking deeply (i.e., a high "need for cognition," Cacioppo and Petty 1982)—the more people value doing so, the less they enjoy spoilers (it might prevent them from developing their very own "theories" prior to experiencing the product).

Producers should thus seek to understand their customers' reaction to spoilers. One way to do so is the Netflix way: the streaming service offers its customers a short online quiz to determine their own "spoiler kind," as well as offering legendary spoilers to enjoy (including the one about the murder of Laura Palmer…).[286] With regard to product characteristics, Leavitt and Christenfeld (2013) show that the complexity of the product determines how much a spoiler can enhance the fluency of the consumption experience, as part of the lower route: if a story is simple and undemanding, spoilers do not increase fluency.

The routes of our framework and the corresponding empirical results help explain the mechanisms underlying the spoiler effect. Please note that all spoiler research so far assumes that audiences have no "built-in" awareness and knowledge of a new product, whereas, in reality, they often know a lot about a new movie, book, or game, particularly when it is a sequel or adaptation. That is why Disney decided to reveal little about of the plot of its STAR WARS sequel THE FORCE AWAKENS—at least not to Western audiences who are so familiar with the saga (Fritz 2015).

Such built-in knowledge doesn't change the basic logic of the spoiler effect. However, it makes it more difficult for the producer to anticipate what kind of information is helpful and what might push the product beyond the "tipping point," reducing anticipation and worsening the liking of the product. In our digital age, information travels instantly around the globe, so that the provision of a target group-specific spoiler has become very difficult. Disney eventually had to recognize this: their trailer for THE FORCE AWAKENS, with story-related information targeted at Japanese and Korean audiences, also became a viral hit in other parts of the world (Fritz 2015).

[286]Find out what "spoiler type" Netflix thinks you are at https://goo.gl/fEv2Fg. But be warned—there might be (*will* be!) spoilers…

For managers, the key challenge based on the insights reported here is to note the different mechanisms and effects that trailers can trigger within consumers and to locate the tipping points at which additional product information would diminish anticipation and impair product evaluation. We are confident that *Entertainment Science* scholars will shed more light on these issues as well.

Now let us put the "what-to-communicate" (and how much of it) question aside—and move on to the different types of entertainment communication, touching on logistical issues such as timing and budgeting, for each. We begin with the communication tool to which entertainment producers usually dedicate the biggest portion of their resources: curtains-up for advertising!

Attracting Consumers via Paid Media: The Role of Advertising

"[Advertising] is the single most discussed and debated issue in Hollywood."
—Terry Press, *as president of CBS Films (quoted in* McClintock *2014)*

Like Mrs. Press states above, advertising is a key concern for entertainment firms when it comes to marketing new movies, TV shows, games, music, and books. Although advertising is used in *every* consumer industry, the particularities of entertainment products lead to some substantial differences when it comes to the mechanics of advertising, including timing and elasticities. In the following, we will first take a quick look at the functions of advertising for entertainment products before diving deeper into what *Entertainment Science* scholars have found regarding advertising's effectiveness, in general and at different points of a product's life cycle.

The Functions of Advertising

How do costly investments in advertising contribute value for an entertainment product? For a new product, advertising plays two roles: (1) it can make consumers aware of the new product, and (2) it can demonstrate the product's quality to them. As advertising is a major tool for the branding of entertainment, it should be no surprise that these roles overlap somewhat with the concepts of brand awareness and brand image, the main functions of branding.

Let us first consider how advertising influences consumer awareness of a product and why this is crucial. One of the best-known illustrations of consumer-decision making describes a consumer's behavior toward a new product as the result of a hierarchical process (e.g., Lavidge and Steiner 1961). According to classic "hierarchy-of-effects" models, advertising can set off a chain of events within the consumer, with the purchase of a product being the culminating event. Becoming aware is the initial—and thus crucial—initiator for the multiple steps; without awareness, the other events, including the product's purchase, cannot happen. But advertising-based awareness can also spread beyond those consumers who received the initial advertising message as ads trigger communication between these receivers and other consumers: the "two-step flow" logic of communication, coined by sociology legends Katz and Lazarsfeld (1955).[287]

Using weekly ad spending and awareness tracking data for 63 movies released in France in 1993 by Columbia Tristar, Fred Zufryden (1996) used an OLS regression to show that consumers' awareness of a film in a given week can be explained almost completely by the level of ad spending and the film's awareness in previous weeks (the R^2 of his model is 0.97). We assume that digitalization has introduced additional awareness sources, but advertising still plays a key role.[288]

Advertising's second role is to convince consumers of an entertainment product's quality. The idea here is that advertising creates and disseminates strong associations that are incorporated into the semantic networks within the mind of the consumer. These associations are essential to move the consumer along to reach the later stages of the decision-making hierarchy towards purchase. Whereas some associations are triggered by the *content* of advertising (e.g., "good" versus "bad" trailers), the mere *amount* of advertising spending for a product can also serve as a quality signal and influence the consumers' quality perceptions. Economists refer to this "signaling" effect as the "money-burning" theory of advertising (e.g., Milgrom and Robert 1986).

Think of the super-expensive Super Bowl advertising for movies and games: in addition to telling consumers that a new film exists and what they can expect from it (i.e., *why* they should watch it!), these ads set the

[287]We will get back to their two-step flow model in the context of our discussion of word-of-mouth effects.

[288]We provide empirical evidence for this in our discussion of antecedents of pre-release buzz for entertainment products in our chapter on "earned" entertainment communication.

advertised product apart from the many others entertainment titles that are not considered "Super Bowl-worthy" by their producers. We don't know of any empirical evidence that separates the content and the signaling effects of advertising for entertainment in particular, but Zhao (2000), based on a general analytical investigation, concludes "that simply 'burning money' is not enough to signal quality.... How the money is burned is also important" (p. 390).

The amount of ad spending for an entertainment product may also serve as a quality signal to other industry actors beyond consumers, such as financial analysts, shareholders, and distributors. When Disney had to set the advertising budget for its STAR WARS movie THE FORCE AWAKENS, it decided to spend a high amount, despite the fact that feverish built-in awareness and anticipation already existed among consumers: spending lower than usual might have risked "drawing undue attention" (Fritz 2015). Joshi and Hanssens (2009), who analyzed the effect of advertising spending levels on studio's stock prices using data from all 200 movies launched by major studios from 1995 to 1998, provide empirical support for a signaling effect of advertising expenditures on investors: higher spending for a film corresponded to higher investor expectations regarding a film's impact on the studio's future net cash flows.

In essence, advertising is an important element of the entertainment marketing mix that influences product success in more than one way. Let us take a look at how *much* should be spent—and *when* it should be spent.

How Much to Spend—and When: Some Introductory Comments on Advertising Budgets and Timing

> "Don't outspend your revenues, but don't underspend your potential."
> —*Former Paramount Pictures executive* Rob Friedman *(quoted in* Squire *2006, p. 290)*

Entertainment producers spend enormous amounts of advertising dollars to make people aware of their newest spectacles and to signal their (high) quality. How has advertising spending for entertainment evolved and how has it been allocated over different media? Figure 11.3 answers these questions for the movie industry and its spendings in the U.S. over the last 15 years. In total, producers these days spend roughly $3.5 billion annually for these

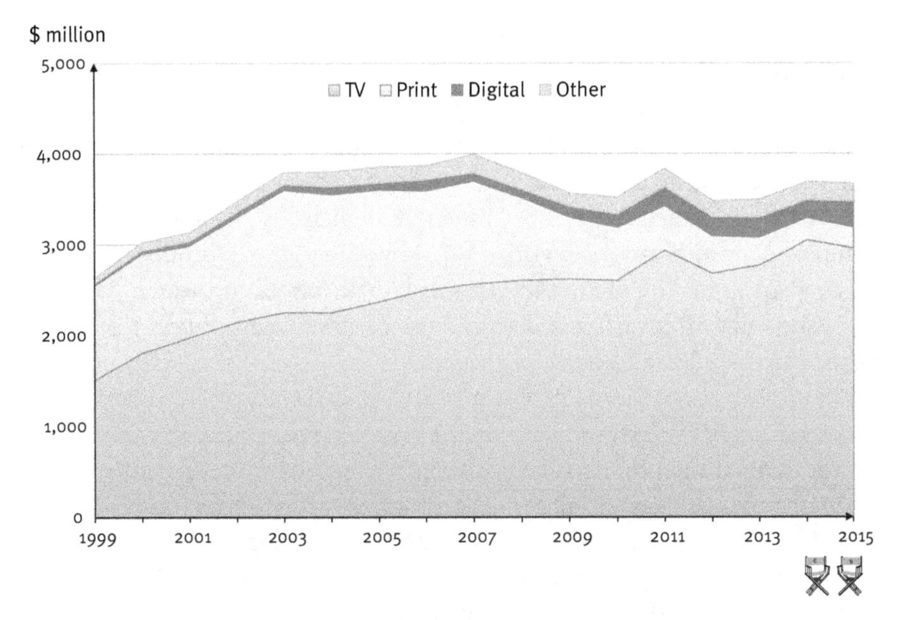

Fig. 11.3 Advertising spendings for movies in the U.S. across media
Notes: Authors' own illustration based on data from Kantar Media. All values shown in the figure are our own estimations which are based on several assumptions; numbers reported here should thus be treated as rough estimates only. The "TV" category includes spendings for network, cable, and syndication, "print" includes magazines and newspapers, and "digital" includes paid search, display/banner, video, and social media, both for stationary and mobile access. Numbers are the raw values for each year (not adjusted for inflation).

media, plus an additional $200 million for other media (mostly billboards and radio). These numbers likely capture about half of what Hollywood studios spend globally for their products (Fritz 2015).

As can be seen in the figure, movie managers in 2015 still assigned the lion's share of their expenditures to TV (which gets about 80% of ad spending for movies, up from 50% in 1999). Managers still believe that, in a highly crowded market, the medium is crucial for reaching the (mass) audience needed for a successful opening of a new product that will have a limited life cycle (see also Fritz 2015). Print, which in 1999 got about 40% of the industry's ad spending, has lost much of the movie studios' love, earning a share of less than 10% in 2015. Digital media have grown from irrelevance to capture nearly 10% of spending, but our data suggests that their role for movie advertising is still substantially smaller than it is for many other industries, where digital is poised to pass TV in terms of advertising revenues (Garrahan 2016).

And at which point in time of an entertainment product's life cycle is the money spent? We have stressed that entertainment products are, because of their hedonic/cultural nature and the industry's high number of new products, faced with a relatively short life cycle, which assigns a key role to the timing of communication. Figure 11.4 shows how movie producers address this challenge. For the movies released in North American theaters in 2012–2014, the bulk of advertising dollars is spent *prior* to the films' release (see also Elberse and Anand 2007).

Specifically, about 76% of film advertising is spent before a new movie is released, another 13% is spent in the week following the movie's release, and the remaining 11% is then spent in the following weeks. This pattern is quite similar for games, although a little less extreme; for the 100 major Xbox 360 games released in the 12 months following October 2011, we found that 52% of advertising is spent before the release, 16% in the week after the release, and the rest afterward (see Marchand et al. 2016). Let us stress that this front-loaded pattern represents a clear contrast to how firms in most other industries allocate advertising budgets over the life cycle of their products—these firms spend the clear majority of advertising only after a product can be purchased by consumers.

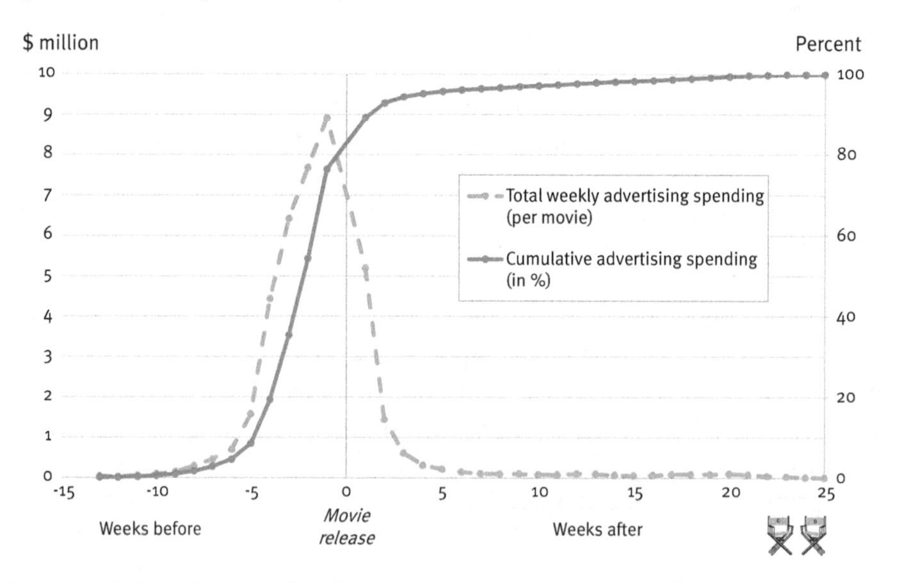

Fig. 11.4 Advertising spending for movies over their theatrical life cycles
Notes: Authors' own illustration based on data from Kantar Media. Numbers are averages for all 442 movies released in North American theaters in 2012–2014 with box-office revenues of $1 million or more.

Several scholars have employed econometric methods to study the effectiveness of advertising for entertainment products, reflecting the critical role of timing by addressing pre-release and post-release advertising separately, and we will structure our discussion in the following accordingly. But before we do so, we will take a look at those studies that investigate entertainment advertising as a whole. Let us also note that determining advertising effectiveness is far from a trivial task, econometrically speaking. The troubling issue is somewhat similar to what we discussed already as the "treatment bias" for different kinds of product factors (such as sequels).

Specifically, advertising spending levels are of an endogenous nature, as they are influenced by managers' expectations about the effectiveness of such spending for a particular product. If a manager does not believe that a new album has the potential to become a major hit, he or she might dedicate less (or even no) advertising to it. Thus, researchers need to find out which part of the success resulted from the advertising itself—and which part resulted from the hit potential of the product (that influenced the advertising spending). Simple correlations between advertising spending levels and product success, as well as unadjusted regression coefficients, are potentially distorted and thus must be treated with care.

In our analysis of different movie distribution channels (Hennig-Thurau et al. 2006), we find that advertising spending is systematically higher for films with stars and/or using a family brand (such as sequels), consistent with the advertising's endogenous nature. Similarly, Prag and Casavant (1994), when using data for 195 films to run an OLS regression with advertising spending as the dependent variable, find that other film characteristics explain almost 70% of the advertising budget. In addition to the involvement of stars, certain genres and the film's budget are among the strongest determinants of advertising spending. Some have argued that producers use a "half-the-production-budget" heuristic when it comes to determining the advertising to spend for an entertainment product (e.g., Quelch et al. 2010).[289] Please keep these intricacies in mind as we now dive into the findings surrounding advertising effects.

[289]Similar rules-of-thumb are at work for entertainment products other than films. For example, book publishers are reported to base their budgeting decision on the existence of a star author and that author's celebrity status or number of previous bestsellers (Shehu et al. 2014).

Some General Insights on the Effectiveness of Entertainment Advertising

Across industries, the average advertising elasticity has been found to be about 0.12—in other words, a 10% increase in ad spending is linked with a sales increase of ($1.10^{.12} = 1.15 =$) 1.5%. So advertising's effect, on average, is positive, but relatively small (Sethuraman et al. 2011). How does advertising for entertainment products perform in comparison? We begin with those studies that do not separate pre- and post-release ad spending and inspect the different forms of entertainment, one by one.

Movies. The key study here is by Bruce et al. (2012), who use advanced econometric tools (such as Kalman smoothing and Markov chains) to study advertising effects for a set of 360 films from 2002 to 2005. Based on parameters they estimate for a dynamic linear model of movie success, they run a number of simulations in which the authors substitute higher and lower spending levels in place of the actual advertising budgets used by a film's producers—doing so allows them to see how spending more or less on advertising would have impacted film success.

Bruce et al., who argue that endogeneity does not harm their results, learn that theatrical advertising budgets should have been higher for two-thirds of the films in their data set (and lower for the remaining third) at the theatrical stage, and for three out of four films at the home video stage. Most insightful for movie marketers are the context factors that they hold responsible for the advantageousness of higher/lower advertising spending. Their findings point to two factors in particular which vary with film type: a film's "wear-in" level (i.e., to what degree does *repeated* advertising influence consumers?) and the film's "forgetting" rate (to what degree does advertising "stick" with consumers?).

Among those films with a higher wear-in rate are science fiction films (perhaps because of their more complex plots and visuals), and a lower "forgetting" rate is associated with the number of professional reviews given for a film (where these reviews seems to prolong awareness among consumers). Thus, for such films, advertising tends to be more effective, making them candidates for higher ad budgets. At the home video stage, Bruce et al. find films that were successful in theaters benefit the most from higher advertising budgets when they are released on video.

Books. Shehu et al. (2014) study a data set of 598 fiction books in Germany and treat advertising as a dummy variable, splitting books into two basic categories (those that were advertised and those that were not).

Doing so enabled the scholars to address the potential endogeneity of advertising with propensity score matching: for each of the 196 books in their data set that were advertised, the researchers identify "twins"—*un*advertised books that were in many other ways equal to those that were advertised. Among those factors that triggered higher ad spending are the appealingness of the title and the book's quality according to professional reviewers.

Correcting for the bias created by these variables reduces the advertising effect on book sales by 41%, but the advertising effect remains quite enormous: advertised books, on average, generated almost twice as much revenue as did their unadvertised twins. But additional analyses by Shehu et al. also show that this "average effect" is fueled by one kind of book only: advertising only makes a noticeable difference for books that do *not* have a star author. In contrast, "star-authored" books reach similar levels of success independent of advertising. We will get back to this in our discussion of advertising contingencies.

Games. And how about advertising for games? Marchand (2016) includes the U.S. advertising budgets in his investigation of the drivers of success for nearly 2,000 console games. He finds that advertising and total sales correlate quite highly (r = 0.58), and his regression with robust standard errors finds an average advertising elasticity of 0.12, exactly matching the cross-industry average reported by Sethuraman et al. (2011). With advertising being only a control variable in this study, Marchand did not apply a bias adjustment or investigate potential contingency effects.

Music. Finally, two studies also link advertising with music success. Papies and van Heerde (2015) study the weekly German record revenues and concert ticket sales of 387 successful, actively touring music artists between 2003 and 2010. In addition to conventional advertising spending for the artists' albums, the scholars' also looked at the amount of airplay that an artist's songs received on radio stations—a key promotional instrument in the music business. The authors estimate separate equations for disc and concert sales with a hierarchical Bayes regression—they correct the endogenous nature of advertising decisions with instrumental variables. They find an advertising elasticity of 0.09 for an artist's record sales (i.e., a 10% increase in ad spending results in an increase of a little less than 1% in record sales in the same week) and a smaller elasticity (0.024) of airplays for the artist's record sales (i.e., a 10% increase in airplay increases record sales by 0.2% in the same week). For the artist's concert ticket sales, airplay affects them too, but (same-week) advertising does not; perhaps concert tickets involve more long-term decision making.

The other study is by Chen et al. (2015) who analyze the impact of ad spending on sales of albums by 616 music artists over a period of 32 weeks in 2008–2009, using a different setting (U.S. sales, using artist ranks from Amazon.com) and a different method (a panel vector autoregression model, or VAR, which addresses endogeneity concerns).[290] Chen et al., while not including airplay, also include owned-media activities in their model (e.g., the artists' activities on the then-popular social media platform MySpace) and also "earned" media in the form of word-of-mouth postings on Amazon.com, as well as several other "success drivers" (such as album price and new releases). Their results suggest that "traditional" advertising affects album sales in the *following* week with an elasticity of about 0.04, which is smaller than what Papies and van Heerde found.[291] We don't know what causes this difference, but the simultaneous consideration of the artist's social media activities might play a role. Let us add that we assume that, because both studies only capture sales that happen in a single week after the ad, the total impact of music advertising (and airplay) might be somewhat higher—it might spill over.

Papies and van Heerde's analysis also gives us an idea how advertising effects might be changing with the growing availability of broadband Internet connections (which enable new streaming models, while also easing illegal access to music). They find that the impact of paid advertising on both record and concert ticket sales is declining—but the importance of radio airplay for record sales grows with improved Internet connections.[292] Papies and van Heerde suspect that this is because airplay, being mainly under the control of the radio station, not the label, takes a "preselection" role for consumers and informs their purchases, which gains importance in the digital age (where the range of choices becomes even larger).

All these results aggregate all the advertising that happens at different points of a product's life cycle. Let's dig a little deeper now and separate out pre- versus post-release advertising, in chronological order.

[290]To be included, artists had to operate a site on the platform MySpace, which was probably more the case for lesser known artists than for superstars; the data set represents about 10% of the total annual advertising spending for music in the U.S.

[291]Interpreting VAR model results is somewhat tricky—elasticities cannot be directly taken from the estimated parameters, but have to be calculated with so-called "impulse response functions." Using the ad spending from the previous week constitutes an exception, with parameters serving as (constant) elasticities.

[292]Papies and van Heerde's results suggest it does not change with regard to concert sales.

The Effectiveness of Pre-Release Advertising

Average Effects

Pre-release marketing activities are particularly prominent for filmed entertainment and video games, so it does not come as a surprise that scholars have focused on these forms of entertainment when investigating pre-release advertising. We start with movies and then broaden our perspective.

The Case of Movies

For movies, findings by *Entertainment Science* scholars on pre-release advertising's effectiveness paint a largely consistent picture. Elasticities for advertising prior to a North American movie release range from 0.30 to 0.40, suggesting that a 10% increase in ad spending corresponds with an average box office increase in the opening week of around 3–4%—a substantially larger impact than the one usually found in other industries.

There are strong indications that part of that effect is of an indirect nature, mediated via the theater owners who, as a kind of "second audience" to advertising, adjust their supply-related decisions accordingly and choose to show a highly advertised film on more screens than one that gets less advertising by its producer. Here are some details from seminal studies:

- Elberse and Eliashberg (2003) report a total elasticity of pre-release advertising of 0.40 for the opening week in North American theaters, using a 3SLS regression for 164 American films in or after 1999 (but not controlling for advertising's endogenous nature);
- Ho et al. (2009), via a GMM regression for 302 movies from 2000 to 2002, also find a total advertising elasticity of about 0.40 on the North American box office. They use the cumulative ad spending in the previous week as an instrument for release-week advertising;
- Gopinath et al. (2013) find a pre-release advertising elasticity for the North American opening weekend of 0.39 when running a two-stage regression approach for 75 movies released in 2004 based on data for different geographic markets in the U.S. They address endogeneity concerns by using a film's production budget as an instrument for advertising spending; and
- Clement et al. (2014), in their study of more than 2,000 films from 2000 to 2010, find a total advertising elasticity of about 0.30 for the North

American opening weekend (via a 3SLS regression). They do not use any bias correction for advertising.

All these results support the idea that spending for advertising prior to a new product's release makes sense. However, because the elasticities are clearly below 1, the effects should not be overestimated. This is also what Elberse and Anand (2007) conclude from studying the impact of advertising for films with data from the virtual Hollywood Stock Exchange. When they link weekly pre-release advertising for 280 movies from 2001 to 2003 with the movies' stock prices (which reflect the traders' revenue expectations) with a dynamic hierarchical linear regression, Elberse and Anand find that a $1 increase in advertising is connected with an expected (total) box office increase of $0.65, on average.

So, does this all mean that film studios spend too much for advertising prior to a new movie's release, an oft-heard claim among Hollywood producers (e.g., Mechanic 2017)? Not necessarily. One has to keep in mind that theaters (for which these effects are measured) are only the first in a series of distribution channels, and studios hope that pre-release advertising not only triggers success in theaters, but also pays off in later channels by contributing to the establishment of a strong entertainment brand. In other words, pre-release theatrical advertising is expected to spill over on sequential home entertainment channels.[293]

But the effectiveness of advertising aside, its high costs certainly suggest careful planning and coordination of advertising with other entertainment decisions. Such care might become even more important in the future, as older advertising studies seem to reveal a declining effectiveness of pre-release advertising over the years. For example, when Basuroy et al. (2006) studied movies from the 1991 to 1993 time period, they find a pre-release advertising elasticity of 0.66 on opening week revenues.

[293]Please see our discussion of the various, and often sequential, entertainment distribution channels. Spillover effects can be expected to be mostly indirect by triggering the success of the film in theaters which then is a major driver of success in subsequent channels. Luan and Sudhir (2010) provide empirical evidence for such an indirect spillover effect for a data set of 526 movies newly released on DVD (from between 2000 and 2003); whereas theatrical advertising spending has no direct statistically significant effect on DVD sales, the movies' box-office results have an elasticity of almost 1 for DVD sales. Their results also point out that advertising at the DVD release is much less effective than theatrical advertising for its respective distribution channel—the average elasticity for DVD advertising is only 0.03 on the release-week sales of the DVD (and drops quickly afterwards). We discuss in much more detail the indirect effect via success later as part of the "earned" communication chapter of this book.

The studies listed above analyzed pre-release advertising, in general, rather than individual media such as print, TV, or digital. [294] Some scholars, though, have looked into the effectiveness of different media prior to a movie's release. We have already mentioned Karray and Debernitz' (2015) work that focuses on trailers prior to the movie's launch. Studying how the release of 140 movie trailers (for films released in 2010 and 2011) influences the success expectations for these films by HSX traders, they find that a trailer that is released before a movie opens is, on average, associated with a $2.2 million increase of the (expected) North American box office. The trailer impact is positive for 90% of the trailers they study and is as high as nearly $8 million for some trailers, with trailer effectiveness being above average for certain types of films (e.g., science-faction/fantasy movies) and release dates (April-May and August-November). The scholars do not empirically distinguish between the impact of trailers in theaters versus trailer views on the Internet (a type of "owned" media), it's not clear which part of the effect can be attributed to advertising (versus "owned" media).

Ho et al. (2009) also look at trailer effects above and beyond general advertising. They focus on trailers that are shown on TV during a very special event: the American Super Bowl. Such trailers deserve particular attention because of their enormous fees; film studios can pay more than $5 million for a single airing (Lieberman and Busch 2016). With a GMM regression analysis, Ho et al. studied whether spending for a Super Bowl ad (which was the case for 19 of their 302 films from 2000 to 2002) made an economic difference at the North American box office—they find that the total elasticity for Super Bowl advertising is significant, but only about 0.02 and thus substantially lower than for conventional TV advertising.

But that doesn't mean Super Bowl advertising is ineffective: elasticities are about *percentage* changes, so that the base value matters strongly. And because *most* films have literally zero ad spending for the Super Bowl, increasing their budget can offer larger returns. Through simulations, Ho et al. show that the average box office increase for an additional Super Bowl spot (ads cost about $2 million at that time) is about $7 million, compared to an average box office increase of $1.7 million if the same money had been

[294]To be precise, Ho et al. use only TV ad spending in their study (because they want to compare its effects with those of Super Bowl advertising on TV—see in the text below). However, because they do not include any other (i.e., non-TV) advertising media in their analysis, the TV spending measure serves as a proxy for ad spending in general, rather than reporting only the specific mechanisms of TV advertising.

added to the *conventional* TV advertising budget. But Super Bowl advertising returns are also highly diminishing—Ho et al. demonstrate that airing a second spot adds less than $1 million in box-office revenues, or 15% of the first spot. But be careful: such simulations are only valid for the conditions under which the data was collected—they do not tell us the value of Super Bowl advertising should many more trailers be aired.

The Case of Other Entertainment Products: TV Series and Video Games

How does pre-release advertising impact entertainment products other than feature films? Lovett and Staelin (2016) look at TV shows, studying the success of the first six episodes of the TV adventure drama series HUMAN TARGET when it was aired in the U.S. in 2010 by Fox. The scholars use survey data from more than 1,000 members of a consumer panel and focus on the exposure to advertising that consumers remembered. Through linear probability models (a type of logistic regression), they find that if a consumer remembered having seen an ad for the show, the probability of watching the show increases by 5%. The authors also included owned and earned communication which assures us that the reported advertising effects are not the result of these alternative information sources.

For video games, we find a pre-release advertising elasticity of 0.13 for a data set that consists of all 100 games released for the Microsoft Xbox 360 console between 2011 and 2012 (Marchand et al. 2016). This result, which is only marginally higher than the elasticity found for general games advertising, refers to the global revenues of a game on its first weekend (games are mostly released on the same day around the globe), using OLS regression and not accounting for advertising endogeneity.

Our own finding is very similar to the one reported by Burmester et al. (2015), who study the link between advertising and the German sales of a much larger set of games: more than 3,000 that were released for consoles or PC between 2004 and 2009. With a fixed-effects panel regression in which they use the number of printed ads as a proxy for ad spending and controlling for advertising endogeneity with a "copula" approach, they find that pre-launch (magazine) advertising affects launch sales with an elasticity of 0.12.

Finally, Xiong and Bharadwaj (2014), for a data set of 673 console games from 2009 to 2010, report even slightly lower elasticities for pre-release advertising (0.05–0.08) from an OLS regression with robust standard errors (but no endogeneity adjustment). The lowest values are found when the pre-release buzz for a new game is included in the model. As advertising can also function also as a major driver of such buzz for new entertainment products, we suspect that the latter variable here "steals" some of the explained variance of game success that should instead be attributed to advertising.

In summary, these results teach us three things: advertising that is done before a game's release is (1) only marginally more effective than game advertising in general, (2) similarly effective as advertising for other (non-entertainment) products, but (3) clearly less effective, on average, than pre-release advertising for new *movies.*

More on Contingencies: Interestingness, Uncertainty, Situational Factors. And Culture?

Average effect sizes are useful because they provide us with a fundamental understanding of how marketing measures usually affect consumers and product success. But entertainment products, of course, differ quite enormously from each other, so it can be insightful to take a closer look to learn about factors that might increase or lower the effectiveness of pre-release advertising for entertainment products. Scholars have shed initial light on a number of such factors, and we already pointed at some of their findings in our previous discussion. But let us take a more systematic look now at such contingencies.

"Interestingness" because of high quality and other factors. Several studies indicate that the effectiveness of pre-release advertising is higher if the advertised product is of "high quality." The logic behind this builds on the "two-step flow" argument of communication, according to which the awareness triggered by advertising can spill over via consumer communication to other consumers who have not seen the original ad. But this will happen much more often if consumers consider the advertised product to be interesting and thus "communication-worthy" (Luan and Sudhir 2010).

Specifically, both Bruce et al. (2012) and Elberse and Anand (2007) have found that movie advertising is more effective for movies that are judged positively by critics; Luan and Sudhir (2010) report a similar effect for the interplay of pre-release advertising and consumers' movie evaluations on initial DVD success. The same logic might also be behind Luan and Sudhir's finding that advertising is more impactful for DVDs that have bonus features: from a consumer's perspective, bonus features might be a (search) quality dimension that help them judge a DVD.

But interestingness might not be limited to quality per se, but may also vary with other product elements, such as having a prominent brand. Basuroy et al. (2006) find that advertising is more effective for sequels than nonsequels. We assume that sequels, with their high built-in familiarity, are more interesting for consumers, which intensifies advertising-based awareness.

But why then do Shehu et al. (2014) find book advertising to be most effective for *less*er known writers, as we reported earlier? Given the critical role of authors for books, it could be that their mere presence assures an awareness-generating treatment by journalists and retailers, so that awareness for star authors is virtually guaranteed even without advertising, and the incremental awareness caused by advertising is small. For books by lesser-known authors, however, advertising is about the only way to make people aware.

Uncertainty. We have argued that, in addition to generating awareness, advertising can also serve as a quality signal on its own and reduce consumer uncertainty. We believe that this is what explains Basuroy et al. (2006) finding that advertising is more effective when there is a lack of consensus among critics about a movie's quality. But do not some features, such as being a sequel, also *reduce* uncertainty for consumers? We argue that the total effect of advertising for an entertainment product is the result of a weighing of the product's interestingness versus uncertainty about its quality. Basuroy et al.'s results suggest that for sequels the "interestingness effect" (which, in the case of sequels, boosts advertising's effectiveness) is stronger than sequels' uncertainty-reduction effect (which would cannibalize advertising effects).[295] A similar argument can be made for a movie's success in previous distribution channels: it amplifies the creation of interest via advertising, but also reduces uncertainty on its own by signaling high quality. Luan and Sudhir's (2010) finding that advertising effectiveness on DVD does *not* benefit from a film's prior theatrical success suggests that the two effects cancel each other out in this case.

The situation. Luan and Sudhir report that advertising is more effective at certain times of the year than others. Specifically, their results suggest that DVD shoppers are more responsive to advertising in "high-demand sea-

[295]But let's keep in mind that Basuroy et al.'s study is based on only 11 sequels.

sons," i.e., certain holidays. Their results point out that timing makes quite a difference; the advertising elasticity is almost twice as high over Christmas, and the specific elasticity for romance movies is even tripled around Valentine's Day.

Culture. We don't know much yet about whether pre-release advertising effectiveness for entertainment varies between cultures, but it seems kind of intuitive, given the differences in media and entertainment usage between countries we have already reported. Further, uncertainty avoidance is a key dimension on which cultures differ, and advertising is a means to address such uncertainty. Consistent with this logic, Clement et al. (2014) detect a much lower pre-release advertising elasticity (<0.10) for German moviegoers: Germans not only go to the movies much less frequently than Americans, in general, but a film requires much more advertising to spur them to go.

The Effectiveness of Post-Release Advertising

Whereas, at least for filmed and programmed entertainment, the majority of advertising takes place before a product's release, spending usually continues afterward. Let us thus take a look at the effectiveness of advertising that takes place *after* a product has been released for the different forms of entertainment.

Movies. Those scholars who have linked advertising spending in the weeks after a movie's release with box-office results have found elasticities for post-release advertising that are even slightly higher than those for pre-release advertising. Specifically, Gopinath et al. (2013) report an advertising elasticity of about 0.50 in the four weeks after the movie's release, and Basuroy et al. (2006) find a weekly post-release advertising elasticity as high as 0.71. Those results should be interpreted with care, however; we have shown that ad spending for movies is, in general, *much* lower after a movie's release which affects the interpretation of elasticities (i.e., the lower *absolute* dollar amounts provide a much smaller initial base). And it is also highly selective, with successful films getting the lion's share of post-release ad spending. Thus, producers should not interpret these results as a call for higher post-release ad spending (but make sure to read our thoughts on such reallocation in the following section).

Moreover, the results from Luan and Sudhir suggest that post-release advertising is largely *ineffective* in later distribution channels. In their study of DVD sales, the authors find that advertising effectiveness declines by 30% per week after release and vanishes completely in the fourth week. Regarding advertising formats in this time frame, Smith and Telang (2016) compared the effectiveness of different digital advertising formats for home entertainment revenues of catalog movies in a series of field experiments. Whereas their findings do not show any differences in ROI between search advertising and banner ads, one type of advertising produced a superior ROI: cookie-based retargeting of users who had shown an interest in a movie via their "online journey." The scholars suggest that managers should pay more attention to this "under-used aspect of online advertising" (p. 3). Managers who want to use consumer journeys to guide ad spending, however, will benefit strongly from a rich understanding of consumer's entertainment-related decision making.[296]

Games. The studies that have looked at post-release advertising effects for games show a somewhat different pattern than those for movies. Burmester et al. (2015) find, in their weekly analysis of German games sales, an elasticity of 0.08 that is one-third smaller than for the pre-release period for a cumulative "stock" measure of advertising. Our analysis of global game sales also shows a lower impact for post-versus pre-release ads (Marchand et al. 2016).

Our findings point at something else though: when estimating advertising effects for different weeks, we find a U-shaped pattern for advertising effectiveness, with an elasticity of 0.06 immediately after the release, complete ineffectiveness in the following weeks, and an elasticity of 0.07 in week 8 (the last one we looked at). Similar to what we have said for movies, however, this latter finding might be affected by a concentration of advertising on successful titles, as well as clearly lower spending levels over time.

Books. Brinja Meiseberg (2016), using a data set of 30,000 books that were already available for purchase, does not look at conventional advertising, but the book samples that are provided to readers. Meiseberg uses an unconditional quantile regression approach to see how the provision of a sample influences the sales rank of a book at Amazon's German site. Controlling for a large number of alternative influences (but not for conventional advertising), she finds that the provision of a sample indeed goes

[296]Our discussion of this process in our entertainment consumption chapter might serve as a good start.

along with higher book sales. The influence is strongest for the lowest selling 20% of books in her data, but it is also significant for the top-selling quantile. Overall, the effect is quite substantial—on average, samples improve a book's sales rank in her data set by between 7% (for higher-ranked books) and 11% (for lower-ranked books).[297]

Music. In the music context, Dewan and Ramaprasad (2014) analyze how airplay influences the sales of music that has already been released. They examine a comprehensive data set comprised of weekly sales from 2006 for about 1,000 songs and 594 corresponding albums. Via a VAR model (which addresses the endogenous nature of airplay—hits are played more often), the scholars analyze "lagged" effects, isolating how the amount of airplay in a given week influences sales in the same week and also in the following weeks.

They find that airplay impacts both song and album sales, and that it does so most strongly in the week it takes place, wearing out shortly afterward. The short-term airplay elasticities are about 0.04, which is a little higher than what Papies and van Heerde (2015) reported. But we have to keep in mind that Dewan and Ramaprasad look at a *specific* song or album instead of airplay effects on an *artist*'s total repertoire of music. Interesting insights also come from an additional analysis of subsets of data: here the authors find that airplay effects are substantially higher for independent songs and albums (which struggle to get airplay at all!) and also for music by artists who do *not* have a high reputation, a finding similar to what Shehu et al. (2014) found for books.

Balancing Advertising Timing Within and Between Sequential Distribution Channels

The comparison of pre-release and post-release advertising effects are informative, but they do not directly address the question of whether allocating advertising budgets differently over time would impact product success. Can *Entertainment Science* teach us something regarding this complicated matter?

---

[297]In Meiseberg's study, the bivariate correlation of −0.39 between the provision of a sample and the book's sales rank is higher than for most other variables such as word of mouth and price, and comparable to the correlation of sales with the TV appearance of a book title.

When Bruce et al. (2012) conduct extensive simulations for their movie data set, they conclude that about 55% of the films in their data set would have benefitted from having allocated more of their advertising budgets *earlier* in the process. They calculate that such a shift would have increased theatrical revenues per film by up to 15%. For the home entertainment stage, their results suggest a similar move: here, 44% of the films would have gained revenue (up to +16%) by spending a larger share of the advertising budget earlier. Figure 11.5 shows the percentage shifts in advertising spending that Bruce et al. recommend for both the theater and the home video stage across their data set of films.

The scholars also offer a glimpse into how product characteristics influence the effectiveness of such intertemporal advertising decisions. With regard to advertising for the theatrical release, they recommend that films that critics rate highly benefit strongly from shifting advertising toward pre-release, a conclusion that is line with the finding that high quality entertainment benefits from early advertising because it triggers communication among consumers (which raises awareness and creates "buzz"). For action films, in contrast, moving a larger share toward later weeks might be preferable, perhaps because of the low uncertainty that consumers perceive for this rather clearly defined type of entertainment.

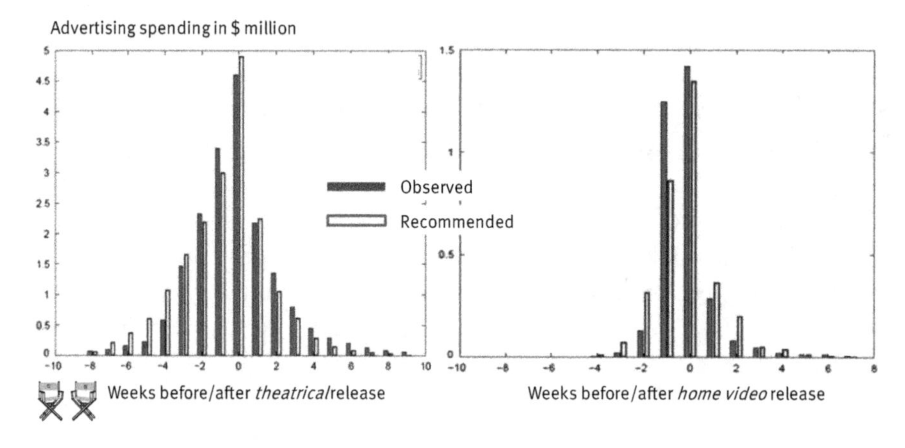

Fig. 11.5 Actual versus "optimal" allocation of movie advertising over time
Source: Reprinted with minor adjustments with permission from Journal of Marketing Research, published by the American Marketing Association, Bruce et al. (2012) Dynamic Effectiveness of Advertising and Word of Mouth in Sequential Distribution of New Products, August 2012, Vol. 49, No. 4, pp. 469–486.

And as entertainment products are often released sequentially through different distribution channels,[298] how should advertising be allocated across these channels? The channel-specific elasticities we reported before provide guidance, but there is also evidence that success in the initial channel spills over to impact product success in later channels via what we call "uninformed cascades."[299] So it seems logical to suggest that advertising in the first channel will, if influential, spur this spillover process, which further enhances the attractiveness of initial-channel advertising over less-powerful sequential-channel advertising.

In summary, what do we learn from the numerous studies into the effectiveness of entertainment advertising covered in this section of the book? We find that early advertising can be highly effective for "high-quality" movies and, to a lesser degree, games; its impact is leveraged as it is capable of setting off communication and buzz cascades for products in which consumers have interest. So, if products are "buzz-worthy," making them well known early in the process makes sense. Although there is no empirical evidence, doing so should also work when a product is highly unique, such that it can instantly trigger chatter and build awareness, essentially "building" the product brand—think of movies such as INCEPTION and AVATAR or games like RED DEAD REDEMPTION. Quality (or its anticipation) will matter here, too.

If a new entertainment product has only limited means to signal strong quality and thus has low communication potential, pre-release advertising will be less effective. Very high amounts of ad spending would be needed to generate sufficient awareness to ensure marketability in this case. But for such products, advertising *after* the release can be impactful, as evidenced by post-release advertising elasticities. Studies suggest that the impact of post-release advertising is affected by quality, too, but experiences play a bigger roles than anticipations for them. For music, receiving airplay is a powerful way to boost sales, particularly for independent labels and unknown artists. In the case that a product is *lacking* quality (both signaled and experienced, in the views of consumers and experts), results suggest that advertising may not help much even after its release.

Empirical research also shows that advertising elasticities clearly vary between the different forms of entertainment, being highest for movies, followed by games and then music. Regarding books, we need to see more empirical evidence before making definite judgments, but paid advertising, as well as the provision of samples as a form of advertising, seems to help product success.

[298]Please see our discussion of this issue in our chapter on entertainment distribution.

[299]We look into this phenomenon more deeply in our discussion of "earned" communication in the next chapter.

But paid communication must no longer be restricted to advertising. Today, advertising needs to be complemented with the "new kid in communication town"—owned channels, and social media, in particular. Let us now explore the role that such channels play for entertainment success—and what we know regarding their effective use.

Attracting (and Keeping) Audiences via Owned Media: Playing Pinball

> "For long periods, [pinball] was widely regarded as a form of gambling, a game of pure chance. [But] anyone who plays today realizes instantly that pinball demands skill: how you can bang the machine's side to change the trajectory of the ball – but not too hard, or else a tilt ends everything; how to trap the ball on a flipper, teeing it up to aim at targets with different scores; how to direct the ball into a slingshot channel… and so on, and so on."
>
> —Cornwell *(2011).*

Back in 2009, then-CEO of Sony Pictures Michael Lynton named the new realities that confront entertainment in the digital era: "[the] Internet with Twitter, Facebook, YouTube or MySpace, but also mobile phones have completely changed how we [as consumers] perceive and understand our environment" (Lynton 2009). With MySpace long gone, replaced by new platforms such as Snapchat (where Mr. Lynton now serves as chairman), his words captured the essence of what is new: entertainment consumers are no longer passive receivers of information, but have adopted a new role as active co-producers of the value they strive for, a role that creates novel expectations and requirements for managers.

The digital space, providing consumers with literally unlimited room for expression and exchanges, as well as multifarious new ways to entertain themselves, offers consumers more power when it comes to dealing with marketers, and consumers have adjusted their behaviors accordingly (Labrecque et al. 2013). In this new world, firms cannot simply speak *to* customers via traditional advertising vehicles as they could in the analogue days. Instead, firms must find ways to *engage* consumers regarding their entertainment products and brands. Internet-based platforms provide entertainment producers with rooms for such engagement to take place—such as Facebook brand pages, Twitter accounts, YouTube channels, and presences on Instagram, Snapchat, etc. These environments are the "owned media" we are talking about in this section—it is here where consumers can be equal participants with firms (or their personnel) in wide-open discussions and interactions.[300]

Although entertainment firms, with their highly involving, emotional, and identity-related products, should be ideal candidates for embracing the potentials of owned media, the road toward owned media has been somewhat rocky for them. Although there are some wonderful examples of how entertainment producers have unlocked the potentials of owned media (we describe them below), the majority of the industry has been rather hesitant to adopt new approaches. They have instead employed the Internet, and owned media platforms, as a purely promotional vehicle—another broadcasting channel instead of making use of the customer's active co-production potentials. As Fritz (2015) described it: "Despite efficient new digital platforms, rarely are producers, executives, and other power players behind a movie willing to try something new that could be blamed for a weak box office performance."

But with the "Nobody-Knows-Anything" mantra as the starting point of our journey toward *Entertainment Science*, this should not really come as a surprise. "Nobody Knows" stands for a risk-averse attitude, implying that nothing can be learned or generalized, and it favors existing models (because with them nobody can be blamed for bad outcomes when one did things the standard way), whereas new approaches and experimentation incur personal responsibility for failure.

Entertainment Science scholars have taken a quite different path—one that provides a way forward for managers. Scholars have been uncovering the "rules" for effective owned-media marketing decisions, which turn out to be paradigmatically different from the rules for traditional media. One key insight is that these new rules resemble those of a chaotic pinball machine, so that marketing activities work best when entertainment marketers consider themselves as "pinball players" when designing owned media strategies, investing in what it takes to become a dedicated digital media "pinball wizard."

In the following, we will first explore more deeply this notion of entertainment marketing via owned media as pinball playing. Then we focus on two of the pinball metaphor's key aspects: (1) finding the "right" content

[300]The term "owned media" itself is actually a little misleading, as producers usually only *rent* the media from the platform providers or use it for free, compensating the platform with advertising spendings. It can be considered a reminder of the early days of such digital meeting places, where the places were usually "brand community" websites hosted by producers themselves. Although such sites still exist, their relevance has fallen far behind those environments provided by platform providers.

that is needed to play successfully (or, to stay true to the metaphor, choosing the "pinball" itself), and (2) moderating the conversations (i.e., handling the "flippers"). Afterward we will inspect what empirical studies tell us about the effectiveness of owned media in entertainment. And we will end the section with a quick look at a non-digital kind of "owned media"—the packaging of physical entertainment, such as a book's or album's cover. But let's play pinball now!

The Pinball Framework of (Entertainment) Communication

Why use a pinball metaphor to understand the digital era? We argue that in the older analogue world without social media, the task of marketing a product was somewhat similar to the activity of bowling—communication, via paid advertising, was linear and one-directional. In a bowling metaphor, consumers are the pins and the ad is the bowling ball that impacts them and causes them to fall (i.e., seeing a new movie or buying a new game). The bowling alley is mass media which transports the ball (ad) to the pins (consumer) (see Hennig-Thurau et al. 2013).

We argue that in the digital world, the bowling metaphor has lost its meaning, and marketing is now better characterized as a pinball game (Hennig-Thurau et al. 2010). In this pinball framework, information about a new product—the "ball"—is introduced, but the consumers are no longer pins that fall down once they have received the information. Instead, they are the various elements of the pinball field, such as bumpers, kickers, and slingshots; these elements are no passive receivers, but actively divert the ball, accelerate or slow it, and shoot it back at the player with high speed.

This reflects the role of consumers in the digital "pinball" environment, who are no longer passive and isolated, but active and interconnected within digital social networks. Consumers can change the intensity and even the meaning of an original message in numerous ways, such as by sharing information, feelings, and experiences with friends via status updates or via reviews posted as videos (Hennig-Thurau et al. 2013). Because such consumer actions happen so fast and can be observed by large numbers of other consumers, a single voice can escalate into massive word-of-mouth and buzz cascades—the equivalent to multi-ball play in a pinball game, if you will.

Unfortunately for entertainment firms, these cascades are not limited to positive information, but can also take the form of negative firestorms (such as when fans do not agree with a casting decision for their hero character;

e.g., Pfeffer et al. 2014, as well as our own work in Hansen et al. 2017). And then there is still traditional mass media. Instead of providing the level path to the consumer as in a bowling alley, mass media now further add to the pinball game's unpredictability by serving as additional bumpers and slingshots that can, through its coverage, multiply individual consumers' social media episodes and provides the basis for even more drastic pinball actions.

Figure 11.6 illustrates the complexities and dynamics of entertainment communication in a pinball environment. Brands stimulate and interact with active and networked consumers, who use and contribute to digital and social media, chatting about the brand, but also "firing back" at it directly. Consumers also get input from (traditional) mass media, which are themselves closely connected with digital and social media, both covering their activities, but also spurring them.

But most importantly, the *goals* of marketing communication need to be different when playing pinball. Whereas bowling was all about making

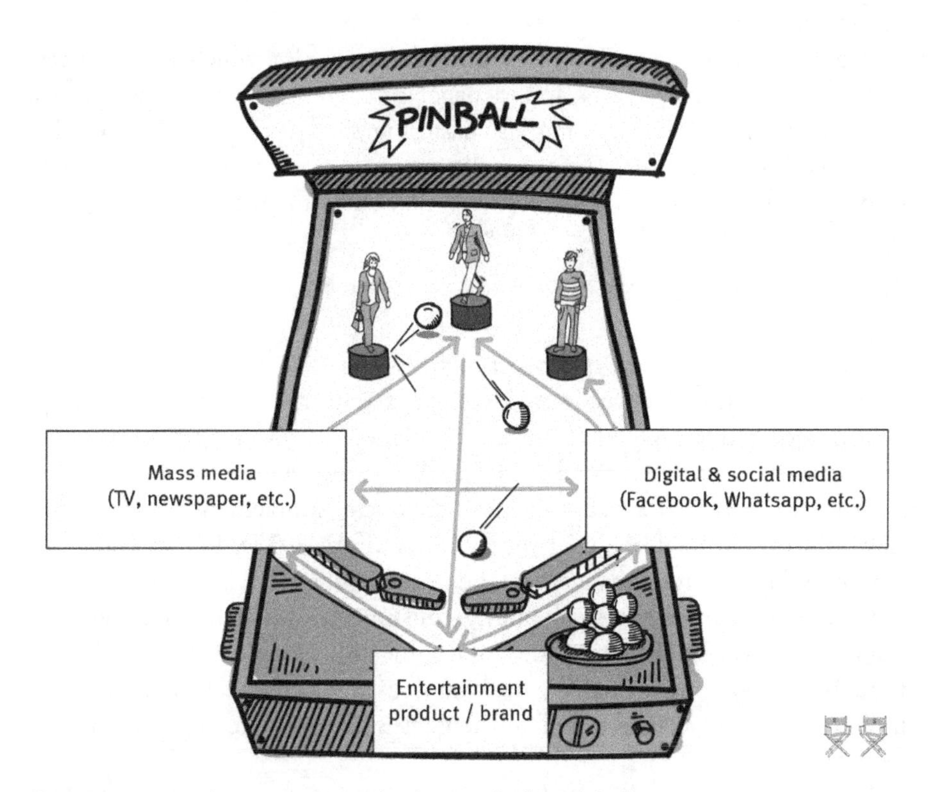

Fig. 11.6 Entertainment communication as playing pinball
Notes: Authors' own illustration based on ideas from Hennig-Thurau et al. (2010). Graphical design by Studio Tense.

people notice and hopefully purchase a new entertainment product, this is not how the pinball machine works. Instead, success in pinball is achieved by stimulating the engagement of consumers and triggering feedback loops (that hopefully are positive) among consumers and media. Scoring points via engagement gets more people involved (creating awareness), but also is the foundation for strong anticipation as expressed in high buzz levels—a key antecedents for success in most entertainment markets.

Scoring pinball points requires excellence in two related fields: (a) in selecting and offering powerful content via owned media platforms that meets these objectives, and (b) in moderating the chatter by the use of specific communicative and organizational practices. Whereas some consider playing pinball a random game, a lottery, a pure gamble (that's why it was banned in the 1920s and 1930s in large parts of the U.S.), it is actually a game of skill: a player's performance in pinball largely depends on his or her ability to manage "deterministic chaos"—to anticipate and react to unpredictable events in a competent manner (see Cornwell's introductory quote to this section).

It is hard to find a major entertainment release for which its producers have not set up a Facebook brand page. But taking a pinball perspective shows that this is not the same as mastering the game. We will now discuss what kind of content (i.e., pinballs) works best in social media and then study the art of moderating interactions (i.e., operating the flippers).

Content that Matters

In pinball times, content has two functions. First, it has to send consumers along a standard decision-making journey, from awareness to buying. Consumers increasingly use producers' owned-media resources on social networks, such as YouTube and Facebook, as important means for getting information about new entertainment products, supplementing what they learn from traditional media channels (see Tedford 2015). But second, content must also trigger engagement because engagement is the source that can make the content go viral among consumers. This changes our understanding of what defines "good" communication content: content has to function in both paid media (where it is used to generate awareness and interest) and owned media (where the goal is to stimulate engagement and feedback).

For stimulating this level of consumer engagement, offering "valuable" content is crucial. This brings us back to the "what-to-communicate" question we discussed earlier—Disney expertly edited their trailers for The Force Awakens in a way that not only generated high awareness and interest, but at the same time invited consumers to fill the social media space with intense speculation regarding the meaning of the content of the trailer (Gallagher 2015). But pinball content is not limited to reediting traditional communication formats such as trailers, as is evidenced on social media pages on a daily basis. Whereas in 2016 the global Facebook page of Marvel offered a rich blend of news, background information about their superheroes, podcasts, and interviews, rival DC Comics at the same point in time basically provided only information about their titles' release dates. Marvel's Facebook page had 18 million fans, while DC's was "liked" by only 2.8 million (Jecke et al. 2015).

So, what exactly makes content valuable in a pinball world? Initial insights come from general scientific endeavors that shed useful light on this evolving issue. For example, based on about 100 qualitative "means-end" interviews (a series of "Why is this important to you?" questions) with users of brand pages on social media, we identified 14 different (and combinable) content practices that provide benefits to consumers on brands' social media sites (Kaczinski et al. 2016). With a follow-up survey of more than 4,000 representative German consumers, we then assessed the value of each type of content for consumers by measuring how often it was mentioned by respondents and linked the content types to consumer behaviors that provide economic contributions to a brand or firm.

We find that the value of content practices differs for customers versus firms, but also between industries. Figure 11.7 shows the Top 10 practices in terms of their value for consumers and also their impact on firm success ("customer engagement value").[301] For media brands (a heterogeneous set of news and entertainment brands/firms in this study), respondents were most often impressed by high-quality and topical content and general (versus product-specific) information. But we found the strongest *economic* impacts for content that is exclusive, shares background insights, and is aesthetic. Also, several content types, such as playful content, are more important in an entertainment context than they are in other industries.

Whereas our study focused on consumers' liking of a brand page, others have put the *virality* of its content on center stage. The work by Jonah Berger and his collaborators, using econometric techniques to explore which

[301]Customer engagement value is a multi-dimensional performance indicator, combining consumers' repurchase and referral intentions, among other contributions. For a more detailed look at the concept, we refer you to the article by Kumar et al. (2010).

Content	*Value for consumers (rank from 1 to 10) – media / all brands & firms*	*Value for firms (rank from 1 to 10) – media / all brands & firms*
Premium quality content	#1 / #3	#6 / #5
Topical content	#2 / #1	#8 / -
General trends and developments	#3 / #2	- / #10
Comedic content	#4 / #8	#10 / #7
Official content	#5 / #10	- / #6
Background insights	#6 / #6	#2 / #3
Diverse content	#7 / #5	- / -
Playful content	#8 / #9	#5 / -
Aesthetic content	#9 / #7	#3 / #4
Education	#10 / -	#4 / #1
Exclusive content	- / -	#1 / #2
Economic incentives	- / -	#7 / #9
Dramatic content	- / -	#9 / #8

Fig. 11.7 Most valuable types of social media content
Notes: Authors'own illustration based on data from Kaczinski et al. (2016). The number before the slash is the importance for media brands and firms only; the number after the slash is the importance rank for all brands and firms. The value for consumers reflects how often a content type was mentioned as valuable by participants; the value for firms is its impact on a multi-item measure of "customer engagement value," estimated through OLS regression. "–" means that a content type was not among the Top 10.

facets of content engage people, is particularly insightful here. In a study of almost 7,000 New York Times articles published during a three-month period in 2008, Berger and Milkman (2012) investigate what makes a newspaper article more likely to be shared with others (via email) by its reader. Using a logistic regression approach (whether an article appeared in the most-emailed list is their dependent variable), they find that it matters for sharing whether the content of an article can provoke consumer emotions.

But we have learned that emotions are a complex, more-dimensional concept. For virality, Berger and Milkman report that content that offers a positive (versus negative) experience gets a general bump in terms of sharing, particularly articles with the potential to spur high-arousal positive emotions (such as awe). But as with entertainment products in general, triggering negative emotions with communicative content is not necessarily a bad thing; news associated with high-arousal negative emotions (anger and anxiety) are also shared more often.[302] In contrast, sad articles ("negative low-arousal")

[302]In Berger and Milkman's study, engagement is most strongly stimulated by anger and awe—a one standard-deviation increase results in a 34% (anger) and 30% (awe) higher probability that content is shared with others. The authors used different methods to measure their drivers of sharing behavior—general emotionality and valence were measured with an automated text mining approach, whereas they used human coders to determine the specific emotional potential of articles.

tend to be shared less by readers. The scholars also find that if a text is considered to be "interesting," "informative," or "surprising," readers' engagement is higher too.

Although stimulating engagement is important, it needs to lead to consumption in the end to warrant entertainment success. Akpinar and Berger (2017) study this link in an advertising context; for 240 online ads they track not only the number of shares by consumers over a six-month period, but also the purchases of the advertised products. In line with Berger's earlier work, they find that emotional appeals (e.g., ads that make strong use of dramatic elements and music) are shared more often—but informative appeals are more effective than solely emotional ones for triggering purchases of the advertised product. Thus, content should integrate emotions and information if the goal is to result in sharing and sales. In some ways, this finding resembles what we have known for offline communication for quite some time: that not everything that arouses consumers (the half-naked model) gets them to buy the product (that he/she is promoting). It is also consistent with our own findings regarding the differential impact of social media content on consumer enjoyment versus spending.

Although entertainment producers have generally been hesitant in their adoption of engagement-targeted communication strategies, some masterful exceptions exist that offer enormous room for learning. Among those entertainment campaigns that triggered enormous engagement and were very successful financially is the one for the low-budget independent horror film THE BLAIR WITCH PROJECT. Considered one of the "best-ever" social media campaigns (regardless of industry affiliation), the campaign was all about content and engagement at a point in time when social media did not even exist (Facebook was founded half a decade later). The rights owners of the film set up a unique website months before its release, creating the illusion that the film would actually be authentic, "lost footage" from the three filmmakers, instead of being a work of fiction. The website provided a timeline of "events" about the "Blair Witch" myth, faux newspaper clippings about the crew's disappearance, police photos of found evidence and their missing car, and interviews with fictional experts (in dedicated MOV format!), such as David Mercer, an anthropology professor from the University of Maryland whose student had discovered a bag that belonged to the missing filmmakers.[303]

[303]At the time of writing this, the original website for the film was still accessible in its historic format: explore it (at your own risk…) via https://goo.gl/2w3gm4.

The website content was accompanied by filmed "mockumentaries" that were aired by regular TV stations before the release and missing-persons leaflets that were handed out at the Sundance Film Festival. Consumers enjoyed the speculation surrounding the film; preceding the film's theatrical release, the website was the most-visited film website of 1999 and among the 50 most-visited sites on the entire Internet. Online forums were full with discussions about the film's mysterious status (e.g., "Re: the answer to if The Blair Witch Project is true!!!;" Harris 2001). The film ended up grossing almost $350 million (in 2017 value) in theaters alone and ranks as one of the most profitable entertainment products of all time.

A more recent entertainment example of using owned-media channels to provide content that works effectively in the pinball environment is the movie Ted, an R-rated comedy about a foul-mouthed teddy bear. About three months before the film's release in the summer of 2012, the newly created Twitter account @WhatTedSaid greeted potential moviegoers in the lead character's dedicated offensive style: "Hello, Twitter. Kindly go f*ck yourself." In the following weeks, the bear sent almost 200 tweets in which he insulted nearly everyone, proving to have a seemingly unlimited repertoire of abusive language. On opening night, he demanded his followers to see his film, tweeting "Here I go, f*cktards! Smoke a fattie and come hang out with me this weekend at your local theater. Or, go to one far away, I don't care."

Such messages, pushed out to the phones of more than half a million consumers, certainly classify as high activation content, particularly as they were consistent in tone with the on-screen persona of the title character. To assure this match, they were worded by Alec Sulkin and Wellesley Wild, the film's screenwriters. The producing studio Universal honored their special social media service with extra salary and provided them with immense latitude; according to the studio's responsible manager, "The parameters were, 'Just go to town'" (Doug Neil, senior vice president of digital marketing, quoted in Dodes 2012). Most tweets were retweeted more than 1,000 times, and consumers' high engagement levels translated into hit status: the film (produced for $50 million and involving only Ted's co-star Mark Wahlberg and creator Seth MacFarlane as brands) greatly exceeded industry expectations and generated theatrical revenues of more than half a billion dollars.[304]

[304]As an aside, the film's producers were later (unsuccessfully) sued for similarities of their teddy bear character and their social media marketing approach, including the wording of some Twitter posts, with a web series that had aired on YouTube three years earlier and its marketing campaign (see Robb 2014).

But Ted also carries one more general insight for entertainment managers: whereas the film's producers used a similar social media approach and the same Twitter account for the sequel, audiences were clearly less enthusiastic the second time. We assume that the approach had worn out its welcome somewhat (i.e., reduced novelty equals fewer sensations) and had diminished potential for activating and engaging consumers. The writers also seemed somewhat less excited this time, as they crafted only 130 tweets over a five-month span, compared to 200 tweets in three months for the parent film. Thus, originality is crucial for stimulating audiences, even for sequels of hit films, or even more so for them. In pinball times, "routine" has a tough job.

Managing Consumer Engagement: Co-Creation and Moderation

> "Don't forget I'm doin' Q&A today at 3 PM EST. Just use #AskTed and will someone remind me? I'm gonna be wicked stoned."
>
> —*Character* Ted *from the movie carrying the same name via Twitter on June 6, 2012*

Playing pinball requires the "right" content, but it needs much more from a marketer than a smooth ball to succeed in this chaotic environment. In pinball marketing times, consumers demand an active role, and marketers have to find ways to deliver. Co-creating value by moderating a conversation with entertainment consumers about brands and their meaning is a potentially powerful approach to address this consumer need. But we also acknowledge that implementing such co-creation is far from trivial—how can you interact with individual consumers when there are millions of them?

The Logic of Co-Creating Entertainment Brand Stories

A brand's image, which is made up of what consumers like about the brand and identify with, is not static, but must change with and adapt to societal and cultural changes. As individuals, we all know this need for constant updating well; think of the way we talk (you don't say 'groovy' anymore, do you?), the information channels (bye-bye MySpace!) and devices (Blackberry anyone?) we use, and how we dress (the painful moment when we are told that we simply cannot wear our favorite suit anymore). The same applies to brands: when the environment changes (and it does all the time), brands have to adjust their appearance and values. In the pinball era, the way brand images are developed must account for consumers' active roles.

A core format for defining brands is through brand stories, which have traditionally been told to consumers via advertising and related communication activities. Ben and Jerry's is all about its origins (the "first ice cream scoop shop in a renovated gas station in Burlington, Vermont"; *BenJerry.com* 2017) and the Disney brand is about the vision of its legendary founder Walt to "make people happy" through entertainment. Each valuable entertainment brand has its own stories to tell.

Such brand stories, with lead characters, a plot, and an emotional outcome that is intended to enable brand attachment, have traditionally been determined solely by managers and then offered to consumers. But today's active consumers question this storytelling monopoly arrangement. They want to share their own brand stories, and the Internet and social media empowers them to do it. Take Britney Spears as an example: When a demo tape for her song HOLD IT AGAINST ME was leaked on the Internet, over 2,000 consumers recorded and uploaded video remixes of the song, with some of them attracting more than 700,000 views (Kaplan and Haenlein 2012).

Marketing scholars Gensler et al. (2013) compare the development of brand stories in pinball times to the assembling and re-configuration of a "brand story puzzle." We illustrate this brand-story-puzzle logic in Fig. 11.8: firm-generated, "official" puzzle pieces (the white elements in the figure) co-exist with consumer-generated, "unofficial" pieces (the orange puzzle pieces). These unofficial pieces are things like consumer reviews and essays, fan-made trailers, recuts, mash-ups, and spoofs. Whereas official puzzle pieces are centrally coordinated, those created by consumers are often heterogeneous, stemming from various sources and offering interpretations that are not necessarily in line with the manager's official narrative.

Gensler et al. recommend that managers should find ways to listen to and integrate some of the "user-generated" brand stories in the overall meaning of the brand, instead of ignoring consumers' actions and insisting on the firm's right to legally define the brand (remember that we have taken a consumer perspective when defining entertainment brands: it matters what consumers think of the brand, not what the firm wants them to do).

This co-creation of brand meaning is illustrated in the figure's right-hand side, where two puzzle pieces originally contributed by consumers have now been turned from orange to white—they have been embraced by the brand's managers and have become part of the brand's official narrative. The figure also shows that such treatment will only happen to a few selected consumer-generated

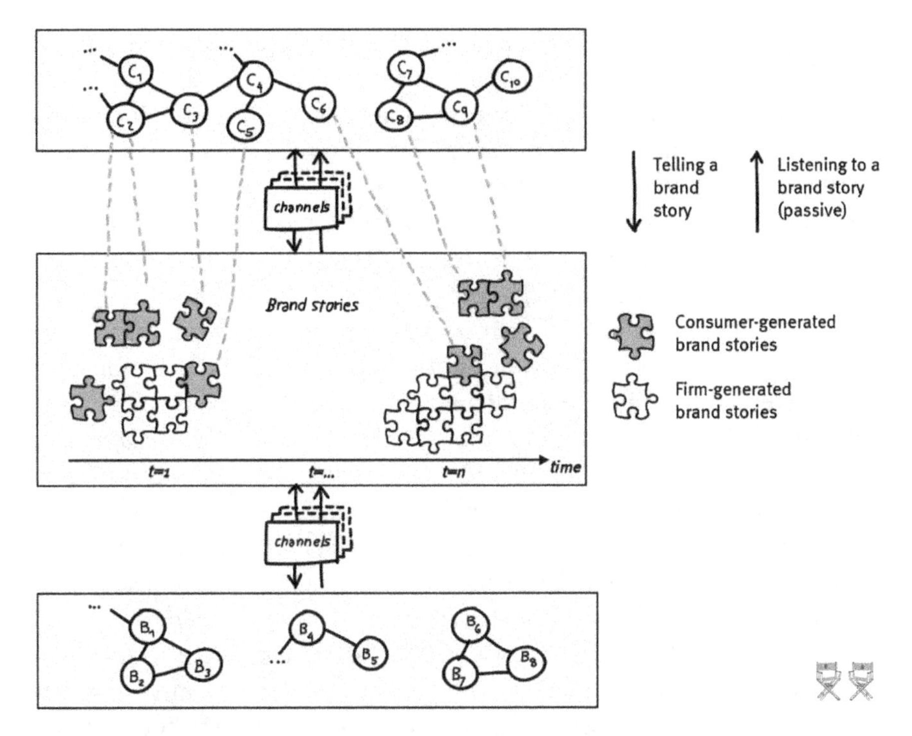

Fig. 11.8 Creating brand stories in pinball times
Notes: Authors' own illustration based on Gensler et al. (2013). Graphical design by Studio Tense.

stories; other orange puzzle pieces have not shared these pieces' destiny, but have remained separate from the brand's historiography (and were hopefully forgotten by consumers).

The key for succeeding in this puzzle task is *moderation*. In addition to developing plans for their brand, managers need to coordinate what is happening in the cacophonic pinball environment, stimulating some kinds of engagement while redirecting, or softening, others. Co-creation must not be confused with offering consumers the driver's seat of an entertainment brand's story and identity. Instead, the manager needs to remain the driver—but he or she should listen to the suggestions that consumers, as passengers, offer and be open to them.

Given that such co-creation of brand meaning implies at least a partial loss of control over one's own brand, and that it is immensely complicated, why should managers nevertheless *voluntarily* embrace such an approach? Here are three reasons:

- *Because the potential gain is immense.* Consumers today *want* to be active and taken seriously. Enabling active consumer involvement via moderation addresses that need. Consumers' thankfulness can set off powerful cascades, often in close connection with their creativity in producing innovative verbal or visual material for free.
- *Because it helps the brand to remain relevant.* Listening to consumers' engagement offers managers of entertainment brands a chance to keep in line with the continuously changing ambitions and desires of a brand's fans, helping to assure the timeliness and relevance of the brand and contributing to its longevity. Given the dynamics of culture, this should be considered essential for any entertainment brand.
- *Because it is happening anyway, and ignoring those who love a brand can only make things worse.* The Internet is full of user-created content for any major entertainment brand, and this content influences the perceptions of millions of consumers, even when managers look away. Several million consumers had seen "fan-made" trailers about The Last Jedi in April 2017 (when no official material was yet available), and almost 3.5 million had watched a fan-made teaser trailer for an Avatar sequel user-titled Return to Pandora, influencing their expectations and anticipation for the actual coming attractions. Ignoring such engagement and lacking a close connection with fans carries the risk of affronting them and making them feel unwanted. This can be counter-productive given the crucial role of "core fans" for the broader adoption of any brand, but particularly for those ones which depend on consumers' emotional attachment—the standard case for hedonic entertainment brands. Alienating one's fan base can become particularly dangerous in times of (brand) crisis, when the support of these diehard fans is needed to change the brand narrative back to the better.

Let us stress that moderation is not the same as the mere stimulation of engagement. It also implies the need to respond to critical articulations from fans. Scholars Parmentier and Fischer (2015) argue that the decay of the TV show America's Next Top Model was caused by consumers' cascading negative word of mouth, creating doppelgänger brands, etc. Fans had "reframed" the brand's identity, with a focus on host's Tyra Banks persona as a "media mogul." This reframed identity then was the basis for consumer "remixing" activities when fans learned about a candidate in the show that violated its rules (such as creating belittling cartoons naming the candidate "Little Miss Cheater," a sarcastic blend of the criticized candidate with the character of the Little Miss Sunshine movie). The scholars argue that the remixing finally led to consumers' "rejection" of the show, when its managers' made a decision regarding which fans were found to be in contrast with what the managers considered the "true" identity of the show. Certainly, the lack of serious moderation efforts and skills by the show's managers was a major factor that contributed to the brand's destabilization process.

We will now name a number of practical examples of actual moderation practices today in entertainment. In addition to demonstrating the approach's potentials through them, we also illustrate some limitations of current implementations and also the pitfalls of underdeveloped pinball playing.

Some Practical Examples of Co-Creating Entertainment Brand Stories with Fans

Active moderation in entertainment is most prominent with musical artists, which, as human brands, occasionally manage their relationships with fans in ways that go beyond "broadcasting" status updates. Singer Britney Spears has relied heavily on the use of social media for managing her brand throughout her career, simultaneously running webpages, YouTube channels, a Twitter account, and a Facebook profile, with some of these channels dating back to 2005 (Kaplan and Haenlein 2012).

Ms. Spears regularly addresses her fans directly through these channels, sometimes directly asking for their responses; for example, her post "Happy Friday people! Am I part of your future?" alone triggered more than 1,000 replies from her fans. But she also actively encourages user-generated content that involves her products. When users created fan-video clips for her leaked song HOLD IT AGAINST ME, she not only stimulated them to do so, but also included links to some of them on her official BritneySpears.com website, recognizing their work and making them part of her official brand narrative (Kaplan and Haenlein 2012).

Other musicians are equally active as moderators of fan conversations (Taylor Swift, for example, sometimes even comments on fans' pages), as are some book authors (notably J.K. Rowling) and film actors (e.g., Russell Crowe, who takes the time to send fans individual birthday greetings, and Kevin Hart, who even has *called* individual fans on Facebook). Whereas such active moderation is less common on the level of "non-human" entertainment brands and products, some interesting examples exist. In the movie context, Marvel stresses the value of interacting with its customers when developing the brand stories of the characters in their Cinematic Universe.

Marvel-CEO Kevin Feige has stated that "the conversation that's taking place around [the casting of actors for Marvel movies] is super-important. … [O]ur upcoming announcements are going to show that we've been listening" (in Fleming 2016). Feige's personal appearance and self-presentation also supports this co-creation approach; rejecting the "suits-suck" image that entertainment managers often carry and, instead, mostly wearing base cap and a sweatshirt, he seeks closeness with the fans and presents himself as a renowned Marvel connoisseur (Jecke et al. 2015).

The producers of the TED movie focused on content provision in general. But they also added some interactive elements such as hosting two live Q&A sessions via Twitter, where fans were able to ask the raunchy teddy bear some equally raunchy questions—our introductory quote for this section gives evidence. The Q&A sessions enabled fans to engage directly with the film's lead character, actually turning their parasocial relationships into two-sided ones, at least for a period of time, and got fans excited about things to come, contributing to the further spreading of the film's hashtag.

The TV series producers of SUITS used an approach that was more directly focused on engaging in conversations with consumers. The 3,400 Twitter followers of the series' Mike Ross received a personal tweet from him in August 2013, asking "Are you a bike or limo to work kind of lawyer?" (Riehl 2014). What makes this interesting is that Mr. Ross is not a real person, but one of the series' characters. The producers' attempt at moderation was kind of halfhearted, however (like the majority of those we came across): although Mike sent out 70,000 replies and now has a follower base of more than 11,000, he meets all the stereotypes of a social bot, instead of a character with whom it pays to engage for fans. He follows only five others himself (all of them are characters from the series…also), has written just 25 tweets, and all his replies are variations of the 25, sent "personally" (but publically) to each of his followers. He would certainly not pass the Turing test for intelligent (human) behavior, this lawyer. Frank Underwood, the American President from Netflix's HOUSE OF CARDS series, is much more effective when it comes to interacting with consumers; at @FrankUnderwood, he has assembled more than 240,000 fans and replies individually to them. The problem here: the character has taken on a life on its own—with the fan-operated account not being affiliated with the series at all…

Quite ambitious—and proficient—were the producers of the DARK KNIGHT movie in their efforts to moderate fan engagement. They developed an alternate reality game that required Batman's fans to master several tasks, some of which required a high level of activity and engagement. In other words, the moderation role was not carried out by humans, but assigned to smart software that provided individual fans with feedback and directions (although several parts of the ambitious game were indeed carried out with "human" support).

The game, in which 10 million people in 75 countries participated (Taylor 2010), was truly multi-media, ranging from "jokerized" $1 bills found at the 2007 Comic-Con fair (which celebrates comics and related forms of entertainment, taking place annually in San Diego), phone numbers written in the sky (and to be called), a lot of online action, real-world scavenger hunts, and 22 actual cell phones stuffed into cakes, to free early-screening IMAX tickets (Lang 2011). Each of these activities was closely tied to the story of the movie which became the second-best selling film of all time. Whereas we are unable to determine the game's exact contribution to DARK KNIGHT's success, the enormous pre-release engagement that the game contributed did not go unnoticed.[305]

We have argued that moderation can be particularly valuable when a brand faces a crisis. But like any other marketing and management action, moderation requires skills, and ill-natured attempts at moderation can escalate criticism. Game producer EA posted that its "intent is to provide players with a sense of pride and accomplishment for unlocking different heroes" in response to early users' revelations that it required enormous amounts of time and money to unlock key STAR WARS characters in STAR WARS BATTLEFRONT II, in addition to the upfront game price of $60.[306] EA's moderation not only became the most down-voted comment in the history of the social media site Reddit (with almost 700,000 user votes; Minotti 2017), but also triggered additional negative feedback by other media—another pinball effect. Although the company later changed its moderation approach, now stating "We hear you loud and clear" in a blog post and (at least temporary) removing any in-game transactions at literally the last minute prior to release (Tassi 2017), it seems that the initial moderation activity still hurt game sales, as well as the company's reputation and financial valuation. Consumers expressed frustration and even petitioned Disney to revoke the license from EA (Kim 2017).

[305]See this book's discussion of the pre-release buzz concept and its contributions to entertainment success in the next chapter.

[306]An illustrative consumer comment was: "Seriously? I paid 80$ to have Vader locked?"

These individual examples provide ad hoc evidence of how skilled moderation might look in entertainment (and what should be avoided). Let us complement these insights with a look at the aggregated statistical evidence that scholars have already assembled regarding the power of pinball playing, despite the approach's youth.

How Effective is Communication Through Owned Media?

Among marketing scholars, quantifying the impact of social media activities on product success is certainly among the hottest topics these days. The results so far are pretty consistent and probably not really surprising, at least not for those who spend a substantial amount of their own time on the Internet and social media platforms: investments in owned media can indeed pay and generate substantial returns. But as with all other elements of marketing, communicating with consumers via social media is not a safe bet—its impact depends on how it is done, and what kind of owned media is used. What do we know so far from empirical research on how social media marketing affects the success of music, TV shows, and movies?

In the field of music, Chen et al. (2015) included two kinds of social media activities in their study of music sales from 616 artists at Amazon, in addition to advertising and user reviews. Those two were automated messages sent by the musician and his or her producer on the platform MySpace (i.e., "friend updates") and personal messages from the artist via the same site ("bulletin board entries"). Both social media messages are "content-only" measures, with no interactive or moderating elements, and the scholars classify them both as "broadcasting." Their panel VAR approach reveals that personal social media messages indeed affect the artists' sales significantly, whereas automated messages have no such effect, with a parameter that is very close to zero.

And how strong is the effect of such personal messages? Chen et al. estimate an elasticity of 0.05, which is slightly higher than the one they find for traditional advertising—a 10% higher number of personal messages should, on average, convert into a sales rank increase of 0.5% without any additional monetary spending (but of course the artist's time is also money, in a certain way). For star musicians (those who also use traditional advertising), personal messages are slightly more impactful, and they are *substantially* higher around the release of a new album. In this case, the elasticity for personal messages is almost 0.20, or four times as high as on average. So sending personal messages to fans via social media around the time of a new product release seems to be a highly rewarding approach.

Other scholars have looked at "owned" social media for TV shows. When Lovett and Staelin (2016) analyzed the drivers of the popularity of TV show HUMAN TARGET among survey respondents, they also asked whether a consumer was engaged in content related to the show on the network's website. Through their linear probability models, the scholars found that doing so had less of an impact than either paid advertising or word of mouth; the probability that a consumer watches an episode of the show increases only by 2%, an effect that is not statistically significant. We do not consider company websites as the kind of media with the highest engagement potential, so this finding should probably not be generalized too much to other used media platforms.

Such interpretation is in line with the results of Gong et al. (2017), who study how a TV show producer's "tweets" (sent via the Chinese Twitter equivalent, Sina Weibo) about its shows affect the shows' viewership among Chinese audiences. The scholars conduct a randomized field experiment in cooperation with a producer of documentaries, whose content is aired by several local TV stations in China. They either sent out a tweet on the day a new show is aired, or did not do so—randomly assigning 98 different shows aired via five stations to one of the two conditions.

And what did they learn? Shows with tweets were indeed watched by more viewers. The average viewing percentage is about two-thirds higher (1.25% versus 0.75%) when the producer sends a tweet about the show to his about 130,000 followers, compared to when the firm sends no such post. When controlling for channel and show characteristics (such as genre) and timing in an OLS regression (with robust standard errors), Gong et al. find that a producer tweet is, on average, associated with a rating increase of 0.6%. The authors also ran some analyses with the *absolute* number of viewers as dependent variable, finding that a posting added about 6,300 viewers, on average. The small scale of these numbers warns us that things could be different for more mainstream content and, as the producer in this study uses only owned social media to promote the content, if producers also use paid channels to promote their showings.

Finally, for a data set of all major movies released in North American theaters in 2012–2014, we examined whether the number of fans a film had attracted on its Facebook page until *three months before its release* affected its box-office results (Kupfer et al. 2018). Using a linear mixed effects model, in which the films' weekly box office in North America serves as the dependent variable and which also included a large number of controls (such as the "brand power" of the film), we find a positive and significant, but relatively decent elasticity of 0.05. In other words, a 10% higher number of such early Facebook fans contributes about 0.5% higher revenues over a film's theatrical life cycle, above and beyond all other factors. Saboo et al. (2016) report a similar link for the number of social media followers of music artists and sales of music. Based on data for 73 weeks on several social media platforms and a control function analysis, their results suggest that the link between fans and music sales might be non-linear: when too many people like an artist on social media, others begin to lose interest in him or her.

And what about the social media actions by those people who are involved in the production of a new entertainment product, such as singers, actors, or directors, that serve as "ingredient brands"? At the heart of our study was a desire to see whether the social media activities of a film's star actor can generate additional returns. Sean Bailey, as president of Walt Disney, revealed that Emma Watson's social media activities were helpful, in that out of the 90 million who viewed the teaser trailer of The Beauty and the Beast, "almost half of them came from one of Emma's vast social media channels. Imagine, 40 million plus views through her social media channels" (quoted in Fleming 2017). But does such activity also affect the bottom-line, bringing in additional revenues? And, if so, how?

Our regression results show that the size and "activity level" of the leading star's Facebook fan base have a decent positive impact on their own, being similar in size to the one of the film's fan base. But we find that what the star *does* with such power potential is the real deal for a producer. On average, we see that a 10% increase in a lead actor's film-related posts leads to a weekly revenue increase of 2.4%, an effect that gets stronger as the actor's fan base is bigger. And it also matters *what* the star posts: content that is authentic, exclusive, and/or persuasive drives box-office revenues notably, whereas posts that contain none of these elements hardly affect movie success. Figure 11.9 reports the dollar value of sending one such post (versus none) to North American moviegoers by a lead actor of a major movie, based on simulations—persuasive star postings bring in about half a million dollars on average.

Film producers thus should not only invest in their film's social media performance; they would also benefit from hiring a star who not only can enchant audiences on the screen, but has also a vivid social media fan base.[307]

[307]One of the controls in our study is "traditional" star power of the actors, which remains significant and important. That's why betting *exclusively* on an actor's social media power would not be a good idea.

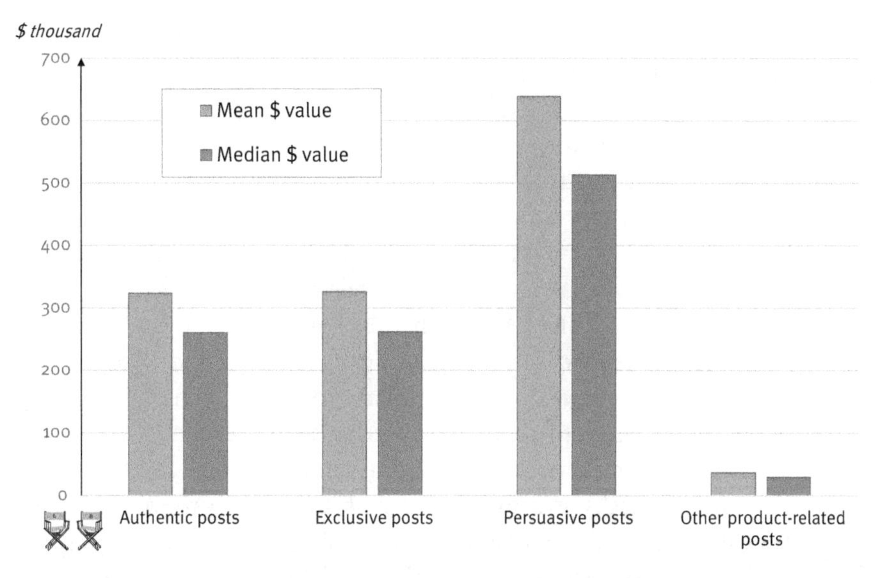

Fig. 11.9 How different kinds of stars' social media activities can help a movie
Notes: Authors' own illustration based on results reported in Kupfer et al. (2018). Numbers are the estimated dollar value of a Facebook post by a movie's lead actor in terms of the resulting change in the North American box office for the film. The top and bottom 5% film-actor combinations were dropped to limit the impact of outliers.

Findings from Gong et al.'s (2017) study indicate that this impact might not even be limited to stars who are personally involved in the making of an entertainment product as ingredients. When the TV producer hired an "influencer" (with millions of followers on his own) to retweet their original tweet about a film, the film's average TV rating increases to 1.44 rating points—about 15% more than for the producer "tweet-only" condition.

A Different Kind of Owned Media: Packaging as a Communication Instrument

Before we wrap up our discussion of owned media and move on to "earned" media, let us take you on a quick detour and look at a very different kind of owned media: a product's packaging. Very much like websites and brand pages, product packaging is in full control of the producer and used, in addition to functional aspects related to logistics, for communicative measures. Lacking the interactive elements of social media, its role in the pinball framework is not one of moderation, though. Instead, the packaging provides a consumer with information about the product to encourage him to buy the product. It is another pinball, if you want, because it can trigger engagement and set off interest cascades. In addition, packages on their own can provide consumers with value.

Several of the general findings we presented earlier on how to communicate a new entertainment product effectively also apply to packaging. However, one particularity is the context in which packaging affects place—consumers react differently to stimuli when they are watching an advertisement, compared to when they are in a shop selecting a book, game, or DVD from a large number of titles. Although we agree with McKay et al. (2012) that this area has remained somewhat under-researched and deserves more attention by (*Entertainment Science*) scholars, some basic empirical insights on packaging exist. Most of them deal with the role of packages in a consumer's decision-making process, and all of them are on books.

So, how are packages processed by entertainment consumers? Reutzel and Gali (1998) use a qualitative, observational approach for studying how children select books; their sample consists of 18 children from either first, third, of fifth school grade. Regarding the book cover as packaging, the scholars find the *design* of the cover to be an important part of the choice process. However, they do not make an attempt to quantify its role or even determine what defines a "good" cover and sets it apart from not-so-good ones. When adult British readers were asked for their purchase motivations for books in a study of book-buying habits (see Buchanan and McKay 2011), respondents similarly self-report books' cover design to be one of the most important drivers of book choice, followed in importance by the "content" displayed on the book.

Two studies avoid the pitfalls of such self-reported importance data and study the role of book covers with actual sales data. Both of them provide further support for the argument that packaging should be treated as an important element of the marketing mix for material entertainment products. Specifically, Meiseberg (2016), in her analysis of what drives the success of about 30,000 books on Amazon's German website, includes a very basic book cover variable—it takes the value of 1 for those books whose covers contain a photo (her measure of a cover's "vividness"), and 0 for those which had no such photo.

Meiseberg's results show that, across all books, photo covers correlate with books' sales ranks quite substantially ($r = -0.33$—so the cover explains about 10% of the variation in sales ranks for the books in her data set, when ignoring potential overlap with other factors). And her regressions corroborate that effect; they suggest that a photo cover is associated with a 3% better sales rank for low-selling titles and with a nearly 7% better rank for best-selling titles. Meiseberg does not account for potential endogeneity though: whereas her results indicate that a vivid cover helps a book sell better, it might also be possible that books with higher sales potential get more costly (and vivid) cover designs.

Whereas the study by Schmidt-Stölting et al. (2011) does not address the latter concern, it facilitates our understanding of the packaging effects in other ways. In their analysis of the success of 1,000+ books in Germany, the scholars measure the "appeal" of a book's cover not with a binary variable, but instead asked four students to rate the "appeal" of each of the book covers (on a 1-to-5 "appealing" scale). They then include a mean appeal score in their SUR analysis of book success, together with several other success drivers.[308] Their results show that, for paperback editions, appeal is associated with book sales—a one-point (or 25%) higher appeal results in 28% higher sales. But this effect is limited to paperbacks, at least in Schmidt-Stölting et al.'s data. For hardcover books they do not find that the appeal of a book's cover will influence success. The authors "blame" advertising: they speculate that for paperbacks, for which usually little or no advertising is spent, the cover serves as the "face" of the book in stores and online, whereas advertising might absorb this effect in the case of hardcovers.[309]

Other areas in entertainment in which packaging can be used as a marketing tool include DVDs and Blu-ray, games, and CDs—as well as vinyl albums (once again). Is the recent renaissance of vinyl a result of pure nostalgia, superior technical quality, or primarily of the value that packaging provides for consumers? Premium packages often go along with other special features added to a product, an issue we return to in the context of versioning as a strategy of entertainment pricing. Across entertainment products, such special packages have established themselves as a niche offering, speaking to the needs of a limited number of highly involved consumers. Overall, it seems pretty clear that the economic role of packaging will tend to diminish as consumers increasingly turn toward digital offers that leave no room for this feat. But for those niche segments who stay loyal to material entertainment offerings, packaging might continue to play a main role.

---

[308]The authors integrate the four ratings by calculating the mean score across raters, which they weigh by each rater's "confidence" in his/her judgment.

[309]To verify their somewhat surprising results, the scholars took an in-depth inspection of the book covers in their data set and found enormous differences in cover design: "The 50 most attractive are densely designed, with vibrant colors, whereas the 50 least attractive are sparsely designed, with a great deal of white space" (Schmidt-Stölting et al. 2011, p. 40).

Concluding Comments

This first of two chapters on entertainment communication decisions focused on those modes of communication through which managers can, more or less, control the message: paid and owned channels.

For paid channels, because the firm is paying a particular media entity (such as a TV channel, a newspaper, or a website) to share a message with that entity's consumers, the firm can specify the form and content of the actual message. Information that a firm shares via paid channels has the goals of creating awareness for the product and building the brand's image. We reviewed the empirical evidence on some of the major decisions facing the entertainment product manager when using paid channels, including how much information to reveal (enough to whet the appetite but not so much as to ruin the meal), along with how much to spend and when to do so. For entertainment products with their short life cycles and movies in particular, pre-release spending has particularly high elasticities.

For owned media (such as a film's Facebook page, Twitter account, or company website), the firm has full control of the initial information they release about their new entertainment product. However, the similarity to paid channels stops there as releasing information via owned media into the digital atmosphere is more like launching a pinball into a pinball machine that is full of bumpers, ramps, and flippers. We show that entertainment producers can benefit strongly from artfully operating the flippers, which offer the opportunity to react and keep the ball alive to spread engagement and anticipation among consumers. Research provides a good understanding about how to most effectively address entertainment consumers via owned media in a pinball world, including the role of content and how to co-create entertainment brand stories. If done properly, entertainment firms can benefit greatly from integrating the creative contributions of their highly-engaged fans, but it certainly requires some managerial courage to relinquish full control about its brands and products.

Let us now move on to the least-controllable category of entertainment communication: the communicative reactions that products "earn" from consumers. We will see that because of the special characteristics of entertainment products, such as their cultural nature (which assigns a public importance to entertainment products, granting them space in newspapers and on websites) and their experience character (which lets consumers value "substitute cues"), such "earned" channels are of particular importance for understanding and managing entertainment success.

References

Akpinar, E., & Berger, J. (2017). Valuable virality. *Journal of Marketing Research, 54*, 318–330.

Appel, M., Richter, T. (2010). Transportation and need for affect in narrative persuasion: A mediated moderation model. *Media Psychology, 13*, 101–135.

Basuroy, S., Kaushik Desai, K., & Talukdar, D. (2006). An empirical investigation of signaling in the motion picture industry. *Journal of Marketing Research, 43*, 287–295.

BenJerry.com (2017). Our history. https://goo.gl/NjSwdH.

Berger, J., & Milkman, K. L. (2012). What makes online content viral? *Journal of Marketing Research, 49*, 192–205.

Bruce, N. I., Zhang Foutz, N., & Kolsarici, C. (2012). Dynamic effectiveness of advertising and word of mouth in sequential distribution of new products. *Journal of Marketing Research, 49*, 469–486.

Buchanan, G., & McKay, D. (2011). In the bookshop: Examining popular search strategies. In *Proceedings of the 11th Annual International ACM/IEEE Joint Conference on Digital Libraries JCDL* (pp. 269–278).

Burmester, A. B., Becker, J. U., van Heerde, H. J., & Clement, M. (2015). The impact of pre- and post-launch publicity and advertising on new product sales. *International Journal of Research in Marketing, 32*, 408–417.

Cacioppo, J. T., & Petty, R. E. (1982). The need for cognition. *Journal of Personality and Social Psychology, 42*, 116–131.

Chen, H., De, P., & Yu Jeffrey, H. (2015). IT-enabled broadcasting in social media: An empirical study of artists' activities and music sales. *Information Systems Research, 26*, 513–531.

Clement, M., Wu, S., & Fischer, M. (2014). Empirical generalization of demand and supply dynamics for movies. *International Journal of Research in Marketing, 31*, 207–223.

Cornwell, R. (2011). DC was never going to get the magic of pinball. *Independent*, June 4, https://goo.gl/uG1ghb.

Das Fernsehlexikon (2008). Das Geheimnis von Twin Peaks. https://goo.gl/jKhBLg.

Dewan, S., & Ramaprasad, J. (2014). Social media, traditional media, and music sales. *MIS Quarterly, 38*, 101–121.

Dodes, R. (2012). Twitter goes to the movies. *The Wall Street Journal*, August 3, https://goo.gl/QLje3S.

Ebert, R. (2000). Movie review for Cast Away. *Chicago Sun-Times*, December 22, https://goo.gl/pM9LB1.

Elberse, A., & Eliashberg, J. (2003). Demand and supply dynamics for sequentially released products in international markets: The case of motion pictures. *Marketing Science, 22*, 329–354.

Elberse, A., & Anand, B. (2007). The effectiveness of pre-release advertising for motion pictures: An empirical investigation using a simulated market. *Information Economics and Policy, 19*, 319–343.

FFA Filmförderungsanstalt (2005). Kinobesucher 2004 – Strukturen und Entwicklungen auf Basis des GfK-Panels, April, https://goo.gl/W4WHDR.

FFA Filmförderungsanstalt (2010). Kinobesucher 2009 – Strukturen und Entwicklungen auf Basis des GfK-Panels, April, https://goo.gl/399mPz.

FFA Filmförderungsanstalt (2015). Kinobesucher 2014 – Strukturen und Entwicklungen auf Basis des GfK-Panels, April, https://goo.gl/kPi6Hf.

Finsterwalder, J., Kuppelwieser, V. G., & de Villiers, M. (2012). The effects of film trailers on shaping consumer expectations in the entertainment industry—A qualitative analysis. *Journal of Retailing and Consumer Services, 19*, 589–595.

Fleming Jr., M. (2016). Kevin Feige on 'Captain America: Civil War' and all things Marvel—Deadline Q&A. *Deadline*, May 6, https://goo.gl/pJo7BX.

Fleming Jr., M. (2017). Sean Bailey on how Disney's live-action division found its 'Beauty And The Beast' Mojo. *Deadline Hollywood*, March 21, https://goo.gl/4aLMxM.

Fritz, B. (2015). 'Star Wars' carries its own marketing weight for Disney. *The Wall Street Journal*, December 7, https://goo.gl/GYqHp8.

Gallagher, D. (2015). Why no Luke in the 'Force Awakens' trailers? Rating the theories. *CNET*, November 3, https://goo.gl/rw7NMr.

Garrahan, M. (2016). Advertising: Facebook and Google build a duopoly. *Financial Times*, June 23, https://goo.gl/ipx36F.

Gensler, S., Völckner, F., Liu-Thompkins, Y, & Wiertz, C. (2013). Managing Brands in the Social Media Environment. *Journal of Interactive Marketing, 27*, 242–256.

Gong, S., Zhang, J., Zhao, P., & Jiang, X. (2017). Tweeting as a marketing tool—Field experiment in the TV industry. *Journal of Marketing Research, 54*, 833–850.

Gopinath, S., Chintagunta, P. K., & Venkataraman, S. (2013). Blogs, advertising, and local-market movie box office performance. *Management Science, 59*, 2635–2654.

Grainge, P., & Johnson, C. (2015). *Promotional screen industries*. London and New York: Routledge.

Hansen, N., Kupfer, A.-K., & Hennig-Thurau, T. (2017). Social media firestorms. Working Paper, Münster University.

Harris, M. (2001). The "Witchcraft" of media manipulation: Pamela and the Blair Witch Project. *The Journal of Popular Culture, 34*, 75–107.

Hennig-Thurau, T., Houston, M. B., & Walsh, G. (2006). The differing roles of success drivers across sequential channels: An application to the motion picture industry. *Journal of the Academy of Marketing Science, 34*, 559–575.

Hennig-Thurau, T., Malthouse, E. C., Friege, C., Gensler, S., Lobschat, L., Rangaswamy, A., et al. (2010). The impact of new media on customer relationships. *Journal of Service Research, 13*, 311–330.

Hennig-Thurau, T., Hofacker, C. F., & Bloching, B. (2013). Marketing the pinball way: Understanding how social media change the generation of value for consumers and companies. *Journal of Interactive Marketing, 27*, 237–241.

Ho, J. Y. C., Dhar, T., & Weinberg, C. B. (2009). Playoff payoff: Super Bowl advertising for movies. *International Journal of Research in Marketing, 26*, 168–179.

Jecke, J., Herger, D., Dederichs, C., & Walk, I. (2015). Der Erfolg des Marvel Cinematic Universe. *Moviepilot*, https://goo.gl/NvWA7o.

Johnson, B. K., & Rosenbaum, J. E. (2015). Spoiler alert: Consequences of narrative spoilers for dimensions of enjoyment, appreciation, and transportation, communication research. *Communication Research, 42*, 1068–1088.

Joshi, A. M., & Hanssens, D. M. (2009). Movie advertising and the stock market valuation of studios: A case of 'Great Expectations?'. *Marketing Science, 28*, 239–250.

Kaczinski, A., Paul, M., Hennig-Thurau, T., & vor dem Esche, J. (2016). Social media marketing: An organizing means-end framework of firm practices and their consumer consequences. Working Paper, Münster University.

Kaplan, A. M., & Haenlein, M. (2012). The Britney spears universe: Social media and viral marketing at its best. *Business Horizons, 55*, 27–31.

Karray, S., & Debernitz, L. (2015). The effectiveness of movie trailer advertising. *International Journal of Advertising, 36*, 368–392.

Katz, E., & Lazarsfeld, P. F. (1955). *Personal influence: The part played by people in the flow of mass communication*. New York: The Free Press.

Kehe, J., & Palmer, K. M. (2013). The gripping, mind-blowing, thrilling evolution of the movie trailer. *Wired*, June 18, https://goo.gl/173sBq.

Kernan, L. (2004). *Coming attractions: Reading american movie trailers*. Austin: University of Texas Press.

Kim, T. (2017). EA's day of reckoning is here after 'Star Wars' game uproar, $3 billion in stock value wiped out. *CNBC*, November 28, https://goo.gl/4dZpFo.

Kinnear, S. (2012). 50 trailers that ruined the movie. *GamesRadar+*, January 9, https://goo.gl/b9hpF1.

Kumar, V. Lerzan, Aksoy, B. D., Venkatesan, R., Wiesel, T., & Tillmanns, S. (2010). Undervalued or overvalued customers: Capturing total customer engagement value. *Journal of Service Research, 13*, 297–310.

Kupfer, A.-K., Pähler vor der Holte, N., Kübler, R., & Hennig-Thurau, T. (2018). The role of the partner brand's social media power in brand alliances. *Journal of Marketing, 82*, 25–44.

Labrecque, L. I., vor dem Esche, J., Mathwick, C., Novak, T. P., & Hofacker, C. F. (2013). Consumer power: Evolution in the digital age. *Journal of Interactive Marketing, 27*, 257–269.

LaFrance, A. (2014). Why classic movies have terrible trailers. *The Atlantic*, January 27, https://goo.gl/Vn8FGJ.

Lang, A. (2011). The 5 most insane alternate reality games. *Cracked*, July 30, https://goo.gl/jcfPPL.

Lavidge, R. J., & Steiner, G. A. (1961). A model for predictive measurements of advertising effectiveness. *Journal of Marketing, 25*, 59–62.

Leavitt, J. D., & Christenfeld, N. J. S. (2011). Story spoilers don't spoil stories. *Psychological Science, 22*, 1152–1154.

Leavitt, J. D., & Christenfeld, N. J. S. (2013). The fluency of spoilers. *Scientific Study of Literature, 3*, 93–104.

Lieberman, D., & Busch, A. (2016). Hollywood tees up movie ads for Super Bowl: TV's most expensive event. *Deadline*, January 29, https://goo.gl/m1zhCu.

Liu, X., Shi, S., Teixera, T., & Wedel, M. (2018). Video content marketing: The making of clips. *Journal of Marketing, 82*, 86–101.

Lovett, M. J., & Staelin, R. (2016). The role of paid, earned, and owned media in building entertainment brands: Reminding, informing, and enhancing enjoyment. *Marketing Science, 35*, 142–157.

Luan, Y. J., & Sudhir, K. (2010). Forecasting marketing-mix responsiveness for new products. *Journal of Marketing Research, 47*, 444–457.

Lynton, M. (2009). Der Produktfluss ist sicher. *Blickpunkt: Film*, https://goo.gl/B39QsV.

Marchand, A. (2016). The power of an installed base to combat lifecycle decline: The case of video games. *International Journal of Research in Marketing, 33*, 140–154.

Marchand, A., Hennig-Thurau, T., & Wiertz, C. (2016). Not all digital word of mouth is created equal: Understanding the respective impact of consumer reviews and microblogs on new product success. *International Journal of Research in Marketing, 34*, 336–354.

McClintock, P. (2014). $200 million and rising: Hollywood struggles with soaring marketing costs. *The Hollywood Reporter*, July 31, https://goo.gl/eoEXYb.

McKay, D., Buchanan, G., Vanderschantz, N., Timpany, C., Jo Cunningham, S., & Hinze, A. (2012). Judging a book by its cover: Interface elements that affect reader selection of ebooks. In *Proceedings of the 24th Australian Computer-Human Interaction Conference* (pp. 381–390). New York: ACM.

Mechanic, B. (2017). Bill Mechanic on movie biz: Big problems & a few suggestions on how to fix them. *Deadline*, March 31, https://goo.gl/6rvu85.

Meiseberg, B. (2016). The effectiveness of e-tailers' communication practices in stimulating sales of niche versus popular products. *Journal of Retailing, 92*, 319–332.

Milgrom, P., & Roberts, J. (1986). Price and advertising signals of product quality. *Journal of Political Economy, 94*, 227–284.

Minotti, M. (2017). EA's defense of Star Wars: Battlefront II is now Reddit's most downvoted comment. *VentureBeat*, November 12, https://goo.gl/GtJMqz.

Papies, D., & van Heerde, H. (2015). How the internet has changed the marketing of entertainment goods: The case of the music industry. Working Paper, Marketing Science Institute.

Parmentier, M.-A., & Fischer, E. (2015). Things fall apart: The dynamics of brand audience dissipation. *Journal of Consumer Research, 41,* 1228–1251.

Pfeffer, J., Zorbach, T., & Carley, K. M. (2014). Understanding online firestorms: Negative word-of-mouth dynamics in social media networks. *Journal of Marketing Communications, 20,* 117–128.

Plumb, A. (2015). The most ludicrous DVD/Blu-ray box sets ever. *Empire,* October 9, https://goo.gl/5n9w1g.

Prag, J., & Casavant, J. (1994). An empirical study of the determinants of revenues and marketing expenditures in the motion picture industry. *Journal of Cultural Economics, 18,* 217–235.

Quelch, J. A., Elberse, A., & Harrington, A. (2010). The passion of the christ (A). *Harvard Business School Case,* 505–025.

Rao, V. R., Abraham (Avri) Ravid, S., Gretz, R. T., Chen, J., Basuroy, S. (2017) The impact of advertising content on movie revenues. *Marketing Letters, 28,* 341–355.

Reutzel, D. R., & Gali, K. (1998). The art of children's book selection: A labyrinth unexplored. *Reading Psychology, 19,* 3–50.

Riehl, K. (2014). Mein Leben als Mensch. *Sueddeutsche Zeitung,* March 18, https://goo.gl/mBrHEo.

Robb, D. (2014). Seth MacFarlane sued over source of 'Ted' idea. *Deadline,* July 15, https://goo.gl/FzeHu7.

Rosenbaum, J. E., & Johnson, B. K. (2016). Who's afraid of spoilers: Need for cognition, need for affect, and narrative selection and enjoyment. *Psychology of Popular Media Culture, 5,* 273–289.

Saboo, A. R., Kumar, V., & Ramani, G. (2016). Evaluating the impact of social media activities on human brand sales. *International Journal of Research in Marketing, 33,* 524–541.

Schmidt-Stölting, C., Blömeke, E., & Clement, M. (2011). Success drivers of fiction books: An empirical analysis of hardcover and paperback editions in Germany. *Journal of Media Economics, 24,* 24–47.

Sethuraman, R., Tellis, G. J., & Briesch, R. A. (2011). How well does advertising work? Generalizations from meta-analysis of brand advertising elasticities. *Journal of Marketing Research, 48,* 457–471.

Shehu, E., Prostka, T., Schmidt-Stölting, C., Clement, M., & Blömeke, E. (2014). The influence of book advertising on sales in the German fiction book market. *Journal of Cultural Economics, 38,* 109–130.

Smith, J. R. (2016). IBM research takes Watson to Hollywood with the first 'Cognitive Movie Trailer'. *IBM THINK Blog,* August 31, https://goo.gl/semSTw.

Smith, M. D., & Telang, R. (2016). Targeting, retargeting, and the effectiveness of search engine advertising: Evidence from randomized field experiments. Draft.

Squire, J. E. (2006). Introduction. In J. E. Squire (Ed.), *The movie business book* (International 3 ed., pp. 1–12). Maidenhead: Open University Press.
Staiger, J. (1990). Announcing wares, winning patrons, voicing ideals: Thinking about the history and theory of film advertising. *Cinema Journal, 29*, 3–31.
Tassi, P. (2017). EA has removed Star Wars Battlefront 2's microtransactions hours before launch. *Forbes*, November 16, https://goo.gl/C8HTYh.
Taylor, V. (2010). The best-ever social media campaigns. *Forbes*, August 17, https://goo.gl/SghPZr.
Tedford, J. (2015). This summer, social ads are the winning ticket for box office success. *MediaPost*, July 22, https://goo.gl/svKJfs.
Xiong, G., & Bharadwaj, S. (2014). Prerelease buzz evolution patterns and new product performance. *Marketing Science, 33*, 401–421.
Yan, D., & Tsang, A. S. L. (2015). The misforecasted spoiler effect: Underlying mechanism and boundary conditions. *Journal of Consumer Psychology, 26*, 81–90.
Zhao, H. (2000). Raising awareness and signaling quality to uninformed consumers: A price-advertising model. *Marketing Science, 19*, 390–396.
Zufryden, F. S. (1996). Linking advertising to box office performance of new film releases: A marketing planning model. *Journal of Advertising Research, 36*, 29–42.

12

Entertainment Communication Decisions, Episode 2: "Earned" Channels

"Earned" media is actually a hodgepodge of quite different kinds of communication about an entertainment product. It encompasses the word of mouth that consumers articulate via various channels, quality signals from consumer "herds" (as reflected by the chart position of a new product—because what sells must be good, right?), and the buzz a product receives on the Internet and elsewhere. Consumers' judgments and behaviors are also the input that automated recommendation engines transform into personalized predictions (i.e., algorithm-generated "earned" media). Beyond consumers, there are other stakeholders of an entertainment product whose evaluations feed "earned" communication channels, such as those critics who judge a new product's quality in their reviews and those who hand out awards for entertainment products that they consider to be of outstanding quality.

The common element of all these kinds of communication is that they cannot be controlled by the producer, or at least to a much lesser degree than paid and owned media. Whether consumers share their opinions about a new product, whether buzz develops, whether critics decide to write reviews (and what they write), and what the members of award committees see in the product, all of these are outside the realm of the producer. Of course, producers *try* to stimulate and steer such communication through targeted advertising, by offering press screenings, or by sending out samples (or by not doing so); we will discuss the effectiveness of such steps in this chapter (along with some more dubious ones). But despite all such efforts, it is the others outside the firm who eventually decide whether an

T. Hennig-Thurau and M. B. Houston, *Entertainment Science*,
https://doi.org/10.1007/978-3-319-89292-4_12

entertainment product "earns" such communication, just as these others decide the valence of such communication. That is the essence of most kinds of "earned" media: if the stakeholders love your product, you will *earn* their praise, but if they do not like it, you might *earn* harsh criticism instead.

In what follows, we will discuss what *Entertainment Science* scholars can tell us about the mechanisms of each kind of earned media and what we know about their respective roles for entertainment product success. We begin with the different ways consumers communicate about entertainment. One important way is definitely word of mouth. We refer to the word of mouth that is exchanged among consumers about entertainment products as "informed cascades": those consumers who talk or write about a product make an explicit "informed" assessment of a entertainment product's quality, based on their own experience with the product. (Please note that this implies that "chatter" about a product by someone who has not yet seen, read, played, or listened to it does *not* fall into our definition of word of mouth. We get back to this in a moment.)

When consumers use the success of a product as their choice criterion, we refer to such herding behavior as "uninformed cascades" because high sales only tell us that a lot of people have bought a product, but not whether they actually *liked* it. The same logic applies to the consumer buzz that exists for an entertainment product prior to its release: none of those who sent the 700,000 tweets about Jurassic World in the week before its release had actually seen the film, nor had any of the 6.6 million who had become a "fan" of the film on Facebook by then. (This includes the "chatter" we have spoken of above, by the way.) We then discuss automated recommendation systems developed from consumer data and, finally, other stakeholders' communication about entertainment: what is it worth to receive a high "Tomatometer" rating, or to win an Oscar?

Informed Cascades: The Power of Word of Mouth

> "[He] only goes to the movies when at least five people in whom he has complete confidence have recommended the film to him as worth seeing."
>
> —*Nobel prize winner* Heinrich Böll *(1963, p. 226) describing his fictitious character* Leo Schnier *in the novel* The Clown *[Translated from the German original by* Leila Vennewitz. *Courtesy of Melville House Publishing]*

Marketers have long considered word of mouth (which we will refer to as WOM on the next pages) to be a powerful source of influence on the behavior of others. And scholars have agreed, after Johan Arndt's (1967) seminal scholarly article added the phenomenon to their theories. However, until recently WOM has been widely considered the "mysterious force" that Arndt (1967, p. 291) named it, rather than something that can be systematically researched and understood. The main problem was that WOM was mostly invisible to the outside observer, as it is shared between consumers in personal conversation, and surveys shed relatively little light on its nature and effects. But with the rise of digitally mediated communication, things have changed quite fundamentally. On the Internet and in social media, WOM is not only visible, but can also be tracked in detail by managers and scholars alike, both on an aggregate and an individual level. As a consequence, WOM has almost overnight become one of the most intensively studied topics in marketing—and *Entertainment Science* in particular.[310]

So, what exactly is WOM after all? Because the concept is so intuitive, an unintended result is a lack of precision when it comes to defining it. Because WOM is all about supposedly objective quality judgments, it is important to limit the concept to personal communication by consumers about a new product *which they have already consumed.* We insist that WOM should not be confused with the behaviors that constitute (pre-release) buzz, as there is a huge difference between WOM's experience-based communication versus chatting about a product which one has not yet consumed. All pre-release communication is necessarily anticipatory and speculative and, thus, lacks the experience component which is so crucial for WOM.[311] It is also this experiential nature which makes the cascades that WOM triggers conceptually different from other cascades that we discuss in this book (De Vany and Lee 2001): only WOM cascades are based on consumers' "true" quality perceptions (which makes them "informed" or "quality-based" cascades), not inferred from the *actions* of other consumers (as is the case for "uninformed" or "action-based" cascades).

[310]In their meta-analyses of WOM effects, You et al. (2015) included 51 empirical articles that link (Internet) WOM with product success, and Rosario et al. (2016) compile close to 100 (!) studies from between 2004 and 2014.

[311]Maybe the words of those who have gained access to a new product prior to its release via illegal sources mark an exception, but we ignore them here—we have dedicated a whole section of our book to their actions in our chapter on entertainment distribution.

And what it the logic that spurs the idea that WOM is influential and powerful? WOM's main "power source" is that it can initiate cascades. Remember the crucial role of personal recommendations in the classic diffusion model by Frank Bass. The recommendations by consumers who have already experienced the product ("Innovators") influence the adoption decisions of other consumers ("Imitators") who can then, after experiencing the product on their own, affect still others. All these recommendations are a central element of the WOM concept.

Let us add that scholars sometimes study the mere amount ("volume") of WOM for a product, in large part because of the ready availability of this type of WOM data from online product reviews posted by consumers on websites, such as Amazon. Such volume information, if studied in isolation, is conceptually different from the *valenced* opinion that triggers informed WOM-based cascades. Volume information does not reveal whether consumers *like* a certain product, only how much *awareness* the product has received among them—it thus falls into the "uninformed" category, very similar to success-related information such as charts. Consequently, our discussion in this section focuses on the valence element of WOM.[312]

In what follows, we will overview what scholars have learned regarding the important, but complex, role of such WOM valence for consumer decisions and entertainment product success. But before we do so, let us take a quick look at the drivers of WOM: what makes a consumer engage in WOM and share his or her thoughts and feelings about entertainment product with friends and others?

What Makes Us Articulate Word of Mouth?

As consumers, we have dozens of consumption experiences every single day, but we only spread WOM about a few of them. Why for some products and not for others? Whereas "inner forces" motivate us to publicly express our experiences with products and services, we have to hold product factors responsible for our selection process for articulating WOM.

Scholars have argued that WOM is "goal-driven." Based in part on an empirical investigation of more than 2,000 online community members we conducted (Hennig-Thurau et al. 2004), the following six basic motivations,

[312]We will get back to volume-related insights in our discussion of "herding" behavior in a few pages.

or psycho-social functions, can be considered as crucial for articulations of WOM regarding entertainment consumption:[313]

- *Impression management.* Consumers engage in WOM to demonstrate their expertise about a product or product category, aiming to enhance their self-worth. Whereas this usually takes the form of positive WOM through which one demonstrates the ability to spend time and money wisely ("This way I can express my joy about a good buy;" Hennig-Thurau et al. 2004), explaining why a popular movie or song does *not* deserve its reputation also falls in this category.
- *Emotion regulation.* Negative entertainment experiences cause negative emotions to accrue, and articulating negative WOM about the responsible product can help a consumer by providing a way to vent those emotions ("I like to get anger off my chest").
- *Concern for other consumers.* Some consumers have a genuine interest in helping others, and WOM is a means to live out such altruistic tendencies. Positive WOM can help others experience joy, while negative WOM can help others avoid hurt or anger ("I want to save others from having the same negative experiences as me").
- *Social bonding.* Whereas the previous motives focus on psychological aspects of a consumer's personality, WOM can also be driven by social motives, serving as a means to address the all-too-human need for social interaction. As consumers, we enjoy talking about a new album, and such conversation can be an integral part of social relationships ("It is fun to communicate this way with other people in the community"). In our study of different Internet platforms, we find social bonding to be the strongest internal driver of writing comments about products, with an effect that is twice as strong as for any other motive.
- *Information acquisition.* WOM can also address consumers' interests in "knowing more." What is the meaning of the TV show's ending last night? How can I master the next level in the new video game? Engaging in WOM can help provide answers to such pressing questions.
- *Persuasion.* And finally, Berger (2014) stresses WOM's role of as a means to convince others to choose one entertainment product (or activity) over alternatives. This is particularly relevant in group consumption settings. When you are going to the movies with friends and want to see the new

[313]See also Berger's (2014) summary of WOM research.

Kevin Costner film instead of the next Avengers episode, praising Mr. Costner's previous efforts might be the way to go.

In addition to such inner drivers, scholars have compiled evidence that WOM activities are also influenced by the characteristics of the object of communication, i.e., the product, or, more specifically, the consumer's perception and evaluation of it. Berger and his colleagues have shed light on such external WOM drivers in a number of studies (Berger and Milkman 2012; Berger and Schwartz 2011). Their findings, which are not specific for entertainment, highlight the following characteristics:

- *Interestingness.* The more consumers find a product "interesting," the higher the probability that they will engage in WOM about it. Naturally, what is "interesting" lies in the eye of the beholder; the term encompasses attributes such as novel, exciting, and unusual. It is closely related to the concept of "involvement," which refers to the importance (or personal relevance) of a product as perceived by a consumer (Jain and Srinivasan 1991). Berger and Schwartz (2011) stress that "interestingness" triggers immediate WOM; when we have seen a movie that has fascinated us, we want to talk about it immediately. Entertainment, per se, scores highly in interestingness, compared to other product categories. That is why we talk (and write) so much about it.
- *Surprisingness.* This concept relates to interestingness, or may be a facet of it: when consuming a product, do consumers perceive it as surprising? Surprise sparks WOM even when controlling for interest (Berger and Milkman 2012)—and it does so both immediately after consuming a product *and* later. So we can expect entertainment experiences that surprise us to get extra conversation, probably one of the reasons why The Sixth Sense was such a big hit movie.
- *Positivity.* Although negative events stick deeply in our memory, we prefer to talk about positive experiences with others. Ratings on Internet forums tend to be largely positive, and the same is true for WOM shared on social media. According to Berger and Schwartz (2011), good products receive more WOM recommendations than bad products receive warnings.
- *Arousal/emotionality.* What affects us emotionally, both in positive ways (e.g., excites us) and negative ways (e.g., scares us or makes us cry) spurs a higher level of WOM than what does not do so. This means that strong emotions associated with an entertainment product not only affect its

success directly (via creating anticipatory/anticipated emotions that lead us to consume a product, or by stimulating us to watch a movie more than once), but also indirectly by attracting *other* consumers via WOM cascades.[314]

The film My Big Fat Greek Wedding addressed several of these aspects and consumers' inner motives. In particular, the emotional reactions of many who had seen the film were quite extreme, and the film thus became a stunning hit mostly through WOM. Its lead actress Nia Vardalos tied WOM to pure chance, quite in line with the "Nobody-Knows-Anything" mantra: "We got lucky. You can't manufacture word of mouth. You can't pay people to tell their 10 cousins" (quoted in Strause 2016).

We kindly object, with all due respect: not only did the film itself provide the content that fueled WOM, but its producers also provided the conditions in which WOM could blossom. They carefully orchestrated cast appearances, packed their screenings, and zoned in on the female audience and the Greek community (e.g., the star traveled extensively making appearances at Greek organizations and bridal shows). All of this, of course, could not *guarantee* that WOM would blossom (it's a probabilistic world after all!), but in entertainment, as in other parts of life, luck favors those who work hard (and know the right things).

Does Word of Mouth Influence Entertainment Product Success? Yes. But It's Complicated

> "If the picture is bad, you might as well shoot everybody coming out of the theater—they will quickly enough kill any film."
>
> —John Friedkin, *former vice president for advertising and promotion at 20th Century Fox (quoted in* Austin *1989, p. 3)*

As noted earlier, empirical studies that address the role of WOM for consumer decision making usually distinguish between the valence of the WOM for a product (such as the average rating at Amazon that a game received from consumers) and the volume of such WOM (i.e., the *number*

[314]See also this book's section on consumer emotions and their role in the sensations-familiarity framework in the entertainment consumption chapter.

of consumer reviews written about the product).[315] We focus on valence here as it is the valence element that captures how consumers judge the quality of a product—the source of informed cascades. We will take a look at average effects across products first, before diving deeper into contingencies that determine how strongly the success of a particular product is affected by WOM.

Average Effects: Word of Mouth (Valence) Matters!

In their meta-analysis of WOM effects on product sales across products and industries, You et al. (2015) find substantial average effects for WOM valence; they also report that this effect tends to be higher for entertainment products than for "others." *Entertainment Science* scholars, taking a more fine-grained look, have provided evidence that WOM valence is influential for all forms of entertainment that we feature in this book, except for music (something we return to when discussing context effects). Here's a summary of what we know about the role of WOM for the success of books, movies, TV shows, and games.

Books. The seminal study on WOM effects is by Chevalier and Mayzlin (2006), who analyze how consumer reviews on the websites of retailers Amazon.com and Barnes and Noble affect a book's *relative* sales rank (the difference in ranks between the two sites).[316] Using a data set of about 1,100 books with at least one consumer review (which combines a random

[315]Let us note that the separation of "valence" and "volume" carries a lot of analytical problems. It mixes the WOM about a product with the product's popularity (which is the source of *un*informed cascades) and also its success (as experiences are required for WOM, more successful products get reviewed more often). Also, valence and volume are systematically inter-related, as the quality of a product is a source of its popularity (the more consumers like a product, the more WOM they will share about it). Scholars have found that WOM valence is a main driver of the amount of WOM for entertainment products; when both "facets" of WOM are included in the same study, WOM volume thus tends to absorb the impact of valence on success (Duan et al. 2008; Karniouchina 2011). Empirical results on WOM valence effects depend strongly on whether a scholar accounts for these problems (e.g., by using instrumental variables for WOM volume and changes in WOM valence ratings over time) and also controls for other drivers of success (such as advertising); they are thus far from consistent across studies (see also for example Forman et al. 2008 and Chintagunta et al. 2010—the latter authors also provide empirical evidence of the consequences of (not) accounting for these aspects). In our coverage of WOM here, we focus on studies that address such challenges in a powerful way.

[316]Note how smart this approach by Judith Chevalier and Dina Mayzlin is: because the books sold at the two sites are identical except for the WOM and some other factors like price (for which the authors control in their analysis), looking at the differences in sales ranks (rather than at absolute sales) eliminates the effect of all product/book characteristics on sales and allows the analytical spotlight to be put on the WOM on the sites.

selection from 1998 to 2002 and bestsellers from 1991 to 2002)[317] and analyzing rank changes at three points in time in 2003 and 2004 with regression analysis, the scholars find that a one-star increase in rating at Amazon corresponds with a 52% increase in rank-difference. Also, if all reviews for a book give it five stars (versus none do so), the book's relative sales rank improves by more than 100%. A similar rise in 1-star ratings has an even stronger sales effect (albeit negative, of course).

Since then, several other scholars have provided additional evidence for the role of WOM valence for books. Sun (2012), applying the same statistical approach to a more recent sample (892 randomly selected books published 2002–2006), also finds that WOM valence is influential. She, though, reports a smaller improvement in relative ranks—a one-star increase at Amazon corresponds with "only" a 21% higher relative sales rank. This result might indicate that the WOM effect weakens in this context, or could simply be the result of the data set (or of changes in Amazon's assortment). For her large set of some 30,000 books, Meiseberg (2016) finds both five- and one-star consumer reviews to influence *absolute* book sales ranks at Amazon. Her effects are again substantial, but also somewhat weaker than the ones reported by Chevalier and Mayzlin: a change from zero to 100% 5-star WOM corresponds with an average improvement in sales ranks of about 52%.

Jabr and Zheng (2014), with a GMM approach, estimate that a one-star improvement in book ratings at Amazon leads to a 26–34% improvement in the book's sales rank. Their data consist of consumer reviews at Amazon for 1,740 randomly selected (non-fictional) books from 2007 to 2009; each book had at least 25 reviews.[318] This is quite similar to Li and Hitt's (2008) finding of a 27% increase in (estimated) sales at Amazon in response to a one-star rise in WOM valence; they estimate this effect for about 2,600 books published in 2000–2004 using a fixed effects regression. And when Schmidt-Stölting et al. (2011) link consumer reviews at Amazon.de to *nationwide* German book sales for their a large data set, they still find a WOM effect—they estimate that, on average, an increase of one star on Amazon.de is associated with 4–7% higher nationwide sales.

When judging the size of these effects, keep in mind that, because WOM is predominantly positive and usually shows relatively little variation between consumers, a one-star change in reviews is quite enormous.

[317]The scholars also conduct analyses with a larger data set of almost 2,400 books, but we focus here on the (more robust) results for those books which had at least one consumer review at the beginning of the investigation period.

[318]The 34% improvement is found when the authors use earlier reviews written by a "WOM giver" as a statistical instrument for that person's WOM valence about a book.

In the Li and Hit data set for example, the average WOM valence is 4, whereas the average deviation from this value is only 0.60—thus, a one-star change would cover about 75% of *all* books in the data set.

Movies. Studies examining movie success that focus on post-release revenues (the time frame in which WOM can actually matter), and thoughtfully account for the role of alternative information sources that are available at this time, also find that experience-based WOM affects product success.

Specifically, Chintagunta et al. (2010) use a data set of 148 movies released in the U.S. around 2004 and analyze box-office results for different geographic areas, not only domestic. Their logic is that WOM spreads over the Internet, so consumers in one location can make use of such information on a movie's release day when the movie is already being shown elsewhere. Using consumer ratings from Yahoo Movies and a large set of controls, such as advertising and distribution, as well as an instrument for the volume of WOM (the two major reasons for biased results in other studies), the scholars estimate a movie's first-day box office in a geographic market, using a GMM approach. They find that a one-unit increase in WOM valence on Yahoo's 1-to-13 rating scale corresponds with a 10% increase in opening-day sales in a specific market.

In a follow-up investigation, the same team of scholars uses a subset of this data for a regression approach in which they study the box office generated during a four-week post-release window (Gopinath et al. 2013). They again account for endogeneity and include several controls. This time the scholars find a WOM valence elasticity for Yahoo consumer ratings of 0.22, which means that a 10% increase in the WOM valence for a film is associated with roughly 2% higher revenues in a movie's first four weeks.

A study by Niraj and Singh (2015) suggests that the role of WOM valence (measured here as a "positivity ratio" of consumer reviews on several websites, portals, and forums) for movie success also exists at the Indian box office. Their investigation, which applies a panel regression to a small data set of 48 Bollywood movies (released in 2010–2011), also provides tentative evidence that the link between WOM valence and the success of movies might not be linear.[319] When they include a squared term of WOM valence, they find that it is negative (and significant). In other words, the value of positive WOM decreases with the number of people who give it—if *everyone* loves a film, that might not necessarily help it. We get back to this idea when discussing the role of the variance in WOM valence in the next section.

[319]Take note that there are certain limitations though—their work does not consider a number of key controls (such as advertising and distribution) and also does not account for the endogeneity of WOM.

TV shows. In their survey panel of TV audiences, Lovett and Staelin (2016) find that remembering WOM is closely linked with TV watching: having been exposed to WOM for a show increases the likelihood that a consumer watches the next episode by 6%. This might not sound like much, but it is 25% higher than the same effect the scholars detected for paid advertising, and almost three times as high as the effect of visits to the broadcaster's website (which was the scholars' measure of "owned" communication).

Games. In our own investigation of 100 Xbox 360 games, we find that WOM valence also makes a difference for games (Marchand et al. 2016). In our 3SLS estimation, we control for several other factors, such as advertising, and also address the endogenous nature of WOM volume. The results teach us that the valence of WOM for a game posted on Amazon.com in the weeks after a game's release influences the game's sales quite strongly; the average valence elasticity in our study is 0.47, suggesting that an improvement of 10% in WOM valence translates into 4.5% higher game sales in the following three weeks.

These results provide clear evidence that WOM matters for entertainment success. But all effects we have reported so far are averages, aggregations over heterogeneous sets of products. Let us now see if contingencies exist that alter the role that WOM plays, as we have shown contingencies to do so for advertising, sequels, stars, and other things in this book. We will shed light on a number of such context factors, ranging from product types, to WOM types, to different groups of consumers. Contingencies will also help us to understand why scholars such as Dewan and Ramaprasad (2012) were unable to find support for an "average" effect for the role of WOM in *music* success. Music has a non-verbal, non-visual character and triggers highly subjective consumer judgments, which will limit the effect of verbal recommendations (and warnings) by other consumers for music, in general. But there are indications in Dewan and Ramaprasad's study that the effect exists at least for some types of music.[320]

The Product Type Matters

We have already shown that certain product characteristics, such as the "interestingness" of a product, influence how much WOM is articulated

[320]In addition and consistent with this logic, we will also see that "uninformed" action-based cascades, which are intuitive to grasp and aggregate the "judgments" of many people, matter a lot for consumers' music choices.

for a product. Scholars also point out that the *impact* that WOM has on product success varies between types of entertainment products. In particular, they argue that WOM valence plays a more prominent role for the success of smaller "independent" products, whereas it tends to be less influential for high-budgeted "commercial" products.

For games, Zhu and Zhang (2010) find, in an analysis of 141 console games (released on both PS2 and Xbox from 2003 to 2005), that an interaction of the valence of WOM posted on the GameStop retail website with a game's "popularity" is strongly negative: WOM valence plays a stronger role for the demand of less-popular games among U.S. consumers.[321] And when Dewan and Ramaprasad (2012) analyze a data set of 1,762 songs that were posted in MP3 format on music blogs in 2006, their 2SLS estimations show no impact of WOM valence (consumer ratings of the songs at Amazon) on song sales for the whole data set. But when they split their data set, the regression coefficient for WOM valence is then clearly higher for niche songs that were ranked below 5,000 than for more highly ranked songs: for niche songs, a one-unit increase in ratings corresponds to a 20% increase in song sales. Neither parameter reaches statistical significance though.

Finally, the book-related results by Meiseberg (2016) are also in line with the prominent role of WOM for niche titles. She finds that the effect for positive WOM (namely 5-star reviews of books on Amazon in her analysis) is highest for the lowest selling quantile (i.e., the least popular products in her data set). The effect is not linear for the other quantiles, so that her results point to an "awareness effect" of positive WOM that makes books with (good) reviews stand out from the vast number of similar titles (that are not written by star authors). WOM tells us that at least *someone* likes them.[322]

Overall, these empirical insights on differing WOM effects for niche and commercial products correspond with industry wisdom among entertainment managers who often consider smaller movies to be much more "WOM sensitive" than bigger productions. It also provides the basis for

---

[321]Zhu and Zhang measure popularity as the above/below average sales of a game in a respective month, compared to all games in the data set in the same period.

[322]We speculate that this effect is further enhanced by Amazon's search engine, which might put products with positive recommendations in a more prominent place when presenting search results to consumers.

entertainment's two main strategic concepts which we will discuss later in this book: the blockbuster concept (for which WOM plays only a marginal role) and the niche concept (for which WOM is essential).[323]

Not All Word of Mouth is Created Equal

> "BRÜNO's box office decline from Friday to Saturday indicates that...[it] could be the first movie defeated by the Twitter effect."
> —Corliss *(2009)*

With the rise of the Internet, WOM has not only become observable, but also increasingly complex and heterogeneous. The empirical studies we listed above all use consumer reviews on Amazon and similar sites, but this particular kind of WOM functions differently than other kinds of personal communication, and thus also affects product success differently. Today, three main kinds of WOM co-exist, each with unique characteristics, and are of particular interest for entertainment managers (Hennig-Thurau et al. 2015). Figure 12.1 overviews what they have in common and what sets them apart.

Traditional (or offline) word of mouth (a.k.a. TWOM) is the face-to-face communication between consumers about a product, based on personal experience with the product. TWOM's main characteristics are that it is shared with an individual consumer or a small group, is transmitted in real time, assumes a personal connection between those who exchange it, and that it enables feedback, combining a "push" element (information whose transmission is initiated by the sender) and a "pull" element (information that is requested by the receiving consumer).

Electronic (or online) word of mouth (or EWOM) is experience-based communication that a consumer makes available to a potentially very large group of anonymous others over the Internet, on forums such as Yahoo Movies or sites such as Amazon. Traditionally the main source for scholars when analyzing WOM effects, EWOM is "pull-only"—we have to actively search for information on a blog or review, the blog does not provide us with information on its own. It is also asynchronous and allows no (or very limited) feedback—but it is often stored for a long time. EWOM also usually offers "summary statistics," such as average ratings and the number of those who have rated a product.

[323]We discuss the two strategic concepts and the role that WOM plays for each of them in detail in our chapter on integrated entertainment marketing.

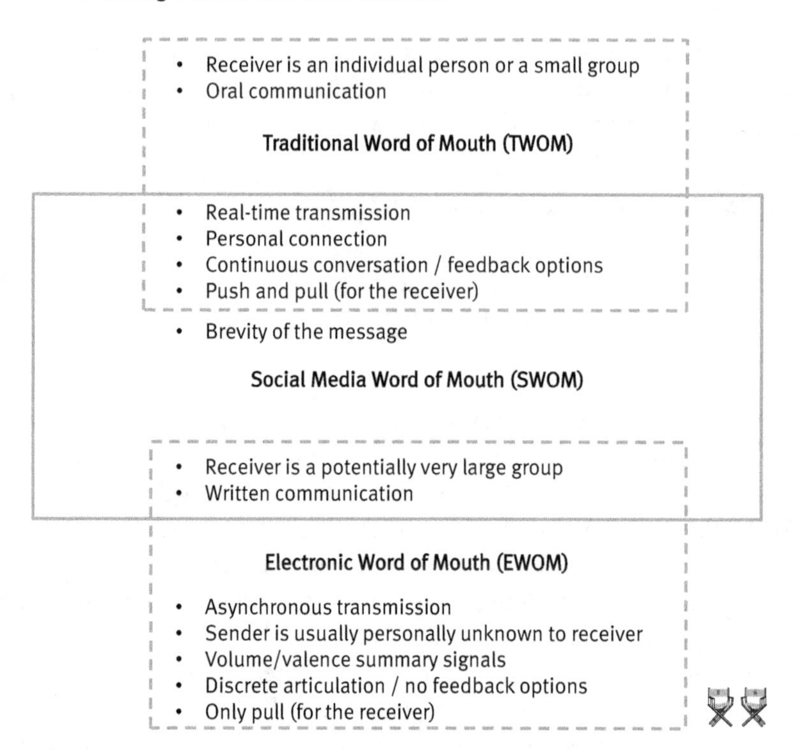

Fig. 12.1 Three major kinds of word of mouth in the digital age
Note: Reprinted with minor modifications with permission by the Journal of the Academy of Marketing Science/Springer from Hennig-Thurau et al. (2015, p. 389).

Finally, *social media (or microblogging) word of mouth* (SWOM) includes statements about a product experience that are broadcast to the sender's social network (select members or all) through a specific web-based service such as Twitter or Facebook. SWOM combines elements of both TWOM and EWOM—like TWOM, it involves a real-time exchange that combines "push" and "pull," a personal connection between sender and receiver, and the chance to provide feedback as part of ongoing exchanges), and just as EWOM, it can reach a very large group of potential receivers. But SWOM is not just a mixture of the two, but adds its own characteristics, namely a medium-specific brevity because of technical or usage-based restraints (nobody reads epic-length rants on Facebook). The unique brew of these elements allows consumers to push their evaluations almost instantly to very large numbers of related others—the basis of the so-called "Twitter effect" that we discuss below.

The distinction of these kinds of WOM is not purely abstract and conceptual, but has manifest consequences for which *Entertainment Science* scholars have compiled initial evidence. In their analysis of movie success, Shyam Gopinath and his colleagues (2013) included a measure of SWOM next to their EWOM data from Yahoo.[324] They find an SWOM-valence elasticity of 0.35, which suggests that a 10% improvement in SWOM valence for a film is linked to about 3% higher revenues in the month after the movie's release. This effect is higher than the effect for EWOM, but more importantly, SWOM and EWOM are both significant influencers of movie success when analyzed *jointly*. The WOM types *carry unique information* and/or *reach unique customer segments*.

In our own study of 100 console games (Marchand et al. 2016), we also included SWOM (expressed in the tweets about each game) in addition to EWOM (via Amazon). Again, the two kinds of WOM work differently: in contrast to the substantial effects of EWOM valence in the weeks after a game's release, the valence of tweets does not significantly link to game sales. Our conclusion: whereas social media communication is best suited to transmit "social" information (such as excitement), consumer reviews are most effective for providing information about a product's performance.[325]

But the crucial time for SWOM is not weeks or even months after a product's release, but way earlier: its real-time character, in conjunction with its ability to push information on the smartphones of large consumer groups, can affect the diffusion of a product *very early* when the impact of both TWOM and EWOM is still systematically limited. In a separate study (Hennig-Thurau et al. 2015), we investigated how SWOM via Twitter affects a film's destiny in the *hours and days* after it has been launched—something journalists and entertainment managers have labeled the "Twitter effect" (see for example the introductory quote by Corliss 2009).

To do so, we collected all four million tweets posted by consumers during the North American opening weekends of 105 wide-release movies between October 2009 and October 2010 and linked them to the movies' daily box office during their initial weekend. Combining the manual coding of 51,000 tweets and machine learning, we separated *evaluative* SWOM from *anticipatory* chatter; Fig. 12.2 shows their respective distribution over the movies' first three days in release. We then ran an OLS regression in which we explained the box office distribution over the first three days (i.e., the drop

[324]Actually, Gopinath et al. do not measure SWOM in its raw form, but data from Google Blogs, which share several features with social media. They determine the valence of blog comments with human coders.

[325]We also probed for interaction effects between the two kinds of WOM, but found none.

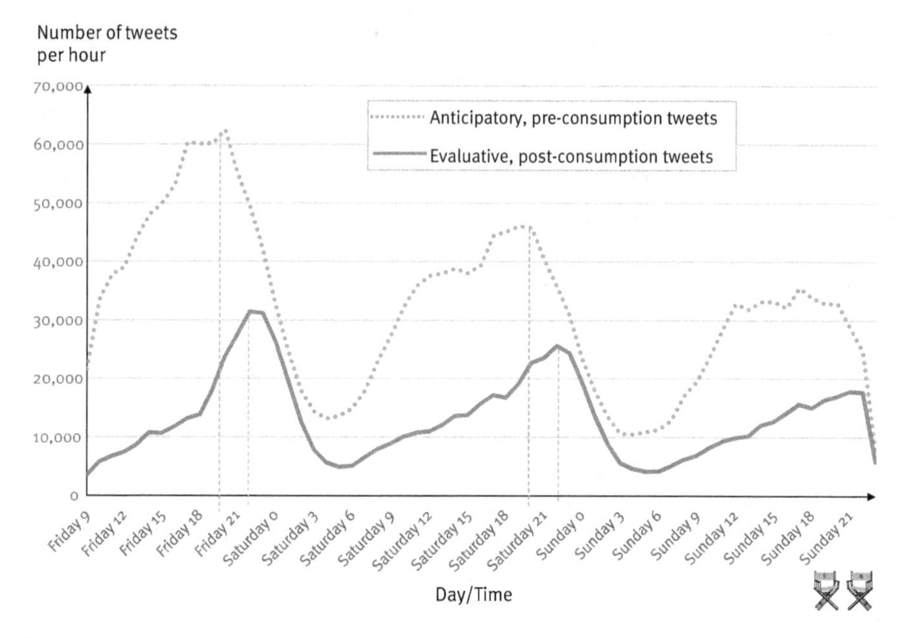

Fig. 12.2 When people tweet about movies during their opening weekend
Note: Reprinted with minor modifications with permission by the Journal of the Academy of Marketing Science/Springer from Hennig-Thurau et al. (2015, p. 382).

or increase on Saturday and Sunday)[326] with (1) the share of moviegoers who tweeted positively valenced comments about a film after having seen a film on its first Friday and (2) the share of opening-day moviegoers who tweeted negatively about their experience; we also included the sheer volume of opening-day SWOM and controlled for multiple other "success factors."[327]

Our results show that the "Twitter effect" does indeed exist. Interestingly, it is much stronger for *negative* tweets that discourage consumers from going to see a film on Saturday and Sunday. The parameter for positive tweets (which are sent *much* more frequently than are negative ones) is only one-tenth of the one for negative tweets and not statistically significant. Simulations show that consequences of negative SWOM can be quite substantial: if the share of negative tweets was 24% for all films (the maximum among the films in our data set), the average opening weekend box office

[326]Specifically, we measured the share of opening weekend revenues a film accrued on the Saturday and Sunday that followed its opening day (Friday).

[327]Those controls included a movie's ad spending and the pre-release buzz it received. Besides, the results remained largely the same when we used the sales as our dependent variable, but such a model carries some additional econometric challenges.

per film would have been 15% lower—the equivalent of $0.5 million. And the *absence* of negative tweets for all films would have increased the average opening weekend box office by more than 4%, or $1 million per film. A film such as the 2010 remake of horror classic NIGHTMARE ON ELM STREET would have lost almost $5 million of its $15.7 million first-weekend revenues if it had a 24% negative-tweets share, but it would have gained about $3 million if negative tweets were completely absent on its opening day.

Whereas the typology of WOM types helps us to understand their respective effects, differences in WOM effects can even exist *within* a single kind of WOM across platforms. This is also what Chevalier and Mayzlin (2006) find in their study of EWOM effects on book sales: EWOM at bn.com had a far weaker effect than EWOM at Amazon.com. A one-star increase at bn.com corresponded with a 16% improvement versus the 52% the scholars estimated for Amazon, and changes in the percentages of 5-star and 1-star ratings are also clearly less strong. So we recommend entertainment managers to pick their WOM data source very carefully.

Not All Word-of-Mouth Givers are Equal, Too…

There is evidence that, for WOM to have an impact, it also matters *who* shares thoughts and feelings about an entertainment product—and what receivers know about him or her.

With regard to personal communication via TWOM, the idea that some consumers are "opinion leaders" that have a stronger influence on others is a central element of Katz and Lazarsfeld's (1955) "two-step flow" theory of communication. Accordingly, opinion leaders are more exposed to media, and their acquired expertise serves as the basis for their influence on less-informed consumers. Other scholars have stressed the critical role of product-specific motivations and involvement for such opinion leadership (such as for cars or video games). Whereas the concept's origin was in politics, Katz and Lazarsfeld themselves provided evidence that opinion leaders exist in different areas of life—including movie-going!

As EWOM usually does not involve a personal connection between giver and receiver, what makes someone an opinion leader in this context? Meiseberg (2016) finds in her large book data set that a review by a "Top 500 reviewer," a status given by Amazon, matters, above and beyond the "normal" WOM valence that exists for a book. Her results do not take into account the valence of the Top reviewer's rating, though, only its mere existence.

In addition, Forman et al. (2008) show that in an environment characterized by anonymity, it can make that EWOM giver's judgment more influential on the receiver's subsequent choices when a receiver learns more about the EWOM giver's expertise and, in turn, builds a sense of closeness with him or her. The scholars learn this from a panel regression of monthly sales ranks of 768 books (which had been a bestseller at Amazon in 2005–2006), which indicates that the more the author of a review at Amazon.com reveals about his or her identity, the more influential the reviews are. The coefficient for such "identity disclosure" suggests that if the percentage of WOM givers for a book who disclose their identity increases by 50%, this would translate into a 15% improvement in sales rank.

The Timing of Word of Mouth

The role of WOM for success also varies over an entertainment product's life cycle. We have reported that the elasticity of EWOM valence for video games is 0.47 in our study (Marchand et al. 2016)—but this effect is actually an average over the whole first 9 weeks of our investigation. A look at distinct time windows shows that the elasticity can differ quite substantially over time. For the EWOM that is articulated 7–9 weeks after a game's release, the valence elasticity is as high as 1.11. Specifically, our "rolling windows" approach, in which we estimated effects for different windows of three subsequent weeks, shows that the effect of EWOM valence *grows* in importance over a product's life cycle. The EWOM parameter only becomes significant about seven weeks after the release.

Why these differences over time? The reliability of WOM valence grows over time with the number of reviews, resulting in more stable averages which offer more powerful insights for consumers. The long-term availability of digital WOM can play to the late strength of this type of information; as time goes on, the availability of alternate information sources (advertising, professional reviews, charts, and buzz) becomes clearly limited, increasing the *relative* importance of EWOM. But we also need to keep in mind that many more copies of a product are sold in the earlier windows, so that higher elasticities do not necessarily mean higher absolute sales volumes as a result of WOM, but only higher *changes* in sales.

Another study that compares the role of EWOM at different points in time is the one by Schmidt-Stölting et al. (2011). In their large-scale analysis of determinants of book sales, the scholars find a smaller effect of EWOM valence for hardcover titles than for paperback versions of the same books (which are usually published several months after the hardcover versions). This result again stresses the role of information availability and the "costs" that accrue for consumers who search for them at a given point in time: because more people have already read a book by the time it is released in paperback, more (and more reliable) WOM should be available about a book compared to when it exists only in hardcover format.

When Consumers Have Different Views: The Role of Word-of-Mouth Variance

Finally, when consumers hear about a new entertainment product's quality from many others, the assessments can differ widely across people. Such variation is not captured by the average WOM valence: for example, a mean score of 4 stars can be the result of ten 4-star ratings, of five 5-star and five 3-star ratings, or of six 5-star, two 4-star, and two 1-star ratings. *Entertainment Science* scholars have also looked whether such variance of WOM valence is informative on its own for consumers, and whether high variance is perceived by consumers as positive or negative information.

In their study of book sales, Jabr and Zheng (2014) find that the variance of Amazon ratings is influential, but in a negative way, with higher variance *reducing* a book's success potential.[328] Sun's (2012) results for books confirm an influence of WOM variance, but also suggest that a more complicated mechanism is at place. Adding an interaction term between WOM valence and variance shows that variance can act in a negative, but also a positive way, depending on a book's WOM valence. If a book's average WOM valence is highly positive, then high variance is not helpful; instead, it discourages consumers from buying a book, as dissonant voices confuse rather than stir interest in a book.

But if the WOM valence for a book is low, a higher level of variance across consumers' Amazon ratings is a positive thing, resulting in more sales. In this constellation, variance might suggest that there is something appealing "hidden" in the book which some reviewers do not see. In Sun's data set, the threshold value is a rating of 4.1, which means that 35% of the books in her data would benefit from more variance—but the majority would suffer from it. Eventually, Karniuchina's (2011) movie-related results point to another potentially positive

[328]Because the independent variables are neither log-transformed nor standardized in Jabr and Zheng's study, their effects cannot be compared.

effect of variance in ratings: variance can stimulate discussions among consumers, contributing to debates and higher interest and buzz.

In summary, we have seen that informed cascades, via WOM, can have a quite substantive impact on the success of entertainment products, and we have laid out the conditions under which it matters more and those under which WOM is less influential. Let us shift the focus now to "uninformed," or "action-based" cascades, where consumers infer quality levels from the actions, rather than the articulated judgments, of other consumers.

Uninformed Cascades: The Power of Herds

> "'Eat sh#t, a hundred billion flies can't be wrong,' the old graffito used to say. 'Follow Stephen, two million tweeters can't be wrong,' I say."
> —*Novelist* Stephen Fry *(2010)*

Even when we, as consumers of entertainment, are not actively talking with others or reading what they have written, we are often still observing what they are doing. What we observe informs our own decision making through what psychologists have named "observational learning" or "social learning" (e.g., Bandura 1977). Through this specific learning mechanism, an action of one person can initiate the actions of others who observe it, who, in turn, affect others, setting into motion so-called "action-based" cascades. Because the information garnered from observation is less rich compared to what can be learned from personal communication, another term for such cascades is "uninformed" cascades. Economists have compared such behavior to that of herds ("herding"), in which members instinctively follow the actions of other herd members without deeper reflection of the logic or wisdom underlying these actions (e.g., Bikhchandani et al. 1998).

In this section, we study two different, but related types of uninformed cascades that are relevant for the success of entertainment products. The first type happens when the adoption of a product by other consumers serves as a signal of the product's quality; here the success of the new product is made visible to others via bestseller or Top Ten lists. This "social proof" increases the desirability of the product to others (through mechanisms we will discuss below) and, thereby, initiate self-enhancing "success-breeds-success" effects. The second type of uninformed cascade takes place even earlier: when the "buzz" about a forthcoming entertainment product is interpreted as a quality signal, enticing others to join the "buzz train" and to eagerly anticipate adopting the product as soon as it is available.

Post-Release Action-Based Cascades: When Entertainment Success Breeds Entertainment Success

Some Words (and Numbers) on the Mechanisms at Work

The information that a product is popular or successful influences its success via two psychological routes. The first mechanism is that popularity biases consumers' perception of the product's quality. Remember that judging the quality of entertainment is notoriously difficult for consumers for several reasons, including the lack of search qualities, and the hedonic character of entertainment which puts emphasis on subjective, holistic aesthetic and artistic achievements rather than objective functional attributes. And artistic taste standards imply that a judgment depends also on one's own, idiosyncratic cultural capital, making the judgment task even more challenging.

Because making a quality judgment is so complicated, an entertainment product's popularity sends a clear and straightforward signal to consumers that is also easy to access, as the media is biased toward hits. The popularity provides "social proof": others must have made the complex choice decision before, so following them should be a good way to get quality. And Lynn et al. (2016) show that the influence of popularity on quality perceptions is not limited to our consumption choices, but even remains *after* we have experienced an entertainment product. When they use experiments to study the reactions of several thousand consumers to downloads of unknown "indie rock" songs, the scholars find that information about a song's popularity (the number of previous downloads by others) impacts the quality rating that consumers give to a song. More popular songs are rated systematically higher by consumers, even though the scholars control for the "inherent quality" of a song, as measured via other consumers' judgments.

But popularity does not affect all of our quality judgments equally: whereas popularity has very little impact on songs with high-quality ratings (i.e., 4 or higher on a "1-hate to 5-love" scale), higher popularity (i.e., ten additional downloads) increases consumers' ratings of a low-quality song (2 or lower on the 1–5 scale) quite remarkably (by 0.25-points). As entertainment consumers, we seem to be confident in our own judgment when we like something, but much less so when we *don't* like it.

The second mechanism through which popularity influences success is a social one: people often want to join the bandwagon. Just as sports fans flock to a winning team (Bayern Munich anyone?!), if an entertainment product is successful, consumers enjoy becoming part of the movement and do not want to miss out. We discuss this "bandwagon effect" in more detail in the context of pre-release cascades based on the buzz for an upcoming entertainment product.

We will now take a closer look at popularity cascades. We will start with an analysis of the impact that popularity information can have on product success within a channel or market and the circumstances that influence this impact. Because entertainment often involves multiple sequential channels and is marketed globally, we then take an inter-channel and inter-market perspective, investigating how popularity in one channel or market can spill over to others. And we will look at moral (and immoral) ways entertainment managers can attempt to harvest the success-breeds-success effect.

The Impact of Popularity on Entertainment Success

Much of what we know today about success-breeds-success effects in entertainment stems from two experiments ran by Salganik et al. (2006). In their inspiring study, the scholars created a number of "artificial markets" in which more than 14,000 consumers were invited to download music (48 songs from 18 bands). In the first experiment, participants were randomly assigned either to a market in which they were provided only with the names of the songs and the bands (the "independent" scenario) or to one of eight markets in which they were also shown the number of previous downloads by other consumers for a song (the "social influence" scenario).

In either scenario, participants could listen to songs and then decide whether or not to download them (for free). All songs started with zero downloads and were presented in random order throughout the experiment. The second experiment was similar—the main difference was that popularity information was presented more prominently; in the "social influence" scenario, the *order* in which songs were presented now reflected the number of their previous downloads.

In essence, the results of their experiments show conclusively that popularity can have a substantial influence on consumers' entertainment choices. Popular songs were *more* popular than less popular songs in all experimental conditions in which downloading information was available, compared to the "independent" scenario (in which no information on a song's popularity existed). And when the downloading information was presented more prominently, the difference in popularity was even higher.[329]

Whereas the download ranks of the songs in the independent, or purely "quality-based" market correlate with those from the social influence markets, they did so far from perfectly—the songs' "performance" took a life on its own based on their early popularity. This is what Panel A of Fig. 12.3 shows by plotting each song's "independent" performance against its performance when popularity information was available. Panel B in the figure shows that the performance dispersion was clearly higher when popularity information was presented in a more prominent way.[330] In other words, the popularity information added unpredictability regarding a song's performance.

The correlational pattern in the figure also stresses that popularity does not fully substitute for the role of a product's quality, but that both coexist and co-determine an entertainment product's economic fate. As Salganik

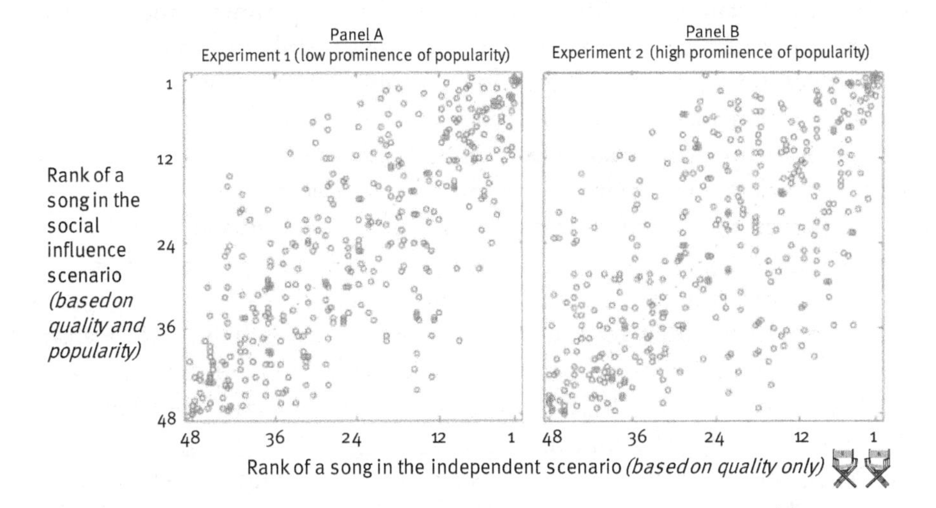

Fig. 12.3 Performance of songs in "artificial music markets"
Notes: Reprinted with minor modifications with permission by the American Association for the Advancement of Science from Salganik et al. (2006, p. 855). The ranks are derived from the number of downloads a song has received in the respective artificial market.

[329]These differences between popular and less-popular songs are statistically reflected in Gini coefficients, which measure the equality between the "market shares" of often and less often downloaded songs. In the first experiment, the Gini coefficient for the independent scenario was 0.25 versus an average of 0.34 for the social influence scenarios (a difference of about 36%). In the second experiment, the Gini coefficients were 0.19 versus 0.50 (+163%, or 4.5 times as high).

[330]The scholars do not report correlations.

et al. argue: "the 'best' songs never do very badly, and the 'worst' songs never do extremely well, but almost any other result is possible" (p. 855).

Keep in mind that this result is based on a lab experiment in which popularity solely develops based on consumers' experiences with the product's quality. Would things look different if popularity develops *before* consumers can actually experience a product? In a follow-up study using the same "artificial market" approach and the same 48 songs, Salganik and Watts (2008) gathered more insights about such popularity that is not based on the actual quality of entertainment. This time they manipulated the songs' *initial* popularity and looked how this "fake" popularity affected the performance of songs among consumers. They found that it indeed also impacts success—low-quality songs became hits solely because of their reported popularity, and the downloads of high-quality songs would suffer greatly due to a lack of initial popularity. We get back to this issue in our next section on pre-release buzz.

There is some positive news though for those who believe in the power of "great entertainment." In Salganik and Watts' experiment, the very best songs in the data set recovered from their low popularity over time. Still, they ended up with fewer downloads compared to when no fake popularity information was provided. In other words, high-quality songs can make up a lack of initial popularity, but only to a certain degree. Further, let us keep in mind that entertainment products usually do not have much time to recover, given the competitive nature of the markets and short product life cycles.

Other scholars have provided additional evidence for the role of popularity using actual sales instead of lab data. Market data certainly add realism, but also carry problems: it is tough outside the lab to demonstrate conclusively that the empirical links from early to later popularity are truly causal (instead of simply reflecting the underlying reasons why a product is successful in the first place). In the music context, Bhattacharjee et al. (2007) analyze albums' survival on the charts, controlling for a number of other variables (such as the artist's star status, the producing label, and the release time). They find that a less-successful debut is associated with a shorter time on the charts—each additional rank at entry alters survival by about 2%. But be aware, however, that the authors do not control for factors like advertising that occurs before release, and we assume that an album's chart entry will often be influenced by these factors (instead of being exogenous).

Similar concerns apply to two studies on success-breeds-success effects for books. Clement et al. (2008) study how the rank of a book in Germany's *Spiegel* bestseller list, and three threshold dummy variables (Top 10, Top

20, and Top 50) affect accumulated weekly sales of 609 novels published in hardcover format in Germany over the course of the 12 months following release. The scholars run OLS and panel regressions in which they also control for a large number of other factors (star status of the author, sequel, the book's quality as rated by readers, and its price—but not advertising). Their results again support the existence of success-breeds-success effect. But they find that the effect is highest in the first week after a book's release (when a higher rank translates into about 4% higher sales) and then deteriorates quickly. After week 4, the specific rank of a book no longer matters—instead, it is whether a book was among the Top 10 or Top 20 in the previous week that affects sales.

We learn that whereas consumers seem to be very attuned to a book's success when it first comes out, and the information is often heightened by media attention and advertising, things change to a more generic perspective, over time. Then it just matters whether a no-longer-new book is a hit or not. This insight is consistent with what Alan Sorensen (2007) reports for an analysis of how being on the New York Times bestseller list affects the weekly sales of 800 books released in 2001 or 2002. In his case, a book's initial appearance on the list adds about 8% in terms of sales, but no significant effect exists in the following weeks.

Logic suggests that popularity should also matter for movies and games. And our own research lends at least tentative empirical support: via structural equation modeling, we find that the long-term theatrical success of a film is highly related to its opening performance, even when controlling for the studio's efforts to produce, promote, and distribute an appealing product (Hennig-Thurau et al. 2006a). For the 331 films in our data set, the results suggest that the short-term success essentially absorbs all marketing measures when it comes to the performance in the following weeks, except for the film's quality itself.

With the pinball marketing environment in place now, we expect the power of success-based cascades to increase. As former Amazon Studios chair Roy Price argued: "If you have one of the top five or ten shows in the marketplace, it means your show is more valuable because it drives conversations and it drive subscriptions" (quoted in Littleton and Holloway 2017). In other words, in the digital era, information about hits will travel farther and faster among consumers, which should let firms profit more than before from their hits.[331]

---

[331]This effect is also reflected in today's shortened diffusion patterns.

We have so far restricted our analysis of success-breeds-success effects to a single channel or market. But entertainment products are usually released in more than one channel or market, and popularity-related information can flow between them. This is what we will look at now.

Success-Breeds-Success Between Channels and Markets

A specific area in which popularity matters is when the success of an entertainment product in one channel (such as movie theaters) or market (such as North America) influences consumers' decision making in a different channel (think DVDs) or market (think Germany). *Entertainment Science* provides some insights into this issue as well. Keep in mind the limitations of real-word data for analyzing success-breeds-success effects—they also apply to most of the studies we mention in the following sections.

Inter-Channel Success-Breeds-Success

Regarding movies, we show for a cross-sectional data set of 331 movies released in theaters and on video during 1999–2001 that the link between theatrical success and video rental success is quite substantial (Hennig-Thurau et al. 2006b). When analyzing a complex structural model with partial least squares, both a film's opening weekend and its performance in theaters in the following weeks are strongly linked with video rental success. The path coefficients are about 0.35 (on a scale from −1 to 1), on par with the impact of advertising and higher than those of all other "success drivers."

Whereas our standardized parameters do not offer elasticities, those from Jozefowicz et al. (2008) do. For a smaller data set that is biased toward theater hits (93 of the 100 highest grossing films in 2001) and few controls, they run two OLS regressions with VHS and DVD revenues as dependent variables. The scholars find that a 10% higher box office corresponds with 5–6% increases in rental revenues for the two home channels—those are large success-breeds-success effects, but with diminishing returns.

Finally, Jordi McKenzie (2010) studied how success at the Australian theatrical box office is linked with DVD sales for a data set of 760 films (from 2004 to 2007). Consistent with our own evidence on rental success, he finds a *very* strong positive correlation between the two channels of 0.88. While correlations are strong across genres and age ratings, they are highest for the biggest films in the data set and for films for which the time gap is neither very short nor very long—success information needs time to spread, but is forgotten after a while. From a simple OLS regression that uses only box office and its squared term as regressors, he derives that the success-breeds-success relationship appears to be nonlinear—the most successful films in theaters experience a disproportionally high bump in DVD sales, compared to less successful films.

Similar inter-channel popularity effects have been reported for other entertainment products. Papies and van Heerde (2015) find that record sales trigger concert revenues, with the effect being part of a feedback loop, as concert tickets also spur higher record sales. In line with our expectation that the digital environment will facilitate such cascades, the scholars show that this effect only exists in recent times with extensive broadband Internet diffusion.

And for books, Sumiko Asai (2015) studies the role of previous hardcover sales on paperback sales for 254 newly released novels which appear on the Japanese Top 200 paperback chart from 2010 to 2013. Using a GMM approach in which she estimates sales and price separately (and controls for several factors—but not advertising and quality), Asai finds an elasticity for hardcover sales of 0.11—on average, a 10% increase in a novel's hardcover sales is linked to 1% higher paperback sales. Schmidt-Stölting et al. (2011) suggest that the role of success in previous channels might be even higher when books which were not hits are also considered. For their large data set of German books, a SUR analysis shows an success-breeds-success effect that is five times as high as the one reported by Asai.

Inter-Market Success-Breeds-Success

Until recently, entertainment media products were often released in their domestic market first and only later in foreign markets. Today, a global "day-and-date" release is nothing unusual for major movies, albums, games, and novels, despite the additional work that is often needed for foreign launches (such as translations, dubbing, etc.). A major advantage of a global release is the synergies in building buzz that producers hope will spread across countries via the Internet. But a potential drawback of this new approach might be the lack of success-breeds-success effects: when a product is released simultaneously in its home market and abroad, there is little time for action-based cascades to develop. What does the evidence say?

Two studies on movies provide evidence that success-breeds-success effects do exist across markets, but these studies also point to problems that occur if the delay between markets is too long. The first study is the often-cited one by Elberse and Eliashberg (2003), in which the two scholars use 3SLS to estimate the effects of a film's success in North America on its foreign reception in France, Germany, Spain, and/or the U.K. For between 127 and 140 films, they find a direct and positive impact of the North American performance on foreign moviegoers in all four European countries. The direct impact of a 10% higher home market success ranges from an increase of 1.6% (in Germany) to as much as 9% (in the U.K.) in their data set. For France, Germany, Spain, the *total* success-breeds-success effect is even higher, as the North American performance significantly impacts the "supply side." Specifically, theater owners in the European markets show films that were more successful in the U.S. on more screens, which also increases the films' success potential with European audiences.

Elberse and Eliashberg's results also teach us that the strength of this success-breeds-success effect is moderated by the time that passes between the two market entries. An interaction between the films' North American box office and the time lag between market entries is strong and negative for the three continental European countries, such that a larger time lag reduces the impact of a film's success in North America for its foreign release. When we analyzed the box-office results of 231 North American films in both their home market and in German theaters (all released between 1998 and 2001), we found the same decay effect for action-based cascades: whereas the correlation between box-office results in the U.S. and Germany was a very high 0.87 for films released in Germany *within 3 months* after their home release, it was only half as strong for films released abroad 6 months or more after the home market release (Hennig-Thurau et al. 2003).

But let us note that both of these studies assumed a linear effect of time—the effect size weakens continuously with every additional month that passes after the home release and across the whole time period. This assumption leaves no room for potential *non-linear* patterns, such as allowing some time for inter-market action-based cascades to build. When we revisited the 1998–2001 data for this book, we find indications of such non-linearity. Take a look at Fig. 12.4: the shared variance between film's North American and German performance follows an inverse U-shape pattern over time, with the highest amount of variance explanation existing not for films which are released simultaneously in both countries, but for films released in Germany 2–3 months after their release in North America. Only then do the success-breeds-success effects start to decline.

These insights must be treated with care, as the published studies do not consider the potential endogeneity of studios' international release decisions. But it seems reasonable that managers systematically market their flops differently than their hits when it comes to international release decisions. But

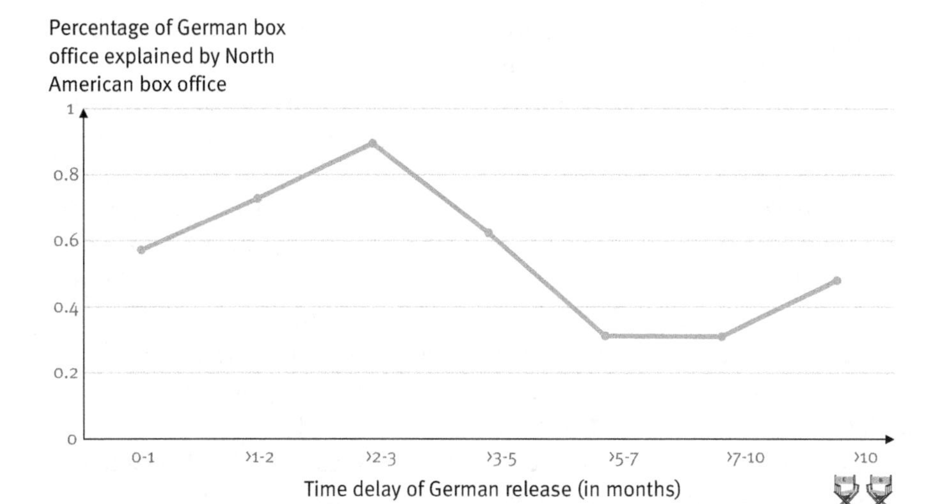

Fig. 12.4 Relationship between films' North American and German box office
Note: Authors' own illustration based on data for 231 U.S. films, as described in Hennig-Thurau et al. (2003).

based on what we see in the data, we are tempted to conclude that delaying the foreign release of what is expected to be a flop in the home market might not be the worst thing: launching at the normally optimal time may enhance the likelihood that the flop information spills over and makes the product dead-on-arrival in the foreign market. But as release decisions and contracts for entertainment products are usually made months in advance, making effective use of inter-market success-breeds-success effects will work best if the producer is able to forecast a new product's success early on—which links our work in this section to the forecasting models we discussed in the context of innovation decisions.

As with all aspects of "earned" communication, managing success-breeds-success effects is a challenge for entertainment managers because of the low controllability of these channels. So, let us take a look at what managers can do (in addition to timing their global releases accordingly) to facilitate such cascades.

Managerial (Mis-)Use of Post-Release Uninformed Cascades

Our findings about the importance and mechanisms of popularity-based action-based cascades for entertainment products carry a number of managerial implications. Some of these fall into the category of "smart" decisions, whereas we consider others to be outright immoral.

But let's start with those we assign to the "smart decisions" category. When Steven Spielberg's movie JAWS was approaching its theatrical release in the summer of 1975, Universal's chairman Lew Wasserman made a tough choice. Although his marketing team had initially arranged for release in a then-unprecedented number of 900 theaters in North America, because of the high demand that theater owners were sensing, Mr. Wasserman dropped more than 400 of those.[332] In a time when box office statistics were not omnipresent and immediate as they are today, Mr. Wasserman's vision was to create long waiting lines which would generate massive news coverage and, in turn, trigger action-based cascades (Shone 2004).

Although it is hard to assert causality in a single case, we know that his decision certainly did not hurt the film, which ran all summer and became the first in history to break the $100 million barrier. In more general terms, Mr. Wasserman's logic was to create images of high demand through an artificial shortening of supply. This is an approach that Hollywood still practices these days when it follows a "limited release" strategy, implying that the showing of a film in carefully selected venues (such as arthouse theaters in art-centric areas) will create high demand signals that are then picked up by the media and, in turn, set off cascades, attracting audiences in other parts of the country and, ideally, the world. And the media today indeed helps by highlighting high "per-screen" averages for limited-release films that pack selected theaters.

Other approaches to exploit the power of success-breeds-success effects might be considered equally clever, but of more questionable morality. A first one was reportedly practiced by management "gurus" Michael Treacy and Fred Wiersema in 1995, when they secretly purchased 50,000 copies of their own book "The Discipline of Market Leaders." They bought them from the stores that are influential for the New York Times bestseller list, helping the book to flourish on the charts, despite mediocre reviews (Bikhchandani et al. 1998). The book then "sold well enough to continue as a bestseller without further demand intervention by the authors" (p. 151).

Such behavior appears to be quite common for novels and music today; it might not be technically illegal, but it certainly generates a fake popularity

[332]See also our more detailed discussion of JAWS as the first incarnation of the blockbuster concept in the integrated entertainment marketing chapter.

that cannot be detected by consumers, who are misled by the numbers. Music manager Tom Silverman, in 2010, argued that music labels have long "hyped the charts" by buying their own songs, and are still doing it. The economics of the digital economy make such behavior even less costly; now the labels only have to cover the 30% share of the digital retailer, such as Apple/iTunes, but not the production costs for the disks. "[I]f they buy 50,000 songs, we're talking $50,000 less 70%, so it would cost about $15,000. For $15,000 in a week, they can buy 50,000 more song downloads, which could drive the record up three or four positions on the chart" (Silverman, quoted in van Buskirk 2010). This success, Mr. Silverman argues in line with the logic of action-based cascades we discussed on the previous pages, makes potential consumers assume that a song is of high quality, which then triggers the "actual" success.

A related, and no less dubious, approach is to tell audiences that a product is a bigger success than it actually is, an approach referred to as "overreporting." Whereas doing so for actual sales would be outright cheating, Malhotra and Helmer (2012) offer empirical evidence that some film studios systematically practice overreporting in publishing box office *estimates*; these are then used by the media for their success-related coverage and Monday morning charts.

Malhotra and Helmer analyze such estimates in combination with actual box office numbers for all 1,000+ wide-released films from 2003 to 2010, finding that overreporting differs strongly between studios.[333] They find that overreporting is particularly strong when the incentives for the studios are the highest—in the release week (versus later), when cascades can be stimulated most effectively, and when competition is high. And if the difference to the second-placed film is less than $10 million by Saturday evening, overreporting for Sunday is highest. Based on these results, the scholars conclude that overestimation is a fact-of-life in the film industry, and that it "is highly unlikely to be due to chance" (Malhotra and Helmer 2012, p. 1411).

[333]In their data set, Malhotra and Helmer find overreporting to be highest for Sony (+10.3% on average) and Summit (+7.9%, now part of Lionsgate), and lowest for Fox, Disney, and DreamWorks (4%, 3.6%, and 1.2%, respectively). Of course, these differences *might* also be attributed to the studios' differing monitoring skills…

Pre-Release Action-Based Cascades: Buzz

> "Failure to create the right buzz beforehand [means] less anticipation and ultimately fewer ticket purchases. ... Greater emphasis has been placed on social media in the hopes of generating the right buzz that will serve to build anticipation, pushing the film into a 'must-see' status."
> —Freedman *(2015)*

Action-based cascades can also be observed *before* the release of a new entertainment product. At this point in time, no information about the actual performance of a product can exist—but consumers can interpret the "buzz" that they observe in the market for a new product as an indicator of the overall anticipation for the product. According to this logic, the observed anticipation itself serves as a signal of the forthcoming product's attractiveness and quality.[334] Building buzz has become a major element in the distribution of new entertainment products, a cornerstone of the now-dominant blockbuster strategy. The idea behind such efforts to stimulate buzz is that pre-release buzz cascades translate into opening success, which might then trigger post-release action-based cascades, as discussed on the previous pages.

Because one cannot measure (let alone manage) something for what no clear understanding exists, let us first clarify what we mean when we say "buzz," a term that is often used in a "happy-go-lucky" way by consumers, journalists, and managers alike. After offering a scholarly definition of the concept, we take a closer look into what *Entertainment Science* scholars have unearthed regarding the drivers of buzz, the concept's link with entertainment success, and finally its mediating role on the road to product success.

So, What Exactly is "Buzz"?

People inside and outside of entertainment have been using the term "buzz" in a pretty superficial way. As a result, we rarely talk about the same thing when discussing buzz, something that has affected the many ways buzz is measured. It has also hindered scientific progress on the topic: there is just no way for *Entertainment Science* scholars to explore a concept and its role for managing entertainment without a common understanding of it. So, what is a useful way to define buzz?

To understand what constitutes buzz, we asked those whose responses to it determine any impact buzz can have on entertainment success: the

[334]See our initial discussion of the buzz concept in the context of (social) consumer behavior.

consumers. Through several depth interviews and focus groups with consumers, we found that new product buzz describes the "aggregation of observable expressions of anticipation by consumers for a forthcoming product" (Houston et al. 2018, p. 349). Let us highlight four core characteristics of buzz that we learned from our exploration:

Buzz is an aggregate-level concept. Buzz is not about what any single person does or says, but it is about what is happening with many consumers at an overall, "macro" level—the totality of things, if you want, not just its individual parts.

Buzz has an anticipatory nature. Buzz is forward-looking, referring to something "to come," not looking back at what has already happened. It is also a positive concept: at least in the context of entertainment, buzz is fueled by people's hopes, not their criticism. "Negative buzz" exists only if consumers are afraid that their hopes will not be met by a new product. Will the AVATAR sequels match the stellar-high standard of the original film? Those studies which include a measure for pre-release *sentiment* of consumer communication find it to be uncorrelated with later success (e.g., Gopinath et al. 2013).

Buzz is multi-behavioral. Consumers' perceptions of the buzz for a product are fed not only by what is said and written about a new product, but also by two other kinds of observable behaviors: people's searches for the new product (as observed in metrics such as IMDb's MovieMeter) and also their "participative" actions (e.g., trailer views on YouTube, the scavenger hunts that took place as part of THE DARK KNIGHT alternate reality game, the re-watching of an earlier movie to prepare for an upcoming sequel, etc.).[335] Figure 12.5 illustrates buzz' multi-behavioral nature and provides examples for each type of behavior.

Buzz is two-dimensional. The amount of buzz, as reflected in "most-popular" rankings on Twitter and other websites, certainly marks a key type of information when consumers judge the buzz for a product. But in addition, buzz involves a more qualitative dimension: *who* is anticipating the new product, in addition to how many? This is the "pervasiveness" dimension of buzz: it tells us to what degree the collective anticipation for a new product is *spread across consumer segments* (see also Fischer et al. 2017). Are only the fan boys excited? Or has the excitement exceeded the niche and reached the mainstream?

[335]In their introspective study, Wohlfeil and Whelan (2008) describe in detail how the anticipation for watching the then-forthcoming movie PRIDE & PREJUDICE shaped the first author's behavior, including participative actions such as buying a new version of the book (which featured the poster as book cover) and buying newspapers which featured articles about the film.

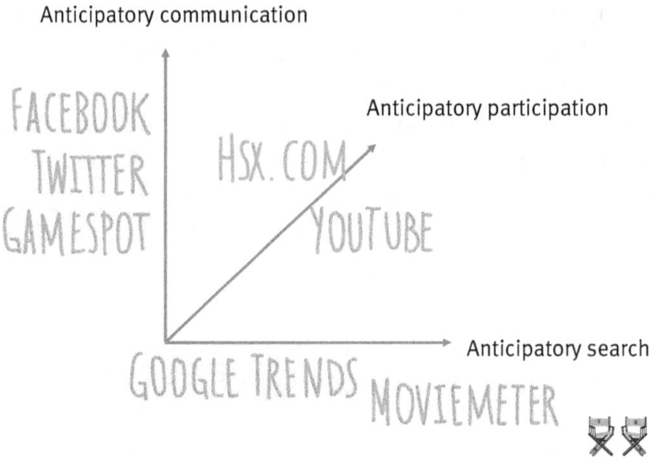

Anticipatory communication
Anticipatory participation
Anticipatory search
FACEBOOK
TWITTER
GAMESPOT
HSX.COM
YOUTUBE
GOOGLE TRENDS
MOVIEMETER

Fig. 12.5 Three general kinds of buzz behaviors (and examples)
Notes: Authors' own illustration based on Houston et al. (2018). Brands are trademarked. With graphical contributions by Studio Tense.

For consumers' evaluation of a coming attraction, niche buzz and mainstream buzz are often separate things.

This two-dimensional character of buzz provides an explanation why some films such as Kick-Ass and Scott Pilgrim Vs. The World flopped at the box office, despite having been strongly hyped by Comic-Con attendees, whereas other films that were surrounded by similar buzz levels at Comic-Con became huge box office hits (such as Avatar and Iron Man).[336] The buzz for the latter films came from fans *and* mainstream consumers (i.e., high pervasiveness), but the former films excited *only* the niche. What did Kaye (2012) say about Scott Pilgrim? "Only nerds like movies about nerds; [leading actor] Michael Cera is not a leading man to the rest of the world."

What do we gain from such an elaborate conceptualization of buzz? Generally speaking, the better someone is able to capture the essence of a concept that influences a product's performance, the more accurate success predictions can be. In the next section, we will first investigate what empirical studies tell us about the link between buzz and success in general, before getting back to the question whether it matters how we conceptualize buzz (and how we measure it).

[336] Iron Man director Jon Favreau explicitly attributed the success of his film to the roaring response of fans at Comic-Con in 2006, telling attendants "It all started here. Nobody cared before you did" (quoted in Horn 2009).

How Buzz is Linked to Product Success

The Link Between Buzz Volume and Product Success

Despite the differences in empirical measures, studies consistently show that buzz is linked to entertainment success. But like with post-release action-based cascades, establishing the causal nature of this link, which is so essential for determining its role in the entertainment marketing mix, is far from trivial.

When Google analysts Panaligan and Chen (2013) reported that they were able to explain about 70% of a movie's opening weekend box office with the search volume for a movie title via Google in the week prior to its release (and that adding the number of theaters and brand status variables increases the variance explained to 92%), they could be sure to get wide attention among entertainment managers. And even more impressive was their proclamation that the search volume for a movie's *trailer* four weeks before release explained 62% of the opening weekend success, 94% when paired with release timing and brand status variables. These result sound too good to be true, and they might be. The authors report very little about the data and method they use, and the same is true for their selection criteria for the small sample of 99 films they analyze—which, as every researcher knows all too well, can have a *huge* impact on the results. We don't know how much of the box office explanation can be attributed to search buzz: the analyses, and particularly the "search-only" ones, suffer from a massive omitted variables bias.[337] And for their "joint" analyses, the authors do not even reveal the share of success that is explained by search buzz, above and beyond the other variables.

Let us thus turn to more rigorous approaches to shed light on the link between buzz and success. In their study of 681 games (all released in 2009 and 2010), Xiong and Bharadwaj (2014) look at how pre-release communication buzz (on a variety of blogs and forums) and search buzz (from Google Trends) affect the games' opening week sales. The scholars conduct separate regressions with robust standard errors for the two buzz behaviors; they also include a large set of controls, including whether a game is a sequel, advertising spending, its genres, and the release timing. They estimate an elasticity of 0.46 for communication buzz (i.e., a 10% higher communication buzz

[337]See the opening chapter in this book for a short introduction to the omitted variables bias.

transfers into 4.5% higher opening week sales)—an effect which is almost nine times higher than the effect of pre-release advertising in their data set.

For movies, results also point to a strong role of buzz. Gopinath et al. (2013), in their investigation of 75 movies, report a strong elasticity of 0.27 for pre-release communication buzz (articulated on blogs) on a film's opening weekend, and Liu (2006), in his OLS analysis of 40 movies, finds an even higher elasticity of 0.59 for pre-release communication on the Yahoo Movies site, which might be somewhat inflated, however, because of a smaller set of controls. Whereas Kim and Hanssens (2017) do not report elasticities, they use search buzz three weeks prior to movie release, along with advertising and distribution efforts, to predict the opening weekend box office of 41 movies. They find that prediction is about 8% (or $6 million) more accurate for the model that includes search buzz versus a model without search buzz.

We, too, looked at movies when we predict the opening weekend box office of 254 movies which were widely released in North America from 2010 and 2011 with partial least squares analysis. We used a multi-behavioral buzz measure that, in line with our conceptualization above, combines information from Twitter for communication, Google for search, and Facebook likes for participation in addition to large set of other "success drivers." We find that this model explains 25% more of the variance in the movies' opening weekend box office than a model that is exactly the same except that it contains no buzz information (Houston et al. 2018).

Scholars have also compiled evidence that buzz matters for TV show episodes, although the effect found in scholarly studies is less radical in this context, probably because episodes, being part of an ongoing series, are less "innovative" compared to a new movie or game. Studying the weekly ratings for the episodes of 30 TV shows when aired in North America between 2008 and 2012, Liu et al. (2016) use a panel GMM estimation approach and find that the number of pre-broadcast tweets increases the explanation of a show episode's rating by 0.5 percentage points, compared to a model that uses only previous ratings for a show.

Finally, only limited evidence exists for the power of buzz for new *music*. The little we know comes from Buli and Hu (2015), employees of music analytics company Next Big Sound, who reported that first-week album sales correlate at 0.70 with views of an artist's Wikipedia page (a measure of search buzz) in the week before release and at 0.25 with YouTube video views (a kind of participation buzz). The authors report Granger causality tests to address the causal nature of these parameters, but there is simply not enough transparence regarding the methods used to judge their findings.

The results we have assembled here give us a solid understanding that buzz is indeed influential for entertainment success; the size of the buzz effects reported by scholars also puts those named by commercial analysts into perspective and gives us an idea of their inflated nature. But the section also shows that quite different measures of buzz are used. Let us now see if we can shed some more light on differences between those measures and the corresponding roles of the buzz behaviors and the pervasiveness dimension of buzz, in addition to buzz' sheer volume.

Differences Between Buzz Measures: Where Behaviors, Pervasiveness, and Content Matter

To understand the success impact of buzz better, studies that compare alternative specifications of buzz can be informative. In their games study, Xiong and Bharadwaj (2014) find a large difference between communication buzz and search buzz. In a separate analysis to their model in which they use communication buzz, they estimate an elasticity of search buzz (via Google) on opening week sales of 0.05, or about just one-tenth of the one by communication buzz.

Results for movies show a different pattern, though. When Kim and Hanssens (2017) report prediction results for their data set of 41 movies for an alternative model which contains communication buzz (via Google blogs), instead of search buzz, they find the search measure to be more effective than the communication measure. And when we estimate alternative models to our multi-behavioral buzz specification (which combines Twitter, Google Trends, and Facebook likes), we also find quite substantial differences between individual buzz behaviors (Houston et al. 2018). When studied in isolation, search buzz explains the most (i.e., has the highest R^2) and also has, along with Twitter, the lowest prediction error (i.e., the smallest MAPE). The number of Facebook likes for a movie, our measure of participation buzz, is clearly less effective in both regards, a finding that is consistent with our results in Kupfer et al. (2018). But none of the buzz measures can, when used in isolation, match the performance of the multi-behavioral buzz specification, which explains, on average, almost 8 percentage points more of movies' opening weekend success.

We also compared the three buzz measures, that are all used by all kinds of consumer segments and thus reflect high-pervasiveness buzz, with low-pervasiveness, niche measures (the number of posts on the fan forum joblo.com for communication buzz, the number of MovieMeter searches on IMDb for search buzz, and the number of page edits by enthusiasts of a film's Wikipedia entry as participation buzz). Using these low-pervasiveness measures for each buzz behavior reduces the explained variance of movie success by 13 percentage points. In other words, at least for our data set of wide-released movies, pervasiveness matters for buzz, and mainstream buzz plays a bigger role than niche buzz.

Finally, Liu et al. (2016) show that the buzz *content* can also matter for the effect that communication buzz has on success. In their analysis of TV series ratings, an in-depth coding of what people chat about via Twitter with regard to a show increases the explanation of the show's rating by 15 percentage points—up to 90%. When buzz features strong doses of positive emotions such as "excited" and "love," the scholars find it to exert a particularly strong effect on success.

In essence, not all buzz is equal, just as we concluded in the case of WOM. Different buzz behaviors produce different insights and suggest different decisions. And although the results for different buzz behaviors and measures vary somewhat between studies and criteria, among the most important insight is that a multi-behavioral buzz measure that captures high pervasiveness is most informative. Understanding this is essential for picking the "right" measures when tracking, and stimulating, the buzz for an entertainment product.

Buzz Cascades and Thresholds

The studies we have covered so far implicitly assume that the volume of buzz has a linear effect on success. But this assumption somewhat conflicts with an argument we made in the consumer behavior chapter of this book. Buzz sends a signal of collective interest to people (the "We-desire") that triggers a cascade, and that at least some consumers only join the cultural bandwagon once they feel that buzz is substantial enough to deserve their interest. In other words, after the buzz for a product has passed a certain threshold, a cascading, non-linear effect should take place.

And studies indeed provide evidence for the existence of such buzz cascades. The results of Xiong and Bharadwaj (2014), for buzz communication about games, show that buzz creates more buzz on its own over the 180 days prior to a game's release, with this self-enhancing effect growing over time. And Kim and Hanssens' (2017) VAR analyses of movies and games point in the same direction; for both forms of entertainment, they find positive "self-elasticities" for both communication buzz (for which they find the effect to become smaller over time) and search buzz (effect becomes stronger).[338] They also show that buzz cascades can exist also between different buzz behaviors; in their data set, communication and search buzz mutually enhance each other in both entertainment contexts, with the effect of communication buzz on search buzz being more than four times as strong as the effect of advertising.

Whereas we now have evidence of buzz cascades' existence, none of these studies has addressed the existence of buzz thresholds that need to be exceeded to set the self-enhancing effect in motion. But Soderstrom et al. (2016) have looked for such thresholds empirically, using buzz data for 309 movies released from 1999 to 2001. Their buzz data come from a professional survey firm that contacts about 300 consumers in the week preceding a particular movie's release, asking consumers whether their friends were talking about the film. In this analysis, the measure of (communication) buzz is the percentage of consumers for which this was the case. To analyze how such buzz is linked with movies' opening weekend success, the scholars combine OLS regressions with a "spline" method, in which separate functions are estimated for different parts of the data set and then the slopes for the parts are compared.

Soderstrom et al.'s findings provide further support for the relevance of (communication) buzz, as adding the buzz measure to a list of controls (such as distribution and stars—though not advertising) increases the explained variance from 44% to 72%. With regard to the cascading effect, their results suggest that a buzz threshold indeed exists, and that once it is passed, the effect of buzz is 3.6 times stronger than before. And where is the threshold located? For the set of movies analyzed by Soderstrom et al., it is when more than 21% of consumers sense the buzz for an upcoming film; this "cutoff" is illustrated in Fig. 12.6. We cannot say whether this insight can be transferred to other data sets and entertainment products, and the authors' findings come with certain limitations. But we learn that for entertainment managers, investments in buzz pay off much more in success once a critical buzz level is passed.

[338]Kim and Hanssens find the self-enhancing effect of "buzz on buzz" to become smaller over time for communication buzz, whereas for search buzz the effect becomes stronger over time until the release.

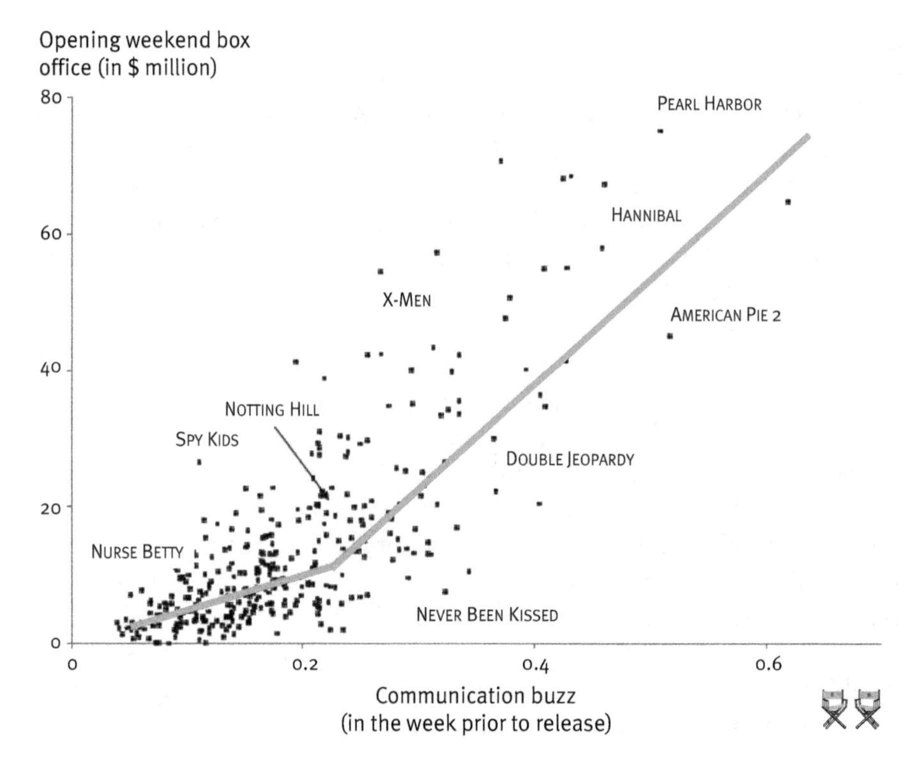

Fig. 12.6 A buzz threshold
Note: Reprinted with some modifications from Soderstrom et al. (2016, p. 925), which is distributed under the terms of the Creative Commons Attribution 4.0 International License (https://goo.gl/n6FkkT).

Buzz Patterns Over Time

Buzz builds over time, and scholars have shown that, beyond its amount at a given point in time and its pervasiveness across groups, there is also value in understanding its *pattern of development*, over time. Foutz and Jank (2010) use functional shape analysis to determine patterns of how the "stock prices" of movies develop over time on the Hollywood Stock Exchange, which, they argue, mirrors the "hype" for a movie as perceived by those consumers who trade movie stocks on HSX. From a data set of 262 films released in North America in 2003–2005, they extract four unique buzz patterns from the movies' stock prices over the 52 weeks prior to their releases: a "higher trading average," a "steeper upward trend," a "last moment hype," and an "early preannouncement hype" pattern.

The two scholars find that all four patterns are linked to higher opening weekend success in their data set. And they show that the patterns might also be used by managers for early forecasting: whereas the average prediction error for a model based on movie characteristics and advertising is about 90% until ten weeks prior to release, Foutz and Jank calculate that the addition of buzz pattern information reduces this error substantially to 35% at 40 weeks before release, and to 20% at ten weeks before release.

Xiong and Bharadwaj (2014) reveal similar insights for their data set of games when they also identify four main functional shapes for pre-release communication buzz, as expressed in blogs and forums before a game becomes available. When the scholars link the shapes with opening week sales in a separate regression model, two pattern are associated with higher sales, both of which had also been found to be influential by Foutz and Jank: the "higher trading average" pattern and "steeper upward trend" pattern. Xiong and Bharadwaj also repeat their analysis with buzz *search* data, finding similar results.

And What Drives Buzz?

Now we have shown what buzz is and that it matters economically. For managing buzz successfully, understanding the factors that influence it is crucial. From the studies that explored variables that drive buzz (or its facets), two major categories of drivers have emerged: the manager's *marketing actions* and the *quality* of the product that the buzz is about. Let us note that most existing studies look at the amount/volume dimension of buzz, but very little is known what determines whether buzz is niche or mainstream.

Marketing actions. Advertising has been shown to be a main factor when it comes to triggering buzz. In Xiong and Bharadwaj's (2014) study of video games, the authors use a panel regression approach to understand what drives buzz communication during the 180 days prior to a game's release. They find that daily *online* advertising has the expected effect on buzz, but it varies over time, with its effect size peaking around 90 days prior to release (but remaining influential until release).

The pattern for TV advertising is similar, but the peak at −90 days is stronger (and the effect weaker closer to release). These patterns might be influenced by current industry practices—a bias that Kim and Hanssens' (2017) longitudinal study of the link between advertising and buzz in a games context counters by using a VAR modeling approach. For a data set of 66 games from 2013 and 2014, they estimate a cumulative advertising elasticity of 0.09 for communication buzz (on blogs) and of 0.06 for search buzz—10% higher advertising spending corresponds with 0.9 and 0.6% higher buzz, respectively. The pattern of results for a data set of 137 movies from around 2009 looks similar, but the effects of advertising on buzz are about twice as high for movies than for games.

Other marketing variables have also been found to influence buzz, with different kinds of brands playing a prominent role. Xiong and Bharadwaj (2014) find that games with a major publisher brand (such as EA) experience more buzz; the publisher brand is particularly effective in the early stages, but becomes small and insignificant about 90 days before the game's release. Craig et al. (2015), in a regression analysis of a sample of 62 wide-released films (from 2008 to 2009), find that buzz (they combine communication and trailer views as a form of participation buzz) is, on average, higher for sequel brands than original films.[339] And Divakaran et al. (2017) find that stars, as human brands, also boost communication buzz (the number of user comments on Fandango, the pre-order fan website about their films) for a data set of 373 movies from 2009 to 2010 which they analyze with partial least squares. Related, Karniouchina's (2011) results suggest that consumers' *chatter* about stars spills over to the chatter about the films in which they appear; she estimates separate equations for the search intensity for films and stars with 3SLS.

Buzz also differs between certain entertainment genres. Xiong and Bharadwaj find much higher buzz for sports games (versus other game genres played on the Xbox360); for movies, Craig et al. find action and horror films to have more buzz, on average, and Liu (2006) reports this to be the case also for action movies and adventure movies. Liu also finds that movies with a rating less restrictive than R produce an average of 150% more communication buzz (on Yahoo Movies).

Quality. Fewer studies have investigated the role of product quality on buzz, but their results suggest that higher quality translates into more buzz, in general. Xiong and Bharadwaj, who measure quality by experts' ratings, find that product quality has an effect on communication buzz, which grows over time, and only loses its impact in the week before the release. In our own partial least squares (PLS) analysis of 254 movies, we also find quality, again measured as experts' quality perceptions, to be positively linked to our multi-behavioral buzz measure (Houston et al. 2018).

But how strong is the (relative) impact of quality on buzz compared to those for the marketing actions we have identified as buzz drivers before?

---

[339]Craig et al.'s result, and those of Divakaran et al. (2017), also suggest a buzz-driving role of the production budget of entertainment products, but here the same concerns apply that we mentioned in our discussion of the production budget's role on product success. Neither study included advertising, whose impact might have been appropriated by the budget variable.

In our study, we also compare the effect size of movie quality on buzz to the effect size of a multi-faceted marketing actions variable (which comprises a film's budget, rating, star, and sequel character). The PLS results show that marketing actions and quality together explain a sizable chunk of buzz (about 35%), and that marketing actions account for a larger part: its impact on buzz is 3.4 times as high as the impact of quality.

The Mediating Role of Buzz

Our analysis on the previous pages has shown that buzz influences the success of entertainment products, but is also affected itself by various marketing actions and the quality of the product–factors that we know influence the market performance of entertainment products on their own.[340] Thus, buzz should not be treated as an exogenous construct, but instead as a mediator between a product's quality and other marketing actions, on one hand, and the product's success on the other (Houston et al. 2018). Figure 12.7 illustrates this mediating role by putting buzz between quality/other marketing actions and success.

Our empirical analyses of this structural model inform us that buzz serves as a *partial*, not a full mediator—both the product's quality and the firm's other marketing efforts still influence consumers directly, aside from buzz. But a substantial share of their impact is transmitted through buzz—in our study, 42% of the effect of quality on movies' opening box office is explained via buzz, and 39% of the effect of other marketing efforts is moderated by the buzz these efforts are able to stimulate (see also Divakaran et al. 2017 for a similar argument).

This mediating role has far-reaching implications for the management of entertainment: because buzz is acknowledged as an important success driver, producers increasingly look for marketing actions that help build buzz. A look at the buzz determinants tells us that this logic favors certain types of products over others. Freedman (2015) states, with regard to the film industry, that producers' attempts to "adapt to [the critical role of buzz] are revealed in the types of films they choose to produce and the manner in which they have amped up their marketing efforts to generate buzz."

In contrast, those products that have lower buzz potential are considered inferior: "It is still possible to start with unfamiliar characters, but only if they can properly be transformed through the magic of social media"

[340]We discuss the role of quality for product success in Chapter 7 of this book, which is dedicated to the entertainment experience.

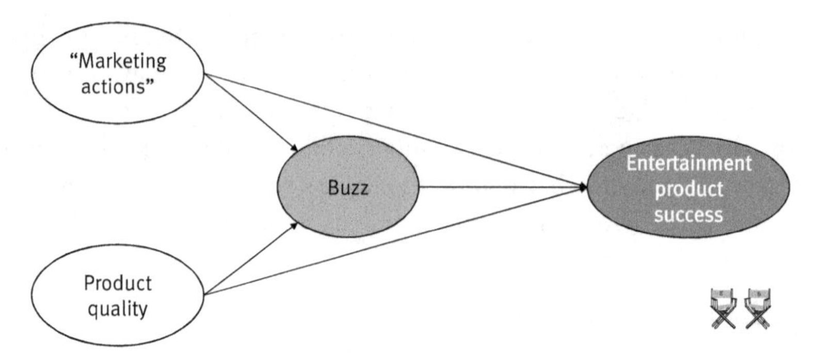

Fig. 12.7 The mediating role of buzz
Note: Authors' own illustration based on ideas in Houston et al. (2018).

(Freedman 2015). We continue this discussion and critically evaluate the industry's adaptation of the powers of buzz in the shape of the "pre-sales" approach to entertainment in the final chapter of this book.

But before we do that, let us inspect other kinds of "earned" media. We begin with what might be considered the most obvious infusion of data analytics in entertainment communication: automated personalized recommendations of entertainment products.

Automated Personalized Recommendations

> "One day we hope to get so good at suggestions that we're able to show you exactly the right film or TV show for your mood when you turn on Netflix."
> —*Netflix-CEO* Reed Hastings *(quoted in The Economist 2017)*

Digitalization has broadened the concept of word of mouth by making the entertainment experiences of other consumers available for us on websites and via social media. It has further facilitated action-based cascades by letting us access success information about entertainment products and the buzz that exists for them in unprecedented ways.

But digital technologies also provide entertainment firms access to new "big" databases and powerful hardware that can process such data and advanced algorithms with enormous speed. The combination of both enables firms to offer personal product recommendations based on the experiences of other consumers who we have never met, and probably will never meet. This is what automated recommender systems do: they apply algorithms to large data sets to generate individualized product recommendations based on the choices of other consumers with similar preferences to our own. Recommenders today exist for all entertainment products that we feature in this book: Pandora and Spotify use them for recommending songs, Netflix and moviepilot.de for guiding us to movies, Gamefly and GameStop develop personalized recommendations for games, and Barnes and Noble suggest which books we should read next. And Amazon uses recommenders for all these products (and for everything else).

Although recommender systems could be considered as a variation of WOM, we discuss them at this point of the book for a reason: unlike WOM and also unlike buzz, recommender systems require the activities of third parties beyond consumers and producers. It is usually an intermediary (such as a retailer like Amazon), a streaming service (such as Spotify or Netflix), or a web platform (such as moviepilot.de or movielense.org) that generates personalized recommendations.[341] For them, recommenders can constitute important strategic resources, or even a "mission-critical technology" (Jannach et al. 2016). Because data technology is highly dynamic, keeping a competitive edge can be quite costly; Netflix is reported to have spent over $150 million as of 2015 just to improve their existing recommendations (*tickld* 2016).

This intermediary involvement is what sets recommender systems apart from the kinds of "earned" media we have discussed so far. Intermediaries, as recommendation providers, blend consumer information with their own skills and ideas. And the latter is what links recommendations with the "earned" media types we will discuss in the following sections, which also involve information from parties other than consumers, namely reviews by critics and awards from (industry) experts.

In the following, we look into the different kinds of recommenders that exist and their respective algorithmic logic; we then discuss selective issues beyond algorithms that also affect the effectiveness of recommenders. Please note that recommendation systems have developed into an academic

[341]One might argue that, with Amazon and Netflix (and, to a certain degree, also Spotify) now also producing entertainment products themselves, their recommender systems fall more into the category of "owned media" than "earned media." However, as their recommenders cover a wide range of products from various sources, with their own productions being only a small fraction of potential recommendations, we consider their recommender systems more part of their value creation role as platform providers than as producers of content. But this discussion raises an interesting issue: at least theoretically, recommender systems could be "owned media," depending on the range of products which are considered. In practice, as we point out in our discussion on the following pages, the value generated by recommenders depends on the (large) number of product alternatives—which presents a serious limitations for the effective use of recommenders as owned media.

discipline on their own, with conferences being held and extensive articles and books being published on the issue. Our discussion does not intend to capture the technical details of this vibrant exploration, but aims to introduce our readers to the fundamental issues of the debate.

How Recommender Systems Work: A Look at Their Algorithmic Logic

The basic idea behind all automated recommender systems is that consumers have individual preferences; thus, person-specific recommendations of entertainment products are more valuable for a consumer than general, average ones. At the same time, the approaches to determine individualized product recommendations differ strongly between systems: whereas collaborative filtering methods use the *judgements or behavior of other consumers* with regard to other products for providing a consumer with suggestions to meet his or her idiosyncratic preferences, content-based recommendations believe the solution for doing so is in the *product and its attributes.*

All recommender systems face a set of challenges against which they are judged. One is the "cold-start" problem: how to deal with new products (for which no consumer ratings yet exist) and new users of the system (who have not rated any products themselves). A second basic problem is referred to as the "serendipity" problem: as captured by the sensations-familiarity framework, consumers value more of the same (the familiarity component of the framework), but they also look for new impressions (the sensations component), i.e., the "serendipity" moment.

Recommenders have a strong bias toward familiar attractions; they generally struggle to provide users with fresh sensations. Some people, such as veteran film journalist Peter Bart, consider this a given and a basis for their aversion against technology: "I wanna try new things!" (Bart 2017 on why he does not like recommenders). As we will show below, different recommender systems address these problems to varying degrees, a fact that has motivated scholars to develop recommendation models that combine the advantages of different systems—so-called "hybrid" recommenders.

Collaborative Filtering: The Matrix Completion Challenge

Collaborative recommenders are all in the "matrix-completion business." Imagine a matrix that includes the ratings of multiple products by multiple users in which not all products have been rated by all users. A recommender algorithm's focal goal in this case is to predict how users would rate those products which they have not yet rated. In other words, how much would a consumer like a movie he has not seen yet?

Figure 12.8 shows such a matrix, which consists of four consumers, or users (Claudia, Mark, Nancy, and Thorsten), and five movies, or items (NOTTING HILL, TERMINATOR 2, PIRATES OF THE CARIBBEAN, THE GOOD THE BAD AND THE UGLY alias GBU, and LOVE ACTUALLY). The rating scale ranges from 1 = "Don't like at all" to 5 = "Like very much." The scale itself does not matter—it could also be from 1 to 10 or binary (0/1). The question marks in the matrix indicate that a user has not yet seen a film, such as Mark has not yet seen GBU. Would he like the film?

The answer depends on the recommender approach used. We will now take a quick look at the two main approaches for generating personalized recommendations via collaborative filtering: user-to-user filtering and item-to-item filtering. Let us note that the "matrix-completion-view-of-the-world" also carries its own definition of what defines a "good" recommender: if you intentionally leave out some of the ratings in the matrix when training the algorithm, the "goodness-of-fit" of the recommender is measured by how accurately the algorithm can predict those left-out ratings.

User-to-User Collaborative Filtering

The logic of user-to-user filtering is pretty straightforward: find other users whose ratings of products are similar to a target user (i.e., have a similar taste) and then predict to what degree this user will like a product he or she has not yet experienced, based on its rating by those "taste neighbors" (e.g., Ekstrand et al. 2010). Doing so requires three major decisions: how the similarity between users is determined, how many neighbors should be considered for the recommendations, and how similarities and ratings are transformed into recommendations.

	NOTTING HILL	TERMINATOR 2	GBU	PIRATES ...	LOVE ACTUALLY
Claudia	5	3	?	4	5
Mark	?	5	?	4	1
Nancy	3	3	1	?	2
Thorsten	5	4	5	1	4

Fig. 12.8 A simple matrix of four consumers' ratings of five films
Notes: Authors' own illustration. Brands are trademarked.

Extensive research has studied potential alternative approaches for user-to-user collaborative filtering (for an overview, see Ekstrand et al. 2010). One popular way is to use the correlation between users as a measure of their respective similarity and to use this similarity measure as a weight when calculating the average of a film's ratings by other users. This approach resembles how consumers integrate the different attributes of a product into an overall attitude rating, according to attitude theory (e.g., Fishbein and Ajzen 1975). And how many neighbors should be considered? Whereas considering too few neighbors is suboptimal because of the artifacts it tends to produce, including too many usually adds too much noise. Empirical research thus suggests that between 20 and 50 "taste neighbors" is a good compromise.

Let us exemplify this by predicting (albeit with fewer-than-ideal neighbors) how much Mark would like GBU, using the data from Fig. 12.8. In this case, only two other users have rated GBU who have rated other films that Mark (M) has also rated. These two can be considered as Mark's taste neighbors: his correlation with Nancy (N) is a perfect 1.0, whereas the one with Thorsten (T) is −0.28. We adjust the rating for Mark's general tendency to judge films positively, as expressed in his average rating $\bar{r}_M$ across all films in the data set. The corresponding equations look as follows (with $r =$ rating and $s =$ similarity/correlation):

$$p_{Mark,GBU} = \bar{r}_M + \frac{s(M,N) \times \left(r_{N,GBU} - \bar{r}_N\right) + s(M,T) \times \left(r_{T,GBU} - \bar{r}_T\right)}{|s(M,N)| + |s(M,T)|}$$

$$p_{Mark,GBU} = 3.33 + \frac{1.0 \times (1 - 2.25) + (-0.28 \times (5 - 3.8))}{1.0 + 0.28}$$

$$p_{Mark,GBU} = 2.09$$

In essence, because Mark's movie taste is (much) closer to that of Nancy (who does not like GBU), the user-to-user filtering algorithm does not recommend GBU to Mark, even though Thorsten considers it a masterpiece. Keep in mind that this is a *very* simple illustrative example—correlations are only stable when many more shared observations exist. In reality, reliable results simply cannot be generated with only two neighbors.

User-to-user recommenders have potential for serendipity in that any surprise discovery made by neighbors has a chance to spread to other users via the algorithm. But even using a large database of users and items, the approach is troubled by the cold-start problem: if new users have not rated a sufficient number of items, the system will not be able to produce any meaningful recommendations for them. However, in practice, an even bigger problem for user-to-user filtering is that it is enormously resource-intensive. Calculating the similarities between the users in a large database requires a large amount of computing time (e.g., for a database with only 100,000 users, 100,000 * 999,999/2 = 49,999,950,000; almost 50 billion similarities have to be calculated). As similarities change with *every* single product rating by a user, this strongly limits the usability of this type of system in many entertainment settings, despite the ever-growing power of computing hardware.

Item-to-Item Collaborative Filtering

A practical alternative for collaborative filtering is item-to-item recommenders—they are used by many retailers and websites, including Amazon and Spotify. Item-to-item, or item-based, recommenders use similarities between rating patterns of *products* (or items), rather than between individual users (Ekstrand et al. 2010). If two items are both liked by one group of users and both disliked by a second group, then the items are considered to be similar for consumers preferences. It is then expected that a user who likes one item but does not know the other will also like the latter (because of its preference-related similarity to the first item).

Based on the illustrative data set from Fig. 12.8, we can calculate product similarities between our five example films. How is this done in such an approach? A robust and often-used measure is cosine similarity, but let us use Pearson correlations once more for simplicity. This results in the item-to-item similarity matrix that is shown in Fig. 12.9.[342] The matrix illustrates that, in our miniature data set, users who like NOTTING HILL also tend to

	NOTTING HILL	TERMINATOR 2	PIRATES ...	LOVE ACTUALLY
NOTTING HILL	1	.50	*cbc*	.95
TERMINATOR 2		1	.00	-.57
PIRATES ...			1	-.28
LOVE ACTUALLY				1

Fig. 12.9 Similarity matrix for four movies based on rating patterns
Notes: Authors' own illustration. Values are pairwise Pearson correlations derived from data in Fig. 12.8 on p. 633. cbc = The correlation cannot be calculated because of a constant value for one film. Brands are trademarked.

[342]Please note that we have to leave GBU out because there are not sufficient observations for it.

like LOVE ACTUALLY, whereas fans of TERMINATOR 2 tend to dislike the latter film.

Let us now use the correlations between products/items to predict how much Mark would like NOTTING HILL (= NH), another movie he has not seen; just like we did above with the user-to-user filtering for GBU. We can use the available similarity information and his ratings of two movies, namely TERMINATOR 2 (T2) and LOVE ACTUALLY (LA). This leads us to the following equations (the notation remains the same as above):

$$p_{Mark,NH} = \frac{s(NH, T2) \times r_{Mark,T2} + s(NH, LA) \times r_{Mark,LA}}{|s(NH, T2)| + |s(NH, LA)|}$$

$$p_{Mark,NH} = \frac{0.50 \times 5 + 0.95 \times 1}{0.50 + 0.95}$$

$$p_{Mark,NH} = 2.38$$

Item-to-item filtering would thus not recommend NOTTING HILL to Mark, mainly as a result of him not liking LOVE ACTUALLY—the movie has a very similar rating pattern as NOTTING HILL by the other users in the database. The recommendation level is still higher for NOTTING HILL than his rating for LOVE ACTUALLY—this is because NOTTING HILL also shows a decent rating overlap with TERMINATOR 2, a film which Mark admires.

A major advantage of item-to-item over user-to-user filtering is that it requires much less calculation time and effort. In theory, when a user changes or adds a rating, it might affect the similarity between items, but in practice, at least in constellations with large user and rating numbers relative to the number of products, similarities between products will be relatively stable. The cold-start problem is also at least partly resolved here; a new user can get solid recommendations after having rated very few products himself, because information about the similarity between products has been generated solely from other users' ratings. (In practice, the system can recommend "similar" products once the user names, or clicks on, one product he or she likes.)

However, the cold-start problem still exists for new *products*, because their similarities with other products are yet to be determined. One approach to overcome this issue is to use so-called "content-based" recommendations, alone or in combination with collaborative approaches in hybrid models—we discuss this in the next section. And whereas user-to-user recommenders provide room for serendipity (because of surprise selections of individual "neighbors"), this is much less the case for item-to-item recommenders, which average ratings between products across consumers.

In summary, collaborative filtering approaches are potentially powerful. They align nicely with consumer behavior theories and can make use of rich insights buried deep in large databases. A major stream of recommender research has been dedicated to improve recommendations by reducing overlap between items as well as users, identifying underlying, latent item dimensions and user segments. Such attempts are often associated with the concept of matrix factorization (singular value decomposition, in particular), a method which was popularized by its use in the algorithm than won the "Netflix Prize," a $1 million award given to the team that increased the prediction accuracy of the former DVD-rental service's recommender algorithm by at least 10% (Gower 2014; see Koren et al. 2009 for a general overview).

In any case, all collaborative filtering approaches aim to complete matrices, and the fewer empty cells a recommender matrix has, the more information can be used for generating user-to-user or item-to-item recommendations. The desire to "fill in the matrix" was a driving force behind Netflix's switch from metric rating data to only asking subscribers to provide "thumps up/down" information. The firm presumes the expected increase in the number of ratings (because of the ease-of-use of the new scale) will overcome the loss of detail associated with switching to a less-granular scale (Goode 2017).

Let us now look at the strength and weaknesses of another type of automated recommender systems which trade information on user preferences against product expertise: content-based recommendations.

Content-Based Recommendations

Content-based recommendation systems share their basic logic with item-to-item collaborative recommenders: they try to suggest new products to a user based on the products he has liked before. But in contrast to item-based recommenders, which identify similar products via the overlap in user ratings, content-based recommenders use the inherent attributes of a product, not other consumers' subjective judgments, as the source of similarity. Products whose attributes match those of products liked by a consumer are recommended, while those with different attributes are not.

The approach combines three key steps (Lops et al. 2011). First, the content of products has to be analyzed and the product categorized accordingly. This is crucial because the attributes used for categorization are the ones that define a product's fit with a consumer's preferences. Second, the system needs to learn about a consumer's preferences regarding the attributes. This is usually done based on the consumer's previous behaviors, expressed either explicitly when rating products or implicitly based on the kinds of products that the consumer has purchased or rented (or at least searched for online). The third step then is matching a product's attributes and the consumer's preferences derived in the previous step to enable the actual provision of recommendations.

Content-Based Recommenders: An Example

To illustrate how content-based recommenders work, let us return one more time to our small exemplary data set of five movies. We have now coded the movies with regard to four attributes: their romantic appeal, level of action, average consumer rating, and runtime. We then transformed each score into a 1–5 scale; Fig. 12.10 shows the results for the five movies.

We once more use Pearson correlations to calculate the similarity between movies (although other criteria could also be used), this time based on their attributes. As we did in the case of item-to-item recommenders, we then combine similarity information for two films which Mark has seen and his ratings for them to determine his predicted rating for NOTTING HILL. For this purpose, we choose LOVE ACTUALLY (the most similar to NOTTING HILL, with an almost perfect correlation of 0.99) and PIRATES OF THE CARIBBEAN (PIR; the most dissimilar film to NOTTING HILL—the correlation between

	Action content	Romantic appeal	Consumer rating	Runtime
NOTTING HILL	1	5	3.50	3.85
TERMINATOR 2	5	1	4.25	4.26
GBU	4	1	4.45	5.00
PIRATES ...	5	2	4.00	4.44
LOVE ACTUALLY	1	5	3.85	4.19

Fig. 12.10 Matrix with attribute ratings for five movies
Notes: Authors' own illustration. Values are our own codings of the five movies, transformed into a 1–5 scale (where 1 is the lowest and 5 the highest level of the respective attribute). Brands are trademarked.

the two films is a strongly negative −0.82). Using the same equation as we did for item-to-item recommenders results in the following[343]:

$$p_{Mark,NH} = \frac{s(NH, LA) \times r_{Mark,LA} + s(NH, PIR) \times r_{Mark,PIR}}{|s(NH, LA)| + |s(NH, PIR)|}$$

$$p_{Mark,NH} = \frac{0.99 \times 1 + 0.82 \times 2}{0.99 + 0.82}$$

$$p_{Mark,NH} = 1.45$$

This approach results in a predicted rating of 1.45, which (again) suggests that recommending NOTTING HILL to Mark might not be a great idea.

Challenges for Content-Based Recommenders: The Critical Role of Attributes

Our approach might look intuitive; it shows that content-based recommenders largely circumvent the "cold-start" problem that recommenders usually deal with. As with item-to-item recommenders, only few consumer reactions are needed to find similar products to the ones a consumer prefers. And, as similarities between products are based on expert codings (rather than consumer reactions) in this case, no cold-start issues hamper the integration of new products. In addition, content-based recommendations can be transparent, because the reasons for suggesting a product can be explicitly provided, something that is hardly possible with collaborative filtering, where the only explanation can be that an item is recommended because either "people like you" have rated it highly or because you like "similar" products.

But our example also points to potential challenges for content-based recommendations, most of which have to do with the selection of attributes and their integration. The usefulness of content-based recommenders is strongly influenced by the attributes which are used to determine similarity among items: Even if you agree that the attributes we selected in our example are relevant for consumers' preferences, it seems similarly problematic to believe that this set of attributes is comprehensive: we discuss *many* others in this book that influence consumers' decision making, but leave them out here.

---

[343]Please note that to account for the negative correlation between Notting HILL and PIRATES OF THE CARIBBEAN, we transformed Mark's evaluation for the latter film from 4 (the second *best* category) to 2 (the second *worst*). By doing so, we assume that a strong *dis*similarity between items has the inverted effect of a strong similarity (i.e., "if a product is highly dissimilar to one I like, I will not like it").

The closer chosen attributes link with consumers' preferences, the more powerful recommendations will be. In our example, we use highly aggregated genre labels such as romantic appeal and action content. Doing so results in predictable ratings: If a consumer has liked one romance movie, he or she will receive recommendations for other films from the same genre. That is why in practice, content-based recommenders often struggle with the serendipity problem, leaning toward satiation—users' lust for sensation is not adequately addressed by them.

But aggregated attributes not only bore us, but they also often simply miss the point why consumers like a particular piece of entertainment. A lot of people like Clint Eastwood's western UNFORGIVEN, not because of a general preference for westerns, but for its star. So recommending a western with John Wayne would only disappoint them. Some of those who like the film because of its star might not be content with other films Mr. Eastwood has directed and appeared in, such as BRONCO BILLY and MILLION DOLLAR BABY. And others may like the film because of its subtle connections to the Italo westerns directed by Sergio Leone and Mr. Eastwood's personae of the Man with No Name in films like A FISTFUL OF DOLLARS—whereas these consumers reject most other westerns and other films by Mr. Eastwood.

In addition to such stable preferences, contingencies also influence what we like (or don't like) at a certain moment: maybe you are among those who can work most effectively while listening to low-key classical film music? Whether a streaming service's content-based recommender can offer a fitting program depends on its ability to separate "low key" soundtracks from others which might be popular, but distract us from our work—take John Williams' spectacular INDIANA JONES theme as an example. In other words, the crucial question is what a streaming service understands when we tell it that we "like" a song we just heard—the value of the information that the service "will play more song like this one" depends on how well it understands *why* we like the song.

To address the issue of fine-grained, contingent preferences, companies such as Netflix and Pandora have invested enormous amounts into a highly differentiated coding of their products. Netflix internally has called its approach "Quantum Theory," which encompasses almost 77,000 unique labels of "micro-genres" for their content. Employees who assign tags to films and series receive a 36-page training document (see Madrigal 2014—a fascinating and highly recommended read, by the way). The firm considers conventional genres, in the words of its vice-president Todd Yelling, to be "just wrappers." It is the micro-genres which are intended to recommend films that are similar to the ones a user loves.

In some ways, this approach even aims to *add* serendipity to content recommenders: Netflix wants to "break these pre-conceived [genre] notions and make it easier for [users] to find stories they'll love, even in seemingly unlikely places" (quoted in *Netflix* 2017). Similarly, music streaming service Pandora has musical experts analyze each song using up to 450 distinct musical facets—their very own "Music Genome Project" (Lasar 2011). By going into this level of detail, the firms hope to minimize the limitations of content-based recommendations while harvesting their strengths.

Let us end this section by naming some further challenges for content-based recommenders. How did we determine the values of the attributes for a given movie in our example? Whereas "objective" information exists for some attributes (such as the films' runtime), this is not the case for others, such as consumer ratings (which we took from the IMDb, but which may differ on other websites) or genre: how romantic is PIRATES OF THE CARIBBEAN *really*? We gave it a 2 on our 1-to-5 romance scale; but what would *you* rate it? It seems also questionable to assume that attributes are of equal importance, and that this attribute importance is the same across consumers—an issue that can be addressed by adding weighting parameters.

Hybrid Approaches: The Best of All Recommender Worlds?

Scholars have also tried to integrate the different recommender approaches, combining their respective strengths while minimizing their limitations. Collaborative filtering suffers, to varying degrees, from the cold-start problem of handling new users and items and the lack of detailed explanation. And content-based recommenders do not make use of the information that is available about the other users in a database and also often have problems in offering fresh sensations.

Popular ways to combine the different approaches include the following (see Burke 2002 for more details):

- *Weighted hybrid recommenders*. The ratings of multiple recommenders for an item are combined via a weighting routine, producing a weighted average recommendation.
- *Switching hybrid recommenders*. The ratings provided to the user come from different recommender systems, which are chosen depending on the characteristics of the situation (e.g., the number of existing ratings for an item or by a user). A challenge here is to determine the "right" switching criteria.
- *Mixed hybrid recommenders*. The user is provided with multiple ratings for a product generated by different recommenders. Such an approach carries the risk of overloading the user with "unnecessary" information. A variation of this approach is to use multiple recommenders, but to

present their respective results not jointly for a specific product, but separately for each recommender (first the top recommendations by content-based recommenders, than by item-to-item filtering etc.). This allows the consumer to decide which recommender to use. Such an approach has been employed by Netflix (Gomez-Uribe and Hunt 2015).

- *Sequential hybrid recommenders.* Different ways have been suggested to combine several recommenders in a sequential order. For example, the output of the initial recommender (such as a content-based approach) is used as input for a second recommender (such as a collaborative filtering approach).

Are such hybrid treatments of recommendations effective? Several studies point to the potential gains of hybrid recommenders; for example, in Marx et al. (2010), we combine content-based and user-to-user collaborative filtering using switching criteria, which increases the fit of the pure content-based recommender by 16% and the fit of the collaborative filter by 10%. But those who offer recommenders always have to weigh the benefits in effectiveness against the higher complexity, longer processing times, and higher costs of such approaches, which limits their usefulness in many practical settings.

So far, we have focused on the algorithms that create recommendations, the backbone of automated recommender systems, if you will. In the following section, we will show that the value of such recommendations for both consumers and those who offer them is influenced also by several other factors.

Recommenders are Way More Than Algorithms: Beyond Matrix Completion

> "Predicting movie ratings accurately is just one aspect of [Netflix's] world-class recommender system."
>
> —Amatriain *and* Basilico *(2012)*

Algorithms are essential for providing powerful recommendations, and the clear majority of scholarly research on recommenders has focused on the development of "better" algorithms, i.e., those that predict user preferences with minimal error. But maximum accuracy at the matrix completion task is far from the only thing that matters for recommenders to be successful (Jannach et al. 2016).

Although it paid a million dollars for it, Netflix has never implemented the algorithm that won the Netflix Prize. The reason is that the firm sees higher gains from other parts of the recommender challenge: "the days when stars were the focus of recommendations [at Netflix] have long passed" (Gomez-Uribe and Hunt 2015).[344] Recommender research and practice have recently turned toward other aspects of recommenders and their implementation at the user interface. Let us name some of those issues that recommender providers in the entertainment world also have to deal with—and the relevant knowledge that scholars have gathered.[345]

Contextualization

Whether we like an entertainment product depends not only on the product itself, as recommenders usually imply. Instead, our liking is influenced by the context in which we use a product. Adomavicius et al. (2011) provide sound arguments that the following specific context dimensions are of particular importance for the enjoyment we derive from an entertainment product—and that taking them into account when making product recommendations can strongly increase the recommendations' usefulness.

Physical context. Our consumption patterns differ with time and place. For example, the music we like to hear when we wake up usually differs from what we like during the day or when exercising. A consumer who generally likes a classical composition might find its recommendation inadequate in the morning hours. Also, whereas a consumer might love epics from David Lean or Paul Thomas Anderson for long-weekend nights, he might prefer much shorter programs for weekday evenings. We know that enjoying the four hour-long LAWRENCE OF ARABIA on a Wednesday evening could be a treat, but that we might pay a high price for the lack of sleep during the Thursday morning's work meeting.

Historical context. Any evaluation of an entertainment product is part of a sequence of consumption acts; there is always a previous novel as there is always a next song. The sequence of events influences a consumer's reaction to a particular product, such as when the similarity of previous experiences triggers satiation with a product. After having experienced five sequels that

[344]Let us add that another reason for Netflix's decision to not implement the "best" algorithm was that the firm changed its business model shortly afterward, from renting DVDs to streaming films and shows. Whereas recommenders have remained crucial for the firm, the new business model introduced a different usage context for recommenders; the now immediate link between recommendations and consumption made other aspects more relevant than prediction accuracy, some of which we discuss in this section.

[345]Jannach et al. (2016) provide a highly informative overview about these non-algorithmic challenges.

satisfy a consumer's desire for familiarity, he or she might have a stronger appetite than before for new sensations from more-original content and would be thankful for being offered corresponding recommendations. Also, dissatisfaction with previous products can lower the expectation level for the next one. The fact that the second and third RAMBO films were such disappointments might provide an explanation for the relatively positive reception of the fourth entry in the RAMBO series.

Emotional context. Managing our mood and emotions is often a motive for consuming entertainment, so the value of recommendations will also vary with our emotional needs and desires at a given point in time. Anger requires a different soundtrack than happiness, sadness demands other films or books than does being in love. Our entertainment choices differ in response to certain emotional states, but standard recommenders have no room for our emotions.

Social context. Are we going to the movies alone or with a friend? Earlier in this book, we discussed how social context influences our decision-making process. Recommenders would benefit from accounting for social context instead of assuming we always act in isolation. Recommending THE SALVATION, a Danish homage to Leone's westerns *dall Italia*, to a dedicated fan of the latter films makes sense in general—but much less so if the person is going on a date with someone who is "allergic" to subtitles and westerns.

Recommendation scholars have suggested considering such context information when designing recommenders, calling for context-aware recommender systems ("CARS"; Adomavicius et al. 2011). How can such contextualization be achieved? One approach that has been suggested is to use contextual filtering: in every situation, recommendations are generated only based on similar situations; all other ratings are filtered out and ignored (i.e., pre-filtering). For example, for someone who wants to watch a movie on Saturday with a group, only those ratings collected for the weekend and/or for group consumption are used for recommending films. Alternatively, results can be filtered post hoc based on context criteria (i.e., post-filtering).

A second approach for "CARS" is contextual modeling. Here, the recommendation function includes contextual information by using statistical techniques, such as decision trees and hierarchical Markov chains. Netflix makes use of this approach, estimating Markov chains. According to Neil Hunt, the firm's former Chief Product Officer, Netflix records "what you have seen first, then what next and what after next. Then we compare this with other people's behavior and calculate 'transition probabilities'" (quoted in Brodnig 2015). Both approaches require extensive data collection, as every cell in the matrix needs to be tied to the specific context in which it was entered. But subscription services such as Netflix and Spotify have access to such data, as they know the timing of when a consumer watched a movie or listened to a song which he then rated.

Also, content-tagging can help (is a song sad, or angry, or happy?), which can then be linked with consumers' moods in a given situation via content-based recommenders, if consumers are willing to reveal their mood. Spotify, for example, has asked listeners for their current mood (Jannach et al. 2016). Another way to make use of content tags is to let listeners self-select mood-adequate titles by offering "context playlists," as Amazon Music does with lists such as "in love" and "relaxed breakfast." Doing so also enables a content-based recommender to address the consumer's desire for "low-key" film music in a working situation; a challenge we have mentioned earlier.

For addressing the social context of a recommendation, scholars have proposed group recommenders that combine the preference data of the different members of a group (O'Connor et al. 2001). Our own suggestion is a two-step approach, where in the first step the individual preferences of each group member are used to generate predicted ratings of potential product alternatives, which are then, in step 2, transformed into collective ratings for the group as a whole, essentially using a weighing mechanism (Hennig-Thurau et al. 2012a).

Does such an approach offer "better" recommendations than a standard one? We tested this approach with two lab experiments with 460 consumers who had to actually watch a recommended film; the recommendations were created with a user-to-user collaborative filtering recommender based on about 4.8 million ratings from a popular German movie website. The results provide evidence that a group recommender can indeed outperform a "single" recommender: the group's satisfaction with the recommended film (the average of both group members) was more than one point higher for the group-recommender condition (on a 0–10 point scale) than when the group watched the film recommended to the "agent" (who made the choice for the group) in isolation. Not surprisingly, most of this effect came from the "partner's" satisfaction with the recommended film.[346]

[346]In the described scenario, the group members had to watch the film recommended by the recommender system. When we allowed the group members to choose freely either to follow the recommendation or to "overrule" it, the group recommender still outperformed the single recommender in terms of group satisfaction with the film. But this effect held only when the group members liked each other and the agent followed the recommender and chose a compromise film over one that maximized his own preferences.

Design and Interaction

Scholars have also highlighted the importance of the usability of a recommender system, as determined by the design of the interface (Jannach et al. 2016). This affects the input of customer information, but also the presentation of recommendations.

Regarding input, the core challenge is to let the users of the system enter ratings easily and intuitively. In the case of "passive" data which derives the consumer's preferences from his or her usage behavior (does a user watch the whole movie? Does he or she listen to the whole song?), no action is required from the consumer, although *a lot* of artificial intelligence is needed to transfer behaviors into preferences. For "active" data, the rating scale matters, as does the input mechanism that must account for consumers' need for convenience (as an example, see Netflix's recent scale change from 5-star to binary; Goode 2017). In this regard, voice-based systems, such as Amazon's Echo device, might provide new ways to increase the usability of recommender systems.

With regard to the presentation of recommendations, it is mostly about how intuitive the results are presented to the user by the system. To address the varying preferences of its consumers for the use of recommenders, Netflix retired it single-recommender approach and substituted it with a collection of different algorithms "all serving different use cases" (Gomez-Uribe and Hunt 2015).

Trust

Many consumers are skeptical about artificial intelligence-powered solutions, so a major hurdle for every recommender system is to earn the trust of its users. Trust involves competence *and* benevolence (e.g., Sirdeshmukh et al. 2002), and both are relevant here. Are the recommended products truly the best for the user, i.e., is the recommender algorithm smart enough to find alternatives that deserve his or her time? One way to demonstrate a recommender's competence is to provide explanations for the recommended titles (e.g., "Titanic is recommended to you because you like big-budget Hollywood movies and films with Leonardo Di Caprio"; see Marx et al. 2010 on the empirical power of such explanations).

And is the platform provider really acting in my best interest, or are recommendations given in a way that primarily benefits the platform? With its increasingly high spending for original productions, Netflix has begun to allocate a disproportional share of its home screen to those films and series, even when recommendation scores are rather low (Bishop 2017). Such a practice carries the risk of threatening consumers' trust in the benevolence of the service's recommender system.

How "Good" are Recommender Systems?

We have seen that the other kinds of "earned" media that provide consumers with information about a product's quality actually influence the consumers' decisions. How about recommender systems? Do they offer consumers insights that they find valuable? Let us take a look at whether the usage of recommenders helps consumers to make "better" decisions and the role of recommendations for such decision making and product success.

A rare study that compares the predictive performance of recommenders to situations in which consumers have no access to them is Krishnan et al. (2008). In this study, the scholars compared the accuracy of film ratings for 14 consumers by a recommender system with the predictions of 50 human raters (who had access to each consumer's past ratings). The recommender, a item-to-item collaborative model of the MovieLense recommender that incorporated 15 million ratings at the time of the study, predicted ratings with a 19% lower average error (MAE) than the human raters—but the human raters' predictions were biased by a number of outlier predictions, and the predictions of the recommender system were closer to actual ratings for less than half of the 14 consumers.

In our own exploration of the performance of group recommenders, we compared the predictions of our user-to-user collaborative filtering approach with a condition in which the "agent" (i.e., the group member who made the choice) lacked access to a recommender system (Hennig-Thurau et al. 2012a). In the latter condition, the agent had to choose a movie based solely on movies' titles, countries of origin, main genre, and a thumbnail poster (with no access to the Internet or word of mouth). We learned that the chosen (and watched) movie was more positively evaluated by the group when group members liked each other and when the agent held a positive attitude toward recommender systems, in general. But for a standard (single) recommender, we found no difference in movie satisfaction: a result which illustrates that the power of recommenders should not be taken for granted across contexts and conditions.

Meiseberg (2016), in her study of book success, did not look at the quality of recommendations or users' satisfaction with recommended products, but their practical impact on decision making: does a recommendation for a product on Amazon's website influence the product's sales at Amazon?

In her analyses of a large book sample (in which she controlled also for a large number of other "success drivers"), she used the sales of all books in which a certain book title was recommended as measure of recommendation intensity for the title. Her results demonstrate that, at least in the context of her study, product recommendations do indeed trigger book sales—and do so to a remarkable degree: correlations are higher than those for word-of-mouth and prices, and comparable to the book title appearing on TV. The recommender variable has the strongest effect for the Top 20 selling books in the sample, followed by the lowest 20% quantile.

This impact of recommenders on product choices is also reflected in Netflix's disclosure that only about 20% of the hours consumers use the service result from searches on the site, while about 80% are inspired by popularity rankings (see our section on "success-breeds-success" effects elsewhere in this chapter) or recommenders (Gomez-Uribe and Hunt 2015). Among those 80%, the share of personalized recommendations that result in the consumer watching a film or series (the "take-rate") is, for attractive titles, up to almost four times higher than for recommendations based on popularity. The two Netflix managers Carlos Gomez-Uribe and Neil Hunt also claim that personalized recommendations at Netflix lead to "lower subscription cancellations rates" (p. 13), but they do not provide empirical details or evidence.

Although recommender systems are operated by a platform, they still use data that are "earned" from consumers, just like word of mouth and action-based cascades. In the final sections of our discussion of entertainment communication, we will now look at the role of other stakeholders of entertainment, namely the professional critics and industry peers (who hand out awards) whose judgments might also influence the success of an entertainment product.

Professional Reviews

> "I heard that some studio insiders want to hold off critic screenings until opening day or cancel them all together."
>
> —Anthony D'Alessandro *(2017) on film managers' reaction to the disappointing opening weekend results for* Baywatch *and the fifth* Pirates of the Caribbean *movie*

> "At his very best, a critic is a cheerleader for films that need support."
>
> —*Film critic* A. A. Dowd *(2015)*

One of the most obvious consequences of entertainment's difference from other products is that media outlets employ journalists to write about them.

It is entertainment's cultural character that makes it so media-worthy, granting films, books, music, and video games space on the pages and websites of prominent newspapers and magazines. From this "media-worthiness" of entertainment, the question arises regarding what role the judgments of professional reviewers play for the commercial success of the products they review.

Whereas this might be an age-old question, it is still highly controversial. Entertainment producers often blame critics' negative comments for the failure of their products—see the introductory quote by Mr. D'Alessandro above. The second quote by Mr. Dowd suggests that critics themselves are much more skeptical regarding the impact of their own work. And they have been that skeptical for quite a while: when we asked Peter Körte, a leading German film critic, what he thinks about the influence of his own writings on audiences, he quoted what French reviewer André Bazin wrote about his occupation in the 1950s: "Film criticism is when you stand on a bridge and spit into a river" (Körte 2009, p. 194).

In this section, we turn to *Entertainment Science* scholars to learn about whether critics influence consumers and, eventually, product success. As with many other questions we have addressed in this book, finding a credible answer econometrically is far from trivial, requiring care and expertise. Critics' judgments, like consumers' word of mouth, overlap with an entertainment product's "true" artistic qualities, so that correlations must not be confused with causal effects. But scholars offer ambitious attempts to isolate such causal effects (i.e., critics as "influencers") from spurious ones (critics as "predictors"), whose main insights we will summarize here. We will then again go beyond "average effects," shedding light on moderators and related issues.

Review(er) Effects on Consumers: The "Influencer Versus Predictor" Controversy

Expert reviews are a source of third-party evaluative information about the quality of entertainment products. Traditionally such information could be accessed by consumers only via newspapers and magazines on a review-by-review basis, and the review literally disappeared the day after it was published. But the Internet has changed the availability and format of reviews. The Internet has created an archive of reviews which, if they are not locked behind a publication's pay wall, can be accessed at any time, independent of their publication date. The Internet has also given us aggregator websites and services such as Rotten Tomatoes (containing reviews of movies and TV shows, founded in 1998) and the more selective Metacritic (for movies, TV shows, games, and music; launched in 1999) that provide consumers with links to individual reviews as well as summary signals, such as mean scores and the number of reviews. Take the movie ALIEN: COVENANT: at the time of this writing, Rotten Tomatoes reports a mean rating of 6.3 (out of 10) and a "Fresh" score of 68% based on 324 different reviews, whereas Metacritic has calculated a weighted average of 65 (out of 100) from 52 reviews. Both aggregators are owned by entertainment conglomerates; Rotten Tomatoes is a joint venture of Universal/Comcast and AT&T/Warner, and Metacritic is part of CBS/National Amusements.

Professional critics have been called "institutional gatekeepers" because they screen the entertainment products that are offered by producers, "winnowing them into a much smaller number of select goods from which everyday consumers then choose" (Hsu 2006a, p. 468). But do expert reviews actually *influence* consumers, and to what extent? When scholars have run regression models with entertainment product success as dependent variable, they often include a product's average review score from critics, as calculated by one of the aggregators.

The estimated parameter in these models is usually positive and significant, which demonstrates that the quality assessments of professional critics go along with a product's commercial performance on average.[347] However, as we have stressed throughout this book, associations are not necessarily causal, and providing evidence that reviewers' tastes are *associated* with what consumers buy should not be confused with evidence for a causal effect by the reviews themselves.

Initial Insights: The Eliashberg and Shugan Study

The first scholars who dived deeply into the issue of review effects were Eliashberg and Shugan (1997): they coined the terms "influencer effect" (for a causal link between critics' reviews and a movie's box office performance) and "predictor effect" (for spurious correlations in which the statistical relationship is not causal, but in which positive reviews *indicate* a film's success because the

[347]Examples of studies that find professional reviews to be positively linked to entertainment product success as part of a regression-type analysis, but do not address the more complex issues we discuss below in their empirical approach, are Clement et al. (2014) and Elberse and Eliashberg (2003) for films, Marchand (2016) and Cox and Kaimann (2015) for games, Clement et al. (2007a) and Schmidt-Stölting et al. (2011) for books, and Lee et al. (2003) for music.

product quality underlying the positive reviews would also drive success).[348] They conducted the first empirical attempt to separate those effects by explaining the weekly North American box office of 56 films released in 1991–1992 (i.e., in the pre-Internet era) with the percentage of positive and negative reviews via regressions. Eliashberg and Shugan found that the regression parameters for positive and for negative reviews are only significant from week 5 onward after a movie's release, but insignificant in the first four weeks.

They considered this finding as support for the dominance of the "predictor effect," because a causal "influencer effect" would be reflected in higher parameters in the earlier weeks (when review information is still fresh, as most reviews are published around the release of a film). But there are alternative explanations which Eliashberg and Shugan's approach does not rule out: think of early audiences often being fan boys who might find reviews less informative than later moviegoers. Also, quality information, the core of any critical review, might have needed more time back then to spread among consumers (as we have shown in our discussion of the timing of word of mouth earlier in this chapter). Basuroy et al. (2003) offer another explanation: maybe the small data set is to be blamed. When they replicated Eliashberg and Shugan's approach for a larger set of films from the same time period, they found significant review parameters for *all* weeks.

What We Have Learned About Review Effects Since Then

From the foundation of this seminal work, other scholars have developed alternative, more sophisticated empirical designs and used even larger data sets in attempts to isolate reviews' spurious "predictor effect" based on movie quality from a truly causal impact.

Also with movies as product category, Reinstein and Snyder (2004) use a "difference-in-differences" approach to circumvent the problem of a spurious correlation. To make use of the *timing* of a review, the authors studied whether films that received a positive review during their opening weekend experienced a higher opening box office than other films (those that were reviewed later and/or received a negative review). The scholars' measure of professional ratings was a single review source: the ratings by prominent

[348]See our discussion of the link between product quality and commercial success in entertainment in Chapter 7 about the "entertainment experience."

U.S. reviewers Gene Siskel and Roger Ebert in their television show SISKEL & EBERT/AT THE MOVIES. Their data set encompasses 609 films released in North American theaters during the long period of time (from 1982 to 1999) that Mr. Siskel and Mr. Ebert's show was on air.

Using an OLS regression (in which they control for distribution size and use movie guide book author Leonard Maltin's much later published ratings as a proxy for the films' "true" artistic quality), they find that films that got "Two thumps up!" from the reviewers during their opening weekend generated an average of 28% higher box office over their first three days than later- and/or negatively reviewed films, despite the fact that the TV show was only aired on the Saturday of a film's opening weekend.[349] Reinstein and Snyder's results indicate that the review effect is non-linear: if only one of the two reviewers gave a positive rating, the effect drops by more than half and becomes insignificant.

One limitation of their study, beyond considering only a single, most-popular, no-longer-existing review source, is that movies of higher quality had a higher chance of being reviewed on their opening weekend by Mr. Siskel and Mr. Ebert, which could have blurred the causal and spurious effects.[350] Whereas Reinstein and Snyder try to reduce the effect distortion by including the "true" quality variable in their regression, we used a different approach when studying the issue ourselves. In an investigation of all 1,370 fictional films that were released in North America from 1998 to 2006 (and generated at least $1 million at the box office), we isolated the unique part of the reviewers' ratings that was not the same as the consumers' quality perceptions (Hennig-Thurau et al. 2012b).

We did this by running a separate ("auxiliary") regression in which we explained the average reviewer rating of a movie (from Metacritic) with the movie's consumer ratings (from Netflix and Yahoo). We then used the residual (the part of the Metascore *not* explained by consumer ratings) to explain movies' box office, along with a large number of "success drivers" including advertising and buzz. In one regression, the opening weekend box

[349]The regression also included one variable that measured whether a film got a positive judgment from at least one of the two reviewers on its release weekend. Let us note that the confidence level for the finding was only 90%, not the usual 95%; this also applies to most of the subsample-related findings in Reinstein and Snyder's study we report later in this section.

[350]The higher "inherent" quality of the early reviews could have erroneously boosted the review effect. The authors empirically show that such a quality effect indeed exists in their data set in a separate probit regression.

office served as the dependent variable; in the other, the dependent variable was the box office generated in the weeks *after* the one in which the movie was released. We find that the unique part of professional reviews indeed impacts product success, but only does so later on in a movie's theatrical run. Specifically, the effect of reviews is not significant for the opening weekend, but positive and significant for the later box office.

And how strong is the latter effect? In our data, a ten-unit increase in professional reviews (e.g., from a Metascore of rating of 60 to 70) corresponds with a 15% increase in box office after the opening week. When including a squared rating parameter in the regression, our findings also corroborate the non-linear nature of review effects suggested by Reinstein and Snyder's results. Figure 12.11 shows that highly positive reviews provide a disproportionate benefit to a film's financial performance.

Whereas all major reviews for movies are published at the same time (around a film's release date), reviews for other entertainment products such as books tend to be published over a longer period. Sorensen and Rasmussen (2004) make use of this characteristic by linking book reviews by the New York Times with sales changes in the weeks following the review's publication.

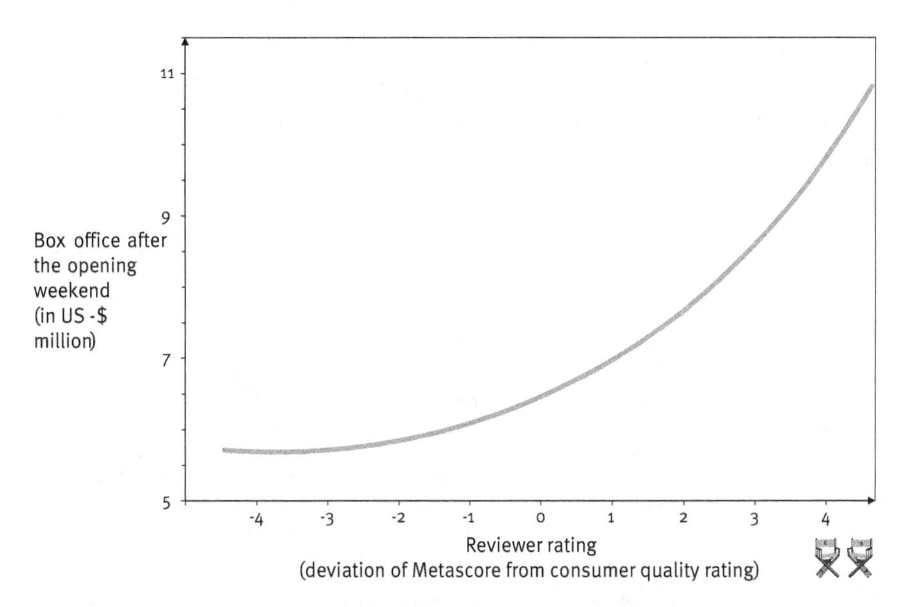

Fig. 12.11 The non-linear relationship between professional reviews and the box office after the opening weekend

Notes: Reprinted with minor modifications with permission by the Journal of Cultural Economics/ Springer from Hennig-Thurau et al. (2012, p. 272). We transformed box office values into raw format from their logarithmized form (which was used in the estimations).

Investigating weekly sales numbers for 175 hardcover books for 2001–2003, the scholars find, via a panel regression in which they control for a book's sales in a previous week (but not for other book variables, such as star author status), that a positive book review in the New York Times causes an enormous 63% increase in next-week sales in the U.S.

Altogether, the skepticism of critics regarding any impact of their writing, as expressed at the beginning of this section by A. A. Dowd and Peter Körte, appears inappropriate. In contrast, entertainment producers' concerns seem justified: at least on average, the ratings of an entertainment product by professional critics have the power to influence its success.[351]

Before we now dig deeper and look beyond average effects, let us address one related issue: can there also be, above and beyond the causal effect of review valence we have looked at here, a publicity (or awareness) effect of professional reviews, in a "all-reviews-are-good-reviews-even-bad-ones" sense? Sorensen and Rasmussen (2004) indeed find that, at least for the set of books they examine, *negative* reviews also have a *positive effect* on sales, which they attribute to the awareness such mentioning raises. But this positive effect is only half as strong than the valence effect and can thus only partially mitigate the negative consequences of being negatively reviewed. For movies, studies which find such a "publicity effect" do not control for the "general" popularity of a film (which exists because of advertising and/or buzz). When we do so, we don't find any success-enhancing effect of the number of reviews written about a film (Hennig-Thurau et al. 2012b).[352]

---

[351]Let us add that this conclusion is also consistent with insights from lab experiments, in which scholars expose consumers to reviews and measure how that impacts them; for example, Wyatt and Badger, as early as 1984, reported a between-subjects design in which consumers who were shown a negative review had a 30% lower interest in seeing the reviewed movie. Whereas such experiments illustrate the value potential of reviews, their contribution is limited by their artificial nature. Media managers are interested in learning about the impact of a specific piece of information in a world in which many other sources of information compete for consumers' attention, whereas lab studies mostly account for only a subset of those alternative sources. In Wyatt and Badger's (1984) study, the professional review was the *sole* information source; in a later study, the authors provided the participants also with information about a film's plot and its stars in a within-subjects experiment with about 200 consumers (Wyatt and Badger 1990). Here they found a 12% increase in interest in a film when shown a positive review, and a 7% decrease when shown a negative one. Separately, the results by Chen et al. (2012) suggest that review effects are not limited to (post-opening weekend) revenues, but can also escalate to the producing studio's stock price—their event study of the professional reviews of 220 movies from 2005/2006 finds, on average, "abnormal returns" of the studio's stock of 12% for above-average reviews and of −24% for below-average reviews. In other words, investors react to critical reviews, and do so stronger to bad reviews than to good reviews.

[352]Please also note that several reviewed books are of a rather "niche" kind, with small advertising budgets, so that awareness effects appear rather more feasible for them than for mainstream-targeted movies or other forms of entertainment (which can be expected to have a higher awareness "baseline."

Toward an Even Richer Understanding of Professional Reviews: Moderators and Mediators of Their Impact

Scholars have also made steps toward a finer-grained understanding of how reviews affect product-level success for entertainment products, not just averages. Entertainment managers and journalists often argue that products differ in their degree of "review-proneness," and we can provide some statistical evidence regarding what kind of products fall into that category. Beyond product characteristics, studies also shed light on consumers, macro-economic conditions, and the effects of reviews on *distributors*.

Product Factors as Moderators

Scholars have analyzed differences in review-proneness for commercial (mainstream-targeted) versus independent (niche-targeted) products, as well as between genres.

Regarding commercial versus independent products,[353] Gemser et al. (2007) argue that audiences of independent products may focus more on the product's artistic achievements (which are what reviews are mainly about), while commercial products will be judged mainly based on other, less artistically ambitious criteria.[354] Also, commercial products, with their attached stars and strong advertising, offer consumers more quality signals (or quasi-search attributes) than independent products do, which should further limit the importance of reviews for them.

Empirical results confirm such a logic. When Reinstein and Snyder (2004) split their data set of films into wide-released and narrow-released ones (in the study of TV reviews by critics Roger Ebert and Gene Siskel), they find that the review effect exists only for movies that were released narrowly. For these films, a "Two thumps up!" is associated with almost 45% higher opening weekend revenues, whereas *no* effect is found for wide releases. In our own study of movie reviews, we also see that professional reviews affect the box office of narrow releases more strongly (Hennig-Thurau et al. 2012b), and Lampel and Shamsie (2000), for a data set of 409 films from 1991 and 1992 which they analyze with OLS regression, report a

[353]We discuss these fundamental product types in our chapter on entertainment business models.

[354]See also our discussion of how the different levels of taste overlap between professional reviewers and mainstream versus niche audiences in the chapter on entertainment product characteristics.

negative interaction effect between a review score and its distribution size on the total North American box office.

In addition, we also find differences for other factors that vary between commercial and independent products: reviews are less influential for films that feature stars, are sequels, and have strong advertising and high buzz. Finally, our results also show that the effect of reviews on later box office is lower for films that open strongly—we assume that early success stimulates action-based cascades, which then crowd out the information offered by professional reviews.

Scholars have also compiled evidence in the context of movies that some genres are affected more strongly by professional reviews than others. Both Reinstein and Snyder (2004) and our own work (in Hennig-Thurau et al. 2012b) support the industry's assumption that dramas are more review-dependent than other genres: Reinstein and Snyder found a 90% boost caused by "Two thumps up!" reviews for this genre. We assume that audiences trust reviewers more when it comes to judging quality as a matter of dialogue than of spectacle; Reinstein and Snyder find no effect of reviews for action films. We find that comedies benefit from a high Metascore even more so than do dramas. Why could this be? Humor is a highly idiosyncratic matter, with the benefits consumers derive from watching a comedy being highly uncertain, so it's the professional reviewers to the rescue for many of us. But not all reviewers provide equally valuable guidance when it comes to deciding what kind of comedy is fun: Reinstein and Snyder do not find a comedy's opening success to be impacted by Mr. Ebert's and Mr. Siskel's ratings.

These insights are a solid start for determining what characterizes a "review-prone" entertainment product. But the product variables studied so far by scholars still encompass quite high levels of heterogeneity—there are all kinds of commercial products, as there are all kinds of dramas. When the movie CHILD 44, a dramatic thriller set in a dystopian Stalin-led Russia, performed poorly at the box office, industry analyst D'Alessandro (2015b) blamed its negative reviews, arguing that "[w]hen you've got spectacularly grim subject matter, you need strong critical support." Are there indeed certain stories and (sub-)genres which depend on reviewers in particular? And which ones? Hopefully, future *Entertainment Science* studies will shed additional light on the matter.

Consumer Factors as Moderators

Different entertainment products appeal to different consumer segments, which might vary in their sensitivity toward professional reviews. One approach to study such consumer factors with secondary data is to code products by their target groups. We have done this for our data set of 1,370 films, using tags by the (no longer available) movie recommender website Jinni.com. The results show that films targeted at families are systematically less impacted by professional reviewers' ratings than films with other target groups; this effect is found only after the opening week.

More detailed light on the issue is shed by a team of Canadian scholars who collected consumer-level data, again in the context of movies. d'Astous and Colbert (2002) surveyed a convenience sample of 120 Canadian student moviegoers and tested a number of hypotheses regarding their usage of professional movie reviews when making movie-going decisions. The scholars' OLS regression results suggest that professional reviews are used more by consumers who (a) have high knowledge about film, (b) are, in general, more susceptible to the social influence of others, and also (c) have lower levels of self-esteem. And the results of a replication study for which the scholars collected data from about 450 consumers in Austria, Colombia, and Italy (d'Astous et al. 2005), propose that these findings can be generalized beyond Canada.[355]

Macro Factors as Moderators

Are there more general factors, beyond characteristics of products and consumers, that alter the impact of professional reviews on the success of entertainment products? Dhar and Weinberg (2015) use a data set that covers almost 1,700 movies over a long time frame—between 1983 and 2009—and test whether the macro-economic conditions at a given time influence the role that professional reviews have for film success.

Using a consumer sentiment index as a measure of the macro-economic conditions and estimating results via a GMM regression,[356] the scholars interpret their results as support for their expectation that consumers are more sensitive toward professional reviews in times of economic crisis (when spending decisions are made more carefully). In contrast, in times in which consumer sentiment is high, they find that people tend to be less critical when making entertainment choices.

---

[355]For their Canadian sample, d'Astous and Colbert (2002) also report a *negative* impact of film involvement on the usage of professional reviews. We suspect that this might be an artifact though based on the concept's overlap with film knowledge—the effect of involvement is *positive* in both settings in which knowledge was not also included in the regression model.

[356]The scholars run separate equations for box office—the "demand side"—and the number of theaters in which movies are shown—the "supply side."

But Dhar and Weinberg also offer a disclaimer: the small size of the interaction parameter they estimate suggests that the change in consumer behavior in less-favorable times is hardly of managerial concern, at least when it comes to the role of reviewers. That is different for the "main effect" of the economic condition; we return to this issue when we investigate the timing of the release of an entertainment product in our distribution chapter.

Distributors as Mediators

Finally, there is initial empirical evidence that professional reviews appear to be read not only by consumers, but also by those in the entertainment value chain who decide which products are offered to consumers, such as movie theater owners. In a study of more than 165,000 exhibition decisions for 788 films (specifically, weekly theater-level choices by theater owners in Quebec, Canada, spanning 2002 to 2011), Legoux et al. (2015) find that professional reviews influence the theater owners' programming decisions.

The scholars use the number of weeks a movie is shown in a theater as the dependent variable in their analysis and estimate a discrete-time survival model in which they control for numerous variables (including a film's performance in the previous week). They find that movies which earned excellent ratings from professional critics had, on average, a 37–66% higher probability of still being shown in the theater in the following week, compared to less positively reviewed films. This effect becomes more pronounced over the life cycle of a film; it is also higher for more successful films.

Now let us conclude our discussion of the effects of professional reviews by taking a look at their relevance for consumers' *quality perceptions*. We stressed earlier in this book that judging the quality of entertainment is not an easy task for consumers—not only prior to experiencing entertainment, but also during and even afterward. We have also shown that consumers' evaluations of entertainment are influenced by its popularity, and there is similar evidence for an effect of professional reviews on consumers' *post*-experience judgments of entertainment products.

This evidence comes from an experiment that Wyatt and Badger (1984) ran. The scholars showed 89 U.S. students professional reviews (that the scholars had manipulated to be positive, negative, or mixed) for the comedy movie THE NATIONAL HEALTH. The researchers found that, on a 77-point multi-attribute liking scale, the average rating for those who had read a positive review before watching the film was 59—a number that was significantly higher than the average liking score for those students who had read a mixed review (51) or a negative review (49).

Remember that the actual film watched was exactly the same in all conditions; only the review read prior to watching the film was different. Overall, in an analysis of variance, the review accounted for a remarkable 11% of the consumers' movie ratings. So it looks that reviewers have an even stronger overall impact than what we have reported in this section so far.

Managerial (Mis-)Use of Professional Reviews

"The producing team of BIG DADDY has produced another winner!"
—*Fictitious film critic* Dave Manning *about the film* THE ANIMAL *(Hoaxes.org 2014)*

Professional reviews are, per our definition, "earned" media. But as with other kinds of "earned" media, that does not prevent entertainment managers from trying to influence the role reviews play for the success of their products. There are two major ways to do so: by influencing the availability of reviews and by combining reviewers' assessments with other, more controllable forms of communication (namely paid and owned media).

A manager's influence on the availability of professional reviews is generally limited as a result of the freedom of the press in a given country. But entertainment managers can decide *when* to let critics access their newest creations. In practice, when producers provide critics access to a new product before its release (a long tradition in the industry for movies, games, songs, or books), they now often contractually prevent reviewers from publishing their reviews prior to a certain date (often the release date or close to it)—establishing an "embargo" (e.g., Brew 2016). Doing so can help prevent reviews from tampering with the producer's own buzz-building plan or to avoid the leaking of unwanted information about the product and its storyline.

A more extreme way to impact the timing of professional reviews is by not providing critics access to the product *at all* prior to its public release. Historically, whereas movie studios have occasionally prevented pre-release access for the reasons we have named above,[357] doing so has been mainly reserved for their *worst* creations in an attempt to prevent negative

[357]For his thriller PSYCHO, Alfred Hitchcock famously not only did not allow press screenings, but also bought copies of the underlying novel to prevent the massive twist to be leaked. And the producers of the eccentric SNAKES ON A PLANE movie also did not offer screenings mainly to add to the "mystery buzz" surrounding it which was considered so essential for the film (Golder 2015).

information from hampering opening week sales (for illustrious examples, see Golder 2015). But there is a catch to this approach: the mere act of withholding a product from critics has become a quality signal itself; products for which no pre-release access is provided are almost never worth their price. For example, when the movie HANSEL & GRETEL: WITCH HUNTERS was not shown to critics, Rotten Tomatoes informed its users (and potential moviegoers) that they'd "love to tell you more about this one, but it doesn't screen for critics until later in the week, which is never a good sign" (*TV Tropes* 2013).

To multiply the power of positive professional reviews, managers often use them as ingredients in their paid and owned media efforts, such as by including snippets from the reviews on posters and in trailers; we discuss the effectiveness of doing so in the context of "good advertising." However, managers sometimes take some liberty with this approach, using the reviewers' critique out of context. The advertisement for the film 27 DRESSES quoted Entertainment Weekly's review of the film: "Katherine Heigl glows!" But those three words were only the first part of a *photo caption* that continued: "…but 27 DRESSES' formulaic romantic comedy stumbles on the way to the altar" (Bialik 2009). A. A. Dowd (2015), the A.V. Club's film editor, dedicated a whole article to a misleading quote on the DVD package for the film ACCIDENTAL LOVE. His article's headline left not much room for doubt: "No, I didn't call your shitty movie a "comedic masterstroke." In these cases and several others, the full quote suggests a very different judgment than the one implied by the advertisement.

Probably the most extreme case of malpractice in the (mis)use of professional reviews was when around 2000, a marketing manager at Sony who had grown up in Ridgefield, Connecticut, created Dave Manning. As a reviewer for The Ridgefield Press, Mr. Manning "wrote" euphoric reviews for four Sony films, quotes from which were featured in print ads (*Hoaxes.org* 2014)—our introductory quote for this section provides an example of Mr. Manning's virtuosity. After a Newsweek reporter found about the true (i.e., nonexistent) nature of Mr. Manning, Sony first argued that their action was justified by "free speech," but in the end agreed to pay $1.5 million to consumers who saw the films in theaters to settle a class-action lawsuit (Phipps 2005).

The downside risk of such an approach would be enormous today; consumers would certainly detect such a farce quickly and soundly destroy it via social media's pinball mechanism. And be aware that *Entertainment Science* teaches us that there is not even an upside potential, as being endorsed by an unknown critic has only very limited potential to drive audiences toward a new product. Remember that Rao et al. (2017) found that it is only blurbs by "top reviewers" (a status that certainly would *not* apply to an unknown critic from an unknown newspaper) that make a difference in advertising effectiveness.

This book has highlighted several far more promising ways to communicate the attractions of a new entertainment product to its target audiences. Regarding professional reviewers, producing a "great" product can still be considered the best assurance. Doing so can have one additional potential benefit: let us now end our discussion of entertainment communication by taking a look at the success potential of winning *awards*.

Awards as Recognitions of Excellence in Entertainment

> "The [Oscar] is important in order to bring people to the movie theater. That's the only principle meaning of any award."
> —*Actor* Javier Bardem *(quoted in Likesuccess.com 2017)*

Some Essentials of Entertainment Awards

The artistic nature of entertainment products is not only reflected in the ubiquitous presence of professional reviews, but also in the existence of awards and the attention that our society devotes to the ceremonies in which the awards are handed out. Awards for entertainment are granted by various institutions specific to each type of entertainment product. Although each award follows its unique rules, the logic is almost always the same: to honor artistic quality and achievements.[358]

For films, the by far most prominent recognitions of artistic excellence are the Oscars awarded by the Academy of Motion Picture Arts and Sciences (AMPAS, an elite group of film creatives and producers); other noteworthy movie awards include the Golden Globes, the Cannes International Film Festival's Palme d'Or, and the British BAFTA awards. For fiction books,

[358]A second type of entertainment award is based on the *commercial* performance of a product (e.g., Gold or Diamond record). In economic terms, such awards function as a visualization of a product's commercial success, similar to charts.

there's nothing bigger than the Nobel Prize in Literature handed out for writers by an expert jury of the Swedish Academy; other widely publicized honors for authors include the Pulitzer Prize, the Man Booker award, and the Newbery Medal (which is given to the most distinguished contribution to children's literature).

For music, the Grammy awards (handed out by the Recording Academy, a peer group of music professionals) stand out; other well-known music prizes include the American Music awards and the MTV Music awards. Game awards include the Game Developers Choice awards (as decided by members of the Game Developers Conference, a peer group of video game developers) and the Game Awards. The most renowned awards for TV productions are the Emmys (with key prizes decided by the Television Academy, another group of production and distribution peers) and, as for movies, the Golden Globes.

Awards and Success: What *Entertainment Science* Can Tell Us About Their Link

Why Determining the Commercial Impact of Awards is Quite a Challenge

The high TV ratings for award shows clearly indicate that we, the people, are interested in what entertainment product experts think are of high quality. When Beck's album MORNING PHASE won the Grammy for best album in February 2015, search volume on Google increased 100 times, to a level that was five times higher than the previous maximum search volume for the album. And when Mario Vargas Llosa was awarded the Nobel Prize for his writing in 2010, people searched ten times more frequently for his 1970s novel AUNT JULIA AND THE SCRIPTWRITER than in the previous months.

However, we also see that search effects vary greatly between products and they are often short term, dropping back to previous levels shortly afterward. And search means consumer interest, but does not equate with commercial success. Learning about the impact of awards on commercial success is a complicated matter though: just like word of mouth, awards can provide information about a product's quality that can be part of an informed cascade among consumers. But as with professional reviews, awards overlap with quality (*and* with professional reviews), which makes measuring the true *causal* contributions of awards a challenging endeavor. A basic requirement for quantifying the effect of awards is to include these other quality variables—otherwise, awards just pick up their share erroneously, and its value becomes inflated.[359]

But determining such a true causal effect is even more complicated in the case of awards, because entertainment products are usually honored months *after* their market release. Thus, in conjunction with entertainment's short life cycles, *reverse* causality is a serious concern: an Oscar given out in February to a movie released in October just cannot impact moviegoers *before* February. In other words, timing is particularly crucial when estimating statistical models for awards.

And there is just one more thing: because of awards' artistic focus, they are usually only given to products which fall into our category of artistic, independent entertainment, but are rarely bestowed upon commercial products. The task thus is not to compare the Oscar-winning, low-budget film MOONLIGHT with Marvel's THE AVENGERS when it comes to the economic success of an Oscar win, but to compare MOONLIGHT to a similar low-budget film that did *not* win the award. When mixing both categories in a joint model, results tend to be biased, with the relatively lesser commercial performance of the Oscar winner being falsely attributed to the award.[360]

Monetizing the Oscars and Other Learnings

So, what then do we know regarding the financial value of awards? The most authoritative study on the matter is by Nelson et al. (2001), who investigate the effect that Oscar wins and nominations in three categories (best picture, best leading actor/actress, best supporting actor/actress) have on a film's box office. Their data set comprises all 131 films that received an Oscar nomination in one of the three categories between 1978 and 1987. The scholars pair the films with a set of another 131 films that received *no* nominations. By comparing apples (Oscar-nominated films) to apples (*similar* films,

[359]Take for example the historic study by Smith and Smith (1986), who, running an OLS regressions to explain the revenues of 600 films released between the 1940 and 1980), report a much higher Oscar parameter than we do for a more recent set of films (Hennig-Thurau et al. 2006b). Whereas Smith and Smith do not include *any* quality controls, we control for CinemaScore ratings *and* IMDb ratings from consumers.

[360]We assume that not addressing this "apples-to-oranges" problem (essentially a selection/endogeneity problem) is the reason why Luan and Sudhir (2010) find a *negative* effect of Oscar nominations on DVD sales, and maybe also why we do not find any significant impact of Oscar wins and nominations on video rental success (Hennig-Thurau et al. 2006b).

except that they garnered no nomination), not to oranges (films which got nominated, but differ systematically), the idea is to avoid the treatment (or "selection") bias we have argued to exist in the context of awards above (i.e., the MOONLIGHT vs. AVENGERS comparison) and which we have already discussed with regard to other success drivers in previous parts of this book.[361]

Nelson and his colleagues then used weekly box-office data to determine, with a fixed-effects regression model, the effect of the award categories on weekly box-office results. They combine this information with the results of a survival function in which they estimate the average number of weeks a film is shown in theaters (assuming that awarded films are shown longer—a supply-sided effect).[362] They find that all three award categories extend a film's run, but only best picture and lead actor categories influence weekly revenues. Using the estimated parameters in a simulation in which they compare an average film that received no Oscar nominations to an equal one that garnered nominations (and wins), their results show that films can benefit from awards (and, to a lesser degree, nominations). But the award effect differs between award categories and also by the time at which the film was released: across categories, in their analysis, a fourth quarter release generates seven times more value from an Oscar than a first quarter release.

In dollar amounts, the "Oscar effect" that Nelson et al. find is quite substantial, as we report in Fig. 12.12: the scholars estimate that a best picture win increases a film's revenues by an average of $16 million (in 1987 value—which translates into $35 million in 2017 currency) for a film that is released in the fourth quarter. While they treat the award categories as

[361]We discuss such treatment/selection bias in the context of entertainment branding. There's a caveat, however, with regard to Nelson et al.'s approach: their sole selection criterion is to pick a film released in the same week as a nominated film, which does not warrant the removal of potential differences in terms of other "success drivers" (think of genres, budgets, or—of particular importance—movie quality). Clement et al. (2007b) conducted a closer investigation of the differences between those films that received an Oscar nomination and those which did not; among others, their results provide evidence that the quality of Oscar nominated films as judged by critics is clearly higher than that for non-nominated films. As a consequence, the results from Nelson et al. we report in this section might still involve a systematic bias between films that were recognized with awards and those that were not; they should thus be considered as "conservative" estimates that mark the lower end of the spectrum.

[362]Our observations of the marketplace suggests that showing a film longer might not be the only supply-sided effect. Theaters also often increase a film's *availability* as a result of Oscar nominations and wins. For example, when THE SHAPE of WATER received 13 nominations in 2018, its number of theaters more than doubled (from 853 to 1,854, or +117%), hand in hand with a 171% increase in box office compared to the pre-nomination weekend. Nelson et al. did not divide their results into supply-sided and demand-sided effects, so we can only report this as anecdotal evidence.

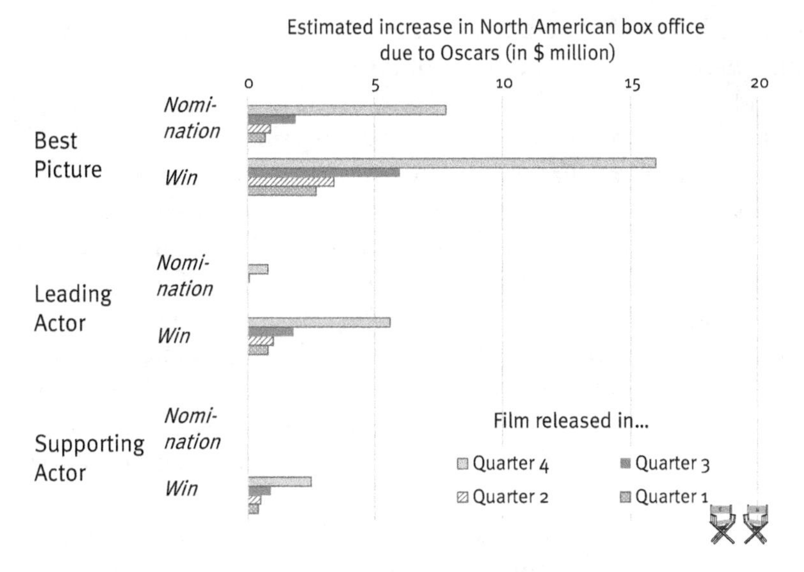

Fig. 12.12 Monetary effects of Oscar wins and nominations
Notes: Authors' own illustration based on results reported by Nelson et al. (2001). The estimates are box office effects which are adjusted for probability of survival. All numbers are in 1987 currency; multiplying them by factor 2.2 approximates their 2017 value.

independent, we expect that negative interactions exist between them—the incremental value of an additional best actor Oscar should be less than if it is the *only* win for a film.

When evaluating these effects, we have to keep in mind that the scholars only look at the success in North American theaters. Movie producers now earn strong returns in channels other than theaters,[363] and we suspect that awards help to gain revenues in them too, either directly or via inter-channel action-based cascades.

With the Oscars being by far the most prominent award in the field of film, to what degree do such effects also exist for other awards? Gemser et al. (2008) study the role of 13 different awards in the financial performance of films, using a data set of around 300 award-winning films that were released between 1997 and 2001 and were still shown in theaters when the award was announced. The scholars run separate OLS regressions for the second and fourth week after the announcement, estimating the effect that awards

[363]For a more detailed discussion of distribution channels for entertainment, see our distribution chapter.

have on both box-office results and on distribution (the number of screens on which a film is shown). They control for the box office and screens prior to the award, as well as several other success drivers (including advertising spending); they do not correct for any potential treatment bias, though. Gemser et al.'s results confirm that the number of total awards won by a film (across categories), and also the number of *different* awards it has won, enhances box-office results and the number of screens allocated to the film. Thus, the economic potential of awards should not be limited to an Oscar win.

Interestingly, the scholars find that *who* picks the winner also matters for any award's impact on audiences: across the different awards, those awards given by experts (such as the Golden Globes) are significantly more influential for audiences and distributors than are awards chosen by industry peers or those picked by consumers. At least in this regard, the Oscars certainly constitute an exception to the (statistical) rule.

We have to admit that we know clearly less about the effects of awards in other fields of entertainment. Future research on music, book, and game awards would help us to better understand their impacts and could serve as the foundation for managerial efforts toward winning an award, which we discuss in the following sub-section. But before we get to this next issue, let us mention one additional thing we indeed do know from scholarly work: the benefits of awards are not limited to *financial* gains.

In a study of the life paths of Oscar winning actors, Redelmeier and Singh (2001) find that Oscar winners live almost four years longer, on average (80 versus 76 years). They derive this insight by comparing the life expectancies of 1,649 Oscar winners and nominees with those of non-awarded same-sex cast members in the same films; birth demographics do not affect the finding. Whereas winning an Oscar more than once tends to add even more life expectancy, being nominated (but ending up not being the winner) does not grant any advantage over those who were not nominated at all. So, whereas an Oscar can't make us immortal, it tends to let us live longer. Why is this? Because, using the authors' words, "success confers a survival advantage" (Redelmeier and Singh 2001, p. 960).

Managerial (Mis-)Use of Awards

We have seen that awards, as a special kind of "earned" media coverage, can grant commercial advantages for entertainment products. As with professional reviews, entertainment managers have developed a number of practices to reach for these advantages: by influencing the chances to win and, if successful, by leveraging those wins.

When it comes to influencing winning probabilities, managers use practices that span the marketing mix. One aspect involves release timing: Nelson et al. (2001) provide evidence that award-winning films that were released (in theaters) later in the year have a *much* higher monetization potential—simply because a later release date increases the chances that a film is still showing in theaters when it receives the award. But, as we discuss in the next chapter, distribution timing is a complex matter, and time periods that are optimal for award potential might be less attractive with regard to other criteria, such as consumer demand and competition. Managers should include award effects in their distribution considerations but balance them with other factors, particularly as winning an award is rare and highly uncertain.

A second popular managerial practice is to dedicate advertising budgets to an award candidate. Very often, studios design campaigns that are targeted at those who choose the award winners. These days, Hollywood's film studios together spend about $150 million for their Oscar campaigns (Kirkham 2015). Follows (2015), in a detailed analysis of such Oscar campaigns, reports that studios spend $5 to $8 million per candidate film; in certain conditions, spending can be twice that amount (e.g., the Weinsteins spent $15 million for SHAKESPEARE IN LOVE's successful Oscar campaign, which equates with almost $23 million in 2017 value). Most of the money flows into targeted advertisements (characterized by "For Your Consideration" notes), followed by producing and sending "screener" versions of their films to members of the AMPAS, along with hosting theatrical screenings. In the film business, more than half of the campaign budget is spent prior to the announcement of the nominations; the remaining budget is then concentrated on those films that managed to achieve a nomination, trying to convert them from nominees into winners (Follows 2015).

Is such spending justified? Putting the campaign costs in relation to Nelson et al.'s revenue estimates can provide a tentative answer. Assuming that an Oscar-winning film released in the fourth quarter returns an additional $20 million to the studio from theaters and, a rough estimate, two times that number from other channels over its lifetime (plus higher brand value that might be harvested by producing sequels or selling the rights),[364]

---

[364]Among Oscar winning films for which sequels (or prequels) were made are ROCKY, THE FRENCH CONNECTION, IN THE HEAT of the NIGHT, THE SILENCE OF THE LAMBS, and THE LORD OF THE RINGS. And THE, GODFATHER of course. We don't argue that the sequels to these films were made *because* of their Oscar wins, but we assume that, based on the value of the Oscar we have shown in this section, they will have benefited from their predecessors' wins. Why have not more award winners been turned into franchises then? We speculate that this might have to do with Oscar voters having shown a preference for drama and sad endings (think: TITANIC!), which affect brand value.

spending up to $20 million on an Oscar campaign would certainly be justified. But, of course, often a film does *not* win. Thus, a portfolio perspective is adequate: over time, a studio must balance its spending in a way that is justified by the number of winners and nominees. The work of some *Entertainment Science* scholars who have tried to demystify the logic behind Oscar nominations and wins might be of help here (e.g., Krauss et al. 2008; Pardoe and Simonton 2008).

Finally, once an entertainment product has managed to win an award, how can this achievement be commercialized? Here, the same basic rules apply as for professional reviews: the nominations and wins are printed on new posters, featured on web pages, added to new trailers, and noted on the packaging of a product (e.g., on the album or DVD cover). And beyond communication, because awards are quality signals (specifically, "substitute cues"), they can also be used as justification for releasing special editions of a product (such as Universal's "Oscar Edition" of films on DVD) as part of a versioning approach. Some movies that had already ended their theatrical run when an award was announced *return* to theaters as a "re-release;" the producer hopes to trigger the interest of those consumers who ignored the film in its original run.

Concluding Comments

We began this chapter by reviewing what insights *Entertainment Science* scholars have compiled regarding the most commonly considered form of earned communication, word of mouth. We distinguished between three types of word of mouth, traditional, social media, and other electronic word of mouth, which today impact entertainment product sales, being more than substitutes for each other. In the weeks and months following an entertainment product's release, the valence of word of mouth influences sales more than advertising and many other "success drivers." This strong effect results from word of mouth being a "substitute cue" for a consumer and involve the sharing of actual people's actual experiences with an entertainment product; we thus refer to the spread of such communication as "informed cascades."

But *un*informed information cascades are also quite influential. We reported that in addition to the desirability that is signaled by high chart rankings ("The top romantic comedy in America!"), consumers' feverish anticipation expressed in pre-release buzz provides a powerful signal; both drive the success of new entertainment products, though at different points in time. Furthermore, automated personal recommender systems process information about consumers' liking of certain products into an information source that is considered as valuable by many consumers. We portrayed the basic approaches that offer recommendations (collaborative filtering, content-based recommenders, and hybrid approaches) and discussed their respective strengths (and limitations).

Whereas recommenders combine consumer data with the actions of those who offer them, we also looked at information from other stakeholders of an entertainment firm that needs to be "earned." Reviewing the empirical evidence, we concluded that professional reviewers do indeed have an impact on product sales, and at least some industry awards such as the Oscar do so, too. We demonstrated how their effect can be monetized from different award categories, offering insights the factors that determine what can be earned by getting nominated or even winning.

This concludes our analysis of entertainment communication. We now turn to distribution: if consumers cannot act on their desire for an entertainment product because it is unavailable or too troublesome to acquire, the product's success will be hindered. In entertainment, the main distribution challenges include finding the right time to release a product, to orchestrate the many distribution channels that exist for entertainment in our digital times, and to deal with illegal competition that takes the form of a pirated version of an entertainment producer's own product.

References

Adomavicius, G., Bamshad, M., Francesco, R., & Tuzhilin, A. (2011). Context-aware recommender systems. *AI Magazine, 32*.

Amatriain, X., & Basilico, J. (2012). Netflix recommendations: Beyond the 5 stars (Part 2). *The Netflix Tech Blog*, June 20, https://goo.gl/N6QNK9.

Arndt, J. (1967). Role of product-related conversations in the diffusion of a new product. *Journal of Marketing, 4*, 291–295.

Asai, S. (2015). Determinants of demand and price for best-selling novels in paperback in Japan. *Journal of Cultural Economics, 40*, 375–392.

Austin, B. A. (1989). *Immediate seating: A look at movie audiences*. California: Wadsworth Pub. Co.

Bandura, A. (1977). *Social learning theory*. Englewood Cliffs, N.J.: Prentice-Hall.

Bart, P. (2017). Peter Bart: Amazon raises bet on movie business, but rivals still baffled about long-term strategy. *Deadline*, March 10, https://goo.gl/t46Vvj.

Basuroy, S., Chatterjee, S., & Abraham Ravid, S. (2003). How critical are critical reviews? The box office effects of film critics, star power, and budgets. *Journal of Marketing, 67*, 103–117.

Berger, J. (2014). Word of mouth and interpersonal communication: A review and directions for future research. *Journal of Consumer Psychology, 24*, 586–607.

Berger, J., & Schwartz, E. M. (2011). What drives immediate and ongoing word of mouth? *Journal of Marketing Research, 48*, 869–880.

Berger, J., & Milkman, K. L. (2012). What makes online content viral? *Journal of Marketing Research, 49*, 192–205.

Bhattacharjee, S., Gopal, R. D., Lertwachara, K., Marsden, J. R., & Telang, R. (2007). The effect of digital sharing technologies on music markets: A survival analysis of albums on ranking charts. *Management Science, 53*, 1359–1374.

Bialik, C. (2009). The best worst blurbs of 2008. *Gelf Magazine*, January 2, https://goo.gl/gvDHhB.

Bikhchandani, S., Hirshleifer, D., & Welch, I. (1998). Learning from the behavior of others: Conformity, fads, and informational cascades. *Journal of Economic Perspectives, 12*, 151–170.

Bishop, B. (2017). How Netflix is trying to rewrite movie marketing with Bright. *The Verge*, December 19, https://goo.gl/VnDjM7.

Böll, H. (1963). *Ansichten eines Clowns*. Cologne: Kiepenheuer & Witsch.

Brew, S. (2016). Movie embargoes: What are they, and why do they matter? *Den of Geek*, July 15, https://goo.gl/Za6igw.

Brodnig, I. (2015). Netflix-Produktchef Neil Hunt: 'Ich weiß das alles über Sie'. *profil*, December 11, https://goo.gl/67ACyu.

Buli, L., & Hu, V. (2015). Data science and the music industry: What social media has to do with record sales. *hypebot*, December 6, https://goo.gl/tET39C.

Burke, R. (2002). Hybrid recommender systems: Survey and experiments. *User Modeling and User-Adapted Interaction, 12*, 331–370.

Chen, Y., Liu, Y., & Zhang, J. (2012). When do third-party product reviews affect firm value and what can firms do? The case of media critics and professional movie reviews. *Journal of Marketing, 76*, 116–134.

Chevalier, J. A., & Mayzlin, D. (2006). The effect of word of mouth on sales: Online book reviews. *Journal of Marketing Research, 43*, 345–354.

Chintagunta, P., Gopinath, S., & Venkataraman, S. (2010). The effects of online user reviews on movie box office performance: Accounting for sequential rollout and aggregation across local markets. *Marketing Science, 29*, 944–957.

Clement, M., Proppe, D., & Rott, A. (2007a). Do critics make bestsellers? Opinion leaders and the success of books. *Journal of Media Economics, 20*, 77–105.

Clement, M., Christensen, B., Albers, S., & Guldner, S. (2007b). Was bringt ein Oscar im Filmgeschäft? Eine empirische Analyse unter Berücksichtigung des Selektionseffekts. *Schmalenbachs Zeitschrift für betriebswirtschaftliche Forschung, 59*, 198–220.

Clement, M., Hille, A., Lucke, B., Schmidt-Stölting, C., & Sambeth, F. (2008). Der Einfluss von Rankings auf den Absatz – Eine empirische Analyse der Wirkung von Bestsellerlisten und Rangpositionen auf den Erfolg von Büchern. *Schmalenbachs Zeitschrift für betriebswirtschaftliche Forschung, 60*, 746–777.

Clement, M., Wu, S., & Fischer, M. (2014). Empirical generalization of demand and supply dynamics for movies. *International Journal of Research in Marketing, 31*, 207–223.

Corliss, R. (2009). Box-office weekend: Brüno a one-day wonder? *Time Magazine*, July 13, https://goo.gl/wYf5v3.

Cox, J., & Kaimann, D. (2015). How do reviews from professional critics interact with other signals of product quality? Evidence from the video game industry. *Journal of Consumer Behavior, 14*, 366–377.

Craig, C. S., Greene, W. H., & Versaci, A. (2015a). E-word of mouth: Early predictor of audience engagement—How pre-release 'E-WOM' drives box-office outcomes of movies. *Journal of Advertising Research, 55*, 62–72.

D'Alessandro, A. (2015b). Tom Hardy Soviet drama 'Child 44' bombs at box office: What the hell happened? *Deadline*, May 1, https://goo.gl/HFaSmh.

D'Alessandro, A. (2017). How 'Pirates' & 'Baywatch' are casualties of summer franchise fatigue at the domestic B.O. *Deadline*, May 29, https://goo.gl/WZJjVP.

D'Astous, A., & Colbert, F. (2002). Moviegoers' consultation of critical reviews: Psychological antecedents and consequences. *International Journal of Arts Management, 5*, 24–35.

D'Astous, A., Carù, A., Koll, O., & Sigué, S. P. (2005). Moviegoers' consultation of film reviews in the search for information: A multi-country study. *International Journal of Arts Management, 7*, 32–45.

De Vany, A., & Lee, C. (2001). Quality signals in information cascades and the dynamics of the distribution of motion picture box office revenues. *Journal of Economic Dynamics & Control, 25*, 593–614.

Dewan, S., & Ramaprasad, J. (2012). Music blogging, online sampling, and the long tail. *Information Systems Research, 23*, 1056–1067.

Dhar, T., & Weinberg, C. B. (2015). Measurement of interactions in non-linear marketing models: The effect of critics' ratings and consumer sentiment on movie demand. *International Journal of Research in Marketing, 33*, 392–408.

Divakaran, P. K. P., Palmer, A., Alsted Søndergaard, H., & Matkovskyy, R. (2017). Pre-launch prediction of market performance for short lifecycle products using online community data. *Journal of Interactive Marketing, 38*, 12–28.

Dowd, A. A. (2015). No, I didn't call your shitty movie a "comedic masterstroke". *A.V. Club*, July 27, https://goo.gl/s3FWnY.

Duan, W., Bin, G., & Whinston, A. B. (2008). The dynamics of online word-of-mouth and product sales—An empirical investigation of the movie industry. *Journal of Retailing, 84*, 233–242.

Ekstrand, M. D., Riedl, J. T., & Konstan, J. A. (2010). Collaborative filtering recommender systems. *Foundations and Trends in Human-Computer Interaction, 4*, 81–173.

Elberse, A., & Eliashberg, J. (2003). Demand and supply dynamics for sequentially released products in international markets: The case of motion pictures. *Marketing Science, 22*, 329–354.

Eliashberg, J., & Shugan, S. M. (1997). Film critics: Influencers or predictors? *Journal of Marketing, 61*, 68–78.

Fischer, S. F., Hammerschmidt, M., & Weiger, W. H. (2017). Signals from the echoverse – the informational value of brand buzz dispersion. *Proceedings of Winter AMA Conference, 28*.

Fishbein, M., & Ajzen, I. (1975). *Belief, attitude, intention, and behavior: An introduction to theory and research*. Reading, MA: Addison-Wesley. [An online version of the book can be found at Mr. Ajzen's website at https://goo.gl/Re6HGP].

Follows, S. (2015). How much do Hollywood campaigns for an Oscar cost? January 12, https://goo.gl/ke5nx4.

Forman, C., Ghose, A., & Wiesenfeld, B. (2008). Examining the relationship between reviews and sales: The role of reviewer identity disclosure in electronic markets. *Information Systems Research, 19*, 291–313.

Foutz, N. Z., & Jank, W. (2010). Prerelease demand forecasting for motion pictures using functional shape analysis of virtual stock markets. *Marketing Science, 29*, 568–579.

Freedman, N. (2015). How social media is changing Hollywood. *Digital America*, April 6, https://goo.gl/ufned5.

Fry, Stephen (2010). Two million reasons to be cheerful, November 30, https://goo.gl/8Ffq5w.

Gemser, G., Van Oostrum, M., & Leenders, M. A. A. M. (2007). The impact of film reviews on the box office performance of art house versus mainstream motion pictures. *Journal of Cultural Economics, 31*, 43–63.

Gemser, G., Leenders, Mark A. A. M., & Wijnberg, N. M. (2008). Why some awards are more effective signals of quality than others: A study of movie awards. *Journal of Management, 34*, 25–54.

Golder, D. (2015). 14 movies that were 'Not Screened For Critics', August 5, https://goo.gl/2WkDhb.

Gomez-Uribe, C. A., & Hunt, N. (2015). The Netflix recommender system: Algorithms, business value, and innovation. *ACM Transactions on Management Information Systems, 6*, 13–19.

Goode, L. (2017). Netflix is ditching five-star ratings in favor of a thumbs-up. *The Verge*, March 16, https://goo.gl/wRWgxY.

Gopinath, S., Chintagunta, P. K., & Venkataraman, S. (2013). Blogs, advertising, and local-market movie box office performance. *Management Science, 59*, 2635–2654.

Gower, S. (2014). Netflix Prize and SVD. Working Paper.

Hennig-Thurau, T., Walsh, G., & Bode, M. (2003). Exporting media products: Understanding the success and failure of Hollywood movies in Germany. Working Paper, Bauhaus-University of Weimar.

Hennig-Thurau, T., Gwinner, K. P., Walsh, G., & Gremler, D. D. (2004). What motivates consumers to articulate themselves on the Internet? *Journal of Interactive Marketing, 18*, 38–52.

Hennig-Thurau, T., Houston, M. B., & Sridhar, S. (2006a). Can good marketing carry a bad product? Evidence from the motion picture industry. *Marketing Letters, 17*, 205–219.

Hennig-Thurau, T., Houston, M. B., & Walsh, G. (2006b). The differing roles of success drivers across sequential channels: An application to the motion picture industry. *Journal of the Academy of Marketing Science, 34*, 559–575.

Hennig-Thurau, T., Marchand, A., & Marx, P. (2012a). Can automated group recommender systems help consumers make better choices? *Journal of Marketing, 76*, 89–109.

Hennig-Thurau, T., Marchand, A., & Hiller, B. (2012b). The relationship between reviewer judgments and motion picture success: Re-analysis and extension. *Journal of Cultural Economics, 36*, 249–283.

Hennig-Thurau, T., Wiertz, C., & Feldhaus, F. (2015). Does Twitter matter? The impact of microblogging word of mouth on consumers' adoption of new movies. *Journal of the Academy of Marketing Science, 43*, 375–394.

Hoaxes.org (2014). Dave Manning. https://goo.gl/tNadxL.

Horn, J. (2009). ComicCon's buzzmakers. *Los Angeles Times*, July 27, https://goo.gl/uAXFGT.

Houston, M. B., Kupfer, A., Hennig-Thurau, T., & Spann, M. (2018). Pre-release new product consumer buzz. *Journal of the Academy of Marketing Science, 46*, 338–360.

Hsu, G. (2006a). Evaluative schemas and the attention of critics in the US film industry. *Industrial & Corporate Change, 15*, 467–496.

Jabr, W., & Zhiqiang, Z. (Eric) (2014). Know yourself and know your enemy: An analysis of firm recommendations and consumer reviews in a competitive environment. *MIS Quarterly, 38*, 635–654.

Jain, K., & Srinivasan, N. (1991). An empirical assessment of multiple operationalizations of involvement. *Advances in Consumer Research, 17*, 594–602.

Jannach, D., Resnick, P., Tuzhilin, A., & Zanker, M. (2016). Recommender systems–beyond matrix completion. *Communications of the ACM, 59*, 94–102.

Jozefowicz, J., Kelley, J., & Brewer, S. (2008). New release: An empirical analysis of VHS/DVD rental success. *Atlantic Economic Journal, 36*, 139–151.

Karniouchina, E. V. (2011). Impact of star and movie buzz on motion picture distribution and box office revenue. *International Journal of Research in Marketing, 28*, 62–74.

Katz, E., & Lazarsfeld, P. F. (1955). *Personal influence: The part played by people in the flow of mass communication*. New York: The Free Press.

Kaye, D. (2012). 4 films that failed to live up to their blockbuster Comic-Con buzz. *Syfy Wire*, December 14, https://goo.gl/baBFuu.

Kim, H., & Hanssens. D. M. (2017). Advertising and word-of-mouth effects on pre-launch consumer interest and initial sales of experience products. *Journal of Interactive Marketing, 37*, 57–74.

Kirkham, E. (2015). Hollywood spends $150 million on Oscar campaigns each year. *The Philadelphia Inquirer*, January 15, https://goo.gl/D7d8s7.

Koren, Y., Bell, R., & Volinsky, C. (2009). Matrix factorization techniques for recommender systems. *Computer, 42*, 30–37.

Körte, P. (2009). Filmkritik ist, wenn man von der Brücke in den Fluss spuckt. In T. Hennig-Thurau & V. Henning (Eds.), *Guru Talk – Die deutsche Filmindustrie im 21. Jahrhundert* (pp. 192–197). Marburg: Schüren.

Krauss, J., Nann, S., Simon, D., Fischbach, K., & Gloor, P. (2008). Predicting movie success and Academy Awards through sentiment and social network analysis. In *Proceedings of 16th European Conference on Information Systems* (pp. 2026–2037). Galway, Ireland.

Krishnan, V., Narayanashetty, P. K., Nathan, M., Davies, R. T., & Konstan, J. A. (2008). Who predicts better?—Results from an online study comparing humans and an online recommender system. In *Proceedings of the 2008 ACM Conference on Recommender Systems* (pp. 211–218). New York: ACM.

Kupfer, A.-K., Pähler vor der Holte, N., Kübler, R., & Hennig-Thurau, T. (2018). The role of the partner brand's social media power in brand alliances. *Journal of Marketing, 82*, 25–44.

Lampel, J., & Shamsie, J. (2000). Critical push: Strategies for creating momentum in the motion picture industry. *Journal of Management, 26*, 233–257.

Lasar, M. (2011). Digging into Pandora's music genome with musicologist Nolan Gasser. *Ars Technica*, January 13, https://goo.gl/yWmG8M.

Lee, J., Boatwright, P., & Kamakura, W. A. (2003). A Bayesian model for prelaunch sales forecasting of recorded music. *Management Science, 49*, 179–196.

Legoux, R., Larocque, D., Laporte, S., Belmati, S., & Boquet, T. (2015). The effect of critical reviews on exhibitors' decisions: Do reviews affect the survival of a movie on screen? *International Journal of Research in Marketing, 33*, 357–374.

Li, X., & Hitt, L. M. (2008). Self-selection and information role of online product reviews. *Information Systems Research, 19*, 456–474.

Likesuccess.com (2017). The award is important in order to bring people to the movie theater. That's the only principle meaning of any award. https://goo.gl/qpYdkm.

Littleton, C., & Holloway, D. (2017). Jeff Bezos mandates programming shift at Amazon Studios. *Variety*, September 8, https://goo.gl/GFtV37.

Liu, Y. (2006). Word of mouth for movies: Its dynamics and impact on box office revenue. *Journal of Marketing, 70*, 74–89.

Liu, X., Singh, P. V., & Srinivasan, K. (2016). A structured analysis of unstructured big data by leveraging cloud computing. *Marketing Science, 35*, 363–388.

Lops, P., de Gemmis, M., & Semeraro, G. (2011). Content-based recommender systems: State of the art and trends. In F. Ricci, L. Rokach, B. Shapira, & P. B. Kantor (Eds.), *Recommender systems handbook* (pp. 73–104). New York: Springer.

Lovett, M. J., & Staelin, R. (2016). The role of paid, earned, and owned media in building entertainment brands: Reminding, informing, and enhancing enjoyment. *Marketing Science, 35*, 142–157.

Luan, Y. J., & Sudhir, K. (2010). Forecasting marketing-mix responsiveness for new products. *Journal of Marketing Research, 47*, 444–457.

Lynn, F. B., Walker, M. H., & Peterson, C. (2016). Is popular more likeable? Choice status by intrinsic appeal in an experimental music market. *Social Psychology Quarterly, 79*, 168–180.

Madrigal, A. C. (2014). How Netflix reverse engineered Hollywood. *The Atlantic*, January 2, https://goo.gl/4frKk3.

Malhotra, N., & Helmer, E. (2012). Inflation in weekend box office estimates. *Applied Economics Letters, 19*, 1411–1415.

Marchand, A. (2016). The power of an installed base to combat lifecycle decline: The case of video games. *International Journal of Research in Marketing, 33*, 140–154.

Marchand, A., Hennig-Thurau, T., & Wiertz, C. (2016). Not all digital word of mouth is created equal: Understanding the respective impact of consumer reviews and microblogs on new product success. *International Journal of Research in Marketing, 34*, 336–354.

Marx, P., Hennig-Thurau, T., & Marchand, A. (2010). Increasing consumers' understanding of recommender results: A preference-based hybrid algorithm with strong explanatory power. In *Proceedings of the fourth ACM Conference on Recommender Systems* (pp. 297–300). New York: ACM.

McKenzie, J. (2010). How do theatrical box office revenues affect DVD retail sales? Australian empirical evidence. *Journal of Cultural Economics, 34*, 159–179.

Meiseberg, B. (2016). The effectiveness of e-tailers' communication practices in stimulating sales of niche versus popular products. *Journal of Retailing, 92*, 319–332.

Nelson, R. A., Donihue, M. R., Waldman, D. M., & Wheaton, G. (2001). What's an Oscar worth? *Economic Inquiry, 39*, 1–16.

Netflix (2017). Decoding the defenders: Netflix unveils the gateway shows that lead to a heroic binge. Press Release, August 22, https://goo.gl/2nmGZf.

Niraj, R., & Singh, J. (2015). Impact of user-generated and professional critics reviews on Bollywood movie success. *Australasian Marketing Journal, 23*, 179–187.

O'Connor, M., Cosley, D., Konstan, J. A., & Riedl, J. (2001). PolyLens: A recommender system for groups of users. In *Proceedings of the 7th European Conference on Computer-Supported Cooperative Work* (pp. 199–218). Dodrecht: Kluwer.

Panaligan, R., & Chen, A. (2013). Quantifying movie magic with Google Search. White Paper, Google.

Papies, D., & van Heerde, H. (2015). How the Internet has changed the marketing of entertainment goods: The case of the music industry. Working Paper, Marketing Science Institute.

Pardoe, I., & Simonton, D. K. (2008). Applying discrete choice models to predict academy award winners. *Journal of the Royal Statistical Society: Series A (Statistics in Society), 171*, 375–394.

Phipps, K. (2005). The ghost of David Manning will have his revenge. *AV Club*, August 3, https://goo.gl/CCSWCW.

Rao, V. R., Ravid, A. S., Gretz, R.T., Chen, J., & Basuroy, S. (2017). The impact of advertising content on movie revenues. *Marketing Letters, 28*, 341–355.

Redelmeier, D. A., & Singh, S. M. (2001). Survival in Academy Award-winning actors and actresses. *Annals of Internal Medicine, 134*, 955–962.

Reinstein, D. A., & Snyder, C. M. (2004). The influence of expert reviews on consumer demand for experience goods: A case study of movie critics. *The Journal of Industrial Economics, 53*, 27–51.

Rosario, A. B., Sotgiu, F., De Valck, K., & Bijmolt, T. H. A. (2016). The effect of electronic word of mouth on sales: A meta-analytic review of platform, product, and metric factors. *Journal of Research in Marketing, 53*, 297–318.

Salganik, M. J., Dodds, P. S., & Watts, D. J. (2006). Experimental study of inequality and unpredictability in an artificial cultural market. *Science, 311*, 854–856.

Salganik, M. J., & Watts, D. J. (2008). Leading the herd astray: An experimental study of self-fulfilling prophecies in an artificial cultural market. *Social Psychology Quarterly, 71*, 338–355.

Schmidt-Stölting, C., Blömeke, E., & Clement, M. (2011). Success drivers of fiction books: An empirical analysis of hardcover and paperback editions in Germany. *Journal of Media Economics, 24*, 24–47.

Shone, T. (2004). *Blockbuster*. New York: Free Press.

Smith, S. P., & Smith, V. K. (1986). Successful movies: A preliminary empirical analysis. *Applied Economics, 18*, 501–507.

Soderstrom, S. B., Uzzi, B., Rucker, D. D., Fowler, J. H., & Diermeier, D. (2016). Timing matters: How social influence affects adoption pre- and post-product release. *Sociological Science, 3*, 915–939.

Sorensen, A. T. (2007). Bestseller lists and product variety: The case of book sales. Working Paper, Stanford GSB Research Paper No. 1878.

Sorensen, A. T., & Rasmussen, S. J. (2004). Is any publicity good publicity? A note on the impact of book reviews. Working Paper, Stanford University.

Sirdeshmukh, D., Singh, J., & Sabol, B. (2002). Consumer trust, value, and loyalty in relational exchanges. *Journal of Marketing, 66*, 15–37.

Strause, J. (2016). How 'My Big Fat Greek Wedding' became an indie phenomenon. *Hollywood Reporter*, March 25, https://goo.gl/RV4prv.

Sun, M. (2012). How does the variance of product ratings matter? *Management Science, 58*, 696–707.

The Economist (2017). How to devise the perfect recommendation algorithm. https://goo.gl/VouerJ.
tickld (2016). 31 fascinating things most people don't know about Netflix. https://goo.gl/8pNw7X.
TV Tropes (2013). Not screened for critics. *TV Tropes*, https://goo.gl/QxQiCs.
Wohlfeil, M., & Whelan, S. (2008). Confessions of a movie-fan: Introspection into a consumer's experiential consumption of 'Pride and Prejudice'. *Proceedings of European ACR Conference* (pp. 137–143).
Wyatt, R. O., & Badger, D. P. (1984). How reviews affect interest in and evaluation of films. *Journalism Quarterly, 61*, 874–878.
Wyatt, R. O., & Badger, D. P. (1990). Effects of information and evaluation in film criticism. *Journalism Quarterly, 67*, 359–368.
Xiong, G., & Bharadwaj, S. (2014). Prerelease buzz evolution patterns and new product performance. *Marketing Science, 33*, 401–421.
You, Y., Vadakkepatt, G. G., & Joshi, A. M. (2015). A meta-analysis of electronic word-of-mouth elasticity. *Journal of Marketing, 79*, 19–39.
Zhu, F., & Zhang, X. M. (2010). Impact of online consumer reviews on sales: The moderating role of product and consumer characteristics. *Journal of Marketing, 74*, 133–148.

13

Entertainment Distribution Decisions

As we noted earlier in this book, entertainment markets are characterized by an abundance of products. This market structure assigns an important role to distribution mechanisms, which function as gatekeepers for entertainment products. Distributors decide which products gain access to consumers and which do not, influencing the success of entertainment products via "supply-side effects." Several scholars have provided empirical evidence of supply-side effects, most of them in the movie context, in which the limited number of available theaters restrict consumers' access to films. Prominent studies include those by Elberse and Eliashberg (2003), Clement et al. (2014), and Karniouchina (2011).[365]

For other entertainment products that are sold or rented via stores or websites, less scholarly research evidence exists, but more should generally

[365]These studies all use large data sets with mainstream films and model the number of theaters in which a film is shown with a separate equation, using the resulting estimates as a "success driver" in the box office equation (instead of the "raw" number of theaters). The distribution elasticities calculated this way are quite consistent. Anita Elberse and Josh Eliashberg estimate a weekly elasticity for the number of theaters in the same week of 0.81 for the North American opening week and of around 1.50 for different European countries; elasticities for the following weeks are around 1, across countries. Michel Clement and his colleagues find elasticities of slightly above 1 for North America and Germany for both the opening week and the next weeks. And Kate Karniouchina estimates a distribution elasticity of slightly below 1 which varies little over the weeks in which the movie is shown in North American theaters. Please note that these elasticities should not be compared with those for product and communication measures, or pricing: whereas product and communication decisions aim at generating demand for entertainment (and price decisions aim at maximizing returns for the producer based on such demand), distribution's role is to ensure that consumer demand for a product can be transformed into revenues. Thus, empirical elasticities of about 1 indicate that markets, in general, work effectively with regard to fulfilling consumer demand for entertainment.

T. Hennig-Thurau and M. B. Houston, *Entertainment Science*,
https://doi.org/10.1007/978-3-319-89292-4_13

be better when it comes to the distribution intensity for entertainment. Intense distribution has been shown to be a key to success for convenience goods (e.g., Coughlan et al. 2006) and durables (e.g., Bucklin et al. 2008 for cars) which are targeted at large consumer segments, just as commercial (and many "independent") entertainment products also are.[366] That is why we consider access to distribution channels to be a critical strategic resource for entertainment companies, by the way. An exception might be in those distribution channels in which "shelf space" is rare and costly, as is the case with physical venue distribution—every empty seat in a movie theater on a Friday night is irredeemably lost. For those channels, if a new product is distributed too widely, such oversupply can threaten the product's availability in the following weeks and months, as theaters drop such films in favor of more lucrative offerings—see our discussion of the niche concept of integrated entertainment marketing.[367]

In this chapter, we look beyond the mere intensity of product availability and which might be the "right" distributors for an entertainment product. Instead, we focus on three other distribution issues that are of key importance for an entertainment product's success. The first is timing the new product release. Timing deserves particular attention because entertainment life cycles are usually short, the consumer motivation for spending money on entertainment is hedonic (the time must be "right" for entertainment fun), and entertainment markets are crowded and competitive. We will discuss both the absolute timing of launching a new product and the timing relative to that of competitors.

Second, because books, music, games, and movies are all information goods and thus can be made available in pure digital format, broadband Internet technology has dramatically increased the number of potential distribution channels for entertainment. Coordinating these channels is tough, particularly with regard to the timing dimension: when should a specific distribution channel be opened and closed? When should the next channel be opened?

[366]To avoid misunderstandings: for some mainly artistic entertainment products, being available via a maximum number of distributors is not the essence, but rather getting access to the "right" ones. But distributors are often the same for commercial and independent products (e.g., Spotify, Amazon), and few producers of artistic products would refuse being successful beyond their core niche target segment, so that in the end, the wider their products are available the better. Things are somewhat different for the theatrical movie channel (and other forms of venue distribution), as we discuss in the text.

[367]In a study of "film survival" at the level of the individual movie theater, Chisholm and Norman (2006) offer rich insights into theater owners' strategic decisions based on 2000/2001 data from 13 movie theaters in the Boston area. Their results provide strong evidence that theaters indeed base their decisions on a film's (relative) performance in the previous week.

Making a specific product available to consumers at different points in time across different channels or formats has a long tradition for books and movies, but channel coordination today is more complex than ever.

And third, again because of its information character, entertainment has been among the industries hit hardest by piracy. We will look at what *Entertainment Science* scholars have to say regarding the effects of piracy, as well as about the effectiveness of anti-piracy measures.

The Timing Challenge: When is the Right Time for an Entertainment Product?

Timing the release of a new entertainment product is a multi-faceted decision. In the following, we discuss both the "isolated" timing of a release and its timing in relation to competitive offerings. Please note that we do *not* discuss the coordination of release timing *between* channels at this point, but have dedicated a separate section to this prevalent issue.

Isolated Timing Effects

Producers of entertainment products have to determine when to enter a market. Even when leaving out an explicit consideration of competitive products, producers have to account for long-term factors (such as the readiness of a market), mid-term timing factors (such as the season of the year), and also short-term aspects (e.g., the day and hour at which a new TV show is aired). Managers have developed heuristics to address these issues, but the evidence does not support all of them. Let's take a look.

Long-Term Timing

> "Hollywood could get used to this recession thing."
> —Cieply *and* Barnes *(2009)*

Long-term timing decisions are situated at the crossroads of distribution and innovation decisions. Developing a new entertainment product is often a time-consuming process that can span years. At the point of greenlighting an idea, producers need to have an eye on multiple factors that, at a much later time, will surround the introduction of the product and influence its reception by consumers. Knowing and understanding these factors allows the producer to vary the speed of the innovation process (to accelerate it or to slow it down) to meet the "right" window in time, but also to completely abandon a project if he or she senses that this window is closed and won't re-open in the foreseeable future. What are these factors that determine a market's readiness for a new entertainment product within a time window? We name technology, infrastructure, cultural trends, and the state of the economy.

Given the important role that *technology* plays in the creation and distribution of many entertainment products, producers have to anticipate the availability of technologies on which the success of their product depends and, if necessary, facilitate the technology's advancement. Remember that James Cameron and his co-workers intentionally delayed the production of their AVATAR movie to personally contribute to the improvement of what they considered to be essential technological advances for putting their vision on film. AVATAR would probably not have become such a gargantuan hit if they had employed premature versions of motion-capturing and 3D, the two key technologies needed for realizing the filmmakers' vision.

With regard to *infrastructure*, take a look at Fig. 13.1: it shows the development of 3D screens in North America, over time. Before 2009, the number of screens that could show a film in digital 3D was too small to make this format a nationwide success; distribution infrastructure was a limiting factor, a price that several early 3D productions (such as Warner's THE POLAR EXPRESS) had to involuntarily pay. The producers of AVATAR timed their release perfectly—2009 was the first year in which sufficient 3D distributive capacity was available in major markets, a development that was at least in part endogenous, being driven by the strong buzz that the film makers were able to create for their film.[368]

Relatedly, ambitious serialized stories, such as HOUSE OF CARDS and THE GET DOWN, have a much higher success potential these days because of technological advances such as broadband Internet connections. It is the wide availability of this technology and its adoption by consumers that allow consumers to indulge in on-demand binge watching and to follow the series' complex, horizontal, episode-spanning storytelling; the infrastructure serves as the foundation of the rise of services such as Netflix and Spotify.

[368]Please see our coverage of empirical findings on the link between 3D technology and product success and the critical role of time in Chapter 8 on entertainment search qualities.

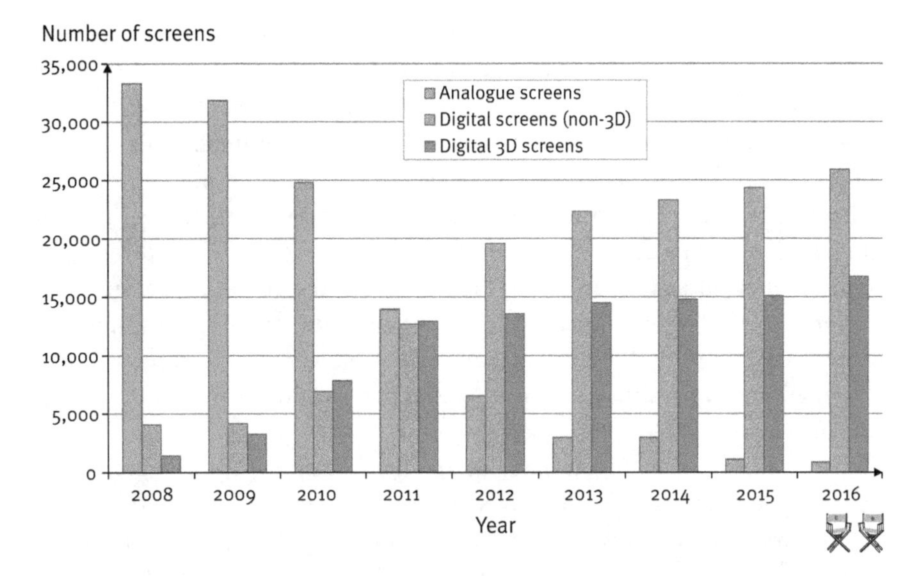

Fig. 13.1 The evolution of analogue, digital, and digital 3D screens in North America
Note: Authors' own illustration based on numbers reported in several annual MPAA publications.

Beyond technology and infrastructure, *cultural trends* can make consumers "ripe" for a product. We have already discussed the *zeitgeist* concept—the idea that the fit of any entertainment product's genres and themes with people's interests and desires may be high or low at a given time. Part of a producer's challenge is to anticipate that *zeitgeist* ahead of time and plan a product's release time accordingly. It could make sense to defer the production or release of a film, game, or book for a number of years, hoping that its topic becomes more *en vogue*. However, postponement can cause substantial conflict with the creatives who are behind the product: if you have spent a year writing a screenplay, you usually hate to see it delayed for economic reasons.

History shows that such patience can indeed pay off though; it may not only help the commercial reception of the product, but also its critical standing. The script for the western movie UNFORGIVEN was written in 1976. It was purchased by Clint Eastwood in 1983, who then delayed transforming it into a film for almost a decade (Schickel 1996). Mr. Eastwood, an artist, did not offer analytical arguments except that he felt "too young" to portray the lead character. But his decision corresponded with the insight that the revisionist nature of the story felt much more timely in the 1990s, when westerns had become a rarity for both for the actor and Hollywood in general. Only in such a modern cultural environment, this film could be appreciated for looking back on the now defunct genre and reflecting on its naïve stereotypes, providing an explanation of its demise, and also as a revision of Mr. Eastwood's own (now long "retired") historic western persona known for its violence. The film became a major commercial hit and also won the Oscar for best picture in 1992.[369]

Finally, the hedonic nature of entertainment and its underlying "pleasure principle" imply that demand for entertainment is also influenced by a society's *economic* conditions. Are entertainment and the economy connected in a counter-cyclical way, with entertainment motives (such as escaping from the everyday life) making entertainment products even more attractive in bleaker economic times, despite consumers having less money at their disposal?

Entertainment Science scholars Dhar and Weinberg (2015) tested this argument with 26 years of weekly North American movie market data, and their results provide empirical evidence for the existence of such a counter-cyclical demand pattern. The nationwide consumer sentiment in the month a movie is released has a negative effect on movies' box office in the scholars' GMM regressions, above and beyond other drivers of movie success (such as distribution and genre—but not advertising). The elasticity of consumer sentiment is −0.32: a 10% decrease in sentiment leads to an increase in movie demand of about 3%, on average. It seems plausible that different kinds of entertainment are affected to varying degrees, as they address more or less crisis-related motives (e.g., are comedies a more intuitive choice than dramas in rough times?). Dhar and Weinberg do not test for such genre differences, so we add this to *Entertainment Science*'s "To Do" list.

Mid-Term Timing

"July 4 Opening Is No Guarantee for Success at Box Office."
—Schwartzel *(2016), after the launch of the movies* T*ARZAN and* T*HE BFG*

So we have seen that distribution timing requires strategic skills from the entertainment manager. But it also demands tactical knowledge to determine the "right" season of the year—one that will maximize a new entertainment product's success potential. As we will show in this section,

[369]UNFORGIVEN's global box office was $159 million (in 1990 dollar value), at a budget of about $14 million.

gaining statistical insights on the matter is challenging, as seasonal revenues are determined not only by fluctuations of consumer demand, but also by managers' heuristics regarding such demand. But before we look at studies that aim to disentangle these effects, let us first take a descriptive glance at seasonal patterns.

Seasonal Patterns

While our need for entertainment is so fundamental to human nature, our ability to consume entertainment is a function of the utilitarian duties we have no choice but to fulfill. Only when we, as employees, students, or *Entertainment Science* writers, have accomplished our day's work, can it be the time for entertainment. These utilitarian duties are unequally distributed over the year. Movie theater owners, for example, often cite the time availability of teenagers and families (in terms of school vacations across the U.S.) as a proxy for a given week's commercial potential—because it facilitates spending time at the movies (Fritz 2017).

Entertainment revenues bear this out. Panel A of Fig. 13.2 reports the North American theatrical box office for each week of the year for the years 2010–2014. The differences between weeks are glaring: the highest revenues are generated in the summer months of June and July, in certain pre-summer weeks, and in the Thanksgiving and Christmas weeks. The first quarter of the year, in contrast, yields the lowest box office numbers, along with the weeks from late August to mid-October—here, over the observation period, average revenues were less than half those of the peak times.

Scholars have provided evidence that such seasonal differences are significantly associated with film success.[370] For example, for our sample of 331 films released in 1999–2001, we find that films released either in the summer or around Christmas generate, on average, a higher box office both during their opening weekend and in the weeks that follow (Hennig-Thurau et al. 2006). Brewer et al. (2009), in an analysis of 466 successful films released in North America between 1997 and 2001, find movies released in summer and at Thanksgiving/Christmas to have about 10% and 7% higher revenues, on average. And Clement et al. (2014), who use an index value for each calendar week based on the week's past theatrical revenues in North

[370]Keep in mind the short life cycles of entertainment products, which, in combination with success-breeds-success effects, often prevent an entertainment product from making up for a disappointing opening.

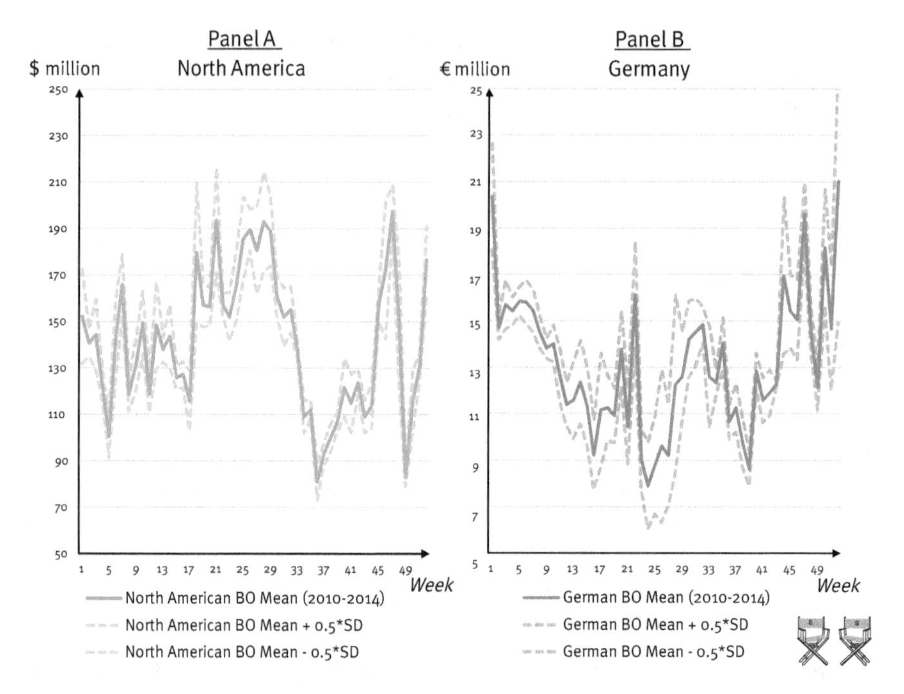

Fig. 13.2 Seasonal box-office revenues for movies: North America and Germany
Notes: Authors' own illustration based on weekly per-movie box office numbers from 2010 to 2014 reported by The Numbers and Blickpunkt:Film. The straight line is the mean value for a week. The dotted lines are the mean for a week multiplied by one standard deviation; they show the variation that the box office in a given week has experienced from 2010 to 2014. Larger deviations from the mean line indicate stronger variation. BO = box office. SD = standard deviation.

America, estimate a "season" elasticity of 0.29 for revenues: a 10% higher average box office in a movie's release week corresponds with approximately 3% higher theatrical revenues for the film. For Germany, the scholars find an even higher elasticity of 0.57.

Seasonality also correlates with the success of other forms of entertainment. Average monthly TV ratings differed by almost 30% over the course of a year between 1998 and 2005 (Hennig-Thurau et al. 2013), and we found that the month in which a film is aired on TV is significantly linked with the film's viewership, even when key film characteristics are also considered (the parameter for month is of a similar size as the one for the film's genre and stars). For music, researchers point to the holiday season as crucial; Bhattacharjee et al. (2007) find that music albums that are released in December enjoy, on average, a 23% longer time on the Top 100 charts than other albums. And for books, Schmidt-Stölting et al. (2011) show that sales differ between seasons in Germany, and that the seasons' commercial appeal differs between hardcover books (for which sales are clearly lower for first quarter releases) and paperback editions (which are most successful when released in the first quarter and least successful when released in the fall).

Seasonal sales patterns not only differ between forms of entertainment and product types, but can also do so between countries. Panel B of Fig. 13.2 illustrates that a seasonal pattern exists for movie attendance in Germany too, but the pattern clearly differs from its North American counterpart. Instead of the summer peak found for North America, we note a summer *dip* in Germany, where June and July are the months with the lowest average movie attendance. Obviously, German and North American consumers decide differently when they have to trade off the cool darkness of a movie theater for sunshine. Managers need to take note of such cultural differences in sales patterns instead of applying an ethnocentric perspective.

Demand-Sided Versus Supply-Sided Effects

But there might be other factors in play beyond simple variations in consumer preferences. Before Steven Spielberg's movie JAWS was released in the June of 1975 in North America, the seasonal pattern of U.S. movie viewership was much more like it is in Germany. Back then, Hollywood studios "viewed the hot weather months as a time when people were too busy traveling or frolicking outdoors and not typically inclined to go to the movies" (Canning 2010, p. 531).[371] So why was JAWS, the expensive inaugural "blockbuster" movie,[372] released in the summer, if it was considered a low-demand period? Not because of a manager's vision, but because of an accident: the movie was scheduled to be released in the winter of 1974, but ran over time and budget, and Universal, the producing studio, was too impatient to wait another 6 months to get reimbursed by moviegoers.

As we now know, the film became a huge hit despite its release date. Or was it a hit *because of* the release date? The key question here is to what degree the differences in sales we observe for an entertainment product are *caused* by the level of customer demand that exists during a certain week or month—versus *supply*-side influences. Think of self-fulfilling effects: if managers release only weak products at a certain time of the year because they *think* demand will be low, consumer demand *will* be low. Not because it

---

[371]See also the statistics provided by Vogel (2015, p. 94).

[372]We discuss the blockbuster concept and Jaws' role for it in our chapter on integrated entertainment marketing.

is intrinsically low, but because of the weakness of the offered products. In econometric terms, this would be a classic case of supply-sided endogeneity.

Separating the demand- and supply-sided factors at play is essential for understanding whether it makes sense to release a product in a week when demand is usually high, or to release the product at a time when demand has been low. Keep in mind that if everyone in the industry uses the same seasonal heuristic, "high-demand" seasons will face heavy competition, whereas there will be much less competition in what is widely regarded as "low-demand" seasons. In fact, there are a number of indicators that point to the existence of such supply-side effects. Figure 13.2 also displays deviations in the box office for the exact same week over the 2010–2014 period, and it is probable that these variations, which are quite strong for several weeks, were caused by the products that were released in the respective periods in a given year. Take the example of what happened when Disney released STAR WARS: THE FORCE AWAKENS at Christmas in 2015: the sequel broke all records and set a new all-time opening weekend record, generating $248 million in North America in its first three days (more than six times the previous record). Isn't it fascinating that the Christmas timing was once more the result of a production delay—the film was originally scheduled to be released in May, just like all six previous STAR WARS films (Child 2013).[373]

Liran Einav (2007) conducted an in-depth investigation of seasonal effects, aimed at the difficult separation of demand-side and supply-side effects. Using a data set of almost 2,000 movies that were released in North American theaters between 1985 and 1999, he estimated a nested logic demand model to separate the part of success that can be attributed to the movie characteristics that determine the commercial appeal of a film (such as the production budget, advertising spending, and genres) from the "true" underlying demand effect. His findings, which he corroborates with several robustness analyses, provide evidence of the existence of supply-side effects—and that their size is managerially relevant. Specifically, he estimates that about one-third of the seasonal variation in movie ticket sales in his data is caused by the kinds of movies that managers choose to show at a given point in time.

In contrast, the other two-thirds can be attributed to differences in "true" demand, according to Einav's results. In other words, supply-sided effects (or endogeneity, technically speaking) amplify the demand effects by about 50%. So consumers' time availability matters, but there is also

---

[373]Completing the picture, you, our readers won't be surprised to hear that the *initial* STAR WARS film was originally slated for a December 1976 release, but had to be postponed to May…

room to release products in seasons that historically have less demand. The industry has recently followed suit and now tends to release its blockbuster productions that have been traditionally considered "summer fare" in a more flexible way, and the sales numbers offer external validity for Einav's findings. Maybe the words of Jeff Goldstein, as president of Warner Bros.' film division, slightly exaggerate the market response, but they capture the essence: "There's no question it's a 12-month calendar now" (quoted in Fritz 2017).

Do such supply-sided effects also exist for other forms of entertainment than movies and other distribution channels than theaters? In general, we assume this to be the case, but the somewhat idiosyncratic nature of theatrical movie distribution patterns reminds us to be careful. Additional scholarly investigation of the matter with other products would certainly be helpful.

Moderating Factors

The impact of seasonal timing on product success also depends on a number of product factors. For example, we have shown that the value of awards can differ depending on release timing; movies that win an Oscar benefit financially from that win when released at the right time of the year. This means that awards moderate the season-success link: for Oscar contenders, the average financial advantage of a fall release will be higher than for movies that have no chance at winning an award. Movie producers thus have to balance these effects with more general demand-sided considerations regarding timing.

Seasonal timing has also been shown to interact with advertising. We have reported earlier that Luan and Sudhir (2010) found that the effectiveness of advertising for a movie on DVD varies strongly with the release timing; consumers are more sensitive to advertising in high demand seasons (i.e., the holidays, Valentine's Day). This means that the economic attractiveness of releasing an entertainment product at such a point in time can be increased by high levels of advertising spending—or, vice versa, there is additional value in releasing a film that is targeted for strong advertising support in a high-demand period.

Short-Term Timing

Finally, for some entertainment products, the short-term timing of release can have an effect on success. This timing dimension applies in particular to products that are part of a linear distribution mechanism where the consumption time is determined by the producer/distributor instead of the consumer, such as series and movies on TV, songs on radio, and movies in theaters.

For TV content, the pioneering work by Goodhardt et al. (1975) and Webster and Wakshlag (1983) stresses the role of “structural” aspects for consumers watching a specific show or film. Among these aspects are the day and time of a TV program’s airing, as well as so-called “inheritance” (or “lead-in”) effects that describe the role of timing relative to the previous program that precedes the program in question. Several studies point to the importance of air time and day. For example, Reddy et al. (1998), in an effort to develop an optimal scheduling algorithm for TV stations, use weekly data for 26 shows aired on a U.S. cable network. They find both certain times and days to be significantly related to ratings, an insight that Wilbur (2008) confirms with data for programs shown during prime time on the major TV networks in May 2003.

Our own PLS analysis of 674 movies shown on nationwide German TV channels gives an estimate of the quite remarkable size of such short-term timing effects (Hennig-Thurau et al. 2013). A measure of the average rating of the day-time combination at which a film was broadcast was the second strongest determinant of a film’s viewership, with only the films’ audience numbers in German theaters exhibiting a higher parameter.

Studies also provide evidence that “inheritance” effects matter for TV programs. A widely used empirical approach here is to study the correlations between a program’s viewership ratings and those of the show that precedes it. Research usually reveals significant associations (e.g., Tiedge and Ksobiech 1986, in a study of almost 1,000 prime-time series aired during 1963–1984, find an average correlation of 0.49). But causality is once more unclear—TV managers may incorporate the effect into their decisions and choose attractive shows to lead consumers into watching other attractive shows. Wilbur (2008), while not fully accounting for the endogeneity bias caused by managers’ decision heuristics, includes a program’s lead-in audience in his regression to explain the program’s market share, controlling for several other program and network characteristics. He finds that the lead-in parameter is significant, as also is a “lead-out” parameter (by which he intends to capture the attractiveness of the subsequent show).

Similar short-term timing effects exist for music played on the radio. Some days (weekdays) and hours (morning and meal times) are more attractive to consumers than others (weekends and evenings). Thus, radio consumption peaks early in the day (in Germany around 7am on weekdays and around 9am on weekends) and declines continuously, dropping to very low levels by the early evening (e.g., Gattringer and Klingler 2014). Lees and Wright (2013) provide evidence that "inheritance" effects also matter in this medium in a study using radio diaries by more than 1,000 consumers in New Zealand; we assume that their findings should hold for most other Western markets.

And moviegoers also have varying preferences regarding different days of the week and the hours at which they prefer to watch a movie in a theater. Figure 13.3 shows that, in 2016, the weekend accounted for about 71% of all box-office revenues. These day/time-related preferences of movie audiences imply that the way theater owners schedule their showings during the day and over the week impacts the success of each film. *Entertainment Science* scholars Jehoshua "Josh" Eliashberg, Charles "Chuck" Weinberg, and Berend Wierenga, together with colleagues, have developed SilverScreener—a complex econometric model that provides theater owners

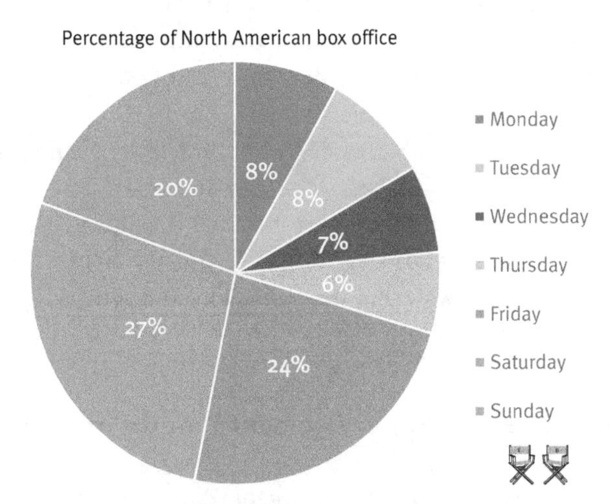

Fig. 13.3 Average box-office revenues per day of the week
Notes: Authors' own illustration based on data from The Numbers. The percentages in the figure refer to the box office generated by the ten most successful films in a given week over the course of the year 2016. For example, the total average weekly box office was $193.7 million, so that 27% equates to about $53 million for an average Saturday (including all holidays).

with scientific support regarding their decision of when to screen a particular movie (e.g., Swami et al. 1999).[374]

Overall, these findings show that short-term timing can influence the success of entertainment products quite substantially, in addition to long-term and mid-term timing factors. But because short-term timing decisions are largely in the hands of distributors, the influence of entertainment *producers* on short-term timing is usually quite limited. The major way to exert some influence might be to do so as part of the production or distribution deal, i.e., when selling a movie to a TV network or leasing it to a theater.[375] If that does not work, then providing the distributor with convincing arguments about the perfect time slot for a product and trusting him or her to implement it might be the only tool remaining for the producer.

Competitive Timing

Our discussion of timing effects so far focuses heavily on demand- and supply-sided effects, but has excluded the role of competition. Despite the hedonic character of entertainment (which limits the products' comparability), managers act under the assumption that competition also affects entertainment success. This is why Hollywood studios have subscribed in unison to the "Competitive Positioning" report assembled by the National Research Group (which determines consumers' awareness of and interest in seeing upcoming films and identifies potential conflicts due to competition; Epstein 2005) and reference the "Feature Release Schedule," as published by Exhibitor Relations.

In the following, we will present scholarly investigations of the extent to which competition influences the performance of entertainment products. We will then look at scholarly attempts to craft strategies for timing the release of a product in a way that accounts for competitive reactions, often drawing on the logic of economic game theory.

How Competition Affects Entertainment Product Success

The majority of empirical scholarly work on competition effects has examined the theatrical performance of movies. Scholars create measures of the competition that existed when a certain film was theatrically released

[374]In addition to the initial study, let us recommend the articles by Eliashberg et al. (2001, 2009) for more details on their approach and its implementation.

[375]As an example of such "deals," Schwartzel (2017) reports the terms that Disney imposed on theater owners for showing The Last Jedi movie.

and then link these measures to the success of the film under scrutiny. Because these measures of competition vary across studies, results not only shed light on the degree to which competition influences movie success, but also highlight the facets of competition that deserve the most managerial attention when making release decisions.

One key facet of competition is the *similarity* of other available films. Studies provide evidence that the presence of highly similar alternatives hurts a new movie's success. Elberse and Eliashberg (2003) find a competition elasticity of −0.22 for the North American box office for the number of films in the Top 25 that are of the same genre *or* the same MPAA rating as a newly released film:[376] a 10% higher competition level reduces the theatrical success of a newly released film by about 2%, on average. In their analysis of three Western European countries, they find that competition is also significant, but less strong, whereas it is more than twice as high in the UK. When Clement et al. (2014) replicated this approach for North America and Germany with newer and larger data sets, they found very similar parameters.

But it matters how such similarity is defined. Ainslie et al. (2005), in their modeling of the diffusion of 825 movies, also find that competition, in terms of films of the same genres and ratings, matters (adding them to the analysis results reduces box office prediction errors by almost 40%). But they point out differential effects of two different kinds of similarity: releasing a new film against movies of the same *genre* hurts opening week sales and reduces the peak of sales over time. Competition in terms of *age ratings* also decreases initial sales, but then the scholars note a "displacement effect"—the audience members lost to same-rating competitors in the opening week flock to the film in subsequent weeks. With ever shorter life cycles, it remains unclear though whether there would be still enough time left today for such a "displacement" effect.

Broader competition measures that also include films that are less similar appear to be less diagnostic. Clement et al. report that the number of *any* other new releases at a movie's opening weekend (weighted with their respective ad spending) correlates with sales only about half as strongly as *similar* new releases do.[377] And Calantone et al. (2010), who use about 3,000 movies released in North American theaters between 1997 and 2004, find the usual negative correlations between same-genre competition and weekly revenues. But those between *different*-genre competition (a *very* low similarity measure) and weekly revenues are basically zero.

[376]Elberse and Eliashberg divide each film by its "age" in terms of previous weeks.

[377]In their empirical models, Clement et al. also find that this (weighted) number of all new films influenced the number of *theaters* allocated to a film by distributors/theater owners.

Clement et al. further include a competition measure that looks at *previously* released films—they find that success barely correlates with the average age of the movies against which the new movie runs. This insight points to the role of *time* for competitive effects. Gutierrez-Navratil et al. (2014) shed more light on this issue in an analysis of 2,811 movies released in five countries (the U.S., United Kingdom, France, Germany and Spain) between 2000 and 2009. They specifically study the role of competition by releases over different weeks, estimating a fixed-effects regression across the different countries. The scholars use a cross-sectional estimation approach (to minimize the endogeneity problem associated with weekly effects of competition) and measure competition as the number of all other movies that were released in a country in a given week, weighting each competitor with its own opening weekend theaters.

Their analysis then includes seven competition measures, ranging from three weeks prior to a film's release to three weeks after its release. For example, when the first IRON MAN movie was released on May 2, 2008 in North America, same-week competition in the area was mainly the parallel release of MADE OF HONOR's 2,729 theaters and the 30 theaters of FUGITIVE PIECES. The previous week's competition (which takes into account that not all consumers interested in a film see it during its opening week) was BABY MAMA, HAROLD AND KUMAR ESCAPE FROM GUANTANAMO BAY, and DECEPTION, which all opened in more than 2,000 North American theaters on April 25. The scholars controlled for a movie's own opening weekend theaters in their analyses, but for no other characteristics of the film in question.

Gutierrez-Navratil et al.'s findings show that the role of competition varies quite strongly with time. Figure 13.4 reports the elasticities per release week, showing that the influence of competition peaks in the release week, and that its impact fades quickly afterward—other releases only play a role in the two weeks prior and two weeks after the release week. Competition's effect is also asymmetric, with post-release actions mattering more than what competitors do before one's film is released.

And Gutierrez-Navratil et al.'s study offers another insight: competition effects appear to be non-linear in a given week. The significant and negative nature of a squared term of the competition measure suggests that the effect of competition decreases with the number of rival products. Gutierrez-Navratil et al.'s elasticity estimate (which takes first-order and squared terms into account) for an average movie peaks at −0.18 in the opening week, quite similar to what we learned from the work of other *Entertainment Science* scholars.

We know much less about competition in other entertainment formats. One noteworthy exception comes from the work by Luan and Sudhir (2010), who study the role of competition for DVD sales. They find that competition also matters in this distribution channel—and that there are,

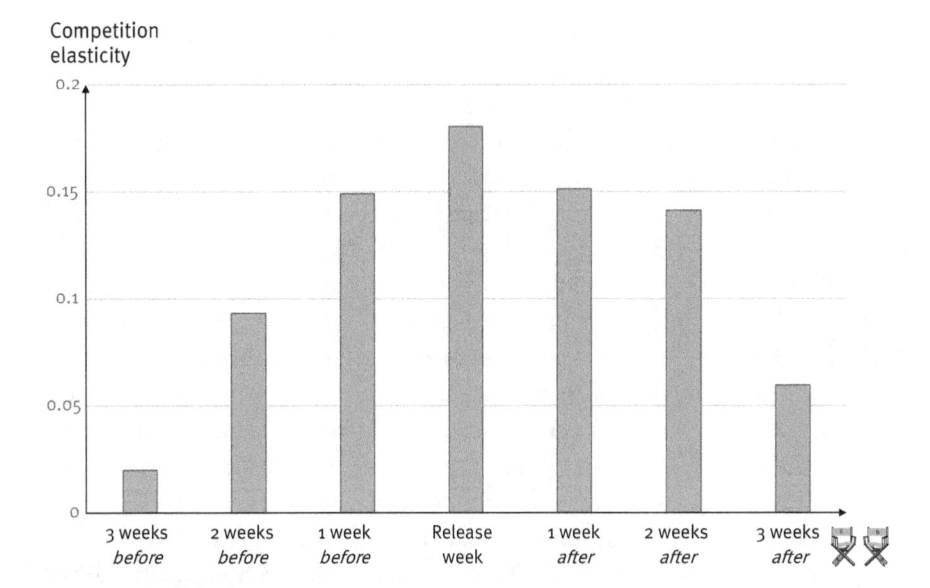

Fig. 13.4 The effect of release, pre-release, and post-release competition for movie success
Notes: Authors' own illustration based on results reported in Gutierrez-Navratil et al. (2014). The numbers are elasticities (first and second order) of the competition in a given week on the total box office of a movie.

in this case, two different sources of competition. The first source comes from other DVD releases (the number of DVD titles released in the next two weeks, weighted by their success in theaters), which are found to have a significant, but small effect: a 10% higher between-DVD competition corresponds with a mere 0.5% decrease in sales.

The second source of competition is much higher and less expected: the authors find that DVD sales are also impacted by what is happening in *theaters* at the same time. Here, a 10% change in theatrical competition (i.e., the box office in the DVD's release week for *all* films shown) reduces the sales of a DVD by 2.2%. Luan and Sudhir's (2010, p. 454) straightforward conclusion: "studios should avoid releasing their DVDs in the same week as box office blockbusters are released."[378]

What is the Right Time to Enter a Competitive Entertainment Market?

Understanding the impact of competitive products is certainly valuable information, as is knowing about the conditions under which substitution—the effect that underlies competition—is most strong. A related, but separate

[378]In a more general way, their findings point at inter-channel cannibalization between theaters and the sales of DVDs.

question is how a producer of entertainment should time the market entry of a new product in a competitive environment, i.e., relative timing.

A number of scholars have tackled this challenge for theatrical movie releases, treating the producer's decision as part of a "game" and drawing from game theory. Game theory is the only economic theory we can think of that Hollywood has ever honored with a leading role in a mainstream movie (and a best picture Oscar!). But be warned: A BEAUTIFUL MIND is not the most reliable source for learning about the theory itself (see, for example, Rey 2008).

An excellent example of timing research using game theory is by Krider and Weinberg (1998), who conduct a game-theoretical analysis based on a "dynamic attraction model." The scholars provide analytical (but not empirical) evidence that in a two-movie constellation in which one movie is strong and the other is weak (in terms of marketability), it is *optimal for both* movies if the weaker one delays its release, compared to both films being released in the same week.

What if both movies are more similar in terms of marketability, with one being only *slightly* more marketable than the other? For this situation, Krider and Weinberg deduce from their analysis that the same release pattern is still optimal for the (marginally) stronger film, but brings suboptimal results for the weaker movie. It is nevertheless likely to occur because the weaker movie has more to lose than the stronger one. Krider and Weinberg argue the situation constitutes what game theory refers to as a "chicken game." They compare it with a situation in which a Volkswagen van and a Greyhound bus are careening toward a head-on collision: whereas both vehicles will lose time when avoiding the other, and although each vehicle swerving would result in a stable situation, the Volkswagen has a higher probability of swerving because its driver is more likely to lose in a collision. Thus "the weaker movie, analogous to the Volkswagen, is most likely to delay its opening" (p. 8)—to swerve, so to speak.[379]

In a follow-up study, Einav (2010) builds on his separation of demand and supply effects on timing decisions for movies and develops a game-theoretical model of competitive distribution timing decisions, empirically analyzing what he calls the "movie release timing game." He uses the same set of films released between 1985 and 1999 as in his earlier study, this time combining it with release data announcements from film producers, as reported by Exhibitor Relations in their "Feature Release Schedule" report.

[379]The scholars also reflect on what can be expected to happen when both movies are *equally* liked by consumers, having similarly high playability. In this case, Krider and Weinberg conclude that it would be best for both movies to open at the same time, and early in the "season," so that their playability potentials can be fully realized. This finding might deserve a second look, though, when taking into account the lesser-than-usually argued differences in consumer demand between different release dates as reported by Einav (2007).

By drawing on his measures of "true" demand per release week (stripped of any supply-side effects), Einav estimates a model which tests whether producers adequately balance the demand at a given time with the substitution caused by competing films. His model assumes a two-player constellation in which one producer makes the first move (setting the release date for his film not knowing the other's release date), and the other then acting afterward, with full information.

His results suggest that producers, in the time frame represented by his data set, tended to overweight the impact of seasonal differences in demand, but underweight the impact of competition. As a result, too many movies were released during holiday weekends. Einav determines that this overweighting is quite substantial: to justify the clustering of releases around peak periods, the inherent demand during those periods would have to be twice as large as he finds it to be in reality. Einav's solution: a more flexible release pattern, with releases shifted away from the most clustered weeks of the year. This has been, as noted above, the direction in which the film industry has been moving.

Changing the (Release) Time

The game-theoretic approach toward competitive timing suggests that, at least under certain conditions, those who set their release dates first have a competitive advantage over those who follow (Einav 2010). So the (film) industry has adopted a routine in which release announcements are often made years in advance, sometimes even before the script or cast of a film exist.[380] A side effect of such early determination of date is that circumstances might require a change of the release, due to delays in the creation of the product or competition-related developments. Among the almost 2,000 films in Einav's (2010) data set, the release date was changed for more than 20% of them. Most of these changes are short-term; only for about 5% of the films was the release date changed by more than five weeks (earlier or later). Announcing and changing release dates is not exclusive to movies, by the way: for example, Brad Thor's publisher switched the release date for his new book, USE OF FORCE, from June 6, 2017, to three weeks later.[381]

[380]As just one example, Fox announced in the summer of 2017 the release dates for four Avatar sequels, none of which had started filming by then (and for most of which, we assume, no screenplay existed). Via the brand's Facebook page, they tell fans, along with industry competitors, that "The journey continues December 18, 2020, December 17, 2021, December 20, 2024 and December 19, 2025!" (*Avatar* 2017).

[381]The publisher argued it was an independent move, not a response to the announcement of star author John Grisham's new novel for the same day (Gamerman 2017).

How are such release date changes perceived by consumers and financial investors? Einav and Ravid (2009) empirically analyzed stock market reactions to release changes, focusing on those changes that affected the release date by at least 60 days (earlier or later). Their data set is a subset of the films that Einav (2007) used; it comprised 302 changes for 260 movies from 25 publically-traded entertainment companies. Using an event-study method, they measured changes in producers' stock prices after a date-change announcement, finding that financial investors, anticipating consumers' responses, react mostly negatively to date changes—regardless of whether a new date is part of the announcement or not. The average negative impact on a firm's value in response to the announcement is $22 million. But there are strong differences between movies: for films with higher production costs, stock market reactions are significantly stronger, probably reflecting the higher losses that might occur for the producer when consumers perceive that the date change indicates a troubled production. The scholars do not investigate whether films with changed dates indeed perform more weakly at the box office—but taking these results and the less-than-usually-argued seasonal variations in consumer demand into account, entertainment producers should balance the pros and cons of very early announcements carefully instead of trying to block a time slate at all costs.

So far, we have only looked at timing decisions that affect the *initial* launch of a new entertainment product. But today, this launch is merely one facet of a new product's distribution, as products are now made available to consumers in various formats across a large number of distribution channels. Orchestrating those formats and channels represents another massive challenge for entertainment producers, with the timing of the market entry for each format being of core concern. We will dive into this topic next.

The Multi-Channel Challenge: Orchestrating the Multiple Formats of Entertainment

> "Television: that's where movies go when they die."
> —*Entertainer* Bob Hope *in 1953 when hosting the first televised Oscars (quoted in* Martinez *2003)*

> "I say to you that the VCR is to the American film producer and the American public as the Boston strangler is to the woman home alone."
> —*Film industry lobbyist* Jack Valenti *in 1982, arguing why the VCR should be illegal in a U.S. Congressional subcommittee hearing (in Committee 1982)*

Many products and services today are distributed via multiple channels. For example, you can buy train tickets at a travel agent, at the train company's physical outlet, via an online site, at a kiosk at the station, or from the conductor on the train. Multi-channel availability is also true for most entertainment products. But something is quite different for entertainment products versus train tickets: the different channels of distribution are not simply gateways to the exact same consumption experience, but they are often linked to a unique entertainment *format*. A train ride remains the same, regardless of through which channel the consumer purchased the ticket. But watching a movie in a theater is not quite the same as watching it on DVD or in streaming format. Each is a different format or mode of consumption for which consumers have differing preferences: some prefer the darkness of the theater, others the convenience of the living room. This is what we have in mind when we talk about distribution channels in the following pages.

Each kind of entertainment content we discuss in this book was originally made available commercially in a single format only: films in theaters, music on vinyl discs,[382] novels on printed paper, and games on machines in mall arcades or bars. Since then, technological developments, and particularly the recent advances in the digital connectivity of consumers, have made alternative formats possible for all those contents. Entertainment has been critical toward new technologies and formats from the very beginning (the introductory quotes in this section bear witness of this quite conservative attitude), but that has not slowed the popularity of the new formats with consumers, usually to the industry's own advantage.

Film producers now make multiple times the revenue from an individual movie compared to what they made when movies where released only in theaters (Epstein 2010), with the majority of revenues now coming from the many channels other than theaters (Friend 2016). Music revenues nearly doubled when the industry made their content available in CD format in the mid-1980s, in addition to via vinyl discs and cassettes (Degusta 2011); they have (finally) started to grow again as firms have begun embracing music streaming (*ifpi* 2017). Book sales have benefited from the availability of digital book formats (Anderson 2016), and games sales have reached unprecedented levels since they were made available for home use by consumers in the early 1980s (*Fandom* 2017). Jon Feltheimer, as Lionsgate executive, illustrated the plurality of formats and its importance as early as 2010: "'Weeds,' our hit show on Showtime, averages about two million viewers an

[382]We know that there was also the shellac disc and the phonautograph, but recorded music *really* became a mass market when vinyl discs were used as a storage medium.

episode but, in addition to more than $100 million in DVD revenue, it has also dominated iTunes charts… is available for streaming on Netflix's Watch Instantly service… and is sold by episode or season on Amazon, Zune, CinemaNow, Movielink and VUDU. Overall, it has already generated nearly five million digital transactions and counting" (quoted in Smith 2010).

Some might argue that the different formats of the same film, game, album, or novel are simply versions of a particular piece of entertainment content whose management we should better discuss in this book's section on "versioning" (see our pricing chapter). However, doing so would gloss over key insights. Although the price might differ between formats, it is not the focal issue. Instead, it is the *timing* that has absorbed the lion's share of the industry's discussion regarding different formats and channels.

Specifically, many entertainment products are rolled out *sequentially* across distribution channels—an approach known as "windowing" in the industry. Such sequential distribution has been at the center of both managerial and scholarly attention, as a result of the proliferation of channels and the presumed far-reaching implications any changes to the "status quo" would have for producers and distributors. In the following, we will first take an analytical look at the state of (sequential) entertainment distribution, before identifying the multiple forces that need to be carefully considered when making channel timing decisions. We will then cover empirical studies of sequential distribution which help explain how changes in the traditional distribution models would affect the industry, and who would win the most (and who might lose). Based on these insights, we will also take a look whether, and how, forms of entertainment which currently do not use sequential distribution could benefit from the approach.

What's to be Considered When Designing the Optimal Channel Mix

A Quick Overview of Entertainment Windows and Underlying Interests

> "Ten years from now, we'll release a film and you'll be able to consume it however you want… Do you want it in a theater? In your home? In your car?"
> —Yair Landau, *as vice chairman and president of Sony Pictures (quoted in* Smith *2005)*

Sequential distribution is a marketing strategy that is designed to maximize a producer's profit by making a product available to consumers in different formats in succession (see also Hennig-Thurau et al. 2007a).

Producers' profit maximization strategies almost never *fully* align with those of the distributors of their products, but joint interests in channel distribution timing of producers and distributors can be substantial. For example, book publishers and retailers have had only limited dispute over the time gap between the release of hardcover and paperback versions of printed books.

But when different formats are offered through *different* distributors, the producer's profit maximization efforts can impact those distributors differently, fueling the flames of conflict. This is why the history of sequential distribution has known some fierce battles. Consider the case of movies: whereas theater owners distribute films in their theatrical format and (only) earn profits from the corresponding revenues, retailers (or rental firms or streaming providers) capture the margins when a film is distributed through them in a home entertainment format: the traditional theater does not get anything when a film is a hit in DVD format. Theater owners thus aim for configurations that maximize theatrical revenues, whereas home entertainment distributors look for a maximization of the revenues that are generated via their respective home entertainment channel.

For movies, sequential distribution was introduced in the mid-1950s when Hollywood studios began to sell the TV rights of their films to individual TV stations (not networks). The initial theater-to-TV "window" spanned multiple years, and the first feature film to be aired in consumers' homes during prime time was MGM's The Wizard of Oz, a full *17 years* after its 1939 theatrical premiere (Dirks n.d.). But even with this delay, theater owners, along with the majority of producers, opposed the alternative movie format because they feared the generic competition between films shown in theaters and on TV; Bob Hope's quote at the beginning of this section speaks volumes about this first channel conflict.[383] The window between a movie's theatrical release and other formats of its presentation has been shrinking continuously ever since.

Today, Hollywood films are released for home consumption three to four months after their theatrical premiere, and frequent efforts are exerted by producers and new "middlemen" who want their own place in the value chain, to shorten this window even further. Recent windowing models reach as far as advocating the simultaneous release of films in theaters and home channels.

[383]For those who want to know more about the complicated historic relationship between Hollywood and TV, which raised the idea of "sequential distribution," we recommend the book by Segrave (1999).

Proposals for this strategy range from "premium VOD" services (such as the one suggested by Sean Parker's The Screening Room, in which consumers would pay $50 for a new film, as well as $150 for a special set-top box; Smith 2016) to simultaneous availability without *any* constraints. This is the recent approach for Netflix movies such as BEAST OF NO NATION, which the streaming firm makes available to its subscribers parallel to the (usually limited) theatrical premiere, and for independent films like MARGIN CALL, which debut on VOD before being offered in theaters (James 2015 and Miller 2012).[384] Each of these models has faced strong headwinds from theater owners who feel that their business model is threatened and have employed measures as far-reaching as boycotts of current or future releases of the studios that participate in efforts to disrupt the established sequence of windowing (e.g., Lang 2015).

In other areas of entertainment, such timing-related conflicts now also exist between producers and other value-chain members. For books, digitalization has enabled the creation of ebooks and e-readers that challenge the sequential hardcover-then-paperback model established in the 1930s, and are favored by digital retailers such as Amazon. Some publishers, most notably Simon and Schuster, a part of CBS/National Amusements, delayed the release of the digital versions of their books in 2008/2009 by up to four months after the hardcover release, in support of printed books and physical retailers, but returned to parallel releasing hardcover and digital versions.

More recently, the format competition for books has shifted from distribution timing to pricing as reflected in publishers' efforts to harmonize the prices for all formats, which reduces the attractiveness of ebooks (that were often available for lower prices, reflecting their lower production costs) for consumers. But the simultaneous release of hardcover and ebook versions has put pressure on the hardcover-paperback window, which, for some titles, is now down to six months, from the former 12-month norm (Bosman 2011). Windowing is also applied to rental offers for books (such as the flat rates offered by Amazon under different programs such as "Prime Reading"), which are designed to protect higher retail margins.[385]

---

[384]In a certain way, such simultaneous release would mean a return to the film industry's channel-related beginnings: on March 10 in 1933, the first motion picture was aired on TV in Los Angeles while it was still shown in theaters. We have to note that THE CROOKED CIRCLE was more of a technological experiment than a business model innovation, with only a handful of consumers owning TVs at that time. It had also premiered half a year earlier in Los Angeles' theaters; the theatrical life cycle was *much* longer by then. See also Novak (2013).

[385]Offering different formats such as retail and rental versions for different prices is a kind of price discrimination, which we discuss in the following chapter.

For games and music, windowing is less prominent these days. Games were traditionally re-released in a separate hardware format, but today, producers focus on offering multiple versions of a game at the same time. Game producers see no advantage in delaying consumers' access for certain formats, so that pure digital formats are released parallel to their packaged equivalents; ailing physical game retailers cannot prevent these simultaneous releases. Most songs and music albums are now available for consumers in the format of their choice, including CD or vinyl, digital download, or stream, either as a single transaction or as part of a subscription.

But as with any other form of entertainment, this simultaneous-release model must not be taken as a given: the music industry has tampered with the idea of distribution windows for quite a while. For example, Bhatia et al. (2001), as Booz Allen Hamilton consultants, suggested a sequence from (high-priced) CDs, to (mid-priced) digital downloads, to (low-priced) digital subscriptions, arguing that this windowing model would "allow labels to protect existing revenue streams and still offer consumers new ways to purchase music" (p. 73). They proposed their model at a point in time when Napster was defanged, and the authors thought that piracy was under control. As we will discuss in a little while, history proved them wrong—and the industry put their windowing ideas aside.

But as we write this about 15 years later, music labels are now beginning to experiment with similar sequential models, although they avoid calling them that. In December 2015, singer Adele released her album 25 on CD and digital download, but not on streaming platforms (Sisario 2015); it was eventually made available (seven months later) in that format on Spotify and other streaming platforms. Whereas Adele named personal reasons for the delayed streaming release,[386] the labels studied the performance of her music closely. The results were stunning: the album sold over eight million copies in its release month and about 18 million in its first six months, during which it was not available for streaming (McIntyre 2016).

These numbers seem to illustrate the benefits of windowing for music releases—although a final judgment would also need to consider the delay's impact on streaming revenues.[387] And, beyond Adele, there is more happening in terms of sequential music distribution. In 2017, Universal imposed

[386]To quote the singer: "It's a bit disposable, streaming." In *Time* (2015).

[387]No streaming numbers have been disclosed for the album, but we saw that it did not make it to the most streamed list after becoming available on Spotify. The final assessment of a decision like Adele's is even more complicated, as we elaborate in the next section: the physical album sold *more* (not fewer) songs in the weeks following its delayed release on streaming (Caulfield 2016).

a general two-week window for its music on the advertising-based streaming of Spotify (Roettgers 2017). In addition, streaming providers make attempts to make some music available *only* through their channel, at least for a certain period, reversing the order that Adele chose—an example is Jay Z's 2017 album 4:44, which was exclusively available on streaming service Tidal for a week before it became accessible in other formats and streaming platforms.[388]

Behind most of the windowing practices (both traditional and newer) are managerial assumptions, which, as we know from discussions with various industry managers, are just that—*assumptions*. The current practices are *not* grounded in reliable empirical data or solid theories of windowing practices. And the limited empirical tests the studios have conducted use designs that are severely flawed from a methodological standpoint. Often the specific content used prevents generalization, such as in the case of Steven Soderbergh's micro-budgeted, very experimental film BUBBLE, which was among the first films made available on home entertainment channels at the same time it premiered in theaters (Risen 2005).

In other cases, effectiveness of the actual windowing strategy is unclear because of boycotts from theaters, as in the case of the sixth PARANORMAL ACTIVITY film, which was part of an "early-VOD" experiment by Paramount (Mendelson 2015). Finally, some of the experiments involve inherently unattractive offerings—we believe that very few people were surprised that offering films for $30 via AT&T's DirecTV 60 (!) days after their release in theaters did not create much excitement among consumers.

Entertainment Science researchers, however, have systematically explored the complex forces that need to be balanced when making multi-format distribution decisions and, acknowledging these forces, also found ways to shed robust empirical light on the intricate issue. But let us first look at the forces that determine the effectiveness of sequential distribution configurations.

A Framework of the Forces that Determine Optimal Windows

The profitability of a distribution model for entertainment products that are sold and/or rented to consumers via multiple channels depends on a number of forces. Figure 13.5 overviews key forces, distinguishing those that are rather abstract (relating to general consumer preferences, "macro-level"

[388]See also our overview of the market for music in our chapter on entertainment business models, in which we name some other exclusive productions by streaming services.

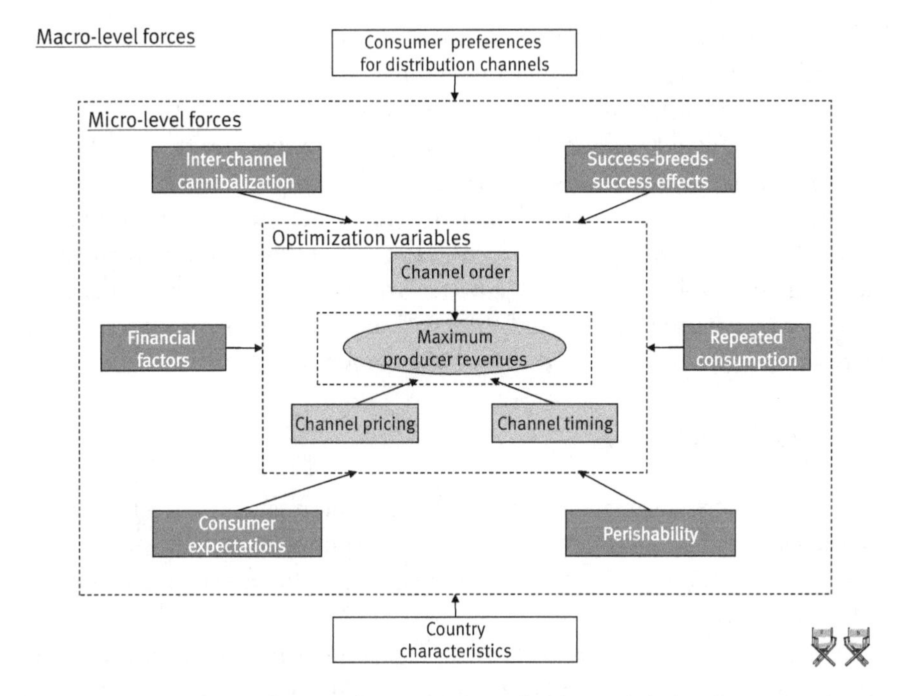

Fig. 13.5 Forces that influence the optimality of a channel design for entertainment producers
Source: Reprinted with minor adjustments with permission from Journal of Marketing, published by the American Marketing Association, Hennig-Thurau et al. (2007c) The Last Picture Show? Timing and Order of Movie Distribution Channels, October 2007, Vol. 71, No. 4, pp. 63–83.

forces), from others that relate to more specific features of channels and their interrelationships (i.e., "micro-level" forces) (Hennig-Thurau et al. 2007a). All these forces are "given" in a certain industry setting and cannot be influenced by a single producer. However, at the figure's center are variables that the producer can indeed determine (or at least heavily influence): the order and timing of the channels, and the pricing within each channel. Because of their active nature, we call these "optimization variables."

In practice, entertainment managers will face other factors when trying to optimize distribution configurations. What about the costs of protests and boycotts by members of the entertainment value chain who oppose a particular channel sequence or configuration, or even generally fight *any* distribution changes? These aspects are matters of channel power, which we separate from our optimization discussion: we consider it more helpful to first determine the raw effects that any distribution configuration would have for producers if *no* conflict occurred, essentially applying the *ceteris paribus* approach.

This then gives us a baseline against which producers can judge the costs of a potential conflict: if I know that I would gain $100 million from a different distribution model, I would have more incentive to fight for it than if I know I would gain only $1 million. But our analysis is by no means a call to seek out such channel conflicts: instead, designing channel configurations in a way that addresses consumer interest better should enable the industry to identify models that benefit *all* parties involved in the creation of entertainment value. But being able to do that requires a better understanding of the mechanisms, which is why we use analytics and theory.[389] Let us now take a closer look at macro- and micro-level forces that we have named in Fig. 13.5.

Macro-Level Forces

Two general, abstract characteristics influence the financial attractiveness of a particular channel configuration for an entertainment producer: consumers' preferences for each channel and, if applying a global perspective, certain country characteristics.

Consumers' channel-related preferences. The more value something offers to consumers, the more those consumers are willing to pay for it. This general truth applies also to entertainment distribution channels—the more customers value a channel, the more are they willing to pay for using it. Entertainment firms have strong normative beliefs regarding what a certain channel is worth: watching a movie in a theater is "superior" to watching it on a TV screen, and experiencing a film on a smartphone is something close to a disgrace. Director Christopher Nolan received applause from many in the movie business when he said about his film DUNKIRK: "[T]he only way to [carry the audience through the film] is through theatrical distribution" (quoted in Busch 2017). Similar arguments are made for music, where vinyl is often considered by managers to be the "true" medium for musical recordings and artists often disrespect streaming—Adele is certainly not alone here. And among publishers and authors, reading a book on a digital device is certainly considered inferior to a "real" printed one.

[389]Our discussion here, as in most of Part II of this book, takes the perspective of an individual producer of entertainment. Let us add that there might also be the need for an aggregate, industry-wide look at distribution configurations. Our analysis of the consumer behavior in entertainment points to the possibility that the amount of emotional and imagery processes might vary with the medium in which an entertainment product is consumed. If movies we watch in a theater have a stronger impact on us than those which we watch on Netflix (because they get more of our attention), the configuration of distribution channels might impact the relevance of filmed entertainment *as a whole*. For example, a shift from theater visits to Netflix views thus might generate higher revenues in the short term, but carry the risk of a reduction of imagery production and, subsequently, importance of movies in the longer term. A similar argument could be made for the music industry for a shift from vinyl and CDs to digital streams of music.

Those industry-insider beliefs are not driven by what consumers actually think and feel about the different channels, but much more so by the history of each medium and considerations of "objective" quality. It is not by accident that theaters, vinyl discs, and hardcover books are the industry's darlings—they were the first widely used formats, and so the artistic status of entertainment is closely tied to them, particularly for those whose personal coming-of-age is tied to their usage (which is the case for most industry leaders these days).[390]

From a scholarly perspective though, it is one thing if *artists* make such normative judgments, but quite another if producers and other managers do. Looking down on a certain channel because of its low "cultural esteem" fails to address changes in consumer preferences. We sense that culture-related arguments often serve as a straw man for reactionary thinking by producers—which is neither adequate nor helpful when it comes to shaping the present and future of entertainment.[391]

In reality, consumers' preferences are quite complex. For movies, some consumers prefer a theatrical experience (Puig 2005 quotes a moviegoer saying "I .. love the mythos of the darkened theater"), whereas others argue that 'there's no place like home,' where "nobody next to you rustles with fast-food and endlessly slurps his empty cola ice-cream, where nobody yatters and kicks you in your back with his feet" (Berlinale chief Dieter Kosslick, quoted in *epd Film* 2017).

Consumers' channel-related preferences are also dynamic. They have been particularly affected by the rise of digital technologies and the new opportunities that come with it. Streaming allows us to "binge watch" (or binge-listen) entertainment content with unprecedented convenience, and new serialized content adds value to this opportunity—and thus the streaming channel. Netflix reports that three out of four consumers who streamed all seven episodes of drama series BREAKING BAD's first season did so in a single session, and that such binge-watching was even higher for seasons 2 and 3 of the show (which had *13* episodes each; Ascharya 2014).[392] Because of these

---

[390]See our discussion of how certain phases in a person's live shape his or her entertainment preferences (e.g., the "reminiscence bump") in our chapter on entertainment product characteristics, which also applies to the channels through which entertainment is experienced.

[391]Focusing on economic effects of channels and configurations is not the same as ignoring artists' opinions about channels. Instead, managers must be aware that their distribution decisions can also influence the "supply side" of the business—the flow of talent that wants to work with them. Thus, there is a certain similarity between the roles of artists and distributors when designing distribution models—their respective attitudes should be considered in conjunction with profitability effects.

[392]Who enjoys binging, and on what content? Schweidel and Moe (2016) are among the few scholars to shed light on the phenomenon so far. Based on an analysis of a very heterogeneous set of TV content from streaming provider Hulu (including game shows, news, and sports), they suggest that binge watching differs according to consumer traits (i.e., watching patterns), as well as with program characteristics (comedies and drama content are more likely to be "binged").

preference dynamics, entertainment producers must regularly update their distribution models. They must do so to strive to achieve profit potentials, but also to avoid threatening the relevance of their content, in general.

If people prefer spending their time at home versus in a theater, restricting a movie's availability to theaters will drive people away from movies to other forms of entertainment. Alternative choices for a person who foregoes a movie might be video games, books, or music; however, they might also turn to completely different forms of entertainment, such as social media (which most consumers today consider as a kind of entertainment on its own; Leggatt 2011) and short videos on YouTube and similar platforms. As Disney-CEO, Bob Iger has pointedly phrased this need to rethink distribution systems: "if we tried to fight [technological change] or slow it down or do anything at all aimed at deterring its impact on our business, we were going to lose that battle. … It's imperative .. that we embrace [Internet distribution channels], .. because [that is] where the consumer is today" (quoted in Ryan 2016).[393]

Country characteristics. We have highlighted differences that exist in the demand for entertainment products between consumers in different cultures and countries more than once in this book. Such differences almost certainly affect consumers' preferences for distribution channels. Some countries, such as France, have a century-old adoration for the institution of movie theaters, whereas in others, such as in many Asian countries, theaters lack this cultural esteem (and history); people there have a stronger inclination to watch films at home.

Related to differences in consumer behavior are infrastructure conditions that affect the quality and convenience of entertainment consumption. Streaming a movie is far less fun without a speedy broadband connection, and so consumers' interest in streaming will be lower under that condition. Convenience is tied to the effort that is required to access a channel, so that a customer's usage of a channel or format will depend on its easy availability, such as access to physical theaters, retail stores, and rental outlets. The same logic applies to the ease of having a CD or DVD shipped to the home, or shopping a printed book: if there is a Barnes and Noble Superstore around the corner, printed books will have a higher commercial outlook compared to when retail availability is lacking.

[393]This forward-looking perspective of its CEO is behind Disney's ambitions to vertically integrate and to become a provider of streaming services itself. Let us note that this does not mean that we wholeheartedly endorse Mr. Iger's *content*-related strategy to address such changing channel preferences—which focuses solely on large-scale, brand-based blockbusters ("My mantra for films is: Make them big and make them great."; quoted in Lieberman 2017).

Other culture/country parameters that need to be considered for the design of distribution configurations are local regulations and laws. Although the design of windowing in the U.S. is fully up to negotiations between market participants, French law prevents any film that has been shown in a theater from being released on VOD until four months have passed. The window for streaming platforms such as Netflix in France, the birthplace of cinema, in 2017 is three *years*. As a consequence of this legislation, Netflix does not release its films in French theaters at all and has experienced clashes with the French film industry (Wilkinson 2017). In Germany, any movie that receives government subsidies for its production faces a legal embargo of six months before it can be released on DVD or VOD, of 12 months for pay-TV, and 18 months for free-TV.[394]

Micro-Level Forces

The macro forces we highlighted impact the financial attractiveness of certain distribution configurations directly (e.g., consumers might switch to other forms of entertainment if their preferred channel is not available), but also indirectly because they influence certain micro-level forces. We will now take a closer look at these forces, each of which also influences the optimality of channel configurations.

Interchannel cannibalization. When the same content is made available in different channels, there is the risk that one channel cannibalizes the other because a "good movie is a good movie, regardless of where it's shown" (film producer Martin Bregman, quoted in Arnold 2005). Interchannel cannibalization provides an argument for postponing the release of a product in a channel that has a lower margin so as not to hurt the higher margins generated in the other (previous) channel. For books, Clerides (2002) offers evidence of such an effect and insight into its size. When he studies sales patterns for about 500 books by an American publisher that were released in both hardcover and paperback format, he finds that when the paperback release is delayed (which happened in three out of four cases), hardcovers accounted for 38% of total book sales, whereas hardcover sales are only 12% of total book sales when both formats are introduced *simultaneously*.

But cannibalization is format-specific. Chen et al. (2017) study consumers' reactions to a delayed release (by between one and eight weeks) of *ebook* versions of 99 book titles as part of a "natural experiment" that happened

[394]For an overview of recent cultural regulations for movies, take a look at MacNab et al. (2017).

when a leading publisher stopped releasing ebook versions of new releases at Amazon in spring 2010. Using a negative binomial panel regression (with weekly book sales as dependent variable), the scholars do not find significant cannibalization between ebooks and print sales. Their results show that ebook sales decline by more than 40% at U.S. book retailers in each week when a title is not immediately available at Amazon, but weekly sales of *print* books at Amazon and other retailers do *not* increase because of the unavailability of the electronic version. In other words, the (un)availability of an ebook format does not make consumers switch to the printed format, suggesting that digital-affine consumers do not really see the formats as substitutes. It's not clear though whether consumers who prefer printed books would react similarly loyal to the printed format.

For movies, scholars provided evidence that, at a time when traditional video rental by Blockbuster and others was still a viable business, video rentals could cannibalize theater attendance (e.g., Frank 1994; Lehmann and Weinberg 2000). They also found that the effect is not one-sided: attractive theater releases can hurt DVD sales in a given week (Luan and Sudhir 2010).

More recent studies show that online channels can cannibalize physical channels. Kumar et al. (2014) investigate how the unavailability of digital versions of films for rental (VOD) and purchase (Electronic Sell Through, or EST) influences DVD rentals. They made use of a "blackout" period for the digital formats (but not the DVD) that existed around 2008–2010 before a film's showing on pay-TV. For a data set of 194 movies aired at that time by the Top 4 pay-cable networks, they find with a SUR analysis that the "blackout" increases DVD sales by around 6% per week.[395] People rent the DVD because a movie is not available in a digital format.

Yu et al. (2017) study the interplay between Netflix and DVD sales, using Netflix's decision to no longer list 128 movies from Paramount, Lionsgate, and MGM in 2015 as a "natural experiment" (the films were picked up by less popular streaming service Hulu instead). A difference-in-difference model finds that from months 2 through 6 after content becomes unavailable at the streaming service, DVD sales of the films increase by 25%, on average. They calculate that this translates into an increase of about $600,000 in retail sales (about 60% of which, or $360,000, would have flowed back to the studio). Splitting the data set showed that availability on Netflix affected only more-recent titles and box office hits; for recent titles, the DVD sales increase averaged about 60%.

[395]The second equation in their model uses DVD sales as dependent variable—we will get to this in a few paragraphs.

Some additional insights on cannibalization from streaming options can be derived from a study by Ananthakrishnan et al. (2016). The scholars do not use actual SVOD consumption data, but use the free availability of episodes of the hit series DOWNTON ABBEY on TV station PBS' website and look how such availability (which can be compared to the "gratis" nature of Netflix content for the service's subscribers) influences the digital EST sales of the series. Using a fixed-effects regression, Ananthakrishnan et al. provide evidence that when PBS made DOWNTON ABBEY episodes available for a limited time frame in 2014, EST sales of those episodes of the series shrank by an average of 8.4%.

Little evidence of cannibalization exists for theatrical visits. Based on a survey of 1,200 consumers, research firm MarketCast reported in early 2016 that close to 25% of the respondents indicated that they would trade a theater visit to see a desired movie for a "$50 Day-and-Date Premium VOD service," if such an option was available (Busch 2016). But we have noted throughout the book the problems associated with such self-reported surveys, and the hypothetical "what-if" nature of the question certainly does not mitigate this concern.

However, for some entertainment formats, there is evidence that they are *not* considered as substitutes by consumers. When Weijters and Goedertier (2016) conduct a latent class analysis of Belgian music consumers, they find that more than half of the respondents are loyal to a single format, using either CD or online formats, but not both, i.e., the formats are not substitutes for them. And when Danaher et al. (2010) use NBC's withdrawal of 5,200 TV series episodes of 75 series such as BATTLESTAR GALACTICA and HEROES from Apple's iTunes store (EST channel) in August 2007 as another natural experiment,[396] they find *no* switching effects for sales of physical DVDs (via Amazon.com). The physical DVD format, which contains whole seasons of a show, is not seen as a substitute for the digital show format (which is sold episode-by-episode) by most consumers. In Kumar et al.'s study of the "blackout" period of digital movie formats, they also look at DVD *sales*—and find them to benefit much less than DVD rentals from the unavailability of digital formats.[397] Obviously, consumers switch more easily when it comes to renting a movie then buying one.

[396]See Barnes (2007) for background information on NBC's content withdrawal. The authors also looked at whether the withdrawal affected *illegal* downloads—see our discussion of piracy later in this chapter.

[397]In a joint model with DVD sales and rentals, the "blackout" effect is 10% compared to the 6% when only DVD rentals are studied. But in an isolated DVD sales equation of the SUR model, the "blackout" parameter is not significant.

Success-breeds-success (SBS) effects. Although cannibalization is generally a negative thing, some channels can also exert a *positive* effect on others. We dedicated a whole section of this book to such action-based cascades, or SBS effects, and we recommend that managers revisit that section before making distribution decisions. In essence, the existence of SBS effects implies that the performance of an entertainment product in an earlier channel can send a signal to consumers about the existence or quality of the product, helping the product's performance in subsequent channels.[398] If a product is released simultaneously in multiple channels, such an effect between channels will be limited or non-existent.

Gong et al. (2015) find a positive spillover effect in a study of two transactional digital channels for movies (i.e., EST purchases and VOD rentals), although it does not stem from a film's success in one channel, but from a marketing measure in the other channel. The scholars conduct a 14-week field experiment to analyze how EST price reductions for 233 "catalogue" studio movies (i.e., excluding the most commercial titles) affect rental revenues. The results of a fixed effects negative binomial regression[399] show that reducing a movie's retail price *increases* rental revenues for the respective film by between 2 and 9%. Gong et al. attribute this to a "positive informational spillover effect" between the channels, i.e., price discounts for sales offers increase awareness of a product in a rental channel.

Repeated consumption. A second positive effect that one channel or format can have on another is that its usage can trigger a desire within the consumer for *subsequent* consumption experiences with the same entertainment product again (and again…; see also Luan 2005). Remember, though, that such repeated consumption is the exception rather than the rule in entertainment because of satiation effects.

A general argument is that rental transactions can trigger purchases, whereas the reverse is much less probable. For music albums, about half of the consumers who buy an album on vinyl have already listened to it through an online channel, such as digital streaming, which suggests that vinyl sales benefit from an album's availability in digital channels (*The Economist* 2017). For movies, Smith and Telang (2009) find that movies' DVD sales can profit from airings of the movies on free-TV. The scholars analyze DVD sales for 522 movies that were aired on over-the-air and free cable TV in the U.S. during

---

[398]Keep in mind that the positive statistical effect of SBS can also work *against* a product—because if the product lacks success, it is then stigmatized as a flop in later channels!

[399]The method accounts for the non-normal distribution of sales and the large number of zero sales in the weekly data.

an eight-month period in 2005–2006. Using a weekly fixed-effects regression, the scholars find a significant increase in DVD sales at Amazon.com that begins in the week a film is aired and lasts for two to three weeks.[400] Thus, watching a film on TV stimulates consumers to buy its DVD, with the effect being substantial: DVD sales increase by up to 120% (for terrestrial broadcasts) and almost 30% (for cable), on average, in the airing week, before declining afterward. Does this mean that movies should be aired earlier to stimulate home entertainment revenues? No—because the results are only valid under the conditions in which the data was gathered, they inform us that DVD sales benefit from TV airings when *several years have passed* since the movies were originally released. Not more, but also not less.

And Kumar et al. (2014), in their study of pay-TV-aired movies, also look at DVD sales. They find that sales increase by 8%, on average, after the airing of a film on a pay-TV channel, an effect the authors label the "broadcast" effect. The authors' explanation is basically the same as the one by Smith and Telang for free-TV broadcasts: buying a DVD allows the consumer to add a movie that was newly discovered on pay-TV to his collection. They find no such repeated consumption effect by TV broadcasts for DVD *rentals*, by the way—in other words, watching a movie on TV, something we consider an act of "renting" in our value creation analysis (see our chapter on entertainment business models), does not stimulate more of the same.

Perishability. The short life cycles of entertainment products imply that products lose attractiveness for consumers as time passes. Scholars such as Frank (1994), Lehmann and Weinberg (2000), and Prasad et al. (2004) all have thus argued that when a producer delays the release of a movie, game, or book in a channel or format, a "wear-out" or "decay" effect takes place. Industry managers have articulated a similar perception, and a decline in consumer interest has been a key argument for shortening channel windows. Bob Chapek, as Disney's home entertainment president, compared a movie to a melting ice cube: "The longer it sits, the smaller it becomes" (quoted in Dutka 2005).

In their empirical investigation of book success in Germany, Schmidt-Stölting et al. (2011) show that the sales of a book's paperback version vary with the weeks that have passed since its hardcover release. Assuming a linear time effect, they find that each week by which the paperback release is delayed, the book's paperback sales shrink by 0.7%. Considering that the mean window length in their study was 78 weeks and the average variation

[400]This effect exists only for movies that were available for purchase at Amazon.com when the airing took place. In the regression, the weeks around the airing of a film were coded 1 and then multiplied with the number of viewers, whereas all other weeks were coded 0.

was 18 weeks, shortening the window by those 18 weeks would result in a sales increase for the paperback of more than 12%. Of course, this does not tell us anything about how such a move would influence hardcover sales through cannibalization.

Consumer expectations. One might argue that producers could develop a profit-maximizing channel configuration based on the other micro-level forces—but open a secondary channel as soon as a specific product has fully exploited the primary one (because then no cannibalization can happen anymore). But such a perspective leaves out consumers' strategic decision making. In the context of films, Prasad et al. (2004) have argued that consumers who would prefer to watch a film in a later channel (such as home entertainment) make their decision whether to watch the film in an earlier channel (e.g., the theater) based on their *expectation* of when the film will become available in their preferred channel.

So, when producers shorten the time between a film's theatrical release and its home entertainment availability (e.g., because the film was a flop), consumers might *strategically defer* their consumption of other movies at theaters—because they expect to soon be able to rent or buy them in the channel that they prefer. Preventing such "anticipatory cannibalization" is a main argument for the current existence of a time lag of several weeks or months in which films are no longer shown in theaters, but are also not available in any other channel for consumers—and thus cannot generate any revenues at all during this "blackout." The tradeoff is that the product's economic value melts during this time period, as we have argued above.

What are the sources for consumers' channel-related expectations? They get information from retailers (who publicize release dates early on),[401] but mainly rely on personal experiences based on "industry standards." These standards are also the reason why theaters do not want producers to shorten windows for single titles, or even to *test* potential effects. Theater owners fear that the mere act of doing so could influence consumer expectations and encourage strategic deferral, with consumers bypassing the theater even when no actual shortening takes place.

Financial factors. In addition to consumers' channel preferences, the economic attractiveness of a specific channel configuration also varies with certain financial parameters. First, because delaying a channel opening means

[401]For example, Amazon sent emails to their German customers in the week before the movie Star Wars: Episode III was released in theaters, inviting them to preorder the DVD for the new movie. Today, the retailer often provides consumers with the opportunity to pre-order the home video format of a film at its theatrical release.

that revenues will flow back later to the producer, the industry-specific discount rate is relevant. The higher the discount rate, the more a producer should be interested in getting his or her money back early, because, all else equal, waiting for a later channel to open means losing more money.

Second, the revenues from different channels flow back to producers at different rates and under different terms. Thus, an economic perspective suggests prioritizing channels in which a producer earns higher shares over those in which others get the lion's share of the money that is generated by the entertainment product. We have glossed over the different ways revenues are shared between producers and distributors in entertainment earlier in this book and in our discussion of fixed payments. These different producer shares need to be considered[402]—even if the shortening of the window between theaters and digital home channels would result in lower total net revenues, such a change *could* still be attractive for producers if their *share* of revenues from digital home channels is substantially higher than that from theatrical box-office revenues.

Valuing Alternative Distribution Models Empirically

The complexity of orchestrating channels results from the parallel existence of all the forces that we discussed in the previous section. In some of the empirical studies we cited, different forces were already intermingled. For example, when Smith and Telang (2009) found a positive effect of a movie's TV airing on its DVD sales due to repeated consumption, this result is essentially a "net effect," because it also incorporates the cannibalization of DVD sales by the previous showing for *some* consumers. And when Kumar et al. (2014) find that a movie's showing on pay TV has no effect on DVD *rentals*, this can also be seen as the result of oppositional forces that neutralize each other: whereas some consumers will probably avoid renting an aired film because they have already seen it (i.e., cannibalization), others may want to revisit it (i.e., repeated consumption).

We have demonstrated that such secondary data-based approaches generate insights for understanding inter-channel effects. They are much less powerful, however, for judging *alternative* channel configurations. The reason is that the variation in channel timing and order in real-world entertainment markets has been very limited, and extrapolations outside of the

[402]Let us note that doing so can be quite demanding if different allocation models have to be compared (such as fixed payments on Netflix versus a revenue share from theaters).

available field data range lack validity in general.[403] Even occasional "natural experiments" (such as the sudden unavailability of some content via a certain channel) only allow selected glimpses, as the changes studied are usually restricted to a single channel and a specific condition.

These problems are avoided by studies that have made the *modeling* of complex channel configurations the focus of their work, using insights from lab experiments to shed light on the effectiveness of channel options that have not been used in industry practice. We combined experimental insights and analytical modeling with data from a representative survey of 1,770 consumers in three countries (the U.S., Germany, and Japan) in an attempt to study simultaneously the various effects at play in complex entertainment channel systems (Hennig-Thurau et al. 2007a).

We focused on movies and studied the four major distribution channels that existed at that point in time: theaters, DVD retail, DVD rental, and EST (i.e., the sales of digital files over the Internet). The general logic of our study followed the INDIANA JONES principle of "Anything Goes": we considered all possible combinations of channel order and timing. At the heart of our study was a conjoint experiment we conducted in 2005, in which we asked the participants to choose between alternative options for consuming a movie they had indicated an interest in watching via one of the four channels listed above. To ensure a high level of realism, the movies from which participants chose were nine actual films that were forthcoming at that time (e.g., X-MEN 3, CHRONICLES OF NARNIA, THE DAVINCI CODE).[404] The consumption options differed with regard to the channel, but also other key characteristics, namely the price, the availability of "extras" (like a "Making Of" featurette), and the language options. The consumers' choices in the experiment were used to extract the personal preferences of each participant regarding the different channels (measured as "conjoint part-worths"), along with his valuation of the other consumption characteristics.

We then combined that preference data with individualized information we had gleaned from the survey to tap the other forces that influence the profitability of a channel configuration (which we discussed above), such as success-breeds-success, cannibalization, and repeated purchase tendencies. For the financial forces of costs of waiting and the revenue allocation of each channel,

[403]And there is also the issue that release dates are often affected by endogeneity: for example, attractive books might have a systematically longer delay before being released in paperback format, which interferes with the measured effects of the delay length on the success of the paperback version.

[404]Each participant rated his or her interest in the films based on their posters, trailers, plots, and casts.

we made assumptions based on real-world observations.[405] Using cutting-edge conjoint techniques, for each of the three countries, we examined all 875 alternative channel configurations and determined which channel an individual consumer would have chosen. When we aggregated this information across all 500+ participants, and calculated how these choices, under the assumptions of our study, would have impacted producer revenues, we learned the following:

- Compared with the standard channel configuration that was in play at the time of our study (a six-month theatrical window followed by a release on DVD retail and rental, and another six-month window until the EST release), producers could increase their revenues substantially via alternative channel configurations. For the three countries we studied, we found configurations that promised between 11.6% and 16.2% higher revenues. In other words, considering alternative distribution models indeed bears the potential for substantial revenue increases for producers of entertainment.
- Optimal distribution configurations for producers varied strongly between countries—and probably still do. The best scenario for a producer in the U.S. was a *simultaneous* release in theaters, DVD rental, and EST,[406] followed by a DVD sales release just three months later. In Fig. 13.6, we report the resulting effects for producers (as well as the different distributors) and contrast them with the results for Germany—where the ideal model was an exclusive theatrical release, followed by a DVD retail release three months later, with another nine-month wait before a film is released on DVD rental and VOD. Consumer preferences regarding channels vary between countries, and channel solutions need to acknowledge such heterogeneity. But we acknowledge that having different distribution patterns in different parts of the world carries its own challenges, given that in today's digital age, entertainment travels fast between continents.
- In *all* the revenue-maximizing configurations for producers, at least one distributor faced shrinking revenues, and sometimes these losses would be substantial. The figure shows that in the simultaneous release scenario in the U.S., theaters would lose more than 40% of their revenues; in the producer-optimal German scenario, theaters would gain, but DVD rental would have shrunk by almost a third.

[405]Specifically, using prevailing industry averages at the time, we assumed that producers would get a 50% share of theater revenues (the remaining 50% remain with the theater owner), 60% of DVD sales revenues (40% remain with the DVD retailer), 40% of DVD rental revenues (and 60% remain with the DVD rental company), and 50% of EST revenues (with the other half remaining with the Internet company).

[406]Please keep in mind that our results reflect a time when consumer preferences for watching movies via the Internet were quite low (below 5% in importance).

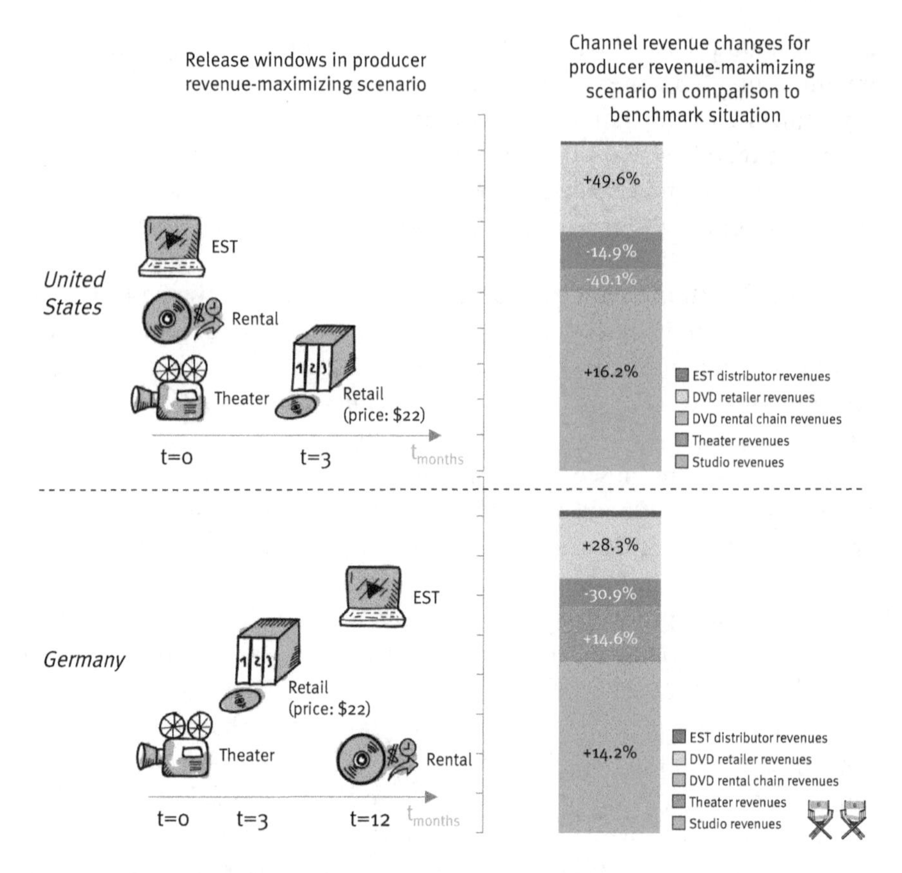

Fig. 13.6 Channel configurations that maximize film producer revenues
Notes: Authors' own illustration based on results reported in Hennig-Thurau et al. (2007a). With graphical contributions by Studio Tense.

- "Win-win" compromise scenarios, in which producers benefit, but no other channel is hurt, also exist. For the U.S., a three-month window between theaters and DVD retail, followed by another three-month delay before a film becomes available on DVD rental and VOD, would increase the producer revenues by 7% (compared to the baseline constellation) without harming theaters. Studying the development of the market over the last decade, it seems to us that the film industry is moving toward the adoption of this type of compromise model. But compromises, by definition, require some players to forgo an even more profitable constellation for a greater good. This implies that the system will experience instability, because who doesn't want a greater slice of the pie?

The main insight from our study is that adjusting channel configurations carries the potential for higher revenues for entertainment producers—and is thus a worthwhile endeavor. At the same time, the implications of changes are extremely multifaceted and cannot be anticipated intuitively, given the complexity of the many forces at play. Given the probably hard-to-reverse consequences, changes should be evaluated thoroughly by producers and their distribution partners, with rigorous analyses.

Whereas our results were robust under the assumptions of the study,[407] our approach leaves out some aspects; the influence of *illegal* channels on a channel configuration's optimality is certainly a notable one. In a partial extension and update of our study, Burmester et al. (2016) added illegal file-sharing to the options a consumer faces when deciding whether, and through which channel, to consume an entertainment product. In addition to theater visits, DVD sales and rentals, and EST they also consider two more legal channels that have gained popularity since we conducted our study: VOD (i.e., digital rentals of movies) and Blu-ray purchases; they also offer the choice between high and low definition for VOD offers. Their revenue model, however, is a simplified version that leaves out several of the key forces we discussed earlier, and the authors take an *industry* perspective (producer plus distributors), not a producer's perspective on channel effects.[408]

When *not* accounting for illegal channels, the scholars find for a sample of about 2,500 German movie consumers that the channel configuration that maximizes revenues for the industry is a parallel release of a film in movie theaters, DVD purchases, and VOD (for a high price of 30 Euro—the equivalent of what the industry terms "premium VOD" these days). But Burmester et al.'s results are biased toward a faster release across channels because they do not account for repeated purchases or SBS effects. And does the inclusion of an illegal channel alter the results much? The scholars' answer is no: although theater visits and DVD purchase prices should be somewhat lower in such an environment (but the channels still should be made available at the same time), and VOD rentals should be introduced one month later, file sharing tends to affect every channel similarly. Burmester et al. also apply their approach to books, facing the same limitations.

[407]For example, when we estimated the actual channel configuration at the time we conducted the study, the results reproduced the "real world" quite accurately, with the estimated percentage of producer revenues resembling actual channel percentages closely (±2 percentage points).

[408]Burmester et al. (2016) calculate each channel's revenues simply by multiplying a channel's price by its "market share" from the channel-preferences determined via conjoint analysis.

In summary, although *Entertainment Science* studies do not provide a final blueprint for managers for the exact design of channel configurations, they do highlight the relevance of considering the topic. And they offer a methodological path through which realistic insights can be generated regarding the effects that changes in the timing and order of channels would have on the producers' and distributors' revenues. Not too much value should be put on the specific *results* of existing studies, as they do not fully reflect the current realities facing entertainment consumers (think of subscription streaming). Thus, general learnings are possible, but updates and extensions are recommended. The approaches developed by scholars should also warn managers to avoid overly simple conclusions: the studies we featured in this section provide us with an idea of how complex channel constellations work.

Before we move on to a more in-depth investigation of the effects of illegal channels (and how they can be dealt with as part of entertainment distribution), let us mention that a related timing problem exists for the coordination of *international* releases of entertainment products. Should a product be released simultaneously across the world (as it is usually done with games and music), or sequentially (as it is often the case with movies and books)? With cannibalization being less of an issue between (most) countries versus between channels ("geoblocking" and languages help managers to limit inter-country substitution), a key element of this problem is the existence of SBS effects between countries—something we already addressed in some detail in the communication chapter of this book.

The Piracy Challenge: How to Deal with Competitors Who Offer One's Own Products for Free

"It's wrong to steal. It hurts other people, and it hurts your own character."
—Steve Jobs *(quoted in The Deadline Team 2013)*

Because entertainment products are information goods, they can be duplicated at relatively low costs. Whereas this carries massive economic advantages for their producers (who can extend supply quickly and globally, realizing enormous economies of scale), it also comes at a price: an incentive for others who do not have the legal right to copy the product to benefit from its duplication. This includes bootleggers *and* consumers, a distinction that is often blurred as the rights of the consumers are somewhat unclear for duplicating an entertainment product.

The concept that the creator of an original idea or creation is granted an exclusive right to use and distribute that idea or creation has a long history; the "Statute of Anne" (which granted the publishers and authors of written works protection from being copied by others), codified by the British parliament in 1709, is usually considered the first copyright law. Since then, producers of entertainment have long complained about copyright violations and have tried to minimize user rights, arguing that duplicates hurt the financial viability of producers and, thus, threaten any creative industry, as a whole. Consumers and their advocates, in contrast, argue that copyright should not restrict consumers' rights of usage (for example, producing backups for your own use of a product you paid for, reselling it, or lending it to a friend). And the discussion of what qualifies as legal "fair use" of copyrighted ideas is quite controversial and ongoing.

In the pre-digital age, in which entertainment products involved a material element and content was stored in analogue form, such as on an audio/video cassette or vinyl disc, producers already ran initiatives to fight duplications. Examples ranged from publicity campaigns such as "Home taping is killing music" in the U.K. in the early 1980s, when dual-cassette players were introduced (Zaleski 2016), to attempts to tax and even prohibit recording technologies (such as the tape recorder and the video recorder). However, in the pre-digital age the problem was of somewhat limited proportions because the analogue format of the content resulted in a loss of quality with every duplication, giving the original a built-in advantage over unauthorized copies.

Things changed fundamentally with the introduction of storage media such as floppy discs and CD-ROMs, CDs, and DVDs, which now contained the entertainment content in a digital format, enabling lossless copying. Further, the Internet and a growing file-sharing infrastructure then popularized the copying of all sorts of entertainment, from music and movies to books and games, in a truly unprecedented way. Peer-to-peer service Napster, founded by then-teenagers Shawn Fanning and Sean Parker and unleashed in May 1999, had 20 million users in less than a year. Complemented by new hardware technology, such as CD burners or MP3 players, downloaded files quickly became an attractive substitute for the purchase of legal entertainment products (Liebowitz 2008). Although Napster itself was short-lived in its original form, the world of entertainment had become forever different.

Napster's heritage is still omnipresent today: statistics show that consumers visited sources of illegal entertainment and software content around 179 billion times globally in 2016 (*MUSO* 2016). In the UK, 25% of those consuming entertainment or software during a three-month period in 2016 accessed at least some of that content via illegal sources (with percentages ranging from 12% for books, to 18% for video games, 20% for music, and 24% for films; *Intellectual Property Office* 2016).

But there are also dynamics at play, and the overall trend seems to point to fewer people being attracted to illegal entertainment sources than in previous years. In the UK, the share of those who consumed illegal audio-video content and music since 2012 shrank by 23 and 13%, respectively.[409] Also, the channels through which piracy is accessed have changed. Sharing entertainment files with other consumers via the Internet is still a substantial resource, but streaming content from unofficial sources has become by far the dominant means of illegal consumption for audio-visual content. Downloading content from filehosting sites and "stream-ripping" (where consumers download content that is intended to be streamed only) have increased in popularity. Across all kinds of content, *MUSO* (2016) reports that about 60% of piracy now happens via streaming, 19% via peer-to-peer file sharing, 17% via downloads, and 4% via ripping. Our discussion of anti-piracy strategies will take these trends into account.

Industry representatives and organizations have published numerous statistics on economic piracy effects, most of which suffer from a biased nature and unrealistic assumptions (such as treating every illegal download or site visit as a lost sale). Instead of helping the industry in its fight against piracy, perhaps such numbers do the industry a disservice as they call the industry's credibility into question by journalists and consumers (see, for example, blogger Masnick 2016). Scholars, driven by academic curiosity versus commercial interests, have also investigated the piracy phenomenon and yielded rich insights regarding piracy. In the following, we first summarize what scholarly studies have found about the link between piracy and the use of legal channels and industry revenues, before we then offer insights on the drivers of entertainment piracy and, relatedly, the effectiveness of anti-piracy measures.

[409]Why then do some argue that piracy is *growing*, instead of shrinking (e.g., Steele 2015)? We speculate that this is mostly because presenting piracy as an essential threat serves their agenda; in a world that is fixated on growth, financial, political, and societal support would be much harder to get for a phenomenon of shrinking importance. Their empirical arguments usually refer not to the number of files pirated, but to the volume of *data* that is exchanged via pirate sites. This volume has indeed been growing—but such growth can be attributed more to the much larger file sizes of today's pirated copies versus those which were shared a decade ago. Just consider the development of storage media since the beginning of digital piracy: whereas early, low-resolution rips of films were between 1 and 3 gigabytes (GB) in size, the availability of high-definition versions increased the file size to about 8 GB, and 4K versions are even around 4.5 times that size.

The Impact of Piracy

Understanding and measuring the effects that illegal sources have on consumers and their spendings for legal forms of entertainment is not trivial, as such illegal sources can have countervailing effects. We have argued, and offered some empirical evidence, that providing consumers access to samples of a product (e.g., trailers or excerpts) can help the product's commercial performance, and also that positive effects can exist between channels, caused by spillover, success-breeds-success, or repeated consumption (e.g., showing a movie on TV may increase purchases in its DVD format). Those who have tried to defend illegal channels with economic arguments have almost always claimed that these effects also apply to piracy.

But, as we will show in the following, empirical evidence shows overwhelmingly that illegal channels predominantly cannibalize legal forms of entertainment, as consumers use pirated versions as a substitute for regular, commercial offerings. This dominance of the "destructive" over the "constructive" effects of piracy is also consistent with the drastic decline of the music industry in the 2000s, which we sketched earlier in this book. But why have other industries not been affected as badly as the music industry? To some degree, this difference may be due to the more-demanding infrastructure challenges associated with transferring audio-visual film and interactive game files over the Internet. It can also be due to the rise of legal alternatives offered by disruptive services such as Netflix (for films) and Amazon (for ebooks), as well as explicit anti-piracy strategies, as we will discuss later in this chapter.

The majority of piracy-related studies have been conducted in music, where Napster & Co. first turned illegal consumption into a mass phenomenon. Broad consensus exists among scholarly studies that music piracy substantially cannibalizes industry revenues, and that piracy is the main single reason for the music industry's shrinkage.[410] Some scholars have used cross-sectional surveys in which they asked about both legal purchases and illegal consumption, and then tried to statistically isolate the effect illegal consumption has on legal purchases. Because music involvement drives both legal *and* illegal behaviors (which tends to inflate the piracy effect due to endogeneity), scholars have used instrumental variables (which should

[410]Let us clarify that "broad consensus" does *not* mean unanimity. The most prominent study which does *not* find a sales-decreasing impact of file sharing is by Oberholzer-Gee and Strumpf (2007). Other scholars, however, have pointed to "important problems with that paper" (Liebowitz 2008, p. 852).

facilitate file sharing, but not affect legal consumption)[411] instead of the raw amount of illegal consumption. Example studies are by Rob and Waldfogel (2006) who, for a survey of 419 university students, estimate that five music album downloads displace the purchase of one album, and by Zentner (2006), who studies a European sample of more than 15,000 consumers and calculates that access to file sharing reduces the probability of music purchases by 30%.

Others have estimated piracy effects based on aggregated secondary data over time. Among those is Stan Liebowitz (2008), who studied the change in cumulative album sales per consumer between 1998 and 2003 in about 100 Nielsen-defined geographic areas. He combined sales data with each area's Internet penetration, as a measure for the amount of file sharing in the area.[412] Liebowitz finds that sales and Internet penetration are strongly linked and estimates that the attractions of the Internet are responsible for an average decline of 1.55 albums sold per consumer in 2003. But how much of that effect can be attributed to *illegal* activities such as file sharing? Comparing album sales with changes of other non-Internet-based media activities (such as TV consumption), Liebowitz estimates that about 76% of that decline (or 1.19 albums) are caused by file sharing.

But, extrapolated, this amount would be more than the *actual* decline in record sales between 1998 and 2003! Liebowitz acknowledges that, pointing out that his results suggest that music revenues would have grown between 3 and 4% each year without file sharing's existence. When he later conducts a review of existing findings (Liebowitz 2016), he concludes that the majority of studies report that "file sharing explains the entire decline" (p. 19) of music sales, and that, when averaging the "raw" numbers from all studies, the share of sales decline to be attributed to file sharing is over 100%.

There is a caveat, however. Those findings are from a period when consumers had *no legal digital alternatives* to file sharing. So, does the threat of piracy persist now that consumers can consume music legally via the Internet, purchasing it on iTunes or subscribing to services such as Spotify? We get back to the role of legal streaming services for the use of illegal channels when discussing anti-piracy measures.[413] Let us add that not all kinds of music might be affected by piracy to the same extent: the findings from

---

[411]An example for such a variable would be the speed of a consumer's Internet connection—at a time when legal versions were not available via the Internet.

[412]For a similar approach using Nielsen's geographic areas, or "DMAs," see Chintagunta et al.'s (2010) study on word-of-mouth effects.

[413]See our discussion of unbundling and its effect on music revenues in our pricing chapter.

Lee (2018), who analyzes weekly bit torrent downloads from an undisclosed "private network" along with sales for about 2,000 albums in 2008 with a GMM panel approach, suggest that cannibalization is strongest for top-tier albums (defined in terms of sales) and weaker for less popular albums for which the creation of awareness via "sampling" might be more valuable than for albums which are widely known already.

For movies, things look quite similar when it comes to piracy effects; robust studies consistently find that piracy hurts industry revenues. When Rob and Waldfogel (2007) collect information in the spring and summer of 2005 from 470 students in two rounds of surveys, they find that the degree to which piracy cannibalizes paid consumption depends on *when* piracy happens—if piracy is the first channel through which a consumer experiences a film, a pirated viewing costs about 80% of the average paid viewing of a film. For their sample, they calculate that paid movie consumption, across legal channels, would have been about 3.5% higher—but only 5% of movie viewings happened via file sharing in their sample, which limits the generalizability of this calculation.[414]

In a study we conducted ourselves in 2006, we ensured generalizability by employing a representative quota sample of 1,075 (German) consumers (Hennig-Thurau et al. 2007b). We surveyed them three times over the course of eight months, tracking their legal and illegal consumption behaviors over the life cycle of 25 movies that were released in theaters and on DVD during that time frame.[415] We asked about both consumption intentions and actual behaviors regarding the movies; when doing so, we carefully avoided any moral or legal judgments. These steps contributed to a much higher, and more realistic, number of "unpaid" consumption experiences than in Rob and Waldfogel's study (i.e., 17% prior to DVD release). To determine whether file sharing had an effect on legal channels, we then ran logistic regressions for each legal channel and movie, in which we included the file sharing intentions and movie characteristics, such as the number of theaters in which a film was released and its quality rating by IMDb users.[416]

So, what did we find? File sharing substantially hurt legal consumption, and did so not only for theaters, but also for DVD sales and DVD rental. Simulations show that in the absence of illegal channels (and corresponding

[414]In addition to the convenience sample, another factor limiting the study's generalizability was that the participants had to recall a period of *three years* of movie watching, which probably caused a potential memory bias.

[415]770 of the respondents participated in all three survey rounds.

[416]Specifically, we used ReLogit regression—a specific variation of logistic regressions which accounts for the large number of zeros (non-viewings) in our data set.

intentions to use them), theater revenues would have been almost 13% higher, DVD sales would have been almost 15% higher, and DVD rental revenues would have gained more than 10.5%. Figure 13.7 also shows the corresponding annual dollar amounts for the German market (in 2007 value), which are substantial. Interestingly, we find that it is not mainly the actual file sharing that cannibalizes revenues—instead, the consumers' *intentions* to watch a free copy of a movie are responsible for the majority of the financial loss attributed to piracy, regardless of whether the consumer actually follows through accessing the illegal copy.

These results are in line with those of Ma et al. (2016), who used secondary data for 533 movies released in North America between 2006 and 2008, and included the time when a pirated version of a movie became available. They build on Eliashberg et al.'s (2000) MOVIEMOD prediction model of movie success,[417] which they adapt for the existence of pirated versions of a film as an alternative for a theater visit. The scholars then calibrate the model parameters with a Markov chain Monte Carlo algorithm.

Their estimations suggest that all movies would have fared better at the box office if illegal formats had not been available, showing a loss of 15% (or about $1.3 billion) per year for the 2006–2008 period and a similarly sized one for 2011–2013. An interesting facet of their approach is that it allows, under the model's assumptions about consumers' decision-making processes, to separate potential positive, demand-enhancing effects of piracy and negative cannibalization effects. When disentangling these effects, Ma et al. find a small positive "sampling" effect of pirated content of 1.5%, which, however, is dominated by the larger number of consumers who prefer the free copy over seeing a movie in theaters.

Games and books have received less scholarly attention regarding how piracy influences industry revenues (see also Watson et al. 2015). We know, though, that game and book markets have developed in ways that limit the destructive effects of piracy, which now seems to be less of a threat for these products (e.g., Depoorter 2014). In contrast to films and music, games are not well suited to be streamed, so they benefit from the recent shift in piracy media from file sharing toward streaming, a change that also reduces piracy of books. In the following section, we demonstrate this shift was not accidental, but at least partly the consequence of strategic industry moves.

---

[417]See our discussion of the MOVIEMOD approach in the innovation management chapter of this book.

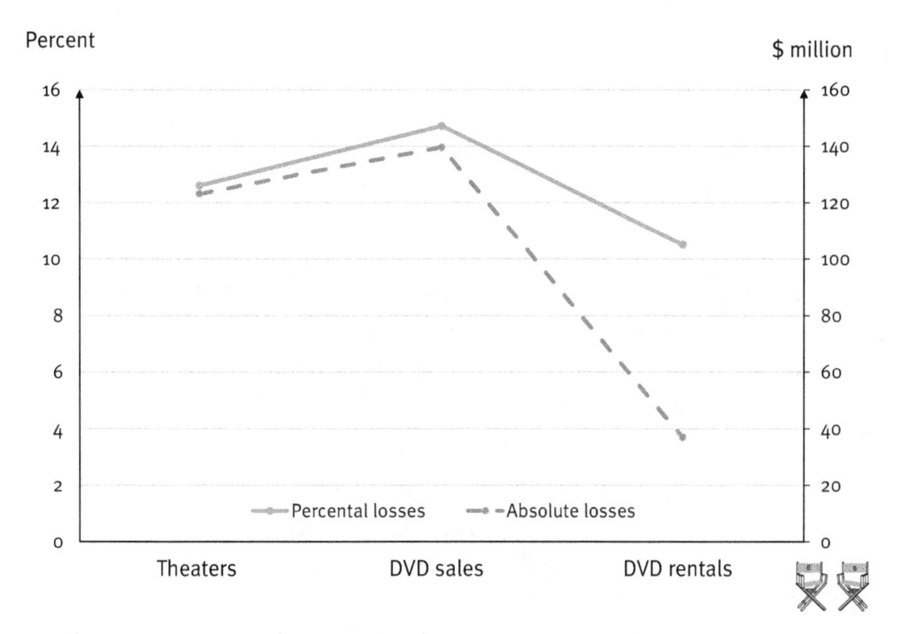

Fig. 13.7 How file sharing impacts German movie revenues
Notes: Authors' own illustration based on results reported in Hennig-Thurau et al. (2007b). Dollar amounts are transformations of the original Euro amounts in the study; they are not adjusted for inflation.

How to Fight Entertainment Piracy

Theories and empirical results can also offer some help for finding ways to counter piracy. Let us provide you with a theory-based, systematic look at the determinants of piracy and distill their relative influences on consumers, as reported by empirical studies. We then discuss the effectiveness of selected anti-piracy strategies for entertainment products, again combining theoretical arguments and empirical insights.

Why and When Do Consumers Prefer the Illegal Copy Over the Original?

One approach to understanding the factors that determine a consumer's decision to engage in piracy is taking an economic utility standpoint. Doing so tells us that legal and illegal formats of any entertainment product are alternatives between which a consumer chooses when making a consumption decision. According to economic utility logic, consumers prefer a pirated format (over the legal one) when they perceive it to offer greater utility.

The first application of a utility perspective to piracy traces back to French scholars Rochelandet and Le Guel (2005), who applied the utility logic to music consumption. They used logistic regression and found that, among other factors, moral costs a consumer faces lower the likelihood of file sharing, whereas his or her Internet skills (which lower the consumer's transaction costs to engage in piracy) increase it. We extended their ideas into a consumer utility framework of piracy, introducing a set of categories that determine consumers' utility perceptions of legal and pirated formats and, subsequently, consumers' piracy behavior (Hennig-Thurau et al. 2007b). The framework consists of four main categories of piracy determinants:

- The *utility* consumers expect to derive from the *legal format*, including both the gross utility and the costs associated with its consumption. The higher the legal format's utility, the less a consumer will be inclined to prefer the illegal copy.
- The *costs* consumers associate with the acquisition and consumption of the *pirated format*. The general logic here is that the higher the costs of the illegal format, the lower its utility and the less a consumer will be inclined to prefer the illegal copy. Higher knowledge about file sharing enables consumers to expend less effort to retrieve illegal copies, thereby making them more attractive.
- The *specific utility of the illegal format*. The higher such utility, the more a consumer will be inclined to prefer the illegal copy.
- Finally, the degree to which consumers consider the legal and the pirated format as *substitutes*. The lesser the perceived degree of substitutability, the less a consumer will be inclined to prefer the illegal copy.

The exact variables in each of the categories will differ somewhat between entertainment forms and products. When buying a CD at an online retailer, other kinds of costs can accrue compared to those for watching a movie in a theater, where one has to find a parking spot and pay for a babysitter at home. Figure 13.8 shows the different specific variables for each of the framework's four categories for the movie context, when the consumer has to choose between going out to the theater to watch a film or accessing a file-sharing version of it (Hennig-Thurau et al. 2007b).

The figure also contains the results of an empirical test we ran for a sample of 813 German consumers. Using partial least squares, we analyzed how each of the proposed drivers of file sharing is associated with consumers

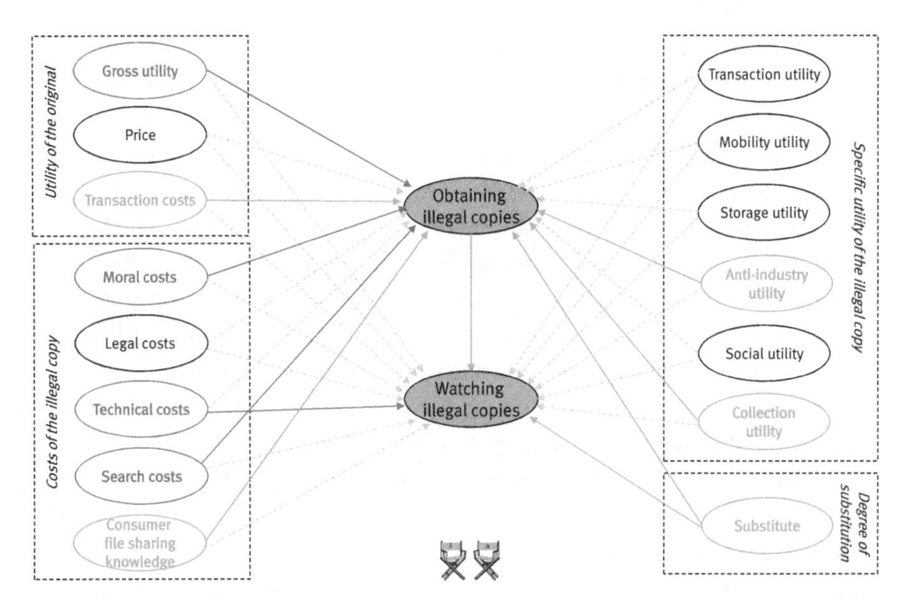

Fig. 13.8 Drivers of movie piracy
Notes: Authors' own illustration based on results reported in Hennig-Thurau et al. (2007b). Green arrows mean that higher values for one variable increase the value of the other variable, whereas red arrows mean that higher values for one variable lower the value of the other. Grey arrows are not statistically significant.

obtaining and watching an illegal movie copy. Data showed that among the features of the theatrical movie, the transaction costs (such as parking and the sitter's pay) facilitate sharing most strongly, followed by the theater visit's utility (which reduces file sharing); the price of the ticket was not significant (probably because it's largely constant for movies).

The costs associated with the illegal copy also matter strongly for our sample: consumers' search costs and their knowledge about file sharing are among the strongest piracy drivers. Moral costs and technical costs (such as the risk of the copy being virus-infected) also determine file sharing. The legal costs, however, do not influence our respondents; we assume that, at least at the time we collected our data, the risk of getting sued was too abstract to prevent consumers from consuming illegal copies. We will get back to this important issue in the next section.

But we also learn that illegal copies can offer benefits that consumers do not see in legal formats, and that these trigger file sharing quite strongly. In our case, the collection utility is a main driver—illegal copies enable consumers to live out their desire to have large repertoires of films at hand. Note that we collected our data before legal digital copies and archives were available, namely subscription services such as Netflix, which provide a consumer with instant access to large numbers of films. And we see that piracy can be an act of protest; the more critical people's attitudes are toward the film industry and its treatment of consumers, the more they are willing to circumvent its legal offers. Finally, the results confirm our argument that the more consumers consider an illegal format as a "full substitute" (versus an imperfect knock-off), the more they tend to prefer it over the legal format.

Overall, our model explains about 20% of file sharing intentions, leaving room for more in-depth studies. Watson et al. (2015) provide a broad review of additional scholarly insights on file-sharing drivers, covering studies on all forms of entertainment, as well as software. From 195 studies, they largely stress the relevance of the categories and variables we discussed above, with few additional explanations being provided. We thus conclude that these determinants will provide a good start for understanding entertainment piracy.

Knowing these factors that influence consumers' decision to engage in piracy is also a prerequisite for developing effective strategies to counter piracy behavior for entertainment products. In the final section of this chapter on distribution decisions in entertainment, let us now explore which anti-piracy strategies are the most promising, and which do not work so well.

Some Thoughts (and Findings) on Anti-Piracy Strategies in Entertainment

Over the last decades, the entertainment industry has developed a number of different approaches to counter illegal consumption, targeting various aspects of the piracy framework we laid out above. *Entertainment Science* scholars have empirically tested several of these strategies' effectiveness, offering feedback which enables the industry to recalibrate their efforts. Here is what scholars have found.

Strategies that increase the attractiveness of legal formats. When digital formats for storing entertainment such as MP3 became available by the late 1990s, the industry was hesitant to use them out of a desire to avoid spoiling the existing value chain of formats and channels. Consequently, fully digital formats of entertainment products were only available illegally, a condition that is widely considered a major reason for the collapse of music sales. With the exception of video games, traditional entertainment producers remained hesitant for many years to foster new digital formats, such as legal downloads and VOD, that would provide consumers with the same utility to which they had become accustomed from illegal competitors.

Thus (and luckily for producers of entertainment *and* for consumers), companies from outside the industry filled the vacuum. Apple provided songs and albums (and later movies) for sale via iTunes from 2001 on, Spotify began to offer a legal music flat rate, and Netflix popularized legal streaming of films and series. All these offers were considered valuable by consumers with regard to price and quality, but they also copied the benefits of the *illegal* offers (e.g., access via the Internet, anytime access to large repertoires in the case of Spotify and Netflix). It was only afterward that the studios embraced such formats, and some still do so only halfheartedly. So, although we describe this rise of new entertainment formats as a "strategy," let us note that it was one that was mostly forced upon the industry's traditional players rather than being actively originated by them.

Some empirical evidence for the effectiveness of fighting piracy by offering *attractive* legal formats has been compiled by scholars. Poort and Weda (2015), in representative Dutch surveys from 2008 and 2012, find indications that the availability of titles, price, and the technical quality of legal offers are crucial factors for getting consumers to switch from illegal to legal formats. More rigorous insights come from Sinha and Mandel (2008) via a series of lab experiments in which they manipulate the features of a legal music web site and then measure participants' "likelihood to pirate." In particular, for a sample of 165 business students, the authors find that a "high functionality" website reduces the students' likelihood to pirate quite substantially (by 27%) compared to a "low functionality" website; even a "medium functionality" website was beneficial, reducing the likelihood of piracy by 17%.

Do you remember Movielink, the Hollywood studios' early failed effort to establish a legal alternative to illegal downloads of movies? The joint venture of five studios, launched in 2002, offered a very limited selection (the initial list was just 175 titles, all of which were already available in rental stores for several months), only rentals, and also restricted watching films to computers, with a focus on security instead of usability (Harmon 2002). Looking at the scholarly findings reported above, we now can be pretty sure why the effort did not resonate more with consumers.

Sinha and Mandel's findings are nicely complemented by an analysis of actual market data by Danaher et al. (2010), who investigated the impact of NBC's removal of a large arsenal of TV series episodes from Apple's iTunes store. In addition to examining the impact of that action on DVD sales of the removed titles, they also studied the amount of piracy for this content. Using a BitTorrent tracker to measure piracy over time, Brett Danaher and his colleagues treated NBC's content removal as a natural experiment and used the content's availability on iTunes for explaining the number of daily illegal downloads. An OLS regression, in which the scholars control for the changes in piracy for other content in the same time period (whose availability on iTunes did *not* change) reveals that piracy of NBC content increased by 11.4% in the two weeks after the closure of the legal format—they measured 48,000 additional illegal downloads per day. And that number is not the upper limit: for certain kinds of content, such as comedy and sci-fi, which appeal mostly to younger, more piracy-affine consumers, the increase is as high as 20%.

We have noted that film studios have begun to consider popular SVOD platforms, and Netflix in particular, as a threat, despite profiting from selling licenses to them. de Matos et al. (2017) indicate that they also benefit from reduced piracy because of these services. Based on a field experiment conducted in an undisclosed country in which a telecommunications firm provides its customers an "SVOD-like" package of movies and series for free, the scholars matched consumer IPs with BitTorrent traffic to study how the legal package influenced consumers' upload and download Torrent traffic. For those consumers whose preferences fit with the movies included in the "SVOD" package, they find that upstreams decrease by almost 50% (downstreams saw less of a decrease, which might be attributed to study limitations, though).[418] But their results also show that content value definitely matters—if the legal content does *not* match a pirate's preferences, no reduction of piracy is found.

And there might even be another positive effect of opening these digital channels that are so popular among consumers: they might help to improve the industry's reputation. Keep in mind that we found that a bad reputation drives piracy, with piracy becoming an act of "revenge." Because of its refusal to open up its models to new technologies, entertainment firms do not hold a sterling reputation among consumers as being customer focused—instead, they have been associated with "friction and customer alienation" (Lasica 2005, p. 4). Making their products available in new formats such as Spotify and Netflix (even if this might have been someone else's idea) might reduce consumers' anti-industry attitudes and, consequently, their use of illegal copies.

Strategies that increase the costs of illegal formats. Opening legal digital channels that meet consumers' expectations for accessing an entertainment product might reduce piracy by switching effects and a potential increase

[418]de Mathos et al. determine the preference fit with the overlap of automatically created recommendations for a user and the inclusion of such recommendations in the "SVOD" package.

in industry reputation. But remember that entertainment choices are also very often about network effects, in which the value a consumer gets from a product increases with the number of other users of a product or the platform on which it is offered. This economic logic also applies directly to file sharing networks: an increase in the attractiveness of legal formats reduces the attractiveness of illegal formats by shrinking the illegal network in terms of the people who make content available, limiting the number of available pirated copies of a product. That, in turn, then increases the consumers' search costs for illegal formats (see also Depoorter 2014).

But there also are other, more active strategies that are targeted at increasing the costs of pirated versions. One of them has been to increase the search costs of pirated copies for consumers *by shortening supply*. Danaher and Smith (2014) investigate how the January 2012 shutdown of the global piracy site Megaupload.com, then a leading platform of piracy activity that was responsible for about a remarkable 4% of global Internet traffic, affected the digital sales and rental revenues of two film studios.[419] The authors measured how Megaupload's share of total Internet traffic varies across countries: they estimate the legal movie revenues in a country in a given week as the result of the country's Megaupload Internet share in that week. An OLS regression provides clear support for the existence of the expected impact: Danaher and Smith calculate that the site's shutdown resulted in 6.5–8.5% higher total digital revenues over the 18 weeks after the shutdown in 12 countries. They find that the size of the effect was comparable for sales and rental transactions.

Increasing consumers' search costs by limiting access to illegal content is also the logic behind "Digital Rights Management" measures, or DRM. In the case of DRM, however, the target is not the platform through which illegal content can be distributed for free, but the consumer who has *paid* to access entertainment content legally. DRM restricts the consumer's own usage of the obtained content to prevent others from accessing the content without paying for it. DRM can either take the form of limiting the number of devices on which a song can be heard or a film can be played (i.e., "individual" DRM) or the consumer's ability to distribute the purchased content to others, such as by limiting the number of shares (i.e., "shared" DRM). Essentially, such behaviors are equated with sharing content with unknown others via file sharing or filehosting sites.

[419]For some background information on the Megaupload case, see, for example, Gross (2012).

We don't want to discuss the industry's moral right to do so here, but instead to see if increasing consumers' search costs via DRM is an effective strategy. The trouble with DRM is that it not only lowers the net utility of any copy for those who have not yet purchased it (because of higher search costs)—but it might also lower the utility of the *legal* format, which we know is a key driver of piracy. Sinha et al. (2010) provide empirical evidence for the problematic nature of DRM for entertainment; they draw on reactance theory (Brehm 1966),[420] arguing that DRM is perceived as a threat to consumers' personal freedom to enjoy music; they also report some descriptive statistics that support this logic.

Based on two lab experiments with 800 and 1,300 students in which they systematically vary individual and shared DRM of music, they conclude that DRM strongly reduces the utility that consumers derive from *purchasing* music. They argue that this finding starkly contrasts with the lack of evidence for DRM's ability to reduce the number of available free copies. The scholars find particularly strong effects for "shared" DRM, the *removal* of which not only increases consumers' willingness to pay for digital music, but also turns consumers from pirates into paying consumers of legal music formats. Individual DRM, in comparison, has neither a positive nor a negative effect on consumers. Based on these results, Sinha et al. conclude that "pursuing a pure DRM-free or significantly relaxed DRM strategy may be optimal" (p. 51).

A different approach for raising the costs of illegal copies for consumers is through increased *technological burdens*. This strategy has been considered a major reason for the reduced piracy for console games. Running illegal copies on a PlayStation or Xbox mandates hardware modifications that require expertise and time and carry the risk of voiding a warranty. The oligopoly of consoles hardware makes it difficult for the consumer to circumvent such restrictions by using third-party technology, something that is much easier to achieve with PCs (for which piracy is a much bigger issue; see Depoorter 2014). This is also the reason why film producers have been so concerned about interfaces like Popcorn Time that have made "accessing pirated content as easy as turning on Netflix" (Thielman 2015).

We have noted that the *legal costs* of using illegal copies are another determinant of piracy. Although we did not find that consumers were influenced by costs in our study from the mid-2000s, several legal approaches have been implemented since then across the world. These legal strategies differ extensively in their specific goals and approaches, so results are difficult to

[420] See also our discussion of reactance theory in the context of (too many) brand placements in our chapter on entertainment business models.

generalize. Danaher et al. (2014) studied the French "Hadopi" law, which passed the French parliament in May 2009. It is a "graduated response system" that considers a three-stage governmental intervention to piracy, ranging from email notification to rigid penalties such as Internet suspension.

The scholars compare how the law influenced legal sales of digital music on iTunes by comparing sales trends in France to a control group of other European countries in which no piracy-related policy changes happened during the same period. They study differences between actual and "normal" revenues (simulated sales in France without the new law which they derived from the other countries). Danaher et al. conclude that digital song and album revenues via iTunes were about 22–25% higher in the two years after the law took effect, the equivalent of nearly 14 million Euro. They also estimate a differential impact for music genres; "younger" and more "piracy-affine" genres like rap and hip-hop experienced an even higher revenue bump of 30% in their models, whereas less "piracy-affine" genres (such as jazz) generated only about 7% more due to the higher legal costs caused by the law.

But let us note that there might be a side effect of increasing legal costs. In their experimental analysis of consumer reactions to file sharing measures, Sinha and Mandel (2008) also manipulate consumers' risks from engaging in illegal consumption. Whereas consumers who tend to avoid risking legal threats have a lower tendency to engage in piracy when the risk is high, no such effect was found for consumers who have a high need for stimulation ("sensation seekers"). Their likelihood to pirate was even about 6% *higher* (although not statistically significant) when legal threats were high!

A final approach to deal with piracy is *strategic (re)targeting*: a strategy that has been successfully practiced by producers of console games. Our previous discussion reported only the *average* importance of piracy determinants across consumers, but the video game industry has built on the assumption that these importances vary between consumer segments. For example, whereas teenagers will be price sensitive (because of small personal budgets) and relatively insensitive toward measures that raise the costs of pirated content, this is not the case for older consumers, who more highly value convenience (and will not engage in piracy if it's inconvenient).

Over the last decade, the games industry has successfully shifted its focus from teens to adults, at least for console games, and drastically reduced piracy for this format. Less tech-savvy adults resist "jailbreaking" their consoles, and instead prefer spending $60 for a game that can be accessed in an easier and less time-consuming way. Now compare this to the film industry, whose products are mainly targeted at teens whose piracy affinity is much more difficult to counter—with a strong intensifying trend toward such segments.[421] In digital times in which entertainment producers cannot avoid competing against illegal copies of their products, putting heavy focus on a consumer segment that tends to prefer "cheap" over "inconvenient" may not turn out to be the most effective approach.

Concluding Comments

The digitalization of entertainment products has dramatically reshaped entertainment distribution. In this chapter, we have shed light on three fundamental distribution-related issues that entertainment managers are dealing with today, based on insights generated by theoretical and empirical studies: the timing of a new product's initial release, the orchestration of the plethora of channels that today exist for most entertainment products, and the threat of illegal offerings.

Scholarly research shows that descriptive statistics on the commercial potential of certain release periods are misleading, as they combine demand-sided influences with supply-sided effects caused by managers themselves—just another case of endogeneity that should be taken into account when deciding when to release a new product. We also pointed to findings by *Entertainment Science* scholars on the role of competition (when does it matter? How should it be defined and measured?) and game-theoretical learning about how to deal with it.

We provided a useful framework for determining channel configurations. The current distribution models are based on decade-old logic, with resistance toward change being the main force behind them. The industry has tampered with new models, but in an ad-hoc way, and we agree with critical voices from within the industry that such power-driven changes could massively harm entertainment, at least in the longer run. We showed that scholars have taken a much more systematic approach by identifying the macro- and micro-level factors that should be considered when valuing alternative distribution models. Approaches that consider these factors indicate that huge gains are available for the producer at the cost of certain distributors, but that also distribution models exist in which producers win, but no distributor loses. The main learning here is that solutions inspired by *Entertainment Science*, that account for the hyper-complex channel relations that exist in today's entertainment world, should be preferred over gut-feeling-based approaches.

---

[421]See also our discussion of where the industry is heading in our chapter on integrated entertainment marketing.

We compiled evidence that piracy remains a major threat that entertainment firms have to deal with, but whose role today should not be exaggerated. We summarized rich insights into the drivers of consumer piracy, which also should guide managers when developing anti-piracy measures. The evidence is pretty clear that increasing the utility (and/or decreasing the costs) of legal copies is a promising path—one that entertainment firms have neglected for a long time in efforts to protect their traditional channels and which they still seem to be embracing only half-heartedly. These strategies also come with the side benefit of enhancing the entertainment firm's reputation, which further decreases consumers' motivations to pirate their content. Strategies that focus on blocking access to pirated content seem to support such efforts, whereas approaches that reduce the value of the legal product by adding restrictions are not recommended.

Having looked at product, promotion (communication), and now place (distribution) issues, we turn to the last "P" of the marketing puzzle, the pricing of entertainment content. Pricing may be the most underdeveloped component of the marketing mix in entertainment, but we will show in the next chapter that good reasons exist to pursue progress in pricing.

References

Ainslie, A., Drèze, X., & Zufryden, F. (2005). Modeling movie life cycles and market share. *Marketing Science, 24*, 508–517.

Ananthakrishnan, U. M., Smith, M. D., & Telang, R. (2016). When streams come true: Estimating the impact of free streaming availability on EST sales. *ICIS 2016 Proceedings*.

Anderson, P. (2016). Glimpses of the US Market: Charts from Nielsen's Kempton Mooney. *Publishing Perspectives*, May 20, https://goo.gl/1Kc8YF.

Arnold, T. K. (2005). Coming back for seconds, thirds.... *USA Today*, September 26, https://goo.gl/C2DjVS.

Ascharya, K. (2014). Binge-viewing is turning Hollywood into Broadway. Here's how. *2machines*, March 19, https://goo.gl/PaM1Lz.

Avatar (2017). Great to be working.... Facebook post on @Avatar site, April 22, https://goo.gl/WXnvaf.

Barnes, B. (2007). NBC will not renew iTunes contract. *The New York Times*, August 31, https://goo.gl/tzHCN5.

Bhatia, G. K., Gay, R. C., & Ross Honey, W. (2001). Windows into the future: How lessons from Hollywood will shape the music industry. *Journal of Interactive Marketing, 17*, 70–80.

Bhattacharjee, S., Gopal, R. D., Lertwachara, K., Marsden, J. R., & Telang, R. (2007). The effect of digital sharing technologies on music markets: A survival analysis of albums on ranking charts. *Management Science, 53*, 1359–1374.
Bosman, J. (2011). Paperback publishers quicken their pace. *The New York Times*, July 26, https://goo.gl/uXZAhW.
Brehm, J. W. (1966). *A theory of psychological reactance*. New York: Academic Press.
Brewer, S. M., Kelley, J. M., & Jozefowicz, J. J. (2009). A blueprint for success in the US film industry. *Applied Economics, 41*, 589–606.
Bucklin, R. E., Siddarth, S., & Silva-Risso, J. M. (2008). Distribution intensity and new car choice. *Journal of Marketing Research, 45*, 473–486.
Burmester, A. B., Eggers, F., Clement, M., & Prostka, T. (2016). Accepting or fighting unlicensed usage: Can firms reduce unlicensed usage by optimizing their timing and pricing strategies? *International Journal of Research Marketing, 33*, 343–356.
Busch, A. (2016). MarketCast study takes a hard look at consumer attitudes toward day-and-date home/theatrical release. *Deadline*, April 13, https://goo.gl/MKjXFL.
Busch, A. (2017). Christopher Nolan shows off 'Dunkirk,' says "The Only Way To Carry You Through" the film is at a theater—CinemaCon. *Deadline*, March 29, https://goo.gl/L6QhGz.
Calantone, R. J., Yeniyurt, S., Townsend, J. D., & Schmidt, J. B. (2010). The effects of competition in short product life-cycle markets: The case of motion pictures. *Journal of Product Innovation Management, 27*, 349–361.
Canning, B. (2010). How Jaws heated up summer movies. In A. B. Block & L. A. Wilson (Eds.), *George Lucas's Blockbusting* (pp. 531–532).
Caulfield, K. (2016). Adele's '25' heading back to top 10 on Billboard 200 after streaming debut. *Billboard*, June 30, https://goo.gl/rvM3G3.
Chen, H., Hu, Y. J., & Smith, M. D. (2017). The impact of ebook distribution on print sales: Analysis of a natural experiment. *Management Science*, forthcoming.
Child, B. (2013). Star Wars Episode VII pushed back to Christmas 2015. *The Guardian*, November 8, https://goo.gl/bApFrt.
Chintagunta, P., Gopinath, S., & Venkataraman, S. (2010). The effects of online user reviews on movie box office performance: Accounting for sequential rollout and aggregation across local markets. *Marketing Science, 29*, 944–957.
Chisholm, D. C., & Norman, G. (2006). When to exit a product: Evidence from the U.S. motion-picture exhibition market. *American Economic Review, 96*, 57–61.
Cieply, M., & Barnes, B. (2009). In downturn, Americans flock to the movies. *The New York Times*, February 28, https://goo.gl/jbgLsE.
Clement, M., Wu, S., & Fischer, M. (2014). Empirical generalization of demand and supply dynamics for movies. *International Journal of Research in Marketing, 31*, 207–223.

Clerides, S. K. (2002). Book value: Intertemporal pricing and quality discrimination in the US market for books. *International Journal of Industrial Organization, 20*, 1385–1408.

Committee (1982). Hearings before the subcommittee on courts, civil liberties, and the administration of justice of the committee on the judiciary house of representatives, Serial No. 97, Part I, April 12, https://goo.gl/tfXKtP.

Coughlan, A. T., Anderson, E., Stern, L. W., & El-Ansary, A. I. (2006). *Marketing channels*. Upper Saddle River: Prentice Hall.

Danaher, B., & Smith, M. D. (2014). Gone in 60 seconds: The impact of the Megaupload shutdown on movie sales. *International Journal of Industrial Organization, 33*, 1–8.

Danaher, B., Dhanasobhon, S., Smith, M. D., & Telang, R. (2010). Converting pirates without cannibalizing purchasers: The impact of digital distribution on physical sales and internet piracy. *Marketing Science, 29*, 1138–1151.

Danaher, B., Smith, M. D., Telang, R., & Chen, S. (2014). The effect of graduated response anti-piracy laws on music sales: Evidence from an event study in France. *Journal of Industrial Economics, 62*, 541–553.

de Matos, M. G., Ferreira, P., & Smith, M. D. (2017). The effect of video-on-demand on piracy: Evidence from a household level randomized experiment. *Management Science*, forthcoming.

Degusta, M. (2011). The REAL death of the music industry. *Business Insider*, February 18, https://goo.gl/2eVt3V.

Depoorter, B. (2014). What happened to video game piracy? *Communications of the ACM, 57*, 33–34.

Dhar, T., & Weinberg, C. B. (2015). Measurement of interactions in non-linear marketing models: The effect of critics' ratings and consumer sentiment on movie demand. *International Journal of Research in Marketing, 33*, 392–408.

Dirks, T. (n.d.). The history of film. *AMC Filmsite*, https://goo.gl/4yfrwo.

Dutka, E. (2005). Battle escalates over the speedy debut of DVDs. *Los Angeles Times*, March 29, https://goo.gl/97HmJF.

Einav, L. (2007). Seasonality in the U.S. motion picture industry. *The Rand Journal of Economics, 38*, 127–145.

Einav, L. (2010). Not all rivals look alike: Estimating an equilibrium model of the release date timing game. *Economic Inquiry, 48*, 369–390.

Einav, L., & Ravid, S. A. (2009). Stock market response to changes in movies' opening dates. *Journal of Cultural Economics, 33*, 311–319.

Elberse, A., & Eliashberg, J. (2003). Demand and supply dynamics for sequentially released products in international markets: The case of motion pictures. *Marketing Science, 22*, 329–354.

Eliashberg, J., Jonker, J., Sawhney, M., & Wierenga, B. (2000). MOVIEMOD: An implementable decision-support system for prerelease market evaluation of motion pictures. *Marketing Science, 19*, 226–243.

Eliashberg, J., Swami, S., Weinberg, C. B., & Wierenga, B. (2001). Implementing and evaluating SilverScreener: A marketing management support system for movie exhibitors. *Interfaces, 31*, 108–127.

Eliashberg, J., Hegie, Q., Ho, J., Huisman, D., Miller, S. J., Swami, S., et al. (2009). Demand-driven scheduling of movies in a multiplex. *International Journal of Research in Marketing, 26*, 75–88.

epd film (2017). E-Mail an… Dieter Kosslick, January 30, https://goo.gl/2BN7r6.

Epstein, E. J. (2005). Hidden persuaders: The secretive research group that helps run the movie business. *Slate*, July 18, https://goo.gl/bjdBRz.

Epstein, E. J. (2010). *The Hollywood economist – The hidden financial reality behind the movies*. Brooklyn: Melville House.

Fandom (2017). Video game industry. *Video Game Sales Wiki*, https://goo.gl/y1XCen.

Frank, B. (1994). Optimal timing of movie releases in ancillary markets: The case of video releases. *Journal of Cultural Economics, 18*, 125–33.

Friend, T. (2016). The mogul of the middle. *The New Yorker*, January 11, https://goo.gl/8hYXxT.

Fritz, B. (2017). The 365 days of summer movies. *Wall Street Journal*, April 27, https://goo.gl/LdkYgB.

Gamerman, E. (2017). The year of the blockbuster. *Wall Street Journal*, May 28, https://goo.gl/R3etWZ.

Gattringer, K., & Klingler, W. (2014). Radio bleibt wichtiger Begleiter im Alltag. *Media Perspektiven, 2014*, 434–447.

Gong, J., Smith, M. D., & Telang, R. (2015). Substitution or promotion? The impact of price discounts on cross-channel sales of digital movies. *Journal of Retailing, 91*, 343–357.

Goodhardt, G. J., Ehrenberg, A. S. C., & Collins, M. A. (1975). *The television audience: Patterns of viewing*. Westmead (UK): Saxon House.

Gross, D. (2012). What's the controversial site Megaupload.com all about? *CNN*, January 21, https://goo.gl/RgzU8H.

Gutierrez-Navratil, F., Fernandez-Blanco, V., Orea, L., & Prieto-Rodriguez, J. (2014). How do your rivals' releasing dates affect your box office? *Journal of Cultural Economics, 38*, 71–84.

Harmon, A. (2002). Movie studios provide link for Internet downloading. *The New York Times*, November 11, https://goo.gl/68ZKvN.

Hennig-Thurau, T., Houston, M. B., & Walsh, G. (2006). The differing roles of success drivers across sequential channels: An application to the motion picture industry. *Journal of the Academy of Marketing Science, 34*, 559–575.

Hennig-Thurau, T., Houston, M. B., & Walsh, G. (2007a). Determinants of motion picture box office and profitability: An interrelationship approach. *Review of Managerial Science, 1*, 65–92.

Hennig-Thurau, T., Henning, V., & Sattler, H. (2007b). Consumer file sharing of motion pictures. *Journal of Marketing, 71*, 1–18.

Hennig-Thurau, T., Henning, V., Sattler, H., Eggers, F., & Houston, M. B. (2007c). The last picture show? Timing and order of movie distribution channels. *Journal of Marketing, 71*, 63–83.
Hennig-Thurau, T., Fuchs, S., & Houston, M. B. (2013). What's a movie worth? Determining the monetary value of motion pictures' TV rights. *International Journal of Arts Management, 15*, 4–20.
Ifpi (2017). Global music report 2017. https://goo.gl/WSPh8U.
Intellectual Property Office. (2016). Online copyright infringement tracker: Latest wave of research Mar 16–May 16, June 20, Newport (UK).
James, C. (2015). 'Beasts of No Nation' hits cinemas and netflix. *The Wall Street Journal*, October 8, https://goo.gl/1cWs85.
Karniouchina, E. V. (2011). Impact of star and movie buzz on motion picture distribution and box office revenue. *International Journal of Research in Marketing, 28*, 62–74.
Krider, R. E., & Weinberg, C. E. (1998). Competitive dynamics and the introduction of new products: The motion picture timing game. *Journal of Marketing Research, 35*, 1–15.
Kumar, A., Smith, M. D., & Telang, R. (2014). Information discovery and the long tail of motion picture content. *MIS Quarterly, 38*, 1057–1078.
Lang, B. (2015). 'Steve Jobs' bombs: What went wrong with the Apple drama. *Variety*, October 25, https://goo.gl/U4wW2d.
Lasica, J. D. (2005). *Darknet: Hollywood's war against the digital generation.* Hoboken: Wiley.
Lee, J. F. (2018). Purchase, pirate, publicize: Private-network music sharing and market album sales. *Information Economics and Policy, 42*, 35–55.
Lees, G., & Wright, M. (2013). Does the duplication of viewing law apply to radio listening? *European Journal of Marketing, 47*, 674–685.
Leggatt, H. (2011). Social media is a form of entertainment, say users. *BizReport*, May 30, https://goo.gl/z6jfDi.
Lehmann, D. R., & Weinberg, C. B. (2000). Sales through sequential distribution channels: An application to movies and videos. *Journal of Marketing, 64*, 18–33.
Lieberman, D. (2017). Marvel and 'Star Wars' films to move from Netflix to new Disney streaming service. *Deadline*, September 7, https://goo.gl/m1R6Qa.
Liebowitz, S. J. (2008). Testing file sharing's impact on music album sales in cities. *Management Science, 54*, 852–859.
Liebowitz, S. J. (2016). How much of the decline in sound recording sales is due to file-sharing? *Journal of Cultural Economics, 40*, 13–28.
Luan, J. Y. (2005). Optimal inter-release timing for sequential releases. Working Paper, Yale School of Management.
Luan, Y. J., & Sudhir, K. (2010). Forecasting marketing-mix responsiveness for new products. *Journal of Marketing Research, 47*, 444–457.
Ma, L., Montgomery, A., & Smith, M. D. (2016). The dual impact of movie piracy on box-office revenue: Cannibalization and promotion. Working Paper, Carnegie Mellon University.

MacNab, G., Niola, G., Blaney, M., Cabeza, E., & Goodfellow, M. (2017). Are much shorter theatrical windows around the corner? *Screendaily*, January 2, https://goo.gl/mhzsqi.

Martinez, A. (2003). Bob Hope, the master of the one-liner, dies at 100. *Los Angeles Times*, July 29, https://goo.gl/pLvGDC.

Masnick, M. (2016). Once again, piracy is destroying the movie industry… to ever more records at the box office. *techdirt,* January 11, https://goo.gl/U71nM1.

McIntyre, H. (2016). Adele's '25' has now sold 15 million copies worldwide. *Forbes*, January 10, https://goo.gl/B4vqRZ.

Mendelson, S. (2015). Theaters are wrong to boycott 'Paranormal Activity: The Ghost Dimension'. *Forbes*, October 16, https://goo.gl/GBNrHh.

Miller, D. (2012). Sundance 2012: The day-and-date success story of 'Margin Call'. *The Hollywood Reporter*, January 18, https://goo.gl/i2LT7E.

MUSO (2016). *MUSO global piracy report 2017.* London: MUSO.

Novak, M. (2013). Friday fun fact: The first movie on TV was in theaters at the time. *Paleofuture*, May 24, https://goo.gl/4nCqtu.

Oberholzer-Gee, F., & Strumpf, K. (2007). The effect of file sharing on record sales: An empirical analysis. *Journal of Political Economy, 115*, 1–42.

Poort, J., & Weda, J. (2015). Elvis is returning to the building: Understanding a decline in unauthorized file sharing. *Journal of Media Economics, 28*, 63–83.

Prasad, A., Bronnenberg, B., & Mahajan, V. (2004). Product entry timing in dual distribution channels: The case of the movie industry. *Review of Marketing Science, 2*, 1–18.

Puig, C. (2005). Movies as you like them. *USA Today*, July 25, https://goo.gl/bXctWu.

Reddy, S. K., Aronson, J. E., & Stam, A. (1998). SPOT: Scheduling programs optimally for television. *Management Science, 44*, 83–102.

Rey, J. (2008). If we all go for the blonde. *Plus magazine*, June 1, https://goo.gl/A4awuk.

Risen, C. (2005). Collapsing the distribution window. *The New York Times Magazine*, December 11, https://goo.gl/m7XRdo.

Rob, R., & Waldfogel, J. (2006). Piracy on the high C's: Music downloading, sales displacement, and social welfare in a sample of college students. *The Journal of Law and Economics, 49*, 29–62.

Rob, R., & Waldfogel, J. (2007). Piracy on the silver screen. *The Journal of Industrial Economics, 55*, 379–395.

Rochelandet, F., & Le Guel, F. (2005). P2P music-sharing networks: Why legal fight against copiers may be inefficient? *Review of Economic Research on Copyright Issues, 2*, 69–82.

Roettgers, J. (2017). Spotify signs Universal Music licensing agreement that includes release windows. *Variety*, April 4, https://goo.gl/18WX6p.

Ryan, G. (2016). Disney CEO Bob Iger talks distribution amid Netflix, Twitter acquisition rumors. *Boston Business Journal*, October 5, https://goo.gl/NEZmkw.

Schickel, R. (1996). *Clint Eastwood: A biography*. New York: Knopf.
Schmidt-Stölting, C., Blömeke, E., & Clement, M. (2011). Success drivers of fiction books: An empirical analysis of hardcover and paperback editions in Germany. *Journal of Media Economics, 24*, 24–47.
Schwartzel, E. (2016). July 4 opening is no guarantee for success at box office. *The Wall Street Journal*, July 4, https://goo.gl/Zo7K3z.
Schwartzel, E. (2017). Disney lays down the law for theaters on 'Star Wars: The Last Jedi'. *The Wall Street Journal*, November 1, https://goo.gl/76AUiG.
Schweidel, D. A., & Moe, W. W. (2016). Binge watching and advertising. *Journal of Marketing, 80*, 1–19.
Segrave, K. (1999). *Movies at home: How Hollywood came to television*. Jefferson: McFarland.
Sinha, R. K., & Mandel, N. (2008). Preventing digital music piracy: The carrot or the stick? *Journal of Marketing, 72*, 1–15.
Sinha, R. K., Machado, F. S., & Sellman, C. (2010). Don't think twice, it's all right: Music piracy and pricing in a DRM free environment. *Journal of Marketing, 74*, 40–54.
Sisario, B. (2015). Adele is said to reject streaming for '25'. *The New York Times*, November 19, https://goo.gl/gouUhT.
Smith, S. (2005). Coming to a theater near you. *Newsweek*, August 7, https://goo.gl/uokEQS.
Smith, N. M. (2010). Jon Feltheimer: Embrace new economic models. *IndieWire*, October 5, https://goo.gl/1At6EV.
Smith, N. M. (2016). Will Sean Parker's screening room hurt or help the film industry? *The Guardian*, April 8, https://goo.gl/xQQbd4.
Smith, M. D., & Telang, R. (2009). Competing with free: The impact of movie broadcasts on DVD sales and Internet piracy. *MIS Quarterly, 33*, 321–338.
Steele, R. (2015). If you think piracy is decreasing, you haven't looked at the data…. *Digital Music News*, July 16, https://goo.gl/TUjzt6.
Swami, S., Eliashberg, J., & Weinberg, C. B. (1999). SilverScreener: A modeling approach to movie screens management. *Marketing Science, 18*, 352–372.
The Deadline Team (2013). Steven Soderbergh's state of cinema talk. *Deadline*, April 30, https://goo.gl/3md7zK.
The Economist (2017). Vinyl gets its grove back, May 20, p. 58.
Thielman, S. (2015). Popcorn Time helps film piracy to live on—even though it technically doesn't exist. *The Guardian*, May 31, https://goo.gl/EYrbVX.
Tiedge, J. T., & Ksobiech, K. J. (1986). The 'lead-in' strategy for prime-time TV: Does it increase the audience? *Journal of Communication, 36*, 51–63.
Time (2015). Adele talks decision to reject streaming her new album. *Time*, December 21, https://goo.gl/bqvyYc.
Vogel, H. L. (2015). *Entertainment industry economics: A guide for financial analysis* (9th ed.). Cambridge: Cambridge University Press.

Watson, S. J., & Zizzo, D. J., & Fleming, P. (2015). Determinants of unlawful file sharing: A scoping review. *PLOS ONE, 10*, 1–23.

Webster, J. G., & Wakshlag, J. J. (1983). A theory of television program choice. *Communication Research, 10*, 430–446.

Weijters, B., & Goedertier, F. (2016). Understanding today's music acquisition mix: A latent class analysis of consumers' combined use of music platforms. *Marketing Letters, 27*, 603–610.

Wilbur, K. C. (2008). A two-sided, empirical model of television advertising and viewing markets. *Marketing Science, 27*, 356–378.

Wilkinson, A. (2017). Cannes 2017: Two vastly different cinema cultures provoke one big Netflix controversy. *Vox*, May 31, https://goo.gl/o83UB1.

Yu, Y., Chen, H., Hung Peng, C., & Chau, P. (2017). The causal effect of video streaming on DVD Sales: Evidence from a natural experiment. Working Paper, The University of Hong Kong.

Zaleski, A. (2016). 35 years ago: The U.K. launches the 'Home Taping Is Killing Music' campaign. *Diffuser*, October 25, https://goo.gl/FdvRuk.

Zentner, A. (2006). Measuring the effect of file sharing on music purchases. *The Journal of Law and Economics, 49*, 63–90.

14

Entertainment Pricing Decisions

"Managing price is an art, not a science…"
—Vogel *(2015, p. 252)*

At the end of the day, consumers spend money only when the value that they expect to receive from a purchase (product or service) exceeds the value of the resources (money and time) that they have to give up in the transaction (Nagle et al. 2011). Judging by that basic axiom, the huge amount of money that is actually spent by consumers for entertainment products suggests that, on the whole, prices for movies, books, songs, and video games are appealing to customers. Happiness does appear to cost "so little," as the cinema marquee slogan of a neighborhood theater near New Orleans in Walker Percy's novel THE MOVIEGOER told its prospect visitors back in the 1950s.

But let's be careful—maybe there is much more to be gained by lower prices, or higher prices, or even by a combination of both. How can an entertainment manager find out? We find Harold Vogel's quote to be an intriguing start to a chapter in a book entitled *Entertainment Science*—Vogel's argument is based on the huge number of highly complex factors that influence the success of *any* pricing strategy for *any* entertainment product, and it is one that has motivated managers to rely on industry norms and rules-of-thumb (the "artistic" route), rather than attempting to determine the ideal price for a new entertainment product using scientific methods. As a result, pricing has been the least explored part of the entertainment marketing mix for quite a while.

T. Hennig-Thurau and M. B. Houston, *Entertainment Science*,
https://doi.org/10.1007/978-3-319-89292-4_14

Whereas we agree that pricing entertainment is indeed a challenge for which no easy algorithmic solution exists, we believe that exploring the knowledge that scholars have compiled regarding pricing is worth the effort. Accordingly, we will explore in this chapter what managers can gain from applying a scientific approach to entertainment pricing. Our starting point are two fundamental questions that entertainment managers face at a strategic level when setting prices:

- Should I stick with the "standard" market price for every movie, book, song, or game I sell—or should I deviate from that price?
- For a single, specific movie, book, song, or game, should I always sell it for the same price—or should I differ its price across audiences or situations?

Because of strong industry norms that carry great inertia, perhaps these questions seem almost irrelevant. But we argue that they do not get the consideration they deserve. Pricing norms for entertainment exist because of the all-too-real challenges of predicting the market reaction to any individual product, i.e., "Nobody Knows Anything." Economists argue that if pricing and product success are "more or less, a total guess," then simply following market norms is not only easy, but may also be rational (McKenzie 2008, p. 173). Whereas some of the norms for pricing entertainment are imposed by external regulations (e.g., the cultural character of books has led to price regulations in several countries such as Germany), most of them result from the industry's acceptance of the Goldman mantra.

Because these norms are so powerful, with market prices often being quite standardized for entertainment products, we had at one time considered *not* including a chapter on pricing in this book. However, because one of our primary goals with *Entertainment Science* is to challenge the Goldman mantra, we decided to tackle pricing anyway—*because* of that industry practice. Theoretical considerations, but also initial empirical findings on the matter, make us believe it is time to take a fresh look at the pricing of entertainment products.

In this chapter, we begin with a review of fundamental pricing theories to articulate how pricing, in an "ideal" entertainment world for which theoretical assumptions hold, should be carried out. Differentiated products offer differential value to customers and those customers should, in turn, be willing to pay different prices across products. For all kinds of consumer products outside of entertainment, consumers' willingness to pay for a specific product differs at least somewhat according to the product's overall appeal. And we are under the distinct impression that some songs/novels/films/

games have more appeal than others and create more value for the listener/reader/watcher/player.

However, in entertainment, variations in prices for individual products have rarely been related to these demand factors. Instead, prices tend to vary by broad categories (new releases versus catalog titles) or by product format (hardcover versus paperback versus Kindle). After a primer on pricing theory, we will explore the pros and cons of differential pricing in its various potential forms, acknowledging insights from innovative empirical studies and settings. These discussions will provide an answer to our first question. We will then turn to our second question, exploring the potential for varying prices for one specific product (or versions of the same) between audience segments and situations. Here, the theoretical concept of price discrimination will guide our analysis of entertainment pricing strategies.

A Primer on Pricing Theory: Customer Value as the True Foundation of Pricing

Let's start with a look at what pricing theory has to say about our first fundamental pricing question, i.e., whether your new entertainment product should simply be priced at the market norm. Many MBA-level treatises on pricing have a central focus on the concept of price *elasticity*—the notion that for most products, a change in price will cause a change in the quantity of the product demanded by a market. They then quickly delve into the derivation of elasticities and discussions of the role of downward sloping demand curves in choosing optimal prices. These topics are certainly critically important to firms and are foundational to making wise pricing decisions.

However, we argue that it is easy to lose sight of what is going on *behind* the elasticity and the demand curve. As Thomas Nagle and his colleagues put it, "[p]rofitable pricing requires looking beneath the demand curve to understand and manage the monetary and psychological value" created by products and that affect the customer's decision of whether to buy an offering at a certain price (Nagle et al. 2011, p. xiii). Scientific labels, like marginal utility and price elasticity, are actually descriptions of very real consumer appraisals, judgments, and behaviors that occur when humans attempt to address their needs through market exchanges and by allocating their limited resources. We, as consumers, turn to the market to purchase products or services that help us preserve our teeth, shelter our bodies, transport us to different locations, or—to entertain us. Because most of us have limited resources to meet all acquisition needs and desires, we make allocation decisions by judging the "value" that a product would provide us with. This value, when converted (consciously or subconsciously) to currency amounts, represents the maximum price (sometimes called the "reservation price") that the customer would be willing to pay for a product.

Because customers differ quite strongly in their needs, priorities, and resource levels, their willingness to pay for a product will also differ across individuals. If the price of the product is higher than the customer's estimation of the value that would be created, the consumer will seek out alternatives that provide the desired benefit at an acceptable price, or do without a solution from the market. As a producer lowers the price of a product or service, this "asking price" will likely fall within the acceptable range of willingness to pay for a larger number of consumers; as the price rises, it will exceed the value created for (and become unacceptable to) more and more consumers. This is why demand curves slope downward and price elasticities generally carry a negative sign.

Panel A of Fig. 14.1 shows the standard course of such a downward-sloping demand function. Panels B and C illustrate constellations in which consumers respond less or more strongly to changes in price. In Panel B, changes in price do not translate to substantial changes of the demand for a

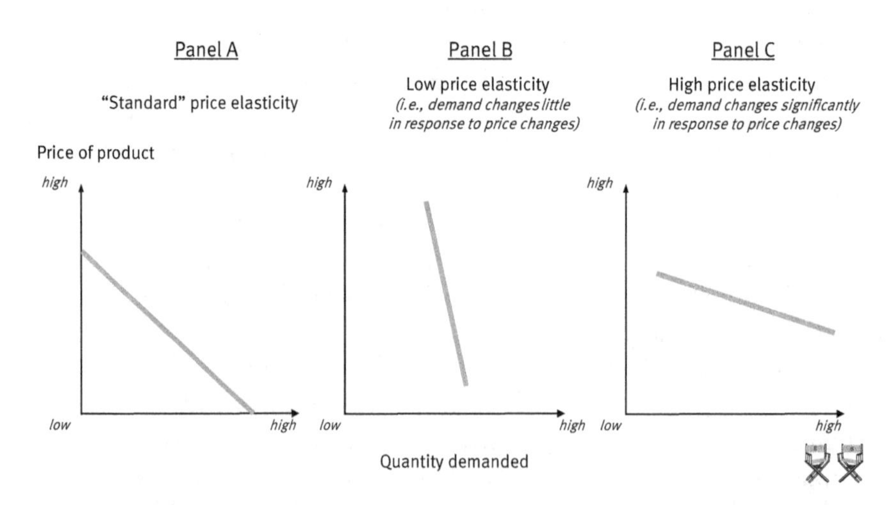

Fig. 14.1 "Standard," low, and high price elasticities
Note: Authors' own illustration.

product—the (absolute) price elasticity is "low."[422] A low price elasticity is typical for products that consumers perceive to be vital and for which substitute solutions are unavailable (their markets are often strongly regulated because of the lack of alternatives—such as electricity or health care in many countries). It is also the case for products that are only liked by a small part of the population and for which others would not pay even a very low price. This could be argued to be the case for at least some media and entertainment products—certain people would just not watch the next Michael Haneke film, regardless the price. Would you? Panel C then describes a market where the product's price elasticity is "high"—small price changes result in large changes in the overall quantity demanded. Such a high price elasticity is usual for products where many highly similar alternatives are available to which consumers can easily switch—something that is the case for most grocery products and news sources. And for standard entertainment offerings?

Demand curves and price elasticities are aggregates of many individual consumers, who often differ strongly in their psychological traits, goals, levels of risk aversion, etc. Often, the demand function is also not linear—whereas Apple's core customers are highly inelastic regarding the prices of iPhones etc. (but maybe only to a certain price point), those who are looking for a good phone other than an Apple product will demonstrate a much higher price elasticity.

The consumer's willingness to pay should be the critical metric for pricing any product if the producer's goal is to maximize revenues or profit. Traditional supply-focused pricing strategies, such as benchmarking against competitors, crafting elegant cost-plus pricing formulas, and setting prices according to firm strategy, can all be "a blueprint for mediocre financial performance" if—at the end of the day—the customer judges that your product or service fails to create the needed value *or* is priced at a level above the worth of the value created (Nagle et al. 2011, p. 2). But to cite one of Hollywood's own creations, IT'S COMPLICATED to price a product according to consumers' willingness to pay, because this willingness to pay is influenced by a large number of factors in addition to the product's inherent characteristics.

Consumers' willingness to pay varies with available alternatives. First, there is specific competition, which we have discussed in the context of

[422]Please note that when we use the terms high or low price elasticities, we always refer to the *absolute* value of the elasticity, leaving out the negative sign (if we won't do so, an elasticity of minus 3 would be much *lower* than one of minus 0.5, which would be counterintuitive).

distribution decisions. The value a consumer anticipates from reading a novel, such as John Grisham's THE ROOSTER BAR, when it is released depends on the value offered by similar novels that contend for the customer's attention at the same point in time. But, as we have shown, competition is not binary; it is continuous and, to a certain degree, generic—as consumers, we are trying to optimize the *total* benefits that we can gain from our *total* pool of resources. And so we make trade-off decisions across many product categories that address many different, more general needs (e.g., "If I read Grisham's novel over the next few days, I cannot watch the new Netflix series that has gotten such great reviews." Or: "If I save the money that I would have spent on the novel, I can afford to buy those new headphones."). All of this influences what the new novel is worth to us and how much we would be willing to spend for it. The market price may be well within the consumer's willingness-to-pay range *in general*, but he or she will not buy it nevertheless.

Costs must also be considered when setting prices based on consumers' willingness to pay. The price for a product should cover the *variable* costs of the product (i.e., those costs that vary relatively directly with the number of units sold, such as the actual materials for a physical product or the supplies to support a service) and provide a reasonable contribution toward covering the *fixed* costs that have been incurred (i.e., those costs that do not vary with the number of products and services produced). We have discussed the specific cost structures for entertainment products as informational products, and particularly those formats that are fully digital, which offer producers enormous flexibility in pricing decisions.

Finally, a firm's market-related strategy will also affect the final price chosen. For example, the goal of deeply penetrating a market would imply a relatively lower price to maximize market share, while the goal of selling first to technophiles implies a high price. Remember that the success of entertainment products often depends on network effects, which can make it reasonable to set a low price when a new product is introduced, even if it does not match the variable costs.

This discussion has presented the essence of pricing according to theory. We will now use these insights to find answers to the two basic questions of this chapter: should all products be priced in a uniform way? And should prices for one specific product be uniform across all conditions? Let's start with the first question, addressing what some consider to be the "pricing puzzle" of entertainment.

The Uniform Versus Differential Pricing Puzzle of Entertainment

> "One of the more perplexing examples of the triumph of convention over rationality is movie theatres, where it costs you as much to see a total dog that's limping its way through its last week of release as it does to see a hugely popular film on opening night."
> —Surowiecki *(2004, p. 99)*

Our review of pricing fundamentals points to the importance of understanding a product's value for consumers and consumers' willingness to pay. Are entertainment products priced according to such a demand-focused approach?

It's hard to think of an entertainment manager who would argue that all books are of equal quality and appeal, all games of equal fun and challenge, all songs of equal ability to move the soul (or body), or all movies of equal impact. Yet, current pricing strategies in each product category do not reflect differences in the value to customers of one book over another, one game over another, one song over another, or one movie over another. Not all prices are equal, but differential pricing occurs mainly across supply factors (a hard copy book costs more to produce than does a paperback) or format characteristics (e.g., a movie in DVD is cheaper than a movie in Blu-ray; a set of songs in digital format versus a physical CD), not demand factors.

Examining over 1,100 books that were released by Yale University Press in both hardcover and paperback, Clerides (2002) finds that "prices are rarely changed over time regardless of demand fluctuations;" he also concludes that "once we control for observable physical characteristics, such as size and binding quality, there is little price variation left." The scholar labels this situation a "puzzle: why do prices depend on cost-related demand shifters but not on 'pure' (not cost-related) demand shifters?" (p. 1386). Orbach and Einav (2007) use the same term when they consider the price of theatrical movie tickets.

In situations in which demand-side characteristics are indeed used as a basis for differential pricing of entertainment, they are applied at a broad category level versus considering quality differences between individual products. For example, new releases are priced more highly than catalog titles (in movies, games, and books; less so in music). But you still pay the same price at the theater to watch the new release of the new hyper-branded STAR WARS blockbuster as you do for the original studio production of THE GREATEST SHOWMAN, as well as for small-budgeted arthouse films, regardless of whether the critics love them (A FANTASTIC WOMAN, for instance) or loath them (such as THE DANCER). Nearly all songs are $0.99 per track on iTunes, and all songs included in Spotify are the same price because they are included in the general subscription fee. Courty (2011) calls this "unpriced quality."

When Orbach and Einav (2007) examine the legalities and economic structure of the movie industry to explain the existence of uniform pricing, they conclude that, other than the regulatory prohibition against vertical distribution arrangements (i.e., studios cannot own theaters), there are no other external constraints. Some industry experts have told us simply that "it has to be this way, because it was always this way." Apart from this being a rather disappointing argument, it is also wrong: in the early days of the movies, differential pricing was standard practice, with ticket prices varying with regard to a film's popularity and stars, among other factors. "For three decades until the 1940s, one theater would have the rights to each movie within a certain zone, and movies received grades (A, B, or C) that corresponded with ticket prices at those theaters. If the rules of the 1920s ruled today, MISSION IMPOSSIBLE might be $15 and YOUNG ADULT [a $12-million comedy-drama] might be $7" (Thompson 2012).

Why then was that established practice abandoned? Orbach and Einav (2007) are among those who consider the Supreme Court decision *United States v. Paramount* in 1948, which prohibited movie studios from being involved in the exhibition business, as the seminal historical event in the return to uniform pricing of movies. Afterward, there could no longer be carefully crafted vertical arrangements that made it easier for a portfolio of products of different grades to be sold by a studio to exhibitors at differing prices (McKenzie 2008). But Thompson (2012) notes that a rudimentary version of differential pricing survived until 1972, as consumers were still charged a premium price to see "event" movies by theaters. The practice was then dropped in conjunction with the release of THE GODFATHER at the behest of the studios, as Pearlstein (2006) argues.[423]

In summary, consumer pricing in entertainment markets is fundamentally different than consumer pricing in nearly every other consumer products category in which higher (versus lower) quality products are able to command higher prices. Uniform pricing exists despite the significant potential for variable pricing for entertainment products, which are information goods with high first-copy costs, but negligible marginal costs for each additional unit sold. So, are there fundamental economic justifications for these differences? Or should uniform pricing be discarded? In the next section, we examine the pros and cons of differentiated pricing for entertainment.

[423]With tongue-in-cheek, Pearlstein calls the fact that this exertion of power happened for Francis Ford Coppola's mafia movie a "coincidence" that would have made a mobster proud.

Theoretical Arguments for Differential Pricing—and Against It

A variety of logics and arguments have been brought forward to explain and/or justify uniform pricing; these need to be contrasted with arguments for differential pricing.

The fundamental argument for uniform prices is that when "Nobody Knows Anything," there is just no basis for charging different prices for different products—because consumers' willingness to pay is unknown a priori (McKenzie 2008). This idea has received support by some scholars—De Vany (2006) argued that "You only know after a movie runs that demand is so high that a premium price could be charged" (p. 626), and Srinivasan (2009) claimed that when "stripped to the barebones, all movies are perceived as the same by customers."

We, however, have stressed our denial of the Goldman mantra from this book's first page on, and this applies also to its use as a legitimation for uniform pricing. First, it would hardly make any sense for a producer to spend hundreds of millions of dollars developing a product if he or she could achieve the same success for much less. But second, our book offers extensive empirical evidence that success predictions for entertainment are possible and informative; studies make use of the rich data that exists for entertainment products in our digital times.[424] As details about a new product's distribution are usually only made available in the days or week prior to the launch, the predictive fit of such models should be rather excellent.

Another key argument that is brought up to justify uniform pricing is that there is just not enough variation in consumers' willingness to pay, the essential requirement for differential prices being meaningful. We have to admit that because entertainment producers have practiced uniform pricing for so long, we simply do not have the (secondary) data to mathematically calculate consumer responses to differential pricing of a specific product (Orbach and Einav 2007). But economic logic shows that setting prices according to consumers' willingness to pay is preferable under most realistic market conditions (Varian 1995), and we have shown that large quality differences exist across products, along with differences in consumer preferences.[425] So we do not see any reason why willingness to pay for entertainment should not vary between products. In fact, we will show in the following section that price elasticities for entertainment, based on other variations, are far from trivial.

[424]See, for example, the section about prediction models in the innovation chapter of this book.

[425]We discuss systematic differences in consumer preferences and the reasons for them in our section on taste, and our review of entertainment markets in the business models chapter shows the heterogeneity of available products and types.

But how would consumers react to differential prices, beyond willingness to pay? Some have noted that consumers would be confused by differential pricing or perceive them to be unfair, as uniform pricing having existed for so long. Orbach and Einav (2007) explain why consumers likely would *not* see differential pricing as unfair. Remember that we, as consumers, are accustomed to paying different prices according to quality in most other product categories. Lower prices could potentially stigmatize products as being of lesser quality, triggering negative feedback effects and turning potential hits into flops merely based on their lower ticket price. But, whereas a lower price might indeed signal a lack of quality to some, others will be less influenced by the quality signal the price would send—remember that uninformed action-based cascades compete with informed cascades. In addition, producers might even benefit from such signals in the case of high-priced films. Would customers rely on prices as quality signals at all, once they have become familiar with their strategic use by producers? We have all learned that the most expensive brand is not necessarily the best one (at least not for *us*) in many other product and service categories, so why shouldn't we be able to learn that when it comes to entertainment?

Other arguments address the challenges of implementing a differential pricing scheme. If platforms would be in charge of setting prices in a way that differ between products, they might have diverging interests and follow different logics/economic models—which might be to the disadvantage of the producer. For instance, if lower prices for some movies would increase admissions, theaters might benefit from additional concessions revenues (which are high margin and are not shared with studios; Davis 2006).[426] But such higher admissions might be obtained at the cost of reducing box-office revenues and thus could hurt the producer. Also, policing costs of differential pricing models could be high. Given the structure of most multi-screen movie theaters today, people might purchase tickets for a low-price movie but, once inside the theater complex, sneak into more expensive movies (Thompson 2012).

These arguments have to be taken seriously—but they are essentially about the *implementation* of the differential pricing model rather than general arguments against differential pricing, per se. The implementation could be influenced by the producer to a large degree; producers might insist on participating in concession revenues, for example. But there might indeed be conditions under which implementing such a pricing model might not

[426]Thompson (2012) reports about a case when in 1970 "some D.C. theaters cut weekday tickets by two-thirds and saw popcorn sales double. That's a huge boost for theaters, since half of theaters' income comes from amenities like popcorn." He provides no details about the occurrence, though.

be economically reasonable—such as when strong distrust exists between producers and distributors and inflates control costs.

In essence, our weighing of arguments regarding the advantages and disadvantages of differential pricing cannot definitively demonstrate the strategy's superiority. But if channel partners can find constructive ways to address implementation challenges, pricing theory suggests that a producer (and the distributor) should gain from offering different products at different prices, at least to some degree (Courty 2011; Orbach and Einav 2007). The conditions for which economists have demonstrated analytically that uniform pricing can be optimal are quite restrictive (some consumer heterogeneity, but not too much; particular assignments of products; e.g., Courty 2011) and do not seem to apply to most entertainment products.

Some scholars have shed additional light on the issue by conducting empirical studies, using contexts which indeed practice variations of differential pricing. But before we report their insights, let us first look how elastic consumers of entertainment products have been found to respond to other kind of price variations than those for one and the same product.

Consumer Reactions to Entertainment Product Prices in General

Can we say something about the elasticity of consumers' responses to price differences in general when it comes to shopping for entertainment? Even in today's market conditions in which products are largely uniformly priced, scholars have compiled evidence that consumers tend to buy more entertainment when prices go down and less when they go up. From where do these variations come, when few systematic differences exist based on consumers' willingness to pay? In addition to timid approaches toward demand-based pricing (fresh albums cost more than catalogue ones; e.g., Mixon and Ressler 2000), some are situational (the discount tickets for Tuesday nights at the movie theater), and others are cost-based (thicker books cost more). We will take a look at findings for the different forms of entertainment.

Books. Clerides (2002) examines sales of 549 paperback books and finds a price elasticity of −3.9 when the paperback is released simultaneously with a hardcover (i.e., a 1% increase in price translates to a 4.5% decrease in demand) and an elasticity of −3.0 for paperbacks that are released sequentially after the hardcover version. Others have found less large elasticities, though still above 1 in absolute terms, indicating a highly elastic price reaction. Chevalier and Mayzlin (2006) estimate an elasticity for book sales at Amazon.com of between −1.5 and −2.1 (and reactions of Amazon sales to price changes at Barnes and Nobles of between 1.8 and 2.7), and Bittlingmayer (1992) reports elasticities ranging from −1.5 to −3.0 for the German book market, probably associated with the government-mandated lack of price competition between retailers.

Music. With music prices being largely uniform these days, the room for studies to explore consumer reactions to price changes is even more limited than for books. Using a sample of 1,457 "best-selling songs" from 286 albums in 2009, Danaher et al. (2014) worked with a major music label and were able to systematically vary the prices of the songs, though only between $0.99 and $1.29. When they put the price change in relation to changes in sales, they found that a 1% increase in the price of a song corresponded with a 0.48% drop in sales, indicating a relatively low elasticity for music. This is also what Chen et al. (2015) report when they examine music sales on Amazon (physical and digital): price elasticity is significant, but relatively low for music. But they find differences between *types* of music: for lesser-known artists, consumer reactions to prices are stronger.[427] Such a difference is well-known by brand managers—strong brands increase loyalty and restrict switching to alternatives, allowing managers to charge higher prices for their products.

Movies—in theaters. Theater prices again show very limited variation. But some scholars have used creative methods to circumvent this limitation when estimating price elasticities. Their results show high consistency: the demand for movies shown in theaters is highly elastic. Davis (2002) cooperated with six U.S. theaters to conduct an experiment. Over a period of three weeks in 1998, the theaters "substantially reduced adult evening admission prices (from $7.75 to $5), and subsequently raised the adult admission price to $8" (p. 82). For his sample of 47 films, Davis finds price elasticities for the six theaters that range from −2.30 to −4.11, which means that a 1% increase in the admission price would result in a 2.5% to nearly 5% decline in demand. Why then were theaters able to raise the ticket price so dramatically (in the U.S., the price increased by 30% from 2007 to 2017)? Because, unlike in Davis' experiment, prices were raised by basically *all* theaters to a similar degree, which left moviegoers with the only alternative of not going at all.

de Roos and McKenzie (2011) use another way to circumvent the problem: they exploit situational price differences. Movie ticket prices are

[427] The scholars' VAR results do not allow us to report any elasticities.

reduced in Australia on Tuesdays. Based on the assumption that the demand for movies is essentially the same on Tuesdays as for regular weekdays, they apply a random coefficients discrete choice model for 314 movies released in 2007 from the Sydney area and find a highly elastic reaction of consumers to the discounting of films—they estimate the median elasticity to be higher than 2.5. And Gazley et al. (2011) use a conjoint analysis approach to determine the price reactions. Using a design that also varied other movie characteristics, such as genre, country of origin, and stars, they test how consumers react to $8 versus $15 ticket prices in a more complex design. Evidence from 225 consumers in New Zealand demonstrate that ticket price is a strong determinant of consumers' choice decisions—its impact is as strong as the participation of their "favorite actor" or "favorite director."

Movies—at home. And there is evidence that consumer demand for movies is also elastic for channels other than theaters. Gong et al. (2015) used an experimental design in which they systematically vary the digital EST prices for 299 catalog movies by a major studio at a digital movie retailer. They find that consumers are highly sensitive to price promotions in that channel. Specifically, their regression results suggest that a $1 drop in a movie's EST price corresponds with increases ranging from 35 to 47% in expected EST sales. Corresponding calculations of elasticities to price show a range from −1.9 to −3.8, or between 20 and 44%, depending upon other promotional activities (such as placement). Finally, Luan and Sudhir (2010), in their study of 526 DVDs, use an approach in which they adjust for retailers' tendency to set lower prices for movies that were hits in theaters (another strategic pricing move); they estimate a price elasticity of about −1.8.

Are there any studies about games and prices? Marchand (2016) shows that prices are positively correlated with the attractions that a game has to offer—although only across very few, very basic pricing levels (e.g., $70, $50, and $30 for console games). He argues that "usually the bestselling video games … are AAA games that offer the highest technological standard currently available, combined with high production budgets, at high retail prices" (p. 147). In his analysis, in which he uses the price of a game only as a control and does not correct for its endogenous nature, he finds a positive association between price and game sales.

We have shown that for those forms of entertainment for which sufficient variation in prices exists or scholars have found ways to overcome the lack of variation, consumer responses are quite elastic to price changes. Let us now complement this fundamental insight with what explorations of differential pricing in entertainment can show us.

What We Know About Consumer Reactions to Differential Pricing for Entertainment Products

Let us start with some anecdotal evidence for movies, Thompson (2012) reports (without offering any details, though) that when tickets in Japan were sold for a premium of 67% for JURASSIC PARK while those for AUSTIN POWERS were discounted by 45%, both experiences were "profitable." We can learn additional insights, however, from scholarly studies.

We have already noted Danaher et al.'s (2014) study with regard to the general price elasticity for music the authors calculated. But the scholars' insights reach well beyond those general elasticities. The systematic design of their study allowed them to see how these price *changes* affect not only each song's sales, but also the sales of albums and the other songs on them. Based on a structural model in which they examine the impact of different scenarios, the scholars recommend that the price of the highest-ranked (i.e., most popular) six songs for a typical album should be $1.29, less popular songs (ranked from 7 to 10) should be priced at $1.09, and the least popular three songs' optimal price is $0.89. The logic behind this "tiered pricing" approach is twofold:

- Danaher et al. find that demand for popular songs is relatively inelastic (increased prices do not reduce sales to the same degree), and
- the price increase for a highly popular song has a positive effect on album sales, whereas price changes of less popular songs do not trigger album sales. The positive utility gained from a lower price of a less popular song is just not enough for consumers to justify purchasing the whole album, despite its relatively higher value.

Another study that tests the success potential of heterogeneous prices is Shiller and Waldfogel (2011) who measured the willingness to pay of about 500 Wharton undergraduates for each of 50 popular songs in 2008 and 2009. They compare uniform pricing with "component pricing" (a different price for each song) and find that the latter increases producer surplus by about 3%. They find that more elaborate price discrimination schemes (bundling, etc.) provide higher payoffs—we will get back to this in a moment.

The richest study in a movie context so far is by Ho et al. (2017), who make use of a rare actual application of differential pricing by movie theaters. They draw on real pricing data from Hong Kong theaters that introduced full-blown differential pricing as a response to the particularly strong effect that home video piracy had on visits to local theaters by the end of the 1990s. The authors obtained daily ticket sales over a three-month period in 2012 from Hong Kong's five leading theater chains and, on that basis, estimated the impact of price differences using a GMM model. Their simulation analyses show that film-specific pricing generated 37% higher admissions and a 24% higher profit for theaters compared to uniform pricing. A large part of this admission and profit increase, however, is captured by the higher prices for 3D versus 2D versions of films in a theater. Nevertheless, an increase in admissions of 24% and a profit increase of almost 8% is generated by forms of differential pricing that reach beyond 3D/2D differences.

It's still somewhat unclear to what extent these Asian findings apply to Western movie audiences, and a true experimental design would be desirable over the use of instrumental variables to get rid of a managerial bias. Recently, the Regal Cinema chain in the U.S. announced that the firm will experiment with premium pricing for blockbusters (and lower prices for flops) in several pilot markets in 2018. And AMC's German subsidiary UCI Kinowelt announced that the chain will from 2018 on employ differential pricing for all its 23 German theaters and 203 screens for at least five years in cooperation with consulting firm Smart Pricer, based on the results of a two-year field test in a limited number of venues. The press release stated that revenues were "above average" and received well by moviegoers, in line with our arguments (*Smart Pricer* 2018). It is our hope that this (and hopefully other field experiments and additional scholarly explorations) will provide scholars with data that can offer generalizable insights.

Let us keep in mind though that differential pricing is only a label for widely varying ways to set prices differently between products, and the approach's effect will vary with the accuracy with which consumer demand and willingness to pay are determined. General popularity and demand must be balanced with information about target groups' specific elasticities. Should arthouse movies be cheaper than blockbusters because demand is lower? Or is the demand tied to a unique segment of consumers who have a high willingness to pay, whereas very few others are willing to watch an Almodóvar movie, regardless the price? In the case of UCI Kinowelt, we found that price variations were quite small and tended to be upward biased, an approach that minimizes conflict with producers. Would larger differences also pay? So the real question is less whether differential pricing for entertainment works, but *how* it should be done.

Price Discrimination: Different Prices for Different Customers (and Products)

We began the chapter by posing two fundamental questions that face the entertainment product manager who is setting pricing strategies. After having discussed whether different products should be priced differently, let us now explore whether the price for *one specific* product should differ across audiences or situations. Our finding that there is substantial elasticity among price variations for entertainment even under conditions of uniform pricing, along with our earlier insight that taste and preferences differ among consumers, points to the usefulness of such price differences—a strategy known by marketers as *price discrimination.*

In its strictest sense, price discrimination refers to the practice of selling the same product to distinct consumers at different prices. But there are other forms of price discrimination in which the product (or product availability) is varied across price levels by firms. In the U.S., the Robinson-Patman Act from 1936 emerged over concerns that price discrimination could harm competition, but it has little practical effect on pricing to consumers. Consumers do not have standing to sue under the act (thus most cases are brought by businesses), and there are three powerful defenses for discriminating certain segments: cost justification (i.e., different costs to serve), meeting competition (i.e., price matching, volume discounts), and changing conditions (in the market or for the firm) (Nagle et al. 2011).

Theory distinguishes between three types of price discrimination around which we will structure this section; we illustrate them in Fig. 14.2. We begin with the "purest" form of price discrimination, or "first-degree" price discrimination, then move to forms in which customers self-select among differing offers that correspond to varying levels of willingness to pay ("second-degree" price discrimination), and conclude with approaches that are familiar ground for most marketers—they involve segmenting markets by manifest or latent customer characteristics ("third-degree" price discrimination).

First-Degree Discrimination: "Perfect" Price Discrimination

First-degree price discrimination (sometimes called "perfect" price discrimination) is making to each customer a separate offer that is priced according to that customer's willingness to pay. Because of the difficulty of knowing each customer's true reservation price and because of the implementation

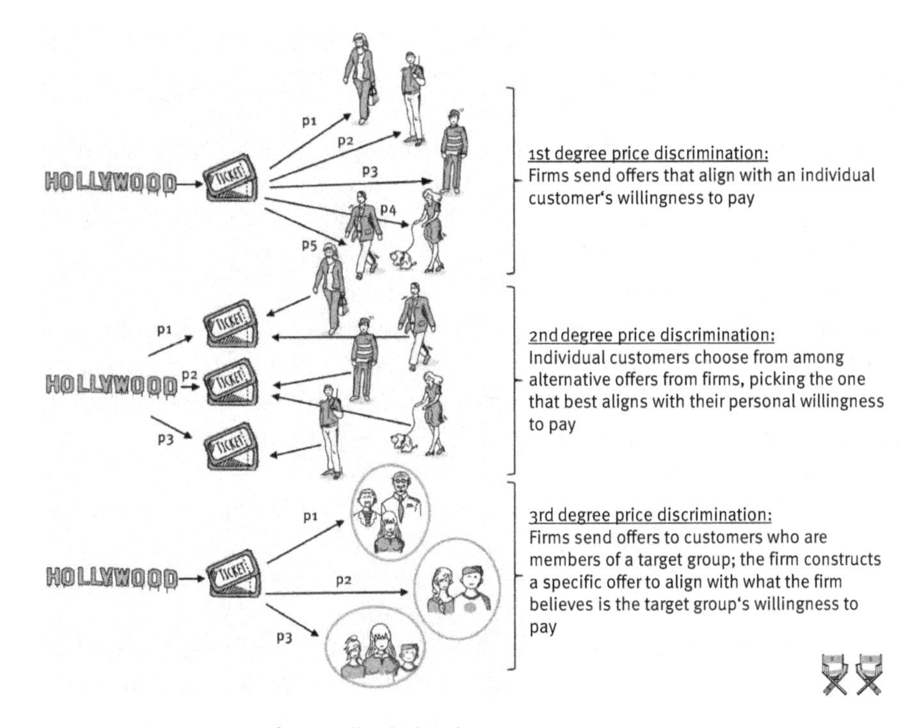

Fig. 14.2 Three types of price discrimination
Notes: Authors' own illustration based on information by Linde (2009). Graphical contributions by Studio Tense.

complexities of managing each offer separately in entertainment, this type of price discrimination is primarily theoretical (e.g., Shiller and Waldfogel 2011).

But entertainment products may actually have an advantage in this regard because of their nature as an information good, and thus entertainment producers have been among the few who experiment with pricing approaches that approximate first-degree price discrimination. Since it is hard to determine an individual consumer's "true" willingness to pay and only then make an offer, an alternative approach is to make an offer that allows the customer to choose the amount he or she *wants* to pay—an approach named "voluntary payments," "participative pricing," or simply "Pay What You Want" (PWYW). A prominent example for such a pricing model was the pre-release of the album In Rainbows by the popular British alternative-rock band Radiohead on their own Internet site in 2007, for which the band asked their fans to name their own price for the digital download.[428]

[428]A screenshot of the website at that time can be found on various sites on the Internet, such as at https://goo.gl/e9KZ7F.

As PWYW essentially lets consumers set the price for an entertainment product, one might think that it just cannot work—won't every rational customer take it for free? But consider the cultural nature of entertainment products and the degree to which consumers identify (and perhaps perceive some form of "relationship") with their favorite musicians, actors, and authors, or the loyalty they feel to a gaming community. In the case of the Radiohead album, 62% of those who downloaded IN RAINBOWS did so without paying anything. But that means that 38% of the people who downloaded the album voluntarily paid some amount: 17% paid up to $4, 6% between $4 and $8, 12% between $8 and $12, and the remaining 4% paid more than $12. The average amount of those who paid *something* was $6, and among all downloaders it was $2.26 (*comScore* 2007).

Since the Radiohead offer, the approach has been used by a number of other musicians (e.g., via the indie-music website Bandcamp.com, which claims to have collected $237 million for artists via PWYW in its ten years of existence), as well as for the pricing of games. For example, the WORLD OF GOO video game brought in over $100,000 in revenues for game developer 2D Boy during a two-week PWYW experiment in 2009 (Groening and Mills 2017), and websites like Humblebundle.com have made PWYW a regular feature for games.

Why do some consumers pay for a product that they can get for free? Kim et al. (2009) argue that voluntary payment models dissolve the distinction between "money-market relations" (in which a monetary value metric regulates the exchange) and "social-market relations" (in which a *social* value metric, based on social norms—such as fairness—regulates the exchange). In short, consumers receive powerful "social utility" that compensates for any monetary utility that is given up by volunteering to pay a non-zero price. Consistent with this line of thought, the scholars found in an experimental study of PWYW that the amount paid by customers was positively influenced by their fairness perceptions and satisfaction, among other factors. Why do you tip the waiter or waitress? The reasons might be similar to those that encourage consumers to pay as part of a PWYW offer.

Other scholars agree. Waskow et al. (2016) state that the main motivation is "the power of social norms, which may outweigh explicit market norms," and Gneezy et al. (2012) suggest that violating these social norms may threaten a customer's self-identity: "voluntary payments may signal a prosocial identity and buyers may tend to avoid to purchase at all when they feel that their [willingness to pay] might be 'too low'." One implication is that firms thus might want to strategically emphasize those social aspects (e.g., concern for seller, norms, reciprocity, equity) via promotional messages to achieve higher PWYW revenues (Groening and Mills 2017).

So, is PWYW a viable strategy? Does it work, and, if so, under what conditions? Gerpott (2016) conducted a meta-analysis of empirical studies conducted on PWYW with data from an entertainment context and a range of other products, from food and beverage (the largest category), to low-priced goods. He concluded that "economic evaluations of economic PWYW outcomes in comparison to those of fixed prices or free giveaways are more often positive than negative" (Gerpott 2016, p. 588), but also noted that most studies only look at short-term effects.

More detailed insights can be derived from those studies that examine only entertainment products. Kim et al. (2009) conduct a field study in a movie theater. For three days (Monday-Wednesday), the management offered cinema tickets under PWYW conditions. The multiplex theater consisted of eight different movie screens that provide seats for 99–355 guests, with a maximum total capacity of almost 1,500 guests. The scholars compared the PWYW revenues with a baseline of regular sales which they calculated from 53 weeks of daily data. Kim et al. found that, on average, consumers paid more than zero, but less than the regular price. On standard days the average PWYW price paid was €4.87 (compared to a "normal" average ticket price of €6.81) and €3.11 on the "discount day" (where the "normal" average price is €4.43). Figure 14.3 shows the distribution of prices paid by customers in their study.

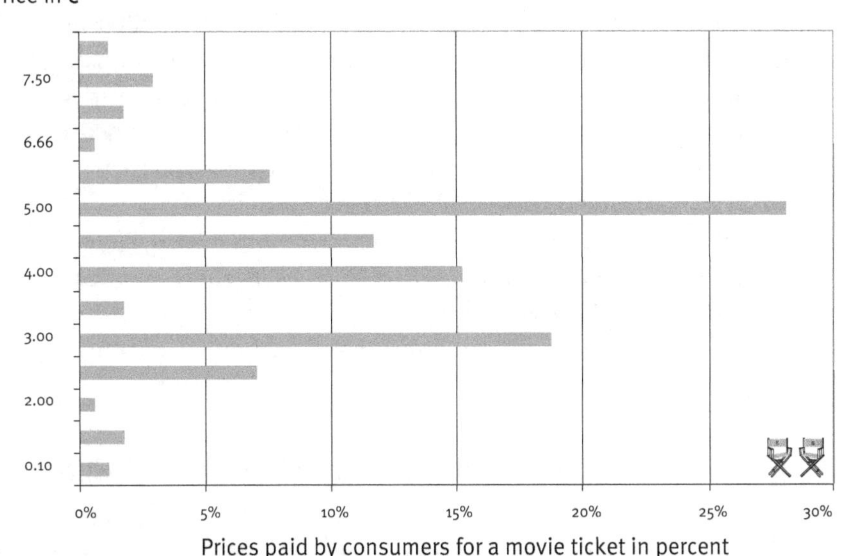

Fig. 14.3 Prices paid by customers under "Pay What You Want" (PWYW) at a movie theater
Source: Reprinted with minor adjustments with permission from Journal of Marketing, published by the American Marketing Association, Kim et al. (2009) Pay What You Want: A New Participative Pricing Mechanism, January 2009, Vol. 73, No. 1, pp. 44–58.

In their experiment, the lower prices did not drive higher volumes of ticket sales, so the overall impact of PWYW on box-office revenues was negative. It would have been interesting to learn whether audiences used some of their savings for spending more at the concession stand, but those revenues were not measured by the scholars. In another entertainment experiment of PWYW, Waskow et al. (2016) analyzed voluntary payments for music albums by 25 participants. Their findings are similar to those by Kim et al., with average prices being higher than zero, but about 20% lower than full payment prices. Interestingly, from studying their participants' neural structures the authors derived that PWYW music purchases triggered stronger neural activity "in brain areas involved in reward-processing" than the control group that was only offered the music for the regular "full" price.

Overall, from the empirical evidence, it is unclear if a PWYW pricing approach can be profitable in real-life settings, at least in terms of short-term generation of voluntary revenues. But there is at least suggestive evidence that the most important value contribution of PWYW may be the creation of attention and consumer buzz. The offer by Radiohead was available for eight weeks only, after which the band released the album via the usual distribution channels, both digitally and in CD (Leeds 2007). And in the first year after its regular release, Radiohead's album sold 3 million times, including 1.75 million CDs (Thompson 2008).

Bourreau et al. (2015) took a closer analytical look at the Radiohead release, using weekly music sales data in the U.S. between 2004 and 2012 to examine the effect of Radiohead's approach on subsequent sales of the band's albums. They concluded that the PWYW offer "had a positive impact on sales revenues, even if one assumes no revenues were obtained directly" through PWYW. Instead, higher-than-usual digital album sales were due to the vast media attention generated by the offer strategy. However, Bourreau et al. find that when Nine Inch Nails decided to provide their new album, THE SLIP, for free, digital album sales were hurt, not increased. So the successful nature of PWYW at least partly depends on whether the pricing approach is able to trigger the interest of consumers and initiates buzz cascades.

Second-Degree Discrimination: Versioning and Bundling

The other two kinds of a price discrimination are based on the realization that a consumer's willingness to pay is hard to observe; it rarely correlates with demographic characteristics, for example. Second-degree discrimination means that variants of a product are offered for different prices. The underlying idea is that a consumer's latent willingness to pay will lead the customer to self-select the appropriate purchase option.

Practical ways to implement second-degree discrimination include varying the price for a product (a) by the quantity purchased (something also known as "non-linear pricing"), (b) by the version of the product chosen (i.e., "versioning"), or (c) by combining a certain product with other products or services, a strategy referred to as "bundling"). Let us take a quick look at each of those three options.

Non-Linear Pricing: Quantity-Based Pricing

Non-linear pricing refers to the practice of charging different prices based on the quantity purchased by the customer. These practices are common in our local supermarket where, for example, we can buy one bag of Lay's Classic Potato Chips for $3.99 or "Two for $6!" In entertainment, McKenzie (2008, p. 95) refers to this exercise as "walking your patron down the demand curve;" he describes the high price of a "small" movie theater popcorn and the small marginal price increases to get the medium and the large: "They aren't so much lowering the marginal price of the additional ounces as they are hiking the price on the first few ounces." This effectively creates a "floor" price for access to popcorn.

But popcorn is an ancillary product. Can firms engage in quantity-based pricing for their primary entertainment products? The entertainment industry's history shows few occasions where producers or distributors have offered different prices based on the quantity of products a consumer accesses, but recent attempts from industry outsiders to set prices according to the consumer's quantity of usage across a *portfolio of products* have turned out to be quite successful, if not disruptive. Keep in mind that this is not exactly the same as offering multiple "copies" of one specific product (like in the potato chips example above), but the information good nature of entertainment deems the latter approach irrelevant—I can watch a movie I have purchased as often as I want anyway.[429]

[429]In some ways, some kinds of versioning (which we discuss below) could be interpreted as a quantity-based pricing strategy—whereas the rented version only allows repeated consumption in a restricted time frame, the purchase version gives us unlimited consumption opportunities. In addition, one might also consider digital rights management measures (DRM) that reduce the number of consumption acts as implementations of quantity-based pricing. But such DRM, with reduced consumption quantity, is usually not offered for a discounted price.

These quantity-based offers take two forms. *Usage-based pricing* consists of different prices depending on the usage levels: Netflix has different subscription prices that determine the number of movies a customer can stream at the same time. The marginal price a consumer pays for access to a higher-usage tier is usually at a lower-per-unit rate than the lower-usage plans. Usage-based pricing shifts the burden of controlling usage to the consumer, but the seller still incurs significant costs for monitoring, billing, and settlement (i.e., higher "transaction costs," Sundarajan 2004, p. 1661).

But it is *fixed-fee pricing* (or subscription) models that have brought so much change to entertainment consumption recently. Such models take the form of one price in exchange for "all you want to listen to" (à la Spotify) or "all you can watch" (à la Netflix and Amazon Prime Video). Even in movie theaters, such "flat rates" are becoming popular; whereas some theaters have experimented with the idea before, it is larger scale services (such as MoviePass) which do not limit consumers' access to a single theater that seem to appeal to substantial numbers of consumers (D'Alessandro 2018).

In fixed-fee pricing, the prices for a single product or consumption act vary because each customer's per-unit price can be calculated as "Price/Number of Units Consumed" (or P/N). Thus, if P is constant across customers, the customer who consumes less (versus more) N pays a higher price per-unit (although the "objective" price he or she pays remains the same). Because of this logic, consumers sense an incentive to consume more (as it reduces their "average" consumption price), so that the strategy makes only sense economically if marginal costs are low—which makes the strategy well suited for entertainment's information-good nature.[430]

Which non-linear pricing scheme is best for producers? Classic findings suggested that usage-based pricing was optimal (e.g., Wilson 1993), but these conclusions were based on economic analyses with some assumptions that are fairly unrealistic for entertainment goods (e.g., pure monopoly, no transaction costs). Relaxing these assumptions, Arun Sundararajan (2004) used sophisticated analytical modeling and concluded (for information goods, in general) that a combination of fixed-fee pricing and usage-based pricing scheme "is always profit improving in the presence of nonzero transaction costs, and there may be markets in which a pure fixed-fee is optimal" (p. 1661). He recommended that early in the life of a category of information good (e.g., when streaming was first introduced), the usage-based/fixed-fee balance should be tipped in favor of fixed-fee, because a low-priced

[430]That is also why it would be less-well suited for non-digital/information-good offers such as concessions in a movie theater.

fixed-fee pricing scheme can be emphasized as a penetration strategy to build consumer usage. High penetration helps to develop market power and adds value for users because of network effects (e.g., via better recommendations).

And findings by Wlömert and Papies (2016) on music streaming services, based on their tracking of the actual behavior of a panel of 2,500 music consumers over more than a year, support the idea that fixed-fee streaming subscriptions can have positive effects in the long run for producer-level and industry-level revenues. But such models also involve other kinds of second-degree price discrimination, namely versioning and bundling—we will get to the scholars' findings when discussing these strategies.

Versioning: Let the Consumer Pick!

Versioning is also a second-degree pricing approach—here a seller offers various "versions" of his or her product to the market. These versions differ by price according to their likely appeal to consumers (and thus their expected higher willingness to pay), i.e., versions with higher quality, more features, and/or extra benefits are priced higher. Consumers then "self-select" the version that they most prefer, paying what they are willing to, instead of dropping out if a product is considered to be too expensive. A classic example from outside of entertainment is the airline practice of offering tiered tickets on the same airplane, such as first-class, business-class, and economy/coach seats, which come with varying space and benefits at different prices.

Versioning is also referred to as "indirect" price discrimination because the alternatives between which the consumer chooses differ not only in terms of price, but also in their quality—which is also why another label for the approach is second-degree *quality* discrimination. The main challenge for the producer is to design the versions so as to induce the consumers to pick an appropriate version—offering a version for everyone, but preventing those who have a high willingness to pay from switching to the cheaper versions.

Versioning is widely used by entertainment firms. Hardcover books, paperbacks, and electronic formats are versions of the same book title that are usually offered at different prices. Theatrical screenings, a DVD, and a VOD/EST stream are versions of the same movie that differ in quality, context, and time of availability. And even within each of these formats, often various versions exist for the consumer to choose from: consider all the different versions of a movie such as BLADE RUNNER, from plain DVD to "Deluxe Collector's Edition," as well as the different "seat versions" sold in European and Asian theaters. Versions of video games are the different formats for which different prices reflect differences in resolution and scale (PC version, console version, mobile version etc.) In addition, players can often choose between a free version that offers limited functionality and a "premium" version with full functionality that provides advantages over those who are playing the game for free—a kind of versioning referred to as "freemium" (we return to the economics of this approach below). And for music albums, a standard version is often complemented with "special" versions that contain additional material and sometimes a unique package.[431]

Empirical findings on versioning effects suggest that producers in entertainment are able to capture significantly different margins with versioning. In the book data set analyzed by Clerides (2002, p. 1395), he finds that "hardcover margins are between 80% and 396% higher than paperback margins," with the difference being statistically significant. And using a data set of more than 5,000 pre-owned games released prior to 2010 and analyzing the determinants of game prices (i.e., an indicator of what consumers were willing to pay for a game) with an OLS regression, Cox (2016) finds that whether a game is a "special edition" is among the main drivers of price variation. The coefficient for the "special edition" variable is 0.39—in other words, special edition versions of games command prices that are, on average, about $\left(e^{0.39} =\right)$ 48% higher than those for standard games. Adding interactions to the analysis reveals that special editions of high-quality titles (those which were positively reviewed by critics) tend to be sold for systematically higher prices than special editions of less highly rated titles, and that the price advantage of special editions tends to decline with the age of a game.

What do we know about "freemium" models, a quite popular kind of versioning? An underlying strategy of the free offer is often to build up the user base of the product and increase its social relevance, essentially harvesting network effects. When the LEGO UNIVERSE MMOG was first launched, only a paid-only version was available, but it failed to attract sufficient user numbers. Lego then switched to free-to-play, which, as intended, attracted additional interest and players, as can be seen by examining the search volume for the game, over time, in Fig. 14.4. But in this case, no noticeable feedback effects and sustainable upward trend could be sustained, and Lego decided to discontinue the game only a few months later.

[431]See also our discussion of packaging in the context of owned entertainment communication.

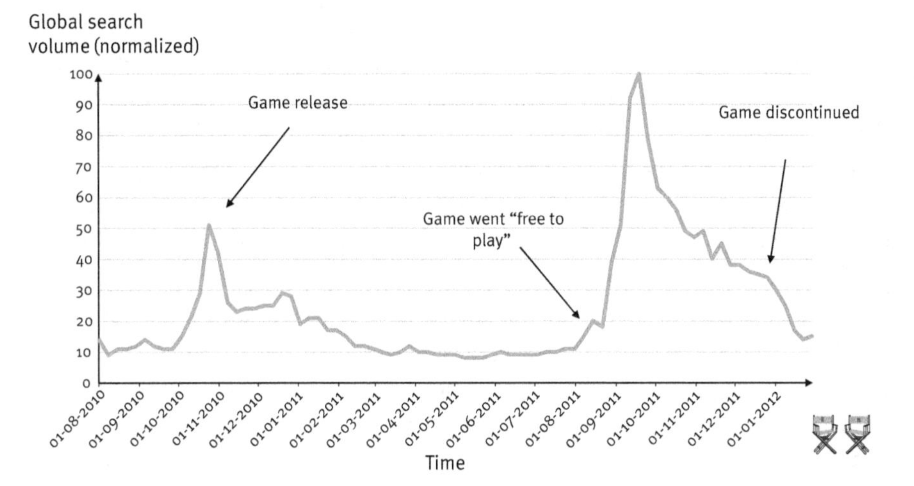

Fig. 14.4 Search volume for LEGO UNIVERSE MMOG over time
Notes: Authors' own illustration based on search data from Google Trends. The measure is the normalized global Google search volume.

For music, Papies et al. (2011) find, in a latent-class, choice-based conjoint experiment, that free song versions (which include advertising in this case) have the *potential* to attract consumers who would otherwise refrain from paying for music at all. Switching between the two versions (from free music with advertising to paid music) does not happen often, though. In their data, ad-sponsored music appears to appeal only to a clear-cut segment that is very price-sensitive and would drop out of the market completely if the only option was to pay. Versioning thus allows the firm to draw revenues (via advertisers) from a segment whose low willingness to pay would otherwise keep them away.

How can a manager determine the right value of the "free" versus the paid version in such models, then? In essence, the net contribution of any user of the "free" version is the incremental advertising revenues he or she generates, plus the potential network value the user creates for those who have subscribed to the paid version. Spotify, for example, can provide their paying users better playlists and recommendations based on the information they get from their "free" users. But one factor still has to be subtracted: the loss in potential subscription fees of any user of the "free" version. If a consumer would subscribe to the paid version if no free version was available, offering him the free version creates "opportunity costs" in the amount of the subscription fee. Entertainment providers thus have to design the value of the "free" version for users in a way that maximizes this net contribution.[432] Let us note that this value can be dynamic—when 70 million people already use the "free" version of a service (as was the case with Spotify in early 2018), the incremental network value of an additional user is less than it was when the service only had one million.

In an analytical model, Wu et al. (2013) examine freemium video games in which the "pay" version is one in which players spend money for in-game purchases, such as "accessory selling." The general idea behind this approach is that the installed base of "free" players generates value via those who switch to the pay version (i.e., buy accessories) and via increasing the attractiveness of the game for existing and new players (i.e., direct network effects of the "installed base" of game players). In the case of the LEGO UNIVERSE game, the development of direct network effects was systematically hampered by the fact that the "free" version included only two levels of the game that were actually free to play; everything else was behind a paywall.

The core challenge in such "accessory-selling" models is to determine the "right" level of attractiveness of the paid features. The scholars show that the advantages of the paid features can induce people to pay—because players want to compete, they invest in accessories that increase their capabilities. They call this the "competitiveness effect." However, highly advantageous paid features can also turn existing and prospective "free" consumers away from the game when free users feel inferior to those who have stocked up their armory with purchased weapons, etc.—the "inferiority effect," as Wu et al. name it.

Managers must try to determine the direct network value of a consumer who plays the "free" version of the game for those who pay for accessories (in addition to his or her own basic switching probability), and then weigh this value against the net effect of the "competitiveness effect" minus the "inferiority effect" of any accessory offer for the user. In this context, it might be helpful to know that Wu et al. report that competitiveness and inferiority effects vary between different types of accessories: whereas weapons can exert a negative network effect on the number of players (because of the "inferiority effect" they cause), this is not the case for decorative accessories. If your avatar looks great in a game, that bothers very few other players.

Let us note that the "inferiority effect" of in-game sales can not only drive free consumers away, but the costs of accessories can also add up to levels

[432]If the *total* net contribution of the "free" version across users is negative, it does not make sense to offer the "free" version at all.

that make the pricing of a game unattractive to those who have *already paid* for it. EA is among those firms who combine fixed fees for a game with the temptations of in-game offers—but the firm had to learn that the "market will only handle so much" (TV host Joe Vargas, quoted in Kim 2017), as part of a major public social media firestorm against its pricing for the game STAR WARS BATTLEFRONT II.[433] Customers calculated that accessing the game's most popular characters would require more money than the upfront payment of $60 and that playing the game successfully without such extra payments would take *years*—or was even not possible at all. These calculations were posted online and then shared widely through the network of gamers. The BATTLEFRONT case shows that using such "pay-to-win" elements can hurt a game's performance and, because they violate the fairness perceptions of several consumers, also negatively influence the company brand's image.[434]

Varian (1997) concludes his summary of research about the versioning of information goods with practical implications. He recommends that (a) producers should design products in a way that make it easy to create versions; (b) it is easier to first create a product with the right quality or right number of features to appeal to the customer segment with high willingness to pay and then downgrade the product or remove features to get the price down to the level to appeal to other segments; and (c) having three versions of a product is better than just two—because consumers may have an "extremeness aversion" (Simonson and Tversky 1992). If you offer two versions (small vs. large; low-speed vs. high-speed), some customers who are indifferent between the two will naturally choose the lower option to avoid being extreme. Adding a third option enables the indifferent or uncertain customer to compromise by selecting the middle option. In practice, according to Varian (1997), the middle option is often identical to the original highest option in the two-option condition (e.g., the new "medium" is the former "large").

Finally, Varian also notes that versioning is particularly effective if "unobservable" consumer characteristics exist that determine his or her willingness to pay to a larger degree than observable characteristics (such as senior or student status) do. Such a constellation makes the setting of segment-specific prices based on observable features (i.e., *third*-degree price discrimination) ineffective.

---

[433]See also discussion of the social media firestorm that resulted from EA's pricing of the game in the context of "pinball" communication.

[434]In the case of BATTLEFRONT and other games, the value of in-game purchases is further reduced by their combination with so-called "loot boxes," which dispense rewards. So even if you pay for Darth Vader, you can't be sure to get him. Lindbergh (2017) provides an anecdotal report about how consumers react to this kind of pricing model.

Bundling: A Special Case of Versioning

A final kind of second-degree ("indirect") price discrimination is bundling, with the bundle of two or more products being akin to another "version" of a product offered to the consumer. Bundling makes sense economically when consumer segments differ in their willingness to pay for individual products, but the seller cannot erect fences to differentiate the price for each segment, so that all consumers, regardless of their willingness to pay, would end up buying the version with the lowest price. "By creating the bundle, the producer can sell at the average willingness to pay, and this will typically be more profitable" (Varian 1995, p. 5).

How does bundling work? Take the example of a live music entertainment venue that serves two fairly distinct segments (Nagle et al. 2011). First, a large, mass "general entertainment segment" loves major headliner acts (e.g., Jay Z or Kenny Chesney) and will pay up to \$60 a ticket for such events, but is less enthusiastic about innovative/cutting-edge, but lesser-known artists—consumers in this segment will generally pay only about \$25 for such acts. In contrast, "music aficionados" absolutely love discovering innovative new artists, while also enjoying headliners; this segment would pay around \$40 a ticket for either type of concert. We illustrate this constellation in Fig. 14.5, with Panel A showing the willingness to pay for the "general entertainment segment" and Panel B for the "music aficionados."

Faced with these customer segments, the concert hall could price concerts separately, charging \$40 for headliners and \$25 for innovative acts. These prices would assure that both segments would attend both events. But this approach leaves a lot of money on the table, potentially threatening the concert halls' economic viability to hire attractive performers: whereas a customer who attends both types of concerts would pay a total of \$65 (i.e., \$40 + \$25) for the two tickets, the total willingness to pay for aficionados is \$80 (i.e., \$40 + \$40) and for general entertainment customers is even \$85 (i.e., \$60 + \$25)—these are the orange bars in the figure. Setting each concert price at the level of the *highest* willingness-to-pay segment (i.e., charging \$60 for headliners and \$40 for innovative new artists) would generate even less revenues, as the general entertainment consumers would buy only tickets for headliners (total spending of the segment: \$60), while music aficionados buy only tickets for new artists (total spending: \$40).

Now consider bundling: if the venue would simultaneously offer a bundle that consists of one concert of each type for \$80, such a bundle aligns with the total willingness to pay of *both* segments and could capture the

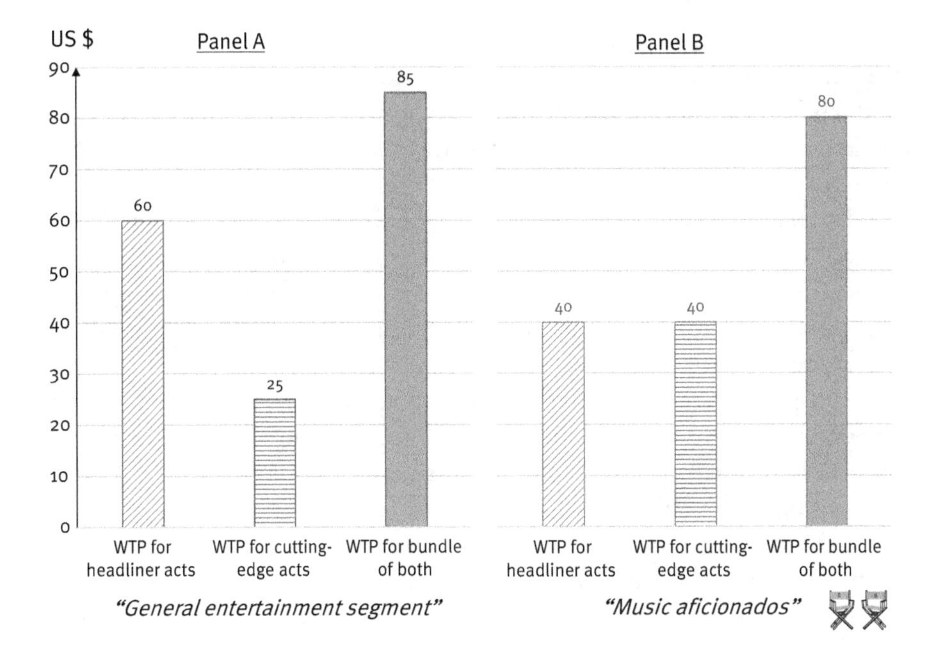

Fig. 14.5 Willingness to pay for concerts by customer segment
Notes: Authors' own illustration based on information in Nagle et al. (2011). WTP in the figure means consumers' "willingness to pay."

entire market, generating total revenues of \$160 versus \$130 or \$100, respectively.[435] In practice, this approach of offering individual products *and* bundles is referred to as "mixed bundling."

Is bundling effective for entertainment products? In entertainment, bundling is often offered by distributors such as the music venue in our example; Spotify offers "bundles" of songs for a monthly price and Netflix and Hulu do the same for movies and TV shows. But entertainment producers also have a long tradition in using bundling—every music album is a bundle of songs that are offered to the customer for a bundle price, with "singles" having mainly served as promotional tools for album sales. In contrast to the mixed bundling approach in our example, music producers have often restricted consumers' choice to the bundle, granting no opportunity to access its elements (except for the notorious hit single on the album).

Such a "pure" bundling approach carries the risk of turning consumers away from buying bundles (and products in general) because they do not

[435]In this simple example, each segment consists of one member only—but multiplying segment revenues with an arbitrary number of segment members does not change the logic of our calculation.

perceive the bundle price as sufficiently attractive; the approach can be lucrative when consumers are forced to buy bundles instead of those individual products they are interested in for a much higher price, but it also drives demand for alternatives ways to access the overpriced content (e.g., if Sean Parker comes along with Napster).

Let's take a look at what scholarly research can tell us about the effectiveness of bundling entertainment. In a foundational study, Bakos and Brynjolfson (1999) demonstrate analytically that, under the assumptions of their study, the low marginal costs of entertainment products provide the opportunity to generate greater sales and profits by offering bundles of entertainment products, compared to when the same products are only sold independently. One reason the scholars name for this advantage is that it is much easier to accurately predict consumers' valuations for a bundle of products than it is to predict valuations for each product individually. This is an important factor because uncertainty about consumers' willingness to pay is "the enemy of effective pricing" (p. 1614), as we have highlighted in our analysis of uniforms versus differential pricing above. They also provide analytical evidence that mixed bundling should be preferred over pure bundling in general.

Hitt and Chen (2005) extended Bakos and Brynjolfson's analysis of bundling's profitability for a kind of bundles in which customers can, for a fixed price, select a defined number of products from a larger pool of options—so-called "customized" bundles. The scholars show that, for low-marginal-cost products such as information goods, customized bundling improves both producer welfare and consumer welfare more than "pure" bundling—and constitutes an economically preferred kind of mixed bundling.

In addition to those analytical investigations, scholars have also shed light on bundling's effectiveness using empirical data from the music industry; the bundling of songs, as we mentioned, has a long tradition. Elberse (2010) determines the effects of unbundling music while controlling for legal and illegal downloading activities. She studies sales of digital versions of individual songs (i.e., "digital tracks") and albums (i.e., "digital albums"), using weekly sales data from 2005 to 2007; she also measures weekly sales of physical albums for *all* titles released by a sample of 224 randomly selected artists. Elberse finds that mixed bundling (i.e., making individual songs available for sale in addition to albums) has *negative* consequences for music revenues—essentially because the differences in margins between individual song sales and album sales is larger than the growth in music demand that results from the unbundling. Piracy plays a large role for her effects; she attributes a reduction of about "one-third of the average weekly mixed-bundle sales … to increased [illegal] digital music downloading activity."

Whereas Elberse examines data from a time of transition, more recent results from Danaher et al. (2014) let us feel less pessimistic regarding unbundling for music. In their study, the scholars systematically varied the prices for a label's music and studied sales of songs and the albums of which they are a part. From simulations, they conclude that an album-only policy (i.e., bundles only) leads to lower revenue compared to the current mixed-bundle practice (in which customers can buy albums or individual songs). Even when album prices were set quite low (e.g., $6.00 instead of $9.99), overall revenues were less when only albums are available.[436] Mixed bundling leads to lower album unit sales, but "the joint revenue from both singles and albums will increase"—mainly because several consumers only enter the market when unbundled offers are available.

Danaher et al.'s simulations also suggest that the mixed-bundle practices that are currently in place might not be optimal. They find that album prices are too high and individual song prices are too low, which tempts customers to simply cherry-pick a few desired songs and not try the other songs as part of the bundle. The "optimal" solution that provided maximum revenue for the record label in their analysis involved an album price of $7.00 and tiered pricing for individual songs ($1.29 for the six most highly rated songs on an album, $1.09 for songs 7–10, and $.89 for the remaining songs)—thus combining bundling (as second-degree price discrimination) with differential pricing, as discussed above.

The logic is that a higher song price deters few customers, but does steer a larger percentage of customers to go ahead and buy the album instead of only the individual song. Even without the differential pricing element, the scholars find a lower album price and higher individual song price to be better than current pricing schemes; the standard $0.99 price-per-song was not economically optimal in *any* condition.[437]

Flat-rate models, as offered by Netflix and Spotify, constitute a special kind of large-scale bundling. In their multi-survey study of more than 2,500 music consumers over the course of a year (from 2012 to 2013), Wlömert

[436]Though album unit sales nearly doubled when only albums are available and the price for them was low, this did not make up for the lack of sales for individual tracks. Keep in mind that marginal costs are zero for digital music.

[437]Papies and van Heerde (2017), using data from 2003 to 2010 (thus spanning the periods studied by both Elberse and Danaher et al.) point at a different negative effect caused by mixed bundling: they find that the availability of individual songs dampens the positive impact live concert sales have on recorded music sales. Many people who go to a concert then want to acquire a digital or physical copy of the music to listen to later—it seems that, when consumers have the option to cherry-pick only the tracks they most want rather than having to buy the full album, customers tend to spend less on recorded music.

and Papies (2016) use the entry of Spotify into the German market as a natural experiment to test how the offering of a massive bundle affects music industry revenue. They find that the bundling by Spotify cannibalizes consumers' other expenditures on music, such as paying for digital downloads and CDs, regardless of whether the bundle is available for a (monthly) subscription fee or "for free" (i.e., advertising-based). But for paid streaming, the *net* effect on label revenue is clearly positive—a finding that is in line with the growth recently noted for *total* music revenues.

For the "free" bundle, though, they find the net effect on revenue to be negative, even when accounting for advertising revenues; the availability of such a bundle has a positive effect on expenditures only for those consumers who were relatively inactive before the adoption, whereas for others it leads to lower overall spending for music. This is consistent with Spotify's revenue sources: whereas three out of four Spotify customers signed up for the "free" (i.e., advertising-based) streaming in 2014, only 9% of Spotify's revenues were generated through this model (see *Mediabiz* 2015). The share of "free" users has declined since then, a trend we link with efforts by Spotify to lower the user value of the "free" version compared to the paid version.

Cross-Subsidization (or Informal Bundling)

> "Every element in the lobby is designed to focus the attention of the customers on its [concession stand] menu board."
>
> —Thomas W. Stephenson, *as CEO of Hollywood Theaters (quoted in* Epstein *2010, p. 33)*

Entertainment firms cross-subsidize some of their products with other products. Such a practice can be considered a special kind of bundling—one in which the bundle is not formally defined by the seller, but is a de facto element of its pricing strategy.[438]

Luan and Sudhir (2010) study the determinants of DVD prices as part of their investigation of advertising effectiveness. They show empirically that retailers discount the DVDs for hit movies at their release and argue that doing so "is consistent with a loss-leader pricing strategy that takes advantage of the release of popular DVDs to boost store traffic" (p. 451f.). In other words, the biggest films serve as "key value items" for retailers whose reduced prices lead consumers to buy other products (also). Kocas et al.

---

[438]Informal bundling shares some similarities with the concept of "customized" bundling, as analyzed by Hitt and Chen (2005).

(2018) make similar observations for books—retailers in the U.S. often price bestsellers at large discounts with the hope that consumers attracted by the deal will purchase other higher-margin goods. Based on the estimation of VAR models with several thousand books at Amazon.com, they find that multi-category retailers (versus pure bookstores) tend to use and benefit from this strategy more often. They have more to gain from subsidizing their bestseller attractions.

McKenzie (2008) argues that a similar cross-subsidization model is standard practice for movie theaters, which make most of their profits from concession sales rather than movie tickets. Popcorn and other concessions have a *very* high profit margin, which suggests that theaters would benefit from subsidizing popcorn et al. by keeping the ticket prices lower than might be optimal if determined in isolation (i.e., without considering the profits from concession sales). By doing so, theaters can gain additional buyers for their high-margin popcorn.

Figure 14.6 illustrates this cross-subsidization approach with 1997 data from Hollywood Theaters, then a 450-screens U.S. theater chain, as reported by Epstein (2010). Although concession revenues were only about

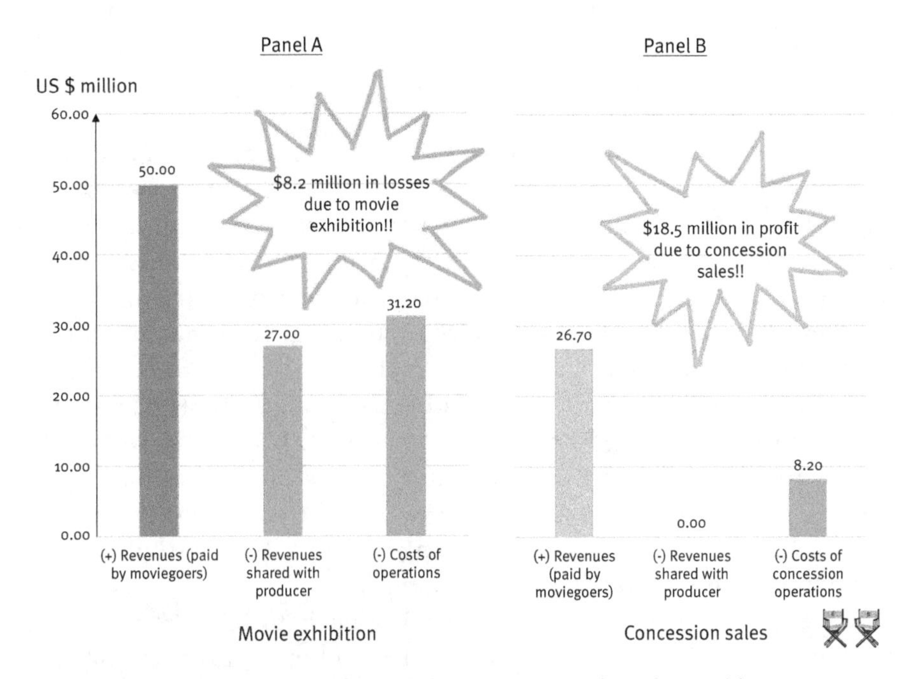

Fig. 14.6 Cross-subsidization of movie theaters. Or: It's the salt, stupid!
Notes: Authors' own illustration based on information reported in Epstein (2010). With graphical contributions by Studio Tense.

half of ticket revenues, the much higher margins for theater owners turn a $8.2 million loss from tickets into a profitable business.

A look at the record-breaking ticket prices in most parts of the world (*inflation-adjusted* prices are now almost 30% higher than 20 years ago!) and the quite high elasticities measured by scholars for movie tickets (that suggest that lower prices would attract *many* additional moviegoers) cause us to doubt that theater owners are currently making full use of the economic potential that cross-subsidization offers. Instead, it seems that they focus too much on revenues from their focal, but low-margin, product.

Third-Degree Discrimination: Segment Pricing

Third-degree price discrimination is a segmentation approach in which the seller charges different prices for a specific product to different groups of customers. These groups can be defined based on measurable (i.e., "manifest") customer characteristics, such as age, region, employment status, and previous purchases. For example, movie theaters frequently offer discounts for students or senior citizens, and Amazon offers special deals to their "Prime" customer segment. In contrast to versioning as "indirect" price discrimination, third-degree discrimination is referred to as "direct" price discrimination (Mortimer 2007)—it is the seller who *directly* assigns customers to different price offers.

The idea underlying this strategy is that the customer groups that are offered the different prices *differ also in their willingness to pay*. Just like second-degree price discrimination, it is an approach to overcome the implementation difficulties of first-degree discrimination. Its effectiveness depends on how closely the segmentation criteria are linked to the underlying willingness to pay. In addition to picking powerful segmentation criteria, challenges for third-degree price discrimination are to erect "fences" that (a) reduce the likelihood of gray market "arbitrage" transactions (e.g., a student buys a movie ticket for $5 and re-sells it to an adult for $7.50 who avoids paying the $10 adult price), and (b) reduce the likelihood of cannibalization (e.g., a price-insensitive person happens to have the afternoon off work and goes to the reduced price matinee showing of ROGUE ONE instead of seeing it at full-price, as she would have done normally).

A special kind of third-degree discrimination is to base the pricing on *latent* characteristics instead of manifest ones and to approach the different segments by adjusting the product price over time. A key distinction here is between product enthusiasts who are less sensitive to the price of a beloved product and other consumers who are more price-sensitive. Assuming that enthusiasts are also eager to access a new entertainment product early, the strategy suggests releasing the product for a "premium" price first (appealing to enthusiasts) and then lowering the price afterward—a price skimming strategy.[439] For example, premium video games are usually priced highly during the first weeks or months of release to capture higher per-unit revenues from enthusiasts.

In this special case, "cannibalization" between segments (i.e., enthusiasts delaying their consumption, waiting for the lower price) not only reduces revenues, but also has a secondary consequence: the reduction in early sales might also hurt the product's success by suppressing success-breeds-success effects. If enough of the fan boys of a new film are price sensitive and know that prices will come down in two weeks, they might skip opening weekend and wait for the price change. The film might then be perceived as a flop by the media and other customers might stay away. Theaters might even stop showing the film based on the small demand, which then further limits the film's success potential.

In addition to games, skimming is common for the home entertainment versions of films.[440] In the case of books, skimming is mostly done through versioning rather than pure life-cycle price management. But *reducing* the price over time is not always adequate—instead, prices should follow the development of consumers' willingness to pay over time in a given context. Scholars have suggested that *increasing* product prices in later stages of the life cycle might be an appropriate strategy for certain music, for example. Mixon and Ressler (2000) argue that, for music, "the typical customer of older CDs is more likely to be a devoted fan of the artist than purchasers of new releases," so that the "demand curve for older material will … be significantly less elastic than that of new material—which corresponds with a higher willingness to pay" (p. 466). In an empirical study of 118 records with OLS regressions, the researchers find tentative support for their arguments.

A second alternative scheme for pricing differently over time is to set prices at different levels at peak versus non-peak demand (times of day or days of week). The logic would be that enough "additional" customers could

[439]Such segmenting by *latent* consumer characteristics somewhat overlaps with the self-selection element of second-degree price discrimination. Whereas the latter practice involves changing the product (the quantity, the version, or by creating a bundle), segmenting according to latent traits means that the exact same product is offered at different prices based on *time*.

[440]As Mortimer (2007) points out, the dynamic part of this pricing approach was necessary in the U.S., as copyright law prevented producers from prohibiting rental firms from renting the cheaper "sell-through" versions to consumers.

be gained to more than offset the lower revenues per transaction and the cannibalization of customers who would have paid full price, but switched to the low-priced time slot. Theater managers in Asia and Europe, having embraced such approaches, obviously have come to a different conclusion regarding their profitability than have the clear majority of their North American colleagues, who predominantly avoid such price variations over time. Who is right? Or are both right, with cultural differences to blame? Similar logic could be applied to offering film or music streaming at different prices at times of the day, week, season, or year in which demand is likely to be particularly high.

Concluding Comments

In this chapter, we have challenged managers (and scholars) to think critically about pricing practices in entertainment—accepting that managing prices requires some artistry, but can genuinely gain from scholarly insights. As with the other areas of entertainment decision making, our discussion has shown that existing *Entertainment Science* studies can shed light on several pressing issues surrounding the pricing of entertainment, while in this dynamic field several issues require additional investigation and testing.

After reviewing pricing theory and comparing it to entertainment practice, our discussion of the pros and cons of differential pricing in entertainment led us conclude that offering different products at different price points in situations in which those products differ in appeal to customers is an idea that deserves closer attention. We are certainly not naïve enough to underestimate the barriers to implementation of such differential pricing; this is a far-from-trivial exercise, given the structure and financial incentives of producers and distributors, combined with the conservatism and traditionalism of many firms in entertainment. We hope that forward-thinking managers and *Entertainment Science* scholars will continue to push the boundaries of pricing strategy.

We also made the case for looking for ways to price an individual product differently, considering three broad categories of price discrimination approaches and illustrating their applications in the context of entertainment. We dedicated particular room to the analysis of different kinds of "freemium" pricing models, a special kind of versioning.

We will now end our analysis of the individual managerial decisions and tackle one last, important quest: how should managers coordinate the different areas of decision making and integrate product, communication, distribution, and pricing decisions? Let us take a look at integrated strategic approaches of entertainment marketing, in general, and the blockbuster and niche concepts, in particular. This analysis will then allow us to end the book with a critical look into current economic developments that we observe in entertainment.

References

Bakos, Y., & Brynjolfsson, E. (1999). Bundling information goods: Pricing, profits, and efficiency. *Management Science, 45*, 1613–1630.

Bittlingmayer, G. (1992). The elasticity of demand for books, resale price maintenance and the Lerner index. *Journal of Institutional and Theoretical Economics, 148*, 588–606.

Bourreau, M., Dogan, P., & Hong, S. (2015). Making money by giving it for free: Radiohead's pre-release strategy for In Rainbows. *Information Economics and Policy, 32*, 77–93.

Chen, H., De, P., & Yu Jeffrey, H. (2015). IT-enabled broadcasting in social media: An empirical study of artists' activities and music sales. *Information Systems Research, 26*, 513–531.

Chevalier, J. A., & Mayzlin, D. (2006). The effect of word of mouth on sales: Online book reviews. *Journal of Marketing Research, 43*, 345–354.

Clerides, S. K. (2002). Book value: Intertemporal pricing and quality discrimination in the US market for books. *International Journal of Industrial Organization, 20*, 1385–1408.

comScore (2007). For Radiohead fans, does 'Free' + 'Download' = 'Freeload'?, November 5, https://goo.gl/amcD3A.

Courty, P. (2011). Unpriced quality. *Economics Letters, 111*, 13–15.

Cox, J. (2016). Play it again, Sam? Versioning in the market for second-hand video game software. *Managerial and Decision Economics, 38*, 526–533.

D'Alessandro, A. (2018). MoviePass jumps past 1.5M subscribers in the post-holiday period at the B.O., January 9, *Deadline*, https://goo.gl/HyNLXe.

Danaher, B., Huang, Y., Smith, M. D., & Telang, R. (2014). An empirical analysis of digital music bundling strategies. *Management Science, 60*, 1413–1433.

Davis, P. (2002). Estimating multi-way error components models with unbalanced data structures. *Journal of Econometrics, 106*, 67–95.

Davis, P. (2006). Spatial competition in retail markets: Movie theaters. *The Rand Journal of Economics, 37*, 964–982.

De Roos, N., & McKenzie, J. (2011). Cheap Tuesdays and the demand for cinema. Working Paper, Sydney University.

De Vany, A. (2006). The movies. In V. A. Ginsburgh & D. Throsby (Eds.), *Handbook of the economics of art and culture* (pp. 615–665). Amsterdam: Elsevier.

Elberse, A. (2010). Bye-bye bundles: The unbundling of music in digital channels. *Journal of Marketing, 74*, 107–123.

Epstein, E. J. (2010). *The Hollywood economist—The hidden financial reality behind the movies*. Brooklyn: MelvilleHouse.

Gazley, A., Clark, G., & Sinha, A. (2011). Understanding preferences for motion pictures. *Journal of Business Research, 64*, 854–861.

Gerpott, T. J. (2016). A review of the empirical literature on pay-what-you-want price setting. *Management & Marketing, 11*, 566–596.

Gneezy, A., Gneezy, U., Riener, G., & Nelson, L. D. (2012). Pay-what-you-want, identity, and self-signaling in markets. *Proceedings of the National Academy of Sciences, 109*, 7236–7240.

Gong, J., Smith, M. D., & Telang R. (2015). Substitution or promotion? The impact of price discounts on cross-channel sales of digital movies. *Journal of Retailing, 91*, 343–357.

Groening, C., & Mills, P. (2017). A guide to pay-what-you-wish pricing from the consumer's viewpoint. *Business Horizons, 60*, 441–445.

Hitt, L. M., & Chen, P. (2005). Bundling with customer self-selection: A simple approach to bundling low-marginal-cost goods. *Management Science, 51*, 1481–1493.

Ho, J., Liang, Y., Weinberg, C., & Yan, J. (2017). Uniform and differential pricing in the movie industry: An empirical analysis. *Journal of Marketing Research*, forthcoming.

Kim, J.-Y., Natter, M., & Spann, M. (2009). Pay what you want: A new participative pricing mechanism. *Journal of Marketing, 73*, 44–58.

Kim, T. (2017). EA's day of reckoning is here after 'Star Wars' game uproar, $3 billion in stock value wiped out. *CNBC*, November 28, https://goo.gl/4dZpFo.

Kocas, C., Pauwels, K., & Bohlmann, J. D. (2018). Pricing best sellers and traffic generators: The role of asymmetric cross-selling. *Journal of Interactive Marketing, 41*, 28–43.

Leeds, J. (2007). In Radiohead price plan, some see a movement. *The New York Times*, October 11, https://goo.gl/nehN5E.

Lindbergh, B. (2017). Battlefront, Reddit: The video game pricing wars that might reshape the industry. *The Ringer*, December 2, https://goo.gl/BJ8tiV.

Linde, F. (2009). Pricing information goods. *Journal of Product & Brand Management, 18*, 379–384.

Luan, Y. J., & Sudhir, K. (2010). Forecasting marketing-mix responsiveness for new products. *Journal of Marketing Research, 47*, 444–457.

Marchand, A. (2016). The power of an installed base to combat lifecycle decline: The case of video games. *International Journal of Research in Marketing, 33*, 140–154.

McKenzie, R. B. (2008). *Why popcorn costs so much at the movies*. Leipzig: Springer.

Mediabiz (2015). Spotify weist Spekulationen um neue Fensterpolitik zurück. *Musikwoche*, December 9, https://goo.gl/c5nFHM.

Mixon Jr., F. G., & Ressler, R. W. (2000). A note on elasticity and price dispersions in the music recording industry. *Review of Industrial Organization, 17*, 465–470.
Mortimer, J. H. (2007). Price discrimination, copyright law, and technological innovation: Evidence from the introduction of DVDs. *Quarterly Journal of Economics, 122*, 1307–1350.
Nagle, T. T., Hogan, J. E., & Zale, J. (2011). *The strategy and tactics of pricing*. New York: Routledge.
Orbach, B. Y., & Einav, L. (2007). Uniform prices for differentiated goods: The case of the movie-theater industry. *International Review of Law and Economics, 27*, 129–153.
Papies, D., & van Heerde, H. (2017). The dynamic interplay between recorded music and live concerts: The role of piracy, unbundling and artist characteristics. *Journal of Marketing, 81*, 67–87.
Papies, D., Eggers, F., & Wlömert, N. (2011). Music for free? How free ad-funded downloads affect consumer choice. *Journal of the Academy of Marketing Science, 39*, 777–794.
Pearlstein, S. (2006). It was better with Bonzo. *Washington Post*, November 24, https://goo.gl/86sVPW.
Shiller, B., & Waldfogel, J. (2011). Music for a song: An empirical look at uniform pricing and its alternatives. *The Journal of Industrial Economics, 59*, 630–660.
Simonson, I., & Tversky, A. (1992). Choice in context: Tradeoff contrast and extremeness aversion. *Journal of Marketing Research, 29*, 281–295.
Sims, D. (2017). Hollywood has a bad-movie problem. *The Atlantic*, July 5, https://goo.gl/BxMfAE.
Smart Pricer (2018). UCI Kinowelt weitet Dynamisches Pricing mit Smart Pricer auf alle Kinos in Deutschland aus, January 8, https://goo.gl/BKVfGX.
Srinivasan, R. (2009). Pricing different movies differently, November 28, https://goo.gl/U3zs7h.
Sundararajan, A. (2004). Nonlinear pricing of information goods. *Management Science, 50*, 1660–1673.
Surowiecki, J. (2004). *The wisdom of crowds*. New York: Random House.
Thompson, D. (2012). Why do all movie tickets cost the same? *The Atlantic*, January 3, https://goo.gl/dF1ohW.
Thompson, P. (2008). Radiohead's In Rainbows successes revealed. *Pitchfork*, October 15, https://goo.gl/vUv355.
Varian, H. R. (1995). Pricing information goods. Working Paper, University of Michigan.
Varian, H. R. (1997). Versioning information goods. Working Paper, University of California, Berkeley.
Vogel, H. L. (2015). *Entertainment industry economics: A guide for financial analysis* (9th ed.). Cambridge: Cambridge University Press.

Waskow, S., Markett, S., Montag, C., Weber, B., & Trautner, P., Kramarz, V., & Reuter, M. (2016). Pay what you want! A pilot study on neural correlates of voluntary payments for music. *Frontiers in Psychology, 7*, 1–10.

Wilson, R. B. (1993). *Nonlinear pricing*. New York: Oxford University Press.

Wlömert, N., & Papies, D. (2016). On-demand streaming services and music industry revenues—Insights from Spotify's market entry. *International Journal of Research in Marketing, 33*, 314–327.

Wu, C.-C., Chen, Y.-J., & Cho, Y.-J. (2013). Nested network effects in online free games with accessory selling. *Journal of Interactive Marketing, 27*, 158–171.

15

Integrated Entertainment Marketing: Creating Blockbusters and Niche Products by Combining Product, Communication, Distribution, and Pricing Decisions

"Friday night is all about the marketing, and Saturday and Sunday are .. about a film's word of mouth."

—*Anonymous Hollywood studio marketing strategist (quoted in* D'Alessandro *2015a)*

Now that we have discussed all the instruments (product, price, place, promotion) that entertainment managers have at hand when marketing their products, from the initial idea to the final release channel and version, let us stress one important point: all of these instruments are part of a broader system. In this system, the effectiveness of each marketing activity is not only determined by the rules we have highlighted so far, but also by the other elements of the system.

If a marketing activity is carried out in a way that "fits" with other actions, enormous synergistic potential can be set free, whereas a lack of fit can drastically curtail any action's effectiveness. We occasionally touched on synergies when we discussed studies that included interaction effects, but those discussions were restricted to the many facets of one particular instrument (such as branding). In this final chapter of *Entertainment Science*, however, we extend this approach and take a holistic look at how product, communication, distribution, and price decisions can be coordinated to maximize the desired results for an entertainment producer.

Entertainment firms have developed two fundamental strategic approaches for integrating the different marketing instruments in a coherent way. We label the first approach the "experience approach" of entertainment; at its core is the creation of high-quality products that have the

T. Hennig-Thurau and M. B. Houston, *Entertainment Science*,
https://doi.org/10.1007/978-3-319-89292-4_15

potential to initiate quality-based cascades. Thus, the experience approach is all about a product's playability, in which the success is driven by informed cascades through which consumers' quality experiences spread.

A major downside of this approach is that it requires that influential consumers and/or experts perceive the product to be great enough to set off such cascades—making this happen is quite a challenging task, given entertainment products' cultural and creative nature. Further, given the scarcity of distribution space in theaters and retail stores and the time required for the news to spread that a new piece of entertainment is "great," a great entertainment product might not be available long enough to enable the number of consumer experiences needed for the cascade to build and have impact.

The second strategic approach aims to avoid these obstacles. Instead of focusing on the quality of the product, it concentrates on creating anticipation for it; the basic idea is to sell the product before it is even released by building high levels of buzz. For this "pre-sales approach," it is not playability that is key, but the product's marketability. And marketability needs to be "factored into the project from the very start of development" (Lewis 2003, p. 65).

Both the experience and the pre-sales approach are each tied closely to a particular integrated marketing concept: the niche concept builds on the experience approach, while the blockbuster concept of entertainment relies on the pre-sales approach. All other strategies are problematic: they are neither fish nor fowl and thus too often result in a product getting "killed in the middle." As Friend (2016) phrases it, a "fifty-million-dollar film costs more to make than a genre film, and nearly as much to market as a tentpole—and, being neither for a sharply defined group nor for everyone, it often can't find a sufficient audience."

In the following, we discuss both integrated marketing concepts and how each makes use of the different marketing instruments. We begin our discussion with the blockbuster concept, as it has become the dominant strategy in entertainment over the last decades, before looking more closely into the economics of the niche concept. We then end our debate by reviewing the state of entertainment regarding the two integrated concepts. We conclude our analysis by offering some high-level reflections on the challenges that we believe result from the present way that marketing and management are executed in the entertainment industry.

The "Pre-Sales Approach": The Blockbuster Concept

Defining the Blockbuster Concept

> "BATMAN was the template for the modern blockbuster. It was based on a pre-existing property. It succeeded as much based on manufactured pre-release hype as any kind of genuine natural excitement or word-of-mouth buzz. It was preceded by a massive saturation marketing campaign that threatened to make its actual release feel anti-climactic. It made its money in a front-loaded manner which made its quality all-but-irrelevant. It made opening weekend king, made the PG-13 rating into the all-purpose general rating for the eventual four-quadrant blockbuster, and kick-started what would eventually be the quick-kill blockbuster."
>
> —Mendelson *(2014) on* Tim Burton's *film BATMAN*

Firms in entertainment have shown quite a bit of flexibility when it comes to defining core concepts and strategies. Regarding the blockbuster concept in particular, people in entertainment have used the term to refer to a variety of different things, including market outcomes (such as when movie director-producer Francis Ford Coppola described blockbuster movies as combinations of "financial success, critical success, great storytelling, and timing;" Coppola 2010, p. VI) and single product characteristics, such as "being exceptionally costly to produce" (Hall and Neale 2010, p. 1).[441] In today's entertainment industry, however, the blockbuster concept de facto describes a holistic strategic effort by producers that combines certain product, communication, distribution, and sometimes also pricing elements[442] to create a high level of pre-release awareness and marketability. This is how we treat the concept in this book. Let's take a closer look at the different elements that, in conjunction, comprise a blockbuster (see also King 2002).

---

[441]What kind of word is "blockbuster" after all? Originally, the term was used by American journalists in the early 1940s to describe massive bombs that were capable of destroying a whole city block. Later this impact was transferred to anything that made a strong public impact—for example, the Chicago Tribune described Broadway managers' reaction to a request to collectively shut down all entertainment as a result of war mobilization "as if a blockbuster had landed on Manhattan." In the context of entertainment, "blockbuster" was first used as a "purely economic term" (Shone 2004, p. 28)—early mentions named theater plays that were so successful that other shows ("on the block") were "busted," and later, all extremely successful products, regardless of their characteristics. The use of the term blockbuster, regardless of an entertainment product's actual market performance, is a more recent phenomenon; it began to gain popularity in the early 2000s.

[442]We discuss the limited pricing decisions for entertainment that are in use today in the previous chapter.

Product elements. The blockbuster concept works best with products that feature a strong brand and/or absorb high production costs. Together, well-known brands and big budgets enable the offering of familiar heroes and stars, as well as intense sensations, such as spectacular historic/futuristic settings, special effects, or technical arrangements. These elements enable the product to command a high level of pre-release awareness and interest.

Another essential product element of blockbusters is a high-concept nature. "High concept" is a term as popular in entertainment as it is fuzzy; in its essence, it refers to products whose value proposition contains "a unique idea that can be conveyed briefly" (i.e., easily communicated), a description that has been attributed to former Disney CEO and high-concept proponent Michael Eisner (see Wyatt 1993). Such easy communicability is critical for the producer to be able to effectively "'pitch' a product to the public" (Wyatt 1993, p. 8), especially in a world that is characterized by heavy generic competition for consumers' attention. In line with this logic, director-producer Steven Spielberg considers it important that one can tell him the idea for a new blockbuster movie in "twenty-five words or less;" if that is the case, he believes, "it's going to make a pretty good movie" (quoted in Wyatt 1993, p. 13).[443]

Communication elements. But having a certain kind of product is not enough. Blockbusters "are also heavily promoted and advertised" (King 2002, p. 50). A large communication budget is spent predominantly before a new blockbuster's release to realize the product's potential, promoting its textual and audio-visual elements to support early awareness and interest. In addition to the amount that is spent, blockbuster communication must align with the product's "high concept" value proposition, transporting it both (audio-)visually and verbally in an "easy-to-grasp" manner.

That is where iconic images are key: just think of the full moon with E.T.'s silhouette flying across it in the basket of his friend Elliott's bike, and the monstrous cracking ALIEN egg in outer space darkness, for example. Figure 15.1 shows the posters for two blockbuster films which capture the films' attractions in an exemplary way: the gargantuan shark below the swimming girl of JAWS (Panel A of the figure) and the tyrannosaurus silhouette of JURASSIC PARK (Panel B). In addition to being easily recognizable and highly memorable, both posters are also self-explanatory, leaving hardly any questions about the sensations that audiences can expect from the films.

[443]The 25-words rule is featured quite prominently in Robert Altman's Hollywood satire THE PLAYER, when executive Griffin Mill is pitched a story—and then criticizes the pitch for exceeding the "magical threshold."

Fig. 15.1 Visualizations of movies' value propositions: JAWS and JURASSIC PARK
Notes: Posters for the movies JAWS (left; © 1975 Universal Pictures; Panel A) and JURASSIC PARK (right; © 1993 Universal City Studios, Inc. and Amblin Entertainment, Inc.; Panel B). Both reprinted with permission from Universal Studios Licensing.

In both cases, the powerful high-concept visualizations are complemented by tag lines that succinctly convey the films' value propositions, in far less than 25 words. JAWS is "[t]he terrifying motion picture from the terrifying No. 1 best seller," stressing the terrifying nature and prominent brand from which the film derives; other slogans for the film warned us to "not go in the water," further highlighting the film's thrill offerings for consumers. And the poster for JURASSIC PARK promises audiences what most of us have dreamed of since our early childhood days: experiencing "[a]n adventure 65 million years in the making." Come to the movies and spend some time with dinosaurs. The material triggered emotions and imagery that caused us to see the film immediately, along with millions of others.

Distribution elements. The blockbuster concept also implies a wide distribution of the product that makes it available immediately to a large number of consumers. This is necessary to harvest the consumers' awareness and interest which, ideally, the product and communication elements have built to a fever pitch at the moment of release.

Tim Burton's BATMAN movie from 1989, to which journalist Scott Mendelson refers in the introductory quote of this section, is an almost perfect incarnation of the blockbuster concept, but it was not its inaugural one. This special honor belongs to Steven Spielberg's film JAWS, which was released by Universal more than 40 years ago, back in 1975. Let us take a quick trip back in time to see how JAWS became the very first modern blockbuster—and how much the marketing of entertainment has changed since the introduction of the blockbuster.

Doing so also illustrates that the blockbuster concept is not "built-in" to entertainment products, as people often argue, but that it was developed by entertainment managers as a strategic response to entertainment products' specific characteristics. We will then analyze the diffusion patterns associated with blockbusters, which point us to the systemic role of pre-release activities and buzz for the blockbuster concept.

"The Monster that Ate Hollywood": A Short Blockbuster History

"[T]hat would all change with JAWS."
—Canning *(2010a, p. 531)*

If you are in in your twenties, perhaps reading this as a student with a particular interest in the business of entertainment, it might be hard to believe that there once was a world of entertainment that was not ruled by blockbusters. But this was indeed the case, and we consider it essential for fully understanding the economic logic of the blockbuster concept to recognize that today's approach by entertainment conglomerates to manage films, books, games, and music is not a "quasi-natural" necessity.

At the occasion of the 25th anniversary of the movie JAWS, PBS crafted a remarkable feature on the film's impact on the film industry and entertainment, in general. The station named JAWS the "monster that ate Hollwood" (*PBS* 2001), referring to the massive, transformative impact the blockbuster concept (which the film introduced) has had on the way entertainment is created and marketed. The blockbuster concept's economic potential made entertainment products much more attractive for investors from outside, as it shifted the focus away from the (artistic) quality of a product to its marketing strategy (to which outside investors could much better relate). Let us take a closer look at the marketing of JAWS, which combined all three elements of the blockbuster concept we have identified above.

Regarding the product element, JAWS was based on a novel by Peter Benchley that was a bestseller of numbing proportions *and* durability. It was on the New York Times bestseller list for 44 weeks, with over 5.5 million copies in print at the time of the film's release (*Time* 1975).[444] The film's producers invested an enormous amount into the film; they paid the novelist a huge $150,000 for the rights (which translates into about $700,000 in today's dollars) and planned to spend $3.5 million for 55 days of filming. Spending then escalated to $12.2 million (i.e., almost $60 million now) and 159 shooting days, one of the highest budgets ever spent at that time (Canning 2010b).[445] And it was also high concept—a summer thrill ride, the first of its kind (Nall 2012).

In terms of communication, JAWS combined public relations efforts with mass advertising. The advertising campaign focused on the "shark-meets-girl" visual motif shown on the poster in Fig. 15.1 and the film's tag lines, highlighting the prominent source material and the thriller elements, which were amplified by John Williams' iconic score ("duh DUN, duh DUN…"). The campaign had a strong pre-release focus; the film's producers and the novel's author appeared on television and radio talk shows as far as eight months prior to the release, and a "massive national advertising blitz" (Shone 2004, p. 27) was unleashed three days before the movie's opening.

The TV spots were shown 50 times in the two nights preceding the release (Nall 2012). In total, Universal spent an enormous $700,000 (close to $3 million today) for its TV campaign in the U.S. alone—something so remarkable that the studio made the campaign itself the subject of ads in trade publications as a signal to theater owners of its high-flying commercial expectations (see Fig. 15.2). By doing so, Universal hoped to hedge supply-sided support for the film, a mechanism whose effectiveness we demonstrated empirically earlier in this book.

Finally, with regard to the distribution element of the blockbuster concept, the film opened simultaneously on 490 screens across North America (*Time* 1975), which at that time, despite the deliberate shortening of supply by its producers, was one of the widest releases in movie history.

JAWS was an instant success, generating an unprecedented opening weekend box office of $7.7 million, and continued its successful run all summer long, becoming the first film to cross the $100 million barrier at the North American

[444]Today, the novel has sold more than 11 million copies (Canning 2010b, p. 573). Interestingly, Benchley's book never made it to the top spot of the New York Times list (that spot was occupied at that time by Richard Adams' wonderful rabbit novel WATERSHIP DOWN).

[445]The immense production problems (which, as we argue in this book's distribution chapter, were responsible for the film's summer release) led the crew to nickname it "Flaws" (Canning 2010b, p. 573).

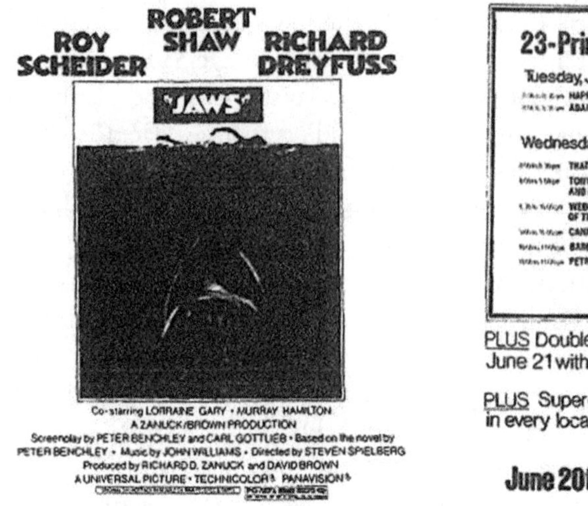

Fig. 15.2 Advertisement for Jaws advertising campaign
Notes: Advertisement for the movie Jaws (© 1975 Universal Pictures) as published in Variety magazine. Reprinted with permission from Universal Studios Licensing.

box office. In theaters alone, it made almost $500 million globally, which translates into more than $2 billion in today's currency; the film still ranks among the ten most successful films of all time in terms of inflation-adjusted box-office revenues (*Boxofficemojo.com* 2017). It was this enormous success that made entertainment firms, first in movies, then also in games, books, and music, aware of the success potential of the blockbuster concept.

We have argued, and provided evidence throughout this book, that entertainment firms are notoriously resistant to changing established patterns—why where Universal's managers so courageous then to experiment with such a *major* asset? They probably had noted the variations of the traditional release model by others in the previous years (Shone 2004); for example, the idea of distributing a film widely was used for some high-budgeted "turkeys" to preempt bad word of mouth (Canning 2010a), but also for The Godfather three years earlier (the film played on 323 screens in its *second* weekend, after only six in its first). Other studios had also used TV advertising before, such as Warner did for a 1971 re-release of the movie Billy Jack—the film had flopped in its original opening, but then earned $40 million in conjunction with the TV ads.

But we speculate that the decisive force might have been Columbia/Sony's action film BREAKOUT, starring Charles Bronson. Although produced for a moderate budget of $1 million, it was released just one month prior to JAWS on a then-unprecedented 1,300 North American screens, accompanied by a nation-wide advertising campaign in the week before the release. It generated about $16 million—and probably convinced Universal to go the "blockbuster route" with JAWS. Nevertheless, by combining all the different ingredients of the blockbuster concept it was JAWS that initiated a disruptive shift in entertainment marketing, turning what were rare and mostly isolated marketing actions into an integrated strategy.

By doing so, Spielberg's film not only changed the way many entertainment products are marketed—but also the kinds of products that are made, consistent with our notion that the product itself is a key ingredient of the blockbuster concept. Let us provide some illustrative empirical evidence how the blockbuster concept has shaped entertainment firms' actions, drawing on data from Hollywood films. Regarding the kind of products that are produced, the adoption of the blockbuster strategy has clearly impacted the presence and utilization of brands for movies.

Panel A of Fig. 15.3 demonstrates that line extensions of movie brands (i.e., sequels and remakes) have traditionally played only a marginal role in the mix of movies produced. The figure also shows how drastically that has changed in conjunction with the development of the blockbuster concept in the mid-1970s. Whereas investments in such branded properties were below 10% in pre-blockbuster times, they exceeded 50% for the first time in 2014. In other words, more than half of the film industry's money now flows into sequels and remakes.[446]

Panel B of Fig. 15.3 shows that the blockbuster concept also has implications for the genres of films that are produced. Genres such as action, science fiction, and animation, which tend to fit the high-concept idea, attract growing resources from Hollywood producers, whereas dramas (whose selling points are less-easily put in a captivating tag line) draw fewer resources. In an analysis of the prevalence of genres among Top-30 films from 1946 to 2013, Rubinson and Mueller (2016) find a rise of "nonrealistic" over "realistic" films since the development of the blockbuster concept. They calculate an increase in action and adventure films of almost 150% (20.6% after 1973 compared to 8.4% before), while film dramas decreased by one-third (from 23.1% to 15.6% after 1973).

[446]For an overview of the prominence of other brands in Hollywood films from 2003-2013, take a look at the database that Elberse and Krasney (2013) have compiled.

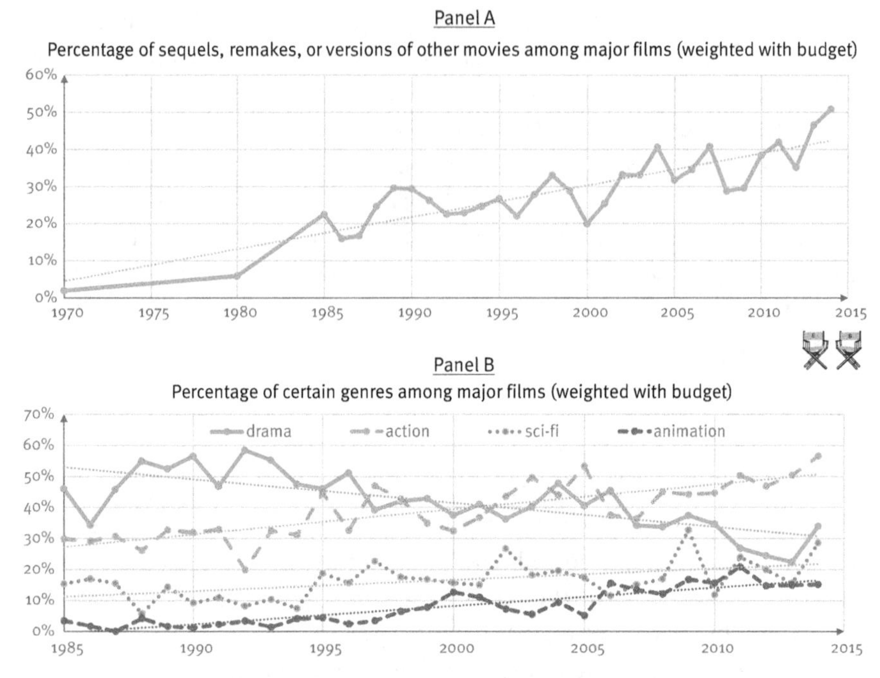

Fig. 15.3 Brands and genres in major Hollywood films over time
Notes: Authors' own illustration based on publically available data from multiple sources including The Numbers (most production budgets) and IMDb (films' genres and line extension status). All percentages are weighted by movies' production budgets. The sample comprises the 100 films with the highest production budget of a given year. The thin lines are estimated linear trends.

Further, Hollywood studios have acknowledged the blockbuster concept's "money nexus" (Stringer 2003) and spend much more for a blockbuster's production and its communication than in earlier years. In Fig. 15.4, Panel A shows that both production budgets and advertising spending have multiplied since 1970, even when accounting for inflation. Whereas the average total costs associated with a Hollywood film in 1970 was about $16 million (in 2014 dollars), the number rose by 130% (to $37 million) within a single decade. And it has more than tripled since then. Further, as Panel B of the figure illustrates, even Jaws' then-breathtaking total costs of $63 million (in today's currency) now look timid compared to the average total costs of Hollywood's ten biggest films today, which devour almost four times the record-breaking number from the earliest blockbuster.[447] Entertainment managers have even coined a term for those "super-blockbusters": they are often referred to as

[447]Please keep in mind that the actual costs of producing and marketing a film are even higher, as these numbers contain only the North American theatrical release. See also our discussion of production and marketing budgets as critical resources for entertainment firms in our chapter on entertainment product characteristics.

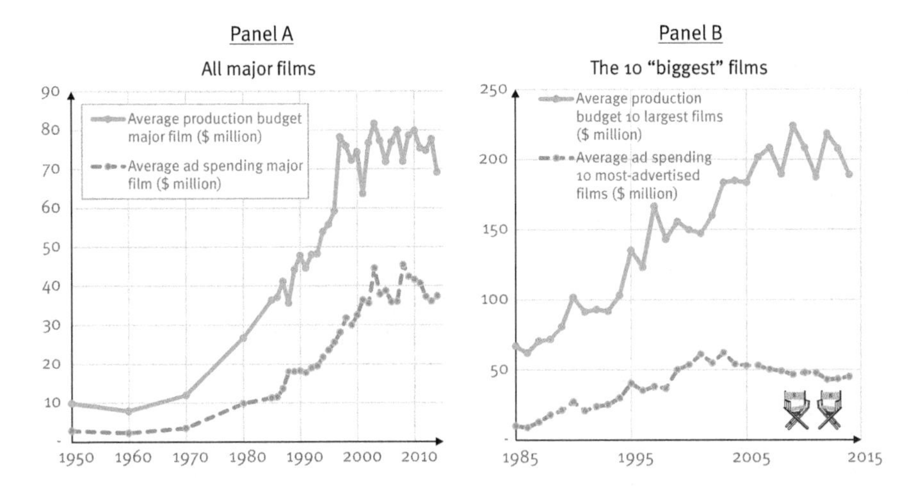

Fig. 15.4 Production and advertising budgets of major Hollywood films over time
Notes: Authors' own illustration based on reports by MPAA and data from The Numbers (production budgets) and Kantar Media (advertising spending), along with other sources. All numbers are our own estimates in 2014 $ million. In Panel A, "major films" refers to all "studio movies" (until 2007) and the 100 largest films of any given year in the following years. The numbers in Panel B are the means for the ten films with the highest production budgets and the ten most-advertised films of a given year, respectively. Digital advertising may not be fully represented in the numbers, which might have contributed to the stable course of the advertising spending function in the figure since the early 2000s.

"tentpoles." This development can also be noted when comparing the different products within a single franchise: the seventh entry in the Star Wars saga, 2015's The Force Awakens, was reportedly made for $306 million, while the initial Star Wars film consumed only $42 million, adjusting for inflation.

The blockbuster concept has also changed Hollywood distribution practices quite fundamentally. Intensive distribution, a central element of the blockbuster concept that aims to make a new product available ubiquitously for consumers, is now the norm when releasing a new entertainment product, at least for Hollywood studios. Figure 15.5 shows that the average number of theaters in which a major film is introduced has sprawled from below 100 in the early 1970s to around 3,000; for the ten highest-budgeted films, the average number of opening weekend theaters now exceeds 4,000. The figure also offers more specific evidence for the impact that the blockbuster concept has had on distribution: whereas the first Stars Wars film was released in only 43 theaters in 1977, the number of opening theaters has grown with every Star Wars episode since. The Force Awakens was released in 4,134 theaters, or almost 100 times as many as A New Hope. What a difference 38 years and a new strategic marketing concept can make in film distribution.

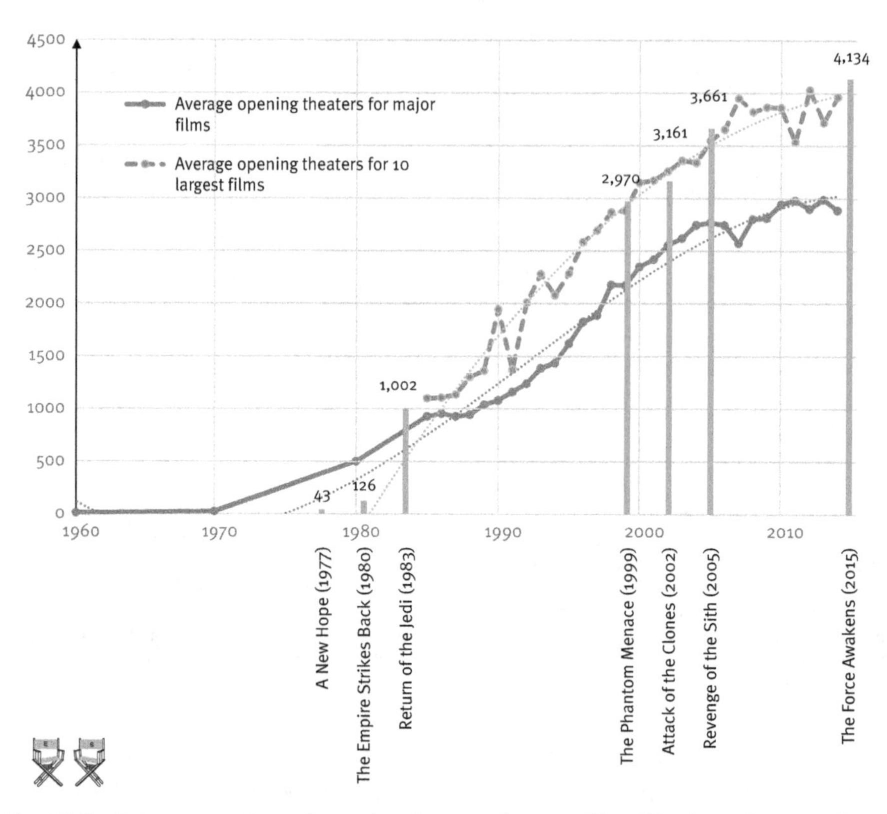

Fig. 15.5 Average number of opening theaters for new (Star Wars) movies over time
Notes: Authors' own illustration based on data from The Numbers and other sources. "Major films" are the 100 largest films of any given year in terms of their production budget. The thin lines are estimated trends. Movie titles are trademarked.

Blockbuster Diffusion: Innovators and Buzz are What It Takes!

> "Kevin, this is a corporate movie. It doesn't matter how good the dialogue is. It's about how many toys we can sell."
> —*Writer-director* Kevin Smith, *recalling an encounter with Warner Bros. executives about a new Superman movie (quoted in* Braund 2013, *p. 182)*

An important implication of the blockbuster concept is that it introduces a substantially different pattern of new product diffusion. Because of their pre-release focus and ubiquitous availability from the very beginning, blockbuster products exhibit a "front-loaded" diffusion pattern. Revenues tend to be highest right at release, followed by an exponential decay in the days, weeks, and months that follow. Blockbuster products can thus generate enormous revenues in a very short period of time. It took the movie AVATAR just 15 days to generate $1 billion in revenues across the globe, and the blockbuster game CALL OF DUTY: MODERN WARFARE 3 needed even one day less to reach that milestone. A side effect is that the inherently short life cycles of entertainment products become even shorter when marketed according to the blockbuster concept, because of the strategic decisions implied by the approach.

Figure 15.6 shows this accelerating effect that the blockbuster concept has on diffusion patterns of entertainment products by comparing the evolution of the cumulative box-office revenues of the first seven STAR WARS episodes, over time, at North American theaters. It shows that the initial film of the series, which used blockbuster marketing only to a moderate degree (just think of the comparably small number of theaters in which the film opened), needed more than 50 days to generate revenues of $200 million (in 2015 dollars). The third film in the series needed about 15 days to reach that level, the sixth only about six days, and the seventh film passed the mark in just 2.5 days.[448]

This acceleration of diffusion for blockbusters is not specific to movies (see also Elberse 2013, p. 64). We find it in our video games data set, where it explains the difference between our own observations and those of Clements and Ohashi (2005) from a decade before. It has been noted for books, where managers refer to blockbuster titles as "rockets" because of their extreme front-loaded nature (März 2011). And it also exists for music—for example, when Asai (2009) contrasts sales patterns in Japan from 2005 with those from 1980, we learn that, in the early phase of the blockbuster age, new singles needed 16 weeks to generate 80% of their total revenues versus only nine weeks a quarter-century later. For albums, 11 weeks had to pass in 1980 until 80% of the total revenues were earned—versus only four in 2005.

Diffusion acceleration is an essential element of the blockbuster concept because it reduces the negative impact that an entertainment product's lack of quality can have on its adoption by consumers. The blockbuster concept focuses on those consumers who decide to adopt a product based on marketing stimuli alone, not influenced by the opinions of other consumers

[448]As we will discuss, the faster return of money brings several advantages for producers. But the time a product needs to return its investments must not be confused with its overall success: the first STAR WARS film still remains the most successful in terms of total North American theatrical revenues, at least when re-releases are considered.

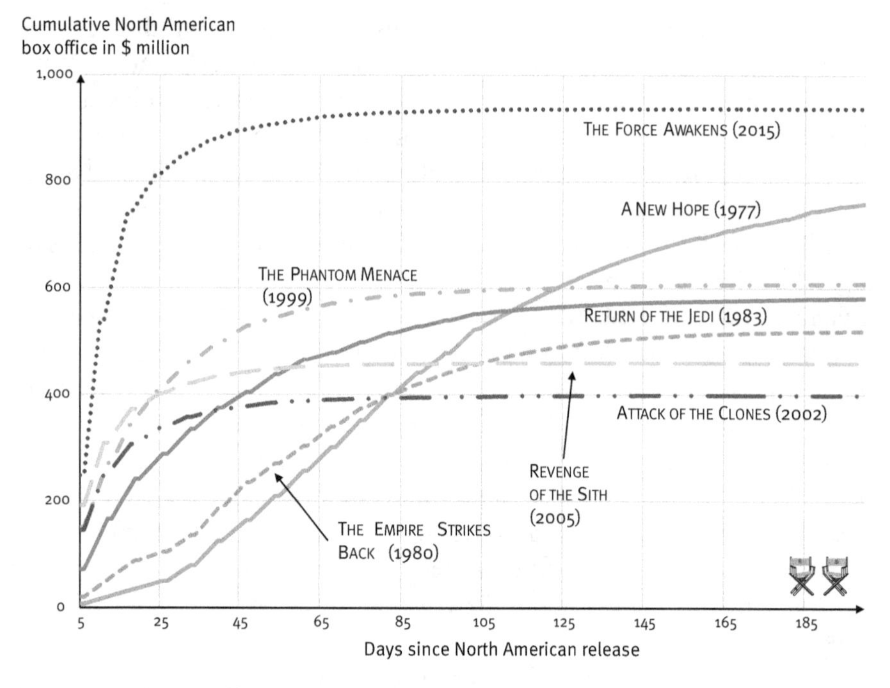

Fig. 15.6 Daily box-office revenues for the first seven STAR WARS films
Notes: Authors' own illustration based on data from The Numbers. All values are inflation-adjusted to 2015 dollars. For periods where data was unavailable, we used logistic regressions to interpolate the missing values (with R^2s of 0.99 and above). Movie titles are trademarked.

(articulated via word of mouth) or experts (via reviews) who have already experienced the product. In other words: those consumers who Frank Bass, in his fundamental diffusion model, named Innovators (see Bass 1969 and our discussion of his model in this book's innovation chapter). Via its integrated mix of marketing instruments, the blockbuster concept aims to create a high innovation parameter, which isolates the new product from criticism and insures a high level of marketability.

Take the example of the movie adaptation of erotic bestseller novel 50 SHADES OF GREY: the film's high innovation parameter and marketability ensured that it generated a North American box office of above $30 million on its very first *day*, and $85 million during its first weekend. Although the film's imitation parameter, and thus its playability, turned out to be quite low (the opening weekend box office ended up as more than half of the film's total North American theatrical receipts), marketability ensured instant success, particularly as the film's production costs were only $40 million. It is this critical role of marketability for blockbuster products that Adam Fogelson, as CEO of STX Entertainment, stressed when he argued that "seventy-five per cent of a [blockbuster] movie's success is due to … its marketability" (quoted in Friend 2016).

If the blockbuster concept is all about Innovators and marketability, buzz is its crucial mechanism. Buzz, which we discussed in depth in the context of "earned" communication forces, is what drove audiences into theaters during the opening days of 50 Shades of Grey: as an anonymous marketing executive stated, there "was a [large] built-in audience for this film with a fervent desire, and they had to see the movie and be part of the conversation during the first weekend" (quoted in D'Alessandro 2015b). For those Innovators, it did not matter that many consumers, as well as professional reviewers felt the film lacked quality.[449]

Because buzz can be influenced by marketers to a much larger degree than the quality of an entertainment product (which depends on the work of artists and consumers' fickle tastes), the blockbuster concept is quite appealing to entertainment producers. Every year a few weeks before the Super Bowl, "some movie producer who thinks he's got a bad movie" asks CBS, the station that airs the Super Bowl, "for two spots that go for [north of $5 million] a spot" (Les Moonves, as CBS CEO, quoted in Lieberman and Busch 2016).

In comparison, the role of the imitation parameter in the Bass model (which corresponds with the playability element of the product) is comparably small for blockbusters. Does this mean that quality is irrelevant for the success of blockbusters? No—a product's quality also matters for blockbusters, and great quality is highly preferred by producers of blockbusters. But it is not a precondition. Low quality means that Imitators are few and playability is small, which we know hurts the *long-term* success of all entertainment products, including blockbusters (e.g., Hennig-Thurau et al. 2006). The key challenge for the management of blockbusters thus is to generate buzz that is so strong that it ensures that a new entertainment product becomes sufficiently successful, even when it turns out that people don't like it very much. If it turns out to be a great product, that's even better.

Let us move on and study the main alternative to the blockbuster concept in terms of integrated marketing strategies: producing entertainment that targets niche segments and becomes successful by exciting them.

[449]The film received a C+CinemaScore rating and an IMDb rating of 4.1 out of 10 from moviegoers, along with a "Tomatometer" score of only 24% from reviewers.

The "Experience Approach": The Niche Concept

The blockbuster concept requires a certain kind of product, one whose appeal can be easily grasped by consumers and can enthuse them upfront. But these conditions are not met by many entertainment projects that feature complex storylines, unconventional rhythms, or lack brand power. What marketing strategy is most promising for these kinds of projects?

The strategic alternative to blockbuster marketing is the niche concept, an approach that places the actual product experience in the limelight. We will now discuss this niche concept and its core characteristics, stressing the role of Imitators among consumers and the role of product quality for the diffusion of niche-marketed entertainment content. Then we take a look at the controversial "long-tail" phenomenon and its role for the success of the niche concept of entertainment marketing.

Defining the Niche Concept

The niche concept abstains from massive financial investments, such as a large production budget and mass advertising, that are so fundamental for blockbuster marketing (see Elberse 2013). Instead, the niche concept involves investing modest sums into the production of a new product that is targeted not to the mass market, but to a specific, clearly defined segment of consumers. There is not a single set of criteria to define those segments, other than the idea that preferences among segment members are somewhat homogeneous. For example, for movies, such segments can be people who love a certain genre (e.g., horror film aficionados) or a certain movie type (think "arthouse drama"); for music it can be a regional group (people from Nashville) or a genre (jazz).

In terms of timing, the niche concept implies that marketing and PR initiatives are mainly carried out once the product has become available, not before. The goal of communication in niche marketing is *not* to create pre-release buzz (there's just not enough marketability potential in the product to warrant pervasive pre-release excitement), but to enable the consumers who experience the product to transport the impact of that experience to others who have yet to consume the product. In other words, communication is about the multiplication of consumer reactions.

Distribution efforts also are very focused in the niche concept; a movie is released in few theaters (which are selected based on the target group), a novel in a small number of handpicked book stores, and a new song in certain record stores or a small set of selected local radio stations. Niche distribution is more difficult for music and games, however, because large shares of sales are now digital, and the Internet is a global platform.

The initial modesty of the niche concept should not be confused with small ambitions. Instead, the idea is to build on the initial reactions of the target group and then expand the customer base, over time. When the target group receives the product positively and there are signs that indicate that others outside the core group are interested, the marketer's task is to support the product's spread with a concerted use of marketing instruments. "[I]f the product takes off—or shows some signs of being on the verge of taking off—will the producer gradually increase the distribution coverage or intensity and support the product with more advertising to further enhance growth" (Elberse 2013, p. 60).

This type of escalation approach has a long tradition in entertainment; it resembles how movies were distributed and promoted in the pre-blockbuster era, starting from the foundation of a few premiere theaters and then expanding over weeks and months throughout the country—and the world.

Niche Diffusion: Imitators and Quality are What It Takes!

If we take a diffusion perspective to better understand niche marketing, we find that the concept's focus is with the Imitators who make their decisions based on social influence from others who have adopted a new product already. As we explained earlier, the number of Imitators is closely tied to an entertainment product's playability. Whereas blockbuster success depends on a product's Innovators and marketability, niche success depends mainly on Imitators and on playability. We have argued that the latter is mostly determined by a product's (experience) quality, which serves as the basis for consumers' word of mouth and critics' reviews.

Let's use the movie My Big Fat Greek Wedding as an example. The movie was filmed for just $5 million and then released in a single North American theater in 2002; it had an advertising budget of just $130,000 prior to release. Figure 15.7 shows how consumer perceptions of high quality combined with adequate distribution and advertising activities result in revenues that grew massively over time. Interest in the film eventually spilled beyond the target group; it became popular among general moviegoers. In the end, the film made more than $240 million (in 2002 value) in North American theaters alone.

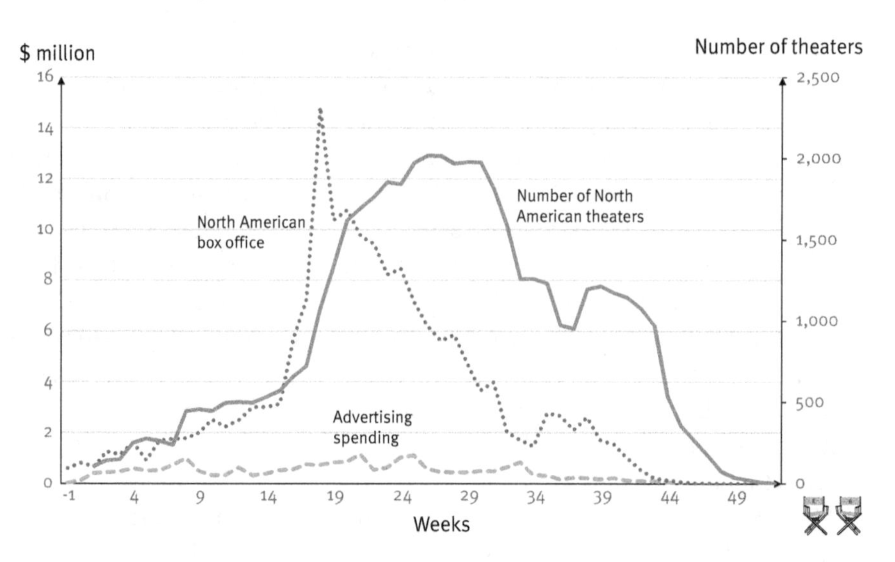

Fig. 15.7 Weekly box office, distribution, and advertising spending for the movie My Big Fat Greek Wedding
Notes: Authors' own illustration based on data from The Numbers (box office and theaters) and Kantar Media (advertising).

A high level of quality thus should be treated as the *conditio sine qua non* for successful niche marketing, and marketers need to allocate their efforts and resources in ways that honor this quality imperative. For example, for a typical blockbuster, the production and release timing decisions are made before the screenplay has been finalized. But this approach would be problematic for niche marketing because time pressure can pave the way for unsatisfactory storylines to be turned into final films. Further, because quality perceptions and preferences differ so strongly among consumers, picking the right target group and approaching it in a competent way is essential for igniting a powerful word-of-mouth cascade (Elberse 2013).

Let us mention two fundamental challenges to be confronted by those who want to engage in niche marketing. The first is the "quality challenge." Because entertainment products are creative, cultural, and hedonic, high quality cannot be ensured, even when all the best practices of innovation management are met by the producer. The other challenge is the "initial awareness challenge." Although the main focus of niche diffusion is on Imitators, it always requires a certain level of initial awareness; if the innovation parameter is zero, there is simply no one who can initiate the diffusion process among Imitators.

This is, for many niche marketers, the greatest difficulty: in today's world of generic hyper-competition, with new entertainment options showing up almost every minute, it is tough to cut through the clutter and gain enough attention to make consumers aware of a new product, prior to its release without deploying an extensive advertising budget. There are exceptions to this rule, but not many. Also, distributors (such as movie theaters and retailers) usually have little patience in such environments; they have to allocate their scarce resources between numerous alternatives, and it is not tenable for many distributors to reserve those resources for a niche product that *might* grow over time (instead of a pouring support into a new offering that is expected to lure in audiences with high marketability).

Finding answers for these challenges, such as through good relations with distributors, will certainly increase the success potential of niche marketing enormously.

The Long Tail: Using the Opportunities of Digital Media

"Embrace niches."
—Anderson *(2004, p. 174)*

The economic potential of the niche concept is closely linked to the so-called "long-tail" phenomenon. The "long tail" is essentially an observation that was first popularized by Chris Anderson, who wrote about it in an article in Wired magazine (which he edited at that time) and later in a blog and a business book (Anderson 2004, 2006). According to Anderson, digitalization gives consumers access to (and brings an increase in demand for) smaller, more dispersed niche products. This development should also apply to entertainment products, increasing niche titles' commercial potential.

Figure 15.8 illustrates the idea of the long tail, comparing the allocation of sales across a repertoire of available entertainment products in pre-digital and digital times. The logic underlying the long-tail phenomenon is that digitalization changes production and distribution in a way that increases the commercial relevance of niche products. Production changes, particularly the lower costs that result from digital production technology, lower the entry barriers for independent products. This should result in an expansion of the already large number of existing products—in other words, extending the tail (visualized by the *horizontal* arrow in the figure).

But it is distribution changes that are fundamental for the phenomenon: they should make the tail thicker (the *vertical* arrow in the figure). Because the Internet drastically reduces the storage costs for entertainment products, products that were previously unavailable for consumers now can be made available. Anderson (2004) argues that in the pre-digital age, Wal-Mart stores could only profitably carry CDs that sold at least 100,000 copies a year. In contrast, Amazon can essentially carry *any* available song and album

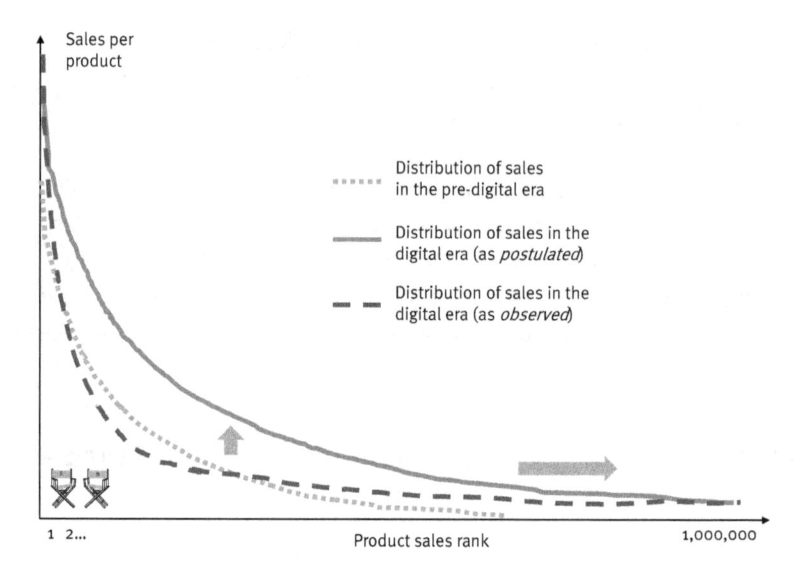

Fig. 15.8 Proposed shift of sales allocation toward niche titles
Notes: Authors' own illustration. Courses of functions are exemplary and not based on actual data.

because storage costs hardly matter for digital retailers and entertainment rental firms. In theory, combining the two kinds of changes should transform the orange dotted sales curve into the solid blue curve in the figure.

Drawing on data from Amazon, music streamer Rhapsody (now the "new" Napster), and then-DVD-rental firm Netflix, Anderson shows that these firms make between 20% and 57% of their revenues from titles that are not available in physical stores. He argues that the market for niche (or "long-tail") products is substantial, if not even "larger than the market for those that are [sold in the average offline store]" (Anderson 2004, p. 174). His assumption is that, in pre-digital times, restrictions prevented us, as consumers, from choosing the products we prefer most, and that we all have highly idiosyncratic preferences that lead us to prefer products that are liked by few others. And recommender technology could help us find what we want out of this enormous array of alternatives.[450]

Empirical research regarding the long-tail phenomenon paints a more complex picture, though. There is widespread agreement that the number of available entertainment products has clearly grown on digital platforms, and also that niche titles attract a substantial share of consumer spending for entertainment. For example, also using data from Amazon, Brynjolfsson et al. (2003) estimate that in 2000 almost 40% of book sales were for titles

[450]See our discussion of recommenders and their role for consumer decision making in our chapter on "earned" entertainment communication.

ranked 100,000 or lower; a more recent follow-up study paints a similar picture (Brynjolfsson et al. 2009). The number of different book titles purchased by Australian consumers increased by 63% in just three years (from 2004 to 2007; *The Economist* 2009). So we, as consumers, obviously make use of the newly available niche titles.[451]

What remains unclear, however, is at whose expense consumers' demand for niche titles comes. Elberse (2008) conducted an extensive analysis of video and music sales (U.S. sales from then-Nielsen's VideoScan and SoundScan), video rentals (from Australian service Quickflix), and music streams (from service Rhapsody) from different periods between 2000 and 2007. Her results suggest that it is not the bestselling titles that suffer; Elberse notes a growing interest in the titles at the top end of the curve. Indeed, across sources, the bestselling products capture the lion's share of the revenues, often in line with (or even exceeding) the 80-20 heuristic referred to as the "Pareto principle" (i.e., 80% of effects come from 20% of cases). And data from TV shows, music, and books consistently show that the share of revenues absorbed by the most successful titles tends to grow (*The Economist* 2009).[452]

Thus, Jeff Bewkes, as Warner CEO, has a point when he states that "[b]oth the hits and the tail are doing well" (quoted in *The Economist* 2009). It appears that the titles that suffer in the digital long-tail environment are the ones *in the middle*, those in-between the megahits and the niche productions. Consumers now increasingly go either for the spectacle (as provided by blockbusters) or for idiosyncratic pleasures (which niche titles can offer), but they increasingly ignore those products that are neither spectacular nor individual enough. This is what the third, purple-dotted line (the "observed" distribution of sales in the digital era) in Fig. 15.8 above illustrates: the line starts higher than the orange one (the biggest hits command more money), but then falls below it and only passes it again toward the end of the tail (niche titles also earn more than before).

Finally, let us clarify that niche products' popularity among consumers must not be confused with a bright outlook for individual producers who engage in niche marketing. Instead, it is the *distributors* (such as Amazon, Google, Apple, Netflix, and Spotify) who benefit most from the increased demand for niche entertainment products. Consumers are attracted to the

---

[451]Their follow-up study also points to certain limitations—Brynjolfsson et al. find that while the tail got longer, consumers were not particularly interested in the far end of the tail, with the slope of the curve becoming steeper for "ultra-niche" titles (those with a rank below 100,000).

[452]For example, No. 1 albums in the UK sold more in 2008 than four years earlier, while all other music sold less. The five most-viewed TV shows in the U.S. clearly lost less audience between 2001 and 2009 than all other shows. And Top-10 bestselling books sold 75% more in the UK in 2008 than a decade earlier, easily beating the trend for all other books. For details, see *The Economist* (2009).

lure of a large number of heterogeneous titles, but only the distributors can offer the desired assortment from which consumers can pick titles that match their idiosyncratic preferences closely.

An individual producer of a single or a small number of niche entertainment products cannot offer such choice. And the costs for consumers to search the works of each individual producer will be prohibitive in most cases, except maybe for powerful niche brands which own a relevant catalogue of titles (such as specialty DVD label Criterion). When it comes to the long-tail phenomenon, "the real business of entertainment is about owning one of the handful of digital platforms that can command consumers' attention" (*The Economist* 2017). As a result, entertainment producers who use the niche concept for marketing their products must not rely on the power of the long-tail phenomenon. Instead, they still must find ways to address the challenges that the niche concept brings, generating sufficient awareness among the target group and distributors, ensuring a high quality outcome, and developing a repertoire of products that stand out enough that consumers actively search for them.

After overviewing the two dominant integrative marketing strategies from which an entertainment producer must choose (if he wants to avoid getting lost "in the middle"), we have almost reached the end of the book. But we simply cannot let you go without taking a short look at the current state of the industry with regard to the use of these two approaches—and the consequences this state might have for the future of entertainment.

Blockbusters Versus Niche Products: Where We Stand Today and the "Too-Much-of-a-Good-Thing" Problem

Living in a Blockbuster World!

> "This industry's in flux, it's run by mucky-mucks pitching tents for tentpoles and chasing Chinese bucks. Opening with lots of zeroes, all we get are superheroes: Spider-Man, Superman, Batman, Jedi Man, Sequel Man, Prequel Man, formulaic scripts!"
>
> —*Actor* Jack Black *at the 87th Academy Awards (quoted by Toronto Mike 2015)*

Today, entertainment is dominated by blockbusters to an unprecedented degree. Major studios and labels are putting all their cards on the big bets, seeming to develop *only* blockbuster productions. Jack Black's sarcastic lyrics capture the concerns of many regarding the dominance of high-budgeted sequels, remakes, bestseller/game/comic adaptations, and toy-inspired productions that now dominate movie theaters.

Many major producers have withdrawn from almost all other production activities, having sold off (or re-positioned) their arthouse affiliations and discarded projects that do not seem to have "blockbuster potential." Film producer Lynda Obst has termed the resulting state of the industry the "new abnormal," a state in which entertainment products can *only* be made in blockbuster format unless they are made completely outside of the major studios and labels. Mrs. Obst (2013) asks whether "the original movie with a good story [can] get made for its own sake in today's Hollywood... Could I get THE FISHER KING made with Terry Gilliam ... and the best script I ever had?" Her answer is: "(Ha! No way!)" (p. 9). That sentiment is similar to what producer Laura Ziskin (of PRETTY WOMAN, among others) replied when she was asked which of her films could get made today ("None of them," she replied; quoted in Obst 2013, p. 9). Comparable arguments have been heard for the other forms of entertainment we cover in this book.

The studios' concentration on blockbusters is as radical as is their interpretation of the concept. Panel A of Fig. 15.9 shows the number of films per year in which American producers invested more than $100 million (in 2014 value; production costs only). The number has increased from one film in 1985 to 31 now; the increase is even more dramatic for films with budgets above $150 and $200 million. This trend takes place in an environment in which major studios and labels are *reducing* the total number of products they release.

The figure's Panel B demonstrates that the film industry is also allocating these investments differently than before, with an even larger portion flowing into big productions. For example, the ten most expensive films absorbed almost one out of four dollars spent for all major productions in 2014, but back in 1985 the corresponding number was only 18%—a change of almost 40%. We also calculated the Gini coefficient as a measure of the inequality (or concentration) in producers' budget allocations across budget categories—the coefficient, which ranges from 0 to 1 with higher values indicating a stronger bias, has increased continuously since 1986, from 0.25 to 0.41 among the 100 most expensive films, and from 0.39 to 0.59 among all major productions.[453]

[453]The increase in inequality of budget allocation over time is almost linear; respective regression functions have R^2 values of 0.84 (for the 100 largest films) and 0.95 (all major films). All numbers we mention here refer to the normalized version of the coefficient (which ranges from 0 to 1).

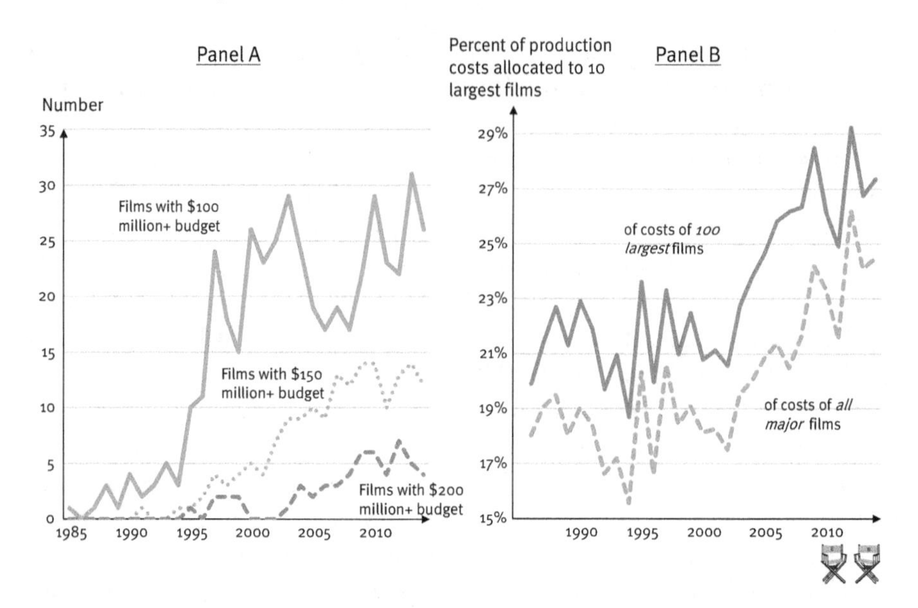

Fig. 15.9 The trend toward larger budgets
Notes: Authors' own illustration based on data from The Numbers and some additional sources. All data is based on 2014 dollars. Includes major films with a total North American box office of $1 million and above (inflation-adjusted) for which budget data was available.

We have pointed to the economic potential of the blockbuster concept and its elements (e.g., strong brands and franchises) at several points of this book. These benefits at least partly explain the industry's shift toward blockbusters. But what is the logic behind the *extreme* incarnations of the blockbuster concept we are observing these days, in Fig. 15.9 above and elsewhere? We hold the mega trends of globalization and digitalization responsible. Globalization has increased the leverage/revenue potential for entertainment blockbusters enormously, with more markets accessible and fewer trade barriers in the way.

And digitalization has further facilitated the exploitation of global markets by enabling global communication via the Internet (essential for building buzz all around the planet) and by massively reducing the costs of rolling out an entertainment product simultaneously and through multiple (digital) channels on a global scale. Today, the moment Netflix uploads a film onto their Amazon-hosted servers, consumers in more than 190 countries can watch it instantly, with minimal marginal costs for Netflix. The logistics are similar for music (via Spotify and iTunes), books (via Amazon), and even for digital movie files sent to theaters worldwide.

Today, most of the money the major (U.S.-based) entertainment producers make comes from outside their home continent. In 1980, a Hollywood studio generated less than 25% of its revenues outside of North America

(Friend 2016)—now international earnings account for up to two-thirds of a studio's revenues (McNary 2016). An immediate consequence is that major producers have the "global consumer" in mind: "[o]ne thing [studios] take into consideration [when deciding about a movie] is the foreign market, obviously" (director Steven Soderbergh, quoted in *The Deadline Team* 2013).

A widely shared assumption among major producers is that the best way to offer enjoyment to heterogeneous audiences is to make the products as spectacular (and thus expensive) as possible and to use brands of global fame; the belief is that higher budgets spent for higher levels of each attraction translate into higher revenues. Although a positive audience evaluation is not the focal element of the blockbuster concept, franchise logic certainly benefits from it, and expensive spectacle has been identified as a global "crowd pleaser." In contrast, less-costly attractions, such as smart dialogue, are considered as less powerful, carrying the risk of not appealing to, confusing, and/or asking too much from at least some of the mainstream global audiences.[454]

Economic results seem to have rewarded managers for their focus on blockbusters: we have cited evidence that the share of revenues captured by the bestselling titles is growing across entertainment-product types. But does this mean that investments in extreme production budgets translate into audience appeal and financial performance? Panel A of Fig. 15.10 illustrates that at the North American box office, the share of revenues generated by the most expensive film productions has indeed grown over the years, along with the higher resources allocated for such productions: the ten largest films now attract a higher share of consumer revenues than in previous years. But we should also note that there is substantial variation in the share of revenues grabbed by the ten most expensive films. And Panel B of the figure casts some additional shadows: it shows that the higher share of costs that is now allocated to the most expensive productions does not translate into a higher revenue share *outside* of North America.

Regardless of these results, the industry considers extremely big budgets and the spectacular attractions they can generate to be the silver bullet to attract large global audiences. The most attractive products can leverage their appeal via digital distribution, so that reaching additional millions of consumers can be done for low marginal costs and helps to secure a

[454]Satterwhite et al. (2016) offer an interesting audio-visual analysis of the music of the Marvel Cinematic Universe, concluding that the producers intentionally avoid intense and memorable musical themes for their films to avoid the risk of confusing or offending audiences. They argue that the films do not use "bad" music—but music that is "bland and inoffensive."

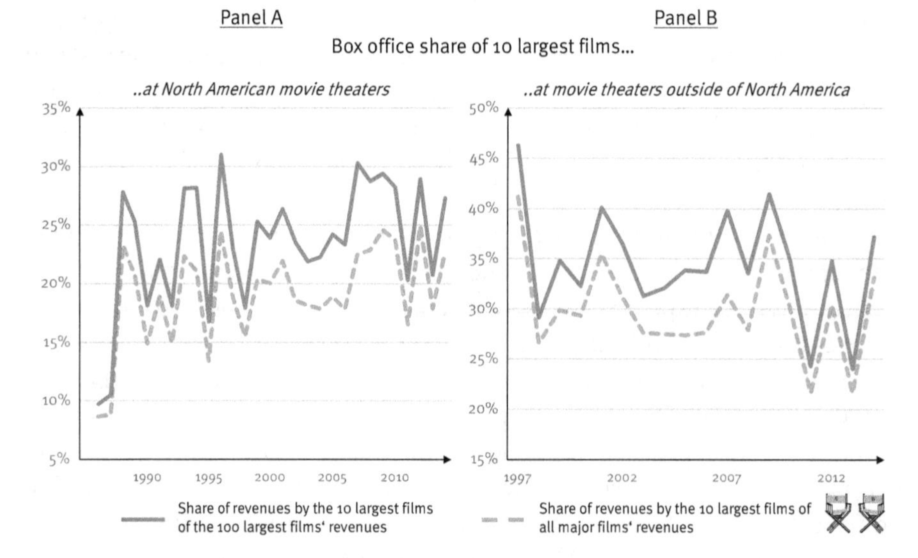

Fig. 15.10 Are the most expensive products the most appealing ones for consumers? Notes: Authors' own illustration based on data from The Numbers and some additional sources. All data is based on 2014 dollars. Major films are all with a total inflation-adjusted North American box office of $1 million and above for which budget data was available.

disproportional share of global *profits*. Take the seventh Star Wars episode, The Force Awakens, as an example: it generated a *profit* of almost $800 million for Disney (Fleming 2016). Follows (2016) estimates that half of the profits in his database "came from the highest paying 6% of movies."

This is how things are today in entertainment: it's a world of extreme blockbusters. With studios and conglomerates being mostly profitable, we could leave you now to conduct your own study, or to produce (or distribute) your own entertainment products. But we cannot do that without first mentioning a systemic risk that we believe entertainment's focus on extreme blockbusters carries with it; a risk that might have the potential to threaten entertainment firms' sustainability and productivity in the longer term. Something that works effectively when applied as an exception, such as the blockbuster concept has done in the past, does not necessarily continue to work effectively when applied as the standard rule.[455]

[455]This insight also offers a final chance to remind us all of the limits of empirical research that uses historical data (in *Entertainment Science*, as well as in other disciplines). The usability of a study is always tied to the relevance of the context in which it was conducted. If the context changes in meaningful ways, past empirical findings lose their predictive power.

The "Too-Much-of-a-Good-Thing" Trap

> "Everybody's optimal strategy is to go aggressively after blockbusters. But when everybody's doing that, it's bad for their collective outcome."
> —*Financial analyst* Doug Creutz *(quoted in* Lieberman *2017)*

Driven by the prospect of low marginal costs for reaching additional consumers in the digital age, the large majority of producers now strive for offering extreme blockbuster products that appeal to a *maximum* number of consumers on a global scale, with higher-than-ever spending per product to assure consumer attention and interest. In addition to offering high levels of familiarity via popular brands, they usually attempt to offer sensations that appeal to diverse consumers by addressing a small set of key, universally applicable motives such as escapism.

The fact that this approach has become the standard brings a number of interconnected challenges for entertainment firms that we believe might, in a few years, threaten the status of some parts of the entertainment business, or even the industry as a whole. These challenges have to do with the increasingly generic character of mass-targeted entertainment products that have emerged from the blockbuster concept being taken to the extreme and becoming the primary focus. Given the challenges' collective nature, it won't be simple, or even possible, for a single manager to counteract them (and it might not even be economically advantageous). But we believe it is essential to share them with you, nevertheless.[456] Here are what we consider as the industry's current main challenges.

The first challenge. We have argued (and provided empirical evidence throughout this book) that consumers' entertainment preferences are diverse. An extreme "blockbuster-only" product portfolio is not able to address such diversity in demand; blockbusters are, by definition, focused on common interests shared by *many* consumers, not the unique interests of smaller groups or individuals. Adding to this challenge is that the blockbuster logic must protect its assets; because key characters are multi-million dollar brands themselves, letting a hero die is simply not an option for a storyline, which further restricts the potential for variety and fresh sensations. Such conformity within an entertainment form (such as movies) may reduce the appeal of that form of entertainment. Signals of such loss

[456]Our discussion here is inspired by what we wrote earlier in Hennig-Thurau (2013). We feel that things haven't appeased since then, but rather worsened. (Although we have been wrong with our expectation of The Lego Movie…)

of interest can already be observed, to a certain degree, for movies, whose blockbuster-only status has been cited as a major reason underlying the growing interest in complex television dramas and immersive video games.

And if the conformity is prevalent in *all* the entertainment forms we feature in this book, then consumers' interest in them may decline altogether. We have already highlighted in our discussion of distribution decisions that, in today's digital world, consumers are not limited to only migrating from one (traditional) form of entertainment to another. Instead, younger consumers see social media as a form of entertainment on its own; the time and money they spend for phones and apps and data gives evidence of this new insurgent to traditional entertainment. And the firms that provide the digital (social) media infrastructure (such as Facebook or Google) are largely different firms than the ones that produce movies, music, or books. One might consider the dealings of traditional and digital (social) media giants another "frenemy" relationship: traditional entertainment producers sell content to the social media firms, but the attractions of social media for consumers are much more than simply watching a film or listening to a song.

Which consumer segments are most affected by the standardization of entertainment content that results from the "over-use" of the blockbuster concept? Every segment whose entertainment preferences diverge from those of the common denominator. This includes the "fan boys," who find their beloved comic, graphic novel, or gaming characters compromised in producers' efforts to make them as widely appealing as possible. Listen how The R-gument(or) (2016), a self-declared movie geek, explains his loss of excitement for movies: "[O]ver the last 5-7 years, most … films have started (to varying degrees) to feel more and more sterile to me. As far as I'm concerned, even the best ones among them are just too 'clean.' Too smooth, too polished, too polite: too perfect. And that perfection and 'cleanness' affects everything. From the language, the clothes and the overall look of the characters to the way the action scenes and fights are designed, heck: even inner conflicts of characters and emotional distress tend to remain well within the viewer's comfort zone. Even dirt looks clean in those movies…" He concedes that many "films are rather expertly made, very entertaining, with usually great artists involved in all aspects of the production; they offer bombastic spectacle, lots of funny one-liners and the latest state of the art special effects. And they bore me to death."

Marketing theory indicates that the loss of fan-boy excitement should be particularly alarming for producers of entertainment, even though fan boys are only a fraction of the total customer base for blockbusters.

Those enthusiasts who, like The R-gument(or), "couldn't wait to see the latest blockbusters" and list big-budget films, such as AVATAR and LORD OF THE RINGS, as among their favorites, fit the profile of what innovation scholars name "lead users": people who are not only enthusiastic about a product category, but whose desires and wishes also foreshadow future preferences of mainstream consumers (von Hippel 1986; Urban and von Hippel 1988). Lead users' frustrations with current marketplace developments should be carefully noted by entertainment producers, as they might indicate a forthcoming trend that could have economic impact.

Let us add that we are not alone in this concern. Some industry managers also see the risks of turning away consumers with a technically perfect, but creatively monotonous product range. For example, legendary Hollywood executive Bill Mechanic voiced concerns that the current blockbuster focus removes the diverse appeal of movies and threatens the movie-going as a whole: "Big is not inherently bad. Small is not inherently good. Good is good and bad is bad. … If theatrical movies disappear in the next decade or so, it would be self-fulfilling prophecy. The world will be a darker place and the culprit isn't new technology, it's [the managers] who didn't think things out" (Mechanic 2017). Recent movies like DEADPOOL and X-MEN spin-off LOGAN, which vary established genre and franchise patterns and were enthusiastically greeted by audiences, are attempts at overcoming this frustration. But they turn away large audience segments with their radicalness, adding violence and "grittiness" (both received an R-rating). In essence this means they violate the blockbuster recipe—and while these two films did so successfully by adding fresh, creative sensations, doing so again would require a continuous stream of artistic achievements, something that introduces risk and is hardly compatible with extreme-blockbuster economics.

The second challenge. We have highlighted the critical role of artistic, creative talent for the production of successful entertainment. An extreme blockbuster-only approach is not very attractive for this group of people, however. Creatives want room for their visions, but blockbusters, franchises, and universes often do not provide such room. We have noted the leading role that Marvel executive Kevin Feige plays in the development of the films in Marvel's cinematic universe; dominance over the creative realm by executives and "committees" is the rule in today's extreme-blockbuster world. If talent objects to the committee's ideas, the talent might be dismissed for "creative differences," as happened to directors Phil Lord and Chris Miller, who had been hired to craft the SOLO movie, a 2018 entry in Disney's STAR WARS franchise (Debruge 2017).

Mass market appeal is essential for a blockbuster, particularly if investments are in the multi-$100-million range, and experimenting with the visions of creatives is a risky option when a single financial flop (caused by the lack of mass-market positioning, not artistic quality) has a seismographic impact on the producer's economic viability. As a consequence, in the short-term, artists migrate to other firms or to other forms of entertainment, where they hope to find better conditions to bring their creative visions to life. This migration has been seen recently with movie directors who decided to become involved in the creation of TV drama series instead of motion pictures. This flow of talent has certainly contributed to the recent popularity of such series, which offer a level of unprecedented artistic quality for this new form of entertainment (e.g., Pähler vor der Holte and Hennig-Thurau 2016). It could contribute to a further deterioration in the artistic quality of traditional entertainment and further weaken their competitiveness against newer digital forms.

The third challenge. The blockbuster economy requires powerful brands that have global appeal to create huge pre-release buzz and attract mass audiences who consume the product upon release because they "cannot wait" for word of mouth (i.e., quality information) from personal friends. But the repertoire of global "uber brands" that can stimulate such reactions is clearly limited. As entertainment producers have shifted nearly all resources into the extension of existing brands, there is hardly anything new coming up that can build the superstar status needed for a blockbuster of extreme proportions. Hollywood is rummaging through all cultural archives these days in the hunt for useful brands, a search that can sometimes trigger surprising results when visionary creatives are involved: think of The Lego Movie, which turned a toy brand into an inspiring piece of narrative storytelling (and made almost $500 million in theaters alone).

But much more often, the hunt produces strange results. The outline of the movie adaptation of the enormously popular non-narrative Fruit Ninja game (in which players, on their smartphones or tablets, slice fruit with a touch screen-controlled blade) goes as follows: "Every couple of 100 years, a comet flies by Earth, leaving in its wake a parasite that descends on a farm and infects the fruit. The infected fruit then search for a human host. The only thing keeping humanity from certain doom is a secret society of ninjas who kill the fruit and rescue the hosts by administering the 'anti–fruit.' The produce–slaying saviors are recruited from the population based on their skill with the Fruit Ninja game" (French 2017). Warner bought the brand, so expect it to be coming soon to a theater near you in dedicated blockbuster format.

In addition, satiation effects (which, as we have shown, are inherent for entertainment consumption) challenge the continuation of even the most powerful brand franchises. In the summer of 2017, several blockbuster extensions of established movie brands performed below expectations (e.g., the fifth TRANSFORMERS film and the eighth entry in the ALIEN saga). Some in the industry attributed the lukewarm results to the films' lack of quality. But with their predecessors having been successful despite similar quality ratings, it seems more reasonable to blame a decaying level of interest in the brands among target audiences. "[O]nce upon a time, Hollywood probably wouldn't have bothered to make a fifth PIRATES [OF THE CARIBBEAN] movie after the fourth entry ... grossed considerably less than the first three, and couldn't even make up its (huge) budget at the domestic box office. Once upon a time, defunct TV properties like BAYWATCH weren't considered ... to be turned into expensive summer tentpoles. Audiences didn't avoid these films because they got bad reviews; they avoided them because they were never interested in them in the first place" (Sims 2017).

The fourth challenge. The blockbuster concept depends on the producer's ability to generate high levels of marketability, i.e., making large numbers of consumers pay for the product before quality information from friends leaks through. But in a time in which nearly everyone is member of one or more digital social networks (even some *Entertainment Science* professors!), the pace of the spread of information among friends continues to accelerate.

Remember that we have presented empirical evidence that sharing negative information about a new film on its opening day among social network members can hurt the film's performance. And aggregation sites such as Rotten Tomatoes offer the same instant information for professional reviews. If this effect accelerates, it could shift people from Innovators to Imitators who only consume a blockbuster if it's "good"—not according to paid and owned media, but "earned" media. This shift carries the risk that blockbuster producers will lose larger shares of their enormous investments if the quality of a product is poor, something that challenges the fundamental economic logic of the blockbuster approach. Brett Ratner, the now-repudiated blockbuster director, producer, and financier, might have been correct when he named such real-time quality information "the destruction of our business" (quoted in Hibberd 2017).[457] Of the *extreme blockbuster business*, to be precise.

[457]Mr. Ratner was speaking of Rotten Tomatoes in particular, but it's easy to see that his critique would also apply to other sources and forms of immediate quality information.

The fifth challenge. Finally, entertainment producers will have to find a way to keep the blockbuster concept effective despite escalating costs for advertising. In an increasingly fragmented media world in which linear TV's reach is deteriorating (e.g., Thompson 2017), creating excitement among a critical number of people for a new entertainment blockbuster prior to its release is becoming increasingly challenging. And expensive: the duopoly of Internet advertisers (Facebook and Google) will ensure that the costs for online advertisements will continue to go up, not down, further threatening the profitability of blockbusters. Escalating digital advertising costs would also accelerate the trend toward even bigger releases, as only they might be able to cut through the clutter and generate needed revenues.

Altogether, the industry is facing the dilemma we label here the "too-much-of-a-good-thing" trap. By turning an occasional strategy into the industry standard, managers create a new environment in which the effectiveness of the strategy itself is no longer certain. The data underlying most empirical findings regarding consumers' adoptions of blockbusters and the economic results were gathered in a context in which such products were the exception rather than the rule; it is unclear whether, and how, findings might change in an "extreme-blockbuster economy." Thus, Epstein (2017) is right when he notes that "by avoiding the inherent risks in greenlighting unproven ideas, [entertainment managers] may be taking the biggest risk of all."

This environmental shift does not mean that the observations we noted regarding the management of blockbusters are invalid, but they will tend to lose grip. When consumers are confronted with branded films only, the advantage that such brands offer will fade as all available alternatives are also big brands. We thus encourage managers to take a closer look at interaction effects rather than main effects. What does not change are the contextual insights: what are the factors and conditions that make a brand particularly powerful?

Does an individual entertainment producer have any options instead of participating in the extreme-blockbuster competition (or by applying the niche concept)? Some are looking to find new, alternative ways to market their products, avoiding the ever-escalating competition for consumer buzz. When singer Beyoncé released her self-titled album BEYONCÉ in December 2013, it simply appeared on iTunes and on CD a few days later. There was no expensive pre-release advertising campaign; instead, the singer announced the release via her Facebook page (Hampp and Lipshutz 2013). Three years later, Beyoncé repeated this approach with her LEMONADE album, and other music stars (such as Rihanna, Drake, and U2) have followed suit. Some observers refer to this approach as a "surprise-release" strategy (Huddleston 2016).

And movie director Steven Soderbergh, after in 2013 declaring his intention to retire from making films (Child 2013), returned four years later with a new "service-deal" marketing strategy, in which he, as the creative, kept full authority over his film LOGAN LUCKY, as well as its advertising and distribution (Guerrasio 2017). He managed to do so by pre-selling the non-theatrical rights for the film and by letting distribution and advertising be handled under the terms of a revenue-sharing deal with a 20-employee start-up, instead of a Hollywood studio.

It is obvious that these approaches are experiments and can only work under specific conditions, if they can work at all (Soderbergh's film opened in over 3,000 theaters in North America, but generated only $8 million on its first weekend). Nevertheless, we are anxious to see what other fresh marketing concepts entertainment managers (or *Entertainment Science* scholars?) might come up with in the future as alternatives to the blockbuster and niche concepts.

Concluding Comments

In this final chapter of *Entertainment Science*, we illustrated that there is value hidden in the way the different elements of the entertainment-marketing mix are coordinated. We introduced two integrated entertainment-marketing strategies, each of which follows a distinct logic: the blockbuster concept and the niche concept. The logic of the blockbuster concept, which combines strong brands, an easily communicable "high-concept" product, massive investments in the product itself and in advertising with intense distribution, is to "pre-sell" new offerings by creating strong pre-release buzz, essentially separating the success of an entertainment product from the quality of its execution (and the risk that is associated with the production of such quality). Its main "selling point" is the familiarity of "household" brands, which it complements with spectacles that appeal to a large number of consumers. The niche concept, in contrast, refrains from the use of large investments and has not a pre-, but a post-release focus. It builds on high quality for a specialized audience and the triggering of informed cascades via word of mouth and the actions of other stakeholders such as professional critics.

We showed that the blockbuster concept has become the dominant one in entertainment, and that it is often taken to quite extreme levels. The combination of digitalization and globalization offers the potential for astronomical rewards, which major producers increasingly target with billion-dollar multi-product franchises like STAR WARS and "universes" such as the one populated by Marvel's superheroes.

And whereas we provide theoretical and empirical support for the blockbuster logic for single products, we are concerned by its dominance—what works as part of a diverse economy, might not do so as the industry standard. A blockbuster-only approach to entertainment faces several challenges that we laid out in this chapter: it carries a systemic risk, potentially driving consumers away from traditional forms of entertainment to new ones such as social media, and thus implies the danger of massive, very expensive failures.

References

Anderson, C. (2004). The long tail. *Wired Magazine*, January 10, https://goo.gl/2NQJUr.

Anderson, C. (2006). *The long tail: Why the future of business is selling less of more.* New York: Hyperion Books.

Asai, S. (2009). Sales patterns of hit music in Japan. *Journal of Media Economics, 22*, 81–101.

Bass, F. M. (1969). A new product growth model for consumer durables. *Management Science, 15*, 215–227.

Boxofficemojo.com (2017). All time box office. Domestic grosses, adjusted for ticket price inflation, July 24, https://goo.gl/hPDwrS.

Braund, S. (2013). *The greatest movies you'll never see: Unseen masterpieces by the world's greatest directors.* London: Cassell.

Brynjolfsson, E., Yu Jeffrey, H., & Smith, M. D. (2003). Consumer surplus in the digital economy: Estimating the value of increased product variety at online booksellers. *Management Science, 49*, 1580–1596.

Brynjolfsson, E., Hu, Y. J., & Smith, M. D. (2009). A longer tail? Estimating the shape of Amazon's sales distribution curve in 2008. In *Workshop on Information Systems and Economics.*

Canning, B. (2010a). How Jaws heated up summer movies. In A. B. Block & L. A. Wilson (Eds.), *George Lucas's Blockbusting* (pp. 531–532).

Canning, B. (2010b). Jaws. In A. B. Block & L. A. Wilson (Eds.), *George Lucas's Blockbusting* (pp. 572–573).

Child, B. (2013). Star Wars Episode VII pushed back to Christmas 2015. *The Guardian*, November 8, https://goo.gl/bApFrt.

Clements, M. T., & Ohashi, H. (2005). Indirect network effects and the product cycle: Video games in the U.S., 1994–2002. *Journal of Industrial Economics, 53*, 515–542.

Coppola, F. (2010). Introduction. In A. B. Block & L. A. Wilson (Eds.), *George Lucas's Blockbusting* (Vols. VI–VIII). New York: HarperCollins books.

D'Alessandro, A. (2015a). Legendary's Michael Mann pic 'Blackhat': What the hell happened? *Deadline*, January 20, https://goo.gl/u7gj4Z.
D'Alessandro, A. (2015b). 'Fifty Shades' lower with $22.26 M; 'Hot Tub 2's tracking off. *Deadline*, February 23, https://goo.gl/PiagvV.
Debruge, P. (2017). Why movies need directors like Phil Lord and Chris Miller more than ever. *Variety*, June 21, https://goo.gl/udAqYg.
Elberse, A. (2008). Should you invest in the long tail? *Harvard Business Review, 86*, 88–96. https://goo.gl/EGtyEJ.
Elberse, A., & Krasney, N. (2013). How Hollywood's original and not-so-original films fare at the box office. https://goo.gl/wfWWdD.
Epstein, A. (2017). 'The Matrix' reboot: It's finally happened. Hollywood has run out of all the ideas. *Quartz*, March 15, https://goo.gl/3oGv18.
Fleming Jr., M. (2016). No. 1 'Star Wars: The Force Awakens'—2015 most valuable movie blockbuster tournament. *Deadline*, March 28, https://goo.gl/Rtj96X.
Follows, S. (2016). How movies make money: $100 m+ Hollywood blockbusters, July 10, https://goo.gl/uYwnJe.
French, A. (2017). How to make a movie out of anything—Even a mindless phone game. *The New York Times*, July 27, https://goo.gl/A3NWGA.
Friend, T. (2016). The mogul of the middle. *The New Yorker*, January 11, https://goo.gl/8hYXxT.
Guerrasio, J. (2017). Steven Soderbergh has a new plan to make Hollywood movies outside the control of big studios. *Business Insider*, June 8, https://goo.gl/Si9mA8.
Hampp, A., & Lipshutz, J. (2013). Beyoncé unexpectedly releases new self-titled 'Visual Album' on iTunes. *Billboard*, December 13, https://goo.gl/mG3NNa.
Hennig-Thurau, T. (2013). Why the 'Buzz Economy' is not a sustainable business model. *Medium*, November 19, https://goo.gl/KJhbpt.
Hennig-Thurau, T., Houston, M. B., & Sridhar, S. (2006). Can good marketing carry a bad product? Evidence from the motion picture industry. *Marketing Letters, 17*, 205–219.
Hibberd, J. (2017). Rotten Tomatoes is 'the destruction of our business,' says director. *Entertainment Weekly*, March 23, https://goo.gl/vnriTx.
Huddleston Jr., T. (2016). Beyoncé's 'Surprise' album wasn't all that surprising. *Fortune*, April 25, https://goo.gl/vUgnts.
King, G. (2002). *New Hollywood cinema*. New York: Columbia University Press.
Lewis, J. (2003). Following the money in America's sunniest company town: some notes on the political economy of the Hollywood blockbuster. In J. Stringer (Ed.), *Movie Blockbusters* (pp. 61–71). London: Routledge.
Lieberman, D. (2017). Hollywood bear: Why media analyst is gloomy about the movies. *Deadline*, March 17, https://goo.gl/HanNbM.
Lieberman, D., & Busch, A. (2016). Hollywood tees up movie ads for Super Bowl: TV's most expensive event. *Deadline Hollywood*, January 29, https://goo.gl/m1zhCu.
März, U. (2011). Bestseller mit Ansage. *Zeit Online,* June 30, https://goo.gl/osWaup.

McNary, D. (2016). Universal, Disney finish 2015 with record worldwide grosses. *Variety*, January 4, https://goo.gl/TR3on2.

Mechanic, B. (2017). Bill Mechanic on movie biz: Big problems & a few suggestions on how to fix them. *Deadline*, March 31, https://goo.gl/6rvu85.

Mendelson, S. (2014). Tim Burton's 'Batman' at 25, and its wonderful, terrible legacy. *Forbes*, June 23, https://goo.gl/93xDpk.

Nall, J. (2012). Open wide: The influence of "Jaws" on modern high concept films. *Student Film Reviews*, May 6, https://goo.gl/oX4yZ3.

Obst, L. (2013). *Sleepless in Hollywood. Tales from the new abnormal in the movie business*. New York: Simon & Schuster.

Pähler vor der Holte, N., & Hennig-Thurau, T. (2016). Das Phänomen Neue Drama-Serien. Working Paper, Department of Marketing and Media Research, Münster University.

PBS (2001). The monster that ate Hollywood. *Frontline*, November 22, https://goo.gl/4DiBo5.

Rubinson, C., & Mueller, J. (2016). Whatever happened to drama? A configurational–comparative analysis of genre trajectory in American cinema, 1946–2013. *The Sociological Quarterly, 57*, 597–627.

Satterwhite, B., Ramos, T., & Zhou, T. (2016). The Marvel symphonic universe. *Every Frame a Painting*, September 12, https://goo.gl/4JDqn3.

Shone, T. (2004). *Blockbuster*. New York: Free Press.

Sims, D. (2017). Hollywood has a bad-movie problem. *The Atlantic*, July 5, https://goo.gl/BxMfAE.

Stringer, J. (2003). *Movie blockbusters*. New York: Routledge.

The Economist (2009). A world of hits, November 26, https://goo.gl/qVMq3v.

The Economist (2017). The battle for consumers' attention, February 9, https://goo.gl/Lyt33S.

The Deadline Team (2013). Steven Soderbergh's state of cinema talk. *Deadline*, April 30, https://goo.gl/3md7zK.

The R-gument(or) (2016). Why Hollywood might want to scale back (a little): A rant against modern tentpole filmmaking, August 26, https://goo.gl/jwS478.

Thompson, D. (2017). TV's ad apocalypse is getting closer. *The Atlantic*, August 10, https://goo.gl/qPZock.

Time (1975). Summer of the shark. *Time Magazine*, June 23, https://goo.gl/oaM6zF.

Toronto Mike (2015). What Jack Black said at the Oscars, February 26, https://goo.gl/ZJ8YY8.

Urban, G. L., & von Hippel, E. (1988). Lead user analyses for the development of new industrial products. *Management Science, 34*, 569–582.

von Hippel, E. (1986). Lead users: A source of novel product concepts. *Management Science, 32*, 791–805.

Wyatt, J. (1993). *High concept: Movies and marketing in Hollywood* (5th ed.). Austin: University of Texas Press.

Now Unlock the Power of *Entertainment Science*!

THIS IS IT.
—*Song by* Michael Jackson.

The management of entertainment has been traditionally dominated by a reliance on "gut feeling" and managerial instinct, a paradigm that has found its epitomization in screenwriter William Goldman's iconic phrase that "Nobody Knows Anything" regarding the commercial performance of entertainment products. This mantra, however, contrasts strongly with more than 35 years of scholarly research that provides ample empirical evidence that consumers' decisions regarding entertainment products and the products' subsequent financial performance follows systematic, non-random patterns. Scholars have studied these patterns with data and econometric algorithms and have identified the rules, or theories, on which they are based.

In this book, we have brought together the vast body of such insights to make a first major step toward what one might call a theory of *Entertainment Science*. As you will have certainly noted over the pages of this book, this theory is far from comprehensive; there a many gaps and empty spaces in explaining what makes an entertainment product successful. And some arguments are much more strongly supported by data than others (and some have yet to be empirically tested at all). But almost *all* theories are works-in-progress, by definition, because there is almost always more to learn about the phenomenon; it is very rare to come across a theory that is definite and final (if there is even one).

T. Hennig-Thurau and M. B. Houston, *Entertainment Science*,
https://doi.org/10.1007/978-3-319-89292-4

The general logic of *Entertainment Science* is that success in entertainment, just as in other fields of life and the economy, follows certain patterns and rules, and that data and algorithms can help us to understand them. The foundation underlying *Entertainment Science* is a probabilistic worldview that sees the world of entertainment as complex and multi-causal. In this complex world, a single factor or use of a single marketing instrument never explains a consumer's reaction or the success of a product with 100% certainty, but only increases the probability that the consumer will react in a certain way or that the entertainment product will be a hit. This is how things are and will always be in any social science, the scholarly field into which *Entertainment Science* falls.

As fragmentary as it is, *Entertainment Science* is not a "pure" theory, but an applied one. It deals with the practical subject of how entertainment products succeed or fail, and we intend it to be of use for managers or others involved in the entertainment industry. *Entertainment Science* makes arguments about what works better (and why), but it make no attempt to oust creativity from entertainment. Instead, the theory of *Entertainment Science* is based on a thorough analysis and understanding of the specific factors that define entertainment and its products, the markets on which entertainment products are traded, and entertainment consumers. Those specifics shape our view of how things work economically in the context of entertainment. We dedicated the first chapter of this book to an in-depth discussion of product characteristics, which included entertainment's hedonic and cultural nature, as well as several other aspects. The chapter also provided linkages to how the different characteristics impact the effectiveness of marketing strategies. In the remaining chapters of the book's first part, we complemented this discussion with an analysis of the characteristics of entertainment markets and the ways consumers make decisions for experiencing entertainment.

The second part of the book was then dedicated to the instruments that can be used by an entertainment manager for marketing an entertainment product—from the creation of the initial idea until the very end of its life cycle. You have noted that our perspective of marketing is not one that is limited to advertising or other promotional actions, but a holistic one which encompasses all activities of an entertainment firm that deal with markets and consumers. This marketing perspective provides room for creativity and artistry, which determine the "experience" quality of any entertainment product—and how audiences and experts react to it. This "experience" quality is essential and thus the first element of the "entertainment marketing mix" we discussed in the book. We reported findings based on cultural theory as well as data and algorithms to better understand what contributes to "great entertainment." But artists can relax; although our insights help define what quality is, there is no formula for the next masterpiece, and we don't expect that to change anytime soon. We argued that the power of algorithms is of much better use for managers when it comes to improving complex business decisions than for creating entertainment itself.

In addition to "experience" quality, we discussed the full spectrum of marketing instruments that entertainment managers have at their disposal. We stressed the power of branding for entertainment success, one of the best-developed areas of *Entertainment Science*, and studied the roles that other (unbranded) features, such as a certain genres, play for a product. We refuted old arguments about the development of new entertainment products and built the case that entertainment innovation can indeed benefit from certain strategies, cultures, structures, and methods that help managers predict the success potential of a new product. Our book did not present a "one-and-only" approach for making predictions; instead, we hope our readers will assess the alternative approaches we present as inspiration for developing an approach that best suits the idiosyncratic needs and resources of his or her firm.

Our book also provided an understanding of how effective different communication approaches are for an entertainment product, encompassing paid media, owned media, and also "earned" media, which includes (negative) word of mouth and professional reviews of (ugly) products. At the time of this writing, there is a lot of chatter in Hollywood that Rotten Tomatoes ruins films (e.g., Rodriguez 2017): by separating correlations from causal effects, *Entertainment Science* will help its reader to judge the "true" effect that professional reviews have and avoid making ad hoc decisions that only worsen an already troubled situation. *Entertainment Science* also adds a scientific layer to the fiery debates regarding (1) changes to established distribution windows and (2) the effects that illegal channels have on entertainment success these days (i.e., the effect of piracy), along with discussing alternative remedies. We showed that it might pay for entertainment managers to reconsider their traditional approach toward pricing all products identically, instead allowing prices to differ between products that differ in attractiveness.

We ended our theory of *Entertainment Science* with a discussion of blockbuster and niche marketing as the two dominant integrative marketing strategies used in practice. Here, we offered a warning for the industry: the application of only fragments of *Entertainment Science* to exploit the economic potential of digitalization and globalization in a way that reduces entertainment to a single type of product, an "extreme blockbuster," poses a threat to entertainment, as a whole industry. It may drive consumers away from films, music, (console) games, and books and toward new rival forms of entertainment, such as social media.

Overall, we intend *Entertainment Science* (the theory and the book) to help its readers to emphasize the value of thinking "scientifically" when making decisions for entertainment products. Doing so improves the power of artistic skills and managerial intuition with theory and data analytics. It enables the reader to leave the "Nobody-Knows-Anything" mantra behind, while also avoiding the "analytics trap," i.e., the naïve and careless use of analytical techniques in a way that ignores the complexities of entertainment and the key role of entertainment products' creative character. "Nobody Knows," but also "theory-free" analytics leave a lot of value on the table.

We hope that we have been able to provide such help, and we wish you good luck in working with, or extending, *Entertainment Science* (the theory). Now it is about time to part ways. Let us leave you by paraphrasing the words of legendary composer Oscar Hammerstein: So long, farewell, auf Wiedersehen, dear reader.

Reference

Rodriguez, A. (2017). The 10 worst summer 2017 movies that Rotten Tomatoes helped destroy. *Quartz*, August 25, https://goo.gl/7CDHTs.

Entertainment Science Scholar Index

T. Hennig-Thurau and M. B. Houston, *Entertainment Science*,
https://doi.org/10.1007/978-3-319-89292-4

T

V

W

Industry Index

T. Hennig-Thurau and M. B. Houston, *Entertainment Science*,
https://doi.org/10.1007/978-3-319-89292-4

C

M

T

Subject Index

T. Hennig-Thurau and M. B. Houston, *Entertainment Science*,
https://doi.org/10.1007/978-3-319-89292-4

K

L

M

N

O

P

T

About the Authors

Dr. Thorsten Hennig-Thurau is Professor of Marketing and holds the Chair for Marketing & Media Research at the University of Münster's prestigious Marketing Center. Prior to joining the University of Münster in 2010, he was Professor at the Bauhaus University of Weimar and served as a part-time Research Professor of Marketing at City University London's Cass Business School for 10 years. His academic work focuses on the entertainment industries and the consequences of digitalization, such as the role of social media for firms and consumers; he teaches entertainment and media marketing classes, along with courses on branding and innovations.

His studies, often conducted together with Dr. Houston, on the valuation of movie rights, the effects of consumer file sharing, the forecasting of movie success, and the role of social media for the adoption of new movies have been published by the world's leading academic journals such as the *Journal of Marketing*; they have also been covered by the leading international media outlets, including America's *Businessweek*, Britain's *Financial Times*, and Germany's *Frankfurter Allgemeine Zeitung*. Dr. Hennig-Thurau has been honored with several research awards, including the Lifetime Award for Published Scholarly Contributions to Motion Picture Industry Studies from the UCLA and the 2015 JAMS Sheth Foundation Best Paper Award for his work on the effect of Twitter chatter on movie success.

The German business magazine *Handelsblatt* has listed Dr. Hennig-Thurau as one of the top 1% of business professors in German-speaking countries in terms of productivity, and the *Frankfurter Allgemeine Zeitung* has named him one of Germany's ten most influential economists. He has

T. Hennig-Thurau and M. B. Houston, *Entertainment Science*,
https://doi.org/10.1007/978-3-319-89292-4

also been President of the Association of Marketing Professors in Germany, the first European member of the Academic Council of the American Marketing Association (AMA), the Director of the Digitalization Think:Lab (a joint initiative with Roland Berger Strategy Consultants at the forefront of the digital revolution), and has co-founded the influential JOURQUAL journal ranking, which he has chaired for 15+ years. He is member of the first DFG-founded Research Unit in the field of marketing and has co-chaired the AMA's Winter Marketing Academic Conference in Las Vegas as well as the Big Data, Big Movies Conference in Berlin in 2016.

He works closely with several leading entertainment companies and loves great novels, video games, and TV series, but most of all he is a serious movie aficionado (or addict, as some say). His all-time favorite piece of entertainment is Sergio Leone's THE GOOD, THE BAD, AND THE UGLY starring Clint Eastwood.

Contact information for Dr. Hennig-Thurau:
Web: http://www.marketingcenter.de/lmm/
Email: thorsten@hennig-thurau.de
Twitter: @ProfTHT

Dr. Mark B. Houston (Ph.D. Arizona State University; M.B.A. University of Missouri; B.S. Southwest Baptist University) is Professor of Marketing at Texas Christian University, where he holds the Eunice and James L. West Chair in Marketing. He is also Visiting Professor of Marketing at University of Münster, and a member of the CSL Research Faculty, Center for Services Leadership, Arizona State University. He previously served as Professor of Marketing at Texas A&M University (where he was also Head of the Marketing Department), University of Missouri-Columbia, where he held the David and Judy O'Neal M.B.A. Professorship, Saint Louis University, and Bowling Green State University.

Dr. Houston's research on marketing and innovation strategy, inter-firm relationship management, and motion picture success has been published in *Marketing Science*, *Journal of Marketing*, *Journal of Marketing Research*, *Journal of Consumer Research*, and *Journal of Financial and Quantitative Analysis*, among others. His awards include TCU's university-wide Deans' Award for Research and Creativity and MU's Distinguished Research Fellowship. As a member of the editorial review boards of six journals, he is Area Editor at *Journal of Marketing*, *Journal of Service Research*, and *Journal of the Academy of Marketing Science*. Dr. Houston co-chaired the 2005 and the 2017 American Marketing Association Summer Educators' Conference, co-chaired the

2010 AMA/Sheth Foundation Doctoral Consortium, and served as President of the AMA's Academic Council (2012–2013).

His research has received press coverage by the *Financial Times*, *BusinessWeek*, *Canadian Business*, the *New York Times*, the *London Observer/Guardian*, *Hollywood Reporter*, *Variety*, and *Screen International*. For his teaching, he has been honored as Outstanding Marketing Teacher by the Academy of Marketing Science and was recognized several times at the university and college levels. He serves on the Board of Directors of the American Marketing Association. He has conducted research, case, consulting, and/or executive education activities with many organizations, including AT&T, Dell, and IBM.

Dr. Houston is an avid reader (science fiction, biographies, and anything by C.S. Lewis) and a huge movie buff, although with decidedly non-critical tastes; his all-time favorites are THREE AMIGOS, TERMINATOR 2: JUDGMENT DAY, and THE PRINCESS BRIDE. He is also a big fan of Stan Lee at Marvel.

Contact information for Dr. Houston:
Web: http://www.neeley.tcu.edu/About_Neeley/Faculty_and_Staff/Houston,_Mark.aspx
Email: m.b.houston@tcu.edu

CPSIA information can be obtained
at www.ICGtesting.com
Printed in the USA
LVHW080535210120
644257LV00004B/272